GEOENVIRONMENTAL ENGINEERING

Geoenvironmental impact management

edited by R. N. Yong and H. R. Thomas

Proceedings of the third conference organized by the British Geotechnical Association and Cardiff School of Engineering, Cardiff University, and held in Edinburgh on 17–19 September 2001

Published by Thomas Telford Publishing, Thomas Telford Ltd, 1 Heron Quay, London E14 4JD.
URL: http://www.t-telford.co.uk

Distributors for Thomas Telford books are
USA: ASCE Press, 1801 Alexander Bell Drive, Reston, VA 20191-4400, USA
Japan: Maruzen Co. Ltd, Book Department, 3–10 Nihonbashi 2-chome, Chuo-ku, Tokyo 103
Australia: DA Books and Journals, 648 Whitehorse Road, Mitcham 3132, Victoria

First published 2001

Organizing Committee

Prof H R Thomas, Cardiff University
Prof R N Yong, Cardiff University
Dr A Al-Tabbaa, Cambridge University
Prof S Jefferis, Surrey University
Dr J M Reid, TRL
Dr D Stewart, Leeds University
Ms C Summers, Cardiff University
Dr P Tedd, BRE
Dr M Winter, TRL

A catalogue record for this book is available from the British Library

ISBN: 0 7227 3033 9

Printed and bound in Great Britain by MPG Books, Bodmin

Preface

This book contains peer reviewed papers presented at the 3rd British Geotechnical Association Geoenvironmental Engineering Conference held at the University of Edinburgh, Edinburgh, Scotland, in September, 2001. The second conference in this series was held two years ago in London at the Institution of Civil Engineers. Issues related to Geoenvironmental Engineering are now well established in importance throughout the world and pollution of the natural environment continues to occur as a result of both industrial activity and natural disasters. Therefore conferences like this have an important role to play in the sharing of knowledge and experience by the international geoenvironmental engineering community.

The main theme of the third conference is Geoenvironmental Impact Management. The papers in this book have been grouped together in eight main sections which reflect the themes which were considered at the conference: Recycling, Reuse, Recovery and Management; Risk Management; Legislation and Regulatory Control; Mine Sites, Tailings Dams, Dredgings and Lagoons; Site Remediation and Land Regeneration; Natural Attenuation, Fate and Transport of Pollutants including Gas; Design, Construction and Operation of Landfill Sites; and Case Histories. The papers in this volume have been submitted by authors from around the world, reflecting the worldwide interest in geoenvironmental engineering.

The conference was organised by the Geoenvironmental Engineering Research Centre at Cardiff University and the British Geotechnical Association. We would like to thank the members of the Organising Committee: Dr A Al-Tabbaa, Professor S Jefferis, Dr J M Reid, Dr D Stewart, Dr P Tedd, Ms C Summers, Dr M Winter and Ms A Reid (Secretary) for their assistance in the paper review process and the organisation of the conference. We would also especially like to thank the Local Organising Committee in Edinburgh (Prof A Barry, Dr W Ferguson, Mr G Paschke, Dr I Pyrah, Ms C Schmolke and Dr M Winter) for their assistance and support.

R N Yong
H R Thomas

Contents

NATURAL ATTENUATION, FATE AND TRANSPORT OF POLLUTANTS INCLUDING GAS

DESIGN, CONSTRUCTION AND OPERATION OF LANDFILL SITES

CASE HISTORIES

Recycling, Reuse, Recovery and Management

The reuse of municipal solid waste incineration aggregates in manufacturing usual concrete

LUSMEILIA AFRIANI[1], JEAN LOUIS QUENEC'H[2], DANIEL LEVACHER[3]

[1]University of Caen, Laboratory of geomorphology, 24 rue des Tilleuls, 14000 Caen, France, afriani@geos.unicaen.fr
[2]Ecole Nationale Supérieure des Ingénieurs des Etudes et Techniques d'Armement, 2 rue François Verny, 29806 Brest Cedex 9, France, quenecje@ensieta.fr
[3]University of Caen, Laboratory of geomorphology, 24 rue des Tilleuls, 14000 Caen, France, daniel.levacher@geos.unicaen.fr

Abstract

Used aggregate residues result from the combustion of refuse. The structure and particle composition are the same as natural aggregates. The chemical composition of the MSWI (Municipal Solid Waste Incineration) is dominated by elements Si, Ca, Al and Fe. These elements are similar to those commonly found in a cement matrix. The characteristics of these materials are defined by testing and measuring the density, the absorption coefficient, the water content of the aggregate, the "Micro-Deval" coefficients and the grain-size distribution. The Dreux-Gorisse method is used in order to obtain the formulation of the composition of a normal and a MSWI concrete. The substitution of natural aggregates is governed by those of the MSWI. Concrete samples are prepared and submitted to compression and tension tests. Mechanical performances at different states are deduced from the tests results.

Key words : concrete, cement, MSWI, solid waste, unconfined compressive strength.

Introduction

MSWI aggregates are solid residues resulting from the incineration of domestic rubbish. They are incinerated at high temperatures reaching more than 1000°C in an incineration plant. Two different types of solid waste result from this incineration: an MSWI aggregate and fly ashes. The MSWI aggregates account for on average 25 to 30 % of incinerated waste and the second type of residue obtained averages from 2 to 5 % of incinerated waste. Fly ashes are not mixed with the MSWI aggregates because they contain high quantities of polluting metals. This solid waste contain different types of materials such as: mixtures of glass, rocks, iron, aluminum, paper, wood and so on. Nowadays, the MSWI materials are mainly used in roadway construction and public works intermixed with bitumen.

The MSWI is chemically characterise by analysis of infra-red spectroscopy [1,8]. The main elements contained in aggregates are: SiO_2 (from 40 to 50 %), Al_2O_3 (from 5 to 15 %), CaO (from 15 to 25 %), Fe_2O_3 (15 %) and other elements include Na, K, Mg and P, with an

Geoenvironmental impact management, Thomas Telford, London, 2001.

aluminosilicate matrix, metal oxides and salts (NaCl, KCl, $CaCl_2$, $CaSO_4$), [2]. The contaminants such as heavy metals, are mainly lead, chromium, zinc and cadmium.

The purpose of this study is to use MSWI as a substitution aggregate in the composition and formulation of concretes. Aggregates used in a concrete formulation can reach 60 to 70% of the total volume of the concrete but they must respect agreements and rules concerning the characteristics. For example, secondary materials include domestic waste, industrial refuse from the steel industry or recycled concrete.

Where MSWI aggregates are used as substituting natural aggregates they must satisfy the physical and mechanical characteristics of a secondary material. These characteristics are defined by French norms (NF.P.18.540,[3]).

These products and their conditions of use in France are defined in a circular issued by the French Ministry of the Environment in May 9, 1994 [4]. As the MSWI aggregates have various performances, this circular specifies strictly their reuse. The main uses specified in this circular are the reuse of secondary materials in the concretes, for earthworks, embankments and parking structures [9]. Some reuses are depending on normalized leaching tests (NF.X.31.210), [5]. MSWI is divided into three categories: V (valorization), M (maturation) and S (storage); only the V and M classes may be adopted for further use.

Before using MSWI aggregates, the rough material is sieved at a diameter of 31 mm and ferrous and non-ferrous metals are removed. In figure 1, can be seen the rough and sieved MSWI aggregates.

a) rough material   b) after sieving

Figure 1 : Municipal solid waste incineration aggregates.

Application of MSWI

Until now, MSWI aggregates have been used primarily in France for road engineering (substrata, roadway foundation) and for filling earthworks. Several public works companies [6] use MSWI aggregates in road engineering and add a bitumen binder for improving the valorization. The treatment with bitumen foam decreases the risk of migration of the polluting elements in the MSWI into the natural environment. This phenomenon has been quantified using leaching tests [4].

Experiments on a roadway were carried out by a public works company in June 1995 and May 1996. This road carries class T3 traffic [6]. In 1997, the construction of a road between Cergy and Roissy (Val d'Oise, France) used a material named "Scormousse" in the construction of the substratum. The adopted thickness was 10 cm and the substratum was covered by 11 cm of bituminous mix. After one year of use under the traffic, this experiment showed a reduction in deflection for the substratum of 35 %. This reduction was linked to a maturation of the material which results in its rigidification with the effect of traffic. MSWI aggregates have also been used in the form of gravel – cement mixtures. The Jean Lefebvre public works company developed and patented formulations using MSWI and hydraulic binders. The product obtained is called Scorcim (SCORies- CIMent) [2]. Scorcim, with a 25cm thickness, covered by 5 cm of bituminous mix used on roadways has been observed for two years. The laboratory tests gave values of tension – bending strength from 0.39 to 0.83 MPa after 90 days. The dry density of Scorcim ranges between 1.82 Mg/m^3 and 1.87 Mg/m^3 and the water content varies from 13 to 15 %.

Experimental methods

Concrete is usually obtained from the mixing of several different components. These components are cement, aggregates (gravels, sand) and water. Concretes are classified according to their resistance to the unconfined compressive strength at 28 days (fc_{28}). This resistance can normally reach 20 to 50 MPa. The value of the tensile resistance is more or less 10 % of the value of compressive one.

The MSWI used in these investigations come from the incineration plant in Colombelles, near the town of Caen in Calvados, France. The incineration residues of domestic waste, collected in April 1999, were used for testing. Natural aggregates (sands ranging between 0-5 mm and gravels ranging between 6-16 mm) produced from the quarry at Mouen, Calvados, France.

The authors have applied the norms published by the French AFNOR in 1992 [3] to evaluate the physical and mechanical characteristics of the aggregates (density, absorption coefficient, water content), their resistance and the workability of the concrete and its grain-size distribution.

The mixing method for normal concrete or MSWI concrete is obtained using the Dreux-Gorisse method [7] which allows determination of the distribution of the aggregates. This hypothesis is used to determine the volumes of materials used in the composition of concrete such as: water, cement, sand and aggregates, given in m^3. This allows the determination of the volumes of water, cement and air from the formulation of the aggregates:

$$Vaggregates = 1 - Vcement - Vwater - Vair \qquad [1]$$

In the experimental part, seven different compositions were made in order to study the influence of the MSWI on a mixed concrete. A concrete with standard aggregates (referenced B0), then a concrete with only MSWI aggregates (referenced B1) were made for comparison. At least 5 other mixtures were made combining natural aggregates with MSWI aggregates. The rates of substitution are: 60 %, 50 %, 35 %, 25 % and 15 % (respectively referenced : B2, B3, B4, B5, B6). 350 kg of cement (CPJ CEM II/A, 32.5 R(L)CP2) are included in the compositions [7].

For each formulation, cylindrical samples were taken with a diameter ϕ of 11 cm, and a height h of 22 cm (slenderness = 2). 5 samples were manufactured for each composition. Subsequently, compressive and tensile tests were run on different days (7, 14 and 28, 56, 210 days). The results obtained for these tests are given in figures 3 and 4.

Results and discussion

The results show clearly differences between sands, natural gravels and the MSWI aggregates concerning the water content values and the absorption coefficient. It is clear that the MSWI aggregates values for water content and the absorption the coefficient are much higher than for the natural aggregates.

This difference may be explained because the crude MSWI aggregates from the incineration oven have a high temperature, and water is used to lower the temperature of the crude aggregate. In addition, the MSWI aggregate is porous and will absorb more water than a natural material. Moreover, the value of volumetric mass of the MSWI aggregates is lower than that of natural aggregates (see table 1).

	Natural sand	Natural gravel	MSWI 0-5 mm	MSWI 0-20 mm
Water content (%)	0.81	1.22	13.68	8.80
Absorption coefficient (%)	0.21	0.70	23.00	16.20
Bulk density (Mg/m^3)	2.59	2.55	2.14	2.23

Table 1 : Water content and absorption coefficient values of natural and MSWI aggregates.

Grain-size distribution measurements have also been undertaken to characterize the MSWI aggregate. They reveal that MSWI aggregates contain 40 % of elements in 0-5 mm (sands), 25 % of elements in 6-10 mm (grits) and 35 % of elements ranging from 10-20 mm (coarse gravels). It can be seen that the MSWI aggregates have almost 10 % of fine particles inferior to 0.4 mm (see below figure 2)

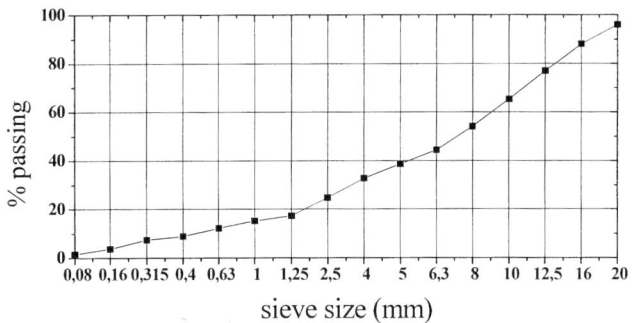

Figure 2 : Grain-size distribution of the MSWI aggregates.

The unconfined compressive strength of B0 concrete gives a value of 38.8 MPa (at 28 days). At 210 days, it is shown that the values notably increase. The concrete B1 has a value at 28 days of 4.2 MPa; it is a weak value. The other unconfined compression strength values are presented in the figure 3. They are inserted between B0 and B1 values.

These experiments show that the MSWI aggregates may be used as substitution aggregate but that the maximal substitution percentage of natural aggregates is 15% of the MSWI aggregates. In this condition the unconfined compressive strength is 20.8 MPa for 350 kg of cement per m^3, and a slump value ranged between 4 to 5.5 cm.

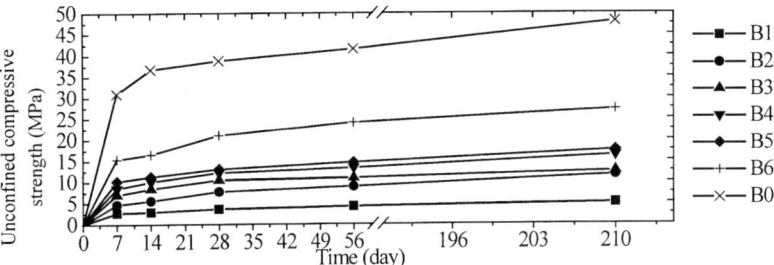

Figure 3 : Compressive strengths of B0, B1 to B6 concretes.

Coutaz [1], obtained a value of 23.5 MPa for a concrete with 50 % of MSWI aggregate, but a different type of cement (CEM I CPA 52.5 PM) was used. A high-performance cement in a normal concrete mix, referenced B20 according to the French norm, was considered. Nectoux *et al.* [8], has also found a maximal value in an unconfined compressive strength of 43 MPa for 15 % of MSWI gravels and of 38 MPa for 25 % of MSWI sands. He also used a cement type CEM 1- CPA 52.5 PM ES and obtained a slump value ranged between 4 and 5 cm.

Figure 4 shows the tensile strength of B0, B1 to B6 concretes (at 7, 14, 28 days). These strengths increase with the age of the concrete and the values for the referenced concrete are always higher than those for the substitution concretes.

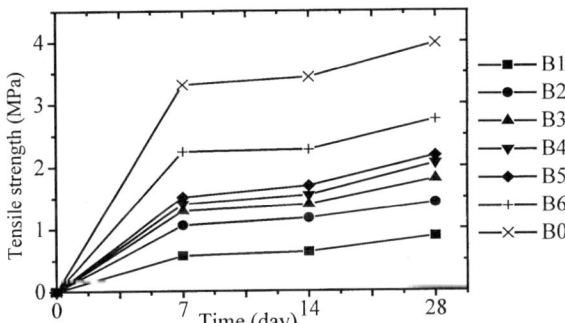

Figure 4 : Tensile resistance for B0, B1 to B6 concretes.

It is noted that the unconfined compressive strength of the concrete increases with time from 28 to 210 days. This phenomenon is only found in concrete composition which contains natural aggregates and MSWI aggregates. The concrete has a very low resistance if it contains 100 % of MSWI aggregates. This kind of composition is not practically acceptable, particularly with a proportion of cement of 350 kg/m³, which is already comparatively higher than for normal concrete.

The lateral surface of the sample of MSWI concrete includes many of cracks, opened pores and swelling. This state has been noted in all MSWI concretes but it is not so important on the B1concrete. Cracks and pores may explain the decrease of values for the unconfined compressive and tensile strengths for concretes. This swelling is due to several causes: first of which is the water absorption by the MSWI aggregates. The second hypothesis concerns the reaction of MSWI with the cement, because aluminum is contained in the MSWI aggregates.

This reaction has been observed in other research studies [1,2]. A chemical treatment in order to neutralize the aluminum reaction can decrease the swelling in MSWI concrete.

Conclusions

The results obtained demonstrate that MSWI aggregates can be partially use as aggregate for concrete. MSWI aggregate could be mixed by substituting a part of gravel and/or sand. The maximum percentage of MSWI aggregates which could be substituted into MSWI concrete is around 10 %; the MSWI concrete will then have an unconfined compressive strength close to 20 MPa at 28 days, see figure 5. For 210 days, this value will increase to more than 27 MPa. This result can be obtained by a cement ratio superior to 350 kg/m^3 (case of cement type: CPJ – CEM II/A 32.5 R), or the use of a superior-type cement. The compressive strength and tensile strength of the MSWI concrete can decrease because there is the reaction of MSWI with the cement while make the MSWI concrete presents a lot of cracks, pores and swelling. Now, MSWI aggregates can also be mixed with cement and bitumen for road construction and for secondary material in concrete composition.

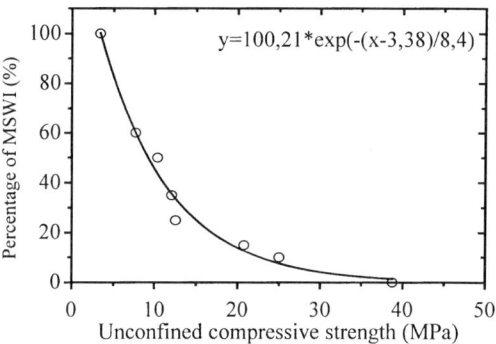

Figure 5: Influence of MSWI substitution level

References

[1] Coutaz L., (1996), "Valorization of the aggregate issued of incinerated domestic wastes", Doctoral thesis, Insa Lyon, France, 230p.

[2] Goacolou H., Oger P., Seigneurie C., Jozon C., Pascual C., Drouadaine I., Troesch O., (1995), "Scorcim for the valorization of the MSWI aggregate in public works", RGRA, n°729, 6p.

[3] AFNOR, (1992), " Building and civil construction, aggregate, French norms", 361p.

[4] Ministry of the Environment, (1994), " Elimination of the MSWI aggregate", Direction of the prevention of pollution and risks – service of the industrial environment, Circular of 9 th may 1994, DPPR/SEI/ BPSEID/ FC/FC n°94-IV-1, 17p.

[5] AFNOR, (1992), " Leaching test", X 31.210, 13p.

[6] Goacolou H., (1997), "Scormousse, a technique for treatment of bottom ash with foamed bitumen", RGRA, n° 757, pp.2-6.

[7] Dreux G., Festa J., (1996), "New guide of concrete and its constituants", Eyrolles Ed., 750p.

[8] Nectoux D., Grandhaie F., Terminaux R., Bonnet S., (1996), "Valorization of the MSWI : Aggregate for concrete", Processes of solidification and stabilisation of waste, pp.395 - 400.

[9] Guide technique for the use of materials, Ile-de-France, (1998), "Municipals solid waste incineration", november, 43p.

Utilization Of Fly Ash For Stabilization/Solidification Of Heavy Metal Contaminated Soils

Dimitris Dermatas*, Director and Xiaoguang Meng, Assoc. Professor, W. M. Keck Geoenvironmental Laboratory, Center for Environmental Engineering, Stevens Institute of Technology, Hoboken, NJ 07030, USA

ABSTRACT

In the present study, fly ash waste materials were tested along with quicklime (CaO) as an effective, yet economic technology to immobilize Pb, as well as Cr^{+3} and Cr^{+6} present in contaminated clayey sand soils. The degree of heavy metal stabilization was evaluated using the Toxicity Characteristic Leaching Procedure (TCLP) as well as other controlled extraction experiments. These leaching test results along with X-ray diffraction (XRD), scanning electron microscope and energy dispersive x-ray (SEM-EDX) analyses were also implemented to elucidate the mechanisms responsible for heavy metal immobilization. Furthermore, the reuse potential of the stabilized waste forms was also investigated by performing unconfined compressive strength (UCS) and swell tests. Addition of fly ash along with quicklime to the contaminated soils effectively reduces heavy metal leachability well below the non-hazardous regulatory limits. Meanwhile, it significantly improves the stress-strain properties of the treated solids, thus allowing their reuse as readily available construction materials. The results presented herein can be applied to the management of incinerator and coal fly ash, boiler slag and flue gas desulfurization wastes.

INTRODUCTION

Stabilization/solidification (S/S) of solid wastes by means of adding cementitious binders, like lime and cement [1,2] is a promising technology to transform hazardous wastes to non-toxic. This is achieved by physically and/or chemically "fixing" the toxic constituents in the waste form, thus reducing their potential release into the environment, to ensure compliance with existing regulatory standards [2]. Moreover, the stabilized wastes may attain adequate stress-strain properties to enable their utilization in construction applications, such as engineering fill [2].

Coarse-grained wastes, stabilized with quicklime, will frequently attain poor geotechnical and environmental properties and may not meet the requirements for reuse in construction applications because of their limited surface area that is available for cementitious reactions. Fly ash, known as pulverized fuel ash (pfa)" in UK, can be added to such coarse-grained wastes to increase the pozzolanic surface area, and hence improve the properties of the waste mixture such as strength, workability, buffering capacity to resist pH changes and heavy metal leachability. Adding fly ash to treat contaminated media would be a cost effective method of waste disposal, since fly ash, by itself, is considered a waste [3]. Pozzolanic reactions will result in the formation of calcium aluminum and calcium silicate hydrate (CAH and CSH) cementitious products, which in turn provide for the physical behavior enhencement of the treated matrix.

Geoenvironmental impact management, Thomas Telford, London, 2001.

Here at Stevens Institute of Technology, an experimental study was initiated to test the use of fly ash in addition to quicklime and sulfate salts, in a stabilization/solidification (S/S) scheme, designed to remediate heavy metal contaminated coarse-grained soils. It was conducted using artificially prepared lead (Pb^{2+}) and chromium (Cr^{+3} and Cr^{+6}) contaminated soils. The addition of sulfates as well as high levels of heavy metal contaminant contents was also pursued to extend the present study results to other waste by-product reuse (incinerator ash, FGD wastes, etc.) [3]. Overall, the main study objectives can be summarized as follows:

1. Immobilize Pb, Cr^{+3} and Cr^{+6} within a solidified matrix to satisfy the TCLP non-hazardous release criteria (5 ppm for all heavy metals studied).
2. Investigate potential reuse of the treatment product based on the strength and swell testing.
3. Elucidate the mechanisms of heavy metal immobilization in the treated solids based on the TCLP and controlled extraction tests results, along with XRD and SEM - EDX analyses.

During our experiments, contaminated soil, stabilized with quicklime, sodium sulfate decahydrate and fly ash was used. The strength and swell, TCLP, solubility and controlled extraction concentrations as well as mineralogical/micromorphological characteristics were monitored, for both treated and untreated specimens. Testing results for Cr^{3+}, Cr^{6+} are not presented in this paper, as only the results pertaining to Pb are presented herein.

QUICKLIME S/S TREATMENT PRINCIPLES

When a significant quantity of lime is added to a soil-fly ash mixture, its pH will reach to approximately 12~13. The solubilities of silica and alumina in the matrix will greatly increase at this high pH level, making them available for reaction with the calcium from lime, fly ash and other constituents to form the cementitious hydrates, like CAH and CSH, which are mainly responsible for the high strength and low swell of the treated solids. Their formation may also assist in heavy metal immobilization through surface sorption, inclusion and physical entrapment.

When the soil and/or groundwater contain sulfates in solution, they may combine with the liberated alumina to form a series of calcium-aluminate-sulfate hydrate compounds, leading to the formation of ettringite, $[Ca_3Al(OH)_6]_2(SO_4)_3 \cdot 26\ H_2O$. The formation of ettringite could lead to significant strength gains of the treated waste forms as well as the immobilization of the heavy metal species [4]. However, it is also a quite expansive mineral when brought in contact with water, and its swelling could lead to catastrophic failures [5]. Even though ettringite formation and effects were extensively studied, only some limited information is presented herein.

EXPERIMENTAL PROCEDURES

During treatment, clay-sand mixes, artificially contaminated using heavy metal salts or oxides, were dry mixed with quicklime (CaO) and sodium sulfate. For the artificial soil mixes, two different types of clay, kaolinite and montmorillonite, were used because they represent the two extremes of layered aluminosilicate surface area attributes and overall behavior. This paper mainly focuses on testing results for kaolinite-sand soil mixes. Fly ash, a non-layered iron aluminosilicate, was added into the contaminated soils to improve their physico-chemical behavior. First, a number of batch-type tests were performed to optimize treatment design. Subsequently, the actual treatment entailed compaction of the dry-mixed soils at optimum water content, and curing of the compacted specimens. Finally, specimens were tested to determine their leaching potential and physical properties, under various testing conditions.

The mechanical and physico-chemical behavior of the compacted specimens formed the basis for evaluating both the degree of heavy metal immobilization and the reuse potential of the treated waste form. Specifically, the effectiveness of the quicklime treatment was evaluated based on the heavy metal leachability, the UCS, and swell results. Meanwhile, XRD and SEM determinations, were used to elucidate the underlying mechanisms of heavy metal immobilization. A schematic representation of the experimental approach was illustrated in Figure 1. Durability, column and monolithic leaching testing results are not presented herein.

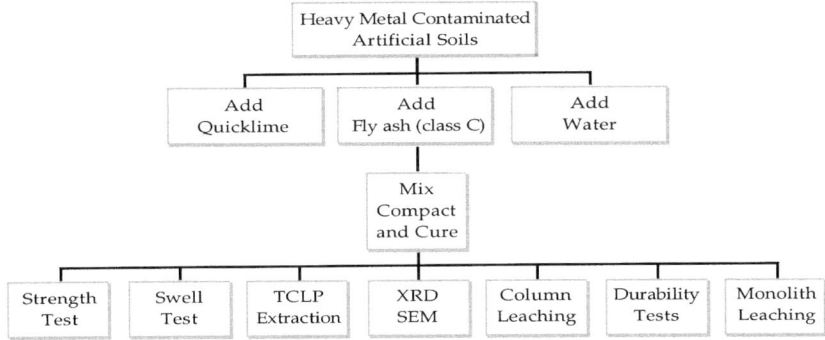

Figure 1:   Outline of the experimental methodology.

Table 1. Characteristics of the materials used during the present study

	COMPOSITION (*% Content*)										
	SiO$_2$	Al$_2$O$_3$	Fe$_2$O$_3$	CaO	MgO	SO$_3$	Na$_2$O	K$_2$O	TiO$_2$	R$_2$O$_3$	Loss on Ignition
Kaolinite	45.70	38.50	0.40	0.20	0.10	-	0.04	0.10	1.40	-	13.60
Montmorillonite	67.20	15.20	1.87	1.92	3.20	-	2.58	0.96	0.16	-	5.70
Quicklime	1.20	-	-	95.40	0.85	0.012	-	-	-	0.75	0.55
Fly ash	34.2	19.3	5.64	25.8	5.07	??	2.04	0.52	-	-	0.11

Coal fly ash (Class C) was used. It basically consists of spherical glassy particles with particle size range between 1 to 6 μm. Over 99% of kaolinite and montmorillonite particles passed through 325 mesh sieve and used as a source of surface area and pozzolanic material in the base soil mixes. Chemical obtained grade CaO (quicklime) powder contains 95% of CaO. The chemical composition of the fly ash, kaolinite, montmorillonite and quicklime are listed in Table 1. Meanwhile, the surface areas of kaolinate, montmorillonite, quick lime and fly ash sample were 66, 760, 40-41.5 and 31(m^2/g), respectively. pH values for these samples, in the same order, were 4-6.5 (20% solids), 7 (10% solids, dist. H$_2$O), 12.9 (50% solids, dist. H$_2$O), and 12.3 (50% solids, dist. H$_2$O), individually. The cation exchange capacity (C.E.C) value of kaolinite was 4.5-5.5 meq./100g, whereas it was 80 for montmorillonite samples. Finally, lead oxide (PbO) was used as the pollutant sources for Pb^{2+}.

Preparation of Artificially Contaminated Soil

Artificial soil specimens composed of kaolinite and fine quartz sand were used for the stabilization/solidification experiments. Mixtures of clay and sand were used to obtain specimens with gradations more comparable to those of naturally occurring soils and to provide materials that could be compacted more easily. In order to avoid dilution effects caused by the addition of fly ash, all contaminants were simultaneously added on a total clay, sand and fly ash weight basis (Table 2). Quicklime (10% by total weight of the solids) and sodium sulfate decahydrate (5% by total weight of the solids) were then added, and following addition of water and mixing, samples were cured for 24 hours. Specimens were then compacted at optimum water content according to ASTM D1557-91 standard [6], and cured at 20°C and 95% relative humidity (RH). Specimen dimensions varied depending on the type of test to be performed. The compacted specimens were cured for 28 days before they were tested for their unconfined compressive strength and vertical swell.

Table 2. Heavy metal contents in the artificial soils

Contaminant source:	$Cr(NO_3)_3$	K_2CrO_4	PbO
Heavy metal species soil concentrations (mg/kg soil)	4,000	4,000	7,000

Strength and Swell Tests
Following specimen compaction and designated curing, specimens were tested for strength and swell based on ASTM D2166-85 [6]. All strength and swell tests were performed on specimen duplicates and average values were used.

Leaching Tests
Following strength testing, samples derived from failed specimens, were tested for their heavy metal leachability using the TCLP test [7]. Pb concentrations in the solution were measured using an ICP atomic emission spectrometer. All TCLP testing was performed on sample duplicates to get average values. Meanwhile, all analyses were performed by using standard additions (spiking) to ensure proper quality control of the reported results. Moreover, other controlled extraction experiments were conducted by following the TCLP test procedure, however, using different acidic solutions. Samples were then filtered and the extract was analyzed. Solubility experiments were also performed for Pb where 377 mg of litharge (PbO) along with 1 gram of quicklime (CaO) were dissolved in 1L TCLP solution (pH=3) and pH was adjusted by adding concentrated HNO_3 or NaOH.

Specimen Designation
Letters in the specimen designation indicate mineralogy, i.e., K: kaolinite, M: montmorillonite, C: class C fly ash, and L: quicklime. Numbers following letters indicate the percent weight of the given attribute. Sand is not included in the specimen designation since its content is always complimentary to the clay or fly ash content on a 100 percent basis. For example the specimen designation K30L10 stands for 30% kaolinite and 70% quartz fine sand at a treatment level of 10% quicklime by weight of the clay-sand soil. Conversely, K30L10S stands for 30% kaolinite and 70% sand at a quicklime treatment level of 10% and a sodium sulfate decahydrate addition of 5%, whereas K5C25L0 stands for 5% kaolinite, 25% fly ash class C and 70% sand at a treatment level of 0% quicklime (untreated).

RESULTS AND DISCUSSIONS
Strength and Swell
Compressive strength results showed that the coarse-grained untreated soil (K5L0) possesses a minimum UCS of only 12.8 kPa, (1 kPa = 6.9 psi). Upon quicklime treatment, a significant strength increase to 144.9 kPa without sulfate addition, and 257.5 kPa in the presence of sulfates was observed. When sulfates were present in clayey soils, the mineral ettringite formed upon quicklime addition (evidenced by XRD), which lead to a significant strength recovery. When the kaolinite clay content was increased to 30% (K30L10), an even higher level of strength enhancement was achieved, probably due to the increased amount of pozzolanic product formation.

Upon fly ash addition to the untreated low clay content, coarse-grained soil mix (K5C25L0), strength increased dramatically to 3830.2 kPa. Fly ash class C, upon exposure to the water of compaction, and due to its high CaO content (~25%), would form cementitious pozzolanic products. Following quicklime treatment (10% by weight), K5C25L10 samples showed a UCS of 6,662.5 kPa without sulfates, and 7,219.7 kPa upon sulfate addition. Clearly, fly ash in the presence of quicklime was a superior strength enhancement agent. Overall, the quicklime/fly ash treatment resulted in almost 1,000 times higher strength levels than untreated soil which was almost comparable to the strength of concrete products.

XRD analyses of K30L10 samples indicated that pozzolanic product formation did take place within the 28-day curing period. Calcium silicate hydrate (CSH) and calcium silicate hydroxide (CSH*) were the main products identified. In addition, unreacted lime was also identified, indicating still on-going pozzolanic reactions following the 28-day cure. XRD analyses of K5C25L10 and K30L10S samples revealed the presence of both ettringite and CSH* pozzolanic products. Moreover, SEM studies confirmed the presence of ettringite in both kaolinite-sand mixes (K30L10S) and quicklime treated fly ash-clay-sand mixes (K5C25L10). The absence of ettringite needle-like crystals and the presence of a book-like fabric, typical for kaolinite clays, were the predominant features of the quicklime treated, no sulfate added, samples (K30L10).

The formation of ettringite may lead to expansion in the presence of water, leading to subsequent failure of the construction application [5]. In the case of K30L10S samples, 9% of vertical swell was observed following sample soaking. However, no development of significant swell (~0.1%) was observed for the ettringite-bearing K5C25L10 specimens, whether additional sulfate was added or not. As previous research has demonstrated [5], in the presence of fly ash, the pozzolanic cementing action may be able to overcome the ettringite-induced swell pressures, so specimens would remain intact upon water exposure. For the K30L10S samples that swelled significantly, additional experiments demonstrated that a barium hydroxide pretreatment, at a molar ratio of at least 0.2, was successful in eliminating ettringite formation and subsequent swell.

Leaching Tests
Lead
Treated solids were tested for their regulatory levels of heavy metal leaching by means of conducting TCLP batch extraction experiments. After 28 days of specimen curing, for untreated cases, the amount of Pb release was 292, 259, 125.6 ppm for sample K5, K30, K5C25, respectively. Whereas, following 10% of quicklime treatment, these values were sharply reduced to 124, 6.4, 1.2ppm, respectively, which indicated that levels of TCLP Pb release were below the regulatory benchmark of 5 ppm only when quicklime and fly ash were added to the contaminated soil. In order to further understand the conditions and mechanisms that lead to Pb immobilization, TCLP-based extraction experiments, using different clay-sand lime treated mixes, were conducted. As can be seen, lead immobilization was ensured if the treatment TCLP pH was kept between 8 and 11 (Figure 2). Moreover, after conducting lead solubility experiments under identical solution conditions, it was concluded that Pb immobilization could not be controlled by Pb solubility in the presence of quicklime treatment, especially at high pH. When pH was greater than about 9, it appeared that a Pb adsorption mechanism was predominant, as Pb release was significantly lower than Pb solubility. Conversely, when the pH was lower than 9, Pb release, even though somewhat influenced by surface adsorption, was mainly solubility-controlled. The addition of fly ash in low clay content quicklime treated solids has an obvious positive effect on Pb immobilization, as indicated by the K5C25L10 release curve shown in Figure 2. More specifically, fly ash addition resulted in further widening the pH range of Pb immobilization from 5 to 13. Similarly, results were also obtained for Cr^{3+} and Cr^{6+} immobilization, but were not presented herein.

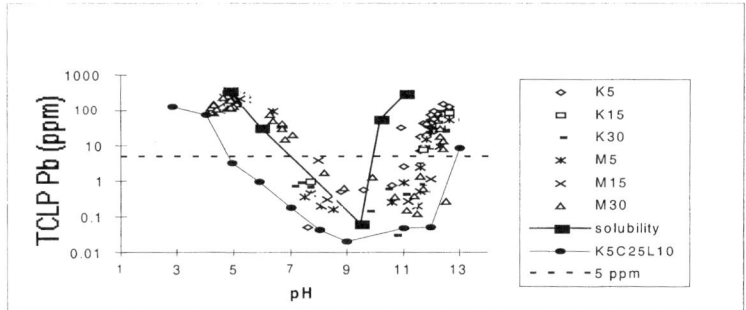

<u>Figure 2:</u>   Summary of TCLP-based, controlled extraction and solubility Pb release results as a function of extract pH for different solid mixes.

CONCLUSIONS AND RECOMMENDATIONS

Fly ash can be effectively used in a variety of construction applications. The utilization of fly ash during S/S treatment of Pb^{2+} contaminated soils, was evaluated by conducting strength, swell, mineralogical/micromorphological and chemical extraction experiments. The addition of fly ash along with quicklime resulted in a high strength, swell-resistant monolithic solid, attaining levels of strength similar to those of concrete products. With respect to heavy metal release, the addition of fly ash was directly responsible for the effective immobilization of Pb. The presence of fly ash increases the immobilization pH region from 5 to 13, thus making treatment more effective.

In summary, the addition of fly ash during the quicklime-sulfate S/S treatment of heavy metal contaminated soils is mainly responsible for their effective immobilization. Fly ash addition also results in improvement of the stress-strain properties of the treated solids, therefore enabling their reuse in construction applications. However, ettringite formation may lead to significant swelling of the treated product in the absence of fly ash. Addition of barium may solve the problem but this has to be researched further. Overall, the results presented here can be safely applied to the management of most kinds of coal burning by-products including fly ashes, FGD wastes, boiler slags, etc.

REFERENCES

[1] Conner, J. R. (1990), *"Chemical Fixation and Solidification of Hazardous Wastes"*, published by Van Nostrand Reinhold, New York, 692 pages.

[2] Dermatas, D. and Meng, X. (1994), *"Stabilization/Solidification (S/S) of Heavy Metal Contaminated Soils by Means of a Quicklime-Based Treatment Approach"*, Stabilization and Solidification of Hazardous, Radioactive, and Mixed Wastes, *ASTM STP 1240*,

[3] Goh, T. C. A. and Joo-Hwa Tay (1993), *"Municipal Solid Waste Incinerator Fly Ash for Geotechnical Applications"*, Journal of Geotechnical Engineering, vol. 119, No. 5, ASCE.

[4] Kumarathasan, P., McCarthy, G. J., Hasset, D. and Plughoeft-Hasset, D.F..(1990), *"Oxyanion Substituted Ettringites: Synthesis and Characterization, and their Potential Role in Immobilization of As, B, Cr, Se and V"*, Materials Research Society, *Symposium Proceedings Vol. <u>178</u>*, pp. 83-104.

[5] Dermatas D., "Ettringite-Induced Swelling in Soils: State-of-the-art", Applied Mechanics Review, Volume 48, Number 10, pp 659-673, October 1995.

[6] American Society for Testing and Materials (1993), *"Annual Book of ASTM Standards" Vol. 4.08, Soil and Rock; Building Stones*, Philadelphia, PA.

[7] S. Environmental Protection Agency (1985), *"Solid Waste Leaching Procedure Manual"*, *SW-924*, U. S. EPA, Cincinnati, OH.

Centrifugal dewatering system by use of rubber elasticity

Dr Ryoichi FUKAGAWA, Prof. Ritsumeikan University, Kusatsu, JAPAN, Kunihiko KAWASHIMA, Graduate Student, Ritsumeikan University, JAPAN, Toyoshige MOHRI, President, Horyo-sangyo Co., Ltd., Osaka, JAPAN, Dr Kazuyoshi TATEYAMA, Associate Prof., Kyoto University, Kyoto, JAPAN and Dr Takeshi KATSUMI, Associate Prof., Ritsumeikan University, JAPAN

INTRODUCTION

A new dewatering system for slurry has been developed with the aim of realization of efficient mud-water treatment. In previous papers[1,2], the new prototype centrifugal dewatering system known as MORIS-1 was reported. However, this machine had the inherent weakness of having a small dewatering capacity. To overcome this fault, MORIS-2 was developed. In the new system, centrifugal force is employed not only for the purpose of densification of mud component but also for separation of mud particles from water. In addition, rubber elasticity is effectively used for the separation of the treated mud and the water. This dewatering system made it possible to achieve a continuous dewatering process without the use of any filtering device. Furthermore the proposed dewatering system neither requires a high initial cost nor a high maintenance cost. Consequently, the system achieves a higher cost performance in comparison with conventional systems. In this paper, the applicability of the system will be discussed based on the results of the experiments carried out with a trial machine on several kinds of slurry materials.

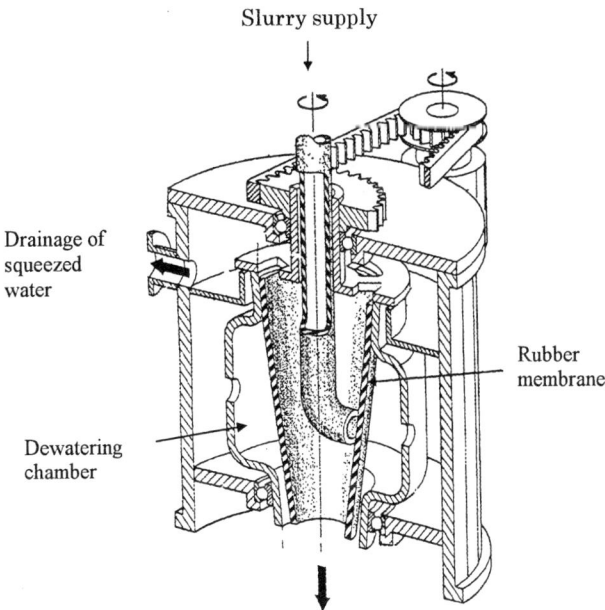

Slurry supply

Drainage of squeezed water

Rubber membrane

Dewatering chamber

Discharge of dewatered soil

Geoenvironmental impact management, Thomas Telford, London, 2001.

NEW CENTRIFUGAL DEWATERING DEVICE (MORIS-2)

The structure of the MORIS-2
Figure 1 shows the inner structure of the MORIS-2.   Mud slurry is supplied from an inlet of the upper part of the dewatering chamber.   The dewatering chamber has outlets for discharge of dewatered soil and squeezed water.   The dewatering chamber is rotated around the central axis by a motor.   The rotation rate is adjustable from 0 to 1700 r.p.m.   The inner diameter of the chamber is about 420 mm, and the height is about 300 mm, so that the space for dewatering is about 20 L (L: Liter).

The dewatering process
Dewatering by use of the new system is carried out as follows (see Figure 2):
a) At first, dewatering chamber is rotated without any load.   In this stage the rubber cylinder is enlarged to cover the inner surface of the cylinder.
b) After the rotation speed reaches a specified value, mud slurry is supplied into the chamber. The slurry piles up on the inner side face of the rubber cylinder due to the centrifugal force.
c) Dewatered mud and squeezed water are separated from the mud slurry. The top end of the chamber has slightly larger outlet than that of the bottom end so that the squeezed water is gradually discharged only from the top end of the chamber.
d) At a specific time in the process cycle, the electric supply causing the rotation is turned off. As the rotation speed gradually slows down, the shape of the rubber cylinder returns to the initial state.   In this stage the squeezed water is first discharged from the upper outlet.
e) When the chamber comes to a halt, the dewatered soil slides down along the surface of the rubber membrane and is discharged through the bottom outlet.

EXPERIMENTS TO TEST MORIS-2 USING TYPICAL SOIL SAMPLES
The effects of soil samples and flocculation agents on the dewatering properties were mainly investigated.

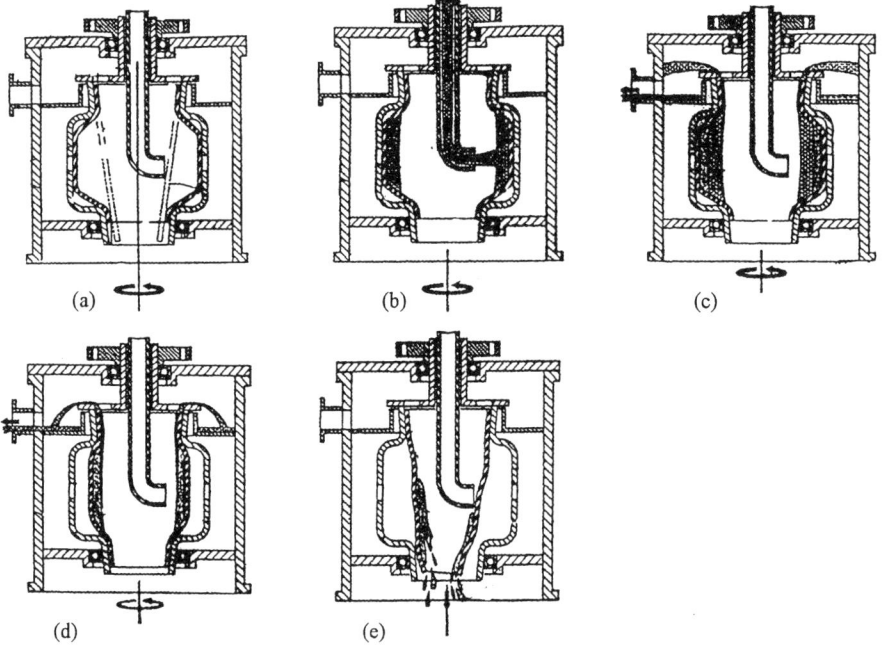

Figure 2. Test procedures

Samples and flocculation agents

5 kinds of soil samples were used, that is Silica Sand No.5, Silica Sand No.8, Soil A, Soil B and Kasaoka Clay. Soil A and B are ordinary garden soils. Kasaoka Clay is a typical kaolin clay. The particle size distribution of the samples is shown in Figure 3 and the physical properties of the samples are listed in Table 1.

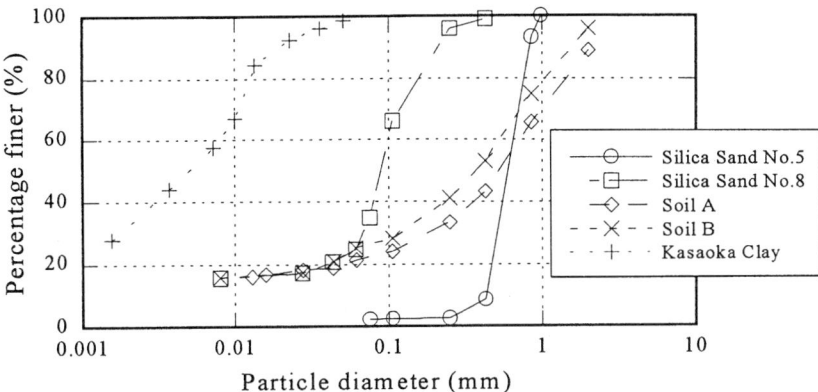

Figure 3. Particle size distribution of the samples

Table 1. Physical properties of samples

Sample	Density (g/cm³)	D₅₀ (mm)	Plastic Limit (%)	Liquid Limit (%)	Plasticity Index
Silica Sand No.5	2.63	0.60	-	-	-
Silica Sand No.8	2.62	0.085	20	34	14
Soil A	2.60	0.50	19	26	7
Soil B	2.56	0.38	20	32	12
Kasaoka Clay	2.67	0.005	21	53	32

2 kinds of flocculation agents were used. Both are inorganic agents which have little effect on water quality. The flocculation agent A can strongly flocculate even under high pH range, whereas the flocculation agent B is ordinarily well used as PAC.

Test procedures

The test procedures are as follows:

a) The initial water contents of the slurry samples are set to 500 %.

b) The chamber rotation is switched on and 26 L of the sample is injected into the dewatering chamber after the rotation speed reaches 1700 r.p.m. (corresponding to 680G). The specified operation times for the dewatering are 2, 4, 6 minutes for Soil A, B and Kasaoka Clay, and only 2 minutes for No.5 and 8 Silica Sand.

c) The squeezed water is sampled during the test and the dewatered soil is sampled after the rotation stops.

d) The test results are evaluated by SS (Suspended Solid) for the squeezed water and water contents for the dewatered soils.

TEST RESULTS AND CONSIDERATIONS

Dewatered soil

Figure 4 shows the effect of the operation time on the water content of the dewatered soil

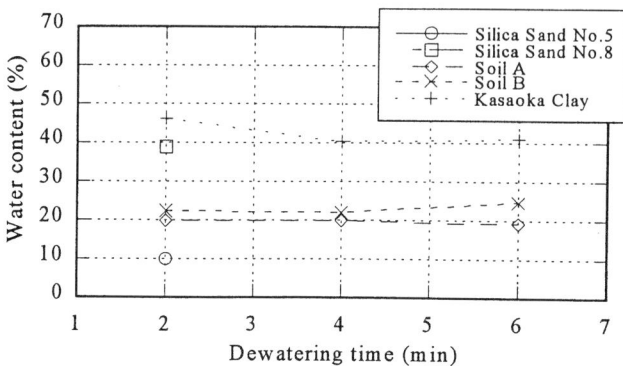

Figure 4. Effect of operation time for dewatering on water content of dewatered soils

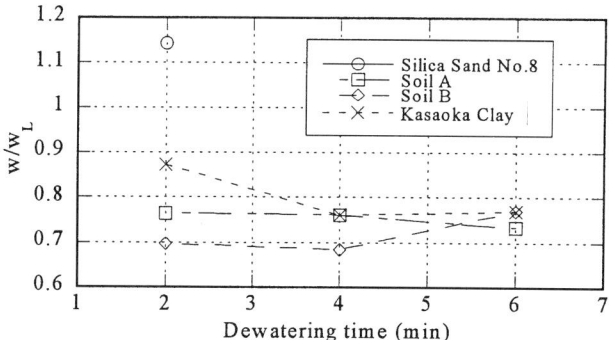

Figure 5. Evaluation of new system based on the comparison with liquid limits

Table 2. Performance of some typical dewatering systems[3]

Dewatering system			Dewatering properties of treated soil	
Classification	Dewatering mechanism	Loading pressure	Water content (%)	Cone index (kPa)
Filter press (Normal type)	Filtering by mud slurry supplying press	500~700 kPa	$(0.8\sim0.9)\times w_L$	100~1000
Filter press (High pressure type)	Filtering by mud slurry supplying pressure and compression	4000 kPa	$(0.6\sim0.8)\times w_L$	1000~3000
Belt press	Mechanical compression	100~150 kPa	$1.0\times w_L$	0~200
Screw decanter	Centrifugal force	500~2000 G	$(1.0\sim1.2)\times w_L$	0

samples.   All the samples have an initial water content of 500 %.   It can be seen that there was no significant change in the water contents during the operation time.   The ratio of water content of the dewatered soil against liquid limit demonstrates the effectiveness of dewatering system.   The result is shown in Figure 5.   This figure shows that each sample was dewatered to a level of 70 – 80 % of w_L.   Table 2 shows the relationship between dewatering systems and the physical properties of the soil samples[3].   Judging from Figure 5 and Table 2, it can be concluded that MORIS-2 has excellent dewatering ability.   It is slightly difficult to dewater from sandy soils by use of MORIS-2 because MORIS-2 has no filter device.   MORIS-2 can compress the sample, but the water in the void can't be fully

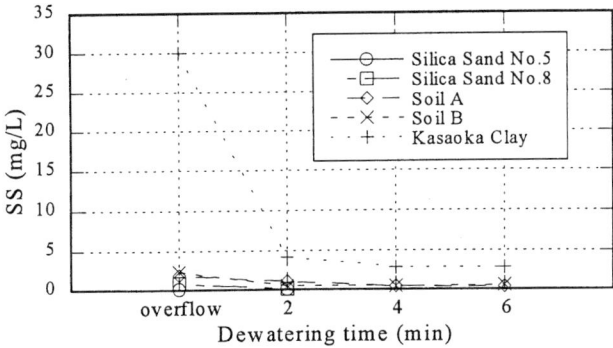

Figure 6. Effect of operation time for dewatering on SS of squeezed water

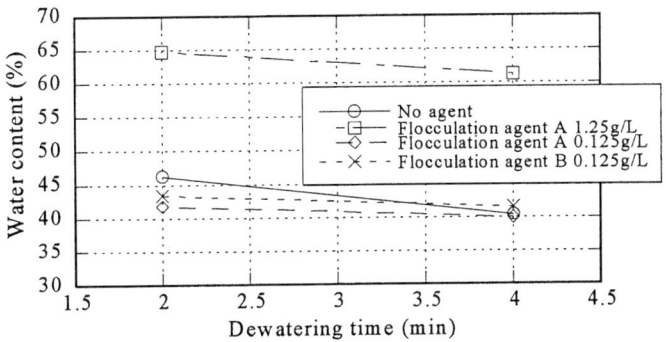

Figure 7. Effect of flocculation agent on water content of dewatered soils

discharged.   Therefore the water content of sandy soils remained comparatively high in Figure 4.

Squeezed water

Figure 6 shows the relationship between SS mg/L and the operation time.   In this figure, "overflow" means the overflow of the water from the top end of the chamber in the initial stages of slurry supply and dewatering (refer to Figure 2 (b) and (c)).   The overflow happens when the slurry supply exceeds the dewatering capacity.   Therefore, the overflowed water may not be fully treated.   The operation time has a great effect on the improvement in SS as can be seen in Figure 6.   According to the Japanese water quality standard, since the average SS of daily value must be lowered 150 mg/L, MORIS-2 gave enough results for each sample. The result obtained from Kasaoka Clay was slightly worse than other samples, because the clay includes much fine particles.

Effect of Flocculation on dewatering

The effect of flocculation on the water content of the dewatered samples is shown in Figure 7. The flocculation agent was mixed with the slurry of Kasaoka Clay.   It can be observed that the usage of flocculation agents has almost no effect on the improvement of the water content. In fact, adding too much flocculation agent worsened the effectiveness of dewatering as shown in Figure 7.   This is because the flocculation agent gathers particles all around, and the strength depends on the volume of the flocculation agents.   The big flocculation results in including much water in it.

The effect of the flocculation agents on SS of the squeezed water is expressed in Figure 8.

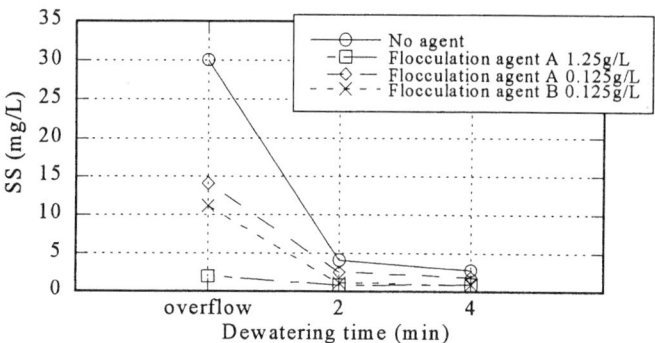

Figure 8. Effect of flocculation agent on SS of squeezed water

It can be seen easily from this figure that the flocculation agents have a great effect on the SS of the treated water, especially when the operation time is relatively short. The volume of the flocculation agent is closely related to the improvement of the SS, but the cost should be considered. Unfortunately the cost performance of the flocculation agents could not be confirmed in this study.

CONCLUSIONS
The main conclusions obtained from this study are as follows:
(1) The water contents of the dewatered soil samples became almost 70 – 80 % of the liquid limit. This means that MORIS-2 has excellent dewatering ability.
(2) The quality of the squeezed water was evaluated by SS. The effect of the operation time on SS was prominent. The SS of the treated water was remarkably low comparing with the Japanese standard.
(3) The effect of flocculation agents was clear in the improvement of SS of the treated water. However, it has almost no significant effect on the water contents of the dewatered samples.

ACKNOWLEDGEMENT
This research is partly sponsored by the Grant-in-Aid for Scientific Research (B:12875088) of the Ministry of Education and Science. We would like to show our appreciation for this support.

REFERENCES
1) Mohri, T., Fukagawa, R., Tateyama, K. and Mori, K.: A new dewatering system for slurry using planetary rotation chambers, Proc. of the 5th Asia-Pacific Regional Conf. Of ISTVS, pp.298-306, 1998.
2) Mohri, T., Fukagawa, R., Tateyama, K. and Mori, K. and Ambassah, N.O.: Slurry dewatering system with planetary rotation chambers, Geotechnics of High Water Content Materials, ASTM STP 1374, pp.279-292, 1999.
3) Katsumi, T., Yamada, M., Ogawa, S. and Kamitani, M.: Geotechnical treatment and effective use of excavated soils and construction waste, Tsuti to Kiso, Vol.45, No.1, pp.55-60, 1997. (in Japanese)

Environmental Suitability Assessment of Incinerator Waste Ashes in Geotechnical Applications

MASASHI KAMON[1], TAKESHI KATSUMI[2], and TORU INUI[1]
[1] Disaster Prevention Research Institute, Kyoto University, Uji, Kyoto, Japan
[2] Department of Civil Engineering, Ritsumeikan University, Kusatsu, Shiga, Japan

INTRODUCTION

The reuse of solid waste products as construction materials (e.g., for filling materials, road base, and embankments) has been encouraged recently because of the lack of disposal sites. In the assessment of the environmental impact of reusing such waste materials due to the toxic substances in them, various leaching procedures have been developed to investigate the release of contaminants which are influenced by many factors including the pH, the redox status, organic substances, etc. Several leaching models, which can account for diffusive leaching or advective-dispersive leaching, have also been presented in many previous research works. However, few methods for predicting the long-term in situ leaching behavior, due to the utilization or disposal of waste materials, have been reported based on availability tests [1] and diffusion-controlled leaching tests [2].

This paper presents some approaches for evaluating the leaching behavior of solid waste materials with some reuse or disposal scenario, considering the results of laboratory leaching tests and specific site conditions. The leaching behavior of municipal solid waste incinerated fly ash (MSW ash) and sewage sludge incinerated ash (SS ash) was characterized through long-term column leaching tests, which could be defined as a proper leaching protocol for the in situ percolation-controlled release from granular materials. Two parametrical analyses were conducted to assess the environmental suitability of solid wastes for geotechnical applications, namely, 1) a subbase was constructed with SS ash and 2) MSW ash is disposed in a landfill. The effectiveness of several measures against the contaminated leachate, such as cement stabilization and the installation of a clay liner, were also examined.

EXPERIMENTS

Materials

Sewage sludge incinerated ash (SS ash) and two types of municipal solid waste incinerated ash (MSW-1 ash and MSW-2 ash) were used for laboratory experiments. The basic properties of these ashes are shown in Table 1.

Table 1. Basic properties of incinerated waste ashes

Incinerated waste ash	SS ash	MSW-1 ash	MSW-2 ash
Particle size distribution (Particle size, mm: fraction, %)	>2.0: 0 0.1-2.0: 27 0.075-0.1: 16 <0.075: 57	>2.0: 1 0.1-2.0: 8 0.075-0.1: 33 <0.075: 18	>2.0: 1 0.1-2.0: 62 0.075-0.1: 28 <0.075: 9
Total content of heavy metal (mg/kg-ash)	T-Cr: 92	Pb: 7600, Cd: 280	Pb: 9800, Cd: 270
Ignition loss (%)	9.3	6.0	11.3

Geoenvironmental impact management, Thomas Telford, London, 2001.

Procedures of the Leaching Tests

Long-term column leaching tests were conducted for cylindrically compacted SS and MSW ashes in molds ($\phi = 51$ mm, $h = 100$ mm) to evaluate the properties of the leaching contaminants from the incinerated waste ashes. The SS ash was prepared by compacting at an approximately optimum water content ($w_{opt} = 58\%$) and then aged for 28 days. The MSW ashes were prepared in two ways, namely, 1) compacted at a 30% water content and 2) stabilized with some additives (Portland cement and coal ash) and then aged for 28 days. The mixing ratio of the additives was 80% MSW ash, 20% Portland cement, and 10% coal ash. Each specimen was permeated with distilled water adjusted to a pH value of 4.0 using nitric acid. The effluent was collected periodically, and the pH values and the electric conductivities were determined. Concentrations of lead (Pb), cadmium (Cd), zinc (Zn), total-chromium (T-Cr), and calcium (Ca) were also analyzed with an Inductive Coupled Plasma Spectrometer (Shimadzu ICPS-8000). In addition, the heavy metal absorption capacity of the soil was evaluated by column-type infiltration tests for the decomposed granite soil (DGS). Each specimen was prepared in a mold with waste ash layers 25, 50, and 100 mm in thickness, respectively, over the 50 mm-thick DGS layer.

Experimental Results

The results of the column leaching tests are presented in Figure 1. Further experimental data have been reported in previous papers [3], but the main results can be briefly summarized as follows:

1) The leaching of T-Cr and Pb and the high alkalinity from the SS ash were investigated. The leaching mass of these contaminants was significantly reduced, however, due to the absorption capacity of the soil.

2) The effect of the cement stabilization on the immobilization of the heavy metal in the MSW ash was examined.   A significant reduction in the leached Cd was achieved regardless of the

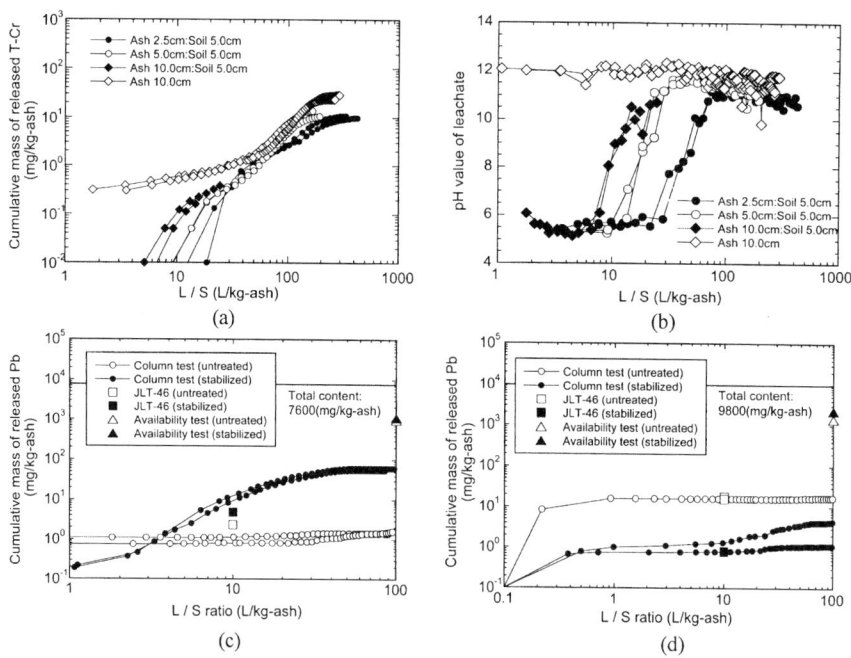

Figure 1.   Results of the column leaching tests: (a) T-Cr leaching from SS ash including the filtration soil layer, (b) pH of discharge from SS ash including the filtration soil layer, (c) Pb leaching from MSW-1 ash, and (d) Pb leaching from MSW-2 ash.

MSW ash properties. However, the immobilization of Pb did not constantly occur, as shown in Figure 1 (c), for the leached mass of Pb largely depended on the pH value of the discharge.

ENVIRONMENTAL IMPACT OF UTILIZATION OR DISPOSAL

Environmental Impact of the Utilization of Sewage Sludge Ash as Base Material

A parametrical analysis was performed in order to discuss the environmental impact of the utilization of SS ash as filling/base material and the design concepts related to the filtration layer, based on the experimental results. The T-Cr and the alkaline migrations from the SS ash were evaluated, considering the effects of the contaminant absorption by the soil on the environment. The cross section for this analysis was assumed as shown in Figure 2. The leachate, including the high alkaline and T-Cr, is generated when the seepage passes through the SS ash layer due to rainfall. However, the leached high alkaline and T-Cr were absorbed while passing through the filtration layer above the groundwater. The suitability of SS ash for construction material is discussed by calculating the amount of T-Cr and the required time for the high alkaline seepage to reach the groundwater. The evaluating method and the assumptions are shown in Table 2.

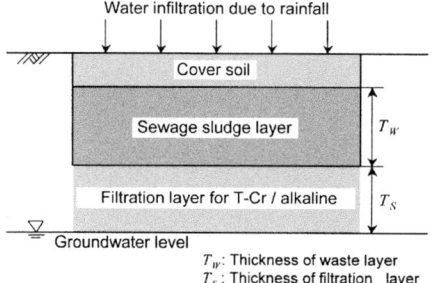

Figure 2.   Cross section for the analysis of T-Cr / alkaline migration in SS ash utilization

Table 2.   Evaluating method for the environmental impact of the SS ash utilization

T-Cr migration	Alkaline migration
$M_C = C_0 \times Q_I - A_C \times \rho_d \times T_S$	$t = (A_A \times \rho_d \times T_S / Q_L) \times \{10^{(pH-14.17)} - 10^{(10-14.17)}\}$

Assumptions:
1) The concentration of leached T-Cr, C_0 mg/L, was simplified to be consistent, i.e., C_0 was the average leaching concentration calculated from the total leached amount of T-Cr and the elapsed time to finish leaching obtained in the column leaching test results shown in Figure 1.
2) Only the vertical seepage flow was considered in the calculations and the soil below the SS ash layer absorbed the alkaline and the T-Cr.
3) The seepage rate into the ground per unit area, Q_L L/m²/year, was assumed to be 1/10, 1/3 or 1/2 of the average annual rainfall in Japan ($L = 1760$ mm/year), parametrically.
4) Leachate from the SS ash exhibited a high pH value, which was assumed to be $pH = 12.0$, and was constantly based on the experimental results.
5) The T-Cr absorption capacity of the soil, A_c mg/kg-soil, was 4.5 mg/kg-soil, which was obtained from the laboratory experiments.
6) Although the leaching of Pb from the SS ash was observed in the column tests, the environmental impact due to the Pb leaching was assumed to be disregarded, because the experimental results shows that all the Pb leached from the SS ash was absorbed by the soil.
Meaning of each symbol: Mc: the cumulative mass of T-Cr to reach the groundwater table per unit area, C_0: the concentration of leached T-Cr, Q_L: the seepage ratio into the ground per unit area, A_C: the T-Cr absorption capacity of the soil, ρ_d: the dry density of the soil (assumed to be 1670 kg/m³), T_S: the thickness of the filtration soil layer, t: the time for leachate of high alkaline over pH=10 to reach the groundwater table, A_A: the alkaline neutralization capacity of the soil (assumed to be 6×10^{-2} mol/kg-soil from the experimental results or 3×10^{-2} mol/kg-soil), and pH: the pH value of the leachate from the SS ash layer.

Figure 3(a) shows the cumulative mass of T-Cr to reach the groundwater with different values for T_S and Q_L. Although the time at which the leakage of T-Cr begins is significantly affected by T_S, even in the severest condition when $T_S = 0.5$ m and $Q_L = L/2$, the leakage of T-Cr would not occur for approximately 45 years, due to absorption by the soil. The leakage of T-Cr also significantly depends on Q_L. T-Cr would be released in 45 years when $Q_L = L/2$, while the release of T-Cr would not occur for about 230 years when $Q_L = L/10$. This indicates that the construction of a cover layer to reduce the seepage flow is very effective for lowering the environmental impact. Figure 3(b) shows the changes in t, affected by T_S, A_A, and Q_L. Calculated t largely depends on T_S proportionally. When $T_S = 4.0$ m, $A_A = 6 \times 10^{-2}$ mol/kg-soil and $Q_L = L/3$, the leakage of alkaline into the groundwater would not occur for 100 years. For cases in which filtration layer consists of various types of soil, the required thickness of the filtration layer, in the design of SS ash utilization, could be approximately estimated according to the equation shown in Table 2 and based on the alkaline neutralization capacities of various soils and the design life.

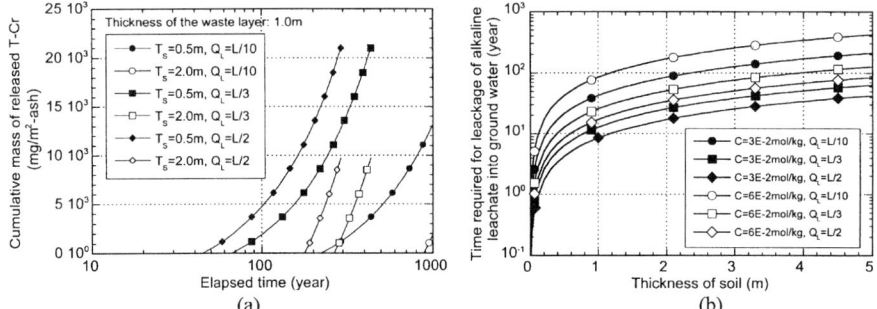

Figure 3.   Estimating the environmental impact of the utilization of sewage sludge ash: (a) cumulative mass of T-Cr and (b) time elapsed for the leakage of the high alkaline leachate

Environmental Impact due to the Disposal of Municipal Solid Waste Ash in Landfill
The environmental impact of the municipal solid waste ash being disposed in landfill sites was also evaluated using two case studies and the experimental results. The simplified cross section for these two analyses is shown in Figure 4(a), which is according to the results of a site investigation for a landfill site in Japan regarding the profile of the soil layers, the groundwater conditions, and change in the leachate water table at the landfill site.

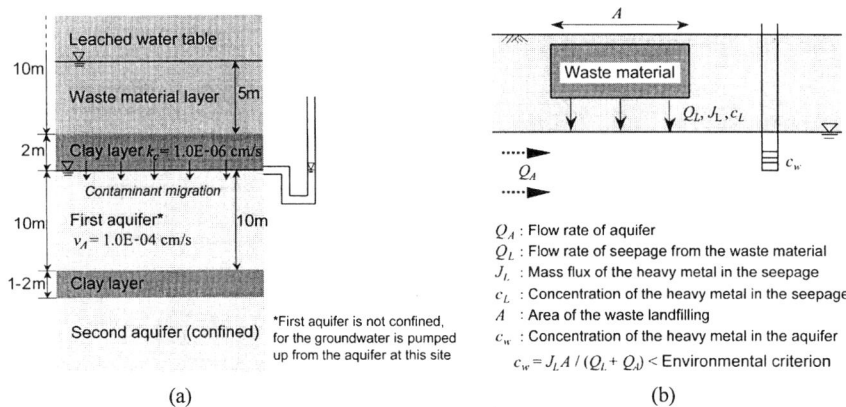

Figure 4.   Evaluation of the environmental impact of municipal solid waste ash disposal: (a) simplified cross section and (b) concept for mass basis environmental impact assessment

Evaluation Based on the Advection-Diffusion Equation

The concentration of leached contaminants from a municipal waste landfill is calculated based on the advection-diffusion equation. The contaminant migration from the waste landfill site is mainly due to the leakage of leachate through the bottom layer, as shown in Figure 4(a). Since clay layers generally have low hydraulic conductivity, it is assumed that the advective and the diffusive transport of contaminants will occur in the clay layer. The one-dimensional advection-diffusion equation, including the absorption effect, can be expressed as shown in Table 3. The analytical solution to this equation and the initial and boundary conditions applied here are also shown in Table 3. In this analysis, changes in the c_0, (i.e., the concentration of contaminants in the leachate from the landfill) versus time is disregarded, however, a more precise evaluation could be conducted by obtaining information on the changes in c_0 according to the elapsed time.

Table 3.   Estimation based on the one-dimensional advection-diffusion equation

Basic equation	Analytical solution	I.C. and B.C.
$\left(1+\dfrac{\rho_d K_p}{n}\right)\dfrac{\partial c}{\partial t} = D\dfrac{\partial^2 c}{\partial x^2} - v\dfrac{\partial c}{\partial x}$	$\dfrac{c(x=L,t)}{c_0} = 0.5\left\{ erfc\left[\dfrac{1-T_R}{2\sqrt{T_R/P_L}}\right] + \exp(P_L)erfc\left[\dfrac{1+T_R}{2\sqrt{T_R/P_L}}\right]\right\}$ $T_R = \dfrac{vt}{RL}$ $\quad$ $P_L = \dfrac{vL}{D}$	$c(0, t) = c_0$ $c(x,0) = 0$ (for $x > 0$) $\partial c\,(\infty, t) / \partial x = 0$

Assumptions for the analytical solution: 1) The clay layer above the groundwater table is generally unsaturated. However, it is assumed that seepage into the clay is a steady state and no suction exists at the bottom layer. This assumption does not result in the underestimation of the leakage rate. 2) The soil properties (e.g., ρ_d, n, K_p, and D) are assumed to be homogeneous and time invariant, and no chemical reactions occur.
Meaning of each symbol: ρ_d: the dry density of the clay, n: the porosity of the clay, K_p: the partition coefficient, $(1+\rho_d K_p / n)$: retardation factor R, c: the contaminant concentration, D: the dispersion coefficient including both the mechanical dispersion and the diffusion, v: the seepage velocity, c_0: the initial concentration of the contaminant, L: the thickness of the clay layer, t: the elapsed time, T_R: the dimensionless time factor, and P_L: the Peclet number (the ratio of advective transport to diffusive transport)

Evaluation Based on the Mass Basis Method

The mass basis method for evaluating the leaching from waste materials has been proposed as an alternative to concentration [4]. The basic concept of the mass basis method is shown in Figure 4(b). In this analysis, the concentration of leached contaminants in the first aquifer, c_A, was evaluated based on the mass flux of contaminants, J_L. Table 4 shows the method for evaluating c_A and J_L. The mass of released contaminants is assumed to be dominated by the volume of seepage represented by the L/S ratio. The cross section and the parameters are shown in Figure 4(a); they are the same as those used in for the previous analysis based on the advection-diffusion equation.

Table 4.   Evaluation based on the mass basis method

(Equation 1) $\quad J_L = -\dfrac{dM_C(t)}{dt}$	(Equation 2) $\quad c_A = J_L \cdot A \cdot \Delta t/(Q_L \cdot A + Q_A)$

$M_C(t)$ was determined from the results of the column leaching tests as follows:

$$LS_{in\,field} = Q_L t/\rho H \qquad M_C(t) = m_C(LS_{in\,labo}) \cdot \rho H \qquad \text{where } LS_{infield} = LS_{inlabo}$$

Meaning of each symbol: $M_C(t)$: the cumulative mass of the heavy metal leached from the unit landfill area at time, t, A: the area of the landfill site, c_A: the contaminant concentration in the downstream aquifer, Q_L: the leakage rate of the leachate from the unit landfill area during interval Δt, Q_A: the rate of horizontal flow in the aquifer during Δt, LS: the liquid to solid ratio, ρ: the dry density of waste, H: the height of the waste layer, $m_C(LS_{in\,labo})$: the cumulative mass of the released contaminants from the unit weight waste at LS_{inlabo} in the column leaching tests

Results and Discussions

The results for the contaminant transport through the clay layer, based on the advection-diffusion equation, are shown in Figure 5(a) for different R and k_c (k_c = 1E-6 cm/s or 2E-7 cm/s). k_c = 1E-6 cm/s is determined by hydraulic conductivity tests on the in situ clay layer, and k_c = 2E-7 cm/s represents the installation of the compacted clay liner. D is assumed to be 2E-6 cm^2/s, derived from the common value for the clay material. Figure 5(a) shows that R affects the transport of the contaminants when k_c is low (2E-7 cm/s), while R has less of an effect when k_c is 1E-6 cm/s. This indicates that a low hydraulic conductivity clay liner (k_c = 2E-7 cm/s) could reduce the release of contaminants. However, c / c_0 reaches 1 with elapsed time regardless of k_c and R. Therefore, the stabilization of wastes and the control of the leachate water table in landfills are also effective in reducing the leaching potential of contaminants.

Figure 5(b) shows the estimated concentration of Pb in the downstream aquifer, c_A, from two different types of MSW ash, influenced by k_c and landfill area, A. The estimated c_A directly reflects the area of the landfill site because of calculating by the mass flux, J_L. The results for the c_A of Pb from the MSW-2 ash indicate little or no Pb release, except in the early years, because the stabilization of the MSW-2 ash has a positive effect on the immobilization of Pb, as shown in Figure 1(d). On the other hand, the c_A from MSW-1 ash, which exhibits less of a Pb stabilization effect, as shown in Figure 1(c), is significantly high. It is indicated that the stabilization in order to immobilize the heavy metal has a great effect on the reduction of the environmental impact. Figure 5(b) also shows that c_A is affected by k_c. A low hydraulic conductivity liner could drastically restrict the mass flux of contaminants, J_L. In addition, the dilution of contaminants with groundwater largely contributes to a lower c_A. This also proves that a clay liner could effectively control the environmental impact from landfill sites.

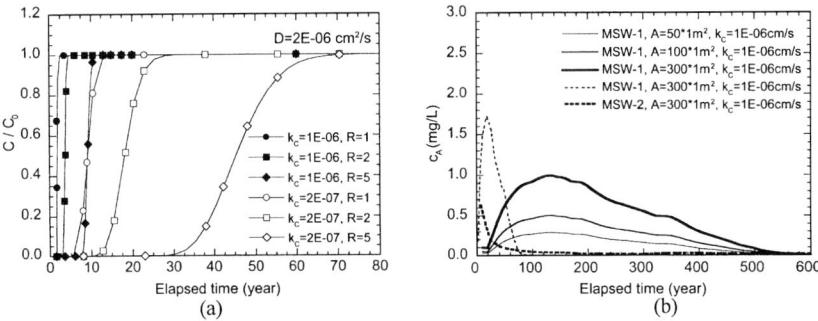

Figure 5.   Results based on (a) the advection-diffusion equation and (b) the mass basis method

CONCLUSION

Simplified methods for evaluating the geo-environmental impact of leachate from reused and disposed waste materials were introduced based on laboratory leaching tests. A discussion on the effectiveness of various measures to reduce the environmental impact (e.g., filtration layers etc.) was also presented using the proposed methods. To promote the efficient reuse or disposal of waste materials, proper design schemes based on leaching tests must be established, in addition to the development of leaching protocols which take into account complex leaching factors.

REFERENCES

[1] van der Sloot, H.A. (1996): Developments in evaluating environmental impact from utilization of bulk inert wastes using laboratory leaching tests and field verification, *Waste Management*, Vol. 16, pp. 65-81.

[2] Kosson, D.S. et al. (1996): An approach for estimation of contaminant release during utilization and disposal of municipal waste combustion, *Journal of Hazardous Materials*, Vol. 47, pp. 43-75.

[3] Kamon, M. et al. (1999): Evaluation of geo-environmental impact due to the leachate from solid wastes, *Annuals of the Disaster Prevention Research Institute, Kyoto Univ.*, No. 42 B-2, pp. 459-469 (in Japanese).

[4] Shackelford, C.D. and Glade, M.J. (1997): Analytical mass leaching model for contaminated soil and soil stabilized waste, *Ground Water*, Vol. 35, No. 2, pp. 233-242.

Applicability of Ferro-Nickel Slag as a Sand Compaction Pile Method Material

M. KITAZUME, Port and Airport Research Institute, Yokosuka, Japan, S. MIYAJIMA, Port and Airport Research Institute, Yokosuka, Japan, and T. YASUDA, Nippon Yakin Kogyo Co., Ltd., Miyazu, Japan

INTRODUCTION

Approximately 2 million tons of ferro-nickel (FN) slag is produced annually in Japan. It is a by-product obtained from the refining process of FN granule from nickel ore to produce stainless steel. Since the FN slag has high particle density, high permeability and a similar particle size distribution to ordinary sand, it is expected to be valuable as a construction material. FN slag has been used in concrete, asphalt and as caisson. However large amounts of FN slag is still dumped in disposal areas. Recently imposed limits to disposal areas means that research effort is required to find new applications for the slag.

The sand compaction pile (SCP) method, in which compacted sand piles are constructed in soft ground, has been widely applied to improve clayey ground in Japan for improvement of bearing capacity of foundation and lateral resistance of sheet wall piles. The current design codes for port and harbor structures in Japan requires that the sand material used for the SCP method should have (1) large shear strength, (2) high permeability, (3) low amount of small particle sand, and (4) easily densified characteristics (Ministry of Transport, 1999). Recently applied limits of the amounts of sand suitable for the method means that investigations to find new materials for the method are required.

In this study, the applicability of the FN slag to SCP material is investigated by centrifuge model tests on lateral pile resistance and on bearing capacity.

PROPERTIES OF FERRO-NICKEL SLAG

The Ferro-nickel (FN) slag used in this study was produced in Miyazu, Kyoto prefecture, whose major chemical composites are SiO_2 of 53.0 %, MgO of 28.8 %, FeO of 9.1 % and CaO of 4.8 %. Table 1 shows the dissolution test results of the slag to show negligible environmental impact to the surrounding area. In the Table, the allowable dissolution values of chemical composites regulated by the Environmental Agency of Japan are listed together. The table shows all the measured dissolution values are within the allowable values.

The grain size distribution tests found that the distribution of the FN slag falls in the range of the suitable sand for SCP method. The Internal friction angle, ϕ_d, of the slag is about 35 degree, which is enough value for SCP method.

Geoenvironmental impact management, Thomas Telford, London, 2001.

Table 1. Dissolution test results on FN slag

Chemical composite		Test result	Allowable value
cadmium compound	*Cd*	< 0.01 mg/ l	< 0.01 mg/ l
cyanide pollution	*CN*	not detective	not detective
organic phosphorus	*P*	not detective	not detective
lead	*Pb*	< 0.001 mg/ l	< 0.01 mg/ l
chromium +6	$Cr\,(VI)$	< 0.05 mg/	< 0.05 mg/
arsenic	*As*	< 0.001 mg/ l	< 0.01 mg/ l
mercury	*Hg*	< 0.0005 mg/ l	< 0.0005 mg/ l
alkyl-mercury	*R-Hg-X*	not detective	not detective
polychlorinated biphenyl	$C_{12}H_{10\text{-}X}Cl_X$	not detective	not detective
copper	*Cu*	< 0.01 mg/ l	< 125 mg/kg
dichloromethane	CH_2Cl_2	< 0.02 mg/ l	< 0.02 mg/ l
carbon tetrachloride	CCl_4	< 0.002 mg/ l	< 0.002 mg/ l
1,2-dichloroethane	$CH_2Cl\,CH_2Cl$	< 0.004 mg/ l	< 0.004 mg/ l
1,1-dichloroethylene	$CH_2\,CCl$	< 0.02 mg/ l	< 0.02 mg/ l
cis-1,2-dichloroethylene	*CHCl CHCl*	< 0.04 mg/ l	< 0.04 mg/ l
1,1,1-trichloroethane	CH_3CCl_3	< 0.1 mg/ l	< 1 mg/ l
1,1,2-trichloroethane	$CH_2ClCHCl_2$	< 0.006 mg/ l	< 0.006 mg/ l
trichloroethylene	CCl_2CHCl	< 0.03 mg/ l	< 0.03 mg/ l
tetrachloroethylene	CCl_2CCl_2	< 0.01 mg/ l	< 0.01 mg/ l
1,3-dichloropropene	$CH_2ClCHCHCl$	< 0.002 mg/ l	< 0.002 mg/ l
thiuram	$C_6H_{12}N_2S_4$	< 0.006 mg/ l	< 0.006 mg/ l
simazine	$C_7H_{12}ClN_5$	< 0.003 mg/ l	< 0.003 mg/ l
thiobencarb	$C_{12}H_{16}ClNOS$	< 0.02 mg/ l	< 0.02 mg/ l
benzene	C_6H_6	< 0.01 mg/ l	< 0.01 mg/ l
selenium	*Se*	< 0.01 mg/ l	< 0.01 mg/ l

CENTRIFUGE MODEL TESTS

Lateral resistance of a pile

The model ground on lateral resistance of a pile is schematically shown in Fig. 1, which is composed of a single pile, normally consolidated clay ground (Kaolin clay) and sand compaction piles (SCP). The model pile with 10 mm in diameter is subjected to horizontal load in a 50 g field by means of the electric motor jack on the specimen box. The centrifuge used in this study was the PHRI (Port and Harbour Research Institute) Mark II geotechnical centrifuge (Kitazume and Miyajima, 1996). As summarized in Table 2, the width and improvement area ratio of the SCP improved area is changed to investigate their effect on the lateral resistance of the pile. Details of the model ground preparation and loading procedure is described by Kitazume et al. (2000).

The horizontal load and displacement curves of the model pile at the ground surface are shown in Fig. 2, in which the displacement is calculated by the measured horizontal displacement at the loading point and the measured moment distribution of the pile. In the figure, the load and the displacement are converted to a prototype scale. The load - displacement curve of the unimproved clay ground is plotted together. The horizontal loads increase

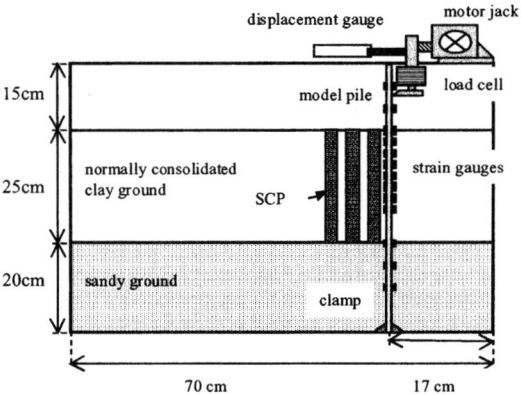

Figure 1. Model ground on lateral resistance of a pile

Table 2. Model test conditions

Test case	sand compaction pile improved ground		
	width and depth	improvement area ratio	material
case 1	10 cm x 25 cm	51 %	FN slag
case 2	10 cm x 25 cm	26 %	FN slag
case 3	10 cm x 25 cm	26 %	Toyoura sand
case 4	-	0 % (unimproved clay ground)	
case 5		100 % (slag ground)	FN slag
case 6		100 % (sand ground)	Toyoura sand

gradually with increasing the horizontal displacement irrespective of the improvement condition. It is found that the horizontal loads of the FN slag ground and sand ground (cases 5 and 6) show the largest load increment among the test cases, and the SCP improved ground is slightly larger than those of the unimproved clay ground. And also found that the horizontal load of the SCP improved ground with the FN slag is almost same order of those of the SCP ground with Toyoura sand.

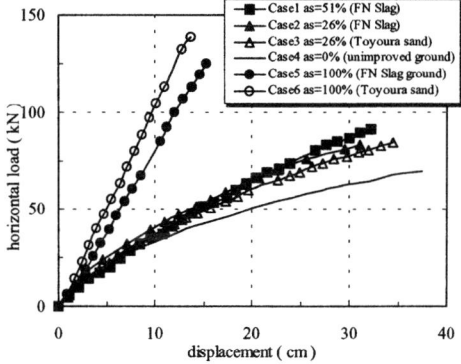

Figure 2. Horizontal load and displacement curves

Figure 3 shows the measured bending moment distributions of the model pile at the horizontal load of about 60 kN in a prototype scale. In the unimproved clay ground (case 4), the moment developing along the pile above the ground surface increases linearly with the decrease of elevation. After reaching a maximum value at about 5 m below the ground surface, the moment decreases with the depth. In the SCP improved grounds (cases 1 to 3), three moment distributions almost coincide each other and show smaller magnitude than that of the unimproved ground. And the moments show a peak value at slightly smaller depth than that of the unimproved ground. The moment distributions in the cases 5 and 6 show a maximum value at about the ground surface. It can be confirmed the magnitude of the moment along the pile decrease with the SCP improvement.

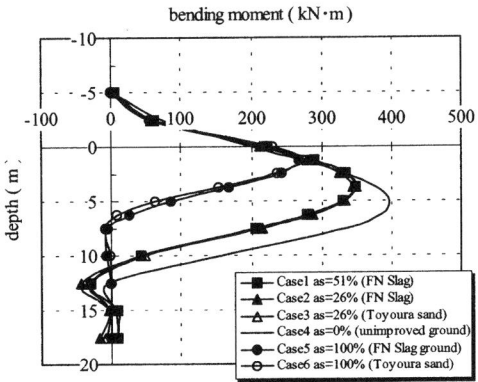

Figure 3. Moment distribution along the pile

The measured moment distribution curves are fitted by polynomial functions and differentiated twice to obtain soil resistance, p and integrated twice to obtain deflection of pile, y at an arbitrary load level. The soil resistance, p can be derived by a following equation;

$$p = kc \cdot y^{0.5}$$

where kc is the coefficient of subgrade reaction (Takahashi and Kasugai, 1987). The method of calculating load and deflection of laterally loaded pile based on this relationship (PHRI method) has been adopted for long by the Japanese Technical Standards for Port and Harbor facilities (Ministry of Transport, 1999) and the validity of the method has been confirmed at a number of practices.

The calculated coefficient of subgrade reaction, kc in all the test cases is plotted against the depth in Fig. 4. It is confirmed that the subgrade reaction for the unimproved clay ground increases almost linearly with the depth, because the undrained shear strength increases linearly. The coefficient of subgrade reaction, kc of the SCP improved ground also increases with the depth and shows larger value than the unimproved clay ground. It is also found that the magnitude of the coefficient of SCP improved ground with the FN slag is almost same as that with Toyoura sand. It can be concluded that the SCP improvement functions to increase the soil resistance and the SCP improvement with the FN slag has the similar effect on the soil resistance to the SCP improvement with sand.

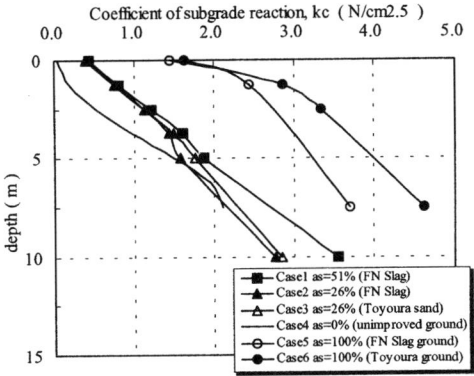

Figure 4. Subgrade reaction distribution along the depth

Bearing capacity test

Another test series was performed to investigate the effect of the SCP improvement with the FN slag on the vertical bearing capacity. The model ground is shown in Fig. 5, in which normally consolidated clay ground partially was improved by SCP. In the test, the model footing is subjected to vertical load to cause ground failure in the 50 g centrifugal acceleration field. Model ground was prepared by the same manner of previous research (Terashi et al. 1990).

Figure 5. Model ground for bearing capacity

The vertical load and the settlement of the footing are shown in Fig. 6. The vertical load in the unimproved ground increases gradually with the footing settlement, and shows no clear peak value. The vertical load of the SCP improved ground also increases with the footing settlement, but its increasing ratio is very high compared with the unimproved ground. The vertical load shows clear bending at the settlement of about 15 cm, and still increases gradually with further settlement. It is also found that the bearing capacity of the SCP improved ground shows large value than that of the unimproved ground. The effect of the SCP improvement with the FN slag is almost same as that with Toyoura sand. The SCP improvement effect is more dominate in the vertical bearing capacity that the horizontal resistance of a pile.

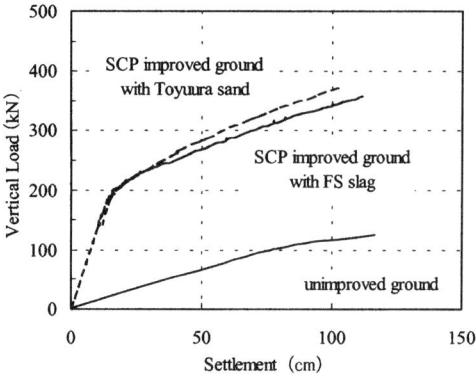

Figure 6. Vertical load - settlement curves

CONCLUDING REMARKS

Two series of centrifuge model tests were conducted on the lateral resistance of a pile and vertical bearing capacity. The model tests show the improvement effect of the SCP ground with the FN slag and Toyoura sand. These tests also demonstrate the improvement effect of the FN slag is almost same as those of Toyoura sand, and the applicability of the FN slag to SCP improvement. The authors are planning to conduct field execution test on the workability and execution ability of the FN slag SCP method to confirm its applicability more.

REFERENCE

Kitazume M. and Miyazima S. (1996) Development of PHRI Mark Ⅱ Geotechnical Centrifuge. *Technical Note of the Port and Harbour Research Institute*, No. 817, 33p.

Kitazume M., Miyajima S. and Yasuda T. (2000) Effect of lateral resistance of a single pile in soft ground improved by SCP. *Proc. of the Symposium on Soil Improvement Techniques*, The Society of Materials Science of Japan (in Japanese).

Ministry of Transport (1999). Design Codes of Port and Harbor Structures, Vol. 1, 460p. (in Japanese).

Takahashi K. and Kasugai Y. (1987) Influence of Pile Width on Lateral Reaction of Sandy Subgrade, *Report of the Port and Harbour Research Institute*, Vol. 26, No. 2, pp.437-462 (in Japanese).

Terashi, M., Kitazume, M. and Minagawa, S. (1990) Bearing Capacity of Improved Ground by Sand Compaction Piles, *Deep Foundation Improvements: Design, Construction, and Testing*, ASTM STP 1089, pp.47-61.

The use of PFA as a Fill Material and the Environment

L. K. A. Sear, PhD, BSc, FICT, Technical Officer, United Kingdom Quality Ash Association, Wolverhampton, UK and R. Coombs, Head of Ash Laboratory, Innogy Plc (National Ash), Drax, UK

INTRODUCTION

PFA (also known as fly ash in many countries) is a by-product from the combustion of pulverised hard coal in electricity power generation. This paper identifies the main potential environmental issues associated with the use of PFA as a fill material. A more detailed analysis of these issues has been produced by the United Kingdom Quality Ash Association. The conclusions are that the issues are generally minor and can be dealt with mainly by good design and site practice.

HANDLING OF PFA FROM THE POWER STATION TO THE SITE

PFA is a fine powder, similar in fineness to cement, which consists of oxides of silica, alumina, iron, calcium and various minor constituents. PFA for fill applications is always supplied moistened with water. This may be from a conditioning plant where water and dry ash are mixed, from stockpiles of previously conditioned material or material recovered from lagoons. Lagoon PFA is deposited as slurry. Once the lagoon is full, the PFA is allowed to drain such that a suitable moisture content to permit compaction is achieved.

The power station must ensure that dust blow does not occur during the recovery and transportation process. This may occur from exposed faces on stockpiles or in lagoons, when loading from conditioning plant or from movements of delivery vehicles. Vehicles should be sheeted and any material adhering to wheels and bodies of the lorries should be removed. Any site roads should be kept clean to prevent the build up of PFA that could become airborne due to lorry movements.

UTLISING THE PFA ON SITE

Design Factors

Account must be taken of the environmental impact of the construction in the design process. This will involve ensuring that there is an adequate drainage layer to prevent capillary rise and saturation of the PFA. Additionally the profile of the PFA should be such to allow efficient run off of rainwater both during and after the construction period. Long term protection of any side slopes is required to prevent build up of run off and subsequent leaching problems. Suitable methods of encouraging the growth of plants and trees or by the use of physical barriers must be designed into the structure.

Geoenvironmental impact management, Thomas Telford, London, 2001.

Although the above measures ensure the PFA is properly engineered they also result in it being effectively isolated from the surrounding environment, reducing the impact of its use in the long term.

Laying and Compaction

The primary considerations when placing PFA is one of minimising dust blow by ensuring that the PFA, when delivered and after compaction, is maintained sufficiently moist. In addition, restrict trafficking to prevent dust being created. Windy conditions result in the greatest risk of the PFA being dried out. Therefore, water-spraying equipment may be needed to remoisten the surface. PFA, which has been accidentally over moistened, can be allowed to dry out by breaking up the surface to encourage evaporation. After a suitable period, the PFA may be reused when the optimum moisture content is achieved.

If material is stockpiled on site, it should be deposited in such a way as to prevent accidental contamination of adjacent watercourses. Vehicles leaving the site need to be in a clean condition and provision of wheel washers or similar may be required. Again, haul roads need to be kept clean.

Summary – using PFA on site

The main environmental risks associated with the process of using PFA as a fill material are primarily ones of common sense when working with fine powders. That is:

- Prevent contamination of adjacent watercourses.

- Prevent dust problems by keeping the material moist.

- Protect the material from wash out by design, e.g. by establishing vegetation or physical barriers.

ENVIRONMENTAL PROPERTIES OF PFA

Physical Properties of PFA - Particle Size and Shape

Because of the way they are produced, PFA particles, particularly those below $50\mu m$, are spherical in shape. As the coal is burnt, the minerals associated with it become molten and form the spherical shape. Because of the rapid cooling experienced by the fine ash particles as they pass out of the furnace, they solidify as an amorphous, glassy material in this shape.

Particles in the coarse silt/fine sand sizes have the potential to become airborne in certain conditions. The particle size also means that compacted PFA has a low permeability, typically 10^{-7} m/s. This means that it is difficult for water to penetrate and because water will only flow through saturated material, this will not occur unless the PFA is placed in areas below water. Experience has shown that if PFA is subjected to heavy rain it is unusual for saturation to affect the surface beyond the top 50mm. Even when saturated there will only be a limited rate of flow through the mass of the material.

Fresh conditioned and stockpile PFA is like a fine-grained soil and it is mainly silt-sized and generally acts like silt. Finer PFA has a silky feel, although a coarser one may feel gritty, they exhibit dilatency and are non-plastic. The suction in partially saturated PFA means it can maintain a vertical face immediately after compaction. This apparent cohesion is augmented in the longer term by chemical bonding.

Chemistry of PFA

Around 60 % to 90 % of PFA is present as an amorphous glassy material composed of silica, alumina and iron oxides, with other metals present in smaller quantities, as shown in Table A. The constituents, apart from the glass, that are of most significance to the properties of PFA are the calcium oxide content (lime) and sulfate content.

If there is sufficient lime present in the PFA then it will result in hardening due to a combination of further crystal formation and reaction between the lime and the glassy material in the PFA (pozzolanic reaction). The high pH is likely to reduce the availability of the trace elements.

Element	Typical range of values for PFA
Silicon (% as SiO_2)	48 – 52
Aluminium (% as Al_2O_3)	24 – 32
Iron (% as Fe_2O_3)	7 – 15
Calcium (% as CaO)	1.8 – 5.3
Magnesium (% as MgO)	1.2 – 2.1
Sodium (% as Na_2O)	0.8 – 1.8
Potassium (% as K_2O)	2.3 – 4.5
Titanium (% as TiO_2)	0.9 – 1.1
Chloride (% as Cl)	0.01 – 0.02*
Loss on ignition (%)	3 – 20
Sulfate (% as SO_3)	0.35 – 1.7
Free calcium oxide (%)	<0.1 – 1.0
Water soluble sulfate (g/L as SO_4) - 2:1 water solid extract	1.3 – 4.0
pH	9 – 12
* Chloride may be up to 0.3 % for PFA conditioned with sea water	

Table A – Typical oxide analysis for PFA

The calcium content of PFA means that most of the sulfate is present as gypsum, which has a limited solubility and will precipitate out in compacted PFA. The sulfate level of lagoon PFA is usually very low because the water/solids ratio used to slurry the PFA means the majority of the sulfate is washed out. Other water-soluble materials are also removed in the process. The sulfate content is typically less than 0.1 g/L.

The sulfate content of PFA means that it cannot be placed within 500 mm of metallic items according to the Department of Transport Specification for Highway Works (SHW). The water-soluble sulfate content of PFA is also sufficiently high to restrict the types of reinforcement that can be used in reinforced earth structures.

Trace Elements in PFA

Typical trace elemental analyses are shown in Table B, which demonstrates that other elements are present in only small quantities, less than 1 % of the total. The values quoted are generally in agreement with other quoted values[i, ii].

Trace element	Typical range of results	Trace element	Typical range of results
Arsenic	4 to 109	Manganese	103 to 1,555
Boron	5 to 310	Molybdenum	3 to 81
Barium	0 to 36,000	Nickel	108 to 583
Cadmium	<1.0* to 4	Phosphorus	372 to 2,818
Chloride	0 to 2,990	Lead	<1* to 976
Cobalt	2 to 115	Antimony	1 to 325
Chromium	97 to 192	Selenium	4 to 162
Copper	119 to 474	Tin	933 to 1,847
Fluoride	0 to 200	Vanadium	292 to 1,339
Mercury	<0.01* to 0.61	Zinc	148 to 918
All expressed as mg/kg * Indicates below the limit of detection			

Table B - Solid phase trace element analysis - Typical ranges from UK sources of PFA.

Leachable Elements in PFA

As discussed above, there is only a small fraction of the constituents that are present on the surface of PFA and that are leachable in water. Typical data obtained from routine analysis are shown in Table C, the extraction is to the German standard DIN 38414-S4[iii] (10:1 water/solids ratio) in this instance.

Typical range of leachable elements for UK PFA (mg/L except pH)			
Aluminium	<0.1* to 9.8	Manganese	<0.1*
Arsenic	<0.1*	Molybdenum	<0.1* to 0.6
Boron	<0.1* to 6	Sodium	12 to 33
Barium	0.2 to 0.4	Nickel	<0.1*
Calcium	15 to 216	Phosphorus	<0.1* to 0.4
Cadmium	<0.1*	Lead	<0.2*
Chloride	1.6 to 17.5	Sulfur	24 to 510
Cobalt	<0.1*	Antimony	<0.01*
Chromium	<0.1*	Selenium	<0.01* to 0.15
Copper	<0.1*	Silicon	0.5 to 1.5
Fluoride	0.2 to 2.3	Tin	<0.1*
Iron	<0.1*	Titanium	<0.1*
Mercury	<0.01*	Vanadium	<0.1* to 0.5
Potassium	1 to 19	Zinc	<0.1*
Magnesium	<0.1* to 3.9	pH	7 to 11.7
Notes: The above data include a seawater-conditioned sample resulting in higher chloride values. The Boron content may also be increased. * Indicates below detection limit.			

Table C – Leachates found using the DIN 38414-S4 method.

From the data it can be seen that the major water-soluble constituents are calcium and sulfur (usually present as sulfate). There are smaller amounts of sodium and potassium, and traces of chloride, magnesium, aluminium and silicon. If it is assumed that all the water soluble calcium, sodium and potassium is present as hydroxide (ignoring the sulfate or chloride) then the total water soluble hydroxide, based on the highest values from Table C, would be 2.1% (m/m). However, calcium hydroxide would make up approximately 2.0%, the others would represent less than 0.1%. In all instances quoted the calcium is very dominant with sodium and potassium present in very small quantities in comparison.

There are only very small amounts of other elements available to leach, the most significant being boron because of its potential for restricting plant growth. However, these can be overcome by the selection of boron tolerant species. Furthermore, there is little significant difference in leachate quality between all UK PFA's, and these are well established.

The leachable fraction of lagoon PFA is lower than for dry or conditioned PFA because a significant amount is removed when the material is sluiced to the lagoon.

Polycyclic Aromatic Hydrocarbons in PFA

Polycyclic aromatic hydrocarbons (PAH's) can result from the incomplete combustion of fuels such as wood, coal and oil. They are widespread in the environment. Metabolic transformations, by aquatic and terrestrial organisms, result in carcinogenic substances[iv]. The most potent PAH's are benzofluoranthenes, benzo[a]pyrene, benz[a]anthracene, dibenzo[a,h]anthracene and indenol[1,2,3-cd]pyrene. Although there has been a significant amount of work on PAH's arising from combustion of coal, most effort has been focussed on airborne particulate matter. PAH's will undergo photo-degradation and are therefore thought to have a limited life span in the atmosphere.

Leaching tests on PFA in accordance with the Environment Agency extraction method[v] have indicated levels of the PAH's benzo[b]fluoranthene, benzo[k]flouranthene, benzo[a]pyrene, benzo[ghi]perylene, fluoranthene and indeno[1,2,3-cd]pyrene to be less than 0.2 µg/L for each species, confirming the above findings that the amount of available PAH on PFA is negligible.

Junk et al[vi] studied levels of PAH's and other organics from stack vapour, stack ash, fly ash and grate ash from Ames power station in the US. Only small amounts were found on both respirable and non-respirable particles. Although there were measurable amounts in the vapour phase, it was noted that if all the vapour were to condense on the particulate matter the amount would still be less than for ambient air particles.

In addition to the above, measurements were made on sluice water carrying fly ash and grate ash to settling ponds. The water used was from an aquifer that had been contaminated by coal tar. The sluice water contained no contaminants above the detection limit of 1ppb, less than was found in the aquifer water. This indicated that the fly ash actually reduced the level of PAH's in the water. This was confirmed by a small trial where water containing 20 to 50ppb of PAH's was mixed with fly ash in a ratio of 10:1. Within 10 minutes the PAH level was reduced to below the detection limit.

Dioxins in PFA

Dioxins are considered to be toxic to humans, as are furans although less so than dioxins. 2,3,7,8-tetra CDD (TCDD) is considered to be the most toxic and therefore the most studied. Dioxins are usually associated with the incomplete combustion of material containing chlorine and as such are commonly associated with the ash from municipal waste incineration, but can be found in small traces in soils. The low chlorine content of coal combined with the high temperatures found in the furnaces of power stations mean that dioxins are unlikely to form and only traces would be expected in the resulting ash. Dioxins are ubiquitous and are present in a wide range of soils and although they can be persistent, they rapidly decay when exposed to light.

Work by the CEGB[vii] in the 1980's examined 18 PFA samples from a range of sources for dioxins from the tetrachlorinated to the octachlorinated. The findings were that the levels were very low, typically less than 25pg/g, with levels of 2,3,7,8-TCDD less than 2pg/g in all but two samples. The only exceptions were samples of PFA from the low NO_x burners at one station (A). It was thought that the low NO_x burners might have had some effect, although the same increase was not observed for samples from another power station utilising similar burners. Although the dioxin levels in the samples from low NO_x burners at station (A) were higher, 210 and 270pg/g, they were still within the range found in soils in the UK; data from unpublished work cited an upper limit in soils of 290pg/g.

Summary – Environmental Properties of PFA
It has been shown that:

- A common sense approach to the use of PFA as a fill material will prevent both atmospheric dust problems and accidental contamination of watercourses.

- There are minimal leachates, including PAH's and dioxins, from PFA.

- The main leachate is calcium sulfate.

- When plants are to be grown directly in the PFA, boron tolerant species should be used with some sources of PFA.

REFERENCES:

[i] Brown J, Ray N J, Ball M, The disposal of pulverised fuel ash in water supply catchment areas, Water Research, Vol. 10. pp 1115 to 1121, Pergammon Press 1976.

[ii] Hoeksema H W, Working conditions for fly ash workers and radiological consequences of living in a fly ash house, Proceedings of ASHTech 84, Second international conference on ash technology and marketing, London, 1984.

[iii] DIN 38414 Part 4, German standard methods for the examination of water, waste water and sludge. Sludge and sediments (group S). Determination of leachability by water.

[iv] Wild S R and Jones J C, Polynuclear aromatic hydrocarbons in the United Kingdom environment: a preliminary source inventory and budget, Environmental Pollution 88 (1995), pp 91-108.

[v] National River Authority, Protocol for a leaching test to assess the leaching potential for soils from contaminated sites, NRA R & D Note 301

[vi] Junk G A, Richard J J and Avery M J, Organic compounds in effluents related to coal combustion, pre-prints of papers – American Chemical Society, Division of Fuel Chemistry, V 30 N 2, pp 171-178, 1985, Published by ACS.

[vii] Freedman A N, The analysis of power station fly ash for the presence of polychlorinated Dibenzo-p-dioxins, CEGB, 1988.

Estimated Recycled Aggregate Quantities in Scotland

M G WINTER[1] and C HENDERSON[2]
[1]TRL Limited, Edinburgh, UK. [2]Formerly ERM Limited, Edinburgh, UK.

ABSTRACT
A survey methodology for the collection of time series data relating to aggregate waste arisings, recycling and landfill in Scotland is described and the quantities estimated from a pilot survey for 1999 are reported. Data is reported for construction and demolition waste, industrial waste and by-products, and bituminous planings and breakout: additional categories are reported within these as appropriate. Of an estimated 10,200kt of arisings, some 5,200kt were landfilled and 5,500kt recycled (the additional 500kt are accounted for by the consumption of existing stocks of industrial wastes and by-products). Data for recycled quantities is reported in terms of the utility of the application to which the aggregates are put. Targets for the recycling of aggregates in the future are also given.

INTRODUCTION
The Government is committed to sustainable development. The Scottish Executive (1994) signalled the role and contribution of recycled aggregates, thus reducing the demand for the production of primary aggregates. By increasing the use of recycled aggregates the need for new quarries should be reduced, with consequential benefits in terms of retaining the countryside in an unspoiled condition. New quarry proposals often generate strong, local opposition, which could be avoided.

Additionally, construction and demolition wastes account for some 50% of controlled waste in Scotland and consequently take significant landfill space. On-site techniques to reuse such wastes can minimise the import of materials. An increase in the recycling and reuse of such wastes could in due course reduce the requirement for landfill. Against this background the Scottish Executive Development Department (SEDD) commissioned research to provide information on the scale and purpose to which recycled aggregates are currently put.

The project comprised two main objectives, namely:
1. The collection of data on recycled aggregates:
   a) Collection and analysis of desk study estimate data.
   b) Development and piloting of a survey system for the collection of time series data.
2. Development of an informed picture of recycling in Scotland:
   a) Current market for aggregates.
   b) Current recycling practice.
   c) Opportunities for and obstacles to the future recycling of aggregates.
   d) The selection of suitable targets for recycled aggregates in Scotland.

This paper reports on the developed survey system and the results of the pilot survey. The data are set within the context of estimates based on a desk study. Targets for future

Geoenvironmental impact management, Thomas Telford, London, 2001.

aggregate recycling are given, as reported by Winter and Henderson (2001), who also report on Objective 2 in full. While the broad categorisation of wastes and the approach to data collection generally follow the proposals presented by Arup (1998), it is important to note that their work took little or no account of the type of application to which recycled aggregates are put. The current work was specifically tailored towards identifying the type of application to which recycled aggregates are put. To this end the economic and environmental utility of the application to which aggregates are recycled is also reported (see also Winter, In Press). For example, bituminous planings and breakout utilities were defined as follows:

Low: Haul road and general fill construction.
Intermediate: Foundation, capping layer and sub-base construction.
High: Hot and cold recycled bituminous road construction.

The utility was surveyed for bituminous planings and breakout, industrial wastes and by-products and estimated for construction and demolition wastes.

SURVEY SYSTEM
Sources of data for aggregate arisings, recycling and landfill were as follows:
- Construction and demolition wastes (from licensed waste disposal sites, waste disposal sites registered exempt, mobile crushing plant).
- Bituminous planings and breakout (from local authorities and the Scottish Executive's agents for trunk road and motorway maintenance).
- Industrial wastes and by-products (from producers and owners of material supplies).

Mobile crushing plant have not been included in the pilot survey process. Instead the number of mobile plant actively engaged in the recycling of aggregates has been identified and the amount of recycling (and the utility of application) estimated from discussions with industry.

Questionnaires were developed for each of the above categories to obtain 1999 data.

PILOT SURVEY METHOD
In the case of bituminous planings and breakout, and industrial wastes and by-products all producers were contacted (i.e., a 100% sample). The sample sizes for Licensed Waste Management sites and Registered Exempt sites were determined for the purposes of the pilot survey, allowing for a reasonable level of null responses.

Licensed and Exempt Sites
Construction and demolition waste data were obtained by surveying selected licensed waste disposal (landfill) sites and sites registered exempt under paragraphs 9, 13 and 19 of the Waste Management Licensing Regulations (Anon, 1994). The former were sites known to accept construction and demolition waste (ERM, 2001) and the latter from SEPA databases (SEPA North (Aberdeen) does not include the registration date in their database and these sites were omitted from the sample). Data were requested from landfill sites for waste handled in 1999.

The questionnaires requested data on types and quantities of material suitable for recycling as aggregate. Data on aggregates incoming to the site, treated at the site (crushed and graded), landfilled by the site (either on or off-site), reused/recycled on the site, reused/recycled off the site, disposed on the site, and other materials (e.g., timber and metals) were requested.

A different approach was required to gather data from exempt sites. A site may be registered exempt for the short duration of a construction project. Thus, some sites were registered exempt by SEPA in 1999 (and handled waste in 1999) but had ceased operating by the time of the survey - March 2000. No attempt was made to correct the data for this effect. However, based upon the number of questionnaires returned with unknown addresses and the likely scale of such operations, it is considered that such sites represent relatively small amounts of material. Similarly, in order to simplify the pilot survey, sites registered prior to 1999 were not actively targeted even though they might have been operational in 1999. To estimate the 1999 waste handled by these sites, data were requested for 1999 and for 2000 onwards. The estimation of waste to be handled annually in future by these sites was thus used as a surrogate for the quantities handled in 1999 by sites registered in previous years.

Mobile Crushing Plant
Information on the number of licensed mobile crushing plant in 1999 was obtained from SEPA. In order to identify those machines utilised for the processing of secondary (as opposed to primary) aggregates SEPA personnel were consulted.

Bituminous Planings and Breakout/Industrial Waste and By-Products
The level of recycled bituminous planings was established from pilot surveys of local authorities, and the Scottish Executive's agents for trunk road and motorway maintenance. Producers of industrial wastes and by-products were targeted by questionnaire. Data were gathered on total arisings, the amounts landfilled and the amounts recycled. All relevant producers and owners in Scotland, including those involved in importing wastes and by-products to Scotland, were contacted. In each case data were gathered on arisings, landfilled and recycled.

RESULTS AND ANALYSIS
Data presented in this paper from the pilot survey have been reported to the nearest 1kt. Assuming that each datum has been rounded up or down, it is reasonable to presume that the systematic errors on each item of reported data is ±0.5kt. The effect on each of the totals presented for arisings, landfill, and recycled can thus be approximated by 100%×0.5kt/mean response. In addition, data are also reported with the associated confidence intervals at the 95% level.

Data from the pilot survey were aggregated to provide pan-Scotland figures. Where sampling of data sources has been undertaken the data was extrapolated on a simple proportional basis. The methods used to undertake this process are explained in the relevant sections.

Construction and Demolition Waste
Licensed Landfill Sites
Some 65 responses were received from the 115 questionnaires sent to licensed landfill sites, corresponding to a response rate of 57%. The number of active landfill sites in Scotland licensed to accept construction and demolition waste in 1999 is not available from the SEPA databases. However, an estimate was made from the databridge project (ERM, 2001) of 214 sites. Based on the questionnaire response rate, a factor of 3.3 (214/65) was applied to the data, to give the extrapolated figures in Table 1.

The total estimated quantity landfilled by all sites was approximately 4,173kt, which is the same as that estimated for 1998 by the Databridge project (ERM, 2001). Of the total estimated quantities landfilled 34% was clean soil, 13% was contaminated soil, 44% was mixed construction and demolition waste, and 8% was contaminated construction and demolition waste; the remainder was asphalt. Of the quantities reused/recycled an estimated 53% comprised clean soil, 22% contaminated soil, 19% mixed construction and demolition waste, and the remainder comprised contaminated construction and demolition and asphalt.

The variations in the data due to the accuracy of reporting are as follows: arising ±2%; landfill ±2.5%; and recycled ±7.5%, respectively. Confidence intervals for the total extrapolated quantities handled by landfill sites are as follows: arising 5,578±2,418kt; landfill 4,173±2,242kt; and recycled 1,405±731kt.

Exempt Sites
The number of responses from the 156 questionnaires sent to exempt sites was 47 (30%). Of these 31 were usable. The total number of exempt sites registered in 1999 in Scotland was 331. Based on the questionnaire response rate, a factor of 10.7 (331/31) was applied to the data, to give the extrapolated figures in Table 1.

An estimated total of 2,025kt of waste was transferred through exempt sites, of which 1,694kt was recycled and 331kt was sent for landfill. Of the total recycled, approximately 74% comprised clean soils, 23% mixed construction and demolition waste, and 3% asphalt.

The variations in the data due to the accuracy of reporting are as follows: arising ±8%; landfill ±50%; and recycled ±10%. The confidence intervals for the total extrapolated quantities handled by registered exempt sites are as follows: arising 2,025±1,167kt; landfill 331±343kt; and recycled 1,694±1,173kt.

Mobile Crushing Plant
Data held by SEPA allowed the number of mobile crushing plant to be identified as 25. Discussions with representatives of the demolition industry indicated that a crusher might process around 100kt pa. However, many licensed mobile crushing plant are engaged in both primary and secondary aggregate processing and are not active for the full year. Mobile crushers on licensed landfill sites were excluded from estimates. An average annual throughput was estimated to be closer to 45kt pa. This figure broadly confirms the findings from an ongoing study for England and Wales. Thus the totals for aggregates recycled using mobile crushing plant were determined from the number of active plant (see Table 1).

Further discussions with industry indicated that almost all material processed was recycled with the bulk being used at low utility, with small amounts at intermediate utility and effectively none at high utility (Table 1). The bulk was used for landscaping and foundations. The methodology adopted for mobile crushing plant is broadly compatible with that adopted in England and Wales (Arup, 1998).

Summary Construction and Demolition Data
The quantities of construction and demolition waste landfilled in the desk study estimate data collection (Winter and Henderson, 2001) and through the pilot survey (Table 1) were remarkably similar, around 4,200kt to 4,500kt. The pilot survey identified around 4,200kt of recycled aggregate while the desk study estimate identified only 3,300kt for licensed and registered exempt sites. Within these quantities, there is a marked difference between the

compositions of wastes estimated from the desk study estimate data and the pilot survey data:

- Landfilled construction & demolition: desk study, 1,260kt; pilot survey, 1,972kt.
- Landfilled soil and rock: desk study, 2,940kt; pilot survey, 1,607kt.
- Recycled construction & demolition: desk study, 1,260; pilot survey, 621kt.
- Recycled soil and rock: desk study, 1,960kt; pilot survey, 2,050kt.

Clearly the nature of the pilot survey and the consequent confidence intervals mean that any conclusions regarding licensed landfill and exempt data must be tentative.

Table 1: Summary construction and demolition pilot survey data for 1999.

Category	Arising	Landfilled	Total Recycled	Recycled		
				Low Utility	Intermediate Utility	High Utility
	kt	kt (percentage of Arising)		kt (percentage of Total Recycled)		
Licensed Landfill Sites	5,578	4,173 (75)	1,405 (25)	1,335 (95)	0 (0)	70 (5)
Registered Exempt Sites	2,025	331 (16)	1,694 (84)	1,609 (95)	0 (0)	85 (5)
Mobile Crushing Plant	1,125	0 (0)	1,125 (100)	1,013 (90)	112 (10)	0 (0)
All Categories	**8,728**	**4,504 (52)**	**4,224 (48)**	**3,957 (94)**	**112 (2)**	**155 (4)**

Bituminous Planings and Breakout

The data returned from the pilot survey of bituminous planings and breakout are summarised in Table 2.

A 100% response was received from the 32 local authorities that manage the non-trunk road network and the nine Premium Units, All-Purpose Units and Concessionaire appointed by the Scottish Executive to manage the trunk road and motorway network.

Table 2: Bituminous planings and breakout pilot survey data for 1999.

Category	Arising	Landfilled	Total Recycled	Recycled		
				Low Utility	Intermediate Utility	High Utility
	kt	kt (percentage of Arising)		kt (percentage of Total Recycled)		
Trunk Roads and Motorways	301	76 (25)	225 (75)	212 (94)	11 (5)	1 (1)
Local Authority Roads	313	62 (20)	251 (80)	143 (57)	74 (30)	34 (13)
All Roads	**613**	**138 (22)**	**476 (78)**	**355 (75)**	**85 (18)**	**35 (7)**

In broad terms the data confirm the impressions formed from the desk study estimate data. By far the majority of bituminous planings and breakout are recycled in low utility applications. A significant proportion is recycled in intermediate utility applications and only a relatively small amount in high utility applications.

The overall proportion of recycling on non-trunk roads compared to trunk roads and motorways is approximately the same. However, local authority activities on the non-trunk roads generally reflect recycling in higher utility applications than on trunk roads and motorways. This appears to be due to the activities of a relatively small number of authorities that are engaged in work to maximise the use of the available resource (e.g., Sinclair and Valentine, 1999).

The other main feature of the data is that the arisings returned from the questionnaires are around 28% in excess of those estimated for the desk study estimate data. Clearly, the pilot survey data should be viewed as more representative of the true level of arisings.

The corresponding variations in the data due to the accuracy of reporting are arising ±3%, landfill ±17% and recycling ±4%, respectively.

Industrial Wastes and By-Products

The data returned from the pilot surveys of industrial wastes and by-products are summarised in Table 3. A 100% response was received from all of the industrial waste and by-products operators contacted.

Table 3: Industrial waste and by-product pilot survey data for 1999.

Industrial Waste or By-Product	Arising	Landfilled	Total Recycled	Recycled		
				Low Utility	Intermediate Utility	High Utility
	kt	kt (percentage of Arising)		kt (percentage of Total Recycled)		
Pulverised Fuel Ash (PFA)	610	439 (72)	170 (28)	48 (28)	59 (34)	64 (38)
Furnace Bottom Ash (FBA)	46	1 (2)	46 (98)	0 (0)	0 (0)	46 (100)
Spent Oil Shale	0	10 (-)	335 (-)	290 (87)	25 (7)	20 (6)
Colliery Spoil	150	85 (57)	65 (43)	65 (100)	0 (0)	0 (0)
Cement Kiln Dust	0	0 (-)	0 (-)	0 (-)	0 (-)	0 (-)
Steel Slag	0	0 (-)	0 (-)	0 (-)	0 (-)	0 (-)
Granulated Ground Blast-furnace Slag (GGBS)	0	0 (-)	90 (-)	0 (0)	0 (0)	90 (100)
Railway Track Ballast	100	19 (19)	81 (81)	64 (80)	16 (20)	0 (0)
All Materials	906	554 (61)	787 (87)	467 (59)	100 (13)	220 (28)

Quantities of recycled industrial wastes approach those of the arisings. This is because there are no arisings for spent oil shale, as the industry is now defunct, or GGBS, as all production is outwith Scotland and the data relates to import of materials. These materials account for around 425kt of recycling with no arising or landfill within Scotland. There were no arisings of cement kiln dust within Scotland as cement production plant use processes that preclude its production. Also no steel slag was imported to Scotland during 1999. While Table 3 indicates that some materials were not imported to Scotland during 1999, imports during 2000 indicate that such materials should be considered by future surveys.

The amounts of industrial waste and by-products landfilled and recycled in Scotland in 1999 were 550kt and 790kt, respectively. Both figures are significantly down on the desk study estimate figures for 1998 (Winter and Henderson, 2001). These discrepancies are entirely due to falls in the arisings of pulverised fuel ash and in the recycling of both pulverised fuel ash and spent oil shale.

In general terms the proportion of recycled materials used in low utility applications remained relatively constant at 59% (cf. 58% for 1998). The level of recycling in intermediate utility applications fell to 13% (from 22%) and that in high utility applications increased to 28% (from 20%). The bulk of this change is due to increases in the high utility application of pulverised fuel ash and furnace bottom ash. It also seems likely that additional opportunities for the recycling of industrial waste and by-products predominantly will be found in low utility applications.

The mean responses for industrial wastes and by-products vary widely according to the type of material. However, the corresponding variations in the data due to the accuracy of reporting are generally ±1% or less, except for cases where very small amounts of material are landfilled (e.g., furnace bottom ash, spent oil shale and railway track ballast).

All Materials
Summary pilot survey data for 1999 are given in Table 4 and illustrated in Figure 1. Although the recycling of materials for which there are no arisings complicates the overall picture it is clear that around half of the aggregate arisings are landfilled and half recycled, albeit at low utilities in the vast majority of instances.

Table 4: Summary pilot survey data for all material categories for 1999.

Category	Arising	Landfilled	Total Recycled	Recycled Low Utility	Recycled Intermediate Utility	Recycled High Utility
	kt	kt (percentage of Arising)		kt (percentage of Total Recycled)		
Construction and demolition waste	8,728	4,504 (52)	4,224 (48)	3,957 (94)	112 (2)	155 (4)
Bituminous planings and breakout	613	138 (22)	476 (78)	355 (75)	85 (18)	35 (7)
Industrial wastes and by-products	906	554 (61)	787 (87)	467 (59)	100 (13)	220 (28)
All Categories	10,247	5,196 (51)	5,486 (54)	4,779 (87)	297 (5)	410 (7)

TARGETS
Aggregate waste arisings in Scotland are estimated at around 10,200kt for 1999 with the consumption of primary aggregates at around 25,000kt to 30,000kt pa in Scotland. It is recognised that the total amount of recycling in any given year is likely to fluctuate with economic activity and that setting a target in terms of an annual tonnage may thus be inappropriate. It is also clear that due to the proportion of Scottish aggregate production that is destined for export markets, the setting of targets in terms of primary aggregate production may also be inappropriate. Similarly setting targets in terms of aggregate demand, as discussed by Kennedy (1999), is unlikely to truly reflect recycling activities in Scotland.

Targets for recycled aggregates in Scotland are thus set as a percentage of primary aggregates destined for consumption in Scotland (production less exports). For 1999 this equates to around 18% (5,500/[35,000-5,000]), based on a production of around 35,000kt, exports of around 5,000kt and recycling of 5,500kt.

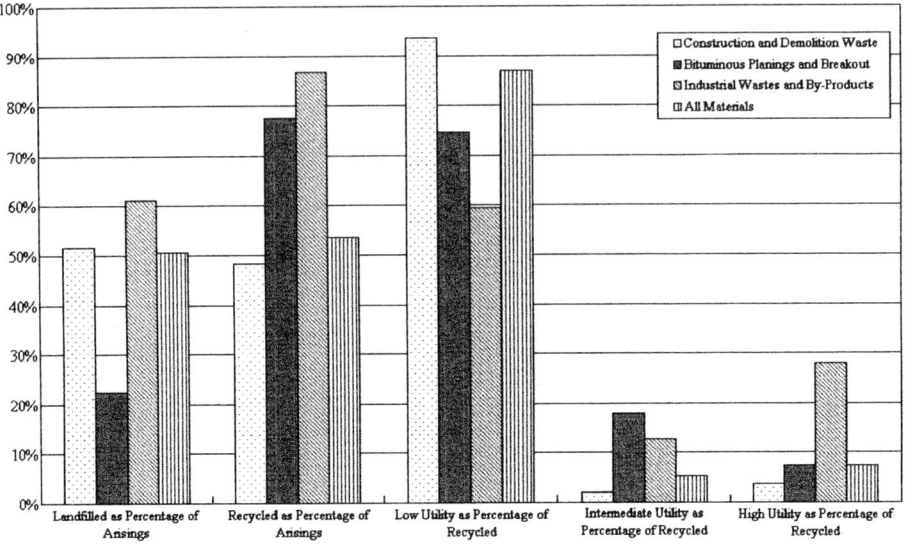

Figure 1: Summary pilot survey data for 1999.

Future surveys are expected to take place every three years. The recycled aggregates market is currently undergoing a period of rapid change and significant innovation and it is clear that the use of recycled aggregates is increasing at a rapid pace, even without targets.

The targets are 20% of production available for consumption in Scotland in 2000, 27% in 2003 and 36% in 2006. This equates with the recycling of 10,700kt of aggregate arisings in 2006; in 1999 5,500kt of aggregate waste arisings were recycled representing 50% of the total. Accordingly, it will be necessary to recycle all of the viable aggregate waste arisings together with a significant increase in the use of stockpiled industrial wastes and by-products, assuming that waste arisings remains constant at the current level (10,200kt).

The above targets are based upon taking the 1999 pilot survey as a baseline. Once the data for the 2000 survey is available the targets should be reassessed to ensure that they remain suitable in terms of the more robust data collected for the full survey. It is recommended that the targets be re-evaluated after each survey in order to ensure that they remain relevant to the trends in the data identified. Targets beyond 2006 should be set once survey data for both 2000 and 2003 is available.

The above targets are set in terms of gross aggregate recycling - no account of utility is taken. It is recommended that as the aggregates recycling industry develops in Scotland that consideration be given to setting future targets in terms of increased utility.

Development of a Contaminated Land Analysis and Risk Assessment Application

S CLEWER, R N YONG, I P ROWLAND and H R THOMAS
Geoenvironmental Research Centre, Cardiff School of Engineering, Cardiff University, Queen's Buildings, PO Box 925, Newport Road, Cardiff, CF24 0YF, Wales.

ABSTRACT
Risk assessment of contaminated land is an important area within the geoenvironmental field. This paper presents a new Windows-based application to address this problem, following the methodology advocated by the Environment Agency. The application has four analysis tiers, and the ongoing development of these tiers is described in this paper.

INTRODUCTION
This paper describes recent developments of a Windows-based application to conduct analysis and risk assessment of contaminated land, following the methodology advocated by the Environment Agency (Marsland and Carey 1999). This involves a system of four levels of analysis sophistication, or tiers, with each tier representing an increasing layer of complexity in terms of the risk analysis of a site.

The most basic analysis tier performs checks of contaminant leachate data against many environmental standards. Other models that use this methodology exist e.g. ConSim, (Environment Agency 1999) but the application being developed involves the implementation of a new fuzzy-logic based method. This is used to enable risk level to be determined from data that is often difficult to interpret.

The method will be used to simulate contaminant flow and perform site visualisation at the highest level of complexity, i.e. tier 4. Full account of uncertainties within site data will also be taken into account. The tier ultimately used in site analysis is determined largely by a cost-benefit analysis, the cost of a more complex tier of analysis being perhaps offset by the reduced cost of remediation indicated by closer study.

The application has been set up to analyse contaminants in both soil and water. Estimated concentrations are then compared with a choice of environmental standards, and this is used to produce an associated level of risk. At the current stage in the development of the application, tiers 1 and 2 are available. Contaminant flow modelling will be considered in subsequent tiers of the application.

TIER 1 ANALYSIS
This relies essentially on a comparison between the soil contaminant concentration and allowed standard values, as selected within the application.

The compliance point, i.e. where the level must lie below the standard value, is the soil zone and the remedial target is set as equivalent to the target concentration. In this tier, processes such as dilution and attenuation that may affect contaminant concentrations along the pathway between the soil and the receptor are not considered.

APPLICATION INPUTS

The basic equation for the soil remediation target concentration (*STC*), the level that the total measured soil concentration should not exceed, can be expressed as (Marsland and Carey, 1999):

$$STC = C_s = C_l \left[K_d + \frac{(\theta_w + \theta_a H)}{\rho} \right]$$

Where C_s = remedial soil target concentration (mg/kg)
$\quad$ C_l = target leachate concentration (mg/l)
$\quad$ K_d = soil water partition coefficient (l/kg)
$\quad$ θ_w = water-filled soil porosity
$\quad$ θ_a = air-filled soil porosity
$\quad$ ρ = bulk density (g/cm^3)
$\quad$ H = Henry's Law constant (dimensionless).

Alternatively, a pore water remediation target (*LTC*) may be applied, if leachate tests or pore water data is available. This is equal to the target leachate concentration, i.e. $LTC = C_l$ (mg/l).

$\qquad$ The value of *STC* is used to produce a distribution of expected standard value, based on the uncertainty distributions for the values entered. This is detailed below.

Sorption and Partition

The total soil contamination, C_s, is distributed within the soil as a leachate component, C_l, and a sorped component, C_{sorb}. The sorption term can be considered in one of two ways within the code. The equation for *STC* above assumes *linear* sorption where $C_{sorb} = K_d\, C_l$. Alternatively, *Langmuir* sorption can be used where $C_{sorb} = k_1\, C_l /(1 + k_2\, C_l)$ (Rudzinski and Everett 1992). Here, k_1 and k_2 are partition coefficients.

$\qquad$ For both sorption types, K_d or k_1 and k_2 can be either input directly or determined based on sample C_l and C_{sorb} values for the soil. If leachate test data are available, these can be input directly to find the sorption factors.

$\qquad$ Partition is determined depending on contaminant classification as organic or inorganic. These factors can be added using the appropriate forms within the application.

Contaminant Selection

The application contains a database of around 200 organic and inorganic contaminants that can be selected, either individually or as a group.

UNCERTAINTY CONSIDERATION

Although the expression for *STC* is straightforward, consideration should be made as to the uncertainties associated with each of the input parameters e.g. soil porosity, bulk density etc. These uncertainties are taken into account within the application by allowing the user to input site data not as a fixed value, but as a value and associated distribution function.

$\qquad$ This uncertainty distribution function, which represents the expected spread in values that are likely to occur in the soil, can be either *Gaussian, triangular* or *uniform* in nature. If a uniform distribution is chosen, the user must enter the upper and lower constraints. For a triangular distribution, midpoint and width are required. For the Gaussian distribution, the inputs are the mean and standard deviation (sd). The option exists, however, for parameters to be entered as single values.

Monte Carlo Generation
Since a set of input parameter distributions are now defined, these can be used to produce realistic data for analysis. To this end, the Monte Carlo (MC) technique is implemented. This is basically a means of repeatedly sampling random events from a given distribution and combining them to calculate an outcome. For each contamination calculation, the MC takes a randomly selected value from the inputted distribution of each variable and then generates an output contaminantion level. For a large number of events, random generation builds up the set of expected data outputs i.e. for a Gaussian, the full distribution will be reproduced if enough samples are taken. By choosing random values from the data, this mimicks the effect of statistical variation in site data. The outcome of this analysis is two sets of data – the expected spread of *STC* within the soil, and the spread of the actual site data (*AC*).

Distribution Analysis
The application performs an analysis on the two distributions, *STC* and *AC*, in order to see the degree to which they overlap. The application calculates where 99%, 95% and 90% of the distribution lies for each of the two distributions. The results are indicated in a contaminant risk assessment report for the two populations. If the *AC* levels exceed the standards for any of these ranges, then this is indicated in the report in red text.

FUZZY LOGIC ANALYSIS
The distribution analysis is augmented by a *fuzzy logic* analysis; this categorises the risk of the *AC* level exceeding the *STC* level, and is complementary to the distribution analysis.

Fuzzy logic is basically a means of modelling a physical system based on rules that relate different physical parameters, rather than complicated theory (Pedrycz 1995). It relies on quantities being defined not exactly (crisply), but within a range of possible values e.g. *fast* could cover all values from, say, 60-80 mph. Crisp values are thus *"fuzzified"* to the fuzzy value *fast*.

A fuzzy logic approach is used to categorise the overlap of distribution functions *STC* and *AC* in order to determine the level of risk for site contamination exceeding the standard value. The mean and sd of both distributions are used to generate a fuzzy set, where the crisp value of the parameter is interpreted as a membership of one or more of a set of areas, or *membership functions*, designed to categorise the value. The application relies on two fuzzy sets: *meanFactor* which is a function depending on the similarity between the two mean distributions, and *sdFactor*, relating the two sds.

These two fuzzy sets are each assigned their appropriate membership functions. For example, some of the membership functions defining *meanFactor* are represented in figure 1. NB: in the fuzzy logic method it is possible for values to appear in more than one membership function. For example, a single value of *meanFactor*, say 1.3, maps onto the membership functions S5 and S4, being inside the ranges of both, with two different levels of membership - being "more" S5 than S4 (see figure 2). In this way, data is fuzzified.

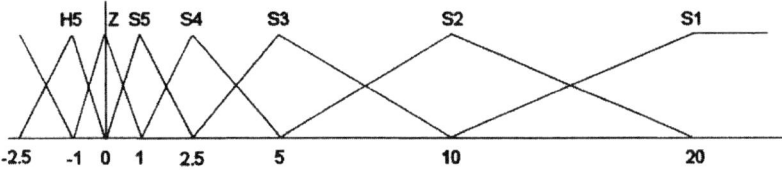

Figure 1: Sample of the membership functions defined for *meanFactor*.

The sets of fuzzified values for *meanFactor* and *sdFactor* are then combined using a rulebase that considers each membership function for both factors. The rules decide how likely it is, given the applicable fuzzy levels, that the site contamination is greater than the distribution of standard value for the site. If it is very likely, then the risk is correspondingly very high. Risk is classified into the following fuzzy sets: *Minimal*; *Low*; *Intermediate*; *High*; *Very High*; *Extremely High*. If any of the last three of these assignments are made, then this is highlighted in red in the contaminant risk assessment report.

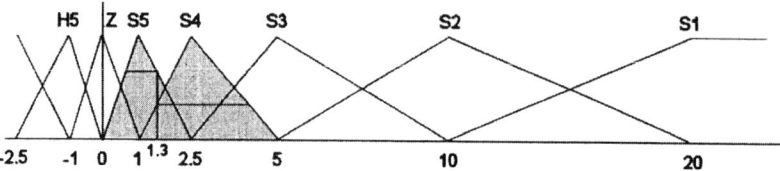

Figure 2: How the crisp value 1.3 is interpreted into the membership functions.

Because of the membership of values within more than one of the membership functions, then more than one rule can operate for a particular value. In this way, uncertainty is taken into account in the evaluation of risk. The outcome of this fuzzy logic analysis should be considered together with the distribution analysis for the user to make an informed judgement regarding the safety of the site with regard to the contaminants considered.

SELECTION OF ENVIRONMENTAL STANDARDS

There are ten environmental standards pertaining to contaminated land within the application. Each contains a set of permitted values for various chemicals. Any or all of these standards can be selected in order to compare allowed levels with the site contaminant levels that are produced by the tier analysis. There is also an option within the application for user-defined contaminants and levels. The defined standards are:

> ICRCL 59/83 Guidelines
> Kelly Indices
> Water Supply (Water Quality) Regulations 1989
> Water Supply (Water Quality) (Scotland) Regulations 2000
> World Health Organisation Guidelines
> Dutch List (Groundwater and Soil)
> EU Shellfish Waters Directive EC79/923/EEC
> EU Bathing Water Quality Directive EC76/160/EEC
> EU Freshwater Fish Directive EC78/659/EEC
> EU Surface Water Directive EC75/440/EEC.

APPLICATION INTERFACE

Once analysis is complete, a graphical results window and risk assessment report window are generated. The default graphics display plots a frequency distribution for the two parameters: generated values for the selected environmental standard target distribution (*STC*) in red and the actual contaminant (*AC*) values in green. This enables a swift visual evaluation of the degree of overlap of the two levels. The generated risk assessment report provides a summary of all soil parameters, environmental standards and contaminants entered. It also indicates the outcome of the distribution and fuzzy logic analyses for each contaminant/standard.

A FUZZY LOGIC RISK ANALYSIS EXAMPLE

The implementation of the fuzzy logic system within tier 1, and its comparison with a standard analysis will now be examined. Consider a soil contaminated with copper, using the *Water Supply (Water Quality) (Scotland) Regulations 2000* as the standard. From this, the acceptable copper level within the soil is 2 mg/l. Using the soil parameters $\rho = 1.65$ g/cm^3, θ_w = 0.05 and $\theta_a = 0.18$, this translates to a concentration of 8006.94 mg/kg, with a standard deviation of 389 mg/kg. An input Gaussian distribution of 11400 mg/kg was used for the site concentration, with a standard deviation of 300 mg/kg. These values are compared graphically in figure 3.

Copper - Water Supply (Water Quality) (Scotland) Regulations 2000 [Mandatory]

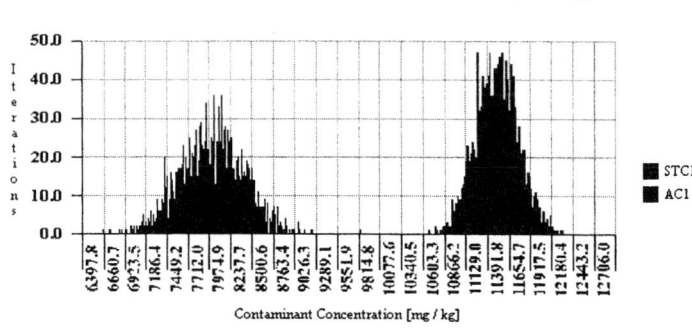

Figure 3: An example of unacceptable values of soil concentration distribution.

It is clear that the entire contaminant distribution lies well above the expected distribution of the standard value. The risk report for these results states:

99% of the STC(1) distribution lies above 7093.45 mg/kg.
99% of the AC(1) distribution lies below 12102.92 mg/kg.
Therefore, overlap of the 99% points of the two distributions is 5009.47 mg/kg (70.6% of STC(1) 99% level).

95% of the STC(1) distribution lies above 7368.40 mg/kg.
95% of the AC(1) distribution lies below 11912.43 mg/kg.
Therefore, overlap of the 95% points of the two distributions is 4544.04 mg/kg (61.7% of STC(1) 95% level).

90% of the STC(1) distribution lies above 7490.66 mg/kg.
90% of the AC(1) distribution lies below 11824.29 mg/kg.
Therefore, overlap of the 90% points of the two distributions is 4333.64 mg/kg (57.9% of STC(1) 90% level).

Fuzzy Logic Analysis
Analysis of the actual and target distributions qualifies the risk of the Actual Concentration, AC(1) exceeding the Soil Target Concentration, STC(1) as:
Extremely High

The first three sets of data above present the degree of overlap between the distribution functions, compared at the 99%, 95% and 90% levels. If there is no overlap, then the actual contaminant can be considered safe. If there is some overlap, then the text is highlighted in red and degree of overlap indicated as a function of *STC* concentration. The fuzzy logic analysis gives an appropriate level of risk for the data – ***extremely high*** – this is the highest risk level the system has.

In cases like the one above, it is self-evident that there is a problem complying with the standards. However, it is often difficult to interpret whether a site level is safe or not.

Consider the data of figure 4, where *Water Supply (Water Quality) Regulations 1989* have been used. The acceptable level is higher (3 mg/l). In this case it is not so clear cut. The analysis produces the following results:

99% of the STC(1) distribution lies above 10567.39 mg/kg.
99% of the AC(1) distribution lies below 12102.92 mg/kg.
Therefore, overlap of the 99% points of the two distributions is 1535.53 mg/kg (14.5% of STC(1) 99% level).

95% of the STC(1) distribution lies above 11001.57 mg/kg.
95% of the AC(1) distribution lies below 11912.43 mg/kg.
Therefore, overlap of the 95% points of the two distributions is 910.86 mg/kg (8.3% of STC(1) 95% level).

90% of the STC(1) distribution lies above 11218.33 mg/kg.
90% of the AC(1) distribution lies below 11824.29 mg/kg.
Therefore, overlap of the 90% points of the two distributions is 605.96 mg/kg (5.4% of STC(1) 90% level).

Fuzzy Logic Analysis
Analysis of the actual and target distributions qualifies the risk of the Actual Concentration, AC(1) exceeding the Soil Target Concentration, STC(1) as:
High/Very High

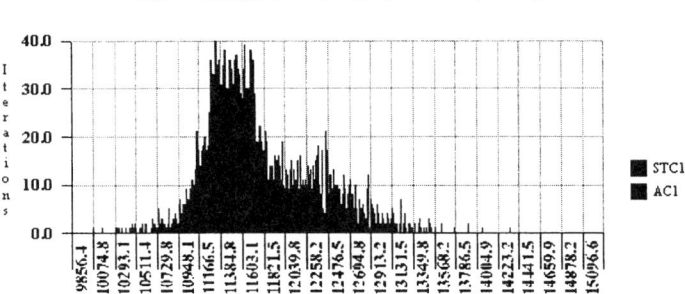

Copper - Water Supply (Water Quality) Regulations 1989 [Guideline]

Figure 4: An example of closely overlapping distributions.

It is at this point that having a means of backing up the analysis with a risk value is very useful. Using the fuzzy logic engine described, this data returns the risk of *AC* exceeding *STC* as **high to very high** i.e. it is reasonable that steps should be taken to reduce the contamination.

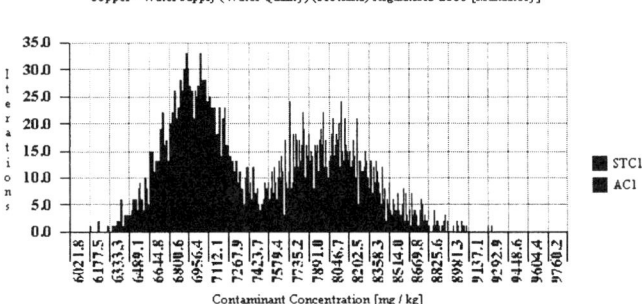

Copper - Water Supply (Water Quality) (Scotland) Regulations 2000 [Mandatory]

Figure 5: An example of soil concentration distribution considered as intermediate risk.

Where there is a smaller degree of overlap, e.g. distributions as in figure 5, the fuzzy logic system recognises this level of risk as *intermediate*, the bulk of the *AC* distribution being below the mean of the standard level.

CONCLUSIONS

The work presented in this paper describes the development of a new application for the analysis of contaminated land and risk assessment. The application takes account of the statistical variation of input values, and employs the use of fuzzy logic to interpret the analysis results and to give an assessment of risk for the site. The development of the full four tiers will produce a comprehensive risk assessment package.

REFERENCES

Environment Agency 1999 *ConSim* application (see http://www.environment-agency.gov.uk/gwcl/publications.htm)

Marsland PA and Carey M A 1999 *Methodology for the Derivation of Remedial Targets for Soil and Groundwater to Protect Water Resources* (Environment Agency)

Pedrycz W 1995 *Fuzzy Sets Engineering* (Boca Raton: CRC Press)

Rudzinski W and Everett D H 1992 *Adsorption of gases on heterogeneous surfaces* (London: Academic Press)

ACKNOWLEDGEMENTS

The work presented has been carried out as a part of a research programme supported by the ERDF. This support is gratefully acknowledged.

Risk to construction materials and services on brownfield sites

S L GARVIN, P TEDD, J P RIDAL and A R J FISHER. Building Research Establishment.

INTRODUCTION

BRE is currently undertaking a project on behalf of DTLR (Department for Transport, Local Government and the Regions) to assess the risks to construction materials on brownfield sites. The project aims to draw together earlier BRE research and information from various sources to produce guidance and inform judgements made in risk assessments of the use of common building materials and products on brownfield sites. The intention is to provide a simple clear guidance document by (a) focusing on the commonly used building materials in contact with the ground and (b) by considering only those contaminants that are likely to be found in significant quantities in the commonly encountered types of contaminated ground.

The long-term performance and durability of construction materials used in contaminated ground may be adversely affected by a wide variety of aggressive chemicals, e.g. acids (both inorganic and organic), alkalis, organic solvents and inorganic salts (sulfates, chlorides etc.) that are potentially highly aggressive. Attack on construction materials is of concern because:

- damage to the materials can lead to weakening of foundation concrete and thereby possibly compromise the stability of the building,

- damage to materials used in services, for example, entry of contaminants into water supplies through damaged, or degraded plastics, or inappropriately protected pipes can lead to health risks, and damage to gas supply pipes may lead to release of gas.

- damage to materials used in works designed to protect humans or the environment from harm due to contamination (e.g. plastic membranes used to protect against gas; cement-bentonite cut-off walls) may compromise their function

The identification of potential contaminant attack on building materials is a part of the overall characterisation of a site and supports the total risk assessment process. There is a need to develop the existing guidance and to present it within a risk assessment framework. There are important factors e.g. pH, groundwater regime and permeability/porosity of the soil which will affect the ability of contaminants to attack building materials.

In recent years more attention has been directed towards the protection of building materials in aggressive environments. A recent survey by BRE indicated that 70% of current building development is on brownfield land. In some cases the type and quality of a building material that is used for a specific environment can be adjusted to suit the aggressive conditions. However, in other cases there may be a limit on the durability that can be expected of the materials, or uncertainty as to the full range of conditions that are likely to be encountered. In these cases it has been suggested that protective coatings could be used for the protection of building materials. BRE report 286 (BRE 1995) details the applicability of certain types of coatings for steel and concrete in a contaminated land environment. Other guidance, such as

Digest SD1 (BRE 2001), recommends the use of coatings for concrete where there are high sulfate concentrations.

For a building material to be attacked by a particular contaminant in the ground depends on the following factors:

- The presence of water or free-phase organic compounds,
- The availability of the contaminant in sufficient concentration and replenishment rate (of the aggressive solution),
- Contact between the contaminant(s) and the building material, and
- The sensitivity of the material to the contaminant (the inherent durability of the material).

However, other factors also influence whether or not deterioration will take place and if this deterioration is likely to occur at a faster rate than would otherwise result (BRE 1994, Garvin 2000). These factors include the following:

- The use of unsuitable materials, for example the wrong choice of cement or aggregates for mortar and concrete.
- Poor specification, production or placing of concrete that leaves the concrete with a more open texture, lack of cover to reinforcement or highly permeable.
- Lack of appropriate protection.

The extensive existing literature is quite general in its application. Barry (1983) identifies the susceptibility of 55 common construction materials to attack from more than 140 generic or specific chemical compounds. Many in the building industry have become confused and seek a simple clear document that brings together the diversity of advice.

FOUNDATIONS
Foundations are primarily required to support the building or structure but could also be acting as a barrier to contaminants. Interaction with contamination could result in loss of strength or change in volume of the foundation material which could cause damage to the building or services as they enter the building. Foundations can be broadly divided into shallow and deep types.

- Shallow foundations include, strip and trench fill, pad and raft.
- Deep foundations include piles, piers and deep basements.

The availability of the contaminants to affect a foundation will depend on the type of foundation. Strip, trench fill and pad foundations are generally less than 1m deep but can be up to 2m deep and occasionally as deep as 4m. Contact with contaminated ground water is less likely for the shallower foundations. Raft foundations are generally built onto a clean engineered sub-base of granular fill, close to the surface, above the water table and therefore the availability of contaminants to the foundation is considerably less than for deeper foundations. The granular fill and/or impermeable membrane would be used as a capillary break within a clean cover system to control upward migration of contaminants. General guidance on the use of cover systems is given in CIRIA SP 124 (Privett et al 1995) and SP 106 (Harris et al 1995).

Piling and penetrative ground improvement methods are commonly used for providing foundations on brownfield sites. Such systems are likely to pass through the site contamination and are more likely to come into contact with any contaminated ground water, therefore the materials used need to be sufficiently durable to resist degradation. Westcott et al, (2001) review piling and penetrative ground improvement methods on land affected by contamination and is largely concerned with creating pathways by which contamination can affect controlled water supplies, although material degradation is an issue. BRE Digest 315 "Choosing piles for new construction" (BRE 1986) reviews the various types of piles and makes reference to maximum durability in aggressive soils and ground water.

The use of stone in vibro stone columns for ground improvement requires particular attention. This method of ground improvement is currently the most common form of engineering treatment of brownfield sites. The specification on vibro stone columns BRE Report 391 (BRE 2000) gives guidance on the type of stone to be used and makes particular reference to unsuitability of limestone in acid ground conditions and the presence of any chemical contaminants being identified. Long term degradation of the stone which forms an integral part of the foundation system could result in unacceptable movements of the building.

Although concrete is the most common foundation material for buildings and the published material available on the performance of concrete and concrete products is extensive, little is specific to contaminated land. In the UK, sulfates in soil and ground water are the most likely to attack concrete. The information in the Industrial Profiles, (NHBC, 2000) indicates that sulfates, sulfide and sulfur are some of the most significant and frequently found contaminants on industrialised land. The effects of sulfate attack can be serious resulting in expansion and softening of the concrete. Another common cause of concrete deterioration is ground water acidity, this sometimes being linked with the presence of sulfate. Guidance on suitable concrete for sulfate contaminated soil is given in BRE Digest SD1 "Concrete in Aggressive ground" and BRE Report BR255 "Performance of building materials in contaminated land".

Risks to Buildings, Building Materials and Services has been the subject of work by BRE for the Environment Agency and DETR in recent years. Environment Agency R&D Technical Report P5 035/TR/01 details a procedural framework for the assessment and management of risks to new buildings, existing buildings and materials used in remedial works (Environment Agency 2001).

Research specific to the effects of soil contaminants on materials in new developments has shown that chemical cocktails affect materials more than single contaminants. This has been particularly noted for concrete where laboratory tests have demonstrated that higher risks exist when sulfate attack is enhanced by acids, heavy metals and organic compounds (Garvin et al 1998, Garvin and Ridal 1999, Garvin and Ridal 2000). In redeveloping contaminated land sites consideration should not be confined to just sulfates and acids when specifying concrete.

The major revision of BRE Digest 363 "Sulphate and Acid Resistance of Concrete in the Ground" which is now BRE Digest SD1 covers site assessment procedures for natural ground and brownfield sites. The revised Digest has been divided into four parts that deal with the specification of new concrete that will be used in contact with aggressive ground.

The four parts are :
1. Assessing the Aggressive Chemical Environment
Part 1 identifies chemicals that are potentially harmful to concrete and describes the procedures for assessing the Aggressive Chemical Environment for Concrete (ACEC) Class in the ground.

2. Specifying concrete and additional protective measures
Part 2 provides general recommendations for the types of cement and aggregate and quality of concrete suitable for use in aggressive chemical conditions.

3. Design Guide for common applications
Part 3 provides design guides for common applications of below ground normal concrete, including foundations, retaining walls and floor slabs for domestic properties.

4. Design guide pre-cast concrete products
Part 4 provides four sets of Design Guides for the specific precast concrete products used for pipeline systems, segmental linings for tunnels and shafts, box culverts and concrete masonry blocks.

Catastrophic concrete foundation failure due to chemical attack from contaminants from brownfield sites is currently unknown. However, a number of instances of damage have been investigated by BRE and other consultants. These are not generally well documented and building owners are often unwilling to release information on work carried out to their buildings. Remedial measures will normally involve replacement of the affected parts of the foundations. This will require excavation of the soil outside the building, with removal of contaminated soil or addition of protective measures or more chemically resistant materials.

Although investigations have often only found minor damage that is not a structural risk to the building, remedial work has been required to ensure the long term life and durability of the building.

SERVICES
It is more likely that problems will arise with services through contaminated land than with the building foundation. Deterioration can lead to safety issues such as permeation of potable water pipes by various hydrocarbons, gas explosions and ignition of combustible materials in the soil from overheating electrical cables. The industries supplying the key services through the ground have issued guidance on the risks of laying them in contaminated ground. The main risks and remedial measures are summarised follows.

Potable Water supply
Comprehensive guidance is given in "Laying potable water pipelines in contaminated ground" (Stephen and Norris, 1994) and in "Pipe Materials selection manual water supply" (Trew et al 1995). These reports provide guidance on site investigation, the range of pipeline materials, their relative merits the adverse effects of various contaminants on different materials.

Organic chemicals which are known to have potential for either affecting or permeating plastic pipes and/or joint sealing rings are:
- Halogenated aliphatic hydrocarbons, eg tetrachlomethane, dichloromethane
- Chlorinated aromatic hydrocarbons, eg. chlorobenzene,

- Aromatic hydrocarbons, eg. benzene, toluene, xylene
- Phenol

Inorganic chemicals which can affect metal and cement pipes are sulfates, sulphides, chlorides, cyanide, alkalis and acids.

Trew et al (1995) provide a simplified assessment of the merits of different pipe materials in situations where no special measures or remedial works have been undertaken. They report that 99.5 % of service pipes (≤50mm) used MDPE in 1992/3 and 81% of all distribution mains pipe (51 – 3000mm) were of thermoplastic materials. Despite the wide application of thermoplastics their use on most contaminated sites is not recommended in the guidance. However, polyethylene systems have been developed that include an aluminium layer to stop the ingress of hydrocarbons. Such systems comply with WRAS (Water Regulation Advisory Service) and BS 6920.

Other than removal of the contamination or routing the pipes around the contaminants, additional protection using impermeable and inert backfill material, protective sleeves and trench drains are recommended. With any services, burying them in inert granular backfill can increase the risk of deterioration if the fill acts as a drain for water borne contaminants.

Electricity supply
Electrical cables are generally protected by plastic sleeves. These sleeves are potentially subject to chemical attack and permeation in similar ways as plastic water pipes (Billing, 1998). Medium and low voltage cables are often laid directly into the ground and are therefore at risk of attack by contaminants. High voltage cables tend to be protected from direct contact with ground as they are buried in clean backfill. The following hazards have been identified as important to electrical cables in contaminated land:

- Degradation of the cable sheathing leading to deformation or permeation.
- Swelling of plastic sheathing leading to deformation or permeation.
- Ignition of combustible materials in the soil due to temperature rises in the soil induced by cable heating.
- Mechanical damage caused by large, sharp or heavy objects in the soil.
- Ageing or softening of cable insulation due to high temperatures in the soil.
- Transport of contaminants, in particular landfill gas, along the cable routes.

The selection of appropriate cable sheathing material is important to the proper long term functioning of the cable. Other factors such as the possibility of underground fires could affect the function of the cable.

Gas Supply
In the UK gas distribution network, the trunk mains are constructed in steel and carry gas from the regional transmission mains to the intermediate distribution mains at 41 bar pressure. Thirty to forty years ago, ductile iron was the preferred material for the lower pressure mains and a small amount of PVC was used. Polyethylene became fully accepted during the 1970s and today the only materials used are MDPE, HDPE and steel. The choice depends on pressure, the proximity to buildings and temperature.

There is no known published information on the effects of contaminants on gas pipelines but the materials used are essentially the same as found in the water supply industry. No case histories of gas pipeline failures solely as a result of contaminants in the ground have so far come to light.

Drainage and sewerage pipes

For the most part, the materials used are the same as found in the water supply industry which are reviewed elsewhere in this report. Permeation of plastics materials used for potable water supply by hydrocarbons, the chief safety concern, is not an issue with waste water and there are no known cases of the total destruction of plastics drainage materials. Traditional materials such as pitch fibre and asbestos cement are at risk but are now infrequently used.

A commonly used material is vitrified clay and guidance on the resistance of vitrified clay pipes, fittings and joints to chemicals in effluents and contaminated ground is available from the Clay Pipe Development Association Ltd, (CPDA, 1999). Vitrified clay pipes, fittings and joints conforming to BS EN 295-1 (BSI, 1991) can safely be laid in ground containing aggressive materials, such as sulphates. Pipes, fittings and joints will resist both effluents and ground water over the pH range 2-12 at average temperatures found in public sewers in the UK of between 14° C and 19°C.

A number of proprietary chemically resistant mortars are available, generally based on furane resin cements, which can be used to produce rigid joints. However, brownfield redevelopments often have structurally poor ground conditions and flexible joints may be used in preference to rigid connections in order to cope with anticipated ground movement. Joint rings and joint materials made from styrene-butadiene rubber (SBR), ethylene-propylene-diene modified rubber (EPDM), nitrile-butadiene rubber (NBR) or polypropylene (PP) are more susceptible to chemical attack and the CPDA document includes comprehensive guidance on the performance of different joint materials in the presence of a wide variety of chemical compounds. If a suitable combination of the jointing materials listed in BS EN 295 cannot be found, normal or extra chemically resistant pipes, as appropriate, to BS 65:1991 (BSI, 1991) may be used with special jointing compounds.

Standard flexible joints can be protected from chemical attack by removing the contaminated ground from around the pipe or by wrapping the joints in protective tape or other suitable material. One type of protection with a high degree of strength and adhesion to the pipe is made from irradiated stretched polyethylene to BS EN 295-4 (BSI, 1995). It is available as a sleeve, which can be placed over the pipes as they are laid, or as a wrap, which can be applied after laying. The resistance of this polyethylene to the chemicals present must be checked before use.

RISK ASSESSMENT AND RISK MANAGEMENT

The long term risks to construction materials in contaminated land need to be addressed in the overall risk assessment when developing a contaminated site for building. Risk assessment should be a systematic and procedural approach to allow the hazards to be identified and then the risks estimated and evaluated. Management of the risks is the essential issue and this will depend on the type of material used and whether it is a service or foundation. For example, for services clean trenches can be used to isolate the receptor and this can also be done for some shallow foundations such as precast concrete or masonry. However, piled foundations will be driven directly into contaminated areas of the ground. In these cases the pile either has to be made of fully resistant materials or be a special design to allow protection.

In dealing with risks to foundations and services it will be necessary to either:

- Remove the risk – by removing the source of the contamination
- Reduce the risk – by protecting the receptor (foundation or service), or by breaking the pathway between the contaminants and the receptors.
- Accept the risk – take no special precautions, which may be suitable if the building is not for habitation or it is a short term building.

The guidance currently being developed will be published in 2002 and it is intended that it will be capable of being used for Part IIA of the Environment Act as it will be based on the Source-Pathway-Receptor model and will include the following phases:

- Hazard Identification
- Hazard Assessment
- Risk Estimation
- Risk Evaluation.

In this way it will be usable and comparative to assessing the risks to human health and the natural environment.

ACKNOWLEDGEMENTS
The project is being funded by the Department for Transport, Local Government and the Regions.

REFERENCES
Barry D L. (1983). Material Durability in Aggressive Ground. Construction Industry Research and Information Association Report 98. London, CIRIA
Billing J W and Greenwood J H (1998). The impact of contaminated land on buried electric cables. ERA Technology. ERA Project 26-01-0280
BRE Digest 315 (1976). Choosing piles for new construction
BRE Digest 362 (1991). Building mortar
BRE Digest 444 (2000). Corrosion of steel in concrete, Part 1 Durability of reinforced concrete structures, Part 2 Investigation and assessment, Part 3 Protection and remediation
BRE Digest SDI (2001). Concrete in aggressive ground, Part 1 Assessing the chemical environment, Part 2 Specifying concrete and additional protective measures, Part 3 Design guide for common applications, Part 4 Design guide for precast products.
BRE Report BR225, Paul V (1994). Performance of Building Materials in Contaminated Land, BRE Report BR255, Construction Research Communications Ltd, Watford
BRE Report BR286, Garvin S L, Paul V. and Uberoi S, (1995). Polymeric Anti-corrosion Coatings for Protection of Materials in Contaminated Land,
BRE Report 391 (2000) Specifying vibro stone columns
BSI 295 (1991). Vitrified clay pipes and fittings and pipe joints for drains and sewers. British Standards Institution. Parts 1, 2 and 3:
BSI 65 (1991): Specification for vitrified clay pipes, fittings and ducts, also flexible mechanical joints for use solely with surface water pipes and fittings. British Standards Institution.

BS EN 295-4: (1995). Vitrified clay pipes and fittings and pipe joints for drains and sewers. Part 4: Requirements for special fittings, adapters and compatible accessories. British Standards Institution

Garvin S L, Hartless R, Smith M, Manchester S, Tedd P (1999). Risks of contaminated land to buildings, building materials and services. A literature review. Environment Agency R&D Technical Report 331.WRc

Garvin S L, Hartless R, Tedd P (1998). Building on contaminated land: The risks, Consoil 98 The 6th International FZK/TNO Conference on Contaminated Soil, Edinburgh, Vol 1, pp641 - 649, Thomas Telford,

Garvin S L, Ridal J P and Halliwell M (1999). Performance of concrete in contaminated land, Extending Performance of Concrete Structures (ed R K Dhir and P A J Tittle), pp 201 - 210, International Congress on Concrete, Dundee,.

Garvin S L and Ridal J P(2000). Risk assessment for building materials in contaminated land, Consoil 2000, 7th International FZK/TNO Conference on Contaminated Soil, Leipzig, pp888-889, Thomas Telford,.

Harris M R, Herbert S M and Smith M A (1995). Remedial treatment for contaminated land. Vol VI Containment and hydraulic measures. CIRIA. Special publication 106.

NHBC/EA (2000). Guidance for the safe development of housing on land affected by contamination. R&D Publication 66.

Privett K D, Matthews S C. and Hodges R A (1996). Barriers, Liners and Cover Systems for Containment and Control of Land Contamination, CIRIA Special Publication 124, 1996

Stephen J and Norris M (1994). Laying potable water pipelines in contaminated ground. Guidance Notes, Water Research Centre Report, FR0448,

Trew J E, Tarbet N K, De Rosa P J, Morris D, Cant J and Olliff J L (1995). Pipe materials selection manual. Water Research Centre Report

Westcott F J, Lean C M B and Cunningham M L (2001).Piling and penetrative ground improvement methods on land affected by contamination: interim guidance on pollution prevention. Environment Agency Report NC/99/73

Landfill Gas Migration Management Through Automatic Monitoring and Control

S R HUNNEYBALL TES Bretby Burton on Trent

SUMMARY
The explosion and asphyxiation hazards associated with the uncontrolled migration of landfill gas are well known. In an attempt to quantify the potential hazard and in order to comply with Waste Management Papers, it is common practice for operators to monitor the levels of methane and carbon dioxide in boreholes placed around the periphery of a landfill site. Of particular interest are the gas concentrations measured in boreholes that are drilled adjacent to vulnerable domestic and industrial properties.

Traditionally, monitoring has been carried out by technicians using portable gas measuring instruments combined with taking samples for subsequent laboratory analysis. The increasing costs associated with manual monitoring together with the limited amount of data provided, have been mirrored by the reduced cost and increasing reliability of continuous automatic gas monitoring systems. Over the last five years, the advantages of automatic monitoring have resulted in a sharp rise in the installation and use of such systems.

Additional advantages are to be gained from the integration of an automatic monitoring system with migration control measures, such as a gas extraction system and flare or emergency gas interception curtain. These measures coupled with the capability to remotely interrogate the monitoring data and status of the monitoring and control systems, provide a powerful tool in the management of landfill gas migration.

This paper describes a typical gas monitoring system and gives examples of the gas control and equipment status monitoring that can be interfaced with it. It also describes how such a system is installed and commissioned on site, highlighting the benefits provided in the successful management of migrating landfill gas.

Finally it provides a case study of a site in Southern England that had a significant specific problem with gas migration near to domestic properties and describes how the installation of an automatic monitoring and control system assisted in the successful management of the problem, plus the design and implementation of a solution.

DESCRIPTION OF GAS MONITORING SYSTEM
The TES Bretby Automatic Landfill Gas Monitoring System (ALGM) is a pumped sequential analysis system capable of extracting gas samples from boreholes, without altering gas equilibrium conditions around the sampling points. The system has the capability of monitoring up to four hundred sample points, although to date the maximum number is sixty.

Geoenvironmental impact management, Thomas Telford, London, 2001.

Figure 1 Installation of Sample Tubing

Installation of the system starts with the digging of a trench for the sample tubing. This is carried out by a JCB or minidigger according to the topography of the site. The trench is normally 500mm wide by 600mm deep. A bed of sand is then put into the bottom of the trench and the tubing laid out in one continuous length as shown in Figure 1. Sometimes multicore tubing is used if a large distance exists between the system cabin and the first sample point. In this case a junction box is used to connect the tubes. Once the tubing is laid a further layer of sand is used to cover it. Marker tape is then laid out on top of the sand to indicate the location of the tubes. The trench is back filled ensuring that no heavy rubble is replaced which could damage the tubing. The next stage is to connect the tubing to individual boreholes. This is carried out by excavating around the borehole casing and feeding the tube under the casing, up through the bentonite seal and connecting to the top of the borehole. A filter is installed at this stage to prevent dust, insects and moisture from entering the sample tubing. Finally the tubes are connected to the system.

When the system is commissioned a gas sample of known concentration is introduced into the sample point end of each tube in turn and analysed by the ALGM. The purpose of this is to ascertain the time taken to draw the gas down each tube into the analysis system. These times referred to as "purge times" are then input into the controlling computer such that analysis of each sample does not commence until the purge time has elapsed. This test also confirms the integrity of the whole sampling system.

Figure 2 ALGM Inside Security Cabin

The monitoring system can either be constructed in a security cabin and then transported to site or transported in modular form and installed into accommodation on site. Inside the cabin each tube is connected to an individual moisture/dust trap and then to a valve. The bank of moisture/dust traps is shown in Figure 2 on the right hand wall of the cabin and the valve cabinet is shown on the right hand side of the rear wall. A vacuum pump shown below the valve cabinet runs continuously after sampling or purging the system with fresh air. When the control computer selects the appropriate valve the sample is passed through a membrane filter, which removes any moisture present in the sample, and into the analysis cabinet, shown to the left of the valve cabinet where it undergoes a further two stages of filtration prior to being analysed. The analyser cabinet contains two infrared gas analysers for methane, one for carbon dioxide and an electrochemical cell for oxygen as specified in Figure 3. The two methane sensors cover low and high concentrations ensuring optimum accuracy over the range of concentrations likely to be measured. Sensors for other gases can be added if required.

Gas Analyser	Analysis Technique	Range (% v/v)	Accuracy (% v/v)
CH_4	Infrared	0 to 5	±0.1
CH_4	Infrared	0 to 100	±2.0
CO_2	Infrared	0 to 30	±0.6
O_2	Electrochemical	0 to 25	±0.75

Figure 3 Analyser Specifications

The analyser cabinet also contains a barometric pressure transducer so that analytical data can be referenced to changing barometric pressure. The system is self-calibrating using a cylinder of gas of known concentration housed within the cabin. The system cabin is fitted with a methane sensor and audible alarm such that if a build-up of gas occurs within the cabin an alarm is raised both locally and remotely using an automatic dial-out alarm facility.

The control computer is a PC based industrial computer specifically designed for maintenance free operation in demanding environments. Although slightly more expensive than conventional, office type PC's, experience has shown that an industrial grade computer is essential for reliable operation over the range of temperatures and humidities typically found inside site cabins. This avoids the need for expensive air conditioning.

The monitoring system is fully configurable such that sampling sequences and intervals, rising and falling alarms and alarm messages can all be pre-set and are customer specific. Results are stored on the control computer's silicon disk and approximately three months data can be stored depending on the number of sampling points. On site results, alarm messages and historical data can be displayed on a VDU in tabular format.

A PC and modem, normally located at the customer's headquarters, can be used to automatically or manually interrogate the system computer and to download and store the analytical data. This facility can be configured to log-on by request or automatically at a pre-set time each day. Should the log-on process not be carried out for several days or weeks the program is configured to download all the readings from the last log-on date. The files are stored in such a manner that they can be imported into several of the data handling packages at present on the market such as Microsoft Excel. This enables customers to continue using their normal reporting formats but with greatly improved efficiency.

The control computer also accepts up to 16 analogue and 24 digital inputs and therefore the system can be used to monitor and control other landfill equipment. The most common example of this is flare monitoring whereby the gas concentration, flow and/or pressure in the gas lines to the flare and the flare temperature are monitored. The results are displayed and logged in the same way as the borehole readings. Examples of other equipment which can be interfaced are leachate level switches, leachate volume flow meters, a weather station and groundwater quality monitors. A novel application is the connection of the ALGM to a landfill gas interception curtain which is described in the case study later in the paper.

Experience to date has indicated that systems have a long-term robustness in what is a hostile operational environment. Although purchase of a system requires a capital outlay the cost of ownership is low which can lead to considerable savings over manual monitoring. This coupled with the improvement in quality and quantity of data when compared with manual surveys makes automatic systems a worthwhile consideration.

CASE STUDY

Bellhouse is an operational quarry owned by a well-known minerals extraction company that is located on the outskirts of Colchester. Between 1981 and 1993 the north-east corner of the quarry, Cells A to D, an area of approximately twelve hectares was filled with domestic, commercial and industrial waste by Essex County Council under licence from the owners. The landfill operations were carried out under The Control of Pollution Act Section 11. In 1993 the site was issued with a Waste Management Licence in accordance with the Environmental Protection Act. At the same time Essex County Council formed ExWaste Ltd a Council wholly owned waste management company which operated the site until 2000. ExWaste ceased tipping waste in 1998 with the completion of Cell D after which the remainder of the site was taken over for tipping by another waste management company.

A gas migration control scheme was installed by ExWaste in the early 1990's which comprised gas extraction boreholes, pipework and flares. A gas utilisation plant was commissioned in early 1998 covering Cells A to C. In addition Cells A and B were capped at this time which resulted in a significant increase in gas migration problems. As a result additional extraction boreholes were installed in order to address these problems and to increase the efficiency of the gas utilisation plant.

In late 1997 and early 1998 high gas readings were detected in the perimeter boreholes near the eastern boundary of the site, approximately thirty metres away from houses. Gas was also found in the service ducts alongside the road. Occasional incidences of spontaneous combustion in the waste were additional hazards. On discovering the problem ExWaste increased the manual gas monitoring from weekly to daily. However the Environment Agency requested ExWaste to produce a Management Plan to overcome the problem. An environmental risk assessment was carried out by environmental consultants whose final report in September 1999 recommended that an interception curtain be installed as well as an automatic monitoring system for the perimeter boreholes.

Well before the final report had been issued, ExWaste arranged for the design and installation of the interception curtain to be carried out. The curtain comprised eighty-five boreholes and vent pipes each two metres apart and divided into five banks. Each bank had a monitoring borehole which would be used to detect migrating landfill gas and lead to the triggering of all or part of the curtain as necessary allowing the gas to escape to atmosphere via the vent pipes. As this work was progressing ExWaste initiated a tendering exercise for the design, supply, installation and commissioning of an automatic landfill gas monitoring system plus a monitoring and control system for the interception curtain.

A summary of the contract specification was as follows:

- An automatic landfill gas monitoring system to be capable of monitoring twenty seven perimeter boreholes and five extraction gas lines at least once per day
- The system to be capable of measuring methane, carbon dioxide and oxygen with an option to measure carbon monoxide because of the spontaneous combustion

- The system to be capable of self calibration, of storing all data generated with the capability for remote interrogation from ExWaste's offices via a modem link
- The system to have sufficient capacity for an additional eight boreholes
- All sample tubing and borehole connections to be installed below ground with tubing runs restricted to the site boundary making the longest run approximately 1500 metres
- Control and monitoring equipment to be installed to give the monitoring system the capability of control and monitoring of the interception curtain
- Five years maintenance.

An order was received from ExWaste on 22 September 1999 for TES to carry out the contract and site work commenced on 4 October. The completed system was handed over to ExWaste during the week commencing 25 October.

The Automatic Landfill Gas Monitoring System installed was a standard TES system with a carbon monoxide sensor expansion fitted. The carbon monoxide data collected on the system have been used to detect and monitor the activity of spontaneous combustion incidences. One possible theory was that the site was being over extracted by the gas utilisation system thus inducing oxygen leakage into the waste. As a result a remote terminal was installed at the companies site office such that the methane, oxygen and carbon monoxide readings could be routinely observed and a suitable gas balance for the site maintained.

Another feature of the Bellhouse ALGM system was a dial-out alarm to automatically notify ExWaste's head office of any alarm conditions. These alarm levels were specified by ExWaste and related to the operation of the interception curtain valves and the gas concentrations in the curtain monitoring boreholes.

A compressed air actuated valve was installed on each of the five interception curtain banks of boreholes, each valve being operated by an intrinsically safe relay fitted locally to the valve which was connected to the ALGM via an underground cable. A sample tube was installed in each of the five monitoring boreholes to monitor for any migrating gas. The ALGM was configured such that any reading above 1.0% methane in any curtain monitoring borehole would trigger the valve for that bank of the interception curtain and send an alarm message to ExWaste indicating that the curtain was open. When all or part of the curtain was open, the migrating gas would be vented to atmosphere via the cowled vent pipe rather than migrate towards the houses. When the methane concentration dropped to below 1.0% the valve would close and a second alarm message would be transmitted.

The system has been successfully demonstrated to the Environment Agency and to the local Parish Council and in the fifteen months of operation the curtain has only been triggered for maintenance or demonstration purposes.

The ALGM is now operated and controlled by Essex County Council who use the daily automatic and manual remote interrogation facility via the modem link to download data for reporting and daily monitoring of the site. There is now no requirement to visit the site solely for gas monitoring and therefore the system has already paid back a significant portion of the capital originally spent.

ACKNOWLEDGEMENT

The author wishes to thank colleagues at TES Bretby for their assistance in producing this paper and Mr J A Scarrow, Managing Director of Environmental Services Group Limited (ESGL), for permission to publish it. The views expressed are those of the author and not necessarily of ESGL or TES Bretby.

Risk assessment and remediation of contaminated land – training material

JOANNE KWAN[1] (CIRIA), DAVID RUDLAND[2] (Halcrow Group Ltd), NICKI NESBIT[3] (Enviros Aspinwall)
[1]CIRIA 6 Storey's Gate, Westminster, London SW1P 3AU, UK
[2]Halcrow Group Ltd Burderop Park, Swindon SN4 0QD, UK
[3]Enviros Aspinwall Walford Manor, Baschurch, Shrewsbury ST4 2HH, UK
Key words: education and training, knowledge transfer, contaminated land, risk assessment, remediation

ABSTRACT

Redevelopment of contaminated land is a substantial business in this country. Over £500 millions are spent every year in the UK on removing contaminants from polluted sites. Part IIA of the Environmental Protection Act 1990, which was implemented in England and Scotland more than 12 months ago, has brought new responsibilities to many construction practitioners.

A recent research carried out by the Construction Industry Research and Information Association (CIRIA) shows that despite large number of construction companies have more than ten years experience in dealing with contaminated land, more than half of their staff are not receiving training on the subject.

Training is needed for consultants, contractors, regulators and other construction professionals, involved in contaminated land schemes, particularly in the following areas:
- the development of objectives for risk assessment
- data collection processes including site investigation and estimating the risks of the site
- identification of contaminated land-related, engineering and management objectives the based on the risk assessment, site specific constraints and legislative requirements
- the development of the remedial strategy
- selection and implementation of the appropriate remediation options including monitoring and aftercare operations.

BACKGROUND

UK construction industry is under growing pressure to build on contaminated land. The government has set targets for increasing the amount of building on previously used sites over the next ten years. Where contamination exists, there are often significant problems to overcome before the land can be redeveloped.

There are also significant changes underway in the legislative regime in the UK. New legislation is in the process of enactment that will significantly change the way in which contaminated land redevelopment projects is carried out in this country. A key piece of legislation is Part IIA of the Environmental Protection Act 1990 which came into force in England and Scotland in 2000.

Geoenvironmental impact management, Thomas Telford, London, 2001.

Whilst many of those to whom responsibility for carrying forward the new regime will be well versed in the procedures and practices on risk assessment and remediation, many will be facing the responsibility for the first time. In particular, local authorities will have the initial responsibility to conduct or instruct assessments for the new regime, however as the regime develops, others will need to have the skills necessary to measure the risk and remove contamination from land. This includes professionals from a wide range of background. (Figure 1).

The Construction Industry Research and Information Association (CIRIA) has carried out a research project to assess the need of training materials for those who:
- have an interest in or specify contaminated land risk management (CLIENTS)
- practise contaminated land risk management (PRACTITIONERS)
- regulate contaminated land (REGULATORS)

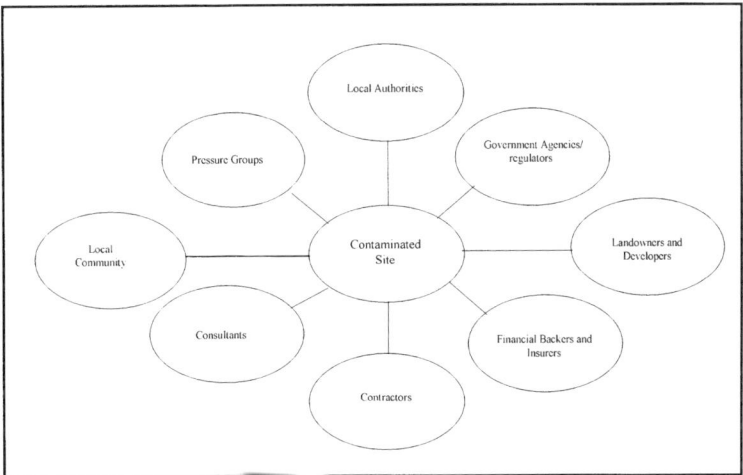

Figure 1 Contaminated land stakeholders

A survey was undertaken using a three-stage approach; a questionnaire, telephone interviews and face to face meetings. The questionnaire was sent out to over 320 selected companies, organisations and individuals. The questionnaire was sent to three main groups of organisations:
- developers, legal, financial and landowners (CLIENTS)
- consultants (PRACTICTIONERS)
- local authorities and environment agencies (CLIENTS, PRACTICTIONERS and REGULATORS)

The result of the survey shows that:
- 83% companies procure services in contaminated land risk assessment and the average length of experience is 12 years.
- 50% of respondents commented that their company does provide training which as reported as a mixture of in-house training, conference attendance and external courses.

This does tend to suggest that whereas many companies might provide services in contaminated land management, and some for many years, training is not always available.

In response to this, CIRIA has commissioned two training packs:
- CIRIA Project RP599 Contaminated Land Risk Assessment – Good Practice Guidance
- CIRIA Project RP601 Remedial Options in Contaminated Land

The aims of the training packs aim to provide sufficient information to allow the user to assess risks due to contaminated land and determine remedial requirements in a rigorous, logical and transparent framework according to recognised good practice at the time of publication. The packs are most suited to those whose duties have recently required them to become involved in different stages of a contaminated land project from other technical fields or who have some knowledge and experience and wish to refine their skills. The packs will also be of benefit to those with experience and who wish to be able to update their abilities in relation to the rapidly changing regulatory regime. The material is primarily intended to be used in a group learning environment, but may also benefit individuals working on their own. In addition, CIRIA has published a third pack, *Environmental Good Practice on Site that* describes the issues that should be addressed, including contaminated land, during construction activities. A feature of these CIRIA training materials is that they provide signposts to existing documents that describe in detail particular aspects of contaminated land management. The users may wish to conduct further studies of their own and each pack will provide directions to sources of information.

These training packs are now available from CIRIA.

What will be in the training materials?

RISK ASSESSMENT
The materials will describe the process by which risk assessment is conducted. The guidance is mindful of the forthcoming *Handbook of Model Procedures for the Management of Contaminated Land, CLR11* to be published by the Department of the Environment, Transport and the Regions. The regulatory regime for the assessment of existing contamination is based on the source-pathway-receptor principle. There are four sub-steps involved in this as follows:
- Hazard identification
- Hazard Assessment
- Risk Estimation
- Risk Evaluation

Each step involves progressive refinement of data. Hazard identification and assessment are the initial data gathering and review stages, more commonly known in the industry as "Phase 1". At this stage the desk studies are undertaken, site history and potential for contamination identified, and the preliminary conceptual model to represent potential sources, receptors and the pathways between them is proposed.

At Risk Estimation stage detailed ground investigations may be undertaken in order to confirm the presence of contamination and to allow estimation of the risks that could be expected under defined conditions of exposure. At Risk Evaluation stage the information is reviewed to decide whether the estimated risks are unacceptable, allowing for any technical uncertainties and other site-specific circumstances. Risk Estimation and Evaluation usually form the stage of assessment known as Phase 2.

Not every risk assessment will need to complete all these stages. It is quite acceptable for the process to be complete at the end of the Hazard Assessment stage, if, for example it has been decided that there are no feasible complete source-pathway-receptor linkages. What is essential for each and every assessment is a transparent decision making process that allows clear concise communication of the risks at each and every stage. Risk communication is a somewhat neglected art and the guidance will discuss this. The result of the risk assessment must always indicate whether the risks associated with the site are acceptable and whether some form of remedial action is required to reduce risks to insignificant levels. This will form the basis upon which the remedial strategy is determined.

REMEDIAL TREATMENT
The material will describe the different stages in the selection of the appropriate remedial options for contaminated sites.

1. Setting of objectives
Remedial action objectives can be:
- contamination-related
- engineering
- management

Contamination-related objectives should take priority in determining the type of remedial action to be undertaken. Remediation, however, will also be influenced by factors other than those arising from the assessment of risks from contamination. These factors include engineering objectives, such as improved stability or ground bearing capacity, and management objectives, such as costs or timescale.

After taking into the account of site specific constraints such as the physical condition of the site, site use, etc., the remedial action objectives should be prioritised into Primary Objectives (those which must be met) and Secondary Objectives (whose which would be desirable to meet). Potential conflicts of these objectives should be resolved early in the project.

2. Review of remediation techniques
A number of remediation techniques are now available and they can be categorised according to the mediums and the contaminants present (Figures 2 to 4). These can be used singularly or in combination to destroy or modify the nature and behaviour of contaminants.

Figure 2 Remedial options for contaminated gases

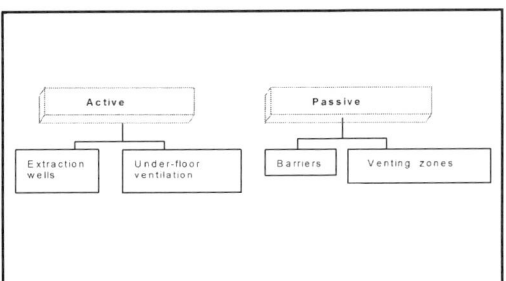

Figure 3 Remedial options for contaminated liquids

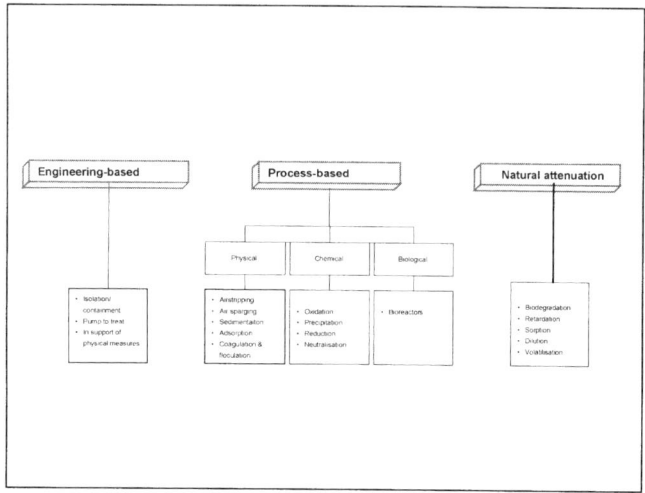

Figure 4 Remedial options for contaminated soil

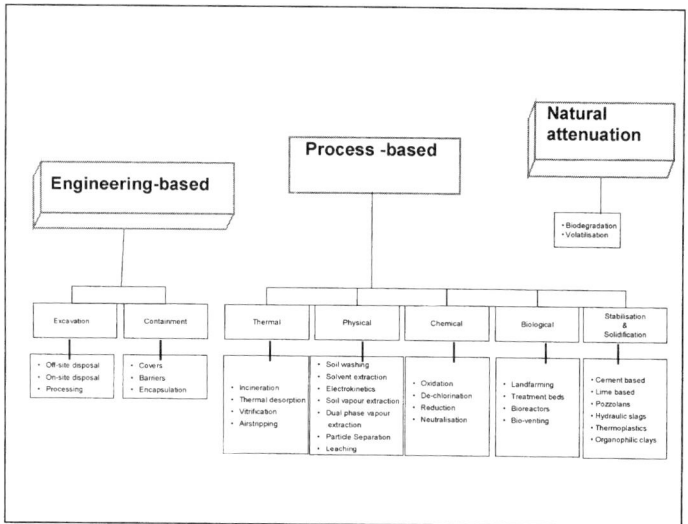

3. Development of remediation strategy

After reviewing the different remedial techniques, it is important to develop a strategy for different zones of the site. The division of zones can be done by depth profile or by lateral profile, media or the suitability of different areas of the site for treatment.

There are a large number of factors that should be considered in the development of strategy (Table 1). It is also important to consider the impact of remediation in one zone or on one pollutants linkage to another.

Table 1 Factors that should be considered in remedial strategy development

• Applicability	• Information requirements
• Effectiveness	• Planning and management needs
• Limitations	• Monitoring needs
• Cost	• Health and safety aspects
• Development status	• Potential environmental impacts
• Availability	• Validation requirements
• Operation requirements	• Post-treatment management needs
• Regulatory requirements	• Local concerns

During the development of the strategy, it is important to consult the following third parties:
• regulators, particularly about the application of the relevant consents.
• developers
• investors/lenders
• insurers
• site neighbours.

4. Selection and implementation of remedial treatment
• Initial selection of the remediation options
In order to select an initial range of options for different zones, overall objectives should be referred to and the options should be compared against the Primary Objectives. This will result in a list of options for each zone. When appropriate, options for different zones should be combined to form alternative site-wide remedial strategies.

• Final selection of the remedial options
The shortlist of strategies for each zone should undergo further detailed assessment. This will involve further evaluation and screening on technical and environmental grounds. Information such as accurate costings and treatability data will be needed.

A final evaluation should be made against Primary and Secondary Objectives before the final option is chosen for each zone.

• Implementation
Validation including a validation report is an essential part of the implementation of any remedial scheme. In the majority of cases it will be needed to demonstrate that the remedial objectives of the project have been met. It may take the form of analysis to confirm that residual contaminant concentrations are less than the required remedial objective, or that containment measures are performing as planned.

Monitoring is also important both during and after treatment. This may include noise, dust, groundwater, surface waters and gas.

The authors and CIRIA wish to acknowledge the work of all those involved in CIRIA research project 601, *Remedial treatment for contaminated land: in-house training material* and 599 *Contaminated land risk assessment – good practice guidance* and the funding from the Partners in Innovation Programme of Department of the Environment, Transport and the Regions, the Environment Agency, SNIFFER, NHBC, English Partnerships and CIRIA Core member organisations, without which none of the work could have been done. © CIRIA, and the authors 2001.

The Use of Risk Ranking Schemes to Assess Potentially Contaminated Sites

M. PENN, A. S. O'BRIEN and A. J. SMALL
Mott MacDonald, 26-30 Wellesley Road, Croydon, CR9 2UL, UK

ABSTRACT
Some existing risk ranking schemes for potentially contaminated land are reviewed and compared to a client specific ranking scheme that was developed to prioritise a large number of sites for further action. The risk ranking schemes are discussed in the context of consistency, sensitivity, the level of available data and their ability to be utilised within an overall management strategy for dealing with risk.

1.0 INTRODUCTION
A risk ranking scheme can potentially provide a systematic framework for prioritising actions, following a preliminary desk study. For landowners and local authorities who may need to assess dozens or hundreds of sites, risk ranking is an important tool for the rapid identification of sites which may need urgent remediation. As noted by the EPA[1]: "If priorities are established based on the greatest opportunities to reduce risk, total risk will be reduced in a more efficient way, lessening threats to both public health and local and global ecosystems." Hence, in this context the ability to consider the effectiveness of remediation options becomes an important aspect of risk ranking within the risk management process, Figure 1.

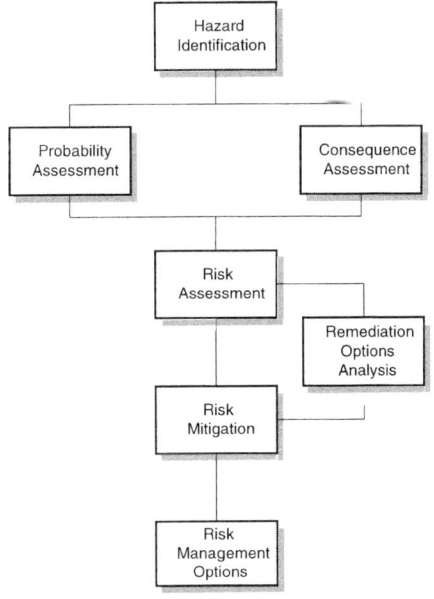

Figure 1. Simplified Flow Chart for Risk Management Process

2.0 RISK ASSESSMENT FOR POTENTIALLY CONTAMINATED LAND

2.1 Qualitative Risk Ranking Assessments
Risks arising from contaminated land are usually assessed on a purely qualitative basis, with sites classified as very high, high, medium, low or very low risk. This approach has the advantage of being simple and relatively cheap to produce. However there are several disadvantages: including a lack of *subjectivity and consistency;* a lack of *transparency;* and the lack of a *framework to assess future changes.* Given the above difficulties, it is worthwhile considering risk on a quantitative basis.

2.2 Prioritisation and Categorisation Procedures
The development of risk ranking schemes is discussed in the Department of Environment CLR No. 6[2]. The report describes a prioritisation

Geoenvironmental impact management, Thomas Telford, London, 2001.

procedure for potentially contaminated sites. The procedure places a site into a priority category between one and four. Category one requires 'urgent action' whilst for category four 'no action is needed'.

2.3   Generalised Risk Ranking Scheme
A general semi-quantitative scheme for estimating environmental risk is presented in Table 1. The scheme requires desk study information to give a score for receptor potential and contamination potential. Only on-site contamination is assessed. The score classifies the site into one of four categories ranging from very low risk to very high risk. It provides a simple and quick assessment and is based on readily available information. It is typical of many risk ranking schemes which are currently in use.

3.0   CLIENT SPECIFIC RISK RANKING SCHEME
A risk ranking scheme was formulated for a large number of potentially contaminated sites owned by one client. The sites all had similar current uses, but the site histories, geological and environmental settings were different. The purpose of the study was to assess the level of risk from current and historical contamination sources both on and off site to nearby receptors. The scheme did not include potential risks to on site workers, since these would be managed by appropriate health and safety systems. The main purpose of the risk ranking was to assess possible third party liabilities.

3.1   Environmental Data
A desk study was compiled for each site and included information available from the client and public sources and site walkover surveys. To maintain consistency between site surveys, a client specific pro-forma was used to record observations. The site manager was also asked to complete a specially developed questionnaire. The desk studies were followed by sampling and testing of surface soils and water from drains and nearby surface water bodies. Intrusive investigations were undertaken at about 30% of the sites. They were selected to cover the range of perceived risk levels, geological settings, geographical locations and adjacent land uses, in order to calibrate the risk ranking model.

3.2   Development of the Risk Ranking Scheme
A semi-quantitative risk ranking scheme was chosen as an objective and consistent approach to determining risk, which would be transparent and auditable by external organisations. It also offered a framework for future re-appraisals, as new information became available. An example of a semi quantitative scheme developed for landfill sites, is given in ref. 3. A summary of the main components of the risk ranking scheme is given in Table 1. It is worth noting that the "sources" scores were cumulative and so reflected the number of potentially contaminating activities as well as the type. An important part of the "receptors" score was a consideration of the overall sensitivity of the surrounding land in order to place the site in the context of its surroundings, in addition to an assessment of specific local receptors. For example, a small housing area surrounded by industrial facilities was judged less sensitive than a purely residential area.

3.3   Sensitivity Analyses
Sensitivity studies were undertaken to assess the integrity and robustness of the overall ranking in order to determine if any one factor or group of factors displayed a disproportionately strong effect on the total risk score. The first type of sensitivity analysis involved one of the sources, pathways or receptors scores being squared or square-rooted. The change in ranking position of each site was assessed for each run.

Generalised Scheme

(A) Receptor Potential	(B) Contamination Potential
Surface Water Proximity	Nature of Use
Surface Water Dilution	Age of Process
Solid Geology Permeability	Nature of Contamination
Superficial / Made Ground Permeability	Nature of Gases
Groundwater Vulnerability	

Risk =>

(a) Present Land Use (A) (a) x (B)
(b) Future Land Use (A) (b) x (B)

Client Specific Scheme

(a) Sources	(b) Pathways	(c) Receptors
(1) On-Site	Surface Run-off	Current Land Use (Downstream)
Current Use (s) And Time Period	Buried River Channels or Ditches	Overall Sensitivity
Historical Use(s) And Time Period	Type of Drainage System (Soakaways, oil interceptors etc)	Aquifer (type and vulnerability)
Storage	Natural Flowpaths (NF) (Permeability, thickness, continuity etc)	Water Abstraction
Evidence		Surface Water
Known Incidents		
(2) Off Site (Upstream Only) Current uses and distance weighting. Historical uses and distance weighting.		

Risk => Σ [(a) (1) x (b) x (c) + (a)(2) x (b) (NF only) x (c)]

Notes
1) Only consider off-site source if less than 250m from site.
2) Only receptors less than 250m from site considered.

Table 1. Overview of Structure of Generalised and Client Specific Schemes

The analysis indicated that regardless of the weighting applied to the scheme, the sites displayed similar risk ranking positions. A second sensitivity analysis was carried out to determine the most influential input factor. The software @RISK was used to determine the influence of a broad range of different input scores, on the magnitude of the output value, i.e. the final risk scores. The chosen simulation model used a multi-variate stepwise regression analysis, which highlighted the most influential factors, Figure 2.

3.4   Application of the Risk Ranking Scheme
Figures 3 and 4 show the scores for on-site sources and overall risk respectively. For the sites considered, Sites A and F had significantly higher overall risk scores than the other sites. Site E had a relatively high on-site sources score, however because there were few pathways for potential contaminant migration the overall risk score was relatively low. The scheme was also used to determine the risk reduction from different types of remedial works. The reduction in the total risk score was used to provide an indication of the cost-effectiveness of each remedial activity at each site. This facilitated the development of a structured, cost effective remediation plan for each of the sites and enabled a hierarchy of future actions and an effective and targeted risk management policy to be developed. It is noteworthy that the risks associated with site A could be significantly reduced with the removal of soakaways from the site, a relatively low cost remedial measure, Figure 4. In contrast, risks at site F could only be significantly reduced with relatively expensive remedial measures (installing cut-off walls).

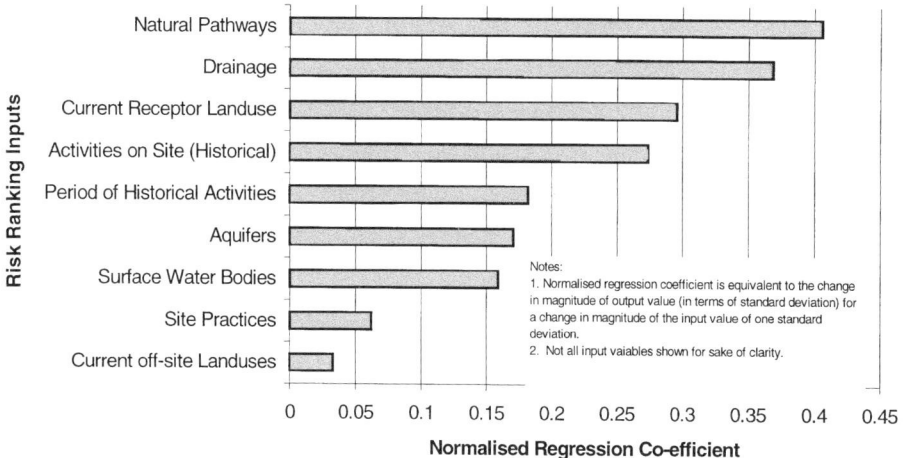

Figure 2. Client Specific Risk Ranking. Result of Multi-Variate Sensitivity Analysis

4.0   COMPARISON OF SCHEMES
The risk ranking schemes described in CLR no. 6, and in section 2.3 above are compared with the more detailed scheme in section 3.0, for seven of the study sites, Table 2. The generalised scheme does highlight significant differences in risk for most of the sites. However, site G attracts a relatively low overall risk score whereas it was subsequently identified as a high risk site. This is partly explained by pathways not being independently considered in the generalised scheme, the probability of a pollutant linkage between sources and receptors cannot be rationally assessed. Because of the simplifications inherent in many generalised schemes the specific circumstances of a site may mean that a generalised risk ranking model may give a misleading indication of the actual level of risk.

Site	Generalised Scheme (Section 2.3)	CLR 6 (Section 2.2)			Specific Scheme (Section 3.0)
		D	SW	GW	
A	330 – very high risk	1	2	1	1st
B	220 – high risk	1	2	2	9th
C	180 – medium risk	1	2	2	14th
D	98 – low risk	2	2	2	19th
E	165 – medium risk	1	2	2	17th
F	390 – very high risk	1	1	2	2nd
G	140 – medium risk	1	2	2	3rd

Note: D = development; SW = surface water; GW = ground water. Ref 2.

Table 2 - Comparison of Risk Ranking Schemes

In contrast the CLR no. 6 approach is inherently a safe assessment methodology in that all the sites (except site D) are categorised as priority 1 sites. Unfortunately, this type of outcome is of little practical benefit to a land owner or local authority who has to prioritise between a large number of 'priority one' sites.

During the study discussed in Section 3.0, an initial qualitative risk assessment was performed based on desk study data. The classification was high, medium or low risk. It was found that the perceived risk allocated after the desk study correlated well with the on-site sources score but not with the overall risk score. This indicated the tendency for visual observations at a site (sources) to strongly influence the result of a qualitative risk assessment, rather than considerations of viable pathways and sensitive receptors.

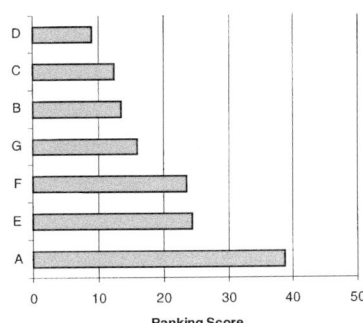

Figure 3.Client Specific Scheme-On Site Sources

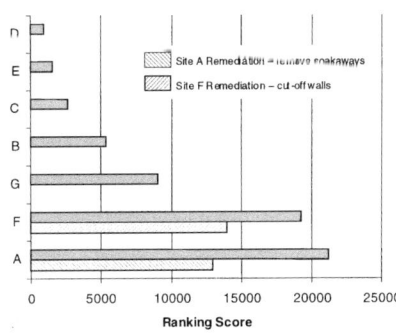

Figure 4.Client Specific Scheme-Total Risk Ranking

5.0   CONCLUSIONS

The paper compares risk based schemes for the ranking of potentially contaminated sites. The schemes have primarily desk study based approaches. The main conclusions from this review are:

(i)   the procedure described in DOE CLR No. 6 will tend to place the majority of sites in old urban industrial areas into a high risk category.

(ii)   care is required in using generalised schemes, since site specific circumstances may lead to unsafe risk assessments. The logic and linkage between different elements of the risk model need to be appropriate to the sites under review.

(iii)   semi-quantitative schemes potentially provide greater objectivity and transparency than qualitative schemes. However, they must be properly calibrated to ensure they are robust and reliable. In addition, clear guidance on scoring procedures is important.

(iv)   client specific schemes can be designed to be used as a powerful tool for overall management of risk across a portfolio of sites, if cost-benefit analyses of remedial measures are incorporated into the scheme.

(v)   a careful balance needs to be maintained between the amount of input data, and the associated cost and time, compared with the reliability and power of a risk ranking model to support judgements and decisions.

Finally, as noted by the Health and Safety Executive, risk assessment techniques can be of most practical value when comparing the cost-effectiveness of different risk reduction methods. Absolute measures of risk (either descriptive or quantitative) are only of significance if the outcome is "no action".

REFERENCES

1)   Environmental Protection Agency (U.S.A.) – "Reducing Risk", 1992.

2)   Department of the Environment (1995) Contaminated Land Research Report, CLR No. 6, Prioritisation and categorisation procedure for sites which may be contaminated, HMSO.

3)   Geraghty and Miller International Inc. (1997) The effects of old landfill sites on groundwater quality - phase I, National Rivers Authority R&D Note 415.

4)   Quantified Risk Assessment: its input to decision making, Health and Safety Executive, 1989.

Remote Sensing Techniques for Representative Environmental Sampling Design

C. SIMEONI, ISPESL/DIPIA; Monteporzio Catone (RM), IT, S. BELLAGAMBA, ISPESL/DIPIA; Roma, IT, A. MARINO, ISPESL/DIPIA; Roma, IT, M. VILLARINI, ISPESL/DIPIA; Monteporzio Catone (RM), IT,G. LUDOVISI, ISPESL/DIPIA; Roma and A. MOCCALDI, ISPESL; Roma

ABSTRACT

The development of a sampling design is basic in environmental risk management plan. The aim of the present research is the environmental sampling design optimisation to monitor herbicides utilised in agriculture, combining remote sensing information whit other qualitative and quantitative data, managed in different format (raster, alphanumeric, vectorial. Remote sensing images can represent terrain surfaces of different nature and origin in a synthetic and uniform way. Using aerial-imaging the proposed method produced a map of soil classification, which allowed locating sources of risk and identifying main human and environmental targets exposed to risk. Then was applied the proximity analysis using information on position of the area and on other priori information. The output of this second step was the identification of homogeneous zones, where samples were collected.

The approach allows using remote sensing to overlap different informative levels in order to collect representative samples in large-size areas and provides a useful tool in the management of environmental risk.

INTRODUCTION

Aim of the work is the environmental sampling design optimisation to monitor herbicides utilised in agriculture, combining remote sensing information whit other qualitative and quantitative data, managed in different format (raster, alphanumeric, vectorial). Sampling was to be made in natural water.

The development of a sampling design is basic in an environmental risk management plan.

The objective of representative sampling is to ensure that a sample or a group of samples accurately characterises site conditions. A representative sampling design that can be able to identify fundamental aspects such as:

- site environmental health
- actual risk sources
- location of human and environmental targets and the level of their exposure to risk.

Furthermore the choice of representative samples ensures the optimisation of analysis costs and sampling time. The time factor is really important for the research. As a matter of fact, the application of herbicides occurs following a specific seasonal pattern related to a phenologic cycle and the pollutant elutes quickly in natural water.

AREA OF STUDY AND AGRICULTURAL PROBLEMS

The site studied is the Sabina area situated north of Rome (Central Italy) near the Tevere river. The landform is hilly and mountainous and the sediments are calcareous and volcanic. The Tevere river valley is a flat area.

The climate is typically Mediterranean characterised by mild winters, raining autumns and dry summers. The mean temperature in the year varies between 13 and 16 degrees. The rain amount varies between 750 and 1500mm in internal zones.

The applied tillage techniques are strongly conditioned by the Politics Guide Line (PAC). They are aimed to maximise output using input like chemical fertilisers, herbicides etc.

Corn and maize has been the studied cultivation. The agricultural techniques required the application of large quantities of herbicides to reduce the weeds presence.

Generally herbicides can be apply into different stages:

1. before seeding;
2. some day after the seeding, to control normal grass infestation;
3. during growth season to control late infestation.

For that concerns the two crops considered (maize and corn) the application occurs only in the second stage. The third phase takes place only if the previously stage is ineffective or in case of non appropriate climatic condition.

METHODOLOGY

Sampling tecniques

Representative sampling approach, recommended to EPA (Environmental Protection Agency- the major American organism of decision-making about environmental pollution), in the "Removal Program – Representative sampling guidance", includes: Judgmental, Random, Stratified random, systematic grid, systematic random, search and transect sampling. A representative sampling plan may combine two or more of these approaches.

Judgmental sampling: is the subjective selection of sampling location in a site, based on historical information, visual inspection and on best professional judgment of the sampling team. This method precludes any statistical interpretation of the sampling results.

Random sampling: is the arbitrary collection of samples within defined boundaries of the area of concern. Random sample locations uses a random selection procedure (random number table).

Stratified random sampling: relies on historical information and prior analytical results (or field screening data) to divide the sampling area into smaller areas called "layers". Each strata is more homogeneous than the site as a whole. Layers can be defined on the base of various factors, including: sampling depth, contaminant concentration levels, and contaminant sources areas. Sample location within each of these strata uses random selection procedures.

Systematic grid sampling involves subdividing the area of concern by using a square or triangular grid and collecting sampling from the nodes.

Systematic random sampling subdivide the area of concern using a square or triangular grid then collects samples from within each cell using random selection procedures.

Search sampling utilises either a systematic grid or systematic random sampling approach.

Transect sampling involves establishing one or more transect lines across the surface of a site.

SAMPLING OBJECTIVE	SAMPLING APPROACH						
	JUDGMENTAL	RANDOM	STRATIFIED RANDOM	SYSTEMATIC GRID	SYSTEMATIC RANDOM	SEARCH	TRANSECT
ESTABLISH THREAT	1	4	3	2[a]	3	3	2
IDENTIFY SOURCES	1	4	2	2[a]	3	2	3
DELINEATE EXTENT OF CONTAMINATION	4	3	3	1[b]	1	1	1
EVALUATE TREATMENT AND DISPOSAL OPTIONS	3	3	1	2	2	4	2
CONFIRM CLEANUP	4	1[c]	3	1[b]	1	1	1[d]

1 - Preferred approach
2 - Acceptable approach
3 - Moderately acceptable approach
4 - Least acceptable approach
a - Should be used with field analytical screening
b – Preferred only where known trends are present
c – Allows for statistical support of cleanup verification if sampling over entire site
d – May be effective with compositing technique if site is presumed to be clean

The arbitrary collection of sampling points requires each sampling point to be selected independently from the location of all other points, and results that all locations within the area of concern have an equal change to be selected. The key to interpreting this probability statements is the assumption that the site is homogeneous with respect to the parameters being monitored. Because sites subjected to environmental monitoring are rarely homogeneous, statistical sampling approaches that provide ways to subdivided the site into more homogeneous areas are favourite. For this, to select sample points, has been used Stratified Random Sampling. This methods allows to select sample points within omogeneous layers The "layers" have been obtained through remote sensing image.

Remote Sensing GIS
Remote sensing images represent in synthetic way land objects having different nature and origin on wide areas with uniformity and measurement homogeneity.
Remote Sensing allowed to obtain an updated classification of land-use.
Land classification was obtain through unsupervised (Isodata, K-means e Fuzzy K-means) and supervised (Maximum Likelihood e Neural Net) algorithm.
Supervised training is closely controlled by the analyst. In this process, you select pixels that represent patterns or land-cover features that you recognise, or that you can identify whit help from other sources, such as aerial photos, ground truth data, or maps. Knowledge of the data, and for the classes desired, is required before classification.
Unsupervised training is more computer-automated. It allows you to specify some parameters which the computer uses to uncover statistical patterns that are inherent in the data. They are simply clusters of pixels with similar spectral characteristics.

In particular, samples outputs from unsupervised classification have been integrated with other thematic information and utilised as input training sites in supervised classification. The results of classification have been validated by direct survey.

The land-use map has been georeferenced assigning geographical coordinates to raster map data. The land-use map has been inserted in a GIS database and integrated with further information like lithology, geological structure, hydrology, hydrogeological characteristics, that were imported in the GIS as layers.

A Digital Elevation Model (DEM) has been realized from interpolation of topographic data (IGM maps).

The EASI-PACE (PCI) software allowed to consider and process information both in raster (like land-use map and DEM) and vectorial (like surface hydrology, isopiestic lines) format.

GIS

The processing of data layers was performed, using the same software through typical GIS operators like indexing, recoding, matrix analysis, proximity analysis. The matrix analysis assigns a different output value to each unique combination of input values. The proximity analysis creates an output layer showing successive zones of proximity (distances) to a specified entity or group of entities. The choice of recoding indexes, weights and matrices was based on references, personal experience and knowledge of the area. In particular the use of matrix analysis instead of simple weighted overlaying allowed a better control on parameters and variables. The recoding and indexing was applied in order to obtain a limited number of classes for each layer (map), for a better understanding of the information.

RESULTS

Applied methodology produced maps related to homogeneous zones. Homogeneity is referred to environmental risk classes related to herbicides transport in natural water. The obtain sample design has been applied for samples collection an analytical study of herbicides occurrence in natural water.

REFERENCES

Brivio P.A., Lechi G.M. & Zilioli E. *"Il telerilevamento da aereo e da satellite"*, Milano. Edizioni Delfino, 1990.

Brivio P.A. & Ziliolo E. *"Il telerilevamento da satellite per lo studio dei rischi ambientali"*, Roma, Edizioni dell'Ulisse, 1995.

Bonciarelli F. *"Coltivazioni erbacee da pieno campo"* Bologna, Edizioni Edagricole, 1990.

Aarts EHL, Korst J. *"Simulated annealing and Boltzman Machines – a Stocastic Approach to Combinatorial Optimization and Neural Computing"*, Wiley, New York. 1990.

Sacks J., Shiller S. *"Spatial designs. In Statistical Decision Theory and Related Topics IV"*, Gupta SS, Berger JO (eds), Springer, New York, 1988. "

Van Groenigen J.W., Pieters G. and Stein A. *"Optimizing Spatial Sampling for Multivariate Contamination in Urban Area"*, Environmetrics, 11, 277-244, 2000.

Decrease of Risk in the Design and Exploitation of Mines

DR E.V. STANIS, Professor, and DR B.I. MASKOVTSEV, Professor
Russian Peoples' Friendship University, Moscow, Russian Federation

INTRODUCTION

The depths and rates of exploitation of coal deposits are constantly increasing and as a result of a rise in rock stresses, the number of rockbursts, instantaneous rock collapses, coal and gas outbursts, continuity disturbances and undermined strata subsidence, damage of the surface, etc. increases.

In a number of cases mine workings cause not only the deformation of buildings and structures but also cave-ins, the appearance on the earth's surface of open fractures, deep throughs and benches. They result in the dewatering or flooding of mine workings and in other technogenic disturbances of the surface retarding it's economic use. To cope with these impacts, considerable expenditures are required. Another result is hazardous water inrushes from pools being undermined. For example, in the field of the Rechnaya mine of the Torezantratsit Industrial Amalgamation (IA) in the Donbass, a funnel-shaped cave-in occurred in 1981, above the air heading of the first eastern face seam h_7. Its diameter was 6 m and its depth was up to 10 m. This was caused by a caving of the roof rocks in the air heading, from which the support had been removed while working the face in retreat. Shear fractures were formed, through which ground water infiltrated into the mine along the boundary of the disturbed rocks of the main roof. Where the heading intersected a shear fracture, the caving into the heading developed to reach the earth surface. Filtered water carried the fallen rocks into the heading and then to mined-out space.

In the working seam l_7 at the Ostraya mine of the Krasnoarmeiskugol IA, the heading cut across a small geological fold at a depth of 160 m. The fold was characterized by increased natural (tectonic) fracturing and rock crushing, and result in a roof caving extending as high as 2 m. Water carrying a considerable amount of sand from the 65 m thick, loosely cemented sandstone layer lying 15 m above seam l_7 began to enter from the zone of collapse. The sand content of the water was 23%. During the failure about 25,000 m^3 of sand was brought into the workings. In a month, a funnel-shaped cave-in 28 m in diameter and 18 m in depth was formed on the surface [1].

Both of these examples demonstrate the need to attach priority to the reliability of mining and geological information. Commonly, the majority of mining and geological parameters are evaluated with a relative error of not less than 30% (in particular the parameters small-sized amplitude tectonic disturbances, thickness of seams and their hypsometry, gas presence, physical-mechanical properties of adjacent strata and mineral).

The significance of geological conditions in the overall reliability of a mine is exhibited by its representation as a composite hierarchic natural technogenic ecological system composed of several subsystems at a different level that play some role in the realization of the principal target, being the extraction of a mineral.

To ensure the safety of underground mine workings, to increase its efficiency, to augment reliability and to decrease costs, studies have been carried out on the geological, mining and technical conditions of the operation of various underground coal mines. Theoretical methods of forecasting and definition of actual reliability of the mining and geological information and the whole mine as natural technogenic system have been developed.

ANALYSIS
Defining the significance of mining and geological conditions n is one of the main tasks in ensuring the reliability of underground mining. A method of application is proposed, in which a mine is described as a composite hierarchic natural .-technogenic ecosystem composed of four subsystems of a different level, each of which plays a definite role in the mining of a useful mineral, in this case coal. Each subsystem is characterized by a weighting coefficient of a significance Z. Thus mining and geological conditions have a role in the abiotic environment of the ecosystem of a mine, which are defined to ensure its reliable operation and the realization of the posed problem. The influence of mining and geological conditions on the reliability of a mine is boosted with the lowering of the level of the subsystem [2].

Where a subsystem is at the first level, its failure will result in coal not being hauled to the surface. The other subsystems can operate normally. The principal ventilation and transport openings comprise this subsystem. The coefficient of significance of the subsystem of the first level Z_1 equals 1.

Where subsystems are at the second level, their failure will result in a reduction in the rate of coal supply to the first level subsystem. Mine workings where coal is conveyed from a group of longwalls, for example, can be described by this subsystem. The coefficients of significance of second level subsystems Z_2 are proportional to the share of extracted coal in the subsystem to the total tonnage of coal extracted.

Third level subsystems include elements whose failure would lead to the loss of coal production from a single longwall. The coefficients of significance of third level subsystems Z_3 are proportional to the share of extracted coal of the third level subsystem to that extracted by the second level subsystem.

As the foundation to the whole system is a subsystem of the lowest fourth level, the failure of which results in the partial or total loss of coal from a given longwall. At this level, the influence of the mining and geological factors on the reliability of the mine is most obvious. At this level a series of original weighting coefficients of reliability is introduced [3]:

α - coefficient of loss of extraction for reasons unrelated to geological conditions;

γ - coefficient of loss of extraction due to inaccurate estimation of the mining and geological factors;

β - slowness coefficient of extraction connected with mining and geological factors.

Then the reliability coefficient of the next highest third level subsystem N_3 will be given by:

$$N_3 = \sum_{i=1}^{n} \alpha_i + \sum_{j=1}^{n} \gamma_j + \sum_{k=1}^{n} \beta_k \qquad (1)$$

where n = 1,2,3, …

The reliability coefficients of the second and first level subsystems N_2 and N_1, respectively, described by serialparallel coupled systems with allowance for the coefficients of significance Z_2 and Z_1 of each subsystem, will be given by:

$$N_2 = \sum_{m=1}^{n} N_{3m} Z_3 + \sum_{b=1}^{n} T_b^3 \qquad (2)$$

$$N_1 = \sum_{m=1}^{n} N_{2m} Z_2 + \sum_{b=1}^{n} T_b^2 \qquad (3)$$

where T^3 and T^2 = coefficients describing a share of loss of extraction of the applicable subsystems due to the failure of a transportation network accounting for coal extraction from subsystems of lower level.

A statistical analysis of reasons for daily failures at all 12 mines of the Donbassantratsit IA shows that more than 78% of dead time results from mining and geological-related factors [4]. Most significant of these are passing through different geologic features (54%) and control of unstable roofs (Figure 1).

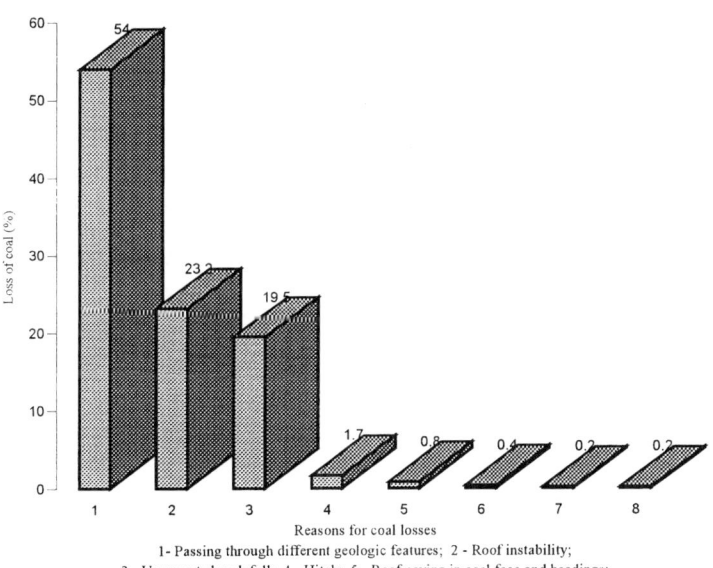

1- Passing through different geologic features; 2 - Roof instability;
3 - Unexpected rock fall; 4 - Hitch; 5 - Roof caving in coal face and headings;
6 - Sudden coal outburst; 7 - Gas pollution; 8 - Floor heave

Figure 1 Mining and geological-related reasons for coal losses

To check this method of estimation of reliability prediction of a mine as a natural technogenic multi-level ecosystem and the definition of the role of mining and geological conditions in forming this reliability, studies have been carried out at the Izvestiya mine (the Donbassantratsit IA) in Central Donbass. The operation of the whole mine, comprising 12 longwalls 9 month period were analysed. The reliability coefficient of the first level subsystem was evaluated using the modified equation (3):

$$N_1^1 = \sum_{m=1} \bar{N}_{m3} Z_3 + \sum_{m=1} \bar{N}_{2m} Z_2 + \sum_{b=1} \bar{T}_b^2 \tag{4}$$

In the analysis, coefficients α_1 (failure of face equipment), α_2 (failure mine face transport for all longwalls), α_3 (organizational dead time), $\gamma' = \gamma + \beta$ (the total dead time and losses of extraction owing to mining and geological conditions) were considered. Also coefficients N_3 and Z_3 were considered (Table 1).

Table 1 Values of coefficients for the Izvestiya mine

Longwall	α_1	α_2	α_3	γ'	Z_3	N_3
1 - western	0.0899	0.0098	0.0005	0.1418	0.1929	0.1759
2 - western	0.1667	0.0085	0	0.2711	0.0075	0.4310
3 - western	0.0174	0	0	0.0073	0.6762	0.0263
4 - western	0.0255	0.0087	0	0.0284	0.3238	0.0715
58' - western	0.0209	0.0026	0	0.0044	0.0913	0.0279
58 - western	0.0103	0.0088	0.0016	0.0044	0.0358	0.0246
54 - western	0.0461	0.0076	0.0015	0.0183	0.0327	0.0735
55 - western	0.1641	0.0069	0	0.0584	0.0704	0.2294
51' - western	0.0315	0.0023	0.0021	0.0051	0.0514	0.0408
12 - north	0.0159	0	0	0.0053	0.0481	0.0212
11 - north	0.0230	0.0015	0.0028	0.0137	0.0400	0.0410
9 - north	0.2178	0	0	0.2709	0.0543	0.4887

RESULTS

The value of the second level coefficient of significance Z_2 was 0.3084, the general coefficient of extraction losses T_2 was 0.0028, and the reliability coefficient of the subsystem N_2 was 0.0357. Based on these coefficients, the reliability coefficient N'_1 (first level subsystem) at the Izvestiya mine was calculated to be 0.1253. As this coefficient is significantly less than 1, the reliability of the mine as a composite natural technogenic system is considered satisfactory. This conclusion was borne out by the fact that the actual extraction of coal during the study period exceeded the norm (over 32,000 tpa). The lowest reliability was found in longwalls 2-western and 9-north (γ' of 0.2711 and 0.2709, respectively). The reliability of these longwalls as third level subsystems is rather low, with reliability coefficients N_3 of 0.4310 and 0.4887, respectively (Table 1). This was borne out by the lower than normal actual extraction in these faces. The analysis of reliability coefficients of the third level subsystems has shown the considerable contribution to this index of mining and geological conditions.

CONCLUSIONS

In relation to coal seams, the main mining and geological factors affecting underground mining by longwall have been identified. These include tectonic disturbance of the coal seam; roof instability; hitch, fluid wash, decomposition and displacement of the coal seam; heightened inflows of water; gas emission from the coal and goaf, outburst hazard; other reasons.

The method of quantitative assessment of the reliability of a mine as composite hierarchic natural technogenic ecosystem incorporating mining and geological conditions, has been described based on the determination of the reliability coefficients of subsystems of different levels.

The values of the reliability coefficients during design must be minimized. With a decrease in the magnitude of coefficient, the reliability of the subsystem will increase.

The influence of mining and geological conditions on the reliability is boosted with a lowering of a level of the subsystem.

Under actual mining conditions there are threshold values for the reliability coefficients, in excess of which the extraction of coal is lower than the norm.

REFERENCES

1. *Airuni A.T., Iofis M.A.* Rock Deformation During Underground Mining of Mineral Deposits. The Mining Sciences in the USSR. Moscow, Nauka Publishers, pp. 42-43, 1985 (in Russian).

2. *Mashkovtsev B.I.* Pattern of ecosystem of a mine. Moscow, TSNIEIUgol, pp. 23-28, 1997 (in Russian).

3. *Stanis E.V., Mashkovtsev B.I.* Reliability prediction of operation of mines on mining-and-geological conditions. Economics of a coal industry, No 5, pp. 16-19, 1995 (in Russian).

4. *Mashkovtsev I.L., Mashkovtsev B.I, Stanis E.V.* Method of control of a unstable roof in longwalls. Pat. of Russian Federation No 2136886, 10.09.1999.

Legislation and Regulatory Control

Environmental Legislation and its Application to the Licensing of Remediation of Contaminated Land.

DR IRENE J ANDERS - CONTAMINATED LAND SPECIALIST
Scottish Environment Protection Agency

INTRODUCTION
Whilst remediation of chemically contaminated land is typically undertaken with a view to improving environmental conditions there are instances where it may have an adverse effect on another part of the environment. In order to minimise such risk legislative controls may be applied to the remediation process.

This paper provides a brief outline of the main pieces of environmental legislation for which SEPA is the regulatory authority, identifies how these may be used to control remediation activities for contaminated land and provides an indication of the typical timescales associated with the application for the various consents, licences and authorisations.

Other regimes such as Planning and Contaminated Land (Part IIA) may be used to determine the extent, nature and standard of remediation required.

THE LEGISLATION
The Control of Pollution Act 1974
The Control of Pollution Act 1974 (as amended) (COPA 1974) is the principal legislation in Scotland controlling discharges of poisonous, noxious or polluting matter to controlled waters. Section 30 of the Act provides the definition of controlled waters, which include:

- relevant territorial waters, extending seaward for three miles from the baseline from which the breadth of the territorial sea adjacent to Scotland is measured

- coastal waters extending from the baselines above as far as the limit of the highest tide or as far as the fresh-water limit of the river or watercourse which adjoins waters within that area

- inland waters, including the waters of any relevant loch or pond and rivers and other watercourses above the fresh-water limit

- groundwaters contained in underground strata, including waters in wells, boreholes and excavations into underground strata

Under COPA 1974 it is an offence to cause or knowingly permit any poisonous, noxious or polluting matter or any solid waste matter to enter controlled waters, unless consented under section 34 of the Act or otherwise licensed or authorised.

Section 46 of COPA 1974 provides SEPA with powers to undertake remediation and prevent or forestall pollution of controlled waters and recover the costs from the person responsible, subject to certain conditions.

Sections 46A to D of COPA 1974, which are yet to be implemented in Scotland, provide powers for SEPA to serve an anti-pollution works notice on a polluter or potential polluter requiring them to carry out works or prevent or remediate water pollution. This provides a more direct means for SEPA to prevent and mitigate pollution than its existing powers to carry out anti-pollution works itself and then recover the costs from the person responsible.

Part I Environmental Protection Act 1990
Part I of the Environmental Protection Act 1990 (EPA 1990) provides the framework for improved control of certain industrial and other processes, to prevent or minimise pollution of the environment. The processes include some activities which may be used as part of remediation such as crushing and screening, incineration and solvent recovery operations. These processes will gradually transfer to regulation under the new Pollution Prevention and Control Regulations.

Part II Environmental Protection Act 1990
Part II of the Environmental Protection Act 1990 (EPA 1990) and the Waste Management Licensing Regulations 1994, (as amended) (WMLR 1994) implement the provisions for licensing the treatment, keeping or disposal of controlled waste and provide details of activities which may be exempt from the licensing regime.

Waste Management Licences can be either Site Licences i.e. specific to the site and can only be surrendered when the regulatory authority is satisfied that the site does not represent a risk to the environment or Mobile Plant Licence i.e. specific to the plant (not the site) and is applicable to any site where that plant is used for the duration of that use, subject to the preparation of a satisfactory working plan for each site.   Mobile plant licences are only applicable to the treatment of waste soils, they do not cover treatment of wastes in liquid form.

Controlled waste is now encompassed by "Directive Waste" which is most simply described as follows:-

In order for a substance or object to be waste it must:-
(a) fall into one of the categories set out in Part II of Schedule 4 to the WMLR 1994 and:
    (i) be discarded, disposed of or got rid of by the holder; or
    (ii) be intended to be discarded, disposed of or got rid of by the holder; or
    (iii) be required to be discarded, disposed of or got rid of by the holder.

Part IIA Environmental Protection Act 1990
Part IIA of EPA 1990 (as amended) provides the framework for identification and remediation of historically contaminated land. Contaminated land is given a specific definition for the purpose of Part IIA as:

"any land which appears to the local authority in whose area it is situated to be in such a condition, by reason of substances in, on or under the land, that -

(a)   Significant harm is being caused or there is a significant possibility of such harm being caused; or

(b)   Pollution of controlled waters is being, or is likely to be, caused".

Guidance on what constitutes significant harm and significant possibility of significant harm are provided in the Statutory Guidance. Pollution of Controlled Waters is defined under Part IIA as "the entry into controlled waters of any poisonous, noxious or polluting matter". Part IIA uses the Polluter Pays Principle and requires remediation of the land and controlled water to a standard suitable for current use.

The regulations prescribe those types of sites for which SEPA is the enforcing authority in respect of securing remediation.

Groundwater Regulations 1998
The Groundwater Regulations 1998 (GWR 1998) complete the implementation of the EC Groundwater Directive 80/68/EEC and supplement Regulation 15 of the Waste Management Licensing Regulations 1994 and existing water pollution legislation.

The Regulations mainly relate to controlling the disposal, or tipping for disposal, of List I or II substances where there might be an indirect discharge of those substances to groundwater, and to controlling other activities which might lead to an indirect discharge of List I substances to groundwater or pollution of groundwater by List II substances.

Regulation 2 excludes from the GWR 1998 any discharge containing quantities of List I/II in such small quantity, and/or concentration as to obviate any present or future danger of deterioration of the quality of the receiving groundwater, i.e. the de-minimius.

Regulations 4 and 5 detail when it is possible to authorise discharges of List I and II substances. Authorisations shall not be granted for direct discharges of List I to groundwater except if investigation reveals that groundwater is permanently unsuitable for use [1]. Authorisations shall only be granted for direct discharges of List II or disposal or tipping of List I or II substances which might lead to indirect discharges of List I or II to groundwater if, after prior investigations, (in line with Regulation 15) conditions are included in the authorisation to ensure that technical precautions are put in place to ensure that there shall be no indirect discharge of List I and all necessary precautions are observed to prevent groundwater pollution by List II.

Regulation 19 may be used to prohibit activities which may result in indirect discharges of List I to groundwater or pollution of groundwater as a result of an indirect discharge of List II.
Offences in respect of the above are treated as contravening 30F of COPA

---

[1] Note, no groundwater in Scotland has been designated as permanently unsuitable for use.

SEPA POLICY

Policy No 19 Groundwater Protection Policy for Scotland

Policy 19 provides a framework for the management and protection of groundwater in Scotland both through SEPA's statutory powers and by consultation with other relevant agencies. The policy also states that where SEPA is of the view that additional powers are necessary, it will apply to government for the provision of such powers.

In addition to that described above, the policy indicates that SEPA will:

- co-operate with local authorities in identifying contaminated land sites posing an actual or potential risk to groundwater.
- recommend to the Planning Authority that it refuses Planning Permission for redevelopment of contaminated sites where water resources could be adversely affected unless it is satisfied that the proposals include effective measures for the protection of groundwater.
- use its powers as the enforcing authority for 'special sites' to secure the protection and remediation of groundwater to an appropriate standard.
- encourage appropriate development of contaminated land so as to mitigate any impact on groundwater.
- encourage planning authorities to ensure that developers control and monitor ground and groundwater contamination during and after development, through planning conditions and/or Section 50 agreement.
- use the Notice provisions of Section 46A of the Control of Pollution Act 1974, as inserted by s22 of Schedule 22 of the Environment Act 1995, whenever appropriate, to require works to be undertaken for the protection or remediation of groundwater.
- refuse to consent the direct or indirect discharge of List I substances to underground strata and will limit the entry of List II substances in accordance with the EC Groundwater Directive (80/68/EEC).
- seek to prevent any direct or indirect discharge into underground strata which may result in pollution of groundwaters and not grant consent to discharge if water resources are judged to be at risk.
- use Prohibition Notices under section 30G of the Control of Pollution Act 1974 to control sewage, trade effluent, and contaminated surface water discharges to land/groundwater where these are perceived to pose a risk to groundwater quality.

APPLICATION OF THE LEGISLATION TO REMEDIATION ACTIVITIES.

Activity	Legislative control	Action and statutory timescale
Treatment of waste soil in - or ex-situ	If waste soil is being treated, either in or ex-situ, then a waste management licence is required. This can be either a site licence or mobile plant licence. A site licence only applies to the actual site licensed and may cover the treatment, disposal or keeping of waste or any combination of these. The conditions applied and surrender provisions will vary depending on the licensed activity. A mobile plant licence applies to the plant and is	Ensure contractor has a mobile plant licence (4 months) and an approved working plan for the site (2 to 4 weeks, non statutory) OR

	transferable to whatever site the plant is used at, subject to preparation and agreement of a working plan for each site. The keeping of waste soil is exempt where it represents temporary storage at the point of production for disposal elsewhere. Any control required is normally met through incorporating the keeping of waste prior to treatment and/or disposal into a mobile plant and/or site licence. Separate control under COPA should be applied to the discharges of effluent from mobile plant, to ensure that effective control is obtained.	Apply for a site licence (4 months)
Disposal of waste matter on-site	Waste soil which is excavated and disposed of on-site is subject to waste management licensing requirements (site licence). Where the waste soil contains substances in Lists I and II of the Groundwater Directive, compliance with Regulation 15 of the WMLR will be required. A waste management licence is also required where contaminated hotspots are discarded by mixing excavated soil with other material and then placed back on site. However, mixing of waste contaminated soil with uncontaminated soil is not a practice that SEPA encourages, as it results in previously uncontaminated soils becoming contaminated	Apply for site licence (4 months)
Disposal of waste matter off-site	Waste soil or other waste which is removed and disposed of off-site is subject to the Duty of Care requirements and should be taken to an appropriately licensed site. The requirements of the Special Waste Regulations may apply where the waste soil constitutes special waste.	Ensure meet Duty of Care requirements. Notify SEPA of Special Waste Consignment (3 days)
Treatment and/or disposal of contaminated groundwater	The treatment of waste water or the disposal of liquid waste requires consent under COPA 1974 unless a waste management licence (site licence) is held. It should be noted that Mobile Plant Licences are not currently applicable to the treatment of waste in liquid form. Treatment is normally encompassed within disposal and so there is generally no requirement for a site licence if a COPA consent is held.	Apply for COPA consent (4 months) unless a site licence is held.
Discharge to controlled waters	Where remedial works result in a discharge to controlled waters, a consent under COPA should be obtained by the operator as a defence to any offence under section 30F. It is an offence under COPA to cause or knowingly permit any trade effluent, poisonous, noxious or polluting matter or	Apply for COPA consent s(34) (4 months)

	any solid waste matter to enter controlled waters, unless consented under section 34 of the Act or otherwise licensed or authorised. Discharges may include the discharge of treated water or the introduction of oxygenating chemicals into water	
Discharge to land	Consent under COPA is not required as a prerequisite to commencing a discharge of trade effluent to land. However, SEPA has powers to serve a prohibition notice should control be required. A prohibition notice may be absolute (requiring the discharger to apply for consent if he wishes to make the discharge) or conditional. Where discharge is made of matter which does not represent trade effluent and which contains substances listed under the Groundwater Directive, control may be required under the Groundwater Regulations 1998.	Notify SEPA Possible licensing under Groundwater Regulations (4 months)
Installation of barriers	The installation of cut-off walls and barriers normally falls outwith waste management licensing requirements, unless contaminants are being discarded through treatment or disposal. Where barriers are formed through pumping of grout or injection of clays containing listed substances into soils, this may constitute an activity subject to the requirements of the Groundwater Regulations 1998	Notify SEPA Possible Licensing under Groundwater Regulations (4 months)
Crushing and screening	The crushing and screening of soil is likely to require authorisation under Part I EPA unless the activity is covered by a waste management licence	WM licence (site or mobile plant) (4 months) OR Authorisation
Incineration of soil or Solvent recovery	The incineration of soil represents a prescribed process and is subject to control under Part I of Environmental Protection Act 1990. Similar measures apply for Solvent recovery	Ensure contractor has authorisation (Part B or IPC) OR Apply for authorisation

For clarification in respect of the above table, contaminated soil is not waste if it is simply left in place: it is only when it is to be discarded that it becomes waste. Regulation 1 of the WMLR 1994 indicates that contaminated materials, substances or products resulting from remedial action with respect to land, which the producer or the person in possession of it discards or intends or is required to discarded constitute Directive Waste. If treatment recovers the contaminants from the soil, then the treated soil ceases to constitute waste. Waste soil which is removed, but from which the contaminants are not recovered, remains as waste.

Piling on land affected by contamination: Environmental impacts, regulatory concerns and effective solutions

F.J. WESTCOTT[1], ERM Ltd, Manchester, UK, J.W.N. SMITH, Environment Agency, Solihull, UK and C.M.B. LEAN, WS Atkins Consultants Ltd, Warrington, UK.

INTRODUCTION

Government policy favours development on previously used land, including 'brownfield' development sites. These sites are often affected by contamination and suffer from poor physical ground conditions. Piling and penetrative ground improvement methods are commonly used to allow structures to be founded. However, environmental regulators are concerned that piling and penetrative ground improvement methods have the potential to create adverse environmental impacts when used on land affected by contamination.

The paper identifies and discusses potential adverse environmental impacts, and regulatory concerns, that could arise from piling or penetrative ground improvement techniques, when used on land affected by contamination. The lack of published research, case studies or monitoring data concerning actual or potential pollution caused by these techniques is noted, and recommendations are made for future study and research.

The lack of knowledge identified makes inevitable the application of the "Precautionary Principle" by environmental regulators, but it has been recognised that a rigid application of this principle may lead to unnecessarily restrictive requirements or prohibitions being placed on designers and developers. The need for guidance for designers and developers on how to consider the potential environmental impacts of piling works has also been recognised.

The paper outlines the decision making framework presented in recent Environment Agency guidance (Environment Agency, 2001) prepared following a research contract with WS Atkins Consultants Ltd. It assists the foundation designer to assess the environmental risks, select an appropriate piling or penetrative ground improvement method, develop mitigation measures, define quality assurance/quality control and monitoring measures, and justify this choice in a format acceptable to the regulator.

PILING METHODS

There is a considerable variety in the materials, design and installation methods used in piling and penetrative ground improvement (BRE, 1986). The following classification has been adopted in this paper. Pile foundations are divided into two basic types, displacement and non-displacement (replacement), and further subdivided according to the methods of construction and installation used. The difference between displacement and non-displacement methods has a potential effect on the environmental impact of the piling method

---

([1]) Formerly of WS Atkins Consultants Ltd, Warrington, UK

Geoenvironmental impact management, Thomas Telford, London, 2001.

used. The large number of variants of these methods developed by an innovative foundation piling industry is noted.

Penetrative ground improvement methods function in a different way from piling methods. Whilst piles are designed to transfer loads through poor ground to a competent founding level, ground improvement techniques aim to improve the bearing capacity or settlement behaviour of the poor ground.

The classification can be further subdivided as summarised in the table below.

Table 1 Types of piling and penetrative ground improvement methods considered in the EA guidance

Generic methodology	Methods that are discussed in EA guidance
Displacement piles	Pre-formed hollow pile
	Pre-formed solid pile (large displacement)
	Pre-formed solid pile (small displacement)
	Displacement cast-in-place pile
Non-displacement piles	Non-displacement cast-in-place pile
	Partially pre-formed pile
	Grout or concrete intruded pile
Penetrative ground improvement methods	Vibro replacement stone column
	Vibro concrete column

POTENTIAL ENVIRONMENTAL IMPACTS

There is evidence from laboratory and mathematical modelling studies that the installation of piles may, under certain circumstances, result in environmental damage. Campbell et al (1984), Hayman et al (1993) and Boutwell et al (2000) report on the potential effects associated with various pile types. Unfortunately, there is currently inadequate field data in the UK to establish the actual environmental impacts at the field scale.

The Environment Agency and others have expressed concern about the potential for foundations and piling to cause, or allow the migration of, pollution into controlled waters, (Agency, 2000; Agency 2001; CIRIA, 1999). The act of causing or knowingly permitting polluting substances to enter controlled waters (including groundwater) is an offence under section 85 of the Water Resources Act 1991. The principal environmental concerns include:

- The creation of preferential flow paths, allowing contaminated groundwater and leachate to move downwards through low permeability layers into underlying aquifers or between permeable horizons in a multi-layered aquifer;
- The breaching of impermeable covers ('caps') by piling or penetrative ground improvement, allowing surface water infiltration into contaminated ground (thus creating leachate) or allowing the escape of landfill or ground gases;
- Contaminated arisings being brought to the surface by piling work, with the risks of subsequent exposure to site workers and residents, run-off into surface waters, and the need for appropriate handling;

- The effects of aggressive ground conditions on materials used in piles, where the secondary effect is to increase the potential for contaminant migration;
- Driving contaminated materials downwards into an aquifer during installation, and;
- Concrete or grout contamination of groundwater and any nearby surface waters.

In the case of piling and ground improvement works, concerns about water protection are likely to be most acute when:

- Mobile contaminants are present on the site and piling could allow them to migrate;
- Piling would breach a low permeability layer or connect two previously discrete aquifers;
- The site overlies a Major or Minor Aquifer, or is within a Source Protection Zone;
- Groundwater is currently of good quality;
- The water table is shallow or likely to be intersected by piles;
- Works are close to surface water and run-off from arisings could pollute these waters.

The overall level of risk is a product of the probability of harm occurring and the consequence of that harm (DETR et al, 2000). For a risk of pollution to exist, there must be a source of contamination (i.e. the contaminated soils or leachate), a receptor (e.g. groundwater in an aquifer) and a pathway between the two. The presence of contaminant sources will normally depend on the past uses of the site, whereas the presence of receptors is essentially defined by the hydrogeological properties of the underlying strata, the proximity to surface water bodies and the use and occupation of the site and its surroundings.

In many instances, the risks to groundwater quality will be the principal concern of Environment Agency officers as described in the Agency's Policy & practice for the protection of groundwater (Agency, 1998). The Agency's response to consultations under the planning process for piling on contaminated sites will be based on the overall level of risk that piling is likely to present, the techniques, any mitigation measures and the quality assurance and control (QA/QC) methods proposed. Where the hydrogeological setting is not sensitive, special precautions or design constraints are unlikely to be necessary. In more sensitive situations the Agency may require a risk assessment to be undertaken and mitigation measures to be incorporated. In the most sensitive situations, the Agency will object to proposals that it considers present an unacceptable risk of pollution.

HAZARD ASSESSMENT

A general summary of the applicability of the generic piling and ground improvement methods, with and without appropriate mitigation measures, against the identified pollution scenarios is given in the table below. This table does not consider structural or geotechnical issues and should be used with care and not in a prescriptive manner as it is not based on site specific considerations. Circumstances may be such that generic methods indicated in this table as being applicable to the pollution scenario are not appropriate to conditions at the site.

The lack of research or field data creates a number of areas of uncertainty. Particular issues which would benefit from further research include the impact of displacement piling on aquitard strata or low permeability cover layers, the effect of densification of the soil caused by displacement, the effect of displacement piling on stiff overconsolidated clayey soils, the phenomenon of frictional drag down of soils, the behaviour of the soil at the interface with a non-displacement pile and the short term effects observed during pile construction.

Table 2 Indicative hazards associated with piling and penetrative ground improvement methods

Pollution scenario	Displace-ment piles	Non-displace-ment piles	Penetrative ground improvement
1: Creation of preferential pathways, through a low permeability layer, to cause contamination of groundwater in an aquifer.	**B-D** (depend-ant on details of method)	**B-C** (depend-ant on details of method)	**D** (stone columns) **B** (VCC)
2: Creation of preferential pathways to allow migration of landfill gas or contaminant vapours to surface.	**B**	**B**	**C** (stone columns) **B** (VCC)
3: Direct contact with contaminated soil arisings which have been brought to the surface.	**A**	**B-C** (depend-ant on contaminant)	**A**
4: Direct contact with contaminated soil or leachate causing degradation of pile materials.	**B-C** (depend-ant on pile materials and contaminants)	**C** (dependant on pile materials and contaminants)	**B-C** (depend-ant on pile materials and contaminants)
5: The driving of solid contaminants down into an aquifer during pile driving.	**B**	**A**	**A**
6: Contamination of groundwater and, subsequently, surface waters by concrete, cement paste or grout.	**A**	**C-D** (depend-ant on details of method)	**A** (stone columns) **D** (VCC)

VCC = vibroreplacement concrete columns

Key:
A: Pollution scenario not likely to be an issue if using this method provided workmanship and QA/QC measures are appropriate.

B: Subject to appropriate workmanship, mitigation and QA/QC measures, to be outlined in the Foundation Works Risk Assessment report and incorporated in the design and contract specification, this method is likely to be acceptable.

C: This method may be considered acceptable, depending on specific type used and subject to appropriate workmanship, mitigation and QA/QC measures, to be outlined in the Foundation Works Risk Assessment report. However a more suitable piling or ground improvement method may be available.

D: This method should normally be avoided on sites where this pollution scenario is likely to be an issue.

RECOMMENDED RISK ASSESSMENT FRAMEWORK
A risk assessment framework has been proposed to provide a robust, effective and transparent decision-making process that allows selection of the appropriate piling method, and mitigation measures if required, when piling on contaminated sites (Figure 1). The framework requires an assessment of risks (based on identification of source-pathway-receptor linkages) associated with the piling at the design stage. Should unacceptable risks be identified and appropriate mitigation measures are not available, a lower risk technique should be adopted.

The risk assessment process will be documented in a Foundation Works Risk Assessment Report, which provides a framework for the designer to select the appropriate piling method and justify this choice, with mitigation if required and appropriate QA/QC measures. It is envisaged that this report would normally be enforced through the planning system, with the Environment Agency acting as consultee on contamination issues.

Figure 1. Risk assessment flowchart

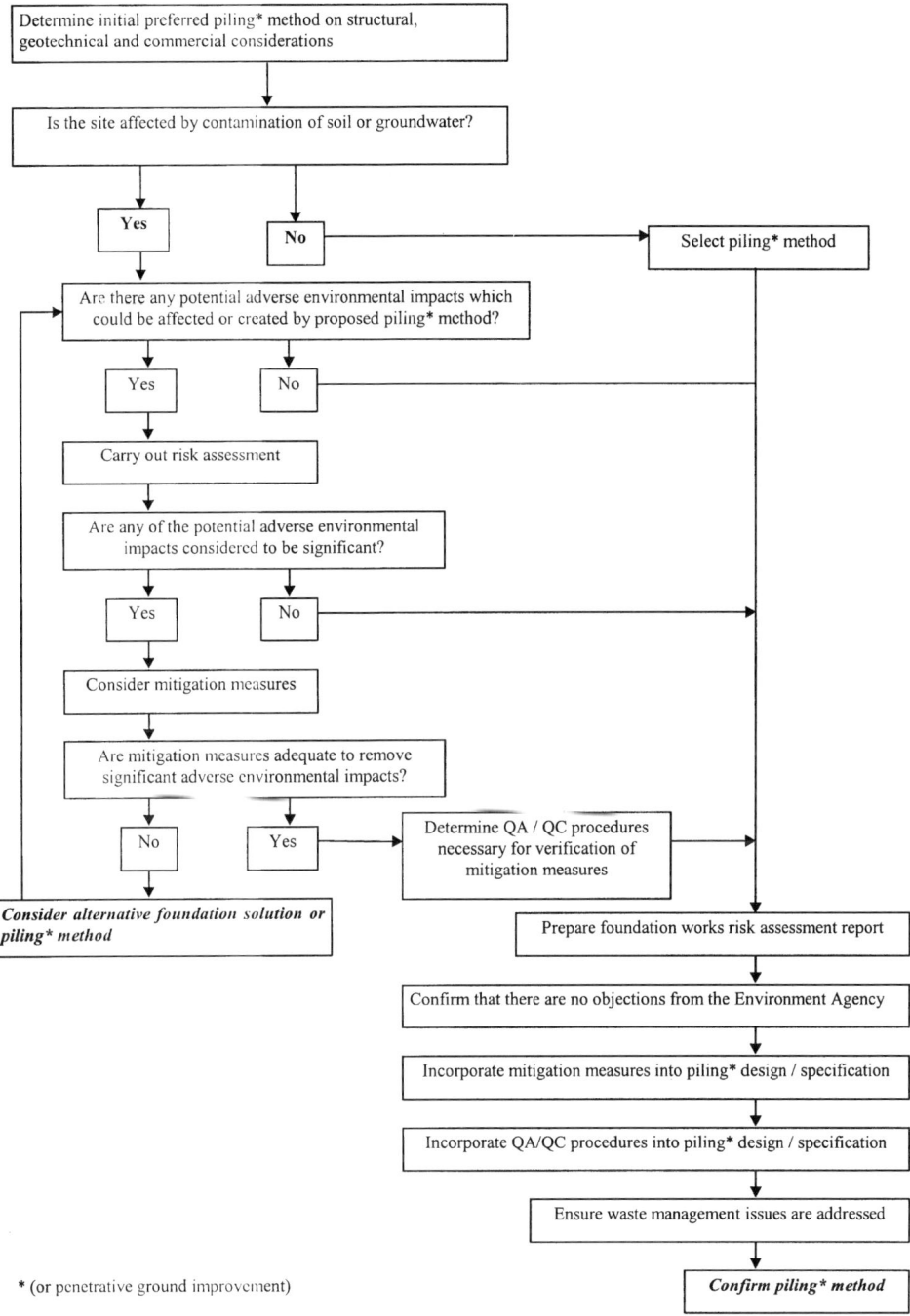

* (or penetrative ground improvement)

In outline, the contents of the Foundation Works Risk Assessment Report may comprise:

- Introduction to site setting and piling requirements;
- Initial selection of preferred piling method;
- Identification of potential adverse environmental impacts;
- Site-specific assessment of the risks to identified receptors in terms of existing problems and new source-pathway-receptor linkages which could be created;
- Identification of changes to preferred method or required mitigation measures;
- Identification of appropriate QA/QC procedures, potentially including establishment of long-term groundwater monitoring schemes to detect detrimental impacts at sites where groundwater resources are considered to be at risk; and
- Justification of selected method with regard to geotechnical, financial and environmental considerations.

This approach recognises that there are a number of factors that determine the choice of piling or ground improvement method for a particular project, including the support requirements (both bearing capacity and settlement) for the structure, the subsoil stratigraphy, the load transfer mechanism from the pile to the ground (friction or end bearing), ease of quality control, effect on adjacent structures, noise and vibration impacts, comparative material and installation prices and speed of installation. The decision framework presented in the guidance document ensures that potential environmental impacts are also considered at all stages of method selection.

REFERENCES
Building Research Establishment (1986). *Choosing Piles for New Construction*, Digest 315.
Boutwell, G.P., Nataraj, M.S. & McManis, K.L. (2000). *Deep foundations on brownfields sites*. PRAGUE 2000, Prague.
Campbell, P.L., Bost, R.C. & Jacobsen, R.W. (1984). *Subsurface organic recovery and contaminant migration simulation*. Proc 4th National Symposium on Aquifer Restoration and Groundwater Monitoring, Columbus, OH.
CIRIA (1999). *Environmental issues in construction: A strategic review*. CIRIA report C510. London.
DETR, Environment Agency & Institute of Environmental Health (2000). *Guidelines for environmental risk assessment and management*. The Stationery Office.
Environment Agency (1998). *Policy & practice for the protection of groundwater*. The Stationery Office.
Environment Agency (2000). *Guidance for the safe development of housing on land affected by contamination*. R&D Publication 66. The Stationery Office.
Environment Agency (2001). *Piling and penetrative ground improvement techniques on land affected by contamination: Guidance on pollution prevention*. National Groundwater & Contaminated Land Centre report NC/99/73, Solihull.
Hayman, J.W., Adams, R.B. & Adams, R.G. (1993). *Foundation piling as a potential conduit for DNAPL migration*. Proc. Air & Waste Management Association meeting, Denver, CO.
Tomlinson, M.J. (1994). *Pile design and construction practice*. 4th Edition. E & FN Spon.

Mine Sites, Tailing Dams, Dredgings and Lagoons

Performance of a Trial Embankment on Hydraulically Placed PFA.

T W Cousens and D I Stewart
School of Civil Engineering
University of Leeds
Leeds, LS2 9JT, UK

ABSTRACT

The paper describes the performance of a 5.3m trial embankment constructed on approximately 45m of hydraulically placed pulverised fuel ash (pfa). It is planned to redevelop the 17 hectare lagoon containing the pfa as a landfill. There is little variation in the particle size distribution of the uniformly graded silt sized pfa over the lagoon. However, the density of the pfa varies with depth with loose material underlying a denser surface layer, in a pattern that probably results from the water level in the lagoon during pfa deposition.

Settlement under the trial embankment was apparently largely complete by the end of the construction period (18 days), with approximately 300mm of settlement under the crest of the embankment. The magnitude of the embankment settlement appeared to be dominated by compression of the loose layers within the deposit.

INTRODUCTION

Much of the electricity generation in the United Kingdom is currently, and has been historically, produced by the combustion of coal. A by product of the process are fine ashes collected by electrostatic precipitation from the flue gases (known as puverised fuel ashes or pfa) and coarser furnace bottom ashes. Some pfa is used as a cement replacement but a large percentage is disposed of by producing a water-based slurry and pumping it into lagoons where settlement occurs. The result is a site that may have a considerable depth of potentially loose, fine-grained material. These areas have potential for development but there are difficulties constructing on the hydraulically deposited pfa.

This paper describes the construction and performance of a trial embankment on a considerable depth of hydraulically placed pfa. The trial embankment was constructed and monitored to provide large-scale settlement data on the behaviour of the pfa. The site is being developed as a landfill site and the trial was part of a programme to predict the settlement of the underlying pfa in order that basal drainage systems could be designed appropriately.

THE SITE

The site consists of two lagoons which were used for the disposal of pfa from an adjacent coal fired power station. Until about 1948 the area was agricultural, although there are

Geoenvironmental impact management, Thomas Telford, London, 2001.

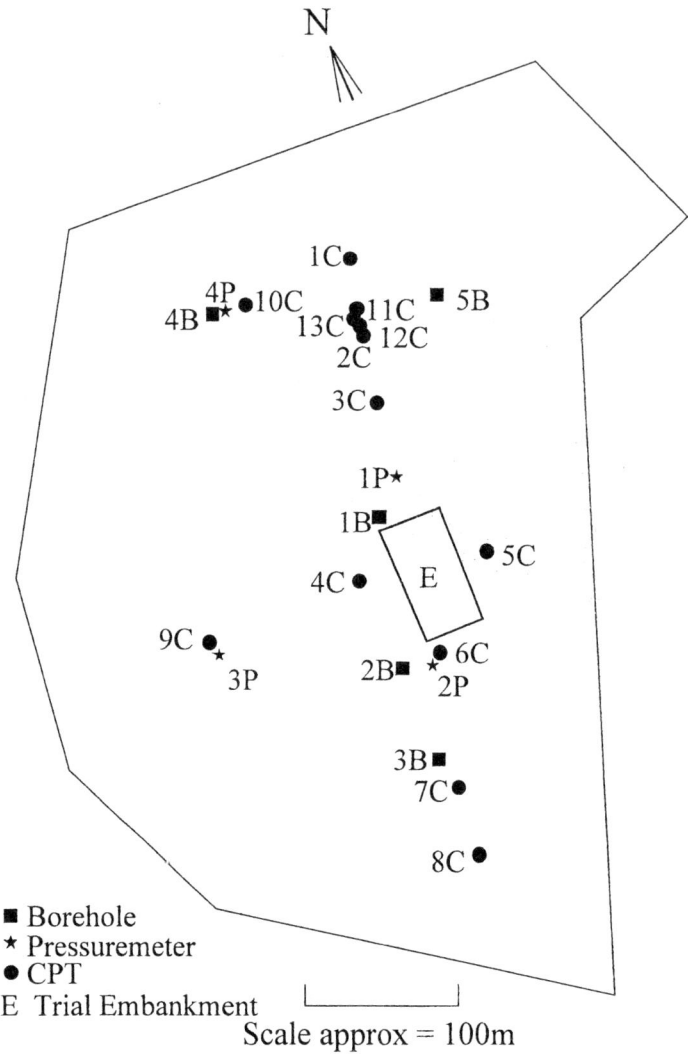

Figure 1. Plan of the pfa lagoon showing the position of the trial embankment and the location of the site investigation

records of coal mining in the area. Opencast coal mining and the extraction of sand and gravel during the 1950's and 60's resulted in large voids. These were partially backfilled with colliery spoil and embankments of the same material were constructed to form lagoons for the disposal of pfa. The pfa was pumped into the lagoons as a water slurry. The pfa was allowed to sediment and excess water was decanted off and disposed of into a nearby river. Pfa disposal took place from 1970 to 1994.

The trail embankment reported in this paper was constructed towards the middle of the larger lagoon on the site (17 hectares in extent). A plan of the lagoon and embankment is shown in Figure 1. Slurry inputs into this lagoon appear to have been largely to its NW corner.

SITE INVESTIGATION

Several previous site investigations have been conducted at the site, but, as part of the study reported here, additional investigations were performed to determine the extent of the pfa and its characteristics. Cone penetrometer (CPT) and pressuremeter testing was carried out at various locations in the lagoon to determine the in-situ behaviour and variability of the pfa. Disturbed and undisturbed samples were collected from five boreholes located around the lagoon both to characterise the pfa and to investigate its spatial variation. Figure 1 shows the locations of the various in-situ tests and boreholes. Laboratory tests included particle size distributions, liquid and plastic limits and one-dimensional compression tests.

At the location of the trial embankment the surface of the pfa is at approximately 25m AOD with a slight fall from north to south. The maximum depth of pfa is about 45m with the ground water level 6m below ground level. The minimum depth of pfa within 40m horizontal distance of the trial embankment is about 20m.

Figures 2 shows CPT data from a location near to the trial embankment as indicated on Figure 1. The data is typical of that obtained from the CPT tests and shows variations in the response of the pfa which is described as varying from firm to very loose. The CPT data also showed the depth of the pfa. Figures 3 and 4 are plots of the relative density of the pfa against depth determined from CPT data. These values were produced using the method described by Meigh (1987), and are based on the cone resistance values and the in-situ vertical stress. Figures 3 and 4 are primarily intended to show patterns in the relative density of the pfa, and the absolute values should be treated with caution. The relative density plot shows an upper denser layer overlying a very loose central layer, above denser material. Other CPTs conducted in the vicinity of the embankment showed a similar pattern for the pfa.

DESCRIPTION OF THE PFA

Figure 5 shows the particle size distribution of a sample taken from a depth of 14.5m in borehole 1B (i.e. near the embankment location). The particle size distributions of most of the samples taken from the site were very similar, and suggest that the pfa is relatively uniform over the site with 5-10% clay sized particles and 60-80% silt sized. This is fairly typical for pfa, which tend to be predominantly silt sized (Cabrera et al., 1984; McLaren and DiGioia, 1987). Occasional thin coarse layers were detected in the pfa, but their extent is unknown, although they appear to be limited.

The average liquid limit of the pfa was 46% (range 38-56%) with an average plastic limit of 42% (range 32-54%). The average plasticity index was 4% with some samples showing no plasticity. The pfa classifies as an inorganic silt with slight plasticity. The in-situ moisture content of the pfa showed a general pattern of a central band with a very high moisture content (55% to 78%) with lower values above and below (38% to 44%). These values suggest loose material, especially in the central band. The in-situ bulk density is estimated as varying between 1.54 and 1.66 Mg/m^3, which correspond to void ratios of 1.1 to 1.6, the latter values corresponding with the soft zone. This voids ratio is quite high; for example, in the extended Casagrande soil classification system void ratio values for silt at maximum dry density at optimum compaction are given as less than 0.7 (Road Research Laboratory, 1952).

Analysis of the pressuremeter data and correlation with the one-dimensional compression tests indicates that the coefficient of volume compressibilty, m_v, decreases with depth from about $0.15m^2/MN$ near the ground surface to about $0.07m^2/MN$ at 30m below ground level.

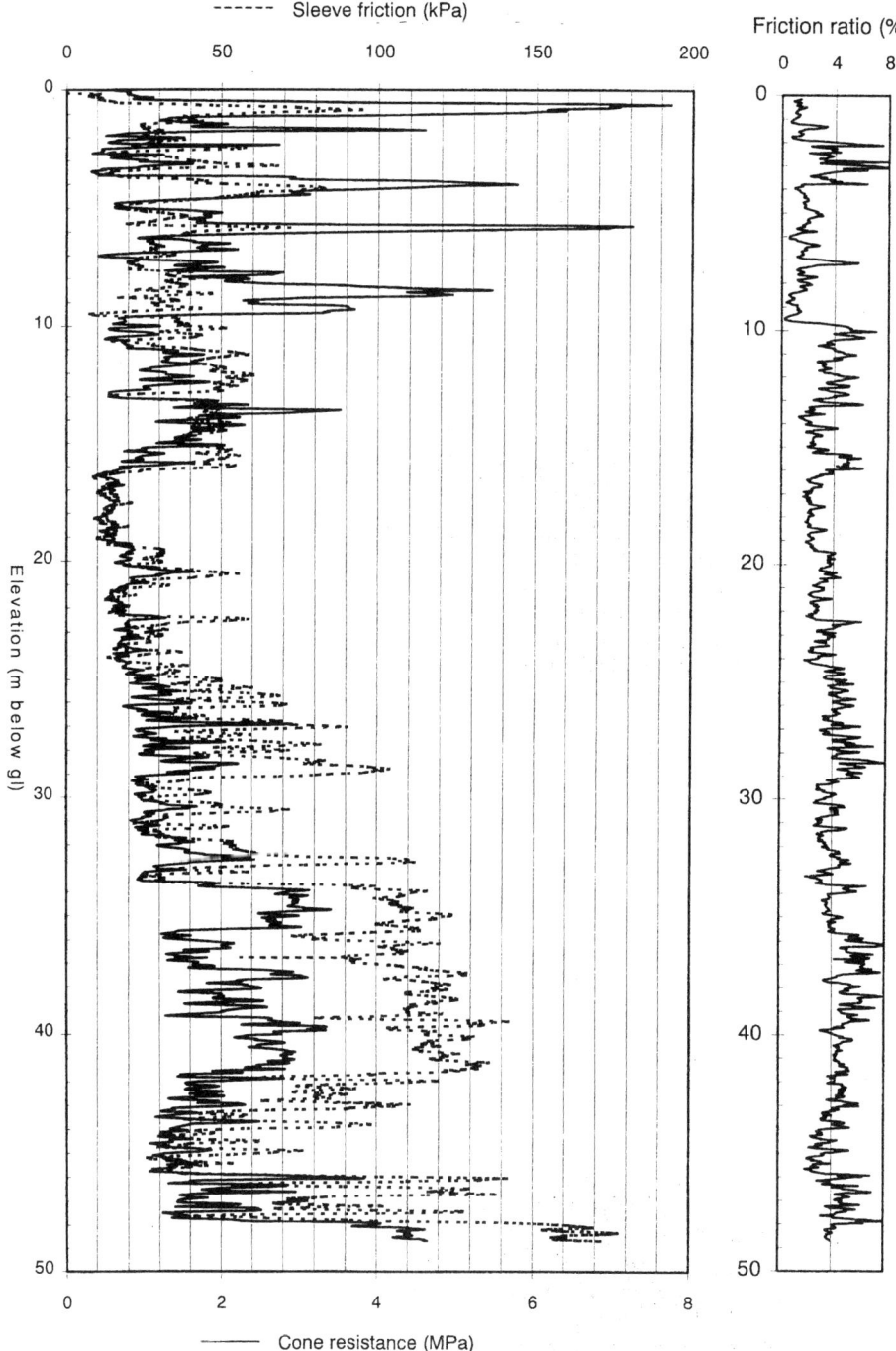

Figure 2. CPT data from location C4 (close to the trial embankment)

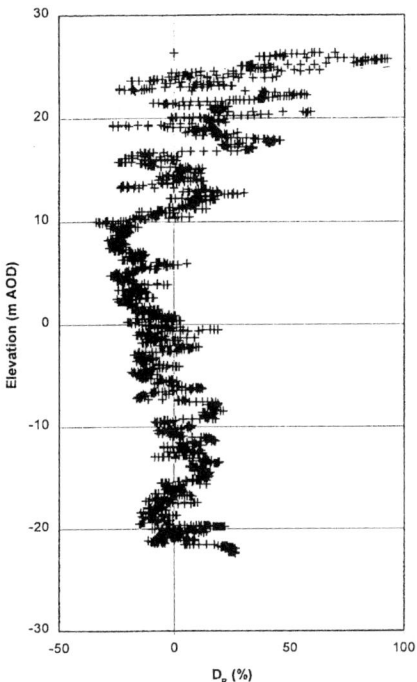

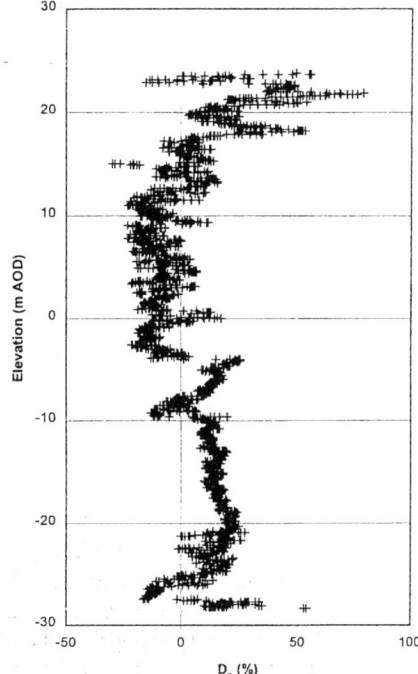

Figure 3. Relative density profile determined from CPT C4

Figure 4. Relative density profile determined from CPT C6

These values of m_v indicate that the pfa has a compressibility comparable to stiff clays (Tomlinson, 1995).

TRIAL EMBANKMENT AND INSTRUMENTATION

The trial embankment had a crest area of about 10 by 50m, a base area of 37 by 71m and a final height of 5.3m. The dimensions of the embankment were chosen so that deformations at the centre section could be considered as approximating to plane strain conditions. The side slopes were about 22° to give reasonably slow changes in the imposed loading on the underlying pfa and ensure stable sides to the embankment. The embankment was constructed on a geogrid overlain with a woven geotextile placed directly onto the pfa (which was cleared of vegetation). The embankment consisted of 0.25m of crushed stone separated by a woven geotextile from 0.5m of colliery spoil, 4m of pfa and 0.5m of colliery spoil. Over the central portion of the embankment a bentonite-impregnated geotextile was placed on the lower layer of spoil. The bottom 0.75m of the central portion of the embankment was representative of the landfill liner that is proposed for the site, and the surface layer of colliery spoil was placed to protect the pfa from erosion by wind and rainwater.

Instrumentation was installed on two planes through the embankment. Both instrumented planes were at right angles to the long axis of the embankment, one at the centre of the embankment with a secondary section five meters away and parallel to it, to act as a back-up in case of damage to the main section (see Figures 6 and 7). The instrumentation comprised inclinometers, magnetic settlement gauges (the gauge positions are not shown in figures 6 and 7 because the deepest gauge is approximately 30m below ground level), hydraulic profile

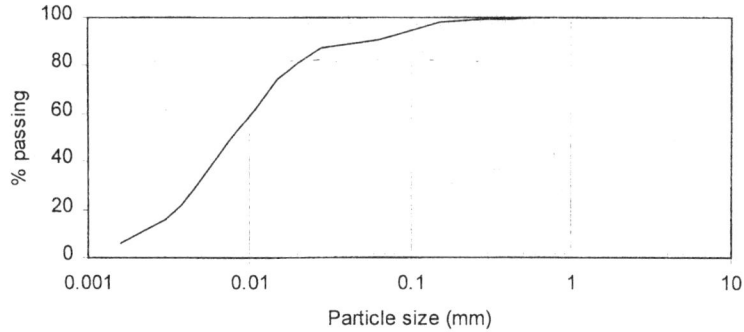

Figure 5: Particle size distribution for the pfa.

gauges and standpipe piezometers. An array of pneumatic piezometers were also installed but gave erratic readings. The elevations of the embankment crest, and of selected points around the embankment, were monitored using a surveying total station. Most of the data reported in this paper are from the magnetic settlement gauges and the surface monitoring points.

CONSTRUCTION SEQUENCE
The instrumentation was installed shortly before the construction of the embankment due to time constraints. After installation all the instrumentation was tested. Construction of the embankment commenced around the 10th September, 1999, (day 4 on Figure 8) and took twelve days to complete. Construction used pfa from the second lagoon and colliery spoil from an embankment. There was one period of heavy rain during construction (day 14). At one stage a large vibrating roller was used but resulted in marked ground vibrations and was abandoned.

PERFORMANCE OF THE EMBANKMENT
Figure 8a shows the average height of the embankment with time. After completion of the embankment the reported data are averaged from ten points on the crest of the embankment. Figure 8b shows average crest settlement after the embankment was complete (i.e. day 16 onwards). Figure 1b indicates that from about day 20, shortly after the embankment was

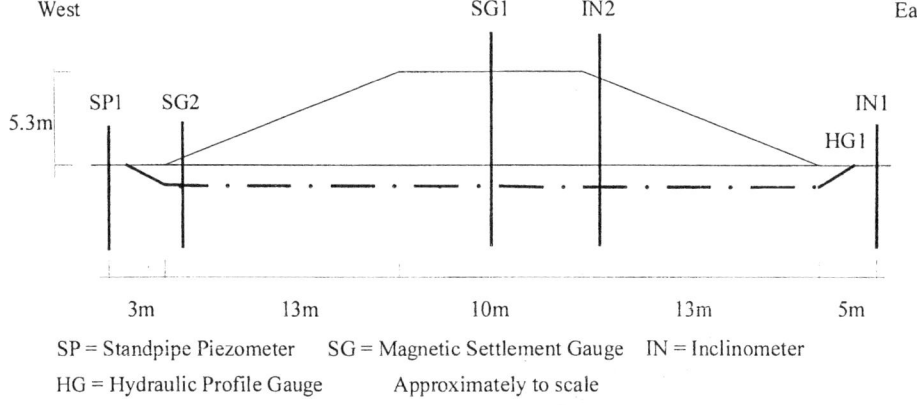

Figure 6. Schematic of the primary instrumented embankment cross-section.

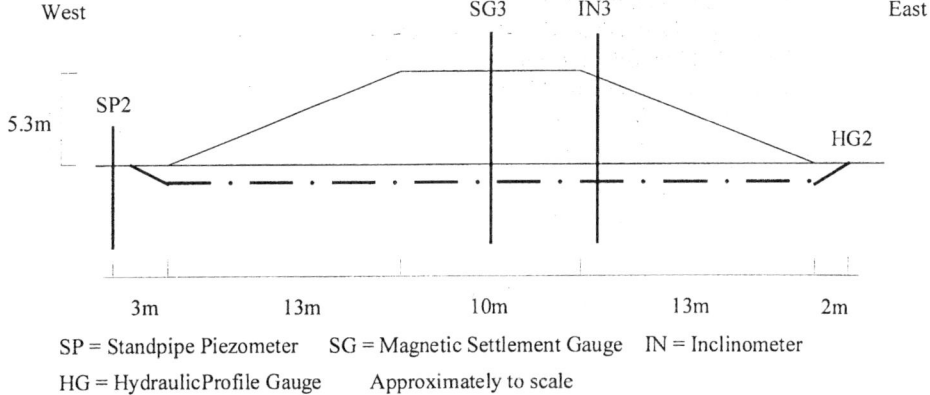

SP = Standpipe Piezometer   SG = Magnetic Settlement Gauge   IN = Inclinometer

HG = HydraulicProfile Gauge   Approximately to scale

Figure 7. Schematic of the secondary instrumented embankment cross-section.

complete, there was no significant settlement of the crest. An additional reading at 150 days confirmed this pattern. The accuracy of individual readings was estimated as +/- 10mm.

Figures 9 and 10 present the magnetic settlement gauge data for gauges 2 and 3 (located at the base of the shoulder and at the midpoint of the embankment respectively). Gauge 1 was damaged during construction of the embankment. The accuracy of the reported elevations is estimated as being +/- 5mm. The data presented in Figure 9 has been corrected for two aberrant events that occurred between days 17 and 18 and between days 21 and 23. In both cases the event consisted of an apparent uniform heave of all the magnets in gauge 2 (the deepest is 30m below original ground level). It is extremely unlikely that such a heave could have resulted from actual soil movements, and may indicate that buckling of the tube occurred. It was noted at about this time that it became more difficult to lower the probe down the tube. This uniform heave has been deducted from the data (indicated by the break in the line). For gauge 3 the settlements of only the two near surface magnets are reported because buckling of the tube prevented access to the deeper magnets.

Figures 9 and 10 indicate that significant settlement of the near surface pfa occurred during embankment construction, with approximately 0.3m of settlement occurring at the gauge 1.6m below the centre of the embankment. The settlement at 1.2m under the shoulder of the embankment was approximately 0.12m. After that time there was a slight settlement of all monitoring points over the next ten days after which the points were essentially stationary.

Figure 11 shows the settlement profile across the embankment in the plane of the main instrumentation array, measured by a hydraulic settlement profile gauge. The profile tube was situated approximately 1m below original ground level. The settlement increased steadily from a very small value under the toe of the embankment slope to a maximum value under the full height of the embankment. There is good agreement between the magnetic settlement gauge and the hydraulic settlement profile gauge data, with a measured long-term settlement using the profile gauge of about 0.1m under the embankment shoulder close to the location of magnetic settlement gauge 2 and 0.3m under the centre of the embankment.

The inclinometer data (not shown) showed only very small lateral movements. These were less than 10mm below 1.5m below original ground level, and less than 50mm above this

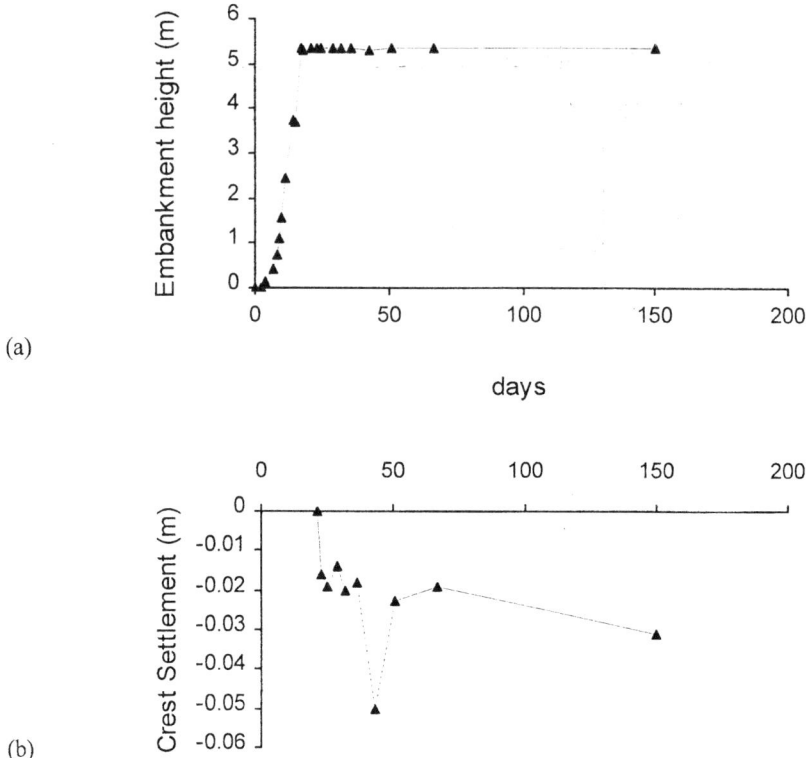

(a)

(b)

Figure 8. Variation in embankment height and crest settlement with time.

where it is thought that the installations may have been affected by plant movement and the placing of fill.

DISCUSSION
The pattern of movements under the trial embankment is complex with a vertical movement of about 300mm beneath the central section. The movement appears to have ceased shortly after the completion of the construction of the embankment (when the maximum increase in vertical effective stress is about 66kPa). The crest of the embankment did not move significantly over 130 days following completion. The overall pattern of surface settlement (observed using the hydraulic profile gauges) was similar to that reported in standard texts (Road Research Laboratory, 1952).

Data from magnetic settlement gauge 2 (below the shoulder of the embankment) suggests that the variation in vertical strain with depth below the embankment was slightly unusual. At the end of the monitoring period, when settlement had effectively ceased, the upper magnet in settlement gauge 2 at a depth of 1.2m below original ground level (approximately 23.6m AOD) settled about 120mm; the magnet at a depth of 4.1m (approximately 20.7m AOD) settled about 100mm; that at a depth of 10.4m (approximately 14.4m AOD) settled about 80mm; whereas the magnet at a depth of 13.0m (approximately 11.8m AOD) settled only 25mm. Thus the vertical strain between at depths between 1.2 and 4.1m below original ground level was about 0.7%, that between 4.1 and 10m was only about 0.3%, whereas

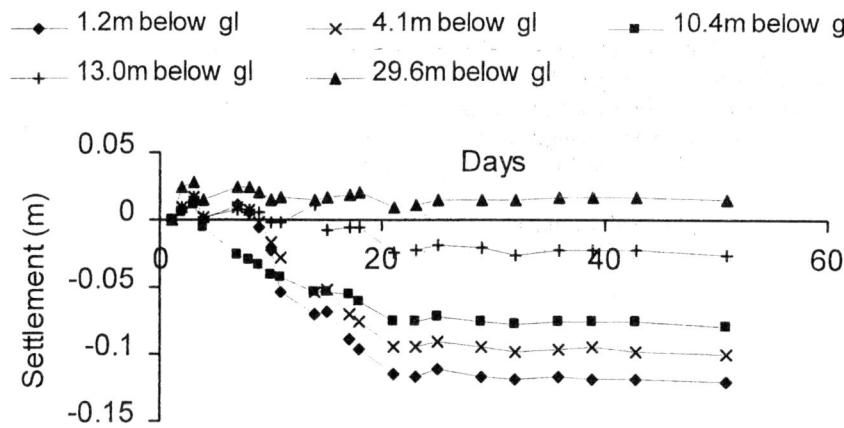

Figure 9.   Settlements beneath the shoulder of the embankment (magnetic settlement gauge 2)

between 10 and 13m below original ground level the vertical strain was about 2.1% (below 13m the vertical strain is about 0.2%).

Data from magnetic settlement gauge 3 (below the centre of the embankment) is compatible with this pattern of vertical strain. At the end of the monitoring period the upper magnet in settlement gauge 3 at 1.6m below original ground level (approximately 23.5m AOD) had settled 300mm while the lower magnet at 4.4m below original ground level (approximately 20.7m AOD) had settled 270mm. This suggests that the pfa between these two gauges underwent a vertical strain of about 1%. At some point below this depth (between about 4 and 8m) the tube within settlement gauge 3 buckled, which may have resulted from the tube passing through a zone of large vertical strains. Without a looser, more compressible zone, the 270mm settlement of the lower magnet would require an average vertical strain of about 0.7% over the entire 45m depth of the pfa (assuming no settlement of the underlying strata).

Analysis of the CPT data suggests that there is a denser surface layer of pfa about 8 to 10m thick (lower level 17 to 15m AOD) overlying very loose pfa. This may be the result of the

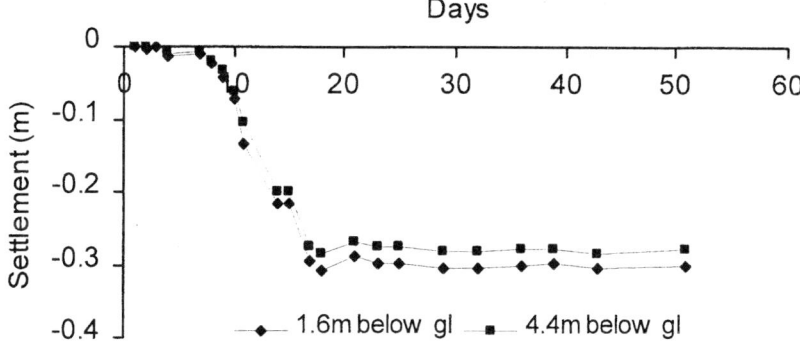

Figure 10.   Settlements beneath the crest of the embankment (magnetic settlement gauge 3)

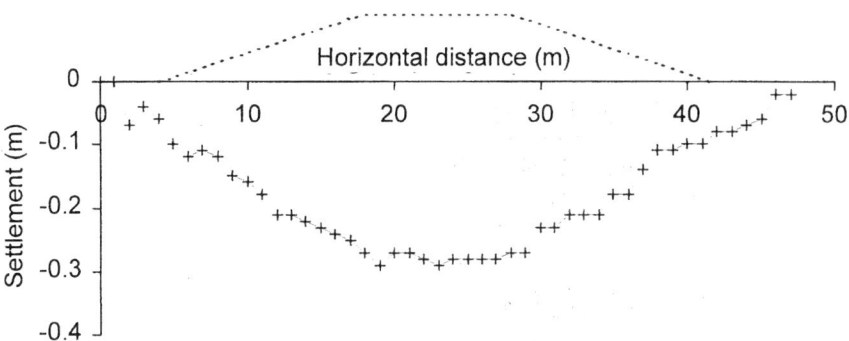

Figure 11. Settlement profile in the plane of the main instrumentation array (HG1, day 27)

depositional history of the pfa and the effect of changing water levels. The water table is currently about 6m below the surface of the pfa (about 19m AOD). The river into which the site drains is about 14.2m AOD as it passes the site. When the slurried pfa was pumped into the lagoon, it is likely that the original open cast void would have filled with water to a level above 14.2m AOD before the excess water from the slurry could overflow into the river. Thus the pfa below about 15m AOD may have settled through water, whereas the pfa above this level probably settled from the slurry as it braided across the surface of the pfa.

CONCLUSIONS
The hydraulically placed pfa, which is a uniformly graded silt sized material, shows significant variations in relative density. These variations are primarily thought to reflect the water level in the lagoon during deposition. Specifically, it is suggested that the pfa that settled through water is looser than the pfa that settled out of water flowing across an exposed pfa surface.

The trial embankment that was constructed on this deep deposit of uniformly graded silt sized pfa reached full settlement shortly after the end of the construction. The shape of the settlement profiles suggests that there was little variation in pfa compressibility in lateral directions. The settlement appeared to be dominated by compression of loose layers within the deposit.

ACKNOWLEDGEMENTS
The authors would like to thank Mr John Ablitt and Biffa Waste Services Ltd. for permission to use the information and data used in this paper.

REFERENCES
Road Research Laboratory (1952). Soil Mechanics for Road Engineers. HMSO, London.
Cabrera, J G, Braim, M and Rawcliffe, J. (1984). The use of pulverised fuel ash for structural fills. Second Int. Conf on Ash Technology and Marketing, London, pp529-533.
McLaren, R J and DiGioia, A M. (1987). The typical engineering properties of fly ash. Geotechnical Special Publication No 13. ASCE, pp683-697.
Meigh, A.C. (1987). Cone penetration testing- methods and interpretation. CIRIA-Butterworths.
Tomlinson, M J. (1995), Foundation Design and Construction. Longman Scientific and Technical, UK.
Somers, N R. (2000). Secondary lines at Gale Common ash disposal scheme. Ground Engineering, **33**, No.8, pp30-34.

The Abandonment of Tailings Lagoons as Environmental Wetland Features – an update (June 2001)

D. R. LAMONT, Health and Safety Executive, UK, J. R. LEEMING, Health and Safety Executive, UK and M.H.K. BRUMBY, UK Coal Mining Ltd, Doncaster, UK

SYNOPSIS.
The paper this is an abbreviated and updated version of a paper first given to the British Dam Society[1] at Bangor, 1998 and subsequently to The Institution of Mining and Metallurgy (Nottinghamshire and South Midlands Branches)[2], which describes an agreement between the Health and Safety Executive (HSE) and the principal operator in the coal mining industry (UK Coal Mining Ltd, formerly RJB Mining (UK) Ltd.) concerning proposals for converting colliery tailings lagoons to environmentally diverse water features as an alternative to overcapping them when restoring spoil heaps. The original paper reviews engineering safety issues in more detail.

INTRODUCTION

Unsurprisingly, the Civil Engineers and Legislators after Aberfan were not immediately exercised by considerations of beneficial land afteruse, of environmental enhancement and of aesthetics. However, over the past thirty years the expectations of the public, Environmental organisations and Minerals Planning Officers have changed. At a number of sites, due to closure of mines, there is significant over-capacity of tipping space. Often there is only a fraction of the soils needed for a good quality agricultural restoration. At a time when good agricultural land is being put to set aside, there is less obvious need to restore land to second rate agricultural use. Conversely, there is increasing demand for sensitively designed restoration schemes, for areas of public access and amenity and for ecological diversity. There is a national and, indeed, international shortage of wetland habitats. Without addressing these aspirations, future Planning Approvals will be increasingly difficult to obtain.

Coal washing and preparation for the market at large coal mines almost invariably produces coal, coarse discard and a tailings fraction, suspended in water. The tailings is predominantly dirt, usually mudstones and shales, and has a particle size range between fine sand, silt & clay (typically of 100% <2mm; 90% <0.5mm and 25% <0.01 mm). Large quantities of this rather problematic material are produced and are normally disposed of by settling out of suspension in lagoons, with cleaned water being recycled.

When a lagoon has filled with tailings, standard practice has been to remove supernatant water, allow the deposits to dry, then to overcap with coarse material. As the tip is developed, a lagoon will finish within the body of the tip. However, overcapping is only possible if sufficient coarse material is available. When a tipping site is nearing capacity, provision is made to stock capping material and to shape the tip to an agreed profile. With the premature closing of a mine, a shortfall in the availability of coarse discard may mean the expensive importation of material to complete the cap. Therefore, proposals to leave parts of lagoons open as water features have in some cases been put forward, with a supporting statement arguing that the resulting feature would have amenity value once the site has been restored. Such features can be constructed by only partial, marginal overcapping of lagoons and by appropriate treatment of adjacent areas.

In the early 1990s, restoration proposals including wetland features on spoil heaps were made at Ledston Luck and Allerton Bywater. British Coal implemented the former and obtained planning permission for the latter. After privatisation, revised restoration proposals incorporating wetland features were submitted for further sites including Askern, Bilsthorpe, Clipstone (Rufford), Gedling (Stoke Bardolph) and Point of Ayr. These proposals have been accepted and (together with Allerton Bywater) have been implemented. In several cases, vegetation is now well established and the sites are becoming extensively colonised by wildlife.

HSE considered the proposals from the mining industry and agreed to it in principle. It was felt that guidance was required on the minimum engineering standards acceptable for safe abandonment and these were subsequently agreed between both parties.

TIP AND LAGOON CONSTRUCTION
In distinction to the absent or weak regulatory framework in some other parts of the international minerals extraction industry, tips and lagoons at British coal mines are constructed in accordance with the standards drawn up following lessons learned from the Aberfan disaster and from intensive subsequent technical investigations (NCB Technical Handbook "Spoil Heaps and Lagoons"[3] and "Codes and Rules" (1971)[4];). Tips are now built on level land and are constructed to enhance the strength of deposited material and to minimise water ingress. They are formed in relatively impermeable compacted layers of maximum thickness 1.5m. The structure is anisotropic and drainage paths are predominantly horizontal. The external flanks are sloped to shed water in a controlled manner and to facilitate restoration.

Lagoon banks are built up from thinner compacted layers, 0.3m maximum, to ensure greater strength and impermeability. Tailings are deposited from varying discharge points around the lagoon and tend to form layers of different permeability. Overall however the vertical permeability of the tailings in the lagoon body is low and is significantly less than its horizontal permeability. The vertical permeability decreases with time due to consolidation of the deposits.

Significantly, both spoil heaps and lagoons are now designed as Civil Engineering structures and are regularly inspected and reported upon by competent and Chartered Civil Engineers.

LEGISLATION AFFECTING TIPS
The Framework
On 12 October 1966, the major tip slide at Aberfan in Glamorgan claimed 144 lives, 116 of them school children, mostly between 7 and 10 years old. The Tribunal of Inquiry held to investigate this disaster determined that to prevent a recurrence and to ensure stability, the construction and maintenance of tips needed strict regulation[5]. This resulted in the enactment of the Mines and Quarries (Tips) Act 1969[6] "the Act", and subsequently the Mines and Quarries (Tips) Regulations 1971[7] "the Regulations".

The Requirements of the Regulations.
The Regulations impose general requirements to ensure that tips are made and kept secure. In particular, operations must not cause an accumulation of water in, under or near the tip which may make the tip insecure and the tip must be efficiently drained. In addition, managers must

appoint a competent person to supervise all tipping operations, maintenance, drainage and security at all tips. Defects and incidents must be recorded in a special book.

Before tipping operations commence, a comprehensive report must be made covering the design of the tip, specifying tipping methods and detailing all matters which may affect tip security, including the topography, geology, hydrology and hydrogeology. Capacity is estimated and site preparation, drainage and fencing specified. A tip plan must be constructed showing this detail together with previous and planned mine workings, water sources and courses and any topographical feature which may affect the security of the intended tip. A geological map of the area with sections of the underlying strata showing significant faulting is also produced.

The Regulations also require further comprehensive civil engineering security reports every two years and a supplementary report as soon as practical after any Dangerous Occurrence, or after any change has been made that might affect the stability of the tip. These inspections record the works that have been carried out since the last report, address any changes in situation, specify remedial or maintenance work which must be carried out and focus on external factors, e.g. underworking, which could affect the stability of the site. The Report must also include an opinion on the present and future stability of the spoil heap. Codes & Rules stipulate that comprehensive reports must be countersigned by a Chartered Civil Engineer and also that a full inspection at monthly intervals is made by the mine mechanical engineer; at three monthly intervals by the mine manager and a six-monthly over inspection by a competent civil engineer.

Tipping must be controlled by Manager's Tipping Rules, which specify not only the technical specification for construction but also the supervision, and the nature and frequency of inspections. Regulations require inspections to be carried out weekly by a competent person appointed to carry them out. This inspection is primarily directed at the drainage of the tip, and "such other inspections as are required by Tipping Rules."

Although both Tipping Rules and periodic comprehensive reports are primarily designed to satisfy the requirements of the Mine & Quarries (Tips) Act and Regulations, it is normal for them also to address the requirements of other legislation and regulations which may apply, including other Health & Safety, Environmental and Planning requirements.

Closed tips, (which are tips no longer in use but still attached to an active mine) generally pose a lesser threat as no tipping operations are being carried out and, if properly maintained, consolidation of the material and the dissipation of elevated pore pressures will increase stability. Nevertheless the size and location of many closed tips precludes any complacency. The tip must continue to be inspected, now at 6-monthly intervals for liquid tips (lagoons) and 12 monthly intervals for solid tips. In addition, a comprehensive civil engineering stability report must be made, now at intervals of 5 years for a liquid tip and 10 years for a solid tip.

Disused tips (when a mine itself is closed) pass into Part II of The M&Q (Tips) Act 1969, which gives Local Authorities (LA's, at County or Metropolitan level) the responsibility of ensuring that the owners of any tip prevent any public danger. The LA is given powers to seek information from the owners, to enter sites to inspect or carry out tests; and to require owners to carry out remedial operations. If the LA believes that any apparent instability constitutes a danger to the public, the LA can carry out the remedial operations itself and recover its expenses from the owner. Again, under the NCB, British Coal and (now) UK Coal Mining Ltd, it is recognised that the most responsible way of managing disused spoil heaps is to treat them as Closed Tips and the inspecting and reporting regime is thus significantly extended.

The reasoning behind all these inspections was to ensure that tips were constructed and maintained in a stable condition. It is worthy of note that since the enactment of the Act and Regulations and following an intensive programme of investigation, analysis and remedial works on tips in the early 1970's, no significant tip instability problems have been reported.

TIP ABANDONMENT
With most tips, the final structure is a domed shape, and the land restored to forestry, amenity or agricultural use. Once a tip site has been vacated by the mine or quarry operator, the Local Authority has a responsibility for ensuring the long term stability of the site, and corrective action as required is taken. It is essential that the mine or quarry owner leaves the site in a stable condition when the site is vacated, regardless of whether or not a water feature is planned.

Two important principles have to be satisfied. These are firstly that, as the LA will have limited resources to examine and maintain the site, any water feature should be maintenance free and secondly, that the water feature when in both its intended state and in any condition of overflow, must not compromise the stability of the tip structure as a whole.

As noted above, one factor in ensuring stability is the elimination of perched water tables from the body of the structure. The intentional leaving of a body of water, perched on a tip above natural ground level, therefore raises long term stability questions which have to be addressed.

ENGINEERING FACTORS INFLUENCING STABILITY ON ABANDONMENT
Various factors which could adversely affect tip stability must be identified and engineering measures taken to counter them. They are related to the existence on the tip of a body of water subject to changes in level due to climatic variations, and the effects that the margins of this water have by wave action and gully cutting during overflow. In countering them robustness and obviating the need for maintenance must be of primary importance. Factors identified and discussed in the 1998 paper are as follows:
1. Water could percolate into the tip and through the lagoon bank causing eventual failure.
2. Not completing the cap could maintain slurry in a wet state with high pore pressure.
3. Water could migrate beneath the edge of the partially completed cap, liquefying it and causing the feature to grow until potentially it is the size of the original lagoon.
4. There is a possibility of seismic activity or blasting causing liquefaction.
5. Wave action may erode the banks.
6. If overtopped, the flank would be washed away
7. Leaving a water feature could be considered to breach Regulation 4(2) of the M&Q (Tips) Regulations 1971:- 'Every active and closed tip shall be efficiently drained.'

The conclusions reached after much thought and investigation were that, in general, the factor of safety against failure in a "wetland feature" were likely to be higher than when the lagoon was in service and full. This is also the case for seismic effects and, in any event, "the liquefaction of lagoon sediments (by) ground shaking by (British) earthquakes, is extremely unlikely "[8]. Of more concern are the issues of wave action and overtopping, which must be addressed in the same way as they would be for any small reservoir. As far as the application of the Regulations is concerned, it can be argued that providing water levels are controlled and there are no unplanned events, the tip can still be regarded as being efficiently drained.

ESSENTIAL ENGINEERING CRITERIA

Following examination of the above concerns, HSE and UK Coal Mining Ltd. agreed on a set of design criteria to ensure long term stability. These form Appendix A to the original paper. The major points of note are:-

1. The embankment shall have been constructed to an appropriate engineering standard.
2. No reduction in embankment width or increase in height of water or lagoon deposits from operational conditions shall be permitted.
3. There shall be no record of instability or significant seepage with the lagoon in reports.
4. The slope of the inner face of the lagoon shall be graded to no steeper than 1 in 5.
5. Crust on deposits should be at least 200mm thick and of 6 kN/m^2 minimum shear strength.
6. Maximum depth of water shall initially be no greater than 1 metre.
7. Scour protection shall be provided around the water's edge by the use of vegetation, by provision of extra soil or stone or by the use of geotextiles.
8. The main spillway shall be a pipe or a channel through the embankment, with vandal resistant intake and outfall. All discharges shall be taken to the foot of the embankment.
9. The spillway shall be constructed to prevent scour and be maintenance free.
10. The auxiliary spillway shall be a broadcrested weir set 100 mm above top water level when the main spillway is in operation at design capacity, shall be of capacity equal to the main spillway and shall prevent the top water level rising to within 500 mm of the crest.
11. The normal operational lagoon design freeboard of 1 metre shall be maintained for the entire embankment (other than the auxiliary spillway) at all times. A design check[9] shall be carried out to ensure that this freeboard is adequate. The embankment including any landscaping fill on the inner face, shall be at least 10 metres wide at the level of the high water mark with the main spillway operating at design capacity.
12. Additional guidance can be found in publications by CIRIA.[10,11]

PUBLIC SAFETY

Any wetland feature in the landscape is a potential hazard to the public. There have been accidents where trespassers have fallen into active lagoons and there is clear evidence at a number of disused spoil heaps that trespassers have for many years been regularly walking or riding motorcycles over uncapped tailings deposits. Such admittedly irresponsible behaviour is not persuasive of the notion that deposits will remain in a state like quicksand in perpetuity!

If a lagoon is to be retained and adapted as a wetland feature with potential for public access, there are certain basic precautions which will minimise risk. All internal slopes should be reduced by the deposit of suitable material, from a typical 1 in 2 to a flatter slope, preferably 1 in 5 or less. This will facilitate any inadvertent trespasser getting out of the lagoon and will also have ecological and aesthetic benefits. Suitable aquatic plants should be established where possible around the margins to increase surface shear strengths by their root systems and to act as a demarcation and barrier to discourage access. Reeds and bulrushes are particularly suitable.

Before any consented access by the public could be contemplated, an adequate crust should have formed on the deposits. shear strengths of the top 200mm around the perimeter should be at least 6kN/m^2. This is very weak but is around the minimum desirable strength to permit normal overcapping operations using a LGP dozer. It is also the kind of strength which can be found between high and low tide levels in a tidal estuary and will support the weight of a pedestrian. Even where the deposits are flooded, this strength should develop in a reasonable period and this should be verified using a vane tester.

Consideration must be given to the provision of fencing, warning notices and life-saving equipment where appropriate. Normally, this would be a temporary requirement until shear strengths are adequate. Subsequently, there is no reason to suppose that a wetland feature would be any more hazardous than any river, pond, lake or seashore.

CONCLUSIONS

Following the announced intention of the mining operator to leave bodies of water in a partially uncapped state on existing tips, work was undertaken to assess the likely hazard and mechanisms of potential failure of the tip structure.

Following this work, it was concluded that providing certain basis design criteria are adhered to, the resulting feature should have no detrimental effect on the stability of the tip structure as a whole. The feature itself should be no more hazardous than any other body of open water to which the public has access. It was concluded that no additional burden will be placed on the LAs by the establishment of these features.

Experience since 1998 has shown that the established wetland features continue to perform as designed from an engineering viewpoint, the overflow arrangements have coped with even the extreme precipitation of November 2000 with ease and minor wave action has apparently reached an equilibrium position due to beaching of marginal erosion material. In most cases aquatic and marginal vegetation is becoming well established and the features have become colonised with a wide variety of wildlife, from insects and invertebrates to heron and swans. It is thus reasonable to hope that a modestly innovative approach to an old problem will, in the fullness of time, lead to an environmental asset.

REFERENCES

1 British Dam Society "The prospect for reservoirs in the 21st century" (Proceedings of the tenth conference of the BDS held at the University of Wales, Bangor on 9-12 September 1998): Ed. Paul Tedd: Thomas Telford, 1998. ISBN 0 7277 2704 4
2 "International Mining and Minerals": January 2001 No.37. ISSN 1461-4715
3 NCB Technical Handbook "Spoil Heaps and Lagoons" (Second Draft Sept. 1970): National Coal Board.
4 NCB (Production) Codes and Rules: Tips. First Draft 1971: National Coal Board.
5 Report of the Tribunal appointed to inquire into the Disaster at Aberfan on October 21st, 1996: H.M.S.O. 1967
6 The Mines and Quarries (Tips) Act 1969 (1969 Chapter 10)
7 The Mines and Quarries (Tips) Regulations 1971 (SI 1971 No.1377)
8 Composition and Engineering Properties of British Colliery Discards: R.K Taylor, N. C.B. 1984.
9 Inst. of Civil Engineers "Floods and reservoir safety" 3rd Ed: Thomas Telford 1996. ISBN 0 7277 2503 3
10 The Construction Industry Research and Information Association "Small embankment reservoirs" Report 161: CIRIA 1996. ISBN 0 86017 461 1
11 The Construction Industry Research and Information Association "Design of reinforced grass waterways", Report 116: CIRIA 1987. ISBN 0 86017 285 6

Advective Transport Through Marine Sediment Caps

HORACE MOO-YOUNG
Assistant Professor, Lehigh University, Bethlehem, PA, USA.

TOMMY MYERS,
RICHARD LEDBETTER
BARBARA TARDY
WIPPAW VAN-ELLIS
Waterways Experiment Station, Vicksburg, MS, USA.

ABSTRACT
The presence of contaminated sediment poses a barrier to essential waterway maintenance and construction in many ports and harbors, which support 95% of U.S. foreign trade. Cost effective solutions to remediate contaminated sediments in waterways need to be applied. Capping is the least expensive remediation alternative available for marine sediments that is unsuitable for open water disposal. Dredged material capping and in situ capping alternatives, however, are not widely used because regulatory agencies are concerned about the potential for contaminant migration through the caps.

This study examines consolidation induced advective contaminant transport in capped sediment utilizing a research centrifuge. Centrifuge modeling simulates the increase the gravitational acceleration (g) of a prototype which is N times larger than the model, where N is gravitational acceleration factor. For contaminant migration, the time of transport in the model is inversely proportional to the square of the acceleration factor in the prototype. In this study, consolidation induced convective transport was modeled for 22 hours at 100-g, which modeled a contaminant migration time of 25 years for a prototype that was 100 times larger than the centrifuge model. A dye tracer was utilized to monitor the advective transport resulting from consolidation induced settlement.

INTRODUCTION
Approximately 523 million cubic meters of sediment must be dredged from waterways and ports each year to maintain the nation's navigation system (with approximately 18-37 million m^3 of contaminated sediment). The US Army Corp of Engineers is responsible for keeping waterways navigable under the Section 10 of the River and Harbor Act of 1899. Operation and production capacity dominated the dredging industry until the 1970s when federal legislation such as the National Environmental Policy Act, Federal Water Pollution Control Act Amendment of 1972, and Clean Water Act of 1977, made environmental compliance an important design consideration for dredging operations. Thus, the presence of contaminants in the sediment has changed the state of practice in the dredging industry (National Research Council (NRC, 1997).

Geoenvironmental impact management, Thomas Telford, London, 2001.

Economics, technical feasibility, and environmental acceptability must be evaluated to determine the most appropriate option. There are three alternatives for the disposal of marine sediment: open water disposal (e.g. sub-aqueous pits), confined disposal facilities, and beneficial use applications. Open water disposal refers to the placement of marine sediment into a water body by a pipeline or release from a barge. Confined disposal involves the placement of marine sediment into large dike region, constructed adjacent to land, in protected waters, harbors, or in open water. Examples of beneficial use applications include wetland creation, beach nourishment, mine reclamation, and land applications (U.S. Army Corp Engineers, 1992). Of the approximately 530 cubic meters of marine sediment, 444 cubic meters are placed in inland, coastal, and estuarine open water sites, confined disposal facilities, or beneficial use sites (Palermo et al., 1998).

When materials are unsuitable for ocean disposal, there are four basic options for remediation of contaminated sediment: containment in-place, treatment in-place, removal and containment, and removal and treatment. Economic considerations make decontamination and upland disposal options unfavorable to many port authorities (Palermo et al., 1998). In-situ capping of sediment and disposal of contaminated sediments in sub-aqueous pits are the least expensive alternative. In situ capping involves placing a layer of clean sand over contaminated sediment (i.e. in-situ). In sub-aqueous pit disposal, contaminated marine sediment is capped with a layer of clean sand, thus reducing the environmental impact of the sediment from the surrounding ecosystem (NRC, 1997).

There is a basic lack of information on the significance of consolidation induced convective (advective) transport of contaminants from contaminated sediment into caps. Estimating the amount of convective contaminant transport due to consolidation is usually not preformed when designing a capping layer. Analysis of consolidation induced convective transport of contaminants into caps could provide major economic and environmental benefits. Economic benefits accrue from avoiding more costly alternatives for dredged material disposal and sediment remediation. Environmental benefits may be accrued from more environmentally protective cap design. In addition to the economic and environmental benefits, there are additional benefits offered by improved analysis capabilities for consolidation induced convective transport of contaminants into cap, which include the maintenance of project schedules and compliance with present and future environmental regulations.

The objective of this research is to evaluate the significance of consolidation induced advective transport of dye from a sediment into and through a cap using a research centrifuge. To accomplish this objective, the following tasks were conducted:
1. Cap and sediment soils were chemically and physically characterized.
2. Settlement of sediment induced by the placement of a cap was monitored on a research centrifuge.
3. A dye tracer study was conducted to monitor the movement of pore water caused by consolidation induced advective transport.

LITERATURE REVIEW
A research centrifuge can be used to model prototype conditions and accelerate flow through porous media. The rationale behind centrifuge modeling is that the centrifugal acceleration induced at the end of a rotating beam may be used to model the earth's gravitational acceleration (g). On the geotechnical centrifuge, a small-scale model, built to a scale of $1/N$, subjected to a centrifugal acceleration N-g, experiences a stress distribution identical to that in a full-scale prototype subjected to gravitational acceleration g on the Earth. In cases of

flow through saturated porous media, the time for advection in the model varies inversely as the square of the acceleration scales, N (Lyndon and Schofield, 1978). For example, at 200-g, a 1-day test using the centrifuge on a model that is 200 times smaller than the prototype represents 109-years of full-scale prototype behavior. Scaling relationships for centrifuge experiments are given by Arulunandan et al. (1988).

Numerous investigators have conducted centrifuge modeling of flow through porous media and consolidation of fine-grained soils (Arulunandan et al., 1988; Cooke and Mitchell, 1991). Researchers have concluded that appropriate scaling laws could be used to model flow processes on a research centrifuge. Numerous researchers have also studied contaminant (i.e. radioactive and organic) transport on the research centrifuge and have verified the utilization of research centrifuge to model environmental contaminant flow problems (Goodings, 1994; Zimmie et al., 1994; Mitchell, 1994; Theriault and Mitchell, 1997).

MATERIALS CHARACTERIZATION

The sediment utilized in this study was a composite of 11 sites in the New York/New Jersey Harbor area collected for the New York Dredged Material Management Plan (NYDMMP). The NYDMMP sediment was analyzed for the geotechnical properties. The dredged sediment contained 33% sand and 66% fines. According to ASTM D-2487, the NYDMMP sediment classifies as sandy organic clay (CH). The initial water content (ASTM procedure D-2974) of the NYDMMP sediment was 113%, and the specific gravity (ASTM procedure D-854) was 2.64. Atterberg limits were conducted on the sediment according to ASTM procedure D-4318. The plasticity index and liquid limits were 39% and 76%, respectively. Table 1 summarizes the geotechnical properties of the sediment.

A silty-sand capping material collected from the Ambrose channel was used in this study. According to ASTM designation D-2487, the sediment classifies as silty sand (SP-SM). The capping material was 93.8% sand and 6.2% fines. The initial water content of the sediment was 29%, and the specific gravity of the sediment was 2.68. Table 1 summarizes the material properties for the capping material.

Table 1. Physical characteristics of sediment and cap

Parameter	ASTM Method	Sediment	Cap
% Sand	D-422	33	94
% Fines	D-422	66	6
Water Content (%)	D-2216	113	29
Organic Content (%)	D-2974	2.6	0.2
Density (pcf) g/cm^3	--	1.4 (88)	1.95 (121)
Specific Gravity	D-845	2.64	2.68
Void Ratio	--	2.98	0.77
Porosity	--	0.75	0.44
Soil Classification	D-2487	CH	SP-SM
Effective Size, D_{10} (mm)		0.004	0.17
Mean Particle Diameter (mm)		.06	0.35
Hydraulic Conductivity (cm/sec)		4×10^{-5}	1×10^{-3}
Plasticity Index (%)	D-4318	39	--
Liquid Limit (%)	D-4318	76	--

Rhodamine WT, a water-soluble, fluorescent dye was used in this study to monitor the movement of pore water through the cap layer. An optimal dye concentration/sediment ratio of 4-mg dye/Kg sediment was established. The initial concentration of the dye in the sediment was 140 ppb.

EQUIPMENT

The research centrifuge is unique in its research applications with capabilities for addressing research needs in physical modeling across a broad range of engineering applications. The research centrifuge at Waterways Experiment Station (WES) is the largest research centrifuge in North America. The research centrifuge has a radius of 6.5-m, and an acceleration range from 10 to 350-g. The maximum payload for the WES centrifuge is 8000-kg at an acceleration of 143-g, and 2000-kg at an acceleration of 350-g.

The modeling box was designed and fabricated from 0.5-inch (1.27-cm) acrylic plastic. The modeling box was 12-inches (30.5-cm) in length, 12-inches (30.5-cm) in width, and 18-inches (45.7-cm) in height. The modeling box was constructed with holes in each side that served as outlets for collecting water samples during the centrifuge tests. Figure 1 is a schematic diagram of the modeling box with the sediment layers and capping layers.

Linear Variable Differential Transducers (LVDTs) were utilized in the centrifuge consolidation studies to measure the vertical settlement of the sediment. An LVDT is an electro-mechanical transducer that produces an electrical output proportional to the displacement of a separate movable core. A special footing was fabricated to allow the LVDT to remain on each surface layer before consolidation was induced. Three LVDT was mounted near the center of the modeling box with its foot placed on the surface of each layer of material (i.e., cap, sediment layers 1 and 2).

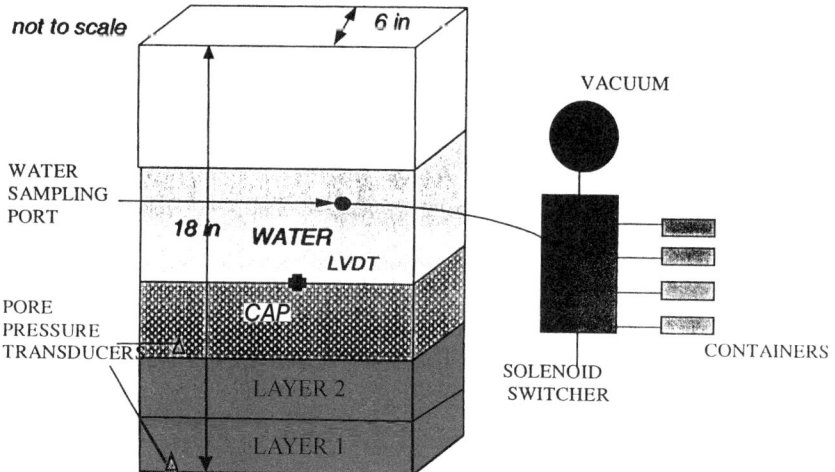

Figure 1. Schematic diagram of modeling box

An overlying water sampling system was fabricated to collect pore water samples at specified time intervals during the centrifuge tests. Four sampling port connectors were made from threaded metal plugs and stainless steel tubing (1/4-inch (0.64-cm) diameter, 6-inches (15.24-cm) in length, and curved to a 120° angle; on one end). One end of the polyethylene tubing was attached to the polyethylene caps of the sampling bottle, and the opposite end of the tubing was inserted in each sampling port about mid-depth below the surface of the overlying water.

Five solenoid valves controlled the vacuum to the sample collection system. Four sampling bottles were connected to separate solenoid valves using polyethylene tubing (1/4-inch (0.64-cm)) and the fifth valve vented the system. The sampling bottles were connected to the inlet side of the solenoid switching box with polyethylene tubing. To determine whether a sample was collecting in the bottles, a current detecting electrode placed on the bottom of each bottle measured voltage changes as water entered the bottles. When water filled the bottle, the data acquisition system showed an increase in the voltage.

Because of the high g-levels attained during the experiments, placement of a conventional laboratory vacuum on the centrifuge was not recommended. Thus, the vacuum used to obtain samples of the overlying water was located in the mechanical room below the centrifuge. The vacuum was connected to the solenoid switching box by placing the vacuum tubing line through the centrifuge's slip ring, which connected instrumentation on the centrifuge basket to the control and mechanical rooms.

Fabrication of a sediment core sampler from 1.9-cm diameter acrylic plastic tubing was accomplished by machining one end of the tubing to a thin edge and using a piston-driven attachment to supply the force needed to push the core sampler through the consolidated sediment and cap layers. Application of a vacuum to the sampling device created a negative pressure on the sample in a manner similar to a piston sampler used in geotechnical engineering. The vacuum applied to the core sampler provided the suction pressure needed to keep the sediment sample in the tube. Sediment cores were capped and stored at 4^0C.

EXPERIMENTAL PROCEDURES
Each centrifuge modeling box was coated with a thin layer of a high viscosity silicone oil (Dow Corning 510) in order to minimize wall effects in the model and to prevent adhesion of the dye to the surface of the acrylic modeling box. The two sediment mixtures and the capping material were placed in separate large polyethylene bags. Loading of the modeling box to the desired sediment height and placement of the cap layer were accomplished by cutting open one corner of the polyethylene bags and slowly squeezing material out of the bag into the modeling box. After placement of the cap material, deionized water was sprayed on the cap in order to minimize void areas in the cap layer, and 0.3-cm of overlying water was placed above the capping layer.

Table 2 lists the testing protocols for the centrifuge consolidation test. Overlying water samples were collected during this study at 5,10, 15, and 20 prototype years (4.5, 9, 13.5 and 18 hours) to monitor concentration changes. Note that two sediment layers were placed into the modeling box where layer 2 contained the dye.

Table 2. Testing protocol for consolidation study

DESCRIPTION	TEST PROTOCOL
Bulk Consolidation Of Sediment (1st Layer)	1. Add sediment to test cell as 1st layer of sediment 2. Consolidate on centrifuge for 26 minutes at 100-g 3. Monitor surface settlement and pore water pressure 4. Remove overlying water 5. Visually measure the magnitude of sediment consolidation
Bulk Consolidation Of Sediment (2nd Layer)	1. Add 4.5-cm layer of premixed dye and sediment as 2nd layer of sediment 2. Consolidate sediment for 26 minutes at 100-g 3. Monitor surface settlement and pore water pressure 4. Visually observe dye breakthrough 5. Remove overlying water ĕ Test dye concentration in overlying water 7. Measure the magnitude of sediment consolidation
Cap/Sediment Consolidation	1. Add a 3-cm layer of capping material, saturate the capping layer with deionized water, and add 0.3-cm of overlying water 2. Centrifuge material for 22.5 hours (25 prototype years) 3. Monitor surface settling and pore water pressure in cap 4. Visually observe dye breakthrough 5. Collect overlying water samples 6. Test dye concentration in laboratory 7. Measure magnitude of sediment and cap consolidation 8. Core sediment, test sediment profile (moisture content) ĕ Core sediment slices on microtome, test slice on a fluorometer (dye test)

RESULTS AND DISCUSSION

The procedure outlined by Arulanandan et al. (1988) was utilized to determine the flow regime where Darcy's law remained valid. For the limiting case when the Reynold's number is equal to one in the model, the maximum increase in the acceleration scale, N_{max}, is given as follows:

$$N_{max} = \frac{v \cdot n}{d \cdot k} \cdot i \tag{1}$$

where v is the kinematic viscosity, d is the effective diameter, n is the porosity, k is the hydraulic conductivity, and i is the hydraulic gradient. Assuming the hydraulic gradient is equal to 1 and the kinematic viscosity is equal to 10^{-6} m^2/s and using the hydraulic conductivity, porosity, and the effective particle size given in Table 1, Darcy's law remains valid as long as the acceleration scale is below 258. Peclet number for these experiments is greater than one, which indicates that the hydrodynamic dispersion is a function of the velocity (the estimated molecular diffusion for the capping material was 1.5×10^{-5} cm^2/sec).

This test was conducted at 100 g, and the centrifuge-model cap was 3-cm thick (300-cm prototype thickness) and the contaminated sediment layer was 9-cm thick (900-cm prototype thickness). At 100-g, 0.875 hours in the centrifuge model is equal to approximately 1 year of

prototype time.   To simulate 25 years in the prototype, centrifuge modeling was conducted
for 22-hours.   Throughout the rest of this paper, the centrifuge data will be presented in
prototype time

The sediment layers were placed into the modeling box at an initial water content of 110%
and were pre-consolidated prior to the placement of the capping layer. Figure 2 shows the
centrifuge prototype settlement curve for test 1 after the placement of the cap.   The data
shown in Figure 2 was obtained from the LVDT at the interface of the sediment and cap. The
total settlement in the model was 2.65-cm (i.e. 2.65-m in the prototype).   Physical
measurements indicated that the average final height of the sediment and cap was 9.4-cm
(i.e., average settlement = 2.6-cm), which is comparable to the average final settlement of the
sediment and cap of 2.65-cm as measured by the LVDTs.

Sediment cores were taken from the modeling box. Cored samples were sectioned utilizing a
microtome to conduct water content analysis. Figure 3 shows the water content and sediment
depth relationship for the cored samples. The capping layer is represented from 0 to 3-cm,
and the sediment layers are represented from 3 to 9.4-cm. Figure 3 indicates that pore water
was advected from the sediment layer through the cap, since the water content of the
sediment decreased from the initial value of 110%.

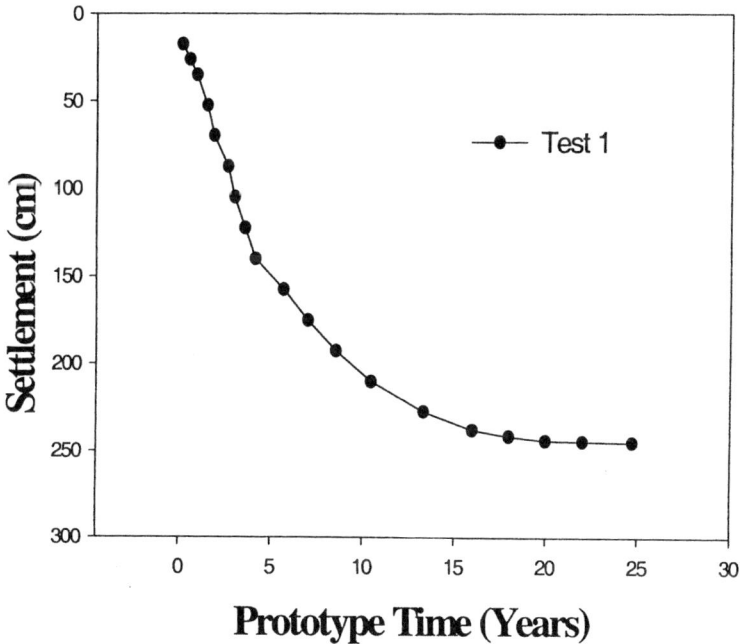

Figure 2. Prototype settlement curve

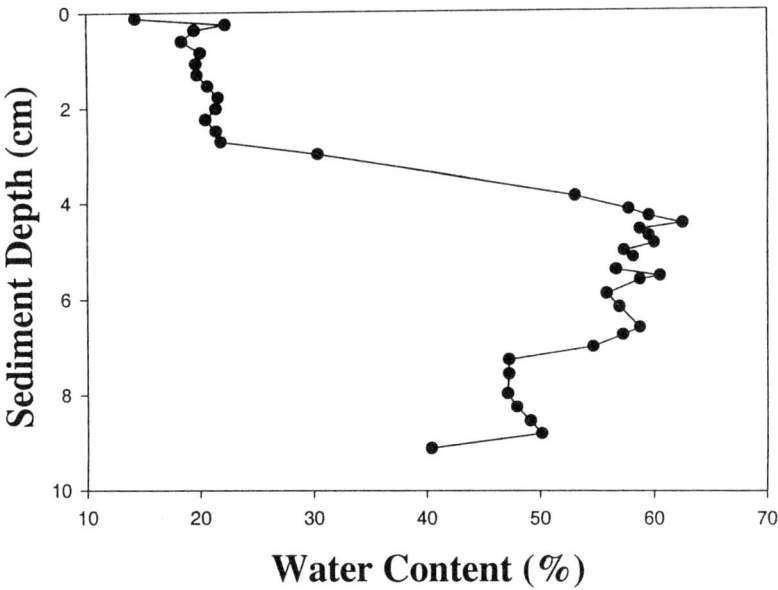

Figure 3. Water content profile for the sediment and cap.

Figure 4a shows the Rhodamine dye concentration in the overlying water. The increase in the dye concentration as time increases indicates that pore water is moving from the sediment layer through the capping material. Furthermore, the instantaneous breakthrough of the dye illustrates that there was no retardation and that advection is the dominant transport process. Figure 4b shows the Rhodamine dye concentration in a cored sample from the centrifuge test. As expected, the dye concentration was much greater in the sediment than in the cap.

SUMMARY AND CONCLUSIONS
The objective of this research was to evaluate the significance of consolidation induced convective transport from the sediment into the cap using a research centrifuge. A centrifuge consolidation test was conducted to estimate the settlement of the sediment and cap. Centrifuge test results illustrate that advection and dispersion were the dominant transport processes. The movement of water during the centrifuge test was illustrated by the transport of dye from the sediment through the cap and into the overlying water. Core samples taken at the end of the centrifuge test also showed that the dye moved from the sediment into the cap and overlying water.

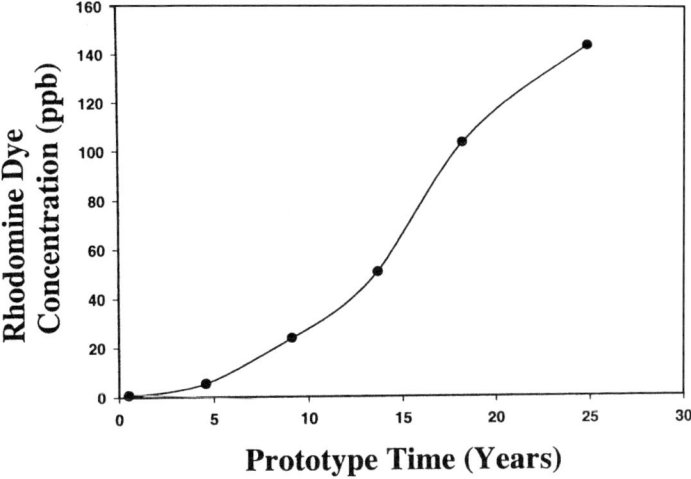

Figure 4a. Rhodamine dye concentration in advected pore water.

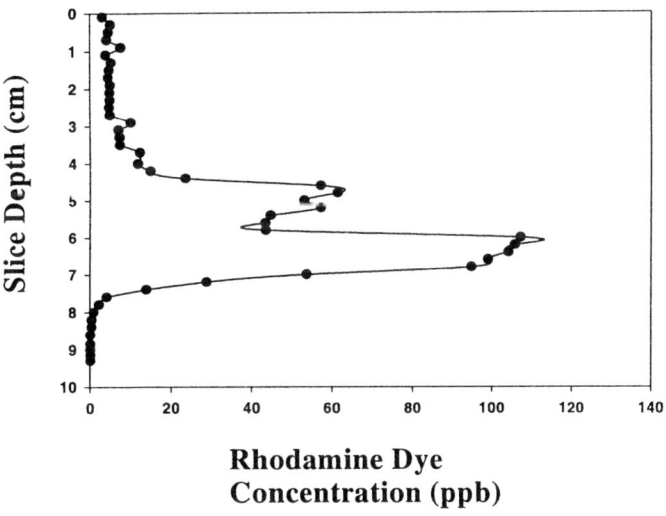

Figure 4b. Rhodamine dye concentration in sediment and cap.

REFERENCES

Arulanandan, K., Thompson, P.Y., Kutter, B. L., Meegoda, N.J., Muraleetharan, K.K., and Yogachandran. C. (1988). "Centrifuge Modeling of Transport Processes for Pollutant in Soils." Journal of Geotechnical Engineering. Vol. 144, pp. 185-205.

Cooke, A.B. and Mitchell, R.J. (1991). "Evaluation of Contaminant Transport in Partially Saturated Soils." Centrifuge 91. H.Y. Ko and F.G. McLean (Ed.). A.A. Balkema: Rotterdam, pp. 503-508.

Goodings, D.J. (1994). "Implications of Changes in Seepage Flow Regimes for Centrifuge Models." Centrifuge 94. C.F. Leung, F.H. Lee, and T.S. Tan (Eds.). Balkema: Rotterdam, pp. 393-398.

Lyndon, A. and Schofield, A.N. (1978). "Centrifuge Model Tests of Lodalen Landslide." Canadian Geotechnical Journal. Vol. 15, pp. 1-13.

Mitchell, R.J. (1994). "Matrix Suction and Diffusive Transport in Centrifuge Models." Canadian Geotechnical Journal. Vol. 31, pp. 357-363.

National Research Council (NRC). (1997). Contaminated Sediment in Ports and Waterways: Cleanup Strategies and Technologies. National Academy Press: Washington, D.C, p. 295.

Palermo, M., Maynord, S., Miller, J., and Reible, D.D. (1998). Guidance for In-Situ Subaqueous Capping of Contaminated Sediment. EPA 905-B96-004. Great Lakes National Program Office, Chicago, IL.

Theriault, J.A., and Mitchell, R.J. (1997). "Use of a Modelling Centrifuge for Testing Clay Liner Compatibility with Permeants." Canadian Geotechnical Journal. Vol. 34, pp. 71-77.

U.S. Army Corp of Engineers and U.S. Environmental Protection Agency. (1992). Evaluating Evironmental Effects of Dredging Material Management Alternatives-A Technical Framework. EPA 842-B-92-008. Washington, D.C.

Zimmie, T.F., Mahmud, M.B., and De, A. (1994), "Acceleration Physical Modeling of Radioactive Waste Migration in Soil," Canadian Geotechnical Journal. Vol. 31, pp. 683-691.

Estimation of Evaporation from Saline Tailings Dams

T.A. NEWSON[1] & M. FAHEY[2]
[1]Lecturer, Department of Civil Engineering, University of Dundee, Scotland, UK.
[2]Assoc. Professor, Geomechanics Group, University of Western Australia, Nedlands, W.A.

ABSTRACT
In the arid regions of Australia, fine-grained slurried wastes produced during mining are typically disposed of in large storages using sub-aerial deposition. Ore processing is conducted using groundwater that can have salinities that approach solution saturation. Precipitation of salts on the tailings surfaces during evaporation leads to the development of thin salt crusts that can significantly reduce the rate of evaporation. Quantification of the actual rate of evaporation from the drying tailings surfaces is important for assessing disposal strategies. Different methods of estimating evaporation were employed at mine sites in Western Australia to provide more information on the drying behaviour of saline tailings. The techniques used are described and typical results from a saline tailings storage are presented.

INTRODUCTION
Fine grained tailings produced in mining operations in arid environments are often deposited as slurries in shallow tailings storage areas using sub-aerial deposition. Sub-aerial exposure after deposition of the tailings leads to desiccation and desaturation under the high evaporative flux. This reduces both the moisture content and compressibility of the material, whilst increasing the density and shear strength. It is possible to maximise benefits from the evaporation, by disposing of tailings in thin layers, which are allowed to consolidate under the effect of evaporation before further filling takes place. With proper management an overconsolidated soil profile can be obtained with a high degree of consolidation using this method. The Western Australian (WA) gold mining industry is situated in arid and semi-arid regions where the net annual pan evaporation rate (E_{pn}) is typically 3-4 m/yr. Since disposal strategies rely heavily on rates of evaporation, it is important to accurately quantify the areal (actual) evaporation rate from the tailings surfaces.

Since the 1980s, mineral extraction in the WA goldfields has required the use of hypersaline water with salinity concentrations (C=mass salt/mass solution) of up to 0.2 (approximately 7 times that of sea water). As a consequence much of the material in tailings storages has high salinities approaching solution saturation (C=0.26). Following the deposition of hypersaline tailings, a surficial salt crust is often seen to develop. Soluble salts accumulate as evaporation removes water from the tailings, forming a thin surface crust (typically <0.5cm thick) as the salts precipitate. Laboratory observations of tailings drying at different salinities have shown that these crusts can drastically reduce (by up to 90%) the evaporation from the tailings surface (Fahey and Fujiyasu, 1994), see Figure 1. It has been found that the factors hindering evaporation from saline soils are the high shortwave reflectivity (albedo) of the salt crust, the vapour density depression due to salinity and the high salt crust resistance to moisture transfer. Since traditional disposal strategies rely on high rates of evaporation, reduced evaporation can affect the drying process and severely affect the behaviour of the storage. This paper discusses methods of estimating areal evaporation from saline tailings, presents results obtained from field studies and discusses the implications of the data presented.

EVAPORATION FROM BARE SOIL

Evaporation involves supplying water with sufficient energy to change it from a liquid to a vapour form, and removing the moist air to allow the process to continue. Many indirect methods of measuring evaporation are based on measuring (or estimating) energy fluxes across the ground surface, with the difference in inward and outward flux being the latent heat of vaporisation. The net radiation energy R_n received at the ground surface is the sum of the incoming and reflected short-wave radiation, and the incoming and outgoing long-wave radiation. The energy balance equation for the ground surface may be written:

$$R_n = H + L_eE + G \qquad (1)$$

where H is the sensible (air convection) heat flux (W/m^2), E is the evaporation rate (kg/m^2.s), L_e is the latent heat of vaporisation (J/kg), R_n is the net (all wavelength) radiation (W/m^2) and G is the soil heat flux (W/m^2). This equation states that the net radiated energy is equal to the sum of: the heat transferred upwards from the ground surface by the air (H); the heat transmitted downwards into the ground (G); and the heat used in vaporising water at the ground surface. Thus, if the radiation and heat flux terms can be measured, the heat used in vaporisation can be determined, and from this the amount of evaporation occurring can be estimated. This is the basis of the 'energy balance' methods of measuring evaporation rate.

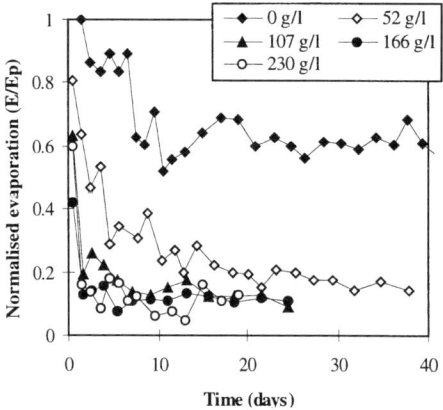

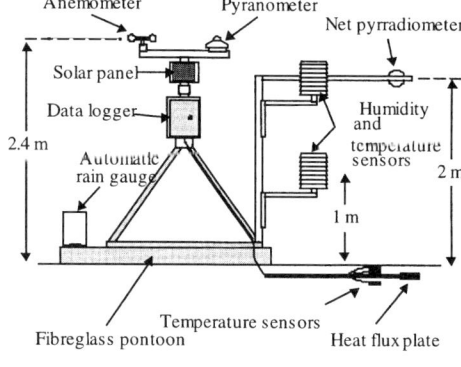

Figure 1: Laboratory tests of evaporation from saline soils (After Fujiyasu and Fahey, 1994)

Figure 2: Layout of Bowen ratio weatherstation

If the amount of water removed by evaporation from the surface of a soil body exceeds the amount being replaced from within the body, then the soil begins to dry. During this drying process (from saturated to residual moisture contents), the areal rate of evaporation from the soil surface, E, can vary from close to that of the potential evaporation, E_p, (stage I) to almost zero (stage III), for desaturated soils (Wilson et al., 1994). The potential evaporation can be defined as the evaporation that would occur from a hypothetically moist surface with radiation absorption and vapour transfer characteristics similar to the area of interest. It is commonly assumed to be equal to be the same as that from a Class 'A' evaporation pan.

Bowen Ratio Method

The fundamental difficulty in quantifying areal evaporation in the environment is the inability to directly measure this process without affecting the process itself. However, there are a number of different methods which can be used to produce reasonably accurate estimates. One of the most popular of these approaches is the Bowen Ratio energy balance method. This effectively partitions the available energy ($R_n - G$) into sensible heat flux (H) and latent heat (L_eE), and expresses this as a ratio β, where:

$$\beta = \frac{H}{L_eE} = \frac{\rho c_{pm}K_h\left(\dfrac{\mathrm{d}T_d}{\mathrm{d}z}\right)}{\rho L_e K_v\left(\dfrac{\mathrm{d}h_s}{\mathrm{d}z}\right)} = \frac{c_{pm}}{L_e}\left(\frac{\mathrm{d}T_d}{\mathrm{d}h_s}\right) = \gamma\frac{\mathrm{d}T_d}{\mathrm{d}h_s} \qquad (2)$$

This equation states that β values may be derived from changes in dry bulb temperature T_d (°C) and in specific humidity h_s (kg/kg) over the same height interval dz, and the thermodynamic value of the psychrometric constant $\gamma = c_{pm}/L_e$, where c_{pm} is the specific heat (J/kg.°C) of moist air (density ρ) at constant pressure. The quantities K_h and K_v (m²/s) are the eddy transfer coefficients of heat and water vapour transport, respectively. In this formulation, these two coefficients are assumed to be equal. Thus, β may be determined by measuring the dry bulb temperature and humidity at two different heights above the surface. The available energy is estimated by measuring the net radiation (R_n) and the soil heat flux G. The areal evaporation E can then be determined from:

$$E = \frac{R_n - G}{L_e(\beta + 1)} \qquad (3)$$

FIELD STUDIES OF EVAPORATION

Methodology

To enable the effects of large scale meteorological phenomenon to be assessed on the evaporation rates from tailings storages a number of field studies were conducted. Both freshwater and saline tailings storages have been investigated. For reasons of brevity only one of the saline sites will be discussed herein, further information on the other sites can be found elsewhere (Fujiyasu et al, 2000; Newson and Fahey, 1998).

The saline site (KT) was equipped with a Bowen ratio weatherstation, which enabled the components of the surface energy balance to be determined and hence the evaporation from the surface to be deduced. In addition, spot measurements of evaporation were made using micro-lysimeters (Boast and Robertson, 1982) and the site was equipped with a Class A evaporation pan. Profiles of moisture content, salinity concentration, suction and undrained shear strength were also determined at different points during the drying process.

The layout of instrumentation for the Bowen ratio weatherstation is illustrated in Figure 2. The instrumentation is mounted on a small fibreglass pontoon so that it can be deployed immediately after completion of filling of the tailings. The weather station instrumentation included: a pyranometer to measure global radiation (R_S) and a net pyrradiometer to measure net radiation (R_n) at a height of 2 m above the tailings surface, an anemometer to measure wind speed at a height of 2 m, a heat flux plate to measure soil heat flux (G) below the surface (10 cm depth), temperature and humidity sensors to measure air dry-bulb temperature (T_d) and relative humidity (h_s) at heights of 1 m and 2 m above the surface (these were located in sensor shelters), temperature sensors to measure soil temperatures at the surface and at depths of 3cm and 7 cm below the surface and an automatic rainfall gauge.

The data were logged continuously by an automatic data logger, with time-averaged values being saved every 15-20 minutes. These data were downloaded onto a portable computer every few weeks. Power for the instrumentation and data logger was provided by battery, charged by a solar panel. Figure 3 shows typical energy flux data measured by the weather station located on the tailings storage over a 3-day period. The components of the surface energy balance equation are shown here: net radiation (R), sensible heat flux (S), soil heat flux (G) and evaporation (L_eE). The soil heat flux is generally negative, implying that heat is being conducted away from the ground surface over most of the day, but some upward flow occurs at night. The net radiation and sensible heat flux vary diurnally with high values occurring during the day. The evaporation rate is determined directly from the latent heat flux term (L_eE), by dividing by the heat of vaporisation.

Field observations of evaporation
The site chosen (KT) was located near Kalgoorlie in Western Australia. The tailings produced are stored in six rectangular cells of approximately 42H. The deposition of tailings can therefore be rotated between the cells on a regular basis to allow significant drying to occur for each layer placed. Layer depths are of the order of 1-2m and are usually left dormant for up to 6 months at a time. The tailings material is hypersaline (C=0.1) and the particle size distribution shows only 5-6% of the fraction is clay sized and the particle size is predominantly in the silt range. The liquid limit (w_ℓ) is a low 32% and the plasticity index (I_p) is only 4. Tailings are deposited at a moisture content of 80%. The field study consisted of monitoring three of the cells throughout a drying cycle. The data shown is from cell #1 between September 1996 and January 1997. The Bowen ratio weatherstation was set up on a pontoon and positioned just off the decant structure to allow access.

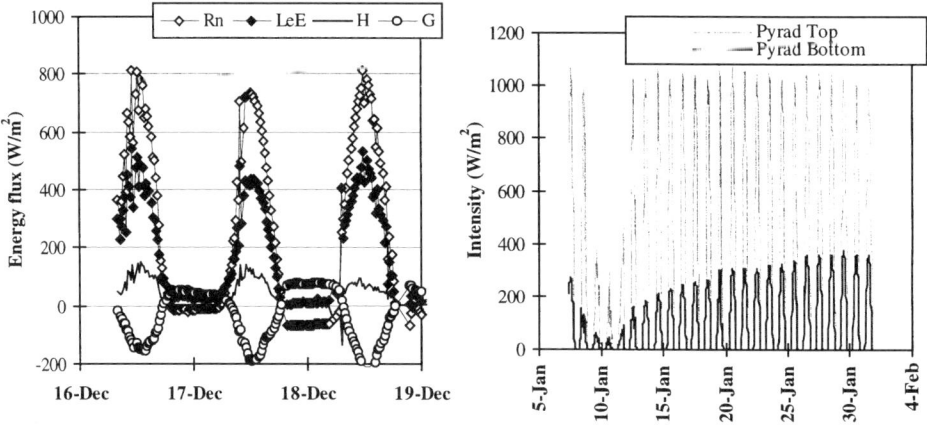

Figure 3: Typical values for the components of the surface energy balance of the evaporating tailings

Figure 4: Incoming radiation flux and surface reflected radiation flux

Over the period, the air temperature varied from 10°C to nearly 40°C, with an average of 20-25°C. As would be expected, the air temperature showed quite marked diurnal variation. Reduced temperatures occurred during periods of rainfall. Similar trends were found for the relative air humidities, which varied from 10% to 100%, with an average of 40-50%. At three centimetres depth, the soil temperature varied between 47°C and 15°C, with an average of 25-30°C. This reading tended to be 5 or more °C greater than the air temperature. The deeper transducer (10 cm) also showed increased measurements compared to the air temperature, but this showed lesser diurnal variations than the near surface transducer. This indicates that the heat stored in the tailings surface appears to be quite significant.

Average windspeeds varied from 0.5 to 10 m/s, with an average of 3-4 m/s. The energy intensities of the net radiation peaked at approximately 700 W/m^2 and soil heat flux at approximately 110 W/m^2. A 'pyrano-albedometer' was also used to take measurements of. the incoming radiation and reflected radiation from the evaporating surface and these are shown in Figure 4. This indicates that the albedo of this material is of the order 0.35-0.4. These values are lower than those measured in the laboratory for salt crusts ($\alpha = 0.75$ for dry crusted soil) and this could be due to the sparse nature of the salt crust that develops on the KT material. This tends to be only 1-2mm thick around the decant areas, compared to 5mm at the edges of the cells. It can be seen that the albedo of the surface changes significantly after the rainfall and then requires about 5-6 days to return back to similar pre-rainfall values.

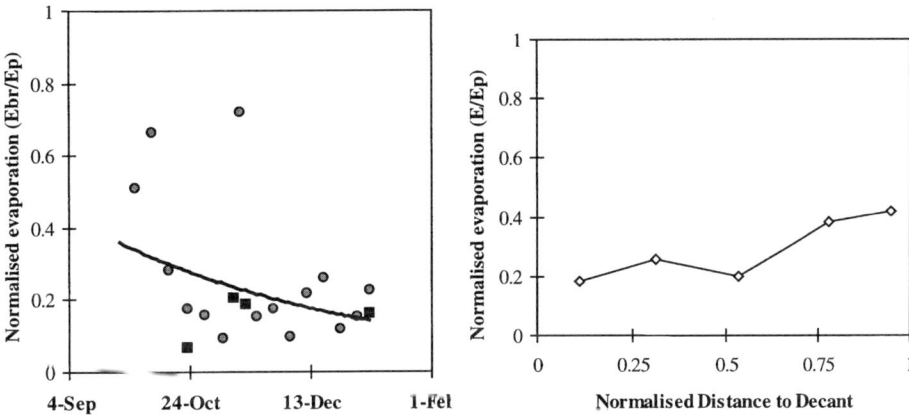

Figure 5: Estimated normalised evaporation using Bowen ratio weatherstation and micro-lysimeter

Figure 6: Spatial variation of evaporation across tailings storage surface

Relative evaporations based on the surface energy balance are shown in Figure 5. Values of Bowen ratio evaporation (E_{br}) have been normalised using the pan evaporation (E_p) data are compared with micro-lysimeter data (shown with square symbols). The weatherstation data has been averaged over weekly periods. The data shows a reduction in relative evaporation from 0.5-0.6 to 0.2-0.3 over the drying period. There are slight increases in the periods that correspond to rainfall events. The micro-lysimeter spot readings correlate reasonably well with these values. It is interesting to note that despite a 'full' salt crust not developing that the reduction in the evaporation is still comparable with the laboratory data. However, the surface evaporation rate does appear to be more susceptible to rainfall than surfaces with heavier salt crusts, which require large rain events to dissolve the crust.

Spatial variation of evaporation
Tailings storages are considered large enough to interact with, and affect the overlying atmosphere. In essence, a wetland environment can be formed over the area of the storage.

The studies conducted with the Bowen ratio weatherstations have provided information regarding single positions on storages. It was felt that information about the spatial variation of evaporation on storages should also be investigated.

Measurements of evaporation from the tailings surface and from the decant pond were made on a number of saline tailings storages. Access was obtained to the area using a hovercraft. Four locations on the surface of the tailings and one on the decant pond were investigated. The measurements were made on a transect from the edge of the storage to the decant structure. At each location, a micro-lysimeters was used to measure the evaporation from the tailings. In addition, a second micro-lysimeter was filled with fresh water, to give a local measurement of the potential rate. Floating lysimeters were used on the decant pond, one filled with pond water and the other with fresh water.

The results are shown in Figure 6; the first three locations on the tailings surface were salt-encrusted, but the fourth had not yet developed a salt crust. The results are plotted as relative evaporation rate versus normalised distance between the wall and the decant (a total distance of 450 m). The results show that the normalised rate in the first three areas was about 0.2, while at the fourth position, the rate was somewhat higher (0.38). The rate obtained from the decant pond was still only 0.4.

CONCLUSION

The field observations show that even before salt crusts are formed, the evaporation rate from saline tailings is well below that of freshwater. Once the salt crust develops and the material begins to desaturate, the relative evaporation can be extremely low (<0.2), which confirms the results found in laboratory tests. The evaporation rates from the decant pond were also found to be extremely low. This result is important since water balance calculations often assume that the evaporation from the decant pond and wet area around it occurs at the potential rate, and this therfore has far reaching consequences for disposal strategies involving these materials. The methods used to estimate the evaporation have proved to be relatively simple and reliable, and have provided accurate estimates of evaporation from the whole of the tailings surface. It is recommended that they be deployed in combination, using the micro-lysimeter method to cover a large area with a number of the devices and extrapolating results from a highly accurate measurement from a Bowen ratio station at a single point, to provide results suitable for accurate water balances to be conducted.

ACKNOWLEDGEMENTS
Part of this work was supported by grants from the Australian Research Council (ARC) and from the Minerals and Energy Research Institute of Western Australia (MERIWA) on behalf of a consortium of gold mining companies. This funding is gratefully acknowledged.

REFERENCES
Boast C.W. and Robertson T.M. 1982. A micro-lysimeter method for determining evaporation from bare soil: description and laboratory evaluation. *Soil Sci. Soc. Am. J.*, 46, 689-696.

Fahey M. and Fujiyasu Y. (1994). The influence of evaporation on the consolidation behaviour of gold tailings. *Proc. 1st Int. Cong. Environ. Geotech.*, Edmonton, 481-486.

Fujiyasu Y., Fahey, M. and Newson, T.A. (2000) Field investigation of evaporation from freshwater tailings. *ASCE Journal of Geotechnical and Geo-environmental Engineering.* 126, 6, 556-568.

Newson. T.A. and Fahey, M. (1998) Saline tailings disposal and decommissioning. *MERIWA Report No. M241*, Vol 1.

Wilson G.W., Fredlund D.G. and Barbour S.L. 1994. Coupled soil atmosphere modelling for soil evaporation. *Can. Geotech. J.*, 31, 151-161.

Tailings in the Environment.

DR A.D.M. PENMAN, Geotechnical Engineering Consultant, Harpenden, AL5 3PW, UK.
Chairman ICOLD Committee on Tailings Dams and Waste Lagoons.

ORIGINS.

A dictionary definition of environment is, "Surroundings, surrounding objects, region, conditions or influences". City dwellers, on their weekend safaris into the countryside, marvel at the beauty of nature, without realising that much of it was created by the efforts of mankind, and give man no credit. When they encounter the site of mining activities they tend to decry as vandalism these efforts of mankind. In those olden days when man had only muscle power and the world supported a smaller population, the battle was between man and nature, with the constant struggle by man to control nature. One of the aims of the Civil Engineer, according to the 1828 Constitution of our Institution, is to direct the Great Sources of Power in Nature for the use and convenience of man. In today's era of political correctness, it is almost as though these roles were reversed. True that with so much power to his elbow and a population exceeding 6 billions, the damage that man can inflict on the world is alarming, but with a population increasing by 90 millions a year, it is going to be an uphill struggle to maintain standards of living.

Mining must be good for man, otherwise it would not be practiced. Fuels, minerals, precious and common metals are all won from the near surface of the world by mining, and are needed by mankind to sustain his level of civilisation. Everyone benefits from readily available metals and raises no complaints so long as the mining is nowhere near their own back yards. But mining must be for the benefit of man, and when it starts killing people not involved in the mining operation itself, perhaps it is time for some checks.

TAILINGS DAMS.

The aspect of mining that we will consider in this short paper is the waste that comes from the tail end of processing plants and its disposal. Many processes produce tailings, from the small volumes of soil washed from farm root crops on their way to the supermarkets to the vast quantities resulting from the beneficiation of the mined ore. In order to extract the desired metals, the ore rock is crushed, milled and ground to silt and sand sizes. After a wet extraction process, the finely ground rock is the waste tailings that is discharged as a pumpable slurry, commonly piped to an impoundment retained by tailings dams. An indication of the volumes involved may be gleaned from a report in the Mining Journal (September 2000) of the development on the Indonesian island Sumbawa of the Batu Hijau mine. The planned production is for an average of 600 000 tonnes/day for the 15 year life of the mine, with the aim of winning 830 tonnes of copper and 0.06 tonnes of gold per day, leaving at least 599 000 tonnes a day of waste fine rock for disposal. An impoundment to store this waste would require capacity to accept more than 3×10^9 tonnes during the expected life of the mine. Predictions of this sort made at the beginning of a new mine are often an underestimate and towards the end, as more ore is discovered, a greater volume is needed requiring the tailings dam to be built higher than provided by the original design, which can lead to difficulties.

FAILURES.

Geotechnical engineering design can produce stable tailings dams if construction, that can extend for more than 15 years, can be carried out to comply with design and remain sufficiently flexible to accommodate changes that may occur. It might be said that the behaviour of tailings dams is fairly well understood, yet failures continue to occur. The list of published failures compiled by the WISE Uranium Project show that during the past 30 years there has been at least one failure every year except 1983, 4, 7 and 1990, seven years with 2 failures, three with 3, one with 4 and 5 failures in both 1994 and 2000, giving an average of 1.7 failures per year during the 30 year period. When a tailings dam fails, the resulting escape of the impounded tailings can cause both human fatalities and considerable environmental damage. Only two examples will be given here, illustrating both the geotechnical and environmental aspects, but in neither case, lost of life.

Aznalcóllar tailings dam.

Mining of the pyrite belt, 45km northwest of Seville, dates back to Roman times. The ore contains zinc, lead and copper. There are two main types, pyretic chalcopyrite (pyroclast) and a polymetallic sulphide (pyrite). In 1960 Andaluza de Piritas SA was formed as part of the Banco Central SA industrial group to acquire and exploit a deposit. Overburden stripping for the beginning of the opencast mine started in 1975, and processing at a rate of 3.5×10^6 tonnes/year began in 1979. Construction of an impoundment to take the tailings was started in 1978 on the flood plane to the west of the river Agrio that flows approximately from north to south. The impoundment is about 2km long and 1.2km wide retained by tailings dams on three sides with a dividing dam as indicated by Fig.1c, to form storage for two types of tailings, those from the pyroclast processing and those from pyrite.

The valley containing the impoundment is underlain by Miocene marl, covered by an alluvium of sand and gravel. The dam was built in stages to suit the volumes of delivered tailings, with a main body composed of waste overburden and rock from the mine and design slopes, upstream and downstream, of 1 on 1.9 and 1 on 1.3 respectively. An upstream earthen layer of low permeability on the upstream slope was connected to a bentonite slurry trench cut-off at the upstream toe, taken through the alluvium into the marl to prevent seepage from the impounded tailings. The mine, bought by Boliden in 1987, continued working until 1996. Since taking over, Boliden had found another ore body called Los Frailes nearby and began working it in 1997, increasing output to 4×10^6 tonnes/year, using the same processing plant and tailings impoundment as before. Prior to the failure, the dams were being raised at about 1m per year.

As so often occurs with tailings dams because of the slow rate of building, construction was not quite as the designer had envisaged and records showed a section as Fig.1b: it is not clear how far up the low permeability layer was carried. Leakage occurred from the main dam towards the river near its junction with the dividing dam and a length of slurry trench wall was constructed from the crest, but leakage continued at a rate, in 1996, of about 100 m^3/hour and a row of 46 relief wells were drilled at the dam toe from the dividing dam towards the north at a spacing of 25 to 30m to pass through the alluvium into the marl, increasing the pumping rate to about 1000 m^3/hour. The impoundment and dams were reviewed by independent consultants and the Spanish government in 1996 and again in 1997. After that inspections were made regularly, the last on 14th April, just 10 days before the failure.

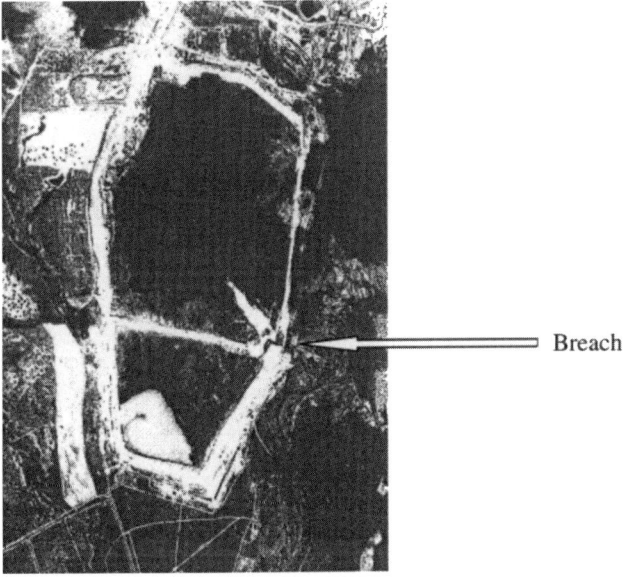

(c)

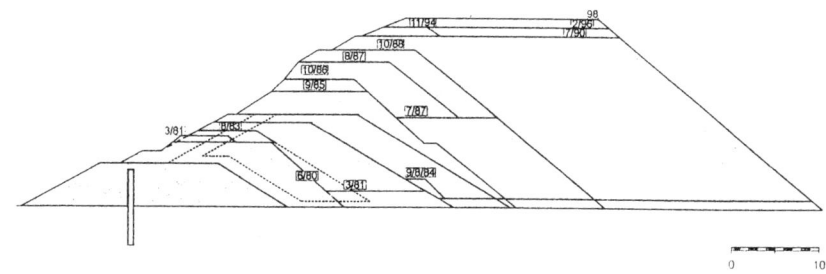

(b)

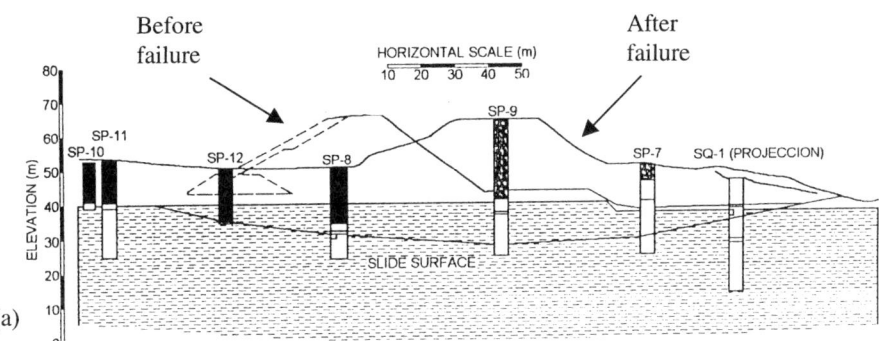

(a)

Figure 1. The Aznalcóllar tailings dam and impoundment.

Failure occurred during the night of 24th/25th April 1998 so there were no eye witnesses. First indications were when a technician noticed that one of the seepage collection pumps was not working. When he went out to investigate, he heard the sound of rushing water, and going further along the dam to see what was causing the noise, saw the failure. It appeared that at the junction with the dividing dam, an intact 600m length of the main dam immediately to the south had swung open like a door hinged at its south end, releasing 6.8×10^6 m^3 of liquefied tailings. It is not clear exactly which part of the main dam gave way first, but the fact that the 600m length remained intact despite an outward movement of about 60m indicated that the dam itself was not unstable. The failure surface appeared to have developed in the marl at a depth of about 14m, as indicated by Fig.1a and it has been suggested that high pore pressures in the marl caused by the weight of the impounded tailings migrated under the dam reducing the effective stresses sufficiently to permit of failure. The tailings had a particularly high bulk density of 28 kN/m^3, the marl was found to be highly fissured and to be very brittle so that it is probable that there was a degree of progressive failure. The situation is unique in that such a mode of failure would not have been predicted during the design stage, and pore pressures were not being measured in the underlying marl. It can be speculated that water leakage through the dam, collected in the relief wells, may have had some destabilizing effect, but this has not been established. The liquefied tailings travelled down the river for 40km covering over 2000ha of agricultural land. Very rapid action was needed to divert the flow to prevent its entry into the Doñana National Park; Europe's largest nature reserve. The clean-up operation, involved mechanically scraping up the escaped tailings and transporting it back to be dumped in part of the opencast mine, continued for more than a year and is said to have cost in excess of US$42.5 millions.

Consolidation of the remaining tailings impoundment will discharge pore water into the alluvium and to prevent this from reaching the river, a slurry trench cut-off wall 2 815 m long has been constructed between the dam toe and the river. It has been taken 1.5 m into the marl and fitted with a 2 mm thick high density polyethylene membrane to attempt a complete seal. During excavation it was found that the alluvium consisted of material ranging from dense sands and gravels to firm and stiff clays, with the depth to the marl ranging from 2 to 9 m. The failed section of the dam has been stabilised and to encapsulate the remaining tailings and provide a long term protection to the surroundings, a four layer cover has been provided. It consists of a fill layer placed over the tailings, a blinding layer, an impermeable layer of Miocene based clay with a design permeability of 1×10^{-9} m/sec, plus a layer of sandy gravel on the surface to protect the clay from desiccation and erosion, to drastically reduce the amount of oxygen and water penetration.

A failure at Baia Mare.

The expanding city of Baia Mare in Romania was beginning to encroach on old mining areas where there were disused impoundments of tailings. Removal of these impoundments and their retaining tailings dams would both release valuable land for city development and allow extraction of remaining metals from the old tailings. The scheme at Baia Mare involved construction of a new impoundment and new efficient processing plant that would accept tailings removed from the old impoundments. Initially three were to be reworked and pipelines were laid out to transmit water from the new impoundment to be used for powerful jets that would cut into the old tailings, producing a slurry that would go to the new processing plant for extraction of remaining metals, with the tailings from it flowing to the new impoundment. The system used the same water going round and round with no interference with the environment.

The site for the new impoundment, well away from the city, was on almost level ground, with its main axis 1.5km long, sloping down only 7m from NE to SW with a width of about 0.6km, as indicated by Fig.2c. An outer perimeter bank 2m high with 1 on 2 side slopes, as shown by Fig.2a, was built from old tailings, and the whole area of about 90ha, lined by HDPE sheet, anchored into the crest of the perimeter bank. Drainage was installed to collect any seepage, that would be pumped back so that there should be no escape of contaminated water into the environment. About 10m inside the perimeter, starter dams were built, also with 1 on 2 side slopes, to heights of about 5.5m along the SW lower edge of the impoundment, tapering down to 2m height about half way along the sides, with the remainder around the NE end of the impoundment, about 2m high. Cyclones mounted along the crest of the SW starter dam and part way along the side starter dams accepted the tailings piped from the new processing plant, discharging the coarser fraction on to the downstream side to fill the space to the parameter dam, and raise the whole dam, with the main volume of fine tailings slurry being discharged into the impoundment. Collected water was discharged into the central decant, drained through a 450mm diameter outfall pipe embedded under the HDPE liner and pumped back to operate the monitoring jets in the first of the old impoundments, 6½ km away, and close to the city.

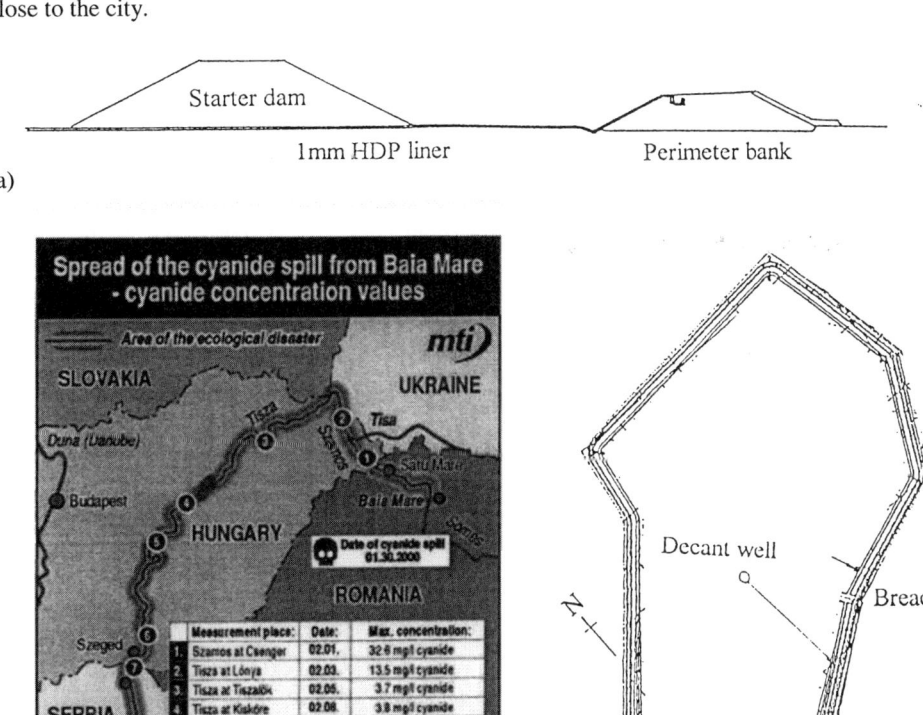

Figure 2. The impoundment at Baia Mare.

Cyanide was used in the new processing plant for the extraction of gold, so that the tailings and water in the new impoundment contained considerable amounts of cyanide. No water should leak from the pipework circuit, although the water used in the cutting jets flowed over the unlined floor of the old impoundment where it could soak into the ground. First discharge into the impoundment was in March 1999, and during the summer everything worked well, particularly during June, July and August when the average evaporation was 142mm per month, although the delivered tailings did not contain quite as much coarse material as had been envisaged and the rate of height increase of the dams was lower than intended.

During the winter, however, conditions became greatly changed. The temperature fell below zero on 20 December and remained low during most of January, freezing the cyclones and producing a layer of ice over the impoundment, which became covered by snow. Tailings from the processing plant was warm enough to keep the operation working, but there was no further height increase for the dams because the cyclones were out of action. Precipitation during September to January averaged 71mm per month and fell as rain and snow on both the whole area of the impoundment but also on the old tailings impoundments that were being worked. This extra water was stored in the impoundment causing the level to rise under the now thick layer of ice and snow.

On 27th January there was a marked change in the weather. The temperature rose above zero and there was a fall of 37mm of rain. The ice and snow covering melted and the dams, half way along the sides of the impoundment, where they were only starter height, were lower than the developing water level. At 22.00 hours on 30th January 2000, a section overtopped, washing out a breach 25m long that allowed the escape of about 100 000m^3 of heavily contaminated water that flowed following the natural slope of the area, towards the river Lapus. This in turn fed into the rivers Somes, Tisa and Danube, as indicated by Fig.2b, eventually discharging into the Black Sea. A very large number of fish were killed with serious consequences for the fishing industry for a time. The Hungarian authorities estimated the total fish kill to have been in excess of one thousand tonnes. Water intakes from the rivers had to be closed until the plume of toxic contaminate had passed and for some time afterwards until the purity of the water could be confirmed. The cyanide plume was measurable at the Danube delta, four weeks later and 2000 km from the spill source.

The concept of a closed system in which none of the process water should escape into the environment should have been excellent, with the new tailings impoundment completely lined with plastic sheeting and provision for the collection of any seepage. Unfortunately no provision had been made for the additional water that would accrue from precipitation, nor had the problems of working at low temperatures been addressed. The scheme was one that could have worked well in the hot and dry conditions found in some parts of Australia and South Africa.

CONCLUSIONS.
Geotechnical knowledge with engineering experience can enable safe tailings dams to be designed and constructed, but the current rate of tailings dam failures that have averaged 1.7 a year during the past 30 years shows that this knowledge and experience has not been applied in every case. A failure can stop production, and clean-up operations, compensation for damage and in some cases death, plus finding new storage for tailings, can be so extremely expensive for a mine that it would be expected that every care would be taken to avoid failure. It would appear that management fail to engage staff able understand tailings dams

sufficiently to be able to detect deficient design or construction procedures, so that dangerous conditions can be allowed to persist until failure occurs. In order to improve the situation the International Commission on Large Dams with the United Nations Environmental Programme are currently publishing an ICOLD Bulletin that gives 221 examples of incidents with tailings dams and discusses causes, with the aim of helping those in charge of tailings dams to understand some of the simple mistakes that continue to occur.

References.

Fernandez Oliva A and Salvi GJ (2000). The use of a slurry cut off wall with HDPE geomembrane to confine the water from the consolidation process of the tailings. *Trans 20th Intn Congress Large Dams, Beijing, vol 4, pp 787-805.*

Report, Cyanide Spill at Baia Mare, UNEP/OCHA Assessment Mission Report, April 2000 available at www.natural-resources.org/environment/BaiaMare

Tailings Dams: Risk of Dangerous Occurrences

DR A.D.M. PENMAN, Geotechnical Engineering Consultant, Harpenden, AL5 3PW, UK.
Chairman ICOLD Committee on Tailings Dams and Waste Lagoons.

ORIGINS

Everything that we have, our trains, planes, cars and the food we eat comes from the one globe on which we live. Looking at the ground beneath our feet we can understand how food grows but it is not so easy to see where metals come from. Ore containing metals looks just like any other rock and to get at the metal, it has to be fragmented. Half a century ago copper ore had to contain more than 4% for it to be worth working, but today ores containing only 0.3 to 0.4% are being mined, so a great deal of the rock is waste. In parts of the Andes, the ore is found at the surface and can be won by normal quarrying techniques. In other parts of the world, for example Poland, a thick ore body was found at depth, is mined with heavy machinery and brought to the surface through vertical shafts. Once obtained, the rock is crushed and ground down to small sizes then passed through a wet process to extract the metal. Reducing to sand sizes required a lot of energy, but some metal was still left, so the trend has been for more intensive milling to reduce the rock to silt sizes of 0.006 to 0.06 mm. Depending on the rock type, some part of it might break down to near the clay size of 0.002 mm, while other parts might resist the milling to come out sand size of 0.1 to 0.5 mm. As an example of the volumes involved, one mine in the Andes mines 34 000 tonnes of rock per day. At an average metal content of 0.3% this would produce about 100 tonnes of metal and leave 33 900 tonnes of waste finely ground rock for disposal. This waste comes from the tail end of the processing plant as a wet slurry that is called tailings.

TAILINGS DISPOSAL

The volumes of tailings are such that it cannot be simply discharged into rivers or the sea, although such methods were used in earlier times. In general the material is now stored in large impoundments, retained by tailings dams. In some cases a valley may be used, and dammed as though forming a water reservoir but with special provision for diverting the river so that the 'reservoir' volume can be used for the tailings, when it is referred to as an 'impoundment'. The dam may be built from borrowed fill as an embankment dam to retain water, and where water is to be stored as well as the tailings, this type is used. But the tailings themselves are nearly all fine rock which should have adequate strength and form a good fill. Preferably when there are sand sizes in the tailings, dams can be built by what has become called the 'upstream' method; so called because during construction, the dam crest moves upstream. A typical section is shown by Fig.1. Construction is begun with a starter bund of borrowed permeable fill and the tailings is discharged from its crest in an upstream direction to begin filling the valley. The aim is to form a gently sloping beach on to which the tailings is discharged uniformly over an area. As the tailings moves slowly down the beach, the larger sizes settle out first, with finer material being deposited further down the beach until the very

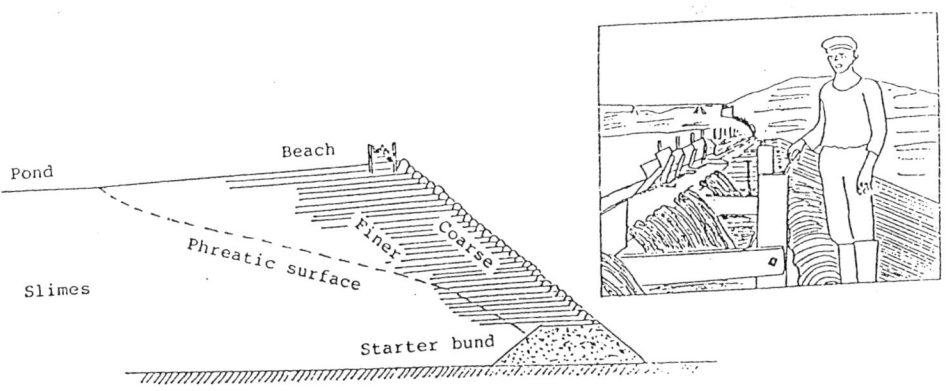

Figure 1. Upstream method of construction. Sketch shows discharge from timber flume on to
beach in 1925. Note flume on trestles bringing tailings from processing plant.

finest, possibly clay size particles go into a pond of water where they settle out and clean
water is discharged through a decant tower, or floating pump barge, to be pumped back to the
processing plant or allowed to flow into the river. Fig.1 depicts the dam of a Spanish copper
mine in 1925. Tailings from the process plant, beyond the left of the insert sketch, flowed
down a timber flume supported on trestles to a distribution tank from which it flowed in a
ditch to the dam crest where it was carried in a timber flume containing outlets on its
downstream side fitted with shutters so that the discharge could be at any position along the
length of the crest. A man with a long handled shovel built up the crest from drier material
and controlled the downstream slope by the amount he placed each addition to the crest
upstream of the last.

The strength of this particulate material is dependant on the prevailing effective stress that is
sustained by its weight and reduced by the pressure of water in its void spaces or pores. To
maintain the stability of the downstream slope it is essential to keep the phreatic surface well
back so that the coarser fraction upstream from the slope surface is free from destabilizing
pore pressures. Tailings is delivered to an impoundment at concentrations of 30 to 40% by
mass and the water in the tailings accumulates on the surface, is lost by drainage and
evaporation but a great deal remains within the pores. Deposition through water leads to a
loose condition well below the critical density. Any disturbance due to earthquake shock or
movement of the retaining dam can lead to development of high pore pressures and
liquefaction of the deposit.

In dry areas a traditional method of dam construction is with the use of paddocks. Paddocks
are formed by constructing small surrounding banks, usually with material from a dug ditch,
that are filled with tailings and left to dry by evaporation, successive paddocks being filled to
form the base of the dam. Dry tailings is dug for banks for the next lift of the paddock and the
processes repeated to form the dam. However constructed, the dams are only raised to keep
up with the rising level of the impoundment that may be no more than one or two metres per
year.

In regions subject to earthquake, it has become usual to build tailings dams by the downstream rather than the upstream method because of their susceptibility to earthquake damage. Instead of a beach, the coarse fraction is separated by centrifugal force in cyclones either on the dam crest or larger units off the dam. This coarse material is placed downstream from the crest, which then moves downstream as height increases.

FAILURES.

When tailings dams fail, the impounded material often liquefies and may flow for considerable distances. In this short paper only two examples of failure will be given, even though, despite our improved understanding of the behaviour of tailings dams, failures continue to occur; during the last three decades, at a rate of about one a year, although in the year 2000 there were four failures.

Stava At 12.23 on 19th July 1985 two tailings dams, one above the other and both built by the upstream method, collapsed. A total quantity of 190,000 m^3 of tailings slurry was released and flowed, initially at a speed of 30 km/hr down into the narrow, steep sided valley of the Rio Stava, demolishing much of the nearby small village of Stava and continued, at increasing speed, estimated to have been 60 km/hr to another small town, Tesero about 4km downstream, at its junction with the Avisio River. The only surviving eye-witness, a holiday maker, had the horrifying experience of watching the disaster from the hillside and saw the hotel where his family were taking lunch being swept away by the torrent of tailings. The damage caused by tailings is very much more than would be caused by the same flood of water, because the tailings are so heavy. Where water could flood a building, tailings can push it over and sweep it along with the flow. This failure caused 269 deaths.

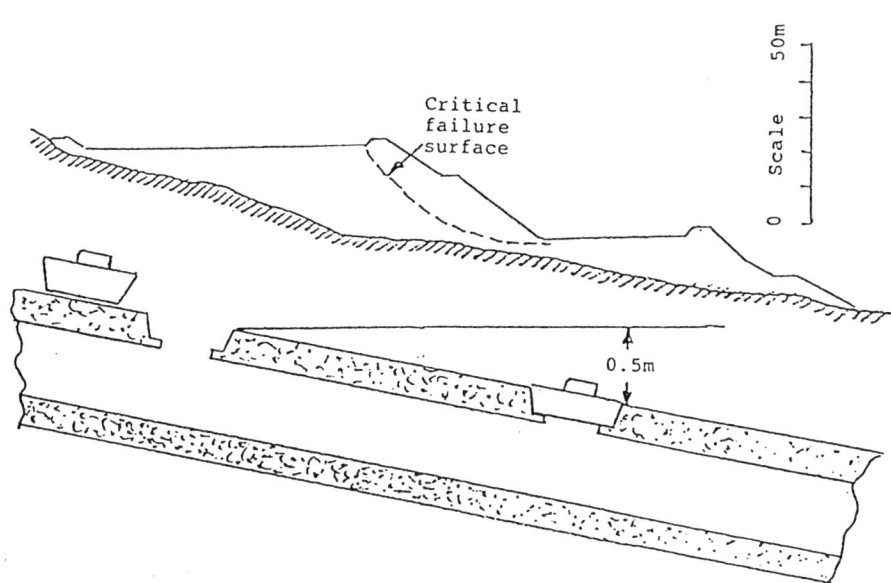

Figure 2. Stava tailings dams. Concrete culvert had coverable openings for decanting.

The tailings dams as indicated by Fig.2, were for a fluoride mine that was begun in 1962 and were sited on a side slope of 1 on 8. The decant was in the form of a concrete culvert laid up the sloping floor, with coverable openings about every 0.5m vertical rise. Water from the pond decanted into the openings, which were covered, one by one, as the level of the tailings rose. The lower dam was built by the upstream method to a slope of 1 on 1.23. When it reached a height of 19m, the second dam was begun at the upstream end of the impoundment and built to a slope of 1 on 1.43. When it reached a height of 19m, further planning consent was required. This was given on the condition that a 5m wide berm was constructed at that level and permission given for the dam to be built to a height of 35m. Construction continued at the same slope of 1 on 1.43 and the failure occurred when it was 29m high. The cause is thought to be due to a combination of blockage and leakage from the culvert under the toe of the upper dam, thereby raising the phreatic surface sufficiently to cause a rotational slip, as indicated by Fig.2. Six months before the failure, a local slide occurred in the lower portion of the upper dam on its right side, in the area where the decant pipes pass underneath the dam, due to freezing the service pipe during a period of intense frost, according to Berti et al (1988). For the next three months water was observed seeping from the area of the slide. A month before the failure, the decant pipe underneath the lower impoundment fractured allowing the free water and liquid mud from the pond to escape towards the Stava river, creating a crater above the point of fracture. A bypass pipe had to be installed through the top portion of the lower dam, and the broken decant pipe blocked to restore use of the system. During this operation the water level in the upper impoundment was lowered as far as possible, then just four days before the failure, both ponds were filled and put back into normal operation. 53 minutes before the failure a power line crossing below the impoundments failed, then only 8 minutes before, a second power line failed. The tailings from the failure reached Tesero about 4 km distance, within a period of 5 to 6 minutes. As a result of this failure, the strict Italian law governing the design and construction of water retaining dams, according to Capuzzo (1990), is being extended to include tailings dams.

Merriespruit The Virginia No 15 tailings dam had been built by the 'paddock' method that is used extensively in South Africa's gold mining industry. It was a long dam encircling and retaining an impoundment of 154 ha holding $260 \times 10^6 m^3$ of gold mine tailings containing cyanide and iron pyrite. The foundation soil was clay and drainage was required under the dams. General experience was that drains were often blocked by iron oxides and other residue. The impoundment formed one of several similar impoundments of the Harmony Gold Mine near Virginia in the Orange Free State. The suburb of Merriespruit containing about 250 houses had been built near the mine in 1956. Virginia No 15 lagoon was begun in 1974 and a straight northern section of the dam nearest the suburb was placed only 300m from the nearest houses. Dam construction and filling of the lagoon continued until March 1993, when the section of the dam closest to the houses was 31m high.

The summer of 1993/4 in the Orange Free State had been particularly wet and on the night of Tuesday 22nd February 1994 there were violent thunderstorms over Virginia and a cloudburst when 40mm of rain fell in a very short time. The water level in the lagoon rose due to direct catchment: there was no stream or other source of water that came into the lagoon, which while operational, had a launder system that removed the transportation water decanted from the tailings slurry that had been delivered into the impoundment. During the early evening at about 19.00, water was found running down the streets and through gardens and an eye witness saw water going over the crest of the dam above the houses. The mining company and contractor were informed, but when their representatives reached the site it was already dark. One of the contractor's men rushed to the decants and found water lapping the top rings but not flowing into the decants. He removed several rings to try to get the water flowing, but the main pool was next to the north dam crest with no direct connection to the decants. At the same time, another contractor's man was near the downstream toe of the dam, and saw blocks of tailings toppling from a recently constructed buttress that had been built against a weak part of the dam. An attempt was made to raise the alarm, but before anyone had been contacted, there was a loud bang, followed by a wave of liquefied tailings that rushed from the impoundment into the town. Cross sections of the dam during early stages and during failure, given by Blight (1997) are shown by Fig.3.

A breach 50m wide formed through the dam, releasing $2.5 \times 10^6 m^3$ tons of tailings that flowed for a distance of 1,960m, covering an area of $520 \times 10^3 m^3$. The flow passed through the suburb where the power of the very heavy liquefied tailings demolished everything in its path, houses, walls, street furniture and cars, carrying people and furniture with it. According to newspaper reports, people already in bed at about 21.00 hours when the mudflow struck, found themselves floating in their beds against the ceiling. 400 survivors spent the night in the Virginia Community Hall, a kilometre away. Hetta Williamson said that her husband had gone back in daylight to their former home and found nothing but the foundations. It is remarkable that only 17 people were killed.

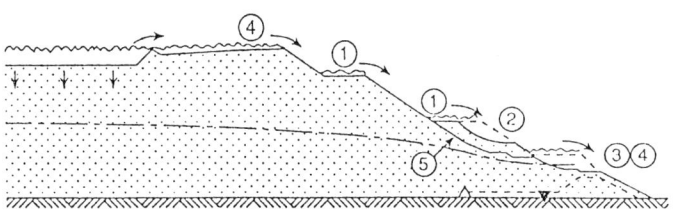

1. Berms overtop after thunderstorm.
2 Loose tailings infill to earlier failures on lower slope erodes.
3. Tailings buttress starts to fail.
4. Pool commences overtopping and erodes slopes and tailings buttress.
5. Unstable lower slope fails and failed material is washed away.

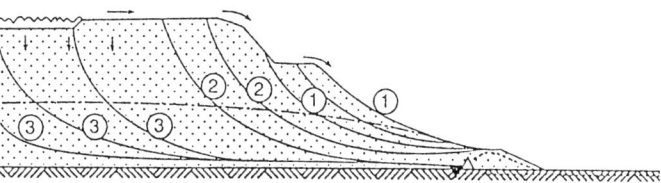

1. Lower slopes fail and are washed away.
2 Domino effect of local slope failures which are washed or flow away.
3. Major slope failures with massive flow of liquid tailings engulfing town.

Figure 3. Section of Virginia No.15 dam. (a) Critical section during early stages and (b) during failure. (Blight 1997).

Apparently this north section of the enclosing dam had been showing signs of distress for several years, with water seeping and causing sloughing near the toe. A drained buttress constructed from compacted tailings had been built against a 90 m long section, but continued sloughing had caused the mine to stop putting the normal flow of tailings into the impoundment more than a year before the failure, i.e. the impoundment had been closed. At that time the freeboard was, according to the contractor, a respectable 1 m. But sloughing at the toe continued, and construction of the buttress was continued. Not long before the failure, slips had occurred in the lower downstream slope just above the buttress. In fact, although the placement of tailings had been stopped, wastewater containing some tailings continued to be placed and the water overflowed into the decants. Unfortunately there formed a sufficient deposit of further tailings to cut off the decants and cause the main pool of water to move towards the crest of the north part of the dam, leaving a freeboard of only 0.3 m, and water was still being pumped into the impoundment from the mill on the night of the failure. Evidence of what had been going on since supposed closure was provided by satellite photography. A Landsat satellite passed over the area every 16 days and the infrared images revealed the positions of the tailings and the water pool. Government Mining Regulations that had come into force in 1976, required a minimum freeboard of 0.5m to be maintained at all times for this type of impoundment, to enable a 1 in 100 year rainstorm to be safely accommodated without causing overtopping. Evidence of the level of tailings in the Virginia No 15 lagoon showed that the tailings had been brought up to within 15cms of crest level prior to abandoning this storage in March 1993. Had the government regulations required inspection of the dam, particularly at closure, the very small freeboard would have been noticed and a further raising of the dam crest enforced to prevent overtopping in the event of a maximum probable precipitation.

ICOLD INVOLVEMENT.

The Civil Engineering profession has became increasingly involved in the design of tailings dams, and mining companies engage dam experts both to assess the stability of many large existing tailings dams and to make designs for new dams. In particular dam engineering expertise is called for to advise on remedial measures when tailings dams, continually being raised, show signs of distress. The International Commission on Large Dams (ICOLD) had not considered tailings dams as dams worthy of consideration and when the World Register of Dams was first being compiled, tailings dams were specifically excluded. By the mid 1970's however there were many tailings dams higher than 100m and following the greater involvement of dam design engineers in the problems of tailings dams, they were included as a subject for discussion by the 12th Congress on Large Dams, held in Mexico City in 1976. So much interest was expressed that ICOLD decided to establish a Committee to consider tailings dams. This Committee on Mine and Industrial Tailings Dams immediately began the task of preparing a Manual on Tailings Dams and Dumps, a Bibliography and, to compliment the World Register of Dams, a World Register of Tailings Dams. These three Bulletins were published in 1982. At that time, there were at least 8 tailings dams higher than 150m and 22 higher than 100m. It was estimated that tailings production exceeded 5×10^9 tons annually, far exceeding the volumes of fill involved in all civil engineering projects.

ICOLD has more than 80 member countries, spread throughout the world, from which the Committees can seek advice. Initially a Committee is limited to about 15 members and each must represent a different country. By seeking representatives from those countries heavily involved in mining, tailings dams and lagoons, the Committee contained within itself a considerable expertise. In addition, when a new Bulletin has reached the stage of a final draft, this is sent to every member country for constructive criticism and a period of at least 6 months is set aside for their replies. Provided that all countries study the draft bulletins in detail and give their considered opinions, the ICOLD Bulletins contain a unique amount of worldwide knowledge. The Committee has continued to look at the problems confronting those involved with tailings dams and has prepared several more Bulletins. These are:-

No.74. 1989. Tailings Dam Safety - Guidelines.

No.97. 1994. Tailings Dams - Design of Drainage.

No.98. 1995. Tailings Dams and Seismicity – Review and Recommendations.

No.101.1995. Tailings Dams – Transport, Placement, Decantation - Review and Recommendations.

No.103. 1996. Tailings Dams and Environment - Review and Recommendations.

No.104. 1996. Monitoring of Tailings Dams - Review and Recommendations.

No.106. 1997. A Guide to Tailings Dams and Impoundments – Design, construction, use and rehabilitation.

Despite all the advice given by these Bulletins, together with text books and many other papers, tailings dams continue to fail. During the last three decades the rate has averaged at one a year. This first full year of the new millennium has excelled itself by providing four failures, two in Romania, one in the far north of Sweden and one in the china-clay mines of Britain. It is not that the safe methods for the design, construction and operation of tailings dams are not understood, but that somehow they are not correctly applied. In an attempt to draw to the attention of mine owners, managers and others associated with tailings dams, some of the silly things that can go wrong and the types of incidents that have caused failures, the Committee has drafted a Bulletin with the title, "Tailings Dams: Risk of dangerous occurrences. *Lessons learnt from practical experience.*" It contains 221 cases of incidents and failures, with general explanations about the types of reasons for these incidents. This latest Bulletin will be published in conjunction with the United Nations Environmental Programme; an organisation that is taking an ever-increasing interest in tailings dams and the damage they can cause to the environment when they misbehave.

REFERENCES.

Berti G, Villa F, Dovera D, Genevois R and Brauns J (1988). The disaster of Stava/Northern Italy. *Proc. Specialty Conf. "Hydraulic fill structures" GT Div/ASCE, Fort Collins, CO, pp 492-510.*

Capuzzo D. (1990). A brief description of the main causes underlying the collapse of Stava tailings dams, Trento - Italy. *Proc. ICOLD Intn. Symp. Safety and Rehabilitation of Tailings Dams, Sydney, Australia, vol 2, pp.64-66.*

Blight GE (1997). Insight into tailings dam failures – technical and management factors. *Proc Int Workshop on Managing the Risk of Tailings Disposal, Stockholm, Sweden, May, pp 17-34.*

Site Remediation and
Land Regeneration

Applicability of Ethanol/bentonite Slurry for Seepage Barriers

MOTOYUKI ASADA and SUMIO HORIUCHI, Institute of Technology, Shimizu Corporation, Tokyo, Japan

ABSTRACT

A high potential of ethanol/bentonite slurry for seepage barriers was reported based on the following test results; (1) using 60% ethanol, 625 kg/m^3 bentonite in a slurry condition can be transported more than 200m using normal grouting pump, (2) the ethanol/bentonite slurry could smoothly fill up the void space. The slurry shows an enough impermeability of k = 4 x 10^{-10} (cm/s).

INTRODUCTION

Bentonites are being used for a wide range of application in the geotechnical and environmental field, such as drilling slurries grouts, clay liners and back-fill & buffer material for radioactive waste containment (JNC 1999). In these bentonite applications, seepage control is the one of the major usage of bentonite, and a higher bentonite content more than 70 kg/m^3 is required for long-term performance. Field preparation of this higher bentonite content mixture is easy by using bentonite powder, but was difficult by the slurry because of the limitation of maximum amount of bentonite suspended in water.

For the example of vertical cutoff walls, the mixed in place (MIP) walls are commonly used in Japan. Cement/bentonite (CB) slurry is the material to construct the MIP walls. This method uses three or five-axis hollow stem augers that churn and mix the soil as the CB slurry is introduced through the base of the augers. TRD method using long chain saws is also widely used to make continuous vertical barriers nowadays. The hydraulic conductivity of the CB-MIP walls, however, is larger than that of the soil-bentonite (SB) walls (Oweis & Khera 1998), and they have no flexibility to the earth deformation. If the MIP walls could be constructed by bentonite slurry, instead of CB slurry, and an enough amount of bentonite could be homogeneously mixed, they could have high flexibility and their hydraulic conductivity could be lowered less than 10^{-7} cm/s. Minimum amount of bentonite required for the impermeable vertical walls, k<1x10^{-7}cm/s, would be more than 70 kg per 1m^3 soil (Kenney et al. 1992). Constructing walls by MIP method, bentonite should be added to soil in a slurry state for homogeneous mixing, however, 20 kg/m^3 of bentonite could be the largest amount by using conventional bentonite/water slurries.

For the other example of pre-compacted blocks of buffer materials, which are one of the most attractive options to be applied, will be of high dense bentonite. It has been pointed out that space between the buffer materials and the walls of the disposal pits and/or that between the buffer materials and the overpacks (Komine et al. 2001). To fill the void space between the bentonite blocks, high dense bentonite slurry are considered to be one of the suitable material to keep the sufficient impermeability of all the radioactive waste containment system.

To increase the bentonite content in the slurry, ethanol substitution for water is one of the effective methods (Asada et al. 2001). Montmorillonite is the major component of bentonite,

Geoenvironmental impact management, Thomas Telford, London, 2001.

and it swells by water intake between the clay interlayers. Natural Na-bentonites, as well as activated Ca-bentonites can adsorb water four to seven times their dry weight and swell to five to eight times their original size (Egloffstein 1995). On this swelling, electrostatic properties of montmorillonite play a key role; the larger the surface potential becomes, the greater montmorillonite shows its swelling property (Horiuchi et al. 2001). Lowering surface potential of montmorillonite by decreasing pore fluid's dielectric constant, swelling can be depressed and thus the bentonite content can be increased. When hydrophilic organic solvents such as ethanol and acetone are used, bentonite swells after its placement by pore solvent substitution to water. Many kinds of solvents are useful for this purpose, however, ethanol is the best because of its moderate properties; easy to be decomposed biochemically and not harmful to underground environment.

Some experimental works were reported on mechanical properties of soil/solvents mixtures (Sridharan & Venkatappa 1973, 1979), however, there is no studies on properties of ethanol/bentonite slurries except our project. The objective of our research is to make clear the execution method of seepage barriers by using ethanol/bentonite slurry. In this paper, a high potential of ethanol/bentonite slurry for grouting is reported based on results of a series of laboratory tests including pumpability and groutability. It recommends the field application of the ethanol/bentonite slurry.

EXPERIMENTS

(1) Materials

Na-Bentonite: Kunigel V1 (Kunimine Industries Co., Ltd. Japan) was selected and used. The density of soil particles ρ_s=2.72(g/cm^3).

Ethanol: Reagent grade of 92.5wt% ethanol was used. Its density at 15 ℃ ρ=0.82(g/cm^3).

Additives: Toyoura sand (average grain diameter D=0.2(mm)) was used for partial substitute for bentonite. The density of soil particles ρ_s=2.64(g/cm^3).

(2) Pumpability test

To check the pumpability of the ethanol/bentonite slurry, the pressure loss during pumping were measured. Fifty meters of 0.5, 1, 2 inches plastic hose were used for the test. Pressure was measured using a bourdon tube type pressure gauge at every 8 meters. A single-axis screw pump, maximum pressure 1MPa, was used for smooth measurement. Table 1 shows the content and viscosity of the slurry tested. Pore ethanol concentration was set to 0, 40, 60, 92.5%, respectively.

Table 1. The content and viscosity of the slurry

Test No.	Slurry content (kg/m^3)				Pore Ethanol Concentration(wt%)	Viscosity (mPa*s)
	Bentonite	Ethanol	Water	Sand		
1-1	136	0	950	0	0.0	860
1-2	112	0	959	0	0.0	93
1-3	97	0	964	0	0.0	35
1-4	129	0	904	129	0.0	870
1-5	118	0	823	353	0.0	970
1-6	108	0	538	541	0.0	1070
2-1	746	596	0	0	92.5	480
2-2	630	630	0	0	92.5	220
3-1	625	438	236	0	60.0	1460
3-2	564	450	244	0	60.0	1290
3-3	512	461	249	0	60.0	260
3-4	509	408	220	255	60.0	1000
3-5	463	372	201	463	60.0	960
3-6	395	316	171	791	60.0	1500
4-1	418	335	438	0	40.0	5050
4-2	349	349	446	0	40.0	980

(3) Model grouting test
To check the applicability of the bentonite/ethanol slurry for grouting or backfilling, No. 3-2 slurry in table 1 was placed underwater using the one-axis screw pump into a filling space shown in figure 1. The filling space, L160cm x W40cm x H10cm in size made of polyacrylate panels with a porous bottom panel, was set 30 degrees underwater. An earth

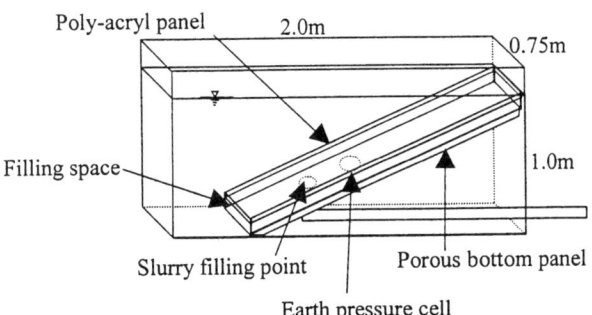

Figure 1. Testing tank

pressure cell was fixed on the bottom panel to measure the swelling pressure during the substitution. Curing for one month, samples were collected to determine the water substitution and the consolidation. Distribution of shear strength and ethanol concentration was also measured.

RESULTS

(1) Pumpability test
All the slurries listed on Table 1 can be transported for 50 m through 1-inch hose. To estimate pressure loss of the slurry, data collected are examined using the Fanning's formula and the Reynolds number; i.e.,

$$\Delta p = 4f(\rho u_a{}^2/2)(L/d) \qquad \text{(Fanning's formula)}$$

$$Re = \rho u_a d/\mu \qquad \text{(Reynolds number)}$$

where Δp = Pressure loss (Pa), f = Pipe friction coefficient, ρ = Density (kg/m^3), u_a = Average velocity (m/s), L = Pipe length (m), d = Pipe inside diameter (m), μ = Viscosity (Pa*s)

Figure 2 shows the relationship between Reynolds number (Re) and friction coefficient (f). The friction loss of ethanol/ bentonite slurry can be expressed by the following equation; f*Re=19.75. This equation is quite close to laminar flow line f*Re=16. As shown in Figure 2, the Reynolds number is the only factor for the friction loss, and factors such as ethanol concentration and

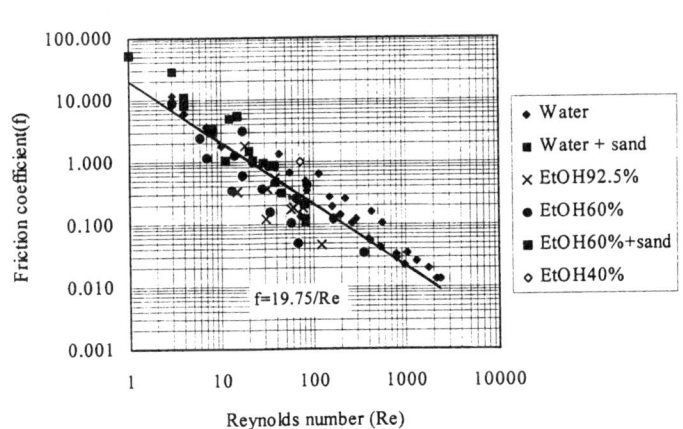

Figure 2. Reynolds number and friction coefficient

sand content have less effect. For the easy field operation, the pumping pressure should be controlled less than 1 MPa for 200m slurry transportation through 2-inch pipe. The results obtained show that maximum amount of bentonite in the slurry can be increased to $136kg/m^3$ for 0% ethanol slurry, $418kg/m^3$ for 40% ethanol slurry, $625kg/m^3$ for 60% ethanol slurry, $746kg/m^3$ for 92.5% ethanol slurry.

(2) Model grouting test

Photo 1 and 2 show typical views during the model test. All the slurry was smoothly placed into the filling space underwater, and no trouble and slope breakdown was observed.

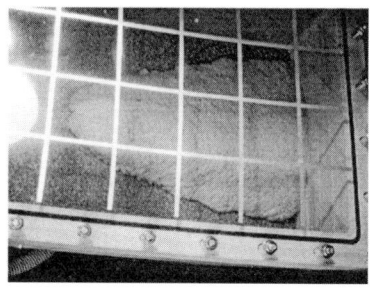

A typical change of swelling pressure is plotted in Figure 3, where a steady increase is confirmed. This pressure increase is caused by water/ethanol exchange, and is expected to continue until fully substituted to water. Almost 1/3 of the ethanol was substituted to water at the end of this test, and swelling pressure can be supposed about 20kPa when all the pore ethanol is exchanged to water.

Photo 1. Slurry placement underwater

Photo 2. Slurry sampling and bane shear strength test

After 30 days curing, the properties of filled slurry were investigated. Figures 4-7 illustrates the distributions of dry density, water content, pore ethanol concentration, and shear strength. Ethanol in the placed slurry moved toward outside of the mass through the permeable bottom panel. Therefore, water content of the bottom part is higher than upper part, and the ethanol concentration decreases as closing to the bottom panel. This phenomenon indicates that pore ethanol is easily substituted to water.

As seen in Figure 7, shear strength increases with water/ethanol exchange, even the water content of the sample increasing. As Sridharan (1979) pointed out, there is a general tendency between strength and dielectric constant of the pore fluid; i.e., the higher the dielectric constant, the higher the strength, and the dielectric constant of ethanol and water at 25 °C are 24.3 and 80.4, respectively. It remains the

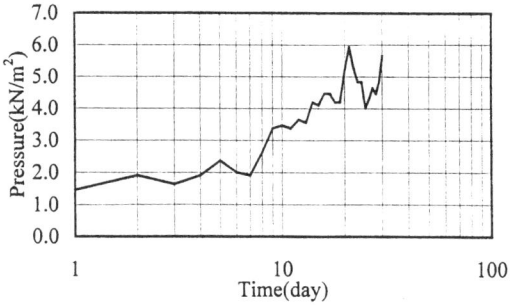

Figure 3. Pressure change of the filling space

further discussion about the effect of density; however, the shear strength of the placed mass shows higher strength when the pore ethanol is substituted to water. From the consolidation test result of the sample at the mass centre, consolidation yield stress p_c=11.8(kN/m^2), and the dry density at p_c ρ_d=0.32(g/cm^3). These results were well fitted to the final pressure of the filling space at Figure 3 and to the dry density at Figure 4.

Figure 8 illustrates the relationship between bentonite dry density and hydraulic conductivity (k) of the placed mass. The sample shows an enough impermeability of k = 4 x 10^{-10} (cm/s). This permeability is well coincident to the data reported for compacted bentonite blocks (Mihara 2000), and thus can be concluded as the ethanol usage has no effect of bentonite properties.

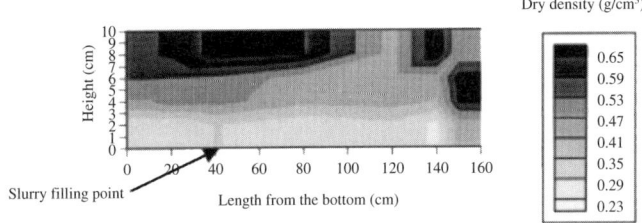

Figure 4. Dry density distribution of the mass vertical section

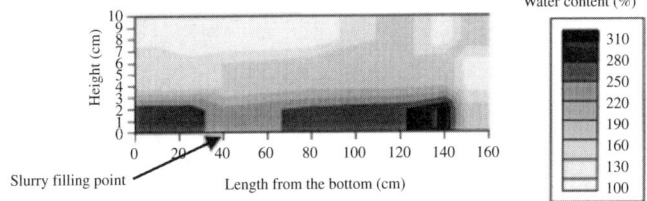

Figure 5. Water content distribution of the mass vertical section

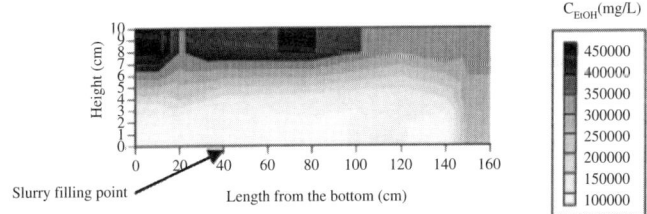

Figure 6. Pore ethanol concentration distribution of the mass vertical section

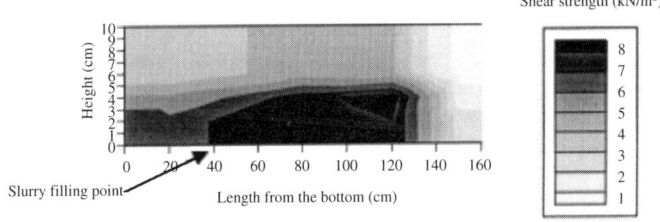

Figure 7. Shear strength distribution of the mass vertical section

CONCLUSIONS

In this paper, a high potential of ethanol/bentonite slurry for grouting was reported based on the following test results; (1) using 60% ethanol, 625 kg/m^3 bentonite in a slurry condition can be transported more than 200m using normal grouting pump, (2) the ethanol/bentonite slurry could smoothly fill up the void space. Shear strength increases up to 6-20 kN/m^2, when the pore ethanol is replaced to water, bentonite in the slurry begins to hydrate, and bentonite shows the swelling potential. The slurry shows an enough impermeability of k = 4 x 10^{-10} (cm/s).

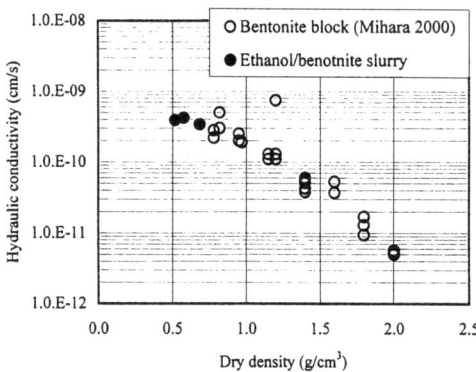

Figure 8. Dry density and hydraulic conductivity

ACKNOWLEDGEMENT

The authors wish to express their gratitude to Ministry of Economy, Trade, and Industry, Japan for their support of this project.

REFERENCES

Asada, M., Ishikawa, A., & Horiuchi, S. (2001): Cutoff wall construction using bentonite/ethanol slurry, Clay Science for Engineering, Rotterdam, A. A. Balkema, pp511-516.

Egloffstein, T. (1995): Properties and test methods to assess bentonite used in Geosynthetic clay liners, Geosynthetic clay liners, Rotterdam, A. A. Balkema, pp51-72.

Horiuchi, S. & Asada, M. (2001): Viscosity and zeta-potential of ethanol/bentonite slurry, 35th Japan National Conference on Geotechnical Engineering, the Japanese Geotechnical Society (in press).

Japan Nuclear Fuel Cycle Development Institute (JNC) (1999): H12 project to establish technical basis for HLW disposal in Japan, JNC TN 1400 1999-022.

Kenney, T. C., van Veen, W. A., Swallow, M. A., & Sungaila, M. A. (1992): Hydraulic conductivity of compacted bentonite-sand mixtures, Can. Geotech. J., Vol.29, pp364-374.

Komine, H., Ogata, N., Takao, H., Nakashima, A., Osada, T., & Ueda, H. (2001): Self-sealing ability of buffer materials containing bentonite for HLW disposal, Clay Science for Engineering, Rotterdam, A. A. Balkema, pp543-551.

Mihara, M. (2000): The comparison concerned with hydraulic conductivities and effective diffusion coefficients for nuclides between Na and Ca bentonite, JNC TN 1340 2000-001, pp 61-68.

Oweis, I. & Khera, R. (1998): Geotechnology of waste management, Boston, PWS Publishing Company, pp258-271.

Sridharan, A. & Venkatappa Rao, G. (1973): Mechanisms controlling volume change of saturated clays and the role of the effective stress concept, Geótechnique 23, No.3, pp359-382.

Sridharan, A. & Venkatappa Rao, G. (1979): Shear strength behaviour of saturated clays and the role of the effective stress concept, Geótechnique 29, No.2, pp177-193.

Immobilisation of contaminated soil using ordinary Portland cement and Hydrofoam

E. M. BENNETT and A. AL-TABBAA
Engineering Department, Cambridge University, Cambridge, UK.

ABSTRACT
The effectiveness of ordinary portland cement (OPC) and a cementitious foam, Hydrofoam (HFR), in the immobilisation of a contaminated sand was investigated. The sand contained two contaminants: an inorganic contaminant, copper sulphate and an organic contaminant, vegetable oil. The effect of the addition of pulverised fuel ash (pfa) to the cements was also considered. The immobilisation effectiveness was assessed using unconfined compressive strength and leachability tests. It was concluded that the contaminants had a stronger retardation effect on the early strength development of OPC-based mixes than on HFR-based mixes. However, the OPC-based mixes retained a greater percentage of both the copper and the vegetable oil.

INTRODUCTION
The techniques of immobilisation, more widely used in the treatment of contaminated waste, are increasingly finding applications in the in-situ immobilisation of contaminated ground (Al-Tabbaa and Evans, 1998). Immobilisation encompasses the physiochemical processes of stabilisation and solidification. Stabilisation minimises the rate of contaminant migration and reduces the toxicity of the waste while solidification increases the waste's strength and decreases its permeability and compressibility (LaGrega et al, 1994). Cement, and predominantly OPC, is the most common principal reagent and is frequently combined with additives such as pulverised fuel ash (pfa). Research has shown that OPC-based stabilisation is best suited to inorganic wastes, especially those containing heavy metals, which are bound in the matrix as insoluble compounds by chemical fixation as well as by physical encapsulation. Conversely, organic contaminants are not easily stabilised (LaGrega et al, 1994).

More recently a number of 'special cements' have been developed with the aim of improving the immobilisation effectiveness of cement-based additives. These use hydraulic cements which hydrate in a different way to Portland cement and contain additives which provide specific properties. This project aims to compare the effectiveness of OPC with HFR, a foamed cement. HFR is characterised by its rapid initial gelation properties and, when set, gives a durable, low density material (Blue Circle Industries, 1995). The comparison of the two cements will consider the immobilisation of two contaminants: an inorganic contaminant, copper sulphate, and an organic contaminant, vegetable oil. There are a large number of variables which may affect the immobilisation effectiveness, this paper investigates the effect of the soil:dry grout and cement:pfa ratios and curing time.

As there are no standard design criteria specific to immobilised contaminated soils, the immobilisation process is validated through a series of physical, chemical and environmental tests based on the U.S. Environmental Protection Agency's specifications usually applied to immobilised waste (LaGrega et al, 1994). Two of these main criteria are investigated in this paper:

(i) UCS: which indicates whether the immobilised material has sufficient strength to support any overburden pressure and has adequate durability and chemical stability. The U. S. Environmental Protection Agency recommends a 28 day UCS value for solidified waste of at least 350kPa (LaGrega et al, 1994).

(ii) Leachability: which determines the quantity of contaminant transferred from the immobilised matrix to a leachant. The leaching tests were carried out in accordance with the NRA Leaching Test for Assessment of Contaminated Land (Lewin et al, 1994). An acceptable concentration of contaminants in the leachate is commonly derived using a multiplier, in the range 50-100, to drinking water standards (Conner, 1993). The Water Supply Regulations (1991) state a maximum allowable concentration of copper as 3mg/l and of mineral oil, in the absence of a more appropriate standard, as 0.01mg/l.

MATERIALS

A medium sand, with particle sizes in the range 300μm - 600μm, was used with a moisture content (using tap water) of 10%. For the copper contaminated samples the water was mixed with copper sulphate, to a concentration of 660mg/l, prior to addition to the sand. Copper is a common groundwater pollutant mainly originating from copper smelting and refining industries with concentrations up to several hundred milligrams per litre on contaminated sites (Fetter, 1993). For the vegetable oil contaminated samples the oil was added directly to the dry sand, to a concentration of 10g/kg of dry soil, and mixed by hand for two minutes before adding the water. Vegetable oil was chosen as a substitute for a mineral oil contaminant which can be present in typical concentrations of up to 2000mg/kg dry soil on sites contaminated by petroleum leaks (Al-Tabbaa and Evans, 1998). Uncontaminated samples were prepared as control samples. All samples were prepared in batches of four.

The grout consisted of cement, water and, for some mixes, pfa. Two cements were used in the experiments: OPC and HFR (Blue Circle Industries, 1995). The grouts consisted of cement:pfa ratios of 1:0, 1:2 or 1:4 and dry grout:water ratios of 1:0.4 (OPC-based grouts) and 1:0.6 (HFR-based grouts). The appropriate solid grout constituents and water were mixed to a slurry prior to their addition to the contaminated sand. The samples were placed in cylindrical moulds (72mm in height, 36mm in diameter) and the moulds gently tapped on the bench to remove any large voids. The batches were covered and allowed to cure at room temperature for one day. The moulds were then removed, the samples wrapped and, if necessary, allowed to continue curing. Samples were cured for 1 day, 7 days or 28 days. The average density of the OPC-based mixes was 2300 kg/m^3 compared to 2025 kg/m^3 for the HFR-based mixes. Table 1 shows the details of the mixes used, both in weight ratios and by percentage weight. Hence the cement content ranged between 3.4 and 19.7% of OPC and 2.8 and 19% for HFR and the pfa content was up to around 15.5% for both cement mixes. For each cement at a constant sand:dry grout ratio there were three different mixes, for example, OPC1 – OPC3 each with a different cement:pfa ratio.

EXPERIMENTAL PROCEDURE

Prior to conducting the UCS test the ends of the samples were made square. The samples were then axially loaded to failure. After testing for UCS the samples were ground and a 10g ± 0.1g subsample placed in a polypropylene screw capped container and 100ml ± 1ml of double deionised water, with a measured pH of around 5.5 ± 0.1, added. After being agitated on an orbital shaker at 100 rpm for 24 hours the contents of the container were filtered through 0.45μm pore size filter paper. The leachate was analysed for copper concentration using an Atomic Absorption Spectrometer. The oil was extracted with a solvent using the partition-gravimetric method (Greenberg et al, 1992). By extracting vegetable oil from samples of known concentration this technique was found to be approximately 81% efficient.

Table 1. Composition of the mixes.

Sample	Sand:dry grout	Cement:pfa	dry: grout:water	sand (%)	cement (%)	pfa (%)	water (%)
OPC1	4·1	1·0	1·0.4	75.9	17.2	0.0	6.9
OPC2	4·1	1·2	1·0.4	75.9	5.7	11.5	6.9
OPC3	4·1	1·4	1·0.4	75.9	3.4	13.8	6.9
OPC4	3.33·1	1·0	1·0.4	72.4	19.7	0.0	7.9
OPC5	3.33·1	1·2	1·0.4	72.4	6.6	13.1	7.9
OPC6	3.33·1	1·4	1·0.4	72.4	3.9	15.8	7.9
HFR1	5·1	1·0	1·0.6	77.4	14.1	0.0	8.5
HFR2	5·1	1·2	1·0.6	77.4	4.7	9.4	8.5
HFR3	5·1	1·4	1·0.6	77.4	2.8	11.3	8.5
HFR4	4·1	1·0	1·0.6	73.3	16.7	0.0	10.0
HFR5	4·1	1·2	1·0.6	73.3	5.6	11.1	10.0
HFR6	4·1	1·4	1·0.6	73.3	3.4	13.3	10.0
HFR7	3.33·1	1·0	1·0.6	69.6	19.0	0.0	11.4
HFR8	3.33·1	1·2	1·0.6	69.6	6.3	12.7	11.4
HFR9	3.33·1	1·4	1·0.6	69.6	3.8	15.2	11.4

RESULTS AND DISCUSSION
The results are presented at 1 day, although all tests were carried out at 7 and 28 days. This is because the two cements exhibited the greatest differences in behaviour after 1 day.

Unconfined compressive strength
The results of the UCS tests on the contaminated and uncontaminated samples at 1 day for the OPC-based and HFR-based mixes are presented in Figures 1 and 2 respectively. The UCS values reported are the average of four values with a maximum margin of error of $\pm$ 23%. The OPC results are for copper only while the HFR results are for both copper and vegetable oil. As expected, the results show that for a constant soil:dry grout ratio the UCS of both uncontaminated and contaminated samples reduced as the pfa:cement ratio increased. The only exceptions are OPC1 to OPC3 mixes which exhibited the opposite. This behaviour is unexpected and may have been caused by improved mixing as the pfa:cement ratio increased.

Figure 1. UCS values of OPC-based mixes.   Figure 2. UCS values of HFR-based mixes.

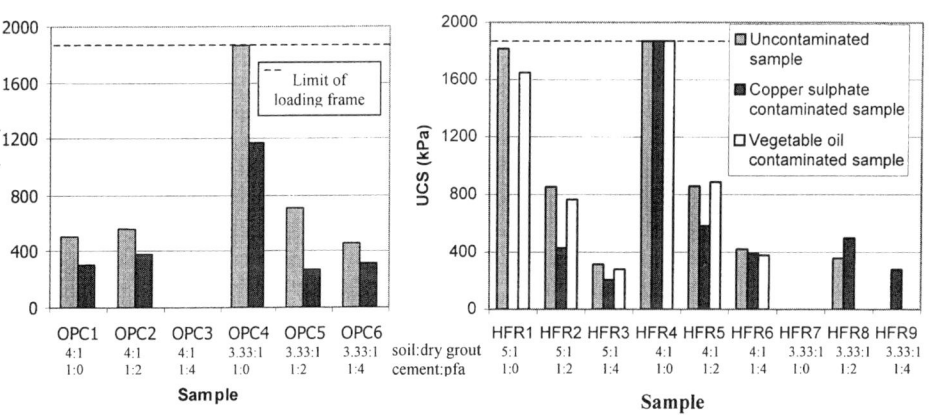

Ordinary Portland Cement-based mixes
Figure 1 shows that all the uncontaminated samples exhibited a higher UCS value than the copper sulphate contaminated samples at 1 day. This result was also observed at 7 days. This indicates that the copper sulphate had a detrimental effect on the early strength development of the mixes. One theory proposed to explain the retardation of early strength development in OPC is the development of a coating on the surface of the cement through which the water must diffuse (Conner, 1993). The results showed that the UCS values at 7 days were approximately 4 times greater than at 1 day for both contaminated and uncontaminated mixes. After 28 days all the OPC samples exhibited UCS values greater than the capacity of the loading frame (~1850kPa). It was, therefore, not possible to determine whether the addition of the contaminants retarded the long term strength development of the grout.

Hydrofoam-based mixes
Figure 2 shows that, in general, the oil contaminated samples produced similar UCS values to those of the uncontaminated samples at 1 day. This result was also observed at 7 days. This indicates that the vegetable oil had a limited detrimental effect on the early strength development of the mixes. After 28 days it was observed that the uncontaminated samples exhibited a slightly greater UCS value than the oil contaminated samples suggesting that the longer term strength development of the mix may have been slightly retarded. Figure 2 shows that the copper sulphate samples generally exhibited lower strengths than the uncontaminated or oil contaminated samples after 1 day. This result was also observed after 7 days. It is postulated that the hydration of the HFR is retarded in a similar manner to that of the OPC. The results showed that the UCS of HFR-based mixes at 7 days was approximately twice that at 1 day. At 28 days the UCS values were generally slightly higher than those at 7 days which is typical for the foamed cement.

Comparison of OPC-based mixes and HFR-based mixes
A comparison of the UCS values for the OPC and HFR-based samples at 1 day shows that for a constant sand:dry grout ratio the copper sulphate had a stronger retardation effect on the hydration and early strength development of OPC than on HFR. However, at 28 days the HFR samples exhibited an UCS value lower than the corresponding OPC sample. It can therefore be concluded that the copper sulphate has less effect on the long term strength development of OPC than on HFR. However, the cement used in HFR is designed to provide high early strength at the expense of later age strength.

UCS value versus design criterion
All the UCS values obtained at 7 and 28 days were higher than 350kPa, hence all samples satisfied the mix design criteria. These results suggest that the percentage of grout added to the soil could be significantly reduced. However, during preliminary experimental work it was concluded that satisfactory results could not be obtained for a dry binder addition of less than 20%. Further experimental work could investigate the effect of increasing the pfa:cement ratio.

Leaching tests
Due to considerable dilution during the leaching test the maximum concentrations in the leachate are significantly less than the initial concentrations in the soil. The maximum possible copper and oil concentrations in the leachate are 6.62mg/l and 1000mg/l respectively. Figure 3 presents the concentration of copper in the leachate of the OPC and HFR-based samples and the concentration of oil in the leachates of HFR-based samples only,

at 1 day. The copper concentrations reported are the average of three values and the oil concentrations the average of two values with a maximum margin of error of ± 10% and ± 21% respectively.

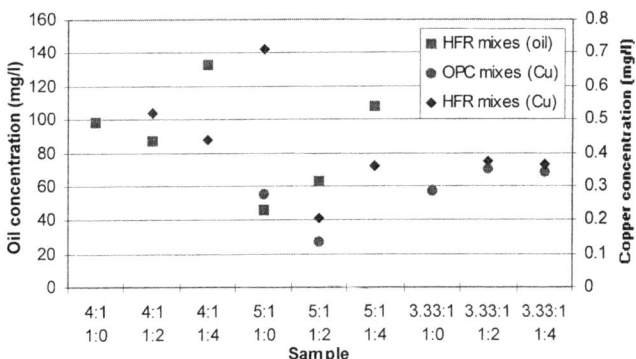

Figure 3. Copper and oil concentrations in the leachate.

Copper concentration in the leachate
Figure 3 shows that the maximum copper concentration in the leachate of the OPC-based mixes at 1 day was 0.35mg/l and in the HFR-based mixes was 0.71mg/l. These are 8.5 and 4.2 times lower than the maximum acceptable drinking water concentration respectively. Over the test period the concentration of copper decreased, with maximum 28 day concentrations of 0.19mg/l and 0.33mg/l for the OPC and HFR-based mixes respectively. The results show that up to 28 days the OPC-based mixes retain a greater percentage of copper than the HFR-based mixes and may be related to the higher porosity (lower density) of the HFR-based mixes.

The results showed that for a constant sand:dry grout ratio the copper concentration was, in general, greatest for the samples containing no pfa. This suggests that, in general, the addition of pfa increases the immobilisation of copper. The level of physical encapsulation might be expected to show a positive correlation with UCS, and hence reduce with increasing pfa:cement ratio. This is contrary to the above result, suggesting that the addition of pfa chemically stabilises the copper. However, the results indicate that the copper concentration does not reduce as the pfa:cement ratio increases from 2:1 to 4:1.

Oil concentration in the leachate
Figure 3 shows that the concentrations of oil in the leachate at 1 day ranged between 46.3mg/l and 132.7mg/l, which are significantly higher than the maximum acceptable drinking water concentration. The results at 7 and 28 days showed that four mixes exhibited a decrease in oil concentration with time while two mixes exhibited an increase in concentration. From these limited results it is not possible to determine whether the effectiveness of the immobilisation improved with time. The results showed that, in general, for a constant soil:dry grout ratio the oil concentration was increased by the replacement of cement by pfa. This suggests that, unlike the copper, the pfa does not increase the effectiveness of immobilisation of oil and that the vegetable oil is retained in the solidified mass purely by physical encapsulation. After 28 days the results showed that the oil concentration was between 2 and 6 times greater in the HFR-based mixes than in the OPC-

based mixes. It was noted earlier that after 28 days the UCS value for oil contaminated samples was greater in the OPC-based mixes, which supports the above theory that the vegetable oil is retained by physical encapsulation.

CONCLUSIONS

All mixes tested obtained a 7 and 28 day UCS value greater than the generally quoted design value of 350kPa. While the copper sulphate resulted in a detrimental effect on the strength development of both cements, the vegetable oil showed a minimal effect on the early strength development of HFR-based grouts. In general, the contaminants had a stronger retardation effect on the hydration and early strength development of the OPC-based mixes than the HFR-based mixes. From a maximum possible copper concentration of 6.62mg/l, the maximum leachate concentration from the OPC-based mixes was 0.35mg/l and from the HFR-based mixes was 0.86 mg/l which are lower than the drinking water standard of 3mg/l. The oil concentrations in the leachate from the HFR-based mixes ranged between 46.3 and 132.7mg/l and hence are significantly higher than the drinking water standard of 0.01 mg/l. In contrast to the result for copper, for a constant sand:dry grout ratio the oil concentration in the leachate increased with increasing pfa:cement ratio. This suggests that the oil is retained in the matrix by physical encapsulation. After 28 days the results showed that HFR offers less effective immobilisation for oils than OPC which may be due to the higher degree of porosity of the stabilised material.

ACKNOWLEDGEMENT

The work presented here was carried out by the first author as part of her final year undergraduate project at Cambridge University. The authors gratefully acknowledge the technical assistance of Steve Chandler and Tim Ablett and the comments made on the paper by Dr David Johnson of Blue Circle Industries.

REFERENCES

Al-Tabbaa, A., Evans, C.W. (1998). Pilot *in situ* auger mixing treatment of a contaminated site. Part 1: treatability study. Proc. Instn Civ. Engrs. Geotech. Engng, 1998, 131, Jan., pp52-59.

Blue Circle Industries PLC. (1995). Hydrofoam - H.F.R.1 Data Sheet.

Conner, J.R. (1993) Chemical Fixation and Solidification of Hazardous Wastes, Van Nostrand Reinhold.

Fetter, C.W. (1993). Contaminant Hydrogeology, Second Edition. Prentice Hall.

Greenberg, A.E., Clescen, L.S. and Eaton, A.D. (1992). Oil and Grease (5520), partition-gravimetric method. Standard Methods for the Examination of Water and Water Waste. American Public Health Association.

LaGrega, M. D., Buckingham, P. L. and Evans, J. C. (1994). Hazardous waste management. McGraw-Hill, New York.

Lewin, K., Bradshaw, K., Blakay, N.C., Turrell, J., Hennings, S.M. and Flavin, R.J. (1994). Leaching Tests for Assessment of Contaminated Land: Interim NRA Guidance, National Rivers Authority, R&D Note 301, Bristol, UK.

Water Supply Regulations. Private Water Supplies Regulations (1991), Doc. No. 2792, HMSO, London, UK.

Long-term durability of in-situ auger-mixed stabilised/ solidified contaminated made ground

N. BOES and A. AL-TABBAA
Engineering Department, Cambridge University, UK.

ABSTRACT
Samples from an in-situ soil-mixed contaminated made ground were tested for their long-term durability performance. Seven soil-grout mixes were used which contained cement (1-6%) and some contained pfa (0-16%), lime (0-0.5%) and bentonite (0-0.5%). The behaviour at 5 years was compared with that at 0.2, 1.2 and 2.3 years. Comparisons were made between seven different mixes applied and between single and overlap mixes. In the light of the results obtained, recommendations as to the applicability of the ASTM test method to the samples tested are presented. Leachability and scanning electron microscopy of the tested samples were also carried out to attempt to relate observed changes in behaviour.

INTRODUCTION
In-situ stabilisation/solidification (S/S) of contaminated ground is a technology which has become increasingly used, particularly in the USA (Day and Ryan, 1995), and has recently emerged in the UK (Evans and Al-Tabbaa, 1999). The treatment is applied to the soil using mixing augers through which a grout is introduced and mixed with the contaminated soil resulting in S/S soil-grout columns. The success of the in-situ treatment methodology relies on the injection and mixing processes of the grout with the soil being thorough and effective so that homogeneous well-mixed soil-grout monolithic columns are produced in which the contaminants are brought into direct contact with the grout. The use of cement-based grouts for the treatment of waste materials containing mainly inorganics and heavy metals has been practised for many years (Conner, 1993), where cement or lime are often used as the stabilising additive. However, the treatment of waste containing organics has been far less successful because some organics interfere with the hydration processes of cement. In these cases special additives are used to provide the link between the organic waste and the cement, examples of which are natural and modified clays (Conner, 1993). For validation purposes, a treatment methodology is expected to produce satisfactory short, medium and long-term performance of the resulting material. Although S/S treatment methodologies have been used in the USA for the past 30 years there is very little information published on the long-term material performance in the field (Kirk, 1996) and concerns about this have recently been addressed (Loxham et al, 1997). The assessment of the treated ground is usually carried out in terms of physical and chemical tests. Since there are no standard design criteria specific to S/S contaminated ground, criteria and standard testing methods used for S/S waste are usually applied. It is the subject of this paper to assess the performance of S/S treated soil in one of these standard tests, namely durability. Durability is required to assess the performance of solidified waste in terms of its ability to resist repeated cycles of extreme weather conditions. This paper looks at the time-related and long-term behaviour of an in-situ S/S made ground from a contaminated site at five years after treatment.

Geoenvironmental impact management, Thomas Telford, London, 2001.

SOIL, CONTAMINANTS AND GROUT

The samples tested originate from a contaminated site in West Drayton near Heathrow Airport treated in 1995 (Al-Tabbaa and Evans, 1998; 2000; Al-Tabbaa et al, 1998). The ground conditions consist of 1.7m of made ground, underlain by 3m of natural sand and gravel deposits which are in turn underlain by London clay to depth. Water was encountered at 2m below ground level. In this paper only the made ground, which is very variable and consisted of clayey sand and sandy clay thin layers containing extraneous materials such as brick, glass and metal, will be considered. Since the turn of the century the site had been contaminated with a wide range of heavy metals and organic compounds as given in Table 1. The soil was treated with seven different cement-based grouts, some of which included pfa, lime and bentonite as detailed in Table 2. The mixes also contain 0.5% of an organophilic clay which was shown not to affect the properties investigated here (Evans 1998). The treatment was applied in the form of a grid of overlapping columns and the samples tested included those from single as well as overlap mixes, the latter containing twice the grout content of the former. The site was cored at 0.15 and 4.5 years after treatment to obtain samples for testing. The 0.15-year samples were left to cure in the laboratory and were tested at 0.2, 1.2 and 2.3 years after treatment. Hence samples were tested at 4 different ages.

Contaminant	Lead	Copper	Nickel	Zinc	Cadmium	Mineral oil
Concentration	2801	1264	105	1589	8.7	1400

Table 1. Contaminants and their concentrations in (mg/kg) in the made ground (Evans, 1998).

Mix	Cement:pfa:lime:bentonite	Water:dry grout	Soil:grout	Soil:dry grout
A	2 : 8 : 0 : 0	0·42 : 1	5 : 1	7 : 1
B	3 : 8 : 0 : 0	0·42 : 1	5 : 1	7 : 1
C	2·5 : 8 : 0·4 : 0	0·42 : 1	5 : 1	7 : 1
D	3 : 8 : 0·1 : 0	0·42 : 1	5 : 1	7 : 1
E	2·5 : 8 : 0·4 : 0	0·42 : 1	3·5 : 1	5 : 1
F	2·5 : 8 : 0·4 : 0	0·30 : 1	3·9 : 1	5 : 1
G	8 : 0 : 0 : 0·8	1·6 : 1	2·8 : 1	7·3 : 1

Table 2. Details of soil-grout mixes applied to the made ground (Evans, 1998).

EXPERIMENTAL PROCEDURE

The durability of the treated made ground was assessed using the ASTM wet-dry and freeze-thaw durability test methods (ASTM, 1995a&b). Each test procedure requires seven samples: three used as test samples, three as control samples and one for moisture content determination. The samples tested were 100-150mm in diameter and 50-100mm high. The control samples were placed in the humidity room at 20°C ± 3°C and 95% relative humidity, the wet-dry test samples in an oven at a temperature of 60°C ± 3°C and the freeze-thaw samples in a freezer at a temperature of −10°C ± 3°C, all covered and for 24 hours. The samples were then removed from the oven, freezer and humidity room and water at room temperature of 20°C ± 3°C was added to the control and wet-dry samples and chilled water at 4°C ± 3°C added to the freeze-thaw samples. All specimens were kept in this immersed state for 23 hours. At the end of each such cycle, the samples were inspected visually for physical deterioration including cracking, fracturing and integrity and the dry mass loss of each sample was determined. The above procedure was repeated for 12 cycles and the cumulative relative average dry mass loss of the test samples in relation to the control samples was calculated. The test was terminated if the corrected cumulative mass loss exceeded 30%

which defined failure. The standard ASTM freeze-thaw test method requires the use of a freezing temperature of –20°C at which most solidified waste materials are usually reported to have failed. In addition, these conditions are severe for the UK and hence a revised freezing temperature of –10°C has been adopted here (Al-Tabbaa and Evans, 2000). Although the justification for the use of severe freezing conditions is that such exposure in the short-term models the exposure to less severe conditions in the longer term, this has not been validated. In order to assess the effect of the durability on the leachability of heavy metals from the test samples, the water used in the durability testing was analysed at the end of each cycle and to examine any microstructural changes in the samples scanning electron microscopy (SEM) analyses was performed.

RESULTS AND DISCUSSION
The samples tested for wet-dry durability survived all 12 cycles with limited mass loss of up to 4% and with very little sign of physical deterioration. In addition, no difference was observed between the different mixes and different age of the samples. Therefore in what follows only the freeze-thaw durability behaviour is discussed.

Physical examination of the freeze-thaw samples
Cracking was observed in most of the freeze-thaw samples, and usually developed starting from the bottom of the sample. In the weakest samples this developed into fracturing with increasingly bigger chunks falling off. The first chunks to fall off were always the extraneous materials. Typical samples before and after durability testing are shown in Figure 1. Figure 1(a) shows a typical sample at the start of a test, the sample in Figure 1(b) is that same sample in Figure 1(a) after losing 7% mass at the end of cycle 8 and the sample in Figure 1(c) is a sample which lost 48% mass during the cycle it failed in.

Figure 1. Typical freeze-thaw durability samples: (a) at the start (left), (b) during (middle) and (c) end (right) of tests.

Time-related freeze-thaw behaviour
The durability test methodology followed for the four difference age samples was different in some cases and hence a direct comparison between all four sets of test results at the end of 12 freeze-thaw cycles is not possible. The samples at 0.2 years were subjected to 6 cycles at –10°C followed by another 6 cycles at –20°C. The samples at 1.2 years were the same samples tested at 0.2 years, and hence had already deteriorated to some degree, and were subjected to 12 cycles at –10°C. The samples at 2.3 and 5 years were identical; tested for the first time and subjected to 12 cycles at –10°C. Hence the only direct comparison that can be made between the four sets of tests is at the end of the first 6 cycles at –10°C and these results are presented in terms of the percentage average cumulative dry mass loss, here after referred to as cumulative mass loss, in Table 3. The table clearly shows that the results at 1.2 years are the worst where most of the samples failed before cycle 6, and this is attributed to their earlier

durability testing and significant deterioration during the cycles at –20°C at 0.2 years. Comparing the other three sets of results shows that generally speaking the samples performed best at 0.2 years with very limited deterioration in all the mixes. The behaviour at 2.3 and 5 years is generally similar with some of the mixes improving and some deteriorating with age. The results show that no particular trend can be observed with time; the time-related performance of mixes C, F and G was similar and the rest was variable.

Long-term freeze-thaw performance

The results in terms of cumulative mass loss at the end of the freeze-thaw test at 0.2, 2.3 and 5 years are presented in Table 4. Comparing the results at 0.2 years in Table 4 with those in Table 3 clearly shows the drastic effect of 6 cycles at –20°C. At 2.3 years. mixes C, E and G performed very well while the remaining mixes failed. At 5 years mixes B, C and G passed the test while the remaining mixes failed. Hence the only mix which performed consistently well is mix C while mixes A, D and F failed in both cases. Mixes E and G deteriorated dramatically while mix B improved between 2.3 and 5 years. Generally speaking there is more deterioration at 5 years compared to 2.3 years. Mixes C, E and F have the same solid grout constituents but mix E has more grout and mix F has a lower water:cement ratio than mix C. The relative performance of mixes C and E suggests that the presence of more grout is detrimental to the durability performance but this does not agree with later observations when comparing single and overlap mixes. Mix D is similar to mix B apart from the presence of a very small lime content. On extrusion of mix D cores it was clear that physically this mix was the weakest and it is possible that the mixing of the grout with the soil during the installation process was not performed effectively. Comparing the results at the end of the test at 0.2 and 5 years shows that a correlation between testing at severe conditions (–20°C) in the short term, six cycles only, and testing at less severe conditions (–10°C) in the long term might not be applicable particularly for mixes B and C.

Age	Mix A	Mix B	Mix C	Mix D	Mix E	Mix F	Mix G
0.2 years	4.7	3.9	3.3	4.9	2.1	3.2	2.4
1.2 years	41.7 (5)	39.1 (3)	32.3 (5)	31.2 (12)	32.2 (3)	31.3 (5)	27.9
2.3 years	22.4	10.5	1.7	20.1	0.1	3.8	0.4
5 years	19.1	3.0	0.7	30.1 (12)	15.9	1.5	1.9

Table 3. Freeze-thaw durability results in percentage cumulative mass loss after 6 cycles at –10°C with the failure cycle number in brackets.

Age	Mix A	Mix B	Mix C	Mix D	Mix E	Mix F	Mix G
0.2 years	33.2 (12)	29.3	26.8	31.8 (12)	13.6	13.7	3.5
2.3 years	36.0 (8)	32.3 (11)	3.6	39.9 (9)	1.3	31.4 (11)	0.7
5 years	36.0 (9)	17.3	8.4	30.1 (6)	31.1 (9)	30.8 (12)	25.1

Table 4. Freeze-thaw durability test results in % cumulative mass loss at the end of the test.

Variability in the freeze-thaw results

Variability in some of the test results was clear even when the samples came from the same core. Figure 2 shows typical results of this observation, in terms of cumulative mass loss of two mixes in parts (a) and (b) where the three samples produced similar and different results respectively. As observed while performing the tests, this variation is mainly attributed to the presence of extraneous materials and to a lesser degree to the quality of the samples which probably reflects variability in the soil mixing process. It is clear from Figure 2(b) that in such cases three samples would not be sufficient to produce a reasonable indication of the freeze-thaw performance. This variability is also considered to be the cause of absence of obvious trends between the different mixes.

Single and overlap mix freeze-thaw behaviour

Overlap columns were constructed for mixes B, C, D and E (denoted as Bo, Co, Do and Eo) and cored only at 4.5 years after treatment. The freeze-thaw durability results in terms of cumulative mass loss at 5 years comparing overlap columns with their single counterparts are presented in Figure 3. Figure 3 shows that the overlap mixes performed better (by 6-8%) than the corresponding single mixes for mixes Bo, Co and Do while Mix Eo showed a much bigger variation. These results show that the increased grout content in the overlap columns improved their resistance to freeze-thaw cycles.

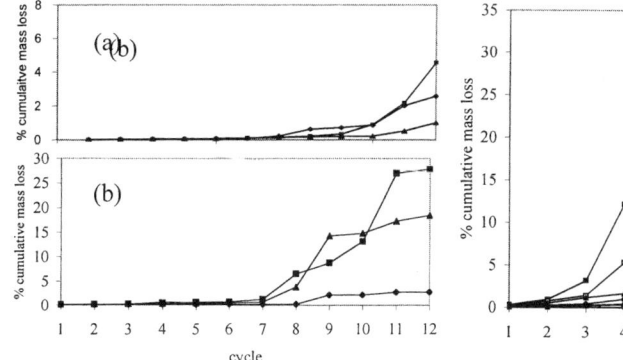
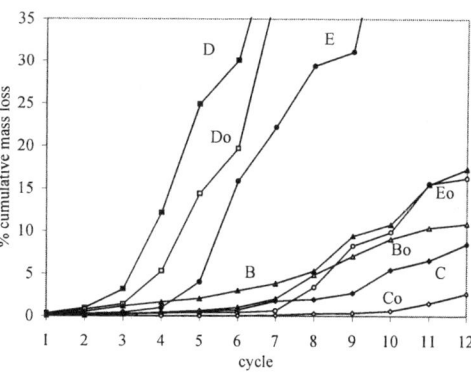

Figure 2. Typical freeze-thaw durability behaviour: (a) consistent and (b) variable.

Figure 3. Freeze-thaw durability results of single and overlap columns at 5 years.

Leachability of freeze-thaw samples

Analyses of copper and nickel in the durability test water showed increased leached concentrations of those contaminants. In addition, the leached concentrations from both the wet-dry and freeze-thaw test samples were always higher than those from the control samples indicating that both weathering conditions have produced some changes to the samples which have affected their leaching behaviour. The concentrations from the freeze-thaw samples were always much higher than those from the wet-dry samples indicating a stronger effect. For the freeze-thaw samples, this is probably caused by the cracks produced by the freeze-thaw process. For the wet-dry samples the reason is not clear but it is possible that the test conditions have weakened the bonding between the heavy metals and the grout matrix. The pH of the solution in the first cycle was around 10 and was higher in the durability samples than the control samples and subsequently gradually reduced to around 8.5 for all samples.

SEM analyses of freeze-thaw samples

Typical SEM micrographs of samples at the end of both the freeze-thaw and wet-dry durability tests and samples which were not subjected to either are shown in Figure 4. These typical samples show very little difference between them which suggests that the durability cycles had a negligible effect on the microstructure of the samples. The wet-dry and control samples show the same level of hydration suggesting that hydration is complete.

CONCLUSIONS

A detailed assessment of the long-term durability behaviour of in-situ stabilised/solidified treated contaminated made ground was carried out. It was found that the behaviour during the first 6 cycles was best at 0.2 years and that the overall durability performance at 5 years was slightly worse than that at 2.3 years. The overlap mixes behaved slightly better than the corresponding single mixes indicating that overlap mixes are not efficient since their grout content is double that of the single mixes. A level of heterogeneity in the results was observed, attributed mainly to the presence of extraneous materials, which indicates that the procedure of the ASTM durability testing of using three test samples is insufficient for such material. The leachability of heavy metals increased during the durability tests which shows that durability does affect leachability performance. For the freeze-thaw samples this is probably caused by the cracks produced during the test. The SEM analysis did not show any obvious changes in the microstructure of the samples.

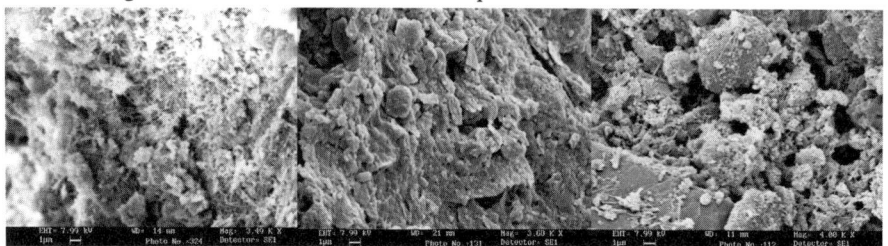

Figure 4. SEM micrographs of mix E before (left) and at end of wet-dry durability test (middle) and freeze-thaw durability test (right)

AKNOWLEDGEMENTS
The authors gratefully acknowledge the financial assistance of EPSRC and the technical assistance of Steve Chandler, Tim Ablett and Alan Heaver.

REFERENCES
Al-Tabbaa A. & Evans C.W. (1998). Pilot in situ auger mixing treatment of a contaminated site Part 1: Treatability study. Proc. Instn Civil Engrs, Geotech. Engng, 131 pp. 52-59.

Al-Tabbaa A., Evans C.W. and Wallace C.J. (1998). Pilot in situ auger mixing treatment of a contaminated site Part 2: site trial. Proc Instn Civil Engrs, Geotech. Engng, 131 pp. 89-95.

Al-Tabbaa A. & Evans C.W. (2000). Pilot in situ auger mixing treatment of a contaminated site Part 3: Time-related performance. Proc Instn Civil Engrs, Geotech Engng, 143 pp. 103-114.

American Society for Testing And Materials Test D4842 (1995a). Standard test method for freezing and thawing test of solid sastes, Vol. 11.04. pp. 148-151.

American Society for Testing And Materials Test D4843 (1995b). Standard test method for wetting and drying test of solid wastes, Vol. 11.04. pp. 152-155.

Conner, J. R. (1993). Chemical fixation and solidification of hazardous waste. Van Nostrand Reinhold, New York.

Day, C. and Ryan C. R. (1995). Containment, stabilisation and treatment of contaminated soils using insitu soil mixing. ASCE Geotechnical Special Publication 46, pp 1349-1365.

Evans C.W. (1998). Studies Related to the In Situ Treatment of Contaminated Ground Using Soil Mix Technology. PhD Thesis, University of Birmingham, England. 317p.

Evans, C W and Al-Tabbaa, A. (1999). Remediation of contaminated ground using soil mix technology: from research to commercialisation. 2nd BGS Geoenvironmental Engineering Conference, London, Thomas Telford, pp. 376-383.

Kirk, D. R. (1996). Summary of U.S. EPA research on solidified/stabilised waste for long-term durability. In Stabilisation and Solidification of Hazardous, Radioactive and Mixes Waste, ASTM, STP1240, Vol. 3, pp 239-250.

Loxham, M., Orr, T. and Jefferis, S.A. (1997). Contaminated land reclamation. In Report of the ISSMFE Technical Committee TC5 on Environmental Geotechnics, pp 113-132.

Laboratory investigation of optimum conditions for sub-surface biofilm barriers

M. J. BROUGH, School of Civil Engineering, Birmingham University, UK,
A. AL-TABBAA, Engineering Department, Cambridge University, UK and R. J. MARTIN, School of Civil Engineering, Birmingham University, UK.

ABSTRACT
This research investigates the feasibility of auger applied bioactive zones in the sub-surface for applications in waste containment and in-situ bioremediation. It is carried out using laboratory-scale investigations focusing on optimisation of the technology in terms of combined clogging and biodegradative potential. Both manually-prepared samples in trixial cells and auger mixed columns in a tank system were used. The results are assessed in terms of permeability reductions.

INTRODUCTION
Proposed clean-up laws make local authorities and the environment agencies duty-bound to locate contaminated sites and require these to be cleaned up by polluters, owners and occupiers. Furthermore, fiscal incentives to encourage brownfield site development have been proposed, with suggested funding coming from taxes on owners of vacant brownfield sites. Therefore there is an overwhelming demand for brownfield remediation technologies. Previous remedial action frequently involved in-situ containment or encapsulation and removal of the contamination. These traditional technologies are being questioned more and more. Removal is seen as transferring unresolved contaminant problems to another site. Encapsulation renders a site safe for immediate use, but future maintenance may be necessary if a change of use of the site occurs or if the encapsulation system fails. Furthermore, stringent regulations regarding the disposal of contaminated material have forced developers to look elsewhere. This has promoted the development of sustainable, low maintenance, low cost applicable remediation technologies that cause minimum disruption to site activities. One example is reactive barriers. Reactive barriers are permeable containment systems that react with contaminated groundwater as it flows through them. Typically, a high permeability trench is constructed with a contaminant specific reactive filling and the groundwater flow directed through it. Although complete contaminant degradation can occur, treatment efficiency is limited by the sub-surface availability of the reactive filling. Therefore, there is the need for a sustainable reactive barrier technology that reacts to the contaminated sub-surface environment. Biotechnology could provide the solution to this problem.

Innovative bioremediation technologies stimulate the natural soil microbial activity, by altering the physical, chemical and biological conditions which often involves the provision of limiting nutrients. Remedial strategies also incorporate the sub-surface inoculation of microbes acclimatized to specific environmental conditions and contaminants (bioaugmentation). This requires the effective application of micro-organisms and associated carrier particles, which can subsequently withstand physical, chemical and biological environmental changes. After bioaugmentation, nutrient stimulation forms a biofilm. A

Geoenvironmental impact management, Thomas Telford, London, 2001.

biofilm is an accumulation of organic and biological material immobilized at a surface. Continued biofilm accumulation changes permeability, porosity and surface roughness of the soil strata. Associated amendments to nutrient or contaminant transport, biodegradation, continued biofilm accumulation and adsorption processes directly influence waste containment and biodegradation potential. Therefore there are many inter-related, physical, chemical and biological, processes which influence a sub-surface biofilm's potential for biodegradation or clogging. If these processes can be controlled, there is the potential for a sub-surface biologically reactive cell with applications in waste containment and bioremediation.

Previous research focused on the accumulation of micro-organisms in porous media and the subsequent effects on permeability. Other work has looked at the sub-surface biodegradation of organic contaminants (bioremediation). However, little research examined the interactions of these processes. Although the work of Brough (1999) considered both aspects, this paper focuses on maximising the clogging potential of compacted and augered soils. The paper starts with the details and results of column tests then moves to practical applications using tank experiments in which a biofilm barrier was installed using soil mixing. Soil mix technology has been used for the treatment of contaminated ground in the past decade or so, mainly in the US, and its main application has been in solidification and stabilisation (Day & Ryan, 1995). Its application in the UK is far more recent (Soudain, 1997) and is receiving considerable interest from industry. Soil mixing is usually carried out using specifically designed mixing auger and its effectiveness relies principally on achieving effective mixing between the soil and the additive. Soil mixing has many advantages in terms of speed of implementation, simplicity of the techniques used, elimination of off site disposal and relatively low cost. This has fuelled interest in new applications for soil mixing; one such novel application is the subject of this paper – that of the installation of a biofilm reactive in-ground barrier.

MATERIALS AND METHODS

Coarse and medium Leighton Buzzard sands with particle sizes of 0.6-1.18mm and 0.3-0.425mm respectively were used as the porous media. To achieve lower permeabilities, the medium sand was supplemented with kaolin clay. Activated sludge was sampled from a wastewater treatment works, stored in covered glass reactors to inhibit algal growth and sparged with air to ensure sufficient dissolved oxygen. Cultures were batch fed on a standard nutrient feed (1.5625g Glucose, 0.3125g KH_2PO_4, 3.1250g bacteriological peptone/litre of distilled water = 5g/l COD). The inoculated sands were compacted into specimen holders and permeated with a nutrient solution. Inoculation was performed during mixing prior or by augering after compaction. Clogging potential was determined by testing permeability prior to inoculation (K_{free}), after inoculation (K_{base}) and after nutrient permeation (K_{min}). Inoculum concentration, initial permeability, clay content, nutrient carbon:nitrogen (C:N) ratio and augering were investigated for their effects on clogging potential. Baseline parameters of the sands prior to inoculation or biofilm treatment are detailed in Table 1. Permeabilities (k_{free}) of clay-amended sands are not quoted for the flow-tank system as inocula and clay particle support media were augered into soils after compaction.

Two experimental rigs were used for permeation experiments as shown in Figures 1 (a) and (b). All systems consisted of an inflow nutrient feed, a specimen holder and an outflow reservoir. The triaxial cell system was used to optimize a biofilm barrier's potential for waste containment and in-situ bioremediation. The tank system was used to simulate and optimise biofilm barrier installation using a model deep soil mixing auger in a confined aquifer. Permeability results are presented with discussion pertinent to biodegradative potential. Permeability was determined by recording permeant flow rate and the resulting pressure head difference between sampling points. Corrections were made for temperature-related changes in water viscosity and density. Chosen flow rates reproduced typical Darcy velocities exhibited near injection wells (Taylor and Jaffe, 1990).

Table 1. Initial void ratio and permeability of compacted sands.

		Mean void ratio (e_{free})	Sample permeability K_{free} x 10^{-3} m/s	Sample permeability K_{free} x 10^{-3} m/s
Test rig			Triaxial system	Tank system
Soil type	Coarse sand	0.56	2.62	2.50
	Medium sand	0.56	0.71	0.80
	Medium sand + 2% clay	0.56	0.0090	n/a
	Medium sand + 8% clay	0.54	0.00063	n/a
	Medium sand + 16% clay	0.45	0.00022	n/a

RESULTS AND DISCUSSION

Effects of inoculum concentration on clogging potential
With increased sludge inoculum concentration, up to around 4g/l, clogging was more severe. This clogging was higher in the medium sand compared to the coarse sand as can be seen in Figure 2. Therefore, sub-surface clogging could be enhanced by increasing sludge inocula concentration or reducing the overall particle size of the sand used. The permeability of the medium sand reduced by nearly one order of magnitude after sludge inoculation. Sludge concentration had a negligible effect on the permeability of the clayey sands due to the clay's overriding effect on initial permeability.

Effect of initial permeability on clogging potential
Inoculated sands were permeated with the standard nutrient feed. Clogging was consistently more severe in the medium sands. Permeability of the coarse and medium sands reduced by average 0.11 and 0.24 orders of magnitude respectively. This was due to improvements in mechanical straining, available surface area and an improved void clogging efficiency. Therefore, increased particle size support media could be used in a biofilm barrier to impede clogging and vice versa.

Effect of clay content on clogging potential
Relatively low clay content supplements had a significant effect on the initial permeability and the nutrient-feed related clogging potential of sands as can be seen in Figure 3. Permeability of the medium sand with 2% clay reduced by one order of magnitude after nutrient permeation. Clogging was enhanced further in the higher clay content sands. The results show that with increased clay content, the biofilm was more efficient at clogging with the 16% clay content causing a 1.5 order of magnitude reduction in permeability. This was due to improvements in mechanical straining processes, biofilm structure and nutrient status. Clay particles acted as inter-particle bridges between micro-organisms, improving nutrient availability, biofilm accumulation and the associated resistance to hydrodynamic shear forces and nutrient limitation. Previous researchers suggested that the permeability of a sand with 17% kaolin reduced by up to three orders of magnitude after biofilm treatment (Turner et al., 1997; Dennis and Turner, 1998).

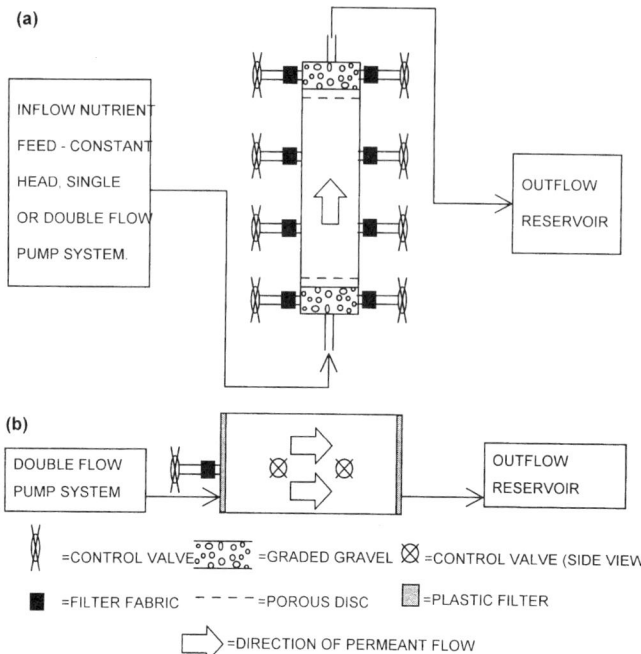

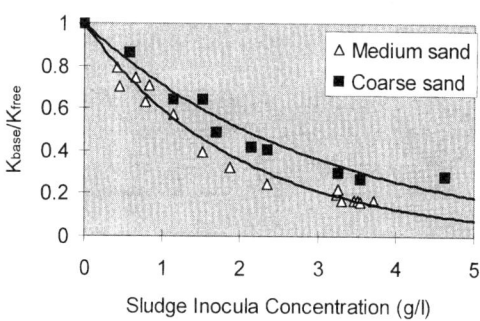

Figure 1. Experimental set-up (a) triaxial cell set-up and (b) tank system.

Figure 2. Clogging of the sands due to sludge inoculation.

Effect of nutrient carbon:nitrogen (C:N) ratio on clogging potential

Previous researchers suggested that micro-organisms subjected to increased C:N ratio feeds exhibited an enhanced extracellular polysaccharide slime production (Wilkinson, 1958). The results are shown in Figure 4 and suggest that the inoculated clay enhanced medium sand, permeated with high C:N feed solutions, exhibited improved clogging, by up to half an order of magnitude, compared with permeation with the standard nutrient solution. Nutrient C:N ratio could therefore be used as a control variable to promote clogging in a biofilm barrier.

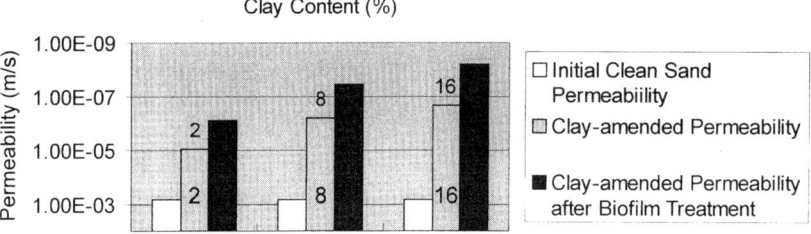

Figure 3. Clogging due to slay supplements and subsequent nutrient permeation.

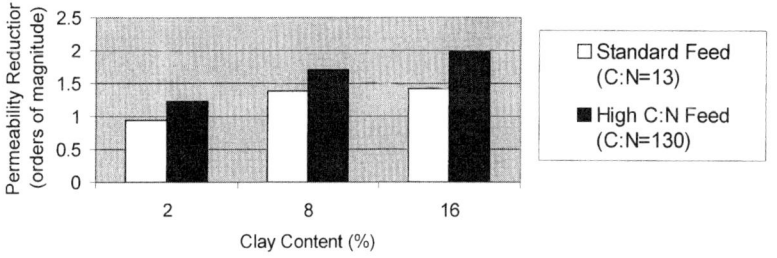

Figure 4. Effect of Increased C:N Ratio upon Clogging Potential.

Effect of augering on clogging potential

Permeabilities of augered soils were consistently higher than the permeabilities of the same soils compacted. These differences were more significant when comparing similar compacted and augered sands that were supplemented with clay. Following nutrient permeation, clogging was consistently less severe in augered soils as can be seen in Table 2. This could be due to heterogeneities in augered soils promoting preferential flow. These localised 'inoculum-free' zones of an augered soil should eventually be colonised with bacteria. As discussed previously, lower clogging in augered soils was more likely related to modified void ratio, density and permeability considerations. Densities of augered inoculated soils were consistently lower than in similar compacted soils. This meant that the void ratio and permeability of augered soils were higher with an associated increased pore volume to 'fill'. Subsequently, accumulated biofilm was less susceptible to mechanical straining processes and not as effective at clogging.

Augering may be advantageous in an in-situ bioremediation technology. Although augered soils exhibited less clogging potential when compared with similar compacted soils, this was not associated with a reduction in accumulated biofilm mass or viability. This has important implications for a biofilm barrier technology. Results suggested bioavailability and biodegradative potential improved in augered soils because clogging was not as severe as biofilm accumulated Brough (1999).

Table 2. Clogging in augered and compacted soils, in terms of orders of magnitude permeability reduction.

Feed	Soil Type (Preparation)	Coarse Sand	Medium Sand	2% Clay	8% Clay	16% Clay
Standard	Compacted	0.23*	0.42*	1.10	1.30	1.60
Standard	Augered	0.4	0.49	0.62	-	1.02
High C:N	Compacted	-	-	1.2	1.4	2.0
High C:N	Augered	-	-	-	-	1.23

* - values are from a constant head system used

CONCLUSIONS

Optimum conditions were identified in the triaxial cell set-up which caused up to one order of magnitude reduction in permeability of the sands tested due to biofilm clogging. This includes increasing the sludge inoculum concentration to 4g/l, the addition of kaolin clay by up to 16%, increasing the C:N ratio in the nutrient solution. Clogging in the augered soils was consistently less severe than that in the compacted soils, and this was consistent with the initial permeability of the former being consistently higher than that of the latter. Perceived bioavailability problems in compacted soils identified the need for a more open-spaced arrangement of soil particles which was achieved by augering. Subsequently, at high clay contents, augering allowed advantages of increased clay content to be realised without encountering permeability-related bioavailability problems. By augering sludge and sludge-clay mixes into the sub-surface, permeability around a site could be controlled to govern contaminated groundwater flow.

REFERENCES

Brough, M.J. Auger applied biofilm barriers a novel waste containment and bioremediation technology. PhD Thesis, 1999, University of Birmingham.

Cunningham, A.B., Bouwer E.J. and Characklis W.G.   Biofilms in porous media. In:Biofilms, Ed. W.G.Characklis K.C.Marshall, John Wiley and Sons Inc., New York, 1990, pp. 697-732.

Day, C. and Ryan C. R. Containment, stabilisation and treatment of contaminated soils using insitu soil mixing. ASCE Geotechnical Special Publication 46, 1995, pp 1349-1365.

Dennis, M.L. and Turner, J.P. Hydraulic conductivity of compacted soil treated with biofilm. Journal of Geotechnical and Geoenvironmental Engineering, 1998, Feb., pp.120-127.

Gavaskar, A.R., Gupta, N., Sass, B.M., Janosy, R.J., O'Sullivan, D. Permeable Barriers for Groundwater Remediation, Battelle Press, Ohio, 1998, pp.55-77.

Soudain, M   Fixed on site. Ground Engineering, 1997, pp 19.

Taylor, S.W., Jaffe P.R. Biofilm growth and the Related Changes in the Physical Properties of a Porous Medium, experimental investigation. Water Resour. Res., 1990, Vol.26, pp.2153-2159.

Turner, J.P., Dennis, M.L., Osman, Y.A., Chase, J. and Bulla, L.A. Biofilm treatment of soil for waste containment and remediation. Proc. Int. Cont. Tech. Conf., Florida, 1997, pp 672-678.

Wilkinson, J.F. The extracellular polysaccharides of bacteria. Bact. Rev., 1958, Vol. 22, pp. 46-73.

Laboratory and full-scale behaviour of soil-mixed columns

B. CHITAMBIRA, DR. A. AL-TABBAA, Engineering Department, Cambridge University, UK and C. W. EVANS, May Gurney (Technical Services) Ltd, Norwich, UK

ABSTRACT
This paper starts with investigating the use of a laboratory-scale auger in the construction of sand-cement and sand-cement-waste columns and assesses the resulting columns in terms of unconfined compressive strength, Young's modulus and load-displacement behaviour. It then investigates and assesses the results from a full-scale site trial. Finally it attempts to correlate the laboratory-scale and full-scale results.

INTRODUCTION
Soil mix technology was originally developed in the 1960s for groundwater cut-off and excavation support and is still widely used in the United States and Japan (Yang, 1994). Its application to the improvement of soft soils then followed and more recently its application to the containment and remediation of contaminated ground has been adopted; initiated in the USA in the late 1980's (Yang, 1994) and has only recently been commercially employed in the UK (Evans and Al-Tabbaa, 1999). Soil mixing is carried out using mixing augers through which a grout is introduced and mixed with the soil resulting in soil-grout columns. It is vital to ensure that the injection and mixing processes of the additives with the soil are thorough and effective so that homogeneous well-mixed soil-additives monolithic columns are produced. Different auger designs and installation techniques are being used depending on the site conditions and depth required. Soil mixing is competitive with most conventional treatment, disposal and remediation methods and has special advantages in reduced health and safety risks, speed of construction, elimination of off-site disposal and low cost. Because of the complexities associated with full-scale trials in terms of site heterogeneities and also cost, full-scale test data is limited and it is convenient to use laboratory-scale model augers to simulate full-scale soil mixing in the laboratory. With the use of model augers it has been possible to investigate many aspects of the applicability of soil mix technology to the improvement of soft ground and the containment and treatment of contaminated ground (Al-Tabbaa and Evans, 1999).

Recent developments in contaminated land treatment have moved from the traditional low permeability physical containment systems to reactive containment systems (low or high permeability) which provide chemical as well as physical containment of the contaminants (Jefferis et al, 1997). The purpose is that contaminant leachates permeating or diffusing through the wall are treated and subsequently treated groundwater emerges. Traditionally cement grouts have been used to provide the low permeability passive containment part of the barrier and materials such as natural or modified clays have been used to provide the reactive containment part (Evans, 1999). New developments have seen the successful use of reactive materials such as activated carbon and iron filings (Gavaskar et al, 1998). More recently the effectiveness of waste materials such as granulated tyre, waste peat and wood shavings have been investigated (Kershaw & Pamukcu, 1997, Ajmal et al. 1998, McKay & Porter, 1997). This paper starts with investigating the strength and stiffness of constructed soil-cement and soil-cement-waste columns and comparing manually-mixed and auger-mixed samples. It then looks at the load bearing capacity of such columns and by presenting results from a full-scale trial, it attempts to correlate laboratory-scale and full-scale results.

MATERIALS AND EXPERIMENTAL PROCEDURES

Medium dense sand with a ϕ value of 37° was used at a moisture content of 10%. Ordinary Portland cement was used as the main additive to which waste material was added. Granulated tyre (4-10mm size), wood shavings and waste fibrous peat were used. A grout of cement and water in a 1:1 ratio was chosen to obtain a flowable mix and a soil to grout ratio of 4:1 was also selected from previous work (Evans 1998). The waste material was added at 5, 2 and 1% of tyre, wood shavings and peat respectively. The column installation was performed using a laboratory-scale auger shown in Figure 1. It consisted of four mixing blades and cutting teeth and is 90mm in diameter and 300mm long. Grout was injected through two diametrically opposite ports positioned under the leading flight to protect them from clogging by the soil. A grout injector was used to pump the grout through an 8mm-diameter rubber tubing via the auger shaft into the soil. The raising and lowering system of the auger was manually-operated while the auger movement was powered by a Parvalux electric motor with one fifth horsepower. A peristaltic flow pump was used to convey the grout through the auger shaft. The columns were installed in drums 300mm in diameter and 420mm high. A 35mm diameter plastic pipe was used to introduce the waste media as a central column, as the particle size was too large to pump through. The auger was then advanced into the soil to the required full length, mixing the sand *in situ* and grout was injected on withdrawal of the auger. Two further mixing cycles were then carried out to ensure homogeneous mixing of the soil and grout. Some columns were then extruded after a period of in-situ curing and others were axially loaded, at 28 days, via a load cell mounted to a loading frame. The loading machine was manually operated at a constant rate of displacement of 2.5mm/min. The load cell was connected to a datalogger and the displacement was measured by means of an LVDT. Samples of the same soil-grout mixes were also prepared manually and tested for their unconfined compressive strength for comparison.

RESULTS AND DISCUSSION

Unconfined compressive strength of the column material

The unconfined compressive strength of the sand-cement and sand-cement-waste mixes was assessed using manually-mixed samples and auger-mixed columns at 28 days. The manually-mixed samples were 100mm in diameter and 200mm high and the results are shown in the second column of Table 1. They show a consistent average UCS value of just over 1100kPa ± 3% and which increased due to the addition of waste. Because of the low percentage of waste material used, it is clear that the waste material acted as reinforcement in the mix. A typical complete column with cross-section is shown in Figure 2. The UCS of the auger-mixed columns was assessed by constructing long columns in the drums used. The proximity of the drum boundary to the column meant that the column material failed rather than the surrounding soil. The columns were 100mm in diameter and around 350-400mm long. The density of the soil-mixed columns and the manually-mixed samples were generally similar.

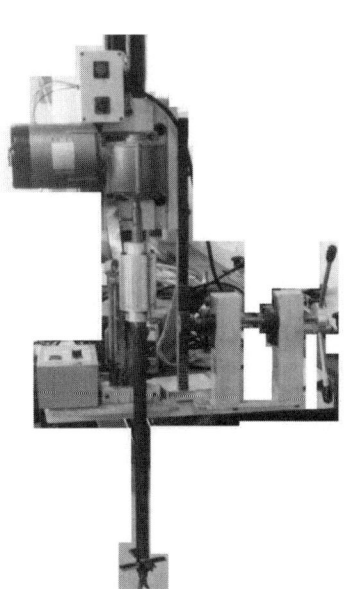

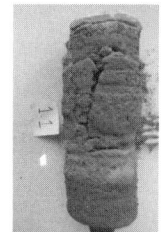

Figure 1. Laboratory auger and   Figure 2. Sand-cement-tyre   Figure 3. Examples of failed
Auger set-up.   column and cross- section.   columns.

The results are shown in column 3 of Table 1. Those UCS values are far more variable than those of the manually-mixed samples. The results for the sand-cement columns were in some cases up to 50% lower than the manually-mixed samples. These values are a reflection of the variability of the grout distribution within the soil-cement columns. The UCS of the sand-cement-waste columns was always much lower than those of the manually-mixed samples and for the case of the tyre and peat was also very variable. This is also a reflection of the difficulty in uniformly mixing the waste material within the soil-mixed columns. It is clear that the variability within the soil-mixed columns meant that the column material failed at the weakest section in the column which shows inconsistent mixing and poor repeatability. This is probably the case because of the relatively low grout content. It should be pointed out here that it was not possible with the laboratory equipment available to monitor the rate of grout injection, something which could be easily achieved on site. However, for laboratory-test purposes more effective mixing should be achieved with a higher grout content. These results agree with other published work showing that the UCS of auger-mixed samples could be much lower than that of the manually-mixed samples (Evans, 1998). Examples of observed failure modes of some of the soil-mixed columns are shown in Figure 3.

Young's modulus, E, values of the columns were also calculated. For the manually-mixed samples these were only calculated for the sand-cement mix and the E values were consistent and ranged between 1.01 and 1.19 x 10^5 kPa. The E values for the auger-mixed columns are shown in the final column of Table 1. The values for the sand-cement columns are the same as those for the manually-mixed samples, hence the compressibility of the mix resulting from the two mixing conditions is consistent. The E values of the columns containing waste material are lower with the values for the mixes containing tyre being 20-70% of the mixes without, and for the mixes containing wood shavings and peat the values were even lower. Hence the compressibility of the columns increased in the presence of the waste material. These low values have design and construction implications which may need to be addressed.

Table 1. UCS and E values for the manually-mixed and auger-mixed columns.

Additive to Cement	UCS of manually mixed samples (kPa)	UCS of auger-mixed columns (kPa)	E of the auger-mixed columns x 10^5 (kPa)
None	1146	637–1146	1.01 - 1.18
Tyre	1273	318–891	0.27 - 0.7
Wood shavings	1528	828–891	0.07 - 0.09
Peat	2037	318–1273	0.07 - 0.09

Figure 4. Typical load-displacement behaviour of the soil-mixed column.

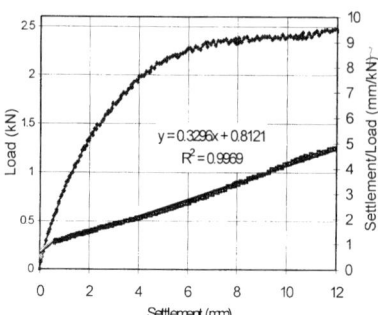

Load-displacement behaviour of the auger-mixed columns

In order to minimise the boundary effects and examine the load-displacement behaviour, shorter columns, 250-300mm long, and with two different diameters of 100mm and 60mm were constructed and tested. A typical load-displacement behaviour of a columns is shown in Figure 4 together with the displacement/load vs displacement graph used in the Chin's method for calculating the failure load (Chin, 1970). Such a plot usually produces two relatively straight line sections from which the shaft capacity (first line) and the total capacity (second line) can be deduced. For end bearing piles only one straight line section would be observed. Chin's method is most appropriate when there is no evidence of structural failure in the material usually indicated by a negative slope. Figure 4 shows as expected that for a pile in sand, since the shaft capacity is only a small proportion of the total capacity, that only one straight line portion was evident giving the total capacity. The initial stiffness of the sand-cement columns, up to one third of the failure load, was found to be around 0.54kN/mm.

Table 2 shows measured failure loads together with predicted loads using theoretical calculation, of shaft and base capacities based on soil properties, and using Chin's method. The results show a general agreement between the three sets of values although in some cases the results are much closer together than in others. It is possible that weaknesses within the soil-mixed column material might have contributed to the variations in the results. The Chin's method results are in some cases higher than the measured failure loads indicating unconservative predictions. Further work is needed to produce more detailed correlations.

Table 2. Measured and predicted failure loads of the laboratory-scale soil-cement columns.

Column details	Diam. (mm)	Height (mm)	Failure load (kN) Measured	Chin	Theoretical
Single	60	280	1.5	1.35	1.71
Single	65	250	2.7	3.03	1.76
Single	70	290	4.0	4.75	2.37
Single	100	300	4.8	6.35	4.99
Double	100	300	10.5	9.6	9.78

Table 3. Measured and predicted failure loads of the site trial piles.

Pile group	Failure load (kN) Chin	Theoretical
1	1111	10860
2	1250	10860
3	1250	14481

Full-scale site trials

Three pile groups of soil-mixed columns were constructed and load-tested, using a pile cap, at Marchington Prison site. All the columns were 4m long and 900mm in diameter. Two pile groups consisted of three piles each arranged in a row at a centre-to-centre spacing of 1.5m (pile group 1) and 1.0m (pile group 2) and the third group consisted of four piles arranged in a square of two by two at a centre-to-centre spacing of 1.5m (pile group 3). The grout mix used was a mixture of cement and pfa in a 3:8 ratio, the soil:grout ratio was 4:1 and the water:dry grout ratio was 0.4:1. The ground conditions consisted of made ground (0-0.8m), sandy clay (0.8-1.6m), dense sandy fine to coarse gravel (1.6m-5.1m) and then siltstone to depth. Groundwater was encountered at a depth of 1.5m below ground level. The average SPT N value in the dense gravel was 32 corresponding to ϕ of $37°$ with no correction for overburden pressure, hence the same value as that used in the laboratory experiments.

Maintained load tests were carried out and the load-displacement results are shown in Figure 5. The load tests were terminated prior to failure because of excessive settlement. Figure 5 shows that pile groups 2 and 3 produced a similar load-displacement behaviour which was better than that produced by pile group 1. This indicates that placing the three piles closer together or increasing the number of piles improved the load-displacement behaviour of the pile group. The theoretically predicted pile group capacities and those predicted using the Chin's method are shown in Table 3. These results are clearly inconsistent with each other with the former being an order of magnitude greater than the latter.

The UCS of 10-day cubes of the site soil-grout mix was found to be 260kPa ±15% (Skanska, 2001). The cube samples had side dimensions of 150mm and an average density of 1900kg/m^3 and were manually mixed from soil-grout material at the top of the columns. This UCS value is surprisingly extremely low despite the average densities achieved. The other surprising observation is that the UCS was the same at 1, 3, 8, 14 and 28 days. This strongly suggests that there was something wrong with the mix, possibly that the grout content was insufficient for the gravel tested or that inadequate mixing was performed, both perhaps producing large voids in the samples. The predicted failure load from Chin's method gave an average bearing pressure at failure of 574kPa ±15%, which is more than twice the UCS of the soil-grout mix samples. This perhaps indicates that better conditions were obtained on site compared to the manually prepared samples and perhaps produced better compaction. Hence the excessive settlements observed in the load test are most likely to be caused by the presence of voids and weak inconsistent soil-grout material within the piles. Further laboratory treatability studies would be required to investigate the very low UCS values and to develop an optimum mix and mixing sequence suitable for the site soil.

The initial stiffness for the individual site piles ranged between 4.7kN/mm for those in pile group 1 and 8kN/mm for those in pile groups 2 and 3. Hence this shows that reducing the column spacing and using more columns has also increased the initial stiffness of the individual piles. To correlate site and laboratory-scale columns, the ratio of the initial stiffness of the two test results is related by the length ratio, hence an initial stiffness of the site piles would be expected to be 13-16 times greater than that of the laboratory-scale columns. Pile group 1 gives a ratio of 9.5 while pile groups 2 and 3, which would be expected to have some pile group effect, give a ratio of 16. Hence the latter is closer to the laboratory results. It is possible that the proximity of the boundary to the piles in the laboratory-scale tests had a effect on the results similar to that produced by a pile group. These results offer possibilities for correlating site and laboratory-scale results and is worth investigating further.

CONCLUSIONS
The laboratory-scale auger was effective in producing sand-cement and sand-cement-waste columns, which were visually homogeneous and with uniform dimensions. However, the UCS results of this material showed a high level of heterogeneity in the distribution of the

grout within the columns, compared to manually-mixed samples. This is probably caused by the relatively small amount of cement used which is 10%. The load-displacement behaviour of the soil-mixed columns produced failure loads which were similar to those prediced using theoretical calculations although the Chin's method results were generally higher and require further investigation. The site trial pile tests showed vast inconsistencies between theoretical and Chin's method predictions for the failure load. They also showed extremely low UCS values which were the same at 1 up to 28 days. These observations are linked and imply that the soil and grout were very poorly mixed producing columns and samples of very weak and inconsistent material. Correlation between the laboratory-scale and full-scale results are encouraging and deserve further investigation.

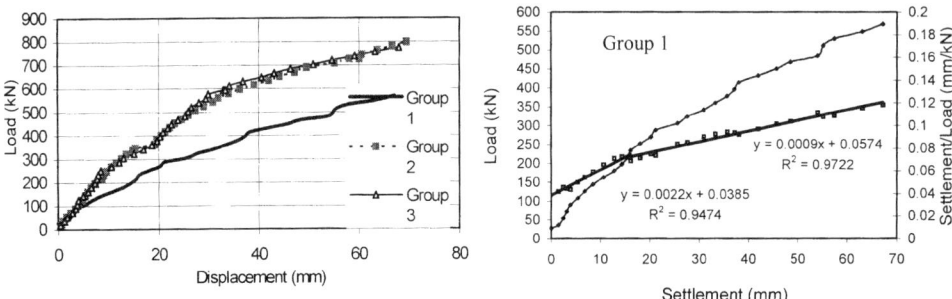

Figure 5. Load-displacement behaviour for the site trial pile tests.

ACKNOWLEDGEMENTS
The authors gratefully acknowledge the techncial assistance of Steve Chandler, Tim Ablett, Martin Touhey, Jim Francis and Si Holder. They also acknowledge Skanska, formerly Kvaerner Technology, for their co-operation with the trial.

REFERENCES
Ajmal, M., Khan, A.H., Ahmad, S, & Ahmad, A. (1998). Role of sawdust in the removal of copper (II) from industrial waste. J. water research, Elsevier, Vol 32, No10: 3085-3091.

Al-Tabbaa, A and Evans, C. W. (1999) Laboratory-scale soil mixing of a contaminated site. Journal of Ground Improvement, Vol 3, No. 3, pp 119-134.

Chin F. K., (1970). Estimation of the ultimate load of piles from tests not carried to failure. Proceedings of the 2nd SE Asian Conference on Soil Engineering, Hong Kong, pp. 81-90.

Evans, C. W (1998). Studies related to the in situ treatment of contaminated ground using soil mix technology. PhD Thesis, Birmingham University, UK.

Evans, C.W. (1999). Cleansing the soil. Preparing the Site, Contract J. Supplement, p. 27, April.

Evans, C. W. & Al-Tabbaa, A. (1999). Remediation of contaminated ground using soil mix technology: from research to commercialisation. 2nd BGS Geoenv. Eng. Conf, TT, pp. 376-383.

Gavaskar A. R., Gupta N., Sass B.M., Janosy R.J. and O'Sullivan D. (1998). Permeable barriers for groundwater remediation. Batelle Press.

Jefferis, S. A., Norris, G. H. & Thomas, A. O. (1997). Development of permeable and low permeability barriers. Proc. Conference Containment Technology, Florida: 817-826.

Kershaw, D.S. & Pamukcu, S. (1997). Ground rubber: reactive permeable barrier sorption media. Proc. Conf. in situ remediation of the geoenvironment. GSP 71. ASCE: 26-40.

McKay, G. & Porter, J. F. (1997). Equilibrium parameters for the sorption of copper, cadmium and zinc ions onto peat. J. Chem. Tech. & Biotech. Vol 69, No 3: 309-320.

Skanska (2001). Personal communication.

Yang, D. S. (1994). The application of soil mix walls in the United States. Geotechnical News, December: 44-47.

Evaluation of leaching tests in contaminated land studies

M..A. CZEREWKO[1], C.C. SMITH[1], J.C. CRIPPS[1], G.M. WILLIAMS[2], A. SMITH[3], A. MIDWOOD[3] and K. SUZUKI[4]

[1] Department of Civil and Structural Engineering, University of Sheffield, Mappin Street, Sheffield, S1 3JD
[2] British Geological Survey, Keyworth, Nottingham NG12 5GG
[3] Macaulay Land Use Research Institute, Craigiebuckler, Aberdeen, AB15 8QH, Scotland
[4] Centre for Analytical Sciences, University of Sheffield, Dainton Building, Sheffield, S3 7HF

ABSTRACT

The study is based on the testing of samples from a number of sites within the Metropolitan Borough of Wolverhampton, encompassing a range of typical historical land uses and industrial wastes. The total and leachable concentrations of heavy metals, metalloids and anions have been determined using a range of standard and non-standard leaching tests allowing comparisons to be drawn between different procedures and the effects of varying leachant composition to simulate different groundwater chemistries. As a result of this investigation the preferred method adopted for the characterisation of leaching potential for sampled material was the three-stage cascade test based on the CEN/TEC 292 test proposal. It has been found that, although a large proportion of the samples contain high concentrations of contaminants, they may not always be leachable and could therefore be classed as chemically inert under ambient environmental situations. The effects of varying the leaching protocol are discussed in the context of the evaluation of the impact of contaminated land on groundwater quality in areas of urban regeneration.

INTRODUCTION

With closure of the traditional industries in many industrial conurbations, groundwater abstraction has declined and in the past few decades groundwater levels have risen by tens of metres. In some areas this is leading to the saturation of contaminated land that has arisen as a legacy of past exploitation of natural resources and associated heavy industries. With the rise in groundwater levels, and changes in land uses, there is potential for the mobilisation of contaminants, including toxic heavy metals, from these materials that could pollute surface waters and also aquifers, where present.

The impact of contaminated land on the urban environment is strongly controlled by the nature of contaminants present and the potential of these to migrate to ground- and surface-water. Both the nature of contaminants and the potential for migration can be evaluated by carrying out leaching tests on representative samples of the various contaminated materials. The present study is based in the Wolverhampton conurbation. Major heavy industries associated with coal abstraction and pig-iron smelting such as coking, manufacture of iron and steel products, tin plating, galvanising and allied gas and chemical production were established in the area during the 19th century. In addition there was a successful brass and copper industry. The combined effects of mining for coal, groundwater extraction and the uncontrolled disposal of industrial wastes have resulted in the presence of extensive deposits of contaminated waste materials and ground, that now lie in an area of rising groundwater. With the saturation of

Geoenvironmental impact management, Thomas Telford, London, 2001.

these materials there is a risk of leaching and migration of contaminants to vulnerable surface-and groundwaters.

The impact of a contaminant on groundwater quality is strongly controlled by its solid and aqueous speciation. In the case of this study this is being assessed by means of detailed chemical characterisation on a suite of samples selected from a range of sites typical of Wolverhampton's industrial past. A groundwater flow model that utilises data provided by the Environment Agency and other sources is being used to predict the consequences of groundwater interaction with contaminated land and the effects on groundwater quality. The leaching experiments are designed to simulate the reaction of contaminated land with a range of water types. These data, with the speciation studies and detailed sample characterisation, are being incorporated into groundwater and contamination transport models which shall be used to assess the impact of groundwater changes, including groundwater rise on contamination mobility within the Wolverhampton conurbation. Integrating the results from the studies involved within this project will improve the risk assessment capabilities and the aim of the project is to help regulators formulate scientifically based strategies to mitigate against the effects of industrial pollution. This paper briefly discusses some of the findings of the characterisation and leaching tests on a selected number of potentially contaminated samples from within the city of Wolverhampton.

The research described forms part of the NERC funded study within the URGENT project titled 'Urban regeneration of coalfields: generic studies of contaminated land and groundwater issues exemplified in Wolverhampton'. The project is a collaborative research involving the University of Sheffield, the British Geological Survey and the Macaulay Land Use Research Institute.

SOIL SAMPLING AND PROCESSING

Sampling sites were selected within Wolverhampton based on historical land use and use of general geochemical data held within the Wolgis database (Bridge et al., 1997). Valuable advice and assistance was also provided by Wolverhampton City Council (WCC) who provided access to over 300 site investigation reports from within the Wolverhampton borough. This assisted in the selection of sites with the potential to provide a range of samples with varying PHE contents typifying the types of soils found within the borough. In total 50 soil samples were collected from 30 sites representing the varied historical and current land use including residential, various industrial, mining, gas works and sewage works.

Soil samples were obtained during 8 site visits spread over a two year period. Sub-surface soil samples were obtained at each site using 'tube' and 'barrel' type hand augers with 150mm, 75mm and 50mm diameter sample heads. Soil samples were obtained as single or incremental samples per locality over a depth profile below topsoil ranging between 0.12 m to 1.70 m below ground level. Sampling at each site was conducted within a sampling grid as recommended by the British Geological Survey for soil sampling (Hooker et al., 1998) which recommends establishing sampling points at each corner and the centre of the grid. Where the sites were exposed such as parks and industrially derelict sites sampling grids of between 5m × 5m and 25m × 25m were adopted. In restricted or sensitive sites, such as road verges or adjacent to structures, reduced sized sampling grids were used.

The material sampled at each site was double bagged using puncture resistant laminated polypropylene sample bag that were labelled and sealed. A mass of between 20 kg to 40 kg of material was obtained at each sample site. The sampling barrels and auger rods were washed clean with de-ionised water and dried between sample points to prevent any cross contamination. The samples were delivered to the laboratory on the day of collection for storage under cool conditions (7 to 10°C). Samples obtained by cable percussive drilling were

also made available by WCC from the Bilston urban village scheme site investigation. In total 18 samples from various horizons across the 200 hectare site were selected for testing.
The samples were generally processed within a day of collection and natural moisture contents determined following the procedures given in BS 1377 (1990). Each bulk sample was transferred to a large HDPE bin and homogenised using a sterile spade for a period of 15 minutes. Following homogenisation a 5 kg quantity of the sample was selected by means of cone and quartering and transferred onto a PVC sample tray and placed in a fan assisted oven set at 40°C for up to 5 days to dry. The dried material was transferred into large air-tight HDPE sample tubs and stored at between 0 and 4°C until required for leaching tests. An additional 4 kg sub-sample of each material was reserved for chemical characterisation. The remainder of each bulk sample was re-sealed and returned to the cool sample storage room. The sample preparation was kept to a necessary minimum. As all the samples contained traces of organic material, drying was necessary to prevent any further anaerobic reaction within the sample. Since representative samples are required for duplicate testing and chemical characterisation sample homogenisation was necessary to provide representative material and therefore unavoidable breakdown of the sample matrix and individual grains would have resulted. Removal of inert material was undertaken by selecting a standard <2 mm size fraction.
In this paper we consider the results of only the initial 24 samples collected which have been characterised and tested. Testing of the remaining 26 samples is almost complete but the data have not yet been compiled or analysed during the writing of this paper. The complete findings including comprehensive charaterisation data shall be presented in future publications.

SAMPLE CHARACTERISATION

The samples were characterised in detail using standard soil testing procedures. The samples were passed through a 2 mm mesh and representative portions of the 2 mm fraction were milled producing finely ground powders that were used for all subsequent chemical analysis. The total elemental contents of each sample were determined by means of aqua regia digestion and ICP-OES analysis of the digest solution, which gives values for the total concentrations for the metals present. This procedure is suited to total heavy metal determination, but silicate minerals are generally resistant to attack by aqua regia therefore the samples were also geochemically characterised using XRF. Total carbon and nitrogen concentrations were determined using instrumental combustion procedures. The mineralogy of each sample was determined and quantified by means of whole rock XRD (see Table 1). The procedure involved analysing spray dried samples which eliminates any preferred orientation of the constituent minerals especially clays. The mineralogy of each sample was quantified using a reference intensity ratio (RIR) procedure based on adding 20 wt% of corundum as an internal standard, the procedure is explained in detail in Hillier (2000).

Mineral group type	Colliery waste fill	Non colliery waste fill
Quartz	18-52 %	55-76 %
Feldspars	9-22 %	16-24 %
Clay minerals	28-53 % -(mainly illite & kaolinite with some mixed-layer illite-smectite).	10-17 % -(Mainly illite and mixed-layer illite-smectite with occasional samples containing small amounts of kaolinite).
Carbonates	0-6 %	0-4 %
Iron minerals	2-7 %	0-19 %

Table 1. Range of mineralogies encountered in the <2mm fraction of the tested material.

The distribution of PHE's in the samples was characterised by sequential extraction following a 4-stage procedure modified from Tessier et al (1979). This consists of a determination of (i)-total exchangable and carbonatic constituents using acetic acid; (ii)- reducible phases bonded to Fe and Mn oxides determined using hydroxylamine hydrochloride; (iii)- oxidisable phases bonded to organic matter and sulphides determined using hydrogen peroxide and ammonium acetate; (iv)- and residual material determined using hot nitric acid. Sequential extraction data permit the estimation of long- term effects on metal mobility, especially where the conditions on a site may change. It is also possible to calculate cation exchange capacities form this data. The samples were also subjected to particle size analysis as means of physical characterisation, each size fraction is currently being subjected to SEM analysis to characterise the physical form and distribution of PHE's and also determinations of specific surface areas, this data shall be presented in future publications.

All the samples collected were subjected to a standard single stage batch extraction based on the Environmental Agency suggested procedure (R & D Note 301, 1994). From the findings of the chemical characterisation and sequential extraction, seven samples were selected for the leaching test investigation. The nature of the fill types can be grouped into 8 categories as seen in Table 2 and suitable samples were identified and selected from 7 of these categories. The samples selected consisted of material having values of PHE's exceeding the ICRCL trigger limits as seen in Table 3. In addition sample UWV:B1 which has low values of PHE's was also selected as a control material for the leaching test investigation.

Nature of made ground	Samples = UWV:	Bedrock
Inert sand, gravel and clay fill	**B1**, B2, B3, B17	Silurian & Triassic
Colliery spoil fill	B13, **B14**, B15, B18	Carboniferous & Permo-Carboniferous
Canal dredging, sewage & chemical waste	B7, **B8**	Triassic
Gas work and ash fill	B4	Permo-Carboniferous
Landfill waste	B10, B11, **B12**, B24	Carboniferous & Triassic
Iron works waste and foundry sand	**B14**, B15, B19, **B20**, B21	Carboniferous
Industrial and building waste including sand and ash	B5, B6, B9, **B16**	Permo-Carboniferous & Triassic
Imported inert industrial fill & ash-residential	B22, **B23**	Permo-Carboniferous

Table 2. Nature of the made-ground and underlying bedrock for the Wolverhampton samples.

LEACHING TEST PROCEDURES

Leaching test procedures are operationally defined, therefore the development of a wide range of leaching tests has resulted which may be unsuitable for certain materials and also undesirable as the data is not comparable. Most leaching tests have been developed for regulatory purposes. Therefore a suite of leaching tests were carried out on selected samples to investigate the leaching behaviour under variable conditions. This enabled the selection of a suitable procedure for the determination of leaching potentials of additional samples under investigation, as it was found that no practical or endorsed leaching procedure specifically for investigation of potentially contaminated soils existed within the UK.

Leaching tests have been in use for a number of decades. They were initially developed to assess the short-term environmental impact of solid waste disposal in landfills following exposure to a leachant such as groundwater, rainwater or open surface water. The leaching tests are generally used to provide an estimate of the potential availability and mobility of metal and non-metal pollutants from contaminated solids by means of elution with various leachants in order to provide information on the behaviour of potential pollutants under environmental conditions. It must be remembered that the tests provide information about constituent concentrations and release from a material under reference testing conditions. Once a material is removed from its natural environment it is impossible to recreate the conditions in the laboratory under which it resided in situ. Basically the tests involve contacting a waste material with a liquid to determine which component of the waste will transfer into solution. There are a large number of research and regulatory test protocols in existence which have generally been developed for the classification and assessment of hazardous waste such as granular or monolithic, solid or radioactive waste and to identify the leachable constituents (EPS 3/HA/7, 1990). The test procedures consist of single stage and single reagent leachate tests for wastes, single stage leachate tests for soils using organic chelators such as EDTA, multi-stage cascade tests and column tests. Short-term leaching effects (<50 years) are usually determined using column or single stage extraction tests, whereas long-term (50-500 years) leaching effects may be determined using multi-stage cascade test procedures (Forstner, 1993). The data are generally expressed as leaching over time and the field concentration estimates are commonly used as input data for groundwater modelling. In leaching tests variations may be introduced by changing test parameters such as leachant composition, method of contact, liquid to solid ratio, contact time and temperature. Therefore a test procedure must be designed with test parameters that are appropriate to the situation and material under investigation. It follows that the majority of test protocols are not suitable for the study of natural soils, or made ground.

Contaminant	Units	Range (& No >)	Trigger Values		Sample ID						
			ICRCL	Kelly	B1	B8	B12	B14	B16	B20	B23
Arsenic	mg/kg	<1-42.3 (3)	10/40	30	<3.4	<1.8	<1.0	7.4	42.3	26.4	15.6
Cadmium	mg/kg	0.07-7.60 (2)	3/15	1	0.122	0.481	1.388	1.446	9.601	1.19	2.036
Chromium	mg/kg	11.45-166.4 (0)	600	N/A	11.45	31.78	70.79	37.27	124.9	166.4	35.58
Copper	mg/kg	10.05-559.7 (7)	130	100	10.05	94.21	559.7	129.8	231.4	275.6	279.8
Lead	mg/kg	16.81-2663 (12)	500	500	18.28	107.3	246.0	2663	1246	248.1	179.1
Nickel	mg/kg	7.80-173.4 (4)	70	20	7.803	22.93	173.4	43.07	100.5	57.1	92.04
Zinc	mg/kg	43.64-6691 (11)	300	250	43.64	220.7	678.7	510.2	4691	828.1	575.1
Iron	%	0.82-42.87 (NA)	NA	N/A	8231	18970	62860	32910	86830	115200	21060
Sulfur	mg/kg	50.03-2171 (7)	1000	N/A	68.31	78.74	2171	1760	1717	3139	551.4

Table 3. Range of PHE contents of soils and selected leaching test samples.

Leaching test procedures may be classed as 'extraction tests' or 'dynamic tests'. Extraction leaching tests involve no leachant renewal and it is assumed that steady-state conditions are achieved by the end of the extraction. This should not be confused with chemical extraction, which is used to quantify the total amount of contaminant available in a sample. Dynamic leaching tests involve leachant renewal therefore maintaining a driving force for leaching. Such conditions are created in cascade tests or column tests from which the temporal release of constituents may be inferred.

In the selection of testing procedures for the present investigation, certain controlling parameters and practical considerations were considered. Although static leaching maintains physical integrity of the sample material the contact periods required to reach steady-state conditions would be so protracted that the method would not be practical. Agitation of the sample-leachate mixture achieves steady-state conditions more rapidly since it improves the solid to liquid contact. However, where agitation is too vigorous, abrasion of individual grains may occur. As this would release otherwise immobile components therefore resulting in overestimates. End over end rotation at a rate of 20-30 rpm was selected which has been shown to cause a minimum of sample abrasion whilst maintaining an effective solid to liquid contact (Bergendahl and Grasso, 1998). The extractions were performed in LDPE extraction vessels with a minimum of headspace.

A range of leachant solutions, ranging from de-ionised water, to simulated leachate solutions and strong acid solutions, are recommended in the existing testing protocols. Generally these leachants are too aggressive and not representative of those present in the urban environment. Therefore it is advantageous to use a standard leachant based on waters such as rainwater, that are liable to be present in the area. Where mild leachant solutions are used the leachate chemistry is controlled by the sample, whereas aggressive leachants give rise to results that are controlled by the leachant rather than the sample. To minimise the exchange with the leaching vessel, plastic should be used when investigating inorganic contaminants.

A liquid to solid (L:S) test ratio is required which is low enough to avoid dilution of the contaminant to below the analytical detection limit and high enough to prevent solubility constraints from limiting the amount of contaminant that can be leached from the waste. Generally ratios of between 1:1 to 100:1 (L:S) are suggested and used. Where low L:S ratios are used the leachate chemistry is controlled by the sample, whereas with high L:S ratios, the leachate chemistry is controlled by the leachant, therefore lower L:S ratios such as 10:1 or 20:1 are preferred.

The contact time influences whether steady-state conditions are achieved. This affects the amount of contaminant released. In extraction tests this is equivalent to the test duration, whereas in dynamic tests this is also controlled by the flow rate for columns and the number of extraction stages for multi-stage batch extraction tests. In extraction tests steady-state conditions are normally achieved within a few hours to a few days. With dynamic tests steady-state conditions may need days or weeks to be achieved as a result of diffusion processes especially in column tests.

Therefore laboratory leachate concentrations are a function of the testing conditions and therefore cannot be normally interpreted as field leachate concentrations. They do, however, enable the leachable constituents to be identifies and the material to be classified.

Three types of test procedures were selected for the investigation, each considered to provide different information. A one stage extraction test which is currently the Environmental Agencies suggested procedure (R & D Note 301, 1994) was a chosen as this is the only procedure specifically recommended for testing soils. A range of leachants was used to investigate the samples which included simulated rainwater, a range of simulated acid rain solutions and a total of 5 natural sampled waters including acid mine water, equilibrated mine water, organic rich acidic surface waters and groundwater.

Two Cascade tests (also referred to as serial batch tests) involving multiple extractions on the same sample were selected. Species detected in solution from leaching tests are dependent on the solubility of species present. Therefore highly soluble species such as Na, K, SO_4^{2-} &c., may swamp a leachate in a single stage extraction. Once highly soluble species have been leached and the buffering capacity of samples exhausted, additional extraction stages can be used to provide information on previously suppressed metal species. A five stage extraction giving a cumulative liquid to solid ratio of 100 was used (Tack et al, 1999) and the simpler CEN/TEC 292- test proposal (Wahlstrom, 1996) which involves a two stage extraction giving a cumulative liquid to solid ratio of 10 was also employed.

Column tests were also performed using standard procedures given in (Fallman and Aurell, 1996). The test provides information on the short- to medium-term behaviour of samples. These tests are useful in providing information on the composition of the first flush (initial leachates), and time to peak of a fresh material such as PFA which is to be placed into a landfill. Column tests are also useful for contaminant attenuation studies. It is generally accepted that they are not readily applied to a regulatory framework because of their lengthy test duration and often poor reproducibility.

RESULTS AND DISCUSSION

It was found that the made ground sampled from the industrial 'eastern' section of Wolverhampton generally consisted of foundry sand, demolition rubble, slag, colliery spoil, and varying amounts of industrial and domestic refuse in a general matrix of sand, gravel and clay. The made ground sampled from the urban and light industrial areas of Wolverhampton which are generally to the east and also north and south of the centre generally consisted of orange sand and gravel deposits with varying amounts of domestic refuse, and occasionally imported foundry sand and colliery spoil. These types of material would normally be considered to comprise potentially contaminated materials.

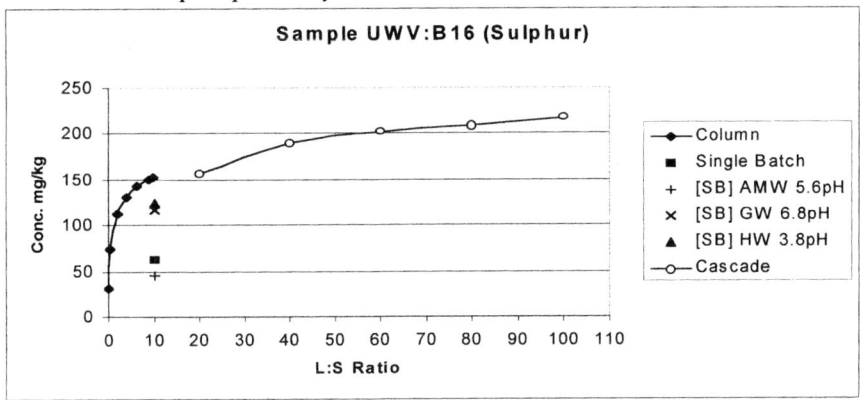

Figure 1. Plot of chemical leaching data for sulphur in sample UWB:B16.

Chemical analysis revealed generally elevated levels of Cu,Pb, Zn,Ni and occasionally Cd and As and elevated levels of S and Fe as seen in Table 3. Leachability testing revealed that the metals within the samples tested were comparatively immobile and as such would not be expected to represent a significant risk to groundwater resources under the range of test conditions investigated. Typical results for total concentrations obtained by aqua regia digestion and ICP-AES determination, total exchangeable content obtained by sequential extraction, and leachable quantities determined by various leaching test procedures for sample UWV:B16 are presented in Table 4. The data shows that of the extraction procedures adopted, the cascade test produced the highest values for available material. As can be seen in Figure 1, which shows data for leachable sulphur from sample UWV:B16, there is a decent fit for results obtained by column testing and sequential batch testing. This is typical of all the samples analysed.   The results obtained by batch testing tend to be lower, although they are within the same order of magnitude, as the other results This is probably a result of

			in mg/kg				
	Concentrations						
Parameter	**Total**	**Exchangable (SE)**	**Batch (10:1) (pH=5.8)**	**Cascade (100:1)**	**Batch AMW (pH=5.6)**	**Batch ASW (pH= 3.9)**	**Column (10:1)**
Cd	9.60	4.747	0.001	5.40	BDL	BDL	0.04
Cr	124.9	0.702	0.17	8.02	0.06	0.06	0.05
Cu	231.4	6.421	0.31	0.93	0.29	0.44	0.27
Fe	86830	4.647	3.30	17.29	-103.18	0.43	1.28

Ni	100.5	8.77	BDL	7.21	BDL	BDL	0.017
Pb	1246	27.19	BDL	0.24	BDL	BDL	0.056
Zn	4691	848.7	BDL	2.49	0.39	BDL	0.33
S	1717	n/a	62.28	216.8	44.8	123.5	152.4
pH	NA	NA	8.02	7.8	7.4	7.5	7.9
Eh	NA	NA	216	297	224	145	215

Table 4. Comparison of results for various extraction procedures for sample UWV:B16.

suppression by more soluble cation and anionic species. Use of artificial leachants and naturally sampled solutions showed very little difference in test results compared with the simulated rain water standard leachant adopted as seen in Table 4. Results for iron using acid mine-water (AMW) show negative values indicating sorption onto the sample. This consists of coal mine waste and foundry waste with a mineralogy consisting of 28% clay minerals and 4% haematite, which are all suitable sorption sites. The sample also contains 6% calcite, which affects its buffering capacity therefore making the material less sensitive to sudden changes in water pH as may result from accidental spillages. As a result of the research the preferred leaching test procedure selected for the work consisted of a cascade test. The procedure is a modification of the two-step CEN TC 292 serial batch test (van der Sloot, 1998) with an addition third step, as shown in Table 5.

Three-step serial batch test
- Step 1 - L:S = 2 : Test duration = 6 hours
- Step 2 – L:S = 2-10 : Test duration = 18 hours
- Step 3 – L:S = 10-20 : Test duration = 24 hours
- Stage n – additional L:S stages may be designed into the procedure to evaluate buffering effects
- Leachants – Simulated rain water (pH 5.6 – 5.8)
- Closed vessel extraction using 1000ml LDPE or HDPE vessel [S = 90g : L = 900ml] – smaller volumes may be used depending on sample amount with a minimum of S = 20g : L = 200ml
- Agitation – End over end tumbling at 20-30 rpm.
- Leachate separation – Centrifuge at 4000rpm for 50 min then filtration using 0.45μm syringe filter
- Measure – pH, Eh, major and minor elements, conductivity (optional) DOC (optional).

Table 5. Test protocol used for testing soils based on CEN TC 292

CONCLUSIONS

The research allows comparisons to be drawn between batch, column and multi-stage leaching methods. The effects of varying leachant composition to simulate different groundwater chemistries has also been explored. These include uncontaminated groundwater and rainwater, acid mine water and landfill leachate with which fills could become saturated in the future. The results for the various methods investigated tended to be within an order of magnitude indicating a generally good level of agreement. A straight forward serial batch test has been selected for use in characterising the leaching potential of potentially contaminated soils in the study. The procedure presents cumulative data of contaminant release and due to its simplicity presents a satisfactory level of repeatability.

The findings of this paper are based on preliminary data and further work which is underway is aimed at verifying a holistic risk assessment procedure based on the proposed test procedures.

Acknowledgements: This research is funded as part of the NERC URGENT project. The authors are grateful to Dr.S.Hillier for undertaking the XRD analysis.

REFERENCES

BERGENDAHL J. and GRASSO D. (1998) Colloid generation during batch leaching tests: mechanics of disaggregation. *Colloid and Surfaces A: Physiochemical and Engineering Aspects,* **135**, pp.193-205.

B.S.1377 (1990) Methods of test for soils for civil engineering purposes. British Standards Institution. London.

BRIDGE, DMcC., BROWN, MJ and HOOKER PJ (1997) Wolverhampton urban environmental survey: an integrated geoscientific case study. British geological Survey Technical Report. **WE/95/49.**

EPS 3/HA/7 (1990) Compendium of waste leaching tests. Wastewater Technology Centre, Environment Canada, p68.

FALLMAN AM. & AURELL B. (1998) Leaching tests for environmental assessment of inorganic substances in wastes, Sweden. *The Science of the Total Environment,* **178**, pp.71-84

FORSTNER U. (1993) Metal speciation-general concepts and applications. *International Journal of Environmental Analytical Chemistry,* **51**, pp.5-23.

HILLIER S. (2000) Accurate quantitative analysis of clay and other minerals in sandstones by XRD: comparison of a Rietveld and a reference intensity ratio (RIR) method and the importance of sample preparation. *Clay Minerals,* 35, pp.291-302.

HOOKER PJ., TRICK JK., STRUTT MH. & FERGUSON AJ. (1998) Soil sampling: a BGS guide. Technical Report **WE/98/30R**, p13.

LEWIN K., BRADSHAW K., BLAKEY NC., TURRELL J., HENNINGS SM. & FLAVIN RJ. (1994) Leaching tests for assessment of contaminated land: *Interim NRA guidance.* **R&D Note 301**, p39.

TACK FMG., SINGH SP. & VERLOO MG. (1999) Leaching behaviour of Cd, Cu, Pb & Zn in surface soils derived from dredged sediments. *Environmental pollution,* **106**, pp.107-114.

VAN DER SLOOT HA. (1998) Quick techniques for evaluating the leaching properties of waste materials: their relation to decisions on utilization and disposal. *Trends in Analytical Chemistry,* **17**, pp.298-310.

WAHLSTROM M. (1996) Nordic recommendation for leaching tests for granular waste materials. *The Science of the Total Environment,* **178**, pp.95-102.

WHITEHEAD TH., ROBERTSON T., POCOCK RW., & DIXON ELL (1928) The country between Wolverhampton and Oakengates. Memoir of the Geological Survey of Great Britain, **Sheet 159.**

Rehabilitation of Contaminated Marine Sediments in Relation to Living Things

M. Fukue, Marine Science and Technology, Tokai University, Japan, Prof., Ph.D
Y. Sato, Ditto, Prof., Dr. Sci
M. Yanai, Graduate student, Tokai University, Japan
M. Nakamura, Ditto,
S. Yamasaki, Aoki Marine Co. Japan

ABSTRACT

This study proposes a rehabilitation technique of marine sediments including the purification of contaminated seawater and sediments. Since the polluted thickness of marine sediments is limited, dredging is one of the most effective techniques in order to clean up the polluted sea bottom. Marine sediments are also useful materials as clay liner in disposal site.

INTRODUCTION

In Japan, some coastal and bay sediments are seriously polluted with heavy metals and other hazardous substances (Fukue et al, 1999, Ohtsuo, 1999). Therefore, it was considered that fishes and shellfishes found were malformed by these substances. A protected animal, horseshoe crab survived for 200 million years is on the verge of extinction in Kasaoka bay, Japan. Now, the Kasaoka bay remains as a water channel, because of reclamation. At present, horseshoe crab has been rarely found in Kasaoka bay. One of the reasons may be that the polluted sediments and seawater in some parts of Kasaoka bay damaged the horseshoe crabs and their embryos (Itow, 1994).

The contamination of sediments causes bio-concentration from sea bottom. It was reported that hazardous substances were more accumulated into man who eats many fishes and shellfishes. For example, man is taking 60 to 70 % of one day intake from fishes and shellfishes (Makiya, 1997). Other hazardous and toxic substances are also involved in the food chain. Under this situation, it is necessary to cut the chains. A root of food chains may be sea bottom. The rehabilitation of sea bottom has to be achieved not for living things there, but also for us.

Therefore, this study investigates the degree of pollution and proposes improvement technique of the sea bottom. For a case study, Kasaoka bay and the surrounding area were selected.

SITE LOCATIONS AND SEDIMENTS

A case study was carried out in Seto Inland Sea in Japan, which is surrounded by Honshu, Kyushu and Shikoku islands of Japan, as shown in Figure.1. The core samples of sediments were obtained from Kasaoka bay, and off the coast of Fukuyama City and Kasaoka City, as shown in Figure. 1. The sites are distinguished as "Bay", "FOff" and "KOff", which are used to describe the distinct site.

The core samples were obtained with a box core sampler. The length of the core was 1.5 meter. The sampling positions are also indicated in Figure. 1.

Geoenvironmental impact management, Thomas Telford, London, 2001.

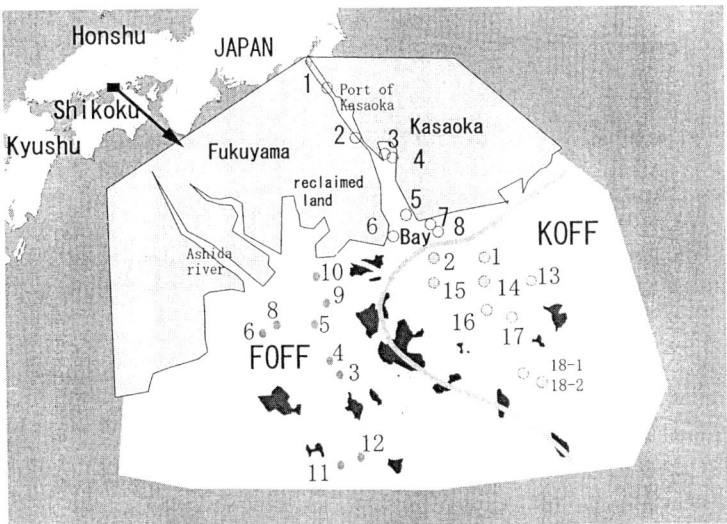

Figure 1   Site locations for sampling.

CONTAMINATION
TBT and TPT

It is understood that organotin compounds are extremely harmful for a creature. In the ocean, organic tin was dissolved from the paint that was used to ships. At present, organic tin is still dissolving from foreign ships yet, although production is stopped in Japan.

TBT (Tributyltin) is one of the dangerous substances in marine environments. Many researchers pointed out that a very low concentration of TBT causes an abnormal generative organ of a kind of spiral shell. For example, female shell has a penis. TPT (Triphenyltin) is also toxic substance.

The concentrations of TBT and TPT in sediments are shown in Figure. 2 and 3, respectively. The figures show that the concentrations are relatively high near the bottom surfaces. The apparent background of TBT is approximately 0.3 μ g/kg and the maximum concentration may exceed 100 μ g/kg for the top sediments in site Bay 1. On the other hand, the apparent background of TPT is 0.1 μ g/kg. The maximum concentration also exceeds 100 μ g/kg at the surface.

The thickness of contaminated sediments with TBT and TPT is around 50 cm. It seems that the contaminated thickness decreases with distance from the coast for offshore sediments. Therefore, as far as TBT and TPT are concerned, it is not difficult to clean up the sea bottom by dredging of surface sediments. This is discussed later.

Heavy metals and others

The vertical and horizontal distribution of heavy metal and other substances are obtained. Examples of Pb concentaration are shown in Figure.4. Figure 4 indicates that with lead, Kasaoka bay is more polluted than the offshore sediments.

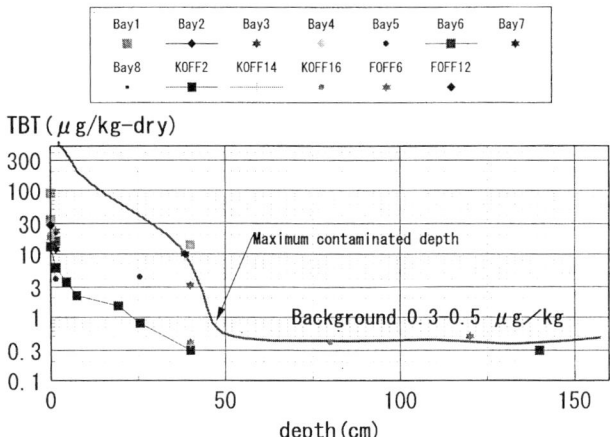

Figure 2   Concentration of TBT in sediments.

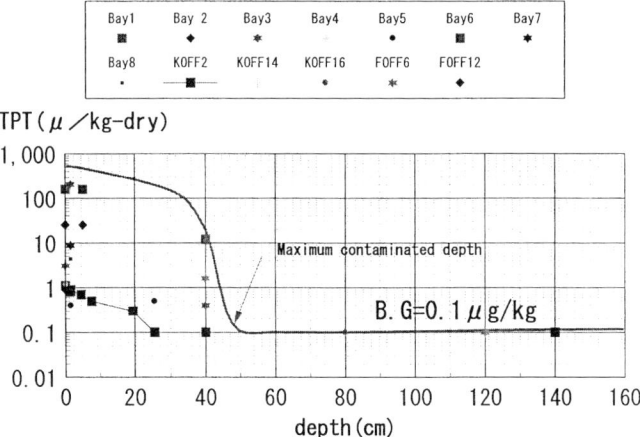

Figure 3   Concentration of TPT in sediments

From Figure 4, the apparent background of Pb for offshore sediments seems to be 20 ppm. Sediments obtained from Kasaoka Bay (Bay 1) and near Fukuyama port (FOff 8) show a relatively high concentration of lead.

Typical background values of heavy metals for various marine sediments are shown in Table 1. The degree of contamination of the sediments shown in Figures 4 and 5 can be evaluated from Table 1. The range shown in Table 1 depends on the number of data.

The concentration profiles of copper are shown in Figure 5. The background value of copper seems to be dependent on sediment type. For example, Bay 3 sediments contain much sand fractions and its background is relatively low, i.e., approximately 10 ppm. With copper, off the coast of Fukuyama is more polluted than Kasaoka bay.

Table 1 Typical background values for marine sediments of Japan.

element	Hg	Cd	As	Pb	Cu
units	mg/kg	mg/kg	mg/kg	mg/kg	mg/kg
Background value	0.05	0.1	10	20	7 - 53
elements	V	Zn	Cr	P	Ti
units	mg/kg	mg/kg	mg/kg	g/kg	g/kg
Background value	45-100	50-150	60	0.23-0.65	1.9-4.2

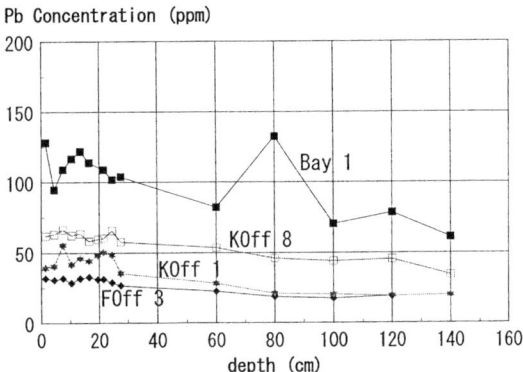

Figure 4   Concentration of lead for various site locations.

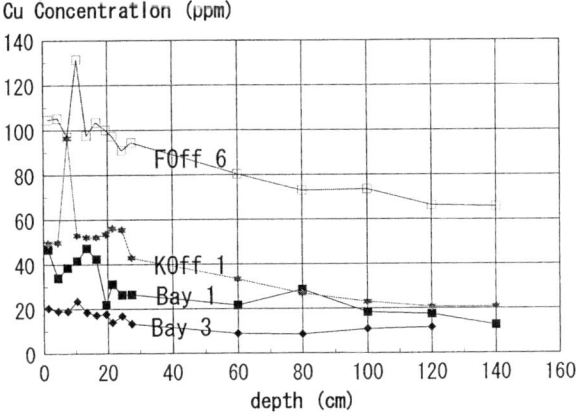

Figure 5   Concentration of copper for various site locations

PROPOSAL OF REMEDIATION

Dredging and disposal

The polluted sea bottom has to be remediated because toxic substances are taken by living things. It may be first step that benthos takes contaminated organic and inorganic particles. For second step, the benthos is taken by larger living things, such as fish, horseshoe crab, crab, etc. This process is explained as a food chain and also bio-concentration.

Calculation from data shows that Japanese are taking a dioxin of approximately 2.5 pg/kg/day from fishes and shellfishes (Makiya, 1997). This amount is more than one half of the total daily intake. It is no doubt that all kind of hazardous substances are also taken from sea products. Herein, rehabilitation of sea bottom is of importance and urgent.

To avoid this peril, it is necessary to break the contacts between living things and contaminated sediments. In order to achieve, the dredging of contaminated sediments is effective. For this purpose, dredging can be achieved by some ways, e.g., grab dredging and pumping up of sediments with seawater. The pumping method requires a treatment of a large amount of muddy seawater.

Dredged sediments can be disposed into a closed sea area, as shown in Figure 6. This is because of no available land in Japan. For disposal sites, two conditions are considered, i.e., wet and dry sites. Dredged sediments can be directly dumped or poured into wet site, as shown in Figure 6, a). In this case, muddy seawater in the closed area has to be filtered to discharge or needs to be treated. On the other hand, solidification and compaction are needed for dry site disposal, as shown in Figure 6, b). Dry site is prepared by pumping out of seawater. Requirement for dredges is "no pollution" by the works.

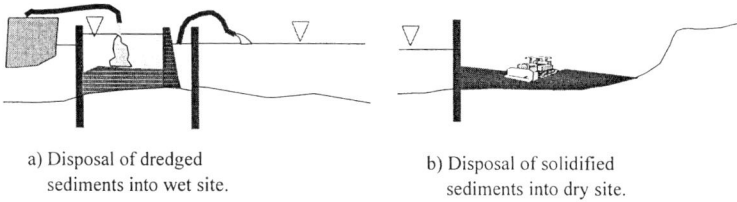

a) Disposal of dredged
sediments into wet site.

b) Disposal of solidified
sediments into dry site.

Figure 6 Closed sites for disposal of dredged contaminated sediments.

Leaching of hazardous substances from marine sediments is very small because sediments have been washed out during deposition (Fukue *et al*, 2001). Thus, the substances remained in sediments may be strongly adsorbed on particles.

However, the change in physical and chemical conditions of sediments may allow for the particles to release hazardous substances adsorbed on them. Therefore, it is better to mix a proper adsorbent with sediments, if sediments contain a high concentration of hazardous substances. The adsorbents are mixed with solidified sediments. A solidification technique of sediments was reported in the literature (Yamasaki *et al*, 1995). Dredged Sediments can be solidified by compression with filter pressers. The solidified sediments or mixtures are compacted into a dry disposal site, as shown in Figure 6 b).

Final goal

The final goal of this study is to create environments of natural purification due to many kinds of living things, such as bacteria, marine algae, benthos, etc. The procedure is shown in Figure 7. The dredging of contaminated sediments and purification of seawater are major works for remediation of marine environments. Purification of seawater is required for some kind of marine algae, e.g. *zostera marina*. Clear seawater is needed for their photosynthesis. To promote natural purification, benthos, marine algae and other living things, such as shellfishes and crabs should be stocked into the marine environments. These living things have a great degradation.

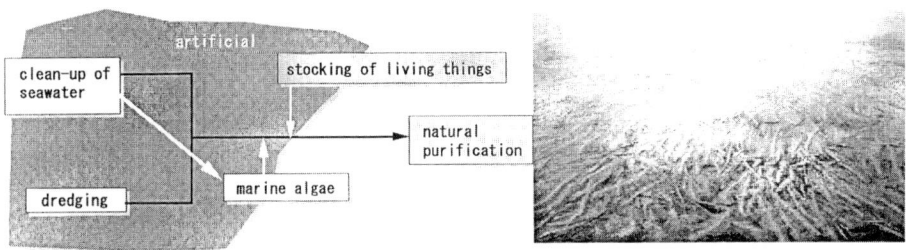

a) flow of artificial and natural purification   b) marine algae

Figure 7   Requirements for recovery of natural purification.

CONCLUDING REMARKS

In order to avoid the bio-concentration of hazardous substance into marine living things, the separation between living things and contaminated sediments is proposed. The rehabilitation of sea bottom can be basically achieved by purification of seawater and dredging of contaminated sediments. The dredged sediments need to be disposed in proper ways. To recover natural purification, stocking of living things is needed.

REFERENCES

Fukue, M., Nakamura, T., Kato, Y. and Yamasaki, S. Degree of pollution for marine sediments, *Engineering Geology* 53, 131-137, 1999.

Fukue, M. Yanai, M., Takami, Y., Kuboshima, S. and Yamasaki, S. Containment, sorption, and desorption of heavy metals for dredged sediments, *Clay Science for Engineering*, Edited by Adachi, K. and Fukue, M., Balkema, pp.389-392, 2001.

Itow, T. The relationships between at mouths of rivers and living things, *Environmental System Research*, No.1, 13-28, 1994 (Text in Japanese).

Makiya, K. Current situation concerning government activities associated with dioxins, *Waste Management Research*, Vol.8, No.4, 279-288, 1997 (Text in Japanese).

Ohtsubo, M. Organotin compounds and their adsorption behabior on sediments, Clay Science 10, 519-539, 1999.

Yamasaki, S., Yasui, S. and Fukue, M., Development of Solidification Technique for dredged sediments, Dredging, Remediation, *Containment for Dredged Contaminated Sediments*, ASTM STP1293, pp.136-144, 1995.

Investigation of the Fundamental Mechanisms Affecting Creep Settlement of Restored Opencast Coal Mine Sites

A K GOODWIN[1], M A O'NEILL[1] & W F ANDERSON[2]
[1] School of Environment & Development, Sheffield Hallam University, UK
[2] Department of Civil & Structural Engineering, University of Sheffield, UK

INTRODUCTION
Increasing environmental and political pressures in the UK have frequently resulted in planning consents for opencast coal operations that demand the controlled restoration of sites on completion of mining. End uses for restored sites can include highways, airports, housing and industrial developments (e.g. Goodwin & Holden, 1993), but the regeneration potential is hampered by uncertainties in the magnitude and timing of post-restoration settlements. These movements are generally divided into relatively short term but often very large inundation strains, and long term creep which, though less rapid, often varies significantly across a site. The nature and magnitude of inundation strains have been discussed widely (e.g. Blanchfield & Anderson, 2000). In contrast, creep has received far less attention. This paper presents interim findings from current collaborative research into the creep behaviour of a compacted opencast backfill by Sheffield Hallam University and the University of Sheffield.

OPENCAST COAL MINING & RESTORATION SPECIFICATIONS THE IN UK
All UK coal is won from strata belonging to the Carboniferous Coal Measures, which typically comprise cyclical deposits of mudstones, siltstones and sandstones. The mudstones usually dominate the Coal Measures, which themselves are sometimes overlain by relatively small quantities of younger deposits such as glacial drift. This complexity of material is amplified by natural spatial variations within a typical opencast coal site, which in the UK may be up to 300 hectares in extent and over 100m deep (Scott Wilson Kirkpatrick, 1995).

Typical stripping ratios of overburden to coal are between 20 and 30, which results in large volumes of spoil being generated and which requires the use of very large plant. Tracked excavators with bucket capacities of $10m^3$ commonly excavate the overburden and load it into dumper trucks capable of carrying over 70 tonnes around the site (see Figure 1). Spoil excavated early in the life of the mine would usually be stockpiled on site, with later spoil either being stockpiled or more commonly being re-used directly as backfill to worked out areas. Stockpiles of now possibly weathered spoil would be re-used as backfill to meet operational requirements, which may be several years after excavation.

The operational necessities mean a large degree of material mixing occurs, which will affect the mechanics of creep. This can be mitigated but not obviated by on site selection of spoil in a controlled restoration zone, where fill is deposited in accordance with a specification. Industrial experience of detailing specifications for opencast restoration works has been accumulating in the UK for over 20 years, and this aspect has been discussed notably by

Charles et al (1998) and Hodgetts et al (1993). However, the discussions to date have largely been based on traditional specifications which presume essentially inert materials that may be classed as either cohesive or non-cohesive.

Figure 1. Large Scale Plant in Use   Figure 2. Freshly Excavated Mudstone

In practice, specification requirements can be difficult to apply rigorously and compacted layers tend to contain a mixture of mudstone, siltstone and sandstone. Though mudstone tends to dominate, the actual mix will vary spatially and with depth depending upon the geology and the operational practices on site. The application of a specification is complicated further by the fact that the dominant mudstones do not fall easily into the cohesive category. Freshly excavated mudstone appears granular (see Figure 2), but weathering reduces it to what classifies as a cohesive fill (Rainbow, 1987).

REVIEW OF PUBLISHED CREEP RECORDS
Figure 3 shows the range of recorded surface settlements at the Ketley Brook site (after Hodgetts et al, 1993) both on linear and log time scales. The type of movement history shown is typical of opencast backfills, and bears out the work of Sowers et al (1965) who found that creep generally follows a log-time relationship of the form:

$$S = \alpha \,(\log t_1 - \log t_2)$$

where S is the strain, expressed as settlement as a percentage of the depth of backfill, that occurred between times t_1 and t_2 after settlement began at time t_0, and α is the creep compression rate parameter. In practice, definition of t_0 is difficult as the lowest layers of fill will have been placed and started to settle before the upper layers are even placed. Sowers et al overcame this difficulty by proposing that t_0 be taken as the time when the fill placed had reached half its ultimate height.

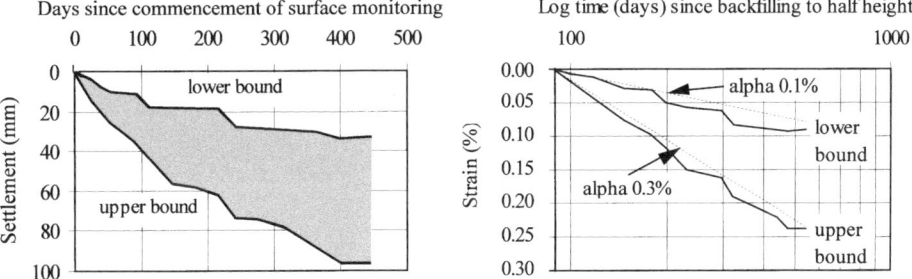

Figure 3. Settlement of Backfill at Ketley Brook Site (after Hodgetts et al, 1993)

The movement records for numerous sites were analysed by Scott Wilson Kirkpatrick as part of a state of the art review for the former British Coal Opencast. The range of α values determined are shown in Figure 4, and these accord with the average α value of 0.2 with a range of ±40% quoted by Hills & Denby (1996) for well compacted fill (usually taken to be fill with air voids less than 10 to 12%). The values vary spatially within and between sites, probably largely due to differences in fill and compaction procedures.

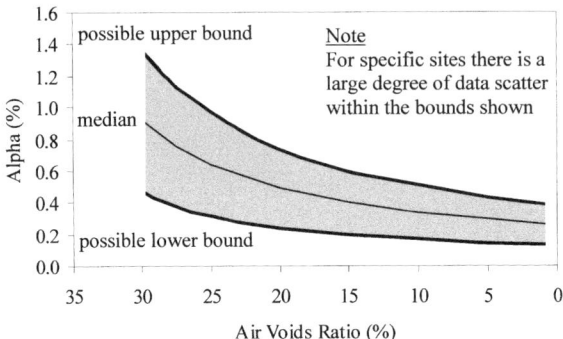

Figure 4. Measured Range of α Values (after Goodwin & Holden, 1993)

CREEP MECHANISMS
Creep settlement is defined as occurring under conditions of constant total stress and moisture content. Hills & Denby (1996) probably represent the consensus view of industry when they state that it results from the gradual re-arrangement of the material fragments due to the crushing of highly stressed contact points. However, there is no direct evidence for this view, and the actual mechanics in the mixed fill commonly produced on an opencast coal site may be far more complex than this. The creep mechanisms that could be at work in an opencast backfill include:

- Variations of particle stiffness and strength with time due to weathering, softening, slaking, and expansion (Bally, 1988)
- Crushing of particle contacts (Sowers et al, 1965)
- Particle splitting and breakage (Marsal, 1973)
- Re-arrangement and / rotation of particles (Hills & Denby, 1996)
- Asperity indentation at contacts and inter-particle ploughing (Scholz & Engelder, 1976)
- Variation of inter-particle friction due to local moisture content variations (Pigeon, 1969)

Other mechanisms, thought to be less important in opencast fills, include inter-particle lubrication by wetted clay and silt sized particles, and viscous yielding of particle contacts.

The relative importance of all these factors probably varies over time as any weathering of the backfill progresses, and as the principal stress transmission paths within the mass change. Moreover, the initial mineralogy of the fill, the relative proportions of mudstone, siltstone and sandstone, the grading of the fill, and the compactive effort, may all be critical factors.

RESEARCH PROGRAMME
A fundamental investigation to assess the relative importance of the differing creep mechanisms is underway. The material to be used is predominantly a grey mudstone sourced from a local opencast site. It is well graded with a maximum particle size of 100mm, has a natural moisture content of between 3% and 4%, and has a specific gravity of 2.73.

The proposed experimental work comprises:
- Investigation of the influence of compaction effort, moisture content and differing amounts of mudstone/siltstone, on particle strength and breakdown, and on fill structure
- Long term (2 years minimum) laboratory creep tests on 600mm diameter specimens of the fill using the apparatus shown in Figure 5
- Short term (4 to 6 months) compressibility tests on 230mm diameter specimens of the same fill but using a scalped grading (particles greater than 100mm removed)
- Short term compressibility tests on 230mm diameter specimens of fill selected to contain differing amounts of mudstone and siltstone
- Direct comparison of site data with laboratory results

The short term compressibility tests will be complemented by the use of x-ray computer tomographic imaging (CT) during the tests, which will allow quantification in three dimensions of particle interactions and changes in fill structure. Details of the apparatus and technique can be found in O'Neill et al (2001).

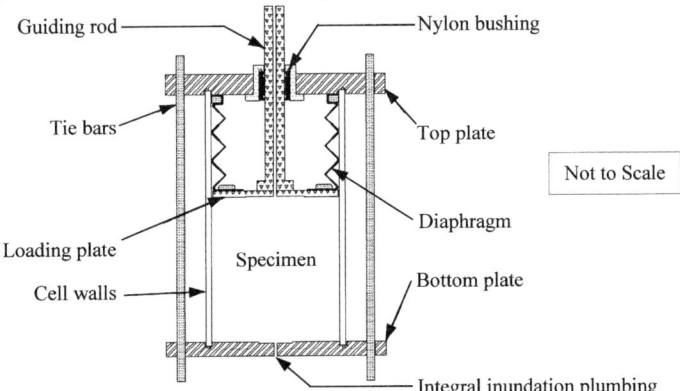

Figure 5. Large Scale Creep Compression Cell

INTERIM RESULTS

The laboratory investigation is in its early stages, and long term creep data is not yet available. Initial short term compression tests on mudstone fill have been completed, and CT scanning has been used to investigate the nature of particulate changes during loading from zero to $800kN/m^2$. Figures 6a and 6b show two 1mm thick CT sections through a 230mm diameter specimen which had been relatively lightly compacted at 2% moisture content to achieve a dry density of $1.78Mg/m^3$. Figures 6c and 6d show approximately the same sections after one dimensional vertical loading had caused 12.5mm compression (3.8% strain).

The images clearly show the larger particles, down to about 1mm nominal diameter, but finer particles appear as a grey mass that is difficult to resolve using the present CT scanner. The observed distribution of larger particles appears to be random, with greater concentrations in some areas than others. Similarly, the voids (black areas on the images) appear random in distribution, and some at least would appear to have formed during compaction, possibly due to arching between large particles above the voids. These observations are significant, as they support the widely held belief that stress is transferred non-uniformly through a particulate medium dependent on its structure. A probabilistic approach to the prediction of behaviour of such materials may present a better avenue for theoretical modelling than discrete physical models of individual particle interactions, and CT imaging is one method that may be used to quantify structural characterisitcs of fill for use in such models.

Interpretation of particulate interactions using two dimensional transverse CT images of the specimens (such as Figure 6) has proved to be difficult. Three dimensional computer reconstructions of the fill are being undertaken, but based on the 2D images alone it has been possible to draw interim conclusions on the particle interactions during settlement of the mudstone fill. The images indicate that the dominant mechanisms at work are local 'collapse' of the fill structure into voids left during compaction, relative inter-particle sliding, and rotation of particles. These mechanisms appear to have been facilitated in part by the presence of some large local voids, but the mere presence of a large void does not mean movements will occur local to that void. For example, comparison of Figures 6a and 6c shows that whilst some voids have changed in shape and size (e.g. voids below and below left of centre) others have not (e.g. above centre by cell wall). Similarly the lower left quadrant indicates that particles have rearranged and moved closer together, whilst in the lower right quadrant the fill structure is virtually unchanged. Overall, the findings support the view that stress transmission and settlements occur non-uniformly within a particulate fill depending on the local structure and its relation to the global structure.

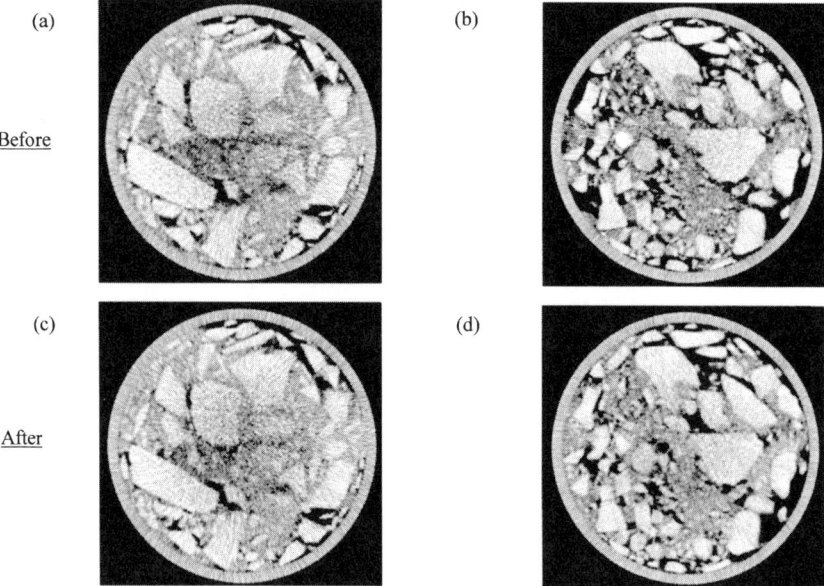

(a)

Before

(b)

(c)

After

(d)

Figure 6. CT Images of Specimen Before and After Compression

Splitting of mudstone particles is discernible on some sections (compare the large particle in the lower left quadrant in Figures 6b and 6d), but this appears at present to be less prevalent and less significant than the other mechanisms. This is despite the expected stress concentrations associated with non-uniform stress transmission. One possible explanation is that local aggregations of small and large particles may be forming that act as a flexible whole to distribute stresses within them to reduce crushing and splitting. This is supported by the occurrence of breakage mainly in sections where voids are relatively more frequent, and where load sharing is less possible

No evidence has been seen thus far of viscous yielding, nor of asperity indentation, but longitudinal sections are being used to assess these aspects further. Similarly, the work has not advanced far enough to comment on the influence of time dependent variations of particle stiffness and strength, nor moisture effects. As the opencast backfill will weather insitu, it is thought likely that the latter factors will have a significant impact on the rate of creep.

CONCLUSIONS

There has been some discussion of possible creep mechanisms in the past, but it has not been possible before to directly to determine their relative importance. A programme of research is underway to rectify this shortcoming, that is making use of both long-term creep tests in a controlled environment and short term compressibility tests. These tests are being backed up and greatly enhanced through the use of computer tomographic imaging to monitor particulate interactions in three dimensions. This relatively new technique in geotechnical engineering offers great potential for further discoveries.

Preliminary research findings have been reported that need further investigation before their general applicability can be assessed. In particular, longitudinal sections and 3D computer reconstructions of the fill are being used to overcome the difficulties of interpreting transverse 2D images. However, the indications from the work to date contradict to some degree consensus opinion that particle crushing is a major mechanism. Rather, local collapse into small voids left by compaction, and relative sliding and rotation of particles, seem to be the dominant factors affecting compression and creep of opencast fills. Particle strength and shape have not been found to be significant, but work is continuing on this and other aspects.

REFERENCES

1. Bally R J (1988), 'Some specific problems of wetted loessial soils in civil engineering', Engineering Geology, **25**:303-324
2. Blanchfield R & Anderson W F (2000), 'Wetting collapse in opencast coalmine backfill', Proceedings of the Institution of Civil Engineers, Geotechnical Engineering, **147**:139-149
3. Charles J A, Skinner H D & Watts K S (1998), 'The specification of fills to support buildings on shallow foundations: the 95% fixation', Ground Engineering, **31**:29-33
4. Goodwin A K & Holden J M W (1993), 'Performance of an engineered fill at Lounge opencast coal site', Proceedings of a Conference on Engineered Fills, Newcastle,413–428
5. Hills C W W & Denby D (1996), 'The prediction of opencast backfill settlement', Proceedings of the Institution of Civil Engineers, Geotechnical Engineering, **119**:167 176
6. Hodgetts S J, Holden J M W, Morgan C S & Adams J N (1993), 'Specifications for and performance of compacted opencast backfills', Proceedings of a Conference on Engineered Fills, Newcastle, 262–280
7. Marsal R J (1973), 'Mechanical properties of rockfill', from "Embankment Dam Engineering - Casagrande Volume", John Wiley & Sons, New York
8. O'Neill M A, Goodwin A K & Anderson W F (2001), 'The use of CT to investigate particle interactions in granular soils under load', 11th Conference of European Union of Geosciences, Speciality Symposium on Applications of Computerised X-Ray Tomography in Geoscience and Related Domains, Strasbourg, France, April 2001,
9. Pigeon Y (1969), 'The compressibility of rockfill', PhD thesis, Imperial College, London
10. Scholz C H & Engelder J T (1976), 'The role of asperity indentation and ploughing in rock friction', International Journal of Rock Mechanics & Mining Sciences, **13**:149-154
11. Scott Wilson Kirkpatrick (1995), 'State of the art review of the compaction of opencast backfill', Report for British Coal Opencast
12. Sowers G F, Williams R C, & Wallace T S (1965), 'Compressibility of broken rock and the settlement of rockfills', Proceedings of the 6th International Conference on Soil Mechanics & Foundation Engineering, Montreal, **2**:561-565

Remediation of Soil with Surfactants in the Form of Foam and Liquid Solutions

CATHERINE N. MULLIGAN AND FARZAD EFTEKHARI
Department of Building, Civil and Environmental Engineering, Concordia University, 1455 de Maisonneuve Blvd. W., Montreal, Canada H3G 1M8

ABSTRACT

An investigation was made into evaluating the capability of surfactants in the form of foam for removing contaminants from the soil. Several surfactants were investigated for their ability to make foam. Two of them, Triton X100 and JBR425 (a rhamnolipid biosurfactant), generated foam with higher quality (99%) and higher stability compared to other surfactants. Triton X-100 and JBR425 were then used to investigate the removal efficiency in soils contaminated with pentachlorophenol (PCP). TritonX100 showed better results in terms of final removal efficiency. TritonX100 (1%) removed 85 and 84 % of PCP from fine sand soil and sandy-silt contaminated with 1000 mg/kg PCP, while these values were 60 and 61 % for JBR425 (1%). The results of this study on a sandy and sandy-silt media found that the foam can be used as a fluid to enhance soil remediation under low pressures compared to other fluids such as liquid surfactant solutions.

INTRODUCTION

Excavation of contaminated soil was once the solution for soil remediation. However because of the high cost of excavation and final disposal of landfills, in addition to lack of available landfill sites, these disposal methods are becoming increasingly less popular (Mann et al., 1993). To decrease costs, various technologies are being developed and implemented for remediation of soils and sediments. In situ treatment of soil is preferable since they are more cost-effective than ex situ processes. However, there are many difficulties with in situ processes.

Two important characteristics of organic contaminants that are the most common among soil pollutants are low solubility in water and high interfacial tension with water. These decrease the efficiency of in situ soil remediation, during flushing and bioremediation processes. To overcome this problem, different surfactants can be used to enhance the rate of remediation. Surfactants enhance organic contaminant recovery through two mechanisms. First, surfactants reduce the interfacial tension between water and contaminants that slows the mobility of the organic components. Therefore surfactants are able to transfer the hydrophobic organic compounds (HOCs) to the mobile phase (Cheah et al., 1998). Secondly, surfactants are capable of forming aggregates known as micelles, thus solubilizing HOCs.

Geoenvironmental impact management, Thomas Telford, London, 2001.

Although surfactants have been effective in removing contaminants, large quantities of these chemicals are required (Mann et al., 1993). One way to reduce usage is to generate foams in the surfactant solutions. If the rate of contaminant removal can be enhanced by foams then the use of surfactants in soil remediation processes can be reduced. Another advantage of using foam in soil treatment processes is that high volumes of air per unit volume of foam are injected into the soil. The foam contains 70-90% air that could enhance remediation of soils contaminated with volatile and semi-volatile compounds by volatilization and by aerobic bioremediation.

Two important characteristics of foams are quality and stability. Quality of foam is used as an indication for the amount of gas in the foam. By definition quality of foam is the ratio of total gas volume per total volume of the foam. Stability is the ability of the foam to resist bubbles breakdown and it is indicated by measuring the time required for half of the liquid in the foam to drain out (Chowdiah et al., 1997). Because foam tends to flow through the soil in a plug flow manner, it penetrates uniformly with reduced channeling effects. This implies that the injected foam is uniformly distributed through the soil, thereby enhancing removal. Due to these potential benefits, we investigated the injection of foam by two different surfactants, a synthetic and a biological surfactant, into soil contaminated with pentachlorophenol (PCP), a toxic, suspected carcinogenic compound used frequently in Canada for the preservation of wood.

MATERIALS AND METHODS
Soil
The soils (purchased from DAUBOIS Inc, Montreal, Quebec, and categorized as fine sand and silty-sandy soil according to the ASTM classification) were used to pack the test column (3.5cm diameter and 15cm long). Both types of soil were washed by water several times and dried in the oven over night before using. The column sample was packed in four layers and each layer was compacted by hitting the column with a wood bar on four different sides. Some characteristics of the soils used in this investigation are summarized in Table 1.

Table 1. Characteristics of two samples used in experiments

Type of Soil	Classification (ASTM)	Hydraulic Conductivity (cm/s)	Specific Gravity
Soil (I)	Fine sand	7×10^{-2}	2.547
Soil (II)	Sandy-silt	6.3×10^{-4}	2.641

Surfactants
Triton X100 ($C_8H_{17}C_6H_4$ (OC_2H_4)$_{10}$ OH, molecular weight (MW)= 646.87g, critical micelle concentration (CMC=1.8×10^{-4} mg/l) and JBR425 ($C_{26}H_{48}O_9$, M.W.= 504g) were chosen among ten different commercial surfactants because they showed better foam-ability compared to the others. Triton X100 is a nonionic surfactant that is obtained by the reaction of octylphenol with ethylene oxide (Union Carbide Chemicals and Plastic Technology Corporation, 1992). It was purchased from SIGMA Chemical Co. JBR425 is an anionic surfactant obtained from Jeneil Biosurfactant.Co. In each trial, surfactant solution was stirred for approximately ten minutes before using.

Experimental setup

Foam was generated by passing the surfactant solutions (1%) and air at the same time through a porous stone. A glass filter was placed on the porous stone to provide uniform distribution of solution/air over the porous stone. The experimental setup is shown schematically in Figure 1. Flow meters A and B control the flow of solution and air, respectively, before injection into the system. The solution is pumped by pump (1) and air tubing is connected to a compressed air system in the lab. The foam column (2.5cm inside diameter and 30cm long) was used to monitor the generation of foam. A valve was installed between foam generator column and soil column to conduct the foam out of the system before its quality becomes constant and uniform, was also used to take foam samples for quality and stability tests. A pressure gage was installed just before the soil column to measure pressure build up in the soil column.

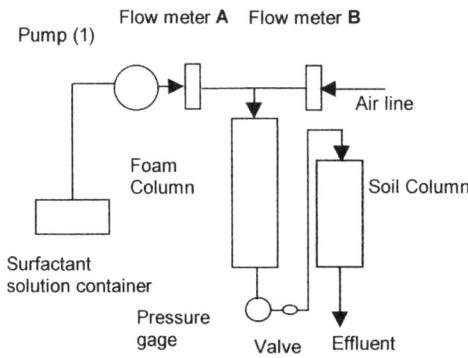

Figure 1 Experimental setup

Experimental procedure

This research consisted of three stages:

Stage 1: Solutions of surfactants (1%) (solutions stirred for 10 min) were prepared and foam generated from each was studied for quality and stability. Fifty ml samples were taken and collapsed. The time required for collapsing half of the foam was used as an indication of stability and the total liquid obtained from the collapsed foam per total volume of the sample (50ml) calculated which indicates the quality of the sample.

Stage 2: Because pressure gradient in soil is one of the most important factors in using foams for soil remediation, the effect of foam quality and foam flow rate on the pressure build up in the soil were investigated in this stage. Solutions of Triton X- 100 (1%) were used to make foam and then foam was injected into the soil sample (packed in four layers) at different flow rates (ranging between 15 ml/min to 35 ml/min) and qualities (ranging between 85% to 98%). In all the experiments, foam flowed through the sand until steady conditions were achieved in terms of exiting foam, before the pressure gradients were measured. It usually took several hours, during which period pressure gradient increased over time.

Stage 3: Both types of soil were then contaminated with PCP. Methanol was used as a solvent for PCP, and a solution of methanol and PCP was added to the soil. 0.5g of PCP was dissolved in methanol and added to 500g (1000mg of PCP/kg of soil) of washed soil and left over night to evaporate the methanol. The pore volume of the compacted column was calculated by measuring the volume of water needed to pump into the column to saturate it (at a pressure gradient of close to zero).

The test column was then saturated and samples of effluent were taken every pore volume passed through the soil column. As effluent from the contaminated soil column was foam also, samples were collected in large covered containers to collapse the foam before transferring to permanent containers. Samples were taken from the soils at the beginning and end of the experiments and PCP was extracted from these soil samples by methanol to measure actual concentration of PCP before and after the experiment.

Sampling and Analysis Procedure
Foam exiting the contaminated column was collected as samples to be analyzed for PCP concentration. Samples were taken for each pore volume passed through the contaminated soil. Foam samples were left for 24 hours to collapse and then transferred to smaller vials.

An ultra violet (UV) method was used to analyze the PCP concentration in the samples. A wavelength of 304nm was chosen to determine the concentration of PCP in samples. Each effluent sample (with known volume) was dried overnight and extracted with methanol (same volume of dried sample) and then absorbance was used to calculate the concentration of PCP. At the end of each test, samples of soil were taken from the column.. Soil samples were weighed and methanol was used as the solvent to extract PCP from soil samples before analysis of PCP concentration.

RESULTS AND DISCUSSION
Foam Characterization Experiments
Characterization of the foam showed that Triton X100 and JBR425 both can be used to generate foam with high quality and stability. Both Triton X100 and JBR425 could form foam with qualities of 99% and stabilities of 6.1 min.

Pressure Buildup in the Soil
The next step was to study the relationship between foam flow rate and pressure in the soil. The pressure was determined for foam generated with 1% solutions of Triton X100 and JBR425, with the type I soil (fine sand) at different flow rates. Pressure in the soil increased from 40 kPa/m to 100 kPa/m as the foam flow rate was increased from 5 ml/min to 35 ml/min. Pressure resulted from the injection of foam generated with these two surfactants, showed the same increasing trend with respect to the foam flow rate, although JBR425 showed slightly lower pressures for all flow rates.

According to the literature (Chowdiah, *et al.*, 1997), pressure should be less than 70 kPa/m to prevent related problems regarding to high pressure. It should be noted that this pressure is different for soils with different characteristics. Therefore, foam flow rates must not exceed more than 10-15 ml/min.

Results regarding the effects of foam quality on the pressure in the soil were obtained. They show that the pressure in the soil drops when the foam quality increases, and also indicate higher pressure build up in the soil for higher foam flow rates. For example, for a foam flow rate equal to 25 ml/min and a foam quality of 92 percent, the pressure was 108 kPa/m. After increasing the quality to 96 percent, this pressure drops to around 68 kPa/m. It is important to note that by increasing the foam quality from 92 to 96 percent, usage of surfactant decreases by an order of five which results in lower consumption of surfactants.

Removal Efficiency

Experiments were then performed to determine the amount of PCP removed by injection of foams of Triton X100 and JBR425. Results obtained from analyzing effluent samples are summarized in Figure 2. Maximum concentration of PCP in the effluent was observed after 6 to 8 pore volumes for fine sand which was about 250 mg/L. The PCP concentration in the effluent was negligible after around 20 pore volumes. For sandy-silt soil, the maximum concentration in the effluent was 200 mg/L.

Flushing of foam shows better results in the sandy soil than for the sandy-silt soil with a lower hydraulic conductivity. It can be seen that sandy-silt soil needs more pore volumes of washing agent to reach the final removal. In Figure 2A, we can see that the final concentration in the effluent foam increases faster in fine sand soil than sandy-silt soil. It seems this technique is more effective in soil with higher porosity. For example in Figure 2A, after five pore volumes, the concentration in the effluent is 100 mg/l for fine sand but only 70 mg/l for sandy-silt soil. The reason may be the time necessary for distribution of foam through the soil and the desorption process that is much longer for soil with a low hydraulic conductivity.

A

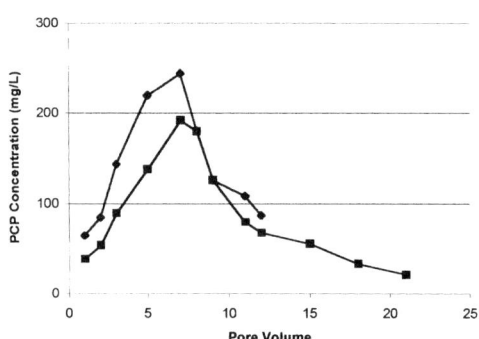

B

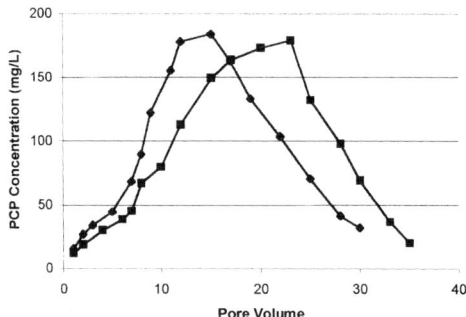

Figure 2. PCP removal by 1% Triton X100 (A) and 1% JBR425 (B) from fine sand (♦) and sandy-silty (■) soil contaminated with 1000 mg/kg.

Flushing contaminated soil experiments was continued with the second type of surfactant (JBR425), which is anionic in nature. A 1% solution was used and foam generated by this surfactant was injected into the contaminated soil. Results from these experiments are summarized in Figure 2B. As this graph shows, the maximum concentration in the effluent was achieved after 14 pore volumes for 1% JBR425 solution for fine sand and 24 pore volumes for sandy-silty soil. This is similar to that obtained for Triton X100 in terms of increasing rate of PCP concentration in the effluent. However, the maximum concentrations were not as high and it took more pore volumes than Triton X-100 to remove the PCP.

For all experiments, mass balance calculations were performed to evaluate the mechanisms of PCP removal. The results are summarized in Table 2. PCP solubilized in the effluent is indicated as mobilized. In all cases, the majority of PCP was removed by the air in the foam . For comparison, the same experimental set up was used in this stage except that the air flow was closed in all experiments. Surfactant solutions of Triton X100 solution were injected into the contaminated soil. Although, the PCP concentrations measured in the foam effluent and liquid effluent experiments were only slightly different, the overall PCP removal was much lower since PCP removal by volatilization in the control experiments was negligible.

Table 2 Mass Balance Calculation Results After Treatment with 1% surfactant foams

Type of Soil	Type of Surfactant Conc.	No. Pore Volume	Level of initial contamination (mg/kg)	Removal by Mobilization (%)	Removal by Volatilization (%)	Total Removal (%)
Fine Sand	Triton X100	12	1000	19	66	85
Sandy-Silt	Triton X100	21	1000	19	57	76
Fine Sand	JBR 425	19	1000	24	36	67
Sandy-Silt	JBR 425	33	1000	22	44	68

CONCLUSIONS
Based on the results of this study, foam-surfactant technology was found to have potential for enhanced remediation of PCP contaminated soils. Pressure gradient in the soil caused by injection of foam was studied and results showed low pressure build up. Since foam contains large amounts of air, it causes high rates of volatilization of volatile compounds and high removal rates. Further studies need to be completed before this technique can be used at large scale. For example, the efficiency of this method for removal of inorganic compounds from the soil, compatibility with bioremediation techniques, and effectiveness of pulsed operation need to be investigated before field study.

REFERENCES
Chowdiah, P., Misra, B.R, Kilbane, J.J., Srivastava, V. J, and Hayes, T. D., (1997), "Foam Propagation through soils for enhanced In-Situ remediation", Journal of Hazardous Materials, 62: 265-280.
Cheah, P. S, Reible, D., Valsaraj, K. T., Constant, D., Walsh, W. and Thibodeaux, L. J., (1998), "Simulation of Soil washing with Surfactants", Journal of Hazardous Materials, 59: 107-122.
Mann, M., Dahlstrom, J.D., Esposito, P., Everett, G., Peterson, G., Traver, R.P (1993), "Innovative Site Remediation Technology, Soil Washing/Soil Flushing", Anderson, W.C., American Academy of Engineers, Annapolis, MD.

The Role and Influence of Clay Fraction of Marly Soils on their Geotechnical and Geoenvironmental Performance

V. R. OUHADI, Asst. Prof., Bu Ali Sina University, IRAN, and R. N. YONG, Distinguished Research Professor, Geoenvironmental Research Centre, Cardiff University, U.K.

ABSTRACT:
This paper focuses attention on some basic aspects of the role and influence of the clay fraction in marly soils (palygorskite and sepiolite) on marl behaviour. The clay fraction of marl is most often dominated by palygorskite and/or sepiolite. It is the performance of these clay soil fractions that are the root cause of many of the problems encountered in the use of such soils as load bearing materials. In this paper some geotechnical and geoenvironmental aspects of marl behaviour are investigated, and a new definition and classification of marly soils is presented. It is concluded that the presence of palygorskite or sepiolite in marly soils will control the geoenvironmental and geotechnical behaviour of such soils.

INTRODUCTION

Marl's uniquely radical behaviour is known to change under dry and wet conditions. In spite of its very high strength in the dry state, its resistance will drop considerably (reportedly by as much as 85%) when saturated. It also exhibits collapsible, dispersive, swelling and low slake durability behaviour when exposed to water. Soil stabilization using cement and lime has often been used to overcome marl-related problems. However, with time, unexpected failures can occur as a result of the formation of expansive minerals called ettringite $\{Ca_6Al_2(SO_4)_3(OH)_{12}, 26H_2O\}$ and thaumasite $\{Ca_6[Si(OH)_6]_2.(CO_3)_2.(SO_4)_2.24H_2O$. The necessary conditions for ettringite formation are (a) the presence of a high pH environment; (b) the presence of soluble alumina ions; (c) the presence of clacium ions provided by stabilizing agents, and (d) the presence of sulfate in the soil pore fluid. Formation of thaumasite needs the presence of soluble silica ions instead of alumina ions (Hunter 1988, Mitchell and Dermatas 1992, Yong and Ouhadi 1997). This paper focuses on an evaluation of the role and performance of palygorskite $\{(Mg,Al)_5 (Si,Al)_8O_{20} (OH)_2 8 H_2O\}$ and sepiolite $\{Mg_4 Al_2 Si_{10} O_{27} 15 H_2O)\}$ in respect to the stability or instability of marly soils. Both minerals are known to have similar open chain-like structures (Grim 1968).

MATERIALS AND METHODS

This study was performed using: (1) a natural unweathered marly soil, representing the marly soils of the southern area of Iran and the northern sector of the Persian Gulf in different states of weathering (this region is highly prone to both natural and artificial slope failures); (2) a pure palygorskite sample extracted from the natural marl; (3) a pure sepiolite sample obtained from Vallecas Spain; and (4) an illitic soil (Domtar sealbond) obtained from Domtar Construction

Materials (Ltd.). Various experiments were performed following procedures and methods given by different investigators (Sheldrick 1984, ASTM 1992, Moore and Reynolds 1989). Pore fluid chemistry analysis in conjunction with the x-ray diffraction method (XRD) was performed to study the chemical and mineralogical aspects of marly soils.

The size fractions of the unweathered marl sample used in this study classified as 55% clay, 41% silt and 4% sand, with compositional features indicating 25.7% carbonate, 0.8% amorphous and 0.6% organic (Ouhadi 1997). The initial soil pH was 8.7 with specific surface area of 74 m^2/g. The sulfate content of the pore fluid of soil was between 580-1300 mequiv kg^{-1}. The soil classified as CL/CH (Yong and Ouhadi 1997). XRD tests indicate that the clay fraction of marl was dominated by palygorskite. The results of double hydrometer testing showed that the percentage of soil dispersion was above 74%. The Sodium Absorption Ratio (SAR) in excess of 50 indicated that the marl was dispersive in performance characteristics (Sherard et al. 1976). As has been previously shown by Yong et al., (1979), the presence of potential determining ions, such as bicarbonate in the marly soil used in this study, not only contribute to the establishment of the surface charge, but also to the dispersive potential of the soil.

ROLE OF MARL COMPONENTS UPON ITS BEHAVIOUR

The Role of Palygorskite on Marl Performance
To investigate the role of palygorskite mineral upon marl stability, several experiments were performed on pure palygorskite extracted from the marly soils (Ouhadi 1997). The physico-chemical tests conducted included the following:

Cation Exchange Capacity – The cation exchange capacity of palygorskite is reported to be in the order of 9-30 meq/100g soil, with a SSA (specific surface area) of about 72 m^2/g using water as the coating agent (Van Olphen and Fripiat 1979). The results from this study using EGME gave a SSA of 192 m^2/g. Using barium chloride as an extractable solution, sodium was determined to be the primary exchangeable cation with an exchangeable sodium percentage (ESP) ratio of 51%. From previous studies, Yong et al. 1996 have shown that soils with ESP greater than 2% are susceptible to spontaneous dispersion in water, and can behave as dispersive clays. Using this as an indicator, it can be reasonably assumed that the presence of palygorskite in the marly soils will render the soil dispersive.

The percentage of swelling for marl soil was determined to be 9.9-10.4% , as shown in Figure 1. It is expected that the swelling is mainly controlled by palygorskite, since it contributes more than 30% of marl mineral composition. Palygorskite was also observed to be a high swelling soil, as shown by the test results in Figure 1. The swelling measurements for the palygorskite obtained using an oedometer showed that after four days, 32% free swelling was obtained.

Figure 2 shows the suction performance of the marly soil sample. This performance of the marly sample, in comparison with an illitic soil, indicates that not only is the water holding capacity of marl higher than the illitic soil, but also that the rate of water removal through pressure application is lower in the marl sample. For instance, from pF 0 to pF 3.14 the water content of the illitic soil decreased 44% while the marly sample decreased by only 11%. This performance can be attributed to the role of the open structure of the palygorskite fraction of marl.

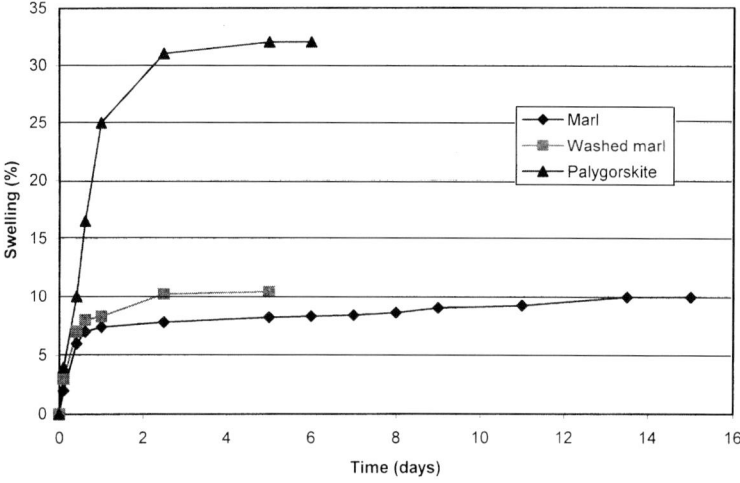

Figure 1. Free-swelling variation of marl and palygorskite samples

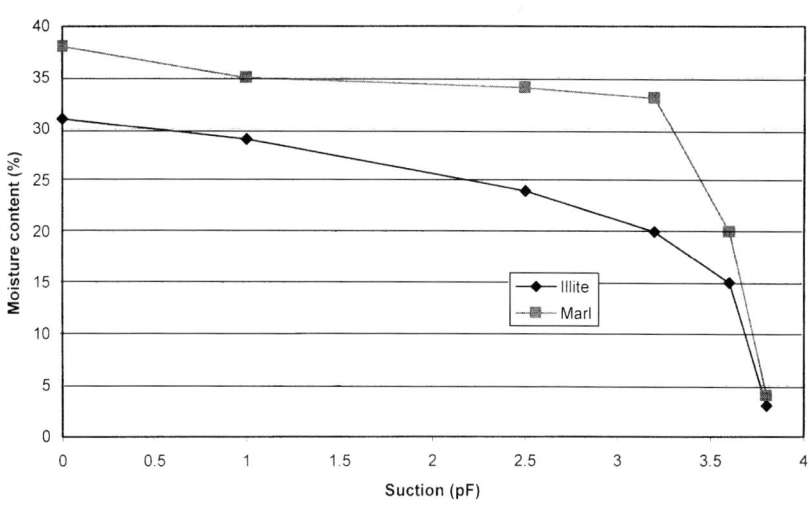

Figure 2. Fluid retention curves for marl and illite

Table 1 presents the results of digestion testing on pure palygorskite. This experiment was performed in the manner outlined by Sheldrick (1984). The Table shows the amount of different elements, including alumina ion, obtained after analysis of supernatant obtained following digestion. Since soluble alumina is known to be one of the main elements contributing to ettringite formation (Mehta et al. 1966, Yong and Ouhadi 1996), it is useful to note from the experimental results reported in Table 1 that the amount of releasable alumina in the palygorskite tested was 578 meq/100 g soil, a quantity which is considered to be quite significant for ettringite formation.

The Role of Sepiolite on Marl Performance

The Atterberg test results for the sepiolite showed a liquid limit of about 500%, a plastic limit of about 200% and a corresponding plasticity index of 300%. The activity coefficient of sepiolite was calculated to be 3. Since clays with activity coefficients > 3 classify as active clays, one can conclude that the sepiolite tested would be considered to be a very active clay. The maximum dry density and optimum water content of the sepiolite were obtained as 0.7 Mg/m^3 and 86% respectively. The CEC of sepiolite, using the barium chloride method, was determined to be 10 meq/100g soil, and the specific surface area obtained using the EGME method was 370 m^2/g. The high SSA and low CEC are typical characteristics for sepiolite and palygorskite -- in contrast with other clay minerals – reflect the open structure of these minerals.

The sepiolite tested showed that after four days, 32% free swelling was obtained. The results of the soil suction tests are given in Figure 3. The high water holding capacity of the sepiolite in contrast that of the illitic soil is evident. At pF 0, the water content of sepiolite is about one order of magnitude larger than the illitic sample. Table 1 also shows the results of the digestion experiment for sepiolite. As can be seen, while sepiolite has a very low amount of alumina in comparison to palygorskite and other clay minerals studied in this research, it has the highest silica ion concentration. Thus, it is not expected that sepiolite will directly affect ettringite development in post stabilization failure. However, it may contribute significantly to the formation of thaumasite.

TOTAL RELEASIBLE ALUMINA AND SILICA OF CLAY MINERALS

In Table 1, the results of digestion tests in terms of total released cations of different clay minerals. The results show that palygorskite releases the largest amount of alumina ions in comparison with the other clay minerals. This accords with the thesis that ettringite formation will likely occur when palygorskite is present in the clay fraction of soil, assuming all necessary conditions for ettringite formation are met.

The results also show that sepiolite releases more silica ions in comparison with the other clay minerals, corresponding to the suggestion that thaumasite formation would be likely when sepiolite is present in the clay fraction of soil, -- once again assuming all necessary conditions for thaumasite formation are met. The expected effects caused by the presence of other ions in the structure of palygorskite, sepiolite and marl on soil behaviour, have been previously addressed by Yong and Ouhadi (1997).

Table 1. The results of digestion testing on different samples

Soil samples	Cations monitored in digestion experiment (meq/100 g soil)					
	Na	K	Mg	Ca	Si	Al
Palygorskite	80±7	51±5	321±21	9±0.8	14±1	578±54
Sepiolite	2±0.2	10±1	1012±90	1±0.1	80±7	9±1
Marl	305±20	12±1	260±20	370±25	23±2	270±20
Illite	8±0.8	16±1	139±10	62±6	18±1	288±20
Kaolinite	5±0.8	0	0.3±0.1	0	16±1	88±7

NEW DEFINITION AND CLASSIFICATION OF MARLY SOILS

Conventional thinking considers marly soils as soils or rocks with 35-65% carbonate with some complementary content of clay (Barth et al. 1939, Pettijohn 1975). However, this criterion cannot be universally applied to many of the marls (rocks) in Britain, the Persian Gulf area, U.S. or other countries. As an example, most of the marls of the Keuper series contain less than 20% carbonate (Bells 1978), while Ontario and Quebec marls are reported to have more than 79% calcite (Guillet 1969). Such soils, according to the classification of clay-lime carbonate mixtures, are marly clays (Barth et al. 1939). It is interesting to note that the general criterion for marly soil classification does not take into account the type of clay fraction in marls. The literature review indicates that usually the clay fraction of marl is dominated by the palygorskite or sepiolite, and that this clay fraction governs marl behaviour. This is the case for the Persian Gulf marl, i.e. it is dominated by the presence of palygorskite (Kassler 1973, Yong et. al 1993, Ouhadi 1997).

In the post-stabilization failures reported in the United States, where lime was used as the stabilization agent, the clay minerals consisted primarily of sepiolite – with minor quantities of montmorilonite and kaolinite. Calcite, gypsum, thenardite and arcanite were also identified (Hunter 1988). The marl in the Virginia region is dominated by the presence of palygorskite. Palygorskite is also the dominant clay fraction in the Russian marl deposits (Ovcharenko 1969). Keuper marl on the other hand also includes sepiolite (Keeling 1956, Davis 1967).

Since palygorskite and sepiolite have chain-like structures, the zeolitic water in the open pores located within their structure can accommodate different ions. This is seen to provide the soils with the capability for ion sorption selectivity. The presence of palygorskite and sepiolite in marly soils can directly contribute to the formation of expansive mineral ettringite and/or thaumasite, respectively. The formation of these can in turn lead to post-stabilization failure of the soils. The presence of palygorskite or sepiolite in marly soils, in combination with carbonates makes the clay-carbonate mixture behave in a fashion quite distinct from other clay-carbonate soil mixtures. It therefore seems appropriate that the current definition of marl, which considers any combination of clay and carbonate (Barth et al. 1939), should be expanded to take into account the type of clay fraction of marl. This is illustrated by means of the triangular diagram shown in Figure 4. The expanded classification criterion has the following advantages: a) It considers the presence of palygorskite or sepiolite minerals a necessary condition for marly soils; b) The suggested classification can be applied to all marly soils classified by Barth et al. (1939), Fookes and Higginbottom (1975), and Pettijohn (1975), and c)By including the percentage of silt or sand in the classification scheme, the expected behaviour of marl in respect to swelling, dispersivity or collapsible performance can be obtained.

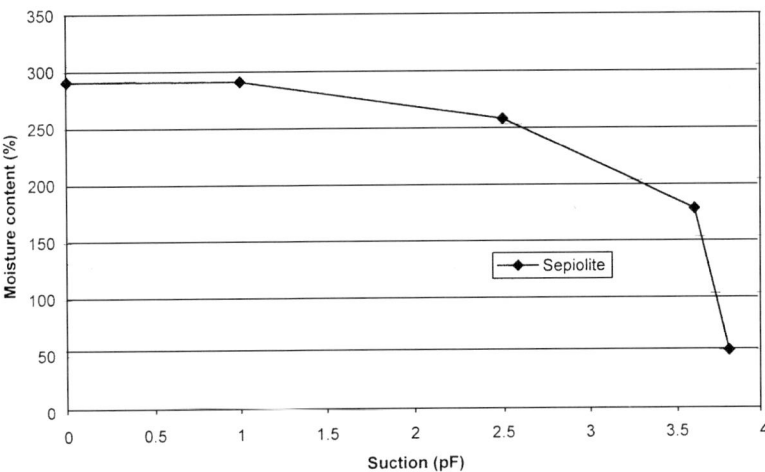

Figure 3. Fluid retention curve for sepiolite sample

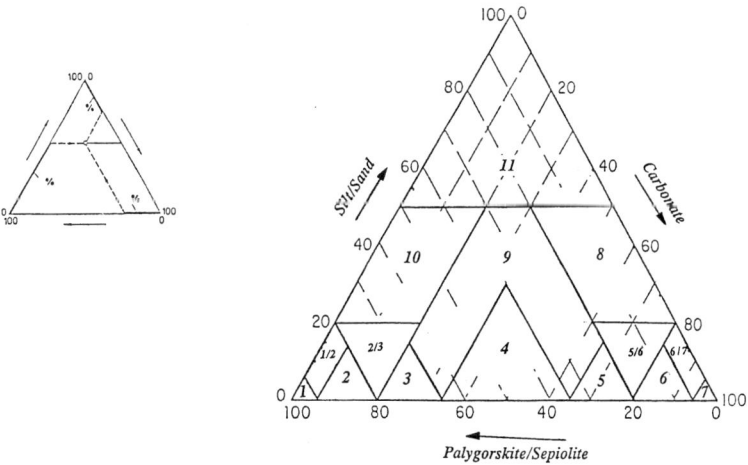

Figure 4. Proposed classification for marly soils

All seven possible combinations of marly soils suggested by Fookes and Higginbittom (1975) are plotted on the same triangular graph in Figure 4. These seven regions are identified as pure clay or mudstone, clay, clayey marl, marl, limey marl, marly limestone and pure limestone, respectively. The base of the triangle includes the suggested classification by Barth et al. (1939) in which the percentage of silt or sand is zero. By considering all combinations of soil fraction including clay, carbonate and silt/sand, all the above different marly soils can be considered in the areas numbered 1 to 7 in Figure 4, respectively. Some of the available information in the literature, including the results of the research conducted by Davis (1967), Hunter (1988), Ismael (1993), Yong et al. (1993), Ouhadi et al. (1993), L.F. Mohammed (1995), which include more than 30 samples, lie in the regions 4 to 9. In all these cases, silt and sand are present. Using the suggested classification scheme and the extensive study conducted by Rollins et al. (1994), regions 8 and 11 (marly silt or marly sand) demonstrate a high possibility of collapsible behaviour. Region No. 10 (silty marl or sandy marl) which has more than 30 percent clay is prone to exhibit swelling (Rollins et al. 1994). Region No. 9 and 4 behave as transition zones between collapsible and swelling performance.

CONCLUDING REMARKS

The significant roles and influence of palygorskite and sepiolite on marl behaviour have been addressed in this paper. Based upon the distinct behaviour patterns of palygorskite and sepiolite, and with the aid of an extensive literature review, a new classification for marly soils has been suggested. This classification scheme focuses attention on the presence of palygorskite or sepiolite as the major fraction of marl. The very noticeable pore fluid holding capacity of marl, which can accommodate sulfate in the zeolitic water of palygorskite/sepiolite and the high alumina content of palygorskite and high silica content of sepiolite, can contribute significantly to the post stabilization failure of marl when lime is used as a stabilizing agent.

REFERENCE

American Society for Testing and Materials, ASTM, 1992, "Annual Book of ASTM Standards", Philadelphia, V.4, 08.

Barth, T.F.W., Correns, C.W., and Eskola, P., 1939, "Die enstehung der gstiene", Sringer, Berlin.

Behnia, K., Ouhadi V.R. and Ghalandarzadeh, A., 1993, " The use of sea-water in stabilization of marly soils with cement and lime" Iranian Journal of Road Eng., No. 26, pp. 54-62.

Bells, F.G., 1978, "Foundation engineering in difficult ground", Billing and Sons Ltd. 598 pp.

Davis, A.G., 1967, "The mineralogy and phase equilibrium of Keuper marl", Q.J. Engng Geol. Vol. 1, pp. 25-38.

Fookes, P. G., and Higginbottom, I.E., 1975, "The classification and description of near-shore carbonate sediments for engineering purposes", Geotechnique,Vol. 25, No.2, pp. 406-411.

Grim, R.E., 1968, "Caly Mineralogy", McGraw Hill International Series in the Earth and Planetary Sciences, 596 pp.

Guillet, G.R., 1969, "Marl in Ontario", Ontario, Dept. of Mines, Industrial Mineral Report 28.

Hunter, D., 1988, "Lime induced heave in sulfate-bearing clay soils", Journal of geotechnical Eng., American Society of Civil Eng. Vol. 114, No.2, pp. B150-167.

Ismael, N.F., 1993, "Laboratory and field leaching tests on coastal salt-bearing soils", Journal of Geotechnical Eng. ASCE, Vol. 119, No. 3, pp. 453-470.

Kassler, P., 1973, "The structural and geomorphic evolution of the Persian Gulf", The Persian Gulf, 11-32, Berlin:Springer-Verlag.

Keeling, P.S., 1956, "Sepiolite at a locality of Keuper marl in the midlands", Min. Mag., Lond., 3, pp. 328-332.

Mehta, P.K., 1966, "Investigation on the hydration products in the system $4CaO.3Al_2O_3.SO_3$-$CaSO_4$-CaO-H_2O", Highway Research Board, Special Report, No. 90, pp. 328-352.

Mitchell, J.K., and Dermatas D., 1992, "Clay soil heave caused by lime-sulfate reactions", Innovations and uses for lime, ASTM STP 1135, D.D. Walker. Jr., T.B. Hardy, D.C. Hoffman, and D.D. Stanley, Eds., ASTM, Philadelphia, pp. 41-64.

Mohammed, L.F., 1995, "Assessment of saline soil stabilization via oil residue and its geo-environmental implications, Ph.D. Thesis, McGill University, Dept. of Civil Eng. and Applied Mechanics.

Moore, D.M., and Reynolds, R.C., 1989, "X-ray diffraction and identification and analysis of clay minerals", Oxford University Press, New York, 332 pp.

Ouhadi, V.R., and Ghalandarzadeh, A., and Behnia, K., 1993, "Engineering characteristics and properties of marly soils", Proceedings of the Second International Seminar on Soil Mechanics and Foundation Eng. of Iran, pp. 36-48.

Ouhadi, V.R., 1997, "The role of marl components and ettringite on the stability of stabilized marl", Ph.D. Thesis, McGill University, Montreal, Canada.

Ovcharenko, F.D., 1969, "The colloidal chemistry of palygorskite", Academy of sciences of the Ukrainian SSR, Institute of General and Inorganic Chemistry, pp. 101.

Pettijohn, F.J., 1975,"Sedimentary rocks", Third edition, Harper and Row, New York, 628 pp.

Rollin, KM, Rollin,R.L., Smith,TD, and Beckwith,G.H.,1994,"Identification and characterization of collapsible gravels", J. of Geotechnical Eng., Vol. 120, No. 3, pp. 528-542.

Sherard, J.L., Dunnigan, L.P., and Decker, R.S., 1976, "Identification and nature of dispersive soils", Jr. Geotechnical Eng. Div. ASCE, Vol. 102, No. GT4.

Van Olphen, H. and Fripiat, J.J., 1979, "Data handbook for clay materials and other non-metallic minerals", Pergamon Press, 346 pp.

Yong, R.N., Sethi, A,J., Ludwig, H.P., and Jorgensen, M.A., 1979, "Interparticle action and rheology of dispersive clays", Journal of the Geotechnical Engineering Division, Proceedings of the American Society of Civil Eng., Vol. 105, No. GT10,

Yong, R. N., Mohammed, L.F., Mohamed , A.M.O., 1993, "Retention and transport of oil residue in soil", ASTM Symp. On analysis of soils contaminated with petroleum constituents.

Yong, R.N., Ouhadi, V.R., 1997, "Reaction factors impacting on instability of bases on natural and lime-stabilized marls", Special Lecture, Keynote Paper, Proceeding of the International Conference on Foundation Failures, pp.87-97, Singapore, Edited by T.W.Hulme and Y.S. Lau.

Web-based support for a Land Regeneration Network

OWEN, D. H., MAHDI, T. A., and THOMAS, H. R.

Geoenvironmental Research Centre, Cardiff School of Engineering, Cardiff University, Queen's Buildings, PO Box 925, Newport Road, Cardiff, CF24 0YF, Wales.

ABSTRACT

The newly formed Land Regeneration Network provides a platform for contaminated land stakeholders in Wales. A web site has been established for the Land Regeneration Network, (www.grc.cf.ac.uk/lrn/) which seeks to aid the network in its core mission to improve effective management and reuse of land assets for all stakeholders in Wales. This paper describes these developments and highlights key features and benefits to contaminated land stakeholders.

INTRODUCTION

With the imminent introduction of the new legal framework for contaminated land in Wales, many stakeholders are likely to be faced with significant clean up costs for their polluted sites. This could affect the profitability of existing companies and deter others from investing in sites that may be contaminated. Consequently, in the later part of 1999, and in conjunction with the Welsh Development Agency, the Geoenvironmental Research Centre (GRC), based at the Cardiff School of Engineering, Cardiff University launched a new initiative for contaminated land stakeholders in Wales. This was aimed at developing greater communication both with and between local small to medium enterprises (SMEs) and stakeholders in contaminated land in order that they may discuss land regeneration issues in an open forum

Today, over one hundred and twenty companies involving over three hundred and twenty individuals within Wales who have an interest in geoenvironmental and land regeneration issues makeup the network. Known as the Land Regeneration Network (LRN) it successfully brings together stakeholders in contaminated land in Wales through regular meetings, evening lectures and seminars.

The Geoenvironmental Research Centre continues to build and maintain the LRN and has recently established a large internet presence for the network (www.grc.cf.ac.uk/lrn/) providing a comprehensive array of information resources which has firmly established itself as a web portal for land regeneration network resources on the internet. This web-based support provided for the network now act as the LRN's main sources for the dissemination of advice and information related to the geoenvironmental area equally to both the specialist and general public. The web site provides the resources for an on-line internet based forum to disseminate state-of-the-art best practices on land regeneration, providing advice and information on regulatory issues, and a source of administrative resource for the network as a whole. This paper describes the design of the web site and highlights its key features and benefits to contaminated land stakeholders in Wales and to the geo-community around the world.

INTRODUCTION TO THE LRN

The Land Regeneration Network was conceived from the conclusions and recommendation of a Technology Audit carried out by the Geoenvironmental Research Centre. The Audit, initiated in April 1999 in liaison with the WDA, involved environmental companies dealing with contaminated land problems in ERDF objective 2b1 Industrial South Wales area. The audit was aimed at establishing the current state of technologies used and scope of work carried out by these companies whilst also identifying problems and barriers that could prevent development of this sector of environmental industry in the region.

From the Audit a number of conclusions were drawn. A definite need was identified for improved networking and interaction between all the stakeholders of contaminated land problems in Wales. Networking would provide opportunities for companies providing services in this sector to introduce themselves and for their Clients to know about them. Additionally greater links with other European and national networks and organisations were required to facilitate the flow of information on funds, regulations and news and events. Links with academia and research sectors also needed strengthening to encourage interactions between industry and research and to facilitate transfer of technology. Finally, the need for regular, informal meetings was identified to discuss regional issues and problems relevant to the contaminated land industry.

These core aspects came together in the very first meeting on Tuesday 7th December 1999, where strong support for the establishment of such a network was given. An advisory board for the LRN was subsequently formed with core representation from the Welsh Development Agency, the National Assembly of Wales, the Environment Agency, the Local Authority, the Countryside Commission for Wales, and the University sectors. Equal representation also exists from large and small consulting engineering companies, land owners, developers, contractors, and property agents within the area. To-date the LRN has grown to a membership of over 320 individual members from a core of approximately 120 local stakeholder organisations.

Networking persists as the core activity of the LRN helping to activate links between Wales and International, European, National and Regional Networks and organisations. These links are employed to improve communication between regional industry to aid other functions of the LRN. The network also provides information to its members about trade missions, exhibitions and conferences and plays an active role by organising such events and devising regular surveys to discover new possible markets. Information about available funds from different European, National and Regional sources can also be provided through the network.

Regulation and guidance on contaminated land problems are still developing and changing. The LRN therefore continually seeks to update its members of new legislation and guidance through its link with the regulators. It provides an opportunity for parties on the receiving end to express their concerns, views and to discuss the regulation and guidance with the decision-makers. The LRN also strives to inform its membership on a whole list of topics relevant to contaminated land industry. Consequently up to date information, including publications, current and potential projects in the region, national and regional events, competitions and awards, case studies, research results, data and reviews are able to be provided.

In the near future the network is also looking to initiate different kinds of studies and explore possible ways of funding in collaboration with industry when the need arises. One of the current issues undertaken by the LRN is the review of, and understanding of current risk assessment standards, procedures and models. The aim being to provide guidelines on the practicality and suitability of currently available methodologies to share and disseminate best practices techniques and to explore new methodologies for risk assessment which could become acceptable for the region and the UK as a whole. Training via workshops, seminars, short and long courses, and evening lectures are additional services that are continuously organised by the network. These courses, being heavily influenced by the potential users are tailored to the memberships needs through direct input from an industrial advisory board to ensure that current topics of concern or interest are continuously targeted.

Some of the events organised specifically by the network in the past year include:

- "The new legal framework for contaminated land", evening lecture in May 2000
- "Risk assessment of contaminated land", evening lecture in September 2000
- "Waste management licensing", evening lecture in January 2001
- "Risk perception as a driver for land regeneration", day seminar in March 2001

Finally, there is a direct link between the contaminated land industry and other related environmental industries, such as waste management, insurance and financial services. The LRN therefore continues to expand, to include these other sectors in order to facilitate interactions between them.

INFORMATION DISEMINATION
The use of the internet was quickly identified by the LRN as an effective mechanism which could act as one of the sources of advice and information provision outlined by the LRN's own objectives. Consequently, and as part of an ERDF program, the Geoenvironmental Research Centre at The Cardiff School of Engineering, started to establish a web site for the LRN. This web site, which can be found at http://www.grc.cf.ac.uk/lrn/ was first launched in October 2000 following six months of research and development.

The site goals were identified from the outset as:
- To facilitate greater awareness of the Land Regeneration Network (LRN);
- To clearly and easily demonstrate the achievements of the LRN;
- To inform the reader of the work of the LRN;
- To act as a notice board for current LRN news and events ;
- To act as one of the sources of advice and information of the LRN;
- To provide a general geoenvironmental information resource area which aids in the provision and dissemination of related geoenvironmental information;
- To provide the resources for an on-line web based forum to disseminate state-of-the-art best practices and information on land regeneration;
- To act as a source of administrative recourses for the network as a whole and its advisory board, - notices of events, meeting, for which minutes, documentation, and registration etc can all become web based;

However, most notably the aim of the new site was to disseminate information relating to the geoenvironmental field equally to both the specialist and the general public. It was also anticipated that the level of technological and internet experience of the typical LRN membership could vary extensively. Consequently the design of the site, in terms of ease of use and flexibility would be as critical to its success as the type and nature of information it contained.

THE STRUCTURE OF THE WEB SITE

The LRN web site consists of more than two hundreds pages and hundreds of dynamic pages that are built on the fly using CGI (Common Gateway Interface) Scripting and Perl. These dynamically built areas of the site utilise large databases for storing, and handling such resources of information as the LRN news, events, conferences and seminars. Equally, dynamically used databases are used for the handling of specific technical and resource drive areas as the LRN directories, standards and contaminants database to name but a few examples. The web site has many thousands of impartially reviewed hyperlinks to other geoenvironmental and related sites to facilitate the dissemination of advice and information related to the geoenvironmental area equally to both the specialist and general public. This area is gaining in recognition as being a large resource or portal to other virtual libraries, and information already housed on the web.

Navigation

A simple, clear and understandable navigation scheme can increase the number of page impressions, boost return visits, and improve the conversion rate. It's a critical aspect of site design that can have a direct effect on the site's success. Consequently an organisational hierarchy, which is simplified to its lowest elements, forms the foundation of the LRN Web site navigation structure, typified by the presence of links to the home page and all of the main topic pages from anywhere in the web site. In designing the LRN site the following navigational strategy was employed.

- A search engine facility is provided on every page in the Website, to help the speedy and easy reach of visitors to required information;
- Uniformity and consistency of navigational tools (same menus "Navbar" and "Sidebar" position, same feel and look);
- Using text menu rather than button or image map, which helps in two ways: a) they are faster to load. b) they can be viewed in browser which don't support images or when the images are turned off for speedy downloads and navigation;
- A link to the home page and to the main topic pages is provided on each page;
- A secondary navigational link is provided which help locate the position of the visitor within the Website and reduce the number of clicks required to go back within a main area.

The Home Page

The home page is the most important navigational tool and critically the most important page of the site as it is the first page seen by the visitors. Equally the home page typically can be a visitor's first impression of the organisation too. Consequently if the home page looks professional, ethical, artistic, appears to have interesting content, and doesn't have any elements that would deter a visitor, then they should theoretically be interested to exploring the organisations web site.

Figure 1: Homepage of the LRN web site [*http://www.grc.cf.ac.uk/lrn*]

It is equally important that the fist page is efficiently downloaded and optimised for all system platforms in order to deter a negative first impression or first experience of both the site and the organisation represented.

The home page should effectively convey to the visitor a number of key details. The most significant aspect is in fact the site's purpose; often (Flanders & Willis, 1996) this is referred to as and *the who, what, when, where, and why.* The home page should also convey what kind of content is contained in the site, and also an effective way for the user to determine how to find that content within the site.

These aspects have been included into the LRN home page as can be seen in Figure 1. Accordingly the home page includes the LRN logo, its main contributors, a search engine to ensure the speedy recovery of information, and text link to the eight core areas of the LRN web site. The links to each of the main areas are provided with a mouse-over event which activates the appearance of text which explains briefly the core contents of that section of the web site.

The Main Pages
All the main pages within the LRN Website have the same look and feel with a top, bottom and side navigation menu. Each main page act as a home page of interconnected sub-web areas within the LRN web site. The side "Navbar" for each area contains links to resources both within and relating to that specific area. The eight main areas of the LRN web site are:

- **News and Events** ~ This houses the latest articles of news, press releases and notices of events relating to the LRN and its subsidiaries
- **About LRN** ~ This section describes the networks missions, goals and objectives. This section also outlines the membership and fundamental historical beginnings of the network.
- **Advisory Board** ~ This area of the web site relates to the advisory board of the Land regeneration network. Details of the boards membership, agendas, notices and minutes of their meetings are also listed there.
- **Information Resource** ~ The information resource area of the LRN web site seeks to become a comprehensive 'one-stop-shop' guide for anything relating to environmental, geotechnical and regeneration purposes.
- **Discussion forum** ~ This section of the web site houses the many discussion forums for dissemination of information, support and help from those members on-line.
- **Contact the network** ~ This aspect gives information on how the network can be contacted through both conventional and new media formats.
- **Feedback** ~ This section relates to online feedback for the variety of LRN activities.
- **Terms of use** ~ This area takes you to the LRN's terms of use and notices.

Subsidiary pages
Finally within each sub-section of the web site there are the subsidiary pages. The consistency in appearance, style, and use of navigation is also maintained here. However, some differences are employed within the core text area of the pages according to the function of the main area, and core content. For example a conference listing in the News and Event area appear in a fixed tabular style for ease of use, and is different in layout from an article about contaminated land in the information resources area of the LRN site.

ACHIEVING THE OBJECTIVES?
A number of measures are used to gauge the success of the LRN web site. However it must be understood that for any web site to succeed it must continue to mature and grow with the demands of the users. Assessment of this site, its contents and functionality is steadily monitored with maintenance and new material continuously being made.

The quantity and pattern of traffic to any web site is one of the key indicators of the success of that web site. In our case the server statistics are continuously employed in a number of ways; this being to monitor the pages visited, origin of visitor, repetition of visit by same visitor, areas of frequent usage to gauge popularity, relevancy of information and potential lack of interest, functionality or requirement. These statistics show that the traffic throughput to the site increased from *500* hits a week in its initial launch month to approximately *8000* hits per week at the current level of service. The statistics also shows great interest from all around the world in the web site, with about 17% of traffic from overseas domains.

Many areas of the site, especially the news, events, conferences and seminars sections are now open to the public to submit information. This not only gives the user the ability to contribute directly to the site but also aids in giving a sense of ownership to the contents and the sites success. This level of user interaction is also a very useful measure of the sites success to promote networking and the effective dissemination of the information both from and to the geo-community. Many of the events publicised by the network are now contributed to the sites either by direct interaction with the online system, or via request for help via e-mail.

The site has also successfully been submitted to many large search engines, portals and information directories. The site also has been reviewed both technically and for content and style. Most notably on the web site "where to Geo" (http://www.wheretogeo.com/) the LRN site has been reviewed and given the highest level for its content (5 Glob out of 5: Awarded to sites with *plenty of original material relating to geotechnical engineering*). For style the LRN site received 4 stars out of five (4 star: Awarded to sites with *tasteful graphics and natural structure*). Equally on a number of large popular search engines, mettasearch engines and directories, such as *yahoo.com* the LRN web site consistently achieves a high rating appearing typically in the top 10 for searches relating to regeneration and the environment.

CONCLUSIONS
The newly formed Land Regeneration Network (LRN) provides a platform for contaminated land stakeholders in Wales. In October of 2000, as part of an ERDF program, the Geoenvironmental Research Centre established a web site for the LRN. The aim of this new web site is to act as the LRN's main sources for the dissemination of advice and information related to the geoenvironmental area equally to both the specialist and general public.

The web site successfully provides resources on land regeneration, providing advice and information on regulatory issues, whilst also acting as a source of administrative resource for the network as a whole. The site is continuously monitored and updated in order to benefit contaminated land stakeholders in Wales.

REFERENCES
1. Vincent Flanders & Michael Willis (1996), Web pages that suck, 266pp, Sybex Inc., San Francisco, USA. See also http://www.webpagesthatsuck.com

ACKNOWLEDGEMENTS
The work presented has been carried out as a part of a research programme supported by the ERDF. The chair of the Land regeneration network is Mr D. Gwyn Griffiths (Welsh Development Agency). The network activities are organised though the Land Regeneration Network secretariat working under the auspices of the Industrial advisory board. This support is gratefully acknowledged.

The remediation of contaminated land in highway works

J M REID and G T CLARK
TRL Limited, Crowthorne, Berkshire, UK

ABSTRACT

A number of techniques are available for the remediation of contaminated land where it is encountered in highway schemes. However, they have not been greatly utilised to date, and the traditional 'dig and dump' approach has been more common. With the legislative regime for contaminated land now in place in the UK, this is likely to change. The factors affecting the use of remedial techniques in highway schemes are reviewed and those most likely to be successful are identified. Changes to the definition of contaminated land in the Specification for Highway Works are presented.

KEYWORDS

Contaminated land, remedial technologies, highways, earthworks, specifications

INTRODUCTION

A number of innovative methods are available for treating contaminated material to render it acceptable for use in highways. However, a review of highway schemes through areas of derelict and contaminated land (Perry, 1994) showed that there was little uptake of innovative techniques. Contaminated material was excavated and removed to landfill, either off-site or to specially constructed tips on site. This is not in line with the principles of sustainable construction, as the contaminated material is simply moved from one location to another, landfill space is used up, valuable natural materials are consumed to replace the excavated material, and a large number of lorry movements are involved, with attendant problems of dust, fuel consumption, congestion and disturbance to residents. The UK Government has adopted a 'suitable for use' policy with regard to contaminated land (Department of the Environment, 1994) and has acted to discourage landfill by the introduction of the landfill tax and the proposed aggregate tax. It is therefore desirable that there is greater uptake of remedial technologies for the treatment of contaminated land in highway schemes. This in agreement with the Government's strategy for sustainable construction (Department of the Environment, Transport and the Regions, 2000b), which encourages recycling over disposal to landfill.

LEGISLATION

A factor that may have limited the use of remedial technologies, not only in highways, was uncertainty over the legislative position with respect to contaminated land. This has now been largely resolved. The primary legislation in dealing with contaminated land is the Environmental Protection Act 1990. It is from the 1990 Act that Statutory Guidance and the Contaminated Land Regulations have been developed to provide a framework for the regulation of land contamination. Part IIA of the 1990 Act was implemented through Statutory Guidance and the Contaminated Land Regulations in April 2000 (Department of the Environment, Transport and the Regions, 2000a). With the implementation of Part IIA, the

Government has put in place a more pro-active regime and legal framework to deal with land which has been contaminated in the past. The new regime provides a means to enforce remediation where the Integrated Pollution Control (IPC) and Waste Management Licensing (WML) regimes may not apply.

For new highways, the requirements for dealing with contaminated land will be agreed with the Environment Agency during scheme development, and may involve the IPC or WML regimes. For existing sites containing contaminated land or on-site waste, Part IIA or the WML regime may apply depending on the extent of the contamination. The remediation of contaminated land is deemed to be a waste management operation, and hence falls under the WML regime. For a number of remedial techniques, the specialised remediation contractor will require a Mobile Plant Licence (MPL). In some cases a full Waste Management Site Licence may be required, if contaminated material remains on site.

The regulatory framework now exists to enforce remediation and clean up of contaminated land in any situation. Where regimes such as IPC or WML are in place on a particular site, they will continue to be used. Part IIA will act to 'mop up' any areas not covered, and to instigate the remediation of 'static' contaminated land, i.e. land that is not undergoing redevelopment. The principles of risk assessment are to be used to assess whether a site is 'contaminated land' under the meaning of the Act, i.e. whether it is causing or poses a significant threat of significant harm to human health or the environment. All these terms are defined in the Statutory Guidance. It is expected that guidance on model procedures for risk assessment and guideline values will be published by DETR and the Environment Agency in 2001. Computer models such as CLEA for human health and ConSim for groundwater may be used to carry out the risk assessment.

REMEDIAL TECHNIQUES FOR HIGHWAY EARTHWORKS
Contaminated land is frequently encountered in highway schemes, particularly in urban areas. The Black Country New Road in the Midlands of England is a typical example; the types of contaminated site encountered included municipal and industrial landfills, sewage works, gasworks, metal works, chemical factories and brickworks (Russell, 1995). A range of remedial techniques will thus be necessary to deal with the wide range of types of contamination. The use of remedial techniques enables the contaminated material to be processed and re-used on site, thus reducing the amount sent to landfill.

The Environment Agency, in line with the legislation on contaminated land will require a risk assessment approach. This may involve the use of computer models. In terms of the CLEA model for human health, highways fall in the 'commercial and industrial' end use category. This is the least sensitive category, as the exposure of the population to the contaminants is extremely low because the materials are covered. Consequently, the limiting values are likely to be much higher than for more sensitive end uses, such as gardens and allotments and only very heavily contaminated materials are likely to exceed them. Pollution of controlled waters - surface water and groundwater - is likely to be the critical factor in the acceptability of remediating contaminated land for highways. The sensitivity of the underlying strata to pollution is thus at least as important as the leachability of the materials. It is likely that a site-specific risk assessment will be required in all cases where controlled waters are at risk.

Remedial works often require a long time scale, to allow for treatability studies, obtaining licences and authorisations, implementation - which may take months or years for some techniques - and validation and long-term monitoring. It may be difficult to reconcile this extended timescale with the short timescale normally associated with highway schemes, particularly Design and Build (DB) and Design, Build, Finance and Operate (DBFO) contracts. Processing contaminated land may be more suitable for traditional contracts or as advance works contracts. This allows the main contract to proceed unhindered by uncertainties as to the timescale and efficiency of the remedial works. On the Black Country New Road, a number of reclamation schemes on areas of contaminated land were carried out as advance contracts (Russell, 1995).

A number of publications on remediation of contaminated land have been released in recent years, including CIRIA reports SP101-112, Remedial Treatment for Contaminated Land (Harris *et al*, 1995), and Martin and Bardos (1996). Techniques include thermal, physical, chemical, biological and stabilisation/solidification methods. Most may be applied both in-situ and ex-situ. A single method is seldom likely to be sufficient; normally a combination of methods will be required. This will be particularly true for old industrial sites with a mixture of contaminants.

In-situ methods are likely to be of limited use on many old industrial sites because:
- the presence of underground pipes, tanks, foundations and structures may interfere with the processes;
- if there is a high water table it may be necessary to dewater the site to enable the process to work satisfactorily, thus increasing the cost;
- it may be necessary to install barriers to prevent migration of contaminants off-site during the remediation works;
- the heterogeneity of made ground and natural strata may limit the effectiveness of techniques developed from laboratory scale tests.

Contaminated groundwater from adjacent sites may recontaminate the cleaned material. This is a potential problem for highways, where the route may pass through extensive areas of contaminated land. Any proposed solutions must take this into account. In all sites, there is a need to keep clean and contaminated waters and materials separate and arrange for treatment of the contaminated waters and materials.

The cost of the various methods will also affect the selection process. Costs can vary widely from site to site even for the same technique, so it is difficult to give clear guidance on this point. A range of prices for each technique is quoted in Harris *et al* (1995) and Martin and Bardos (1996), but the range for each technique is generally greater than the difference between techniques. Cairney (1996) quotes a financial break-even reclamation cost of about £30 per cubic metre of soil as a criterion for acceptance of remediation process on development sites in the UK private sector. As more experience of the application of the various techniques in UK conditions is gained, it is likely that the relative costs will become clearer. However, it is important to emphasise that a specific strategy should be developed and costed for each site, based on those techniques that will achieve the required standard of clean up, not on those which are cheapest or most familiar.

A further factor to be considered is whether a Waste Management Mobile Plant Licence or a Site Licence will be required for the particular remediation scheme proposed. Decisions will be made on a site-specific basis and it is therefore important to open discussions with the Environment Agency at as early a stage in scheme development as possible. The Environment Agency have recently produced guidance on land contamination and remediation licensing (Environment Agency, 2001).

The cost of remediation may not be the most critical factor; additional costs due to remediation may be more than offset by increased ease of working and resulting savings by avoiding disposal of contaminated material to landfill. Unless the remediation costs are extremely high, the limiting factors are likely to be uncertainty about the reliability and timescale of the techniques and difficulties in obtaining approvals from the regulatory authorities. Thus techniques which are familiar, reliable, rapid, relatively cheap and which yield a product which can be re-used in the works are most likely to find favour with contractors.

On the basis of the information summarised in the preceding sections, remedial techniques have been classified in terms of the likelihood of their finding widespread acceptance in highway schemes. The results are shown in Table 1. A more detailed discussion of individual techniques and their suitability for use in highway earthworks is given in Reid and Clark (2001). Table 1 can be used as a guide to potential techniques that might be utilised on highway schemes, but the list is not exhaustive and the categories should not be regarded as fixed. Technologies are developing all the time, and methods, which are regarded as unlikely to be suitable at present, may become more attractive in the future. Specialised advice should always be taken when considering options for remedial treatment. This should include consultation with the Environment Agency as well as specialised environmental consultants and remediation contractors. At present, the technologies that have greatest application to highways appear to be:

- Solidification/stabilisation with lime/cement (ex-situ and in-situ)
- Ex-situ bioremediation
- Soil washing
- Dry sieving/mechanical sorting
- Vacuum extraction (soil vapour extraction)
- Hydraulic treatment of groundwater (pump and treat)

There are few examples of the use of remedial technologies in highways in the UK. The use of lime and pulverised fuel ash (pfa) to treat lightly contaminated silt dredgings on the A13: Thames Avenue to Wennington scheme is described by Nettleton *et al* (1996). This operation proved very successful. Approximately 100,000 m^3 of the silt was treated to produce around 150,000 m^3 of lightweight fill, which was used in embankments. The operation illustrated the importance of site trials as part of the development of the treatment process. The site was revisited in 1998 and some of the treated material was sampled. Tests showed that the geotechnical properties had not changed significantly, and were still within the requirements of the specification (Reid and Clark, 2001). No significant leaching of contaminants appeared to have occurred, although the pH of the material had dropped to 8.2 from a value of 10.7 at the time of construction (Reid and Clark, 2001). The use of dry sieving to obtain acceptable general fill from an old landfill on the line of the A12 Hackney to M11 Link Road, Contract 1 is also described by Reid and Clark (2001).

SPECIFICATION

The Specification for Highway Works (MCHW 1), which is used within the contract documents for major highways in the UK, is written and structured to allow the maximum use of materials. Within the MCHW 1, contaminated materials are defined as either Class U1 or Class U2 depending on their degree of contamination. There is a need to update these definitions in line with the definition of contaminated land in Part IIA of the Environmental Protection Act 1990 and to include the principles of risk assessment. The revised definitions have been trialed on a major highway scheme in England with satisfactory results.

Table 1 Classification of remedial techniques for application to highway schemes

Technique	Comments
Widespread application to highway schemes	
Solidification/stabilisation with lime/cement (ex-situ and in-situ)	Ease of execution, rapid rate of reaction, ability to immobilise contaminants, yield 'value added' product, relatively low cost.
Ex-situ bioremediation	Satisfactory, low cost and environmentally friendly way of destroying organic contamination. Widely used on redevelopment sites in the UK in recent years, methods developing to shorten the timescale of treatment and extend the range of contaminants that can be treated.
Soil washing	Applicable to a wide range of soils, plant and expertise available in America and Europe, minimises waste product and/or prepares an even feedstock for further processing, yield of 'value added' clean granular material, relatively low cost.
Dry sieving/mechanical sorting	Likely to be used as a cheap alternative to soil washing and to minimise quantity of material going to landfill.
Vacuum extraction (soil vapour extraction)	Established technique for dealing with spillages of organic chemicals, can allow subsequent excavation or treatment of contaminated material, is generally of relatively low cost.
Hydraulic treatment of groundwater (pump and treat)	Likely to be required on schemes where contaminated groundwater is present.
Possible application to highway schemes	
Stabilisation with organic binders	Uses familiar technology, can yield high value end product suitable for use in road pavements.
Thermal desorption (ex-situ and in-situ)	Deals with organic contamination, possibly in conjunction with soil washing for ex-situ applications.
Unlikely to find application in highway schemes	
Electro-remediation (ex-situ and in-situ)/solvent extraction	Can be used for fine-grained soils and sludges with inorganic contamination; slow, little experience at site scale.
In-situ vitrification	Use for securely immobilising inorganic contaminants such as asbestos, as an alternative to excavation and disposal to landfill.
Chemical methods (ex-situ and in-situ)	Reactions too uncertain and possibility of producing toxic by-products too great, especially for in-situ applications.
Incineration	Concern over toxic by-products (e.g. dioxins), process alters soil significantly, likely to require sending material off site to a central processing unit, high cost.

In-situ soil washing/leaching	Limited field applicability, especially in made ground, risk of introducing or spreading contamination to groundwater.
In-situ biological methods	Results too uncertain given generally low ground temperatures in the UK and difficulty of adequate mixing of reagents and soil, timescale likely to be extended.

Amendments to the existing Specification for Highway Works (MCHW 1) and Notes for Guidance (MCHW 2) are shown in *italics* in the following text.

#601 Classification, Definitions and Uses of Earthworks Materials
General Classification
2   Unacceptable material Class U1 shall be:

i)   material which does not comply with the permitted constituents and material properties of Table 6/1 and Appendix 6/1 for acceptable material;

ii)   material, or constituents of materials, composed of the following unless otherwise described in Appendix 6/1:

a)   peat, materials from swamps, marshes and bogs;

b)   logs, stumps and perishable material;

c)   materials in a frozen condition;

d)   clay having a liquid limit determined in accordance with BS1377: Part 2, exceeding 90 or plasticity index determined in accordance with BS1377: Part 2, exceeding 65;

e)   material susceptible to spontaneous combustion except unburnt colliery spoil complying with sub-Clause 15 of this Clause;

f)   contaminated materials, including controlled wastes (as defined in the Environmental Protection Act 1990), whose level of contamination is above that given in Appendix 6/14 [limiting values for pollution of controlled waters] or Appendix 6/15 [limiting values for harm to human health and the environment], and excluding all special wastes (as defined in the Special Waste Regulations 1996) and radioactive wastes (as defined in the Radioactive Substances Act 1993).

3   Unacceptable material Class U2 shall be:

i)   special waste (as defined in the Special Waste Regulations 1996) and radioactive wastes (as defined in the Radioactive Substances Act 1993).

NG 601 Classification, Definition and Uses of Earthworks Materials and Table 6/1: Acceptable Earthworks Materials: Classification and Compaction Requirements
8   The definition of contaminated materials in Class U1 is based on the concept of risk assessment and is in accordance with the definition of contaminated land in Section 78A(2), (5) and (6) of the Environmental Protection Act 1990 and associated statutory guidance.

9   A site specific risk assessment should be undertaken for each earthwork section, as the degree of exposure to living organisms or the hydrogeological conditions can vary significantly within a scheme, leading to different limiting values in different sections. However, appropriate generic guideline values, which are based on a risk assessment model, may be used as default values.

10   For general fills, the limiting values for harm to human health should normally be based on the 'commercial/industrial' end use category of guideline values, as there is a very low risk of exposure to the public from any contaminants in the fill. For landscaping fills, considerations of phytotoxicity will be important. Where slopes are to be returned to

agricultural use, the limiting values should be based on the 'residential with gardens' end use.
The appropriate category should be decided for each section or sub-section of the scheme.

11   Details of the limiting values adopted and explanations of their derivation should be
given in Appendix 6/14 and Appendix 6/15.

12   Materials, which would be classified as Class U1 because of contamination using
generic guideline values, may be rendered acceptable by remedial techniques such as
stabilisation with cement or lime. The contaminant levels are not changed by the stabilisation
process, and remain above the generic guideline values, but their ability to migrate is
reduced. A site specific risk assessment must be carried out to demonstrate whether the risk to
human health and living organisms, and of pollution of controlled waters, is acceptable
before the remediated materials can be reclassified as acceptable fill materials.

NG SAMPLE APPENDIX 6/14: POLLUTION OF CONTROLLED WATERS

[Note to compiler. This should include:]
1   Limits on the amount of contaminants in a material above which there is a significant
possibility that controlled waters (surface water and groundwater) will be polluted.
2   An explanation of the derivation of the limits (e.g. generic guideline values for given
soil conditions, or values derived from site specific risk assessment quoting relevant input
parameters and methods).

NG SAMPLE APPENDIX 6/15: LEVEL OF CONTAMINATION OF MATERIAL

[Note to compiler. This should include:]
1   Limits on the amount of contaminants in a material which, if exceeded, will lead to a
significant possibility of significant harm to human health or the environment.
2   An explanation of the derivation of the limits (e.g. generic guideline values for given
soil conditions, or values derived from site specific risk assessment quoting relevant input
parameters and methods).

CONCLUSIONS

There are a number of methods for the remediation of contaminated land, which are suitable
for use on highway schemes. Experience to date has been limited but has generally been
satisfactory, provided the design and execution of the work was carried out to a high standard.
There is now a comprehensive framework of legislation surrounding the issue of
contaminated land and its remediation. It is possible to negotiate this successfully if
consultations are held with the relevant authorities from an early stage in the development of
a scheme. It is important to allow sufficient time to carry out the necessary laboratory and
field scale trials, and to obtain the necessary licences and authorisations. For this reason, it
may be preferable to carry out remediation works as advance contracts, which will then allow
the main works to be carried out more efficiently.

The definition of contaminated land in the Specification for Highway Works has been revised
to bring it into line with the definition of contaminated land in Part IIA of the Environmental
Protection Act 1990. This introduces the concept of risk assessment when setting limiting
values for contaminants in the Specification. Feedback on the proposed definitions would be
welcome, pending a revision to the Specification for Highway Works.

ACKNOWLEDGEMENTS

The work described in this paper was carried out under a research contract for the Highways
Agency. The permission of the Agency to publish the paper is gratefully acknowledged. The
views expressed represent those of the authors and not necessarily those of the Agency.

REFERENCES

Cairney T (1996). Dry sieving – a cost effective reclamation solution in many metal & gas contaminated granular soils. Proceedings of the Fourth International Conference on Polluted and Marginal Land: Re-use of Contaminated Land and Landfills, held at Brunel University, London 2-4 July 1996. (Ed M C Forde). Edinburgh, Engineering Technics Press, pp 249-251.

Department of the Environment (1994). Framework for Contaminated Land: Outcome of the Government's Policy Review and Conclusions from the Consultation Paper 'Paying for our past'. London, Department of the Environment.

Department of the Environment, Transport and the Regions (2000a). Environmental Protection Act 1990: Part IIA. Contaminated Land. DETR Circular 02/2000. London, The Stationery Office.

Department of the Environment, Transport and the Regions (2000b). Building a Better Quality of Life: A Strategy for more Sustainable Construction. London, Department of the Environment, Transport and the Regions.

Environment Agency (2001). Guidance on the Application of Waste Management Licensing to Land Contamination Remediation Activities, Version 2.0. Environment Agency web site www.environment-agency.gov.uk.

Harris M R, S M Herbert and M A Smith (1995). Remedial treatment for contaminated land, Volumes I to XII. CIRIA Special Publications SP101-112. London, Construction Industry Research and Information Association.

Manual of Contract Documents for Highway Works. London, The Stationery Office.
    Volume 1: Specification for Highway Works
    Volume 2: Notes for Guidance on the Specification for Highway Works

Martin I and P Bardos (1996). A review of full scale treatment technologies for the remediation of contaminated soil. Richmond, Surrey, EPP Publications.

Nettleton A, I Robertson and J H Smith (1996). Treatment of silt using lime and pfa to form embankment fill for the new A13. Lime stabilisation (Ed C D F Rogers, S Glendinning and N Dixon). London, Thomas Telford, pp 159-175.

Perry J (1994). Case studies of highway schemes through areas of contaminated land. Unpublished TRL Report PR/GE/16/94. Crowthorne, Transport Research Laboratory.

Reid J M and G T Clark (2001). The processing of contaminated land in highway works. TRL Report 489. Crowthorne, Transport Research Laboratory.

Russell L (Ed.) (1995). Black Country New Road. Commemorative Supplement to New Civil Engineer, November 1995. London, Emap Construct.

The Stationery Office (1990). The Environmental Protection Act 1990. London, The Stationery Office.

The Stationery Office (2000). Contaminated Land (England) Regulations. SI No 227. London, The Stationery Office.

Building on brownfield sites – the highwall problem

H D Skinner and J A Charles
Building Research Establishment Ltd, Watford, UK

INTRODUCTION

When brownfield sites are developed, the main emphasis is often on the hazards to the human population and the natural environment from pollution and contamination. However, there may be physical hazards for the built environment which impose significant restraints on the economic redevelopment of a brownfield site.

Opencast coal mining has been carried out over extensive areas of Great Britain and, generally, the ground has been restored for agriculture. Although such sites may not be regarded as brownfield land, there can be problems for building development associated with the previous use of the site. A particular hazard for building development on restored opencast mining sites is the steep excavation slope termed the "highwall" which may emerge at ground surface or may be buried under a substantial depth of fill. The economic feasibility of building on many restored opencast mining sites, and other types of brownfield site where excavations have been infilled, will be dependent on the proportion of the site from which buildings have to be excluded because changes in depth of fill over the highwall give the potential for unacceptable differential settlement.

Where building development takes place on an infilled excavation the received wisdom is not to build over the edge of the excavation, thereby ensuring that no building is founded partly on fill and partly on natural ground. This recommendation only deals with part of the much larger problem which arises wherever there is a significant variation in depth of fill at the location of a building. A general description of the hazard posed by a variable depth of fill has been given by Skinner and Charles (1999). The most acute problems are commonly associated with small buildings for which deep foundations are not an economically viable solution. It is usually feasible to provide a stiff raft foundation which will prevent distortion of the building due to differential settlement and which will resist horizontal tensile forces. The principal interest then lies in the surface settlement profile over the highwall and the delineation of the zone from which buildings should be excluded because tilt is unacceptably large.

This paper describes experimental work in the BRE 4m deep test pit, which has examined the effect of a buried highwall on the pattern of ground movements in a fill. Displacement and tilt measured in this large-scale model have been compared with a finite element analysis, using linear elastic isotropic parameters to represent the fill. Settlement over highwalls outcropping at the ground surface has been monitored at two restored opencast mining sites. At one site the backfill was end-tipped and at the other the upper third of the fill was heavily compacted.

On the basis of the experimental, field and analytical data, some general conclusions concerning building exclusion zones are put forward.

LARGE-SCALE MODEL HIGHWALL

A 2m high model highwall with a face angle of 60 degrees was constructed towards one end of the BRE 6m × 4m × 4m deep watertight reinforced concrete test pit, as shown diagrammatically in Figure 1. The faces of the wall consisted of double sheets of 20mm plywood bonded together and attached to a scaffold frame which was supported by three columns of concrete blocks to ensure adequate stiffness. Horizontal battens were fixed at 300mm intervals onto the outer faces of the plywood to create a rough interface with the fill.

A granular drainage layer was formed in the base of the test pit to facilitate uniform upwards inundation of the fill. A 4m depth of colliery spoil fill was placed in 250mm layers with little compaction in order to ensure that there was substantial collapse potential. About 15% of the fill consists of silt and clay size particles, with 25% of particles greater than 32mm. Compaction tests and one-dimensional collapse tests on this material have been described by Skinner et al (1999). The drainage layer was separated from the fill by a geotextile. Settlement points were installed during fill placement to measure the settlement of the fill at the level of the top of the wall. Three lines of surface movement points were installed in the top layer of the fill to measure vertical and horizontal movements. Systematic inundation of the fill then commenced raising the water level up through the fill in stages of 1m per day and monitoring movements at each stage. The water level was then lowered in stages back to the base of the fill during one day.

Four stages of the test have given useful data on the displacements over the highwall:
- Settlement under self-weight following completion of fill placing
- Collapse settlement on inundation of the bottom one metre depth of fill
- Collapse settlement on inundation of the bottom two metres of fill
- Settlement due to the increase in effective vertical stress when the water level (WL) was lowered from the level of the top of the highwall to the base of the test pit

The settlements measured at points between 1m and 2m from the edges of the test pit were sufficiently similar to demonstrate that there were no significant boundary effects. The settlement measured at the end of the pit furthest from the highwall can be characterised as *far-field* settlement as it is unaffected by the presence of the wall. Far-field settlements at both ground surface and at the level of the top of the wall are listed in Table 1 for the four stages of the test. The measured settlement profiles for these stages are shown in Figure 1 for each of the three lines of surface movement points.

Horizontal and vertical displacements of the centre-line movement points during inundation of the bottom 2m layers showed that, in conjunction with large vertical movements, large horizontal movements occurred. There was horizontal compression of 3% above the base of the wall and the significantly larger horizontal extension above the top of the wall resulted in cracks in the fill. These horizontal strains occurred in a situation where vertical compression of about 12% was observed.

Table 1: Far field settlement in mm at different stages of the test

	Creep under self weight	Inundation: rise in WL from 4m to 3m depth	Inundation: rise in WL from 4m to 2m depth	Drawdown: fall of WL from 2m to 4m depth
Surface	21	105	218	12
2m depth	14	103	178	6

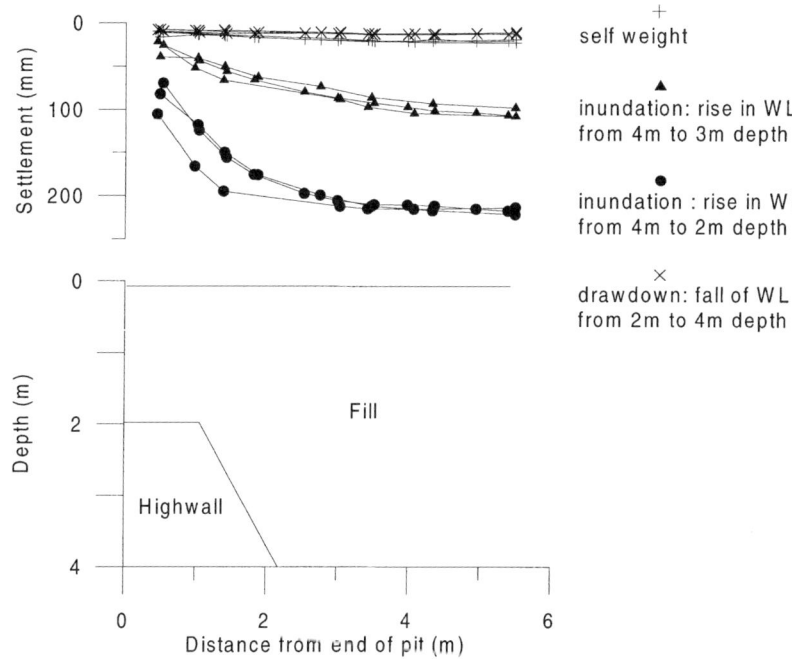

Figure 1: Measured surface settlement profiles over large-scale model highwall

The surface settlement profiles have been normalised by, firstly, subtracting the far-field compression measured in the upper layers of fill above the level of the top of the highwall and then, secondly, dividing the resulting corrected settlement (s) by the maximum corrected settlement (s_{max}). The normalised profiles for the four stages of the test are plotted in Figure 2. This normalising process facilitates a comparison of the data in the different stages of the test on a scale of 0 to 1 and reflects the situation in which there is no far-field compression in the fill above the level of the top of the highwall. Tilts have been normalised by dividing them by the far-field vertical compression in the fill below the top of the highwall.

The normalised results of a finite element analysis, using linear elastic isotropic parameters to represent the fill, are shown in Figure 2 along with the range of normalised measured settlements from the experimental work for the 2m deep inundation and the self-weight settlements. Tilts are shown in addition to settlements. While the shape of the settlement and tilt profiles are broadly similar, the analysis shows the settlement profile at a greater distance from the highwall than do the measurements.

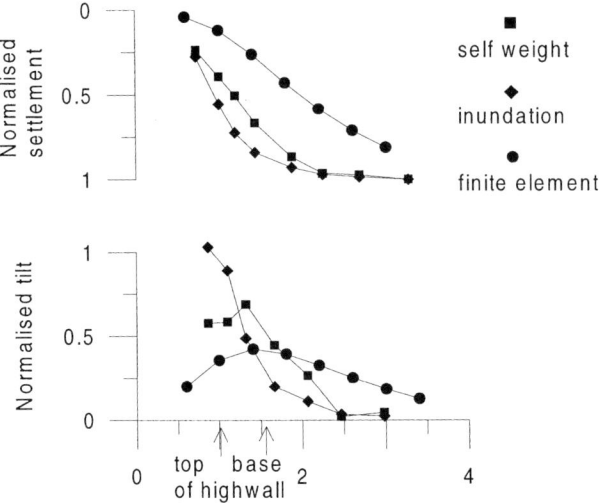

Distance from end of pit / height of highwall

Figure 2: Measured and predicted normalised surface settlement profiles over large-scale model highwall

The finite element analysis showed very small horizontal movements. This represents a major difference between the observed and the predicted deformations.

FIELD STUDIES

Field monitoring of ground surface movements has been carried out at two restored opencast coal mining sites.

At Orgreave opencast mining site, near Sheffield, settlement has been monitored over a highwall, with an angle of $38°$. The creep settlement of the end-tipped mudstone rockfill which has occurred between May 1999 and June 2000 is shown in Figure 3. It would seem that a ledge close to the top of the highwall has caused a local reversal in the settlement. The presence of an overburden heap prevented settlement observations being made at a distance from the toe of the highwall. Tilts of up to 1/250 have been calculated from the measured surface settlement profile.

The backfill at Blindwells opencast mining site, near Edinburgh, consists of mudstone, siltstone and sandstone. It was placed by dragline and face shovel in 1982. To reduce the settlement of a 1.4 km section of the Tranent by-pass, built across the site in 1985, the top 16 m of the backfill was systematically compacted (Charles and Burford, 1987). Figure 4 shows settlement over the highwall, which has an angle of $43°$, measured between February 1988 and June 1995. The compaction of the upper zone of fill has effectively moved the zone affected by differential settlement some distance from the toe of the wall. Doubtless, compaction greatly reduced the potential for vertical compression and, hence, the magnitude of the differential settlement.

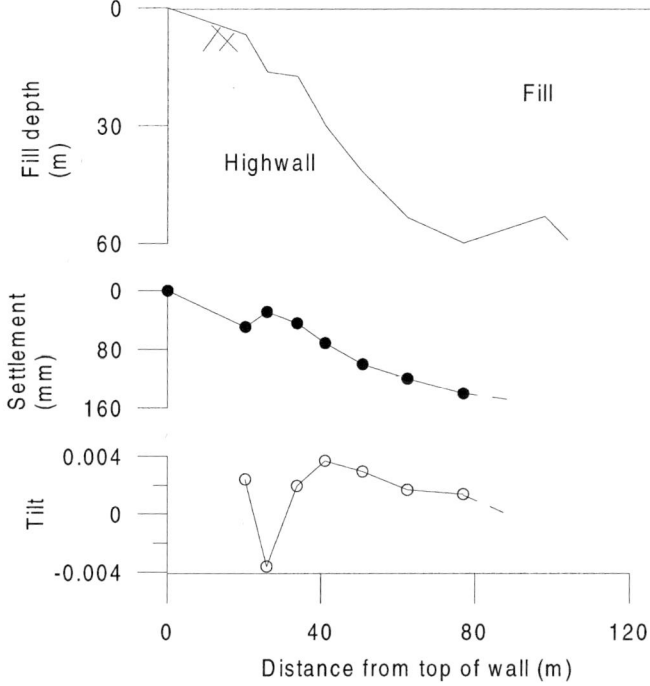

Figure 3: Settlement at Orgreave opencast mining site

The two case histories demonstrate the manner in which details of the geometry of the highwall and the method of fill placement can substantially affect the settlement profile.

DISCUSSION AND CONCLUSIONS

The displacements measured in the test pit have provided experimental data against which finite element analyses can be calibrated. The shapes of the settlement profiles at different stages of the test are similar, but the location of the profile in relation to the position of the highwall is a little different for the larger movements induced by the inundation of the full depth of the highwall. A finite element analysis show a broadly similar profile, but with larger differential settlements beyond the base of the highwall.

In field situations it is unlikely that the properties of the fill will be accurately known or that the fill will be homogeneous; the geometry of the highwall may not be known with any degree of precision. The two case histories of field measurements show how details of the highwall geometry or fill placement can substantially affect the settlement profile. There is little practical value, therefore, in refining the analysis or using a more sophisticated soil model to achieve precise agreement with the experimental observations. In a field situation, the fill adjacent to the highwall will be less well compacted than the general body of the fill thereby introducing a zone of fill with low stiffness at the fill/highwall interface and producing a settlement profile that closely follows the profile of the highwall. This factor would have been present in the large-scale model test, but was not replicated in the finite element analysis. A simple model based on an assumption of a linear variation of tilt with distance is being developed by BRE (Charles and Skinner, 2001).

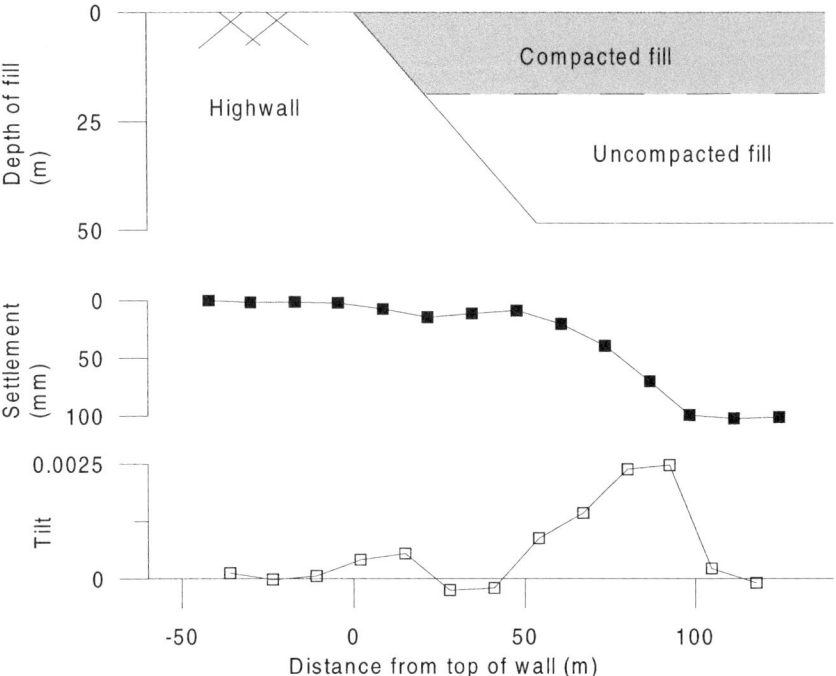

Figure 4: Settlement at Blindwells opencast mining site

Horizontal movements have been measured in the large scale model which are much larger than those predicted in the finite element analyses. The induced horizontal extension strains above the top of the highwall could cause major damage to buildings unless foundations are designed which are strong enough to withstand the tensile forces.

REFERENCES

Charles J A and Burford D (1987). Settlement and groundwater in opencast mining backfills. Proceedings of 9th European Conference on Soil Mechanics and Foundation Engineering, Dublin, vol 1, pp 289-292.

Charles J A and Skinner H D (2001). The delineation of building exclusion zones over highwalls. Submitted for publication in Ground Engineering.

Skinner H D, Charles J A and Watts K S (1999). Ground deformations and stress redistribution due to a reduction in volume of zones of soil at depth. Geotechnique, vol 49, no 1, February, pp 111-126.

Skinner H D and Charles J A (1999). Problems associated with building on a variable depth of fill. Ground Engineering, vol 32, no 7, July, pp 32-35.

ACKNOWLEDGEMENTS

The research work described in this paper has been carried out for DETR under the Building Regulations Framework Agreement and Environment Business Plan.

Bioremediation – Black Art or Green Con?

I. M. SUMMERSGILL
General Manager, VHE Technology Ltd, Barnsley, UK

SYNOPSIS

Bioremediation is now commonly used for the treatment of former industrial and brownfield land, predominantly for contamination with hydrocarbon compounds. Methods used include ex-situ landfarming, biopiling, windrows, bioreactor vessels and in-situ injection; several specialist contractors exist and flourish in the civil engineering market.

The Author has undertaken several ex-situ projects, as Consultant, Contractor and as third-party Verifier. Some have involved open discussion of the methods involved; others have claimed commercial sensitivity/exclusivity for the processes. Monitoring results from three such projects are discussed herein and other research is reviewed in brief. The similarity in the test results shown here masks a wide variability in applied technique, and it is argued that the efficacy of 'special' additives may be a smokescreen to hide the more fundamental natural processes that constitute effective ex-situ bioremediation.

THE MARKETPLACE

Most commercially-available processes involve the addition of a bacterial culture or other substrate, then the irrigation and/or aeration of the soil by physical turning. A mystique has perhaps grown up around the various techniques and processes, and the additives used, fuelled by the specialist operators themselves in many cases. The most common process used in the UK tends to be the (field) adaptation of composting processes, using man-high windrows or tilled rows of soil. On the Continent, it is more common for work to be undertaken in centralised treatment 'factories'. In the USA, the use of sealed vessel 'bio-reactors' is more common *(Alexander, 1999)*, especially for the treatment of recalcitrant compounds. Materials such as 'bio-gels' can be injected directly into the affected soils.

The UK market for ex-situ bioremediation has slowly developed from virtually zero usage in the late 1980's to perhaps 4-12% of the remediation market by the late 1990's *(Petts et al, 2000; Kean, 2000 & Swannell, 1999)*. The market has been polarised between specific specialist contractors and direct 'research' work by problem holders and biotechnological companies. The specialist Contractors have emerged as both home-grown and subsidiary offshoots of continental operators, and there have been several recent entrants to the market as the use of bioremediation becomes more acceptable or understood. Availability of proprietary 'bugs'/inoculants on the market is a relatively new sales phenomenon.

At the same time, it has not gone unnoticed that bioremediation and adaptation goes on in Nature without human intervention. Indeed, the disastrous intervention of Mankind's marine oil spills in Alaska (Exxon Valdez), Pembrokeshire (Sea Empress) and Brittany has unwittingly led to a plethora of research into natural bio-adaptation *(Swannell, 1999; Madsen, 1991 & Lessard et al, 1995)*, and the efficacy of aeration and phytoremediation.

Geoenvironmental impact management, Thomas Telford, London, 2001.

UK RESEARCH

From such research and practical experience has come a better appreciation of the processes involved in chemical conversion, and clues as to whether the bio-process is essentially a 'black art' or not. Publication of acceptable data from bioremediation projects in the UK has been limited to simple representations of contaminant decline with time *(Kean, 2000)*; speciation and detailed analysis remains mostly unpublished or university research. In respect of publicly-monitored projects, the BioWise initiative has published details of one project on a former coke works in 1999 *(DTI, 2000)*, it is envisaged that the CL:AIRE organisation will review its first ex-situ project this year, and localised bio-trials have been carried out on projects by several Development Agencies in the North of England.

One organisation, BG Technology (now Advantica) has already carried out in-house research and provided a brief overview, with some 'surprising' conclusions *(Jones, 1999)*. They were primarily concerned with the effectiveness of biotreatment on PAH (coal tar byproducts) rather than the hydrocarbon contaminants monitored in most ex-situ projects. BG Technology set up a series of experiments using a mobile, steel-tank landfarm with the same basic contaminated soil in it, but carried out fertilisation, inoculation, co-solvent addition, peroxide mixing and also co-metabolism with an orange 'product'. Their results indicated that the greatest influence on biodegradation of PAH was the <u>management</u> of the landfarming, <u>not</u> the additives. The soil turning, irrigation and fertiliser regime gave 65% PAH removal, whilst the specialist additives only resulted in 58-60% removal; volatilisation was not significant for PAH, even with forced aeration.

This latter research would seem to confirm that there are other issues at work during such remediation, not least the activity of the natural ('in-house') bacteria. In the field, it has long been tacitly recognised that the mixing and volatilisation that takes place as a result of soil movement itself, plus the inherent inaccuracy of sampling/testing, could form a large part of the apparent biodegradation seen as a reduction in original contaminant level. Until this aspect is researched in full detail at field size, the suspicion will remain that the process is rather more of a 'green' con than a scientific biotreatment.

FIELD EXAMPLES

Three examples are now provided to consider this hypothesis. They relate to projects carried out over the past three years, and are part of a suite of data from the author's experience. No apology is made for the simplicity of the data and presentation, as it is the aim to draw conclusions from their similarity, not to challenge the detail of the site analysis. The first example was a specialist contract 'trial' and the subject of significant sampling and analysis (significant in that those costs constituted 25% of the whole sub-project); the next incorporated a specific product (USEPA inoculant) in an unsophisticated use; and the third project is included to illustrate unintended 'natural' processes in action.

BioPile - London

The first example relates to the treatment of some 6000 cum of soils from a former oil terminal site, using an ex-situ Biopile with forced air extraction and moisture control. The chief contaminant of the soil was hydrocarbons and particularly the lubricant oil end of the range; the soil was mostly reworked Fill material with a range of constituents from brick to silty clay. The range of initial Total TPH contamination was generally 5000-20,000 ppm, averaging 12,000 ppm for C6-C36. The initial C6-C10 average was 250 ppm and C11-C20 averaged around 2000 ppm (range 500 – 5000). The set targets for bioremediation over a 13-week period were 5000 ppm for Total TPH with 1000 ppm maximum for C6-C20.

The 6000cum was set out in two long biopiles but these were individually designated in 100cum zones; a covering LDPE liner prevented undue volatilisation or rain infiltration. Sampling occurred in each zone, as it was turned/treated, so individual lots of 100cum could be monitored. Initial results are shown in Figure 1, which indicates that by 6-7 weeks (at the time of a second turning of the first soil, but only shortly after deposition of the last 100cum) there was a reduction in peak concentrations but a general averaging around 5000 ppm. This suggested at the time that the aeration process was not wholly effective, and that the narrowing range of TPH results reflected 'mixing' soils as a result of biopile placement.

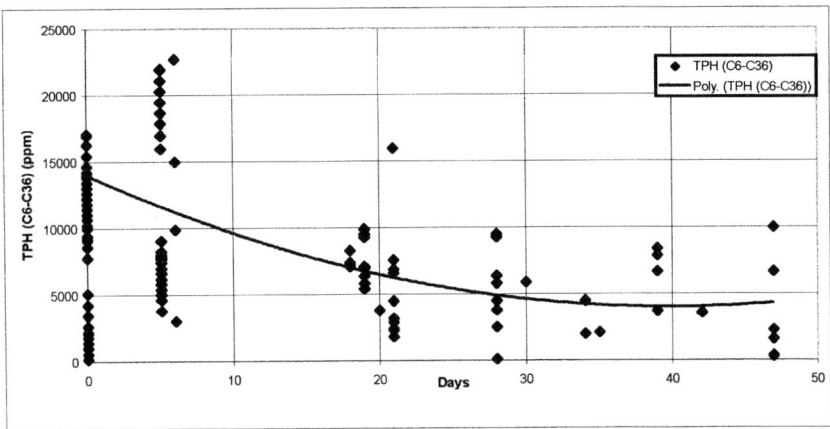

Figure 1: Biopile Degradation (C6-C36), Weeks 0-7

Examination of the specific test results for the C6-C10, C11-C20, C21-C36 ranges however indicated that the 'heavier' range had only commenced partial degradation. Continuation of the aeration and turning for 13 weeks resulted in satisfactory achievement of the targets, with an average of 300ppm for C6-C20 and 1500ppm (range 500-3500) for Total TPH. The following Figure 2 indicates the percentage reduction from an original concentration for each zone in the biopile, showing clearly that any biotreatment of the heavier fraction may not commence immediately. And that there is still considerable scatter after 3 months.

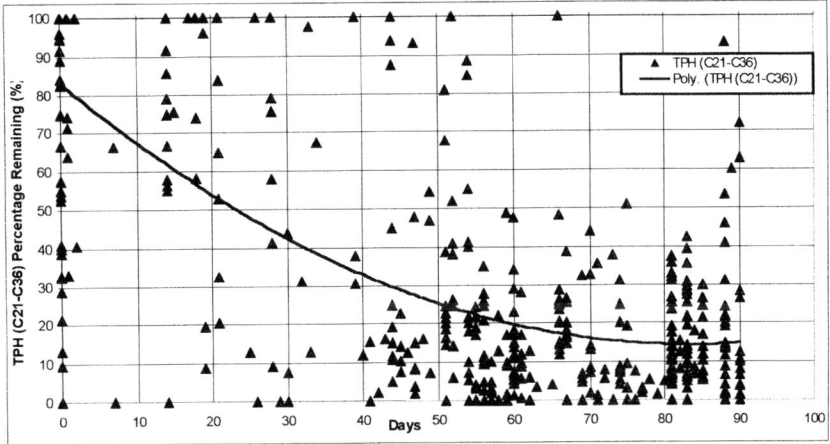

Figure 2: Biopile TPH Percentage Remaining (C21-C36), Weeks 0-13.

It has become standard practice to set an absolute or 95% confidence target for bioremediation, but this plethora of results perhaps indicates how the <u>average</u> final target figure will need to be set far below any absolute (or 95%) target in order to achieve 'compliance'. In this writer's experience, test results follow a skewed, normalised distribution and the peak value is frequently less than half the 95percentile; it is statistically naïve to set an absolute limit. The 'black art' involved here is an understanding of mathematical processes, <u>not</u> biological activity; for engineers, this should be second nature.

Landfarm – Midlands

This second example relates to a fairly simple process, spreading of a thickness of soil (in this case, 300-500mm depth) on land and rotavating it on a fortnightly basis. The project requirement was to remediate a volume of diesel-impacted clayey soil from 4000mg/kg to below 1000mg/kg, in order for it to be retained on site as infill material. In the timescale available (6-8 weeks) it was considered that the 'natural' aeration process created by such landfarming needed to be speeded up and the soil was augmented by a bacterial inoculant. A proprietary material, USEPA approved, was identified and applied by spray one week after soil spreading; a second application after rotavation at 3 weeks was intended but proved unnecessary. An initial application of nutrient, at spreading, was also used to enhance the 'in-house' bugs and to ensure some activity along with optimal moisture level.

There was limited testing on this project, only 8 samples being taken per visit. The results for Diesel range organics (DRO) and Total TPH are shown on Figure 3. It is self-evident that the material swiftly degraded under the combined action of aeration, phytoremediation and enhanced/augmented bacterial activity. Some 75% of degradation occurred in the first month, and the remainder in the second. It should also be evident that the starting concentration was significantly below that expected from site testing/investigation, perhaps evidence of averaging down by mixing and/or the effects of excavation and transportation, albeit only a short distance.

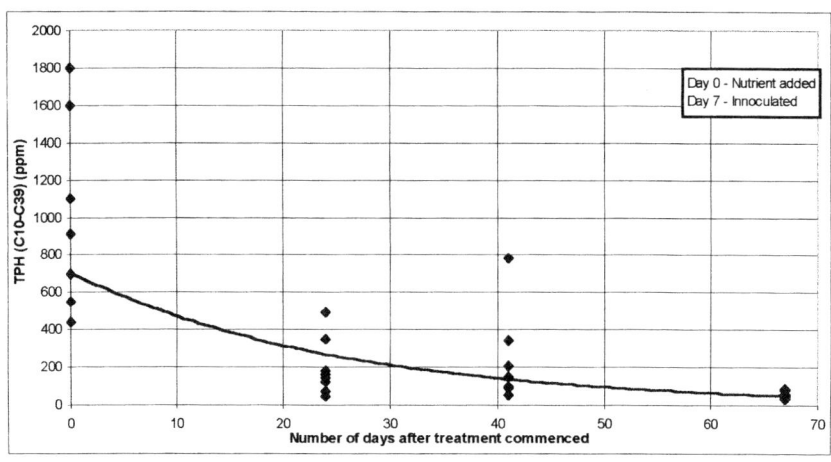

Figure 3: Landfarm TPH Degradation (C10-C39) (ppm).

Unfortunately, insufficient funds were available to 'test' the imported bacteria against the incumbents on an untreated/control section of land, as contractually the risk of a longer degradation period was not acceptable. This does indicate, however, that such sustainable practices could be applied to many sites. The significant degradation (75-80% in 3-6 weeks, and ~95% after 9 weeks) does suggest that careful nurturing of the soil can be an alternative to landfill; in this case, the vast proportion of the contaminants were the less volatile DRO's.   Understanding of the soil toxicity and the adaptation of the inoculant for its target contaminants were possibly an example of a 'black art' in practice here.

'Natural' Degradation – Estuarial Site

This third example relates to material dug from an old scrapyard adjacent to an estuary. The material was generally scrap car pieces (engine blocks included) in a sandy/clayey matrix; they had been deposited in a shallow pit for the past 20 years at least. As part of a larger contract, the soil was stockpiled and tested for off-site disposal, where it was found that both the TPH and some other contaminant constituents exceeded the levels expected from site investigation. The assigned Special Waste tip would only accept material below 10,000 ppm TPH and, due to other contaminants, imposed a weekly volumetric constraint.

The Contractor took steps to seek permission to send the worst material out of the County, and to segregate that material which could be sent to local tip. Minor contractual and administrative delays meant that the soil heap, roughly 2000cum, remained on site in the height of summer, and was actually relocated twice due to other site works. Fortunately, the site was remote from public housing and its near neighbour was a Sewage Works.

It became evident, as samples were taken weekly to accompany Disposal Notes, that the TPH concentrations still declined with time (Fig. 4) and particularly after heap relocation. A decision to translocate the 'worst' material into a separate stockpile was taken after eight weeks, and this again gave results that indicated a 30-50% degradation had taken place in the three weeks after movement. In the end, all the soil was disposed to the local waste tip as it eventually fell below the 'gate limit' for Mineral Oils.

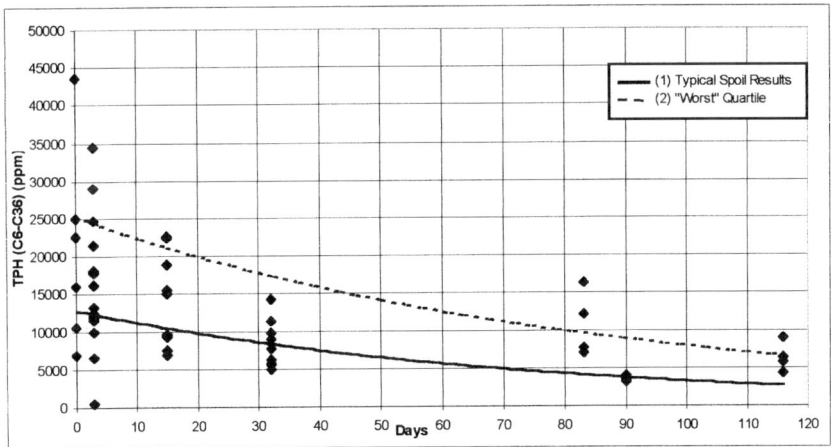

Figure 4: "Natural" TPH Degradation (C6-C36) (ppm).

It is suggested that this material exhibited a natural degradation partly through aeration of a loose stockpile (it having been compacted in a pit before), but also through sunlight and temperature degradation because of its open exposure. Evidentially, the surface soils changed colour in the stockpile when exposed, which was a strong indicator of aerobic activity.   Whether indigenous microbes existed within the car parts, soil and waste oil, and were stimulated into life by the conditions, will not be definitely proven.   What is the case with this example is that no 'magic' ingredient was mixed into the matrix, yet two near-typical bioremediation degradation 'curves' were the result.

SUMMARY/CONCLUSIONS

As more demonstrations and trials are reported, the mystique that has grown up around bioremediation will hopefully be dispelled, or at least de-mystified.   These examples, and others that remain unpublished, should illustrate that Man is only assisting, not controlling, Nature in these activities.   It seems apparent that a contaminated soil matrix, long held in apparent stasis in the ground, can rapidly degrade into less polluting elements when dug up and exposed to air and sunlight.   What is not fully proven yet is the optimal treatment process, and whether there is a 'special skill' that can be applied to greatly accelerate the natural activity, or whether a 'manufactured' compound is required to augment the reaction.

The first example gives several clues as to the progressive degradation of compounds by aeration, starting with the most volatile; the main skill here was management of soil condition.   The second example is unfortunately not scientifically rigorous, but tends to show that an additive can speed up a natural process to a significant degree; thus the use of an inoculum may not be the 'green con' that many observers report.   The third example is perhaps the most intriguing, in that no 'black art' or added ingredient has been used, yet over three months a 60% degradation has occurred 'naturally'.   The Author is left with the impression that there are several 'commercial' ways to speed up ex-situ bioremediation, but that the process can and will occur naturally at a significant rate once materials are removed from their (compacted, sometimes saturated and anaerobic) resting place in the ground.

ACKNOWLEDGEMENTS

The Author would like to acknowledge the various site/office colleagues at VHE and WSAtkins who assiduously recorded, compiled and assisted in preparation of the data.   The conclusions and comments herein remain, however, resolutely those of the Writer and are not necessarily those of his employers.   The research of Dr. Steve Jones at BG Technology, now unfortunately no longer with us, and his colleagues at Loughborough, into the detailed aspects of biological treatability is gratefully acknowledged and respected.

REFERENCES

Alexander, M. - *Biodegradation and Bioremediation*, 2nd Edn. 1999 (Chap. 17).   Academic Press, New York.
Department of Trade and Industry, UK (DTI) Bio-Wise Initiative. - *Case Study 1 – Cutting the cost of coaltar clean-up.*   Crown Copyright Leaflet, DTI, April 2000.
Jones, S. – *What are the Myths and Truths of Land-Farm Treatment of Contamination?*   Article (page 7) in Local Authority Waste & Environment, Vol7 Issue 11 November 1999.
Kean, A. – *Rapid exsitu Bioremediation in the UK: the development of a commercial success.*   Pp 241-5 in Land Contamination and Remediation Journal Vol 8, Number 3, July 2000
Lessard, R. Wilkinson, J. Prince, R. Bragg, J. Clark, J. & Atlas, R. – Pages 207-225 in *Bioremediation of Pollutants in Soil and Water (Schepart, ed.).*   ASTM Publications, Philadelphia, 1995.
Madsen, E.L. – Pages 1662-1673 in *Environmental Science and Technology* Journal. Vol 25, 1991.
Petts, J. Rivett, M. & Butler,B. – *Survey of Remedial Techniques for Land Contamination in England and Wales.*   Environment Agency R&D Technical Report P401.   Bristol, 2000.
Swannell, R. – *Benefits of Bioremediation.*   Article in Industrial Environmental Management, Vol. 9 Issue 8, January 1999.

Impact Management During Remediation of a Former Landfill in an Urban Environment

PAUL TAUNTON
Principal Geoenvironmental Engineer, ENSR International Limited, Chepstow, Gwent
RUPERT ADAMS
Principal Environmental Health Officer, Vale Royal Borough Council, Winsford, Cheshire

SYNOPSIS

An old unlined landfill, containing mixed industrial waste, was sited in an area now surrounded by housing, overlying a highly permeable minor aquifer. A remediation strategy based on complete removal of the waste to engineered landfills was agreed with the regulators. Before the remediation commenced, systems were set up to monitor and control the impact of the remediation work on the environment. Additionally, channels of liaison were established with nearby residents. During the course of the work, odour became a matter of concern to the wider public and led to adverse press reports regarding the site. A broader residents' liaison group was elected, and involved in the assessment and management of the odour issue. The paper details both the site environmental controls and the provisions for public liaison employed to bring the remediation to a successful completion.

SITE DESCRIPTION & HISTORY

During the 1940's and 1950's a sand pit was excavated in a semi-rural area in Cheshire. In this area the glacial sands are underlain by mudstones of the Mercia Mudstone Group at some 50m depth. The pit was situated immediately behind a row of Victorian houses along a main road, and reached a maximum depth of 15m, eventually covering a site area of about 2ha. During the 1950's and 1960's the area became increasingly urbanised as housing estates were developed adjoining the eastern and northern boundaries of the site.

It is thought that backfilling of the pit with industrial waste commenced about 1960. Records held by Vale Royal Borough Council show approval of planning applications for the disposal of "factory waste, refuse to be dry and not of a noxious odour" throughout the 1960's. Examination of Council records has revealed that, even then, concerns were frequently raised by residents regarding the type of waste being deposited and its possible off-site impact. Recorded complaints include those relating to a major fire, chemical drums left on the site surface, noise and dust. Disposal of "soil and dry industrial waste" continued until about 1973, when the pit had been filled to just below the original ground level.

Subsequently a workshop and bungalow were constructed on the southern margin of the backfilled area and the site used as a transport depot, until its closure in early 1999. To the west of the site lies an open field which had been scheduled for housing but where development was restricted because of the proximity of the landfill.

Geoenvironmental impact management, Thomas Telford, London, 2001.

DEVELOPMENT OF REMEDIATION PROPOSALS

From 1977 onwards, varied schemes were put forward for the redevelopment of the site. In late 1997, when the reassessment of existing information commenced as part of the present scheme, eighteen site investigation reports were available, relating to earlier proposals. To allow the development of realistic remediation proposals, further studies relating to waste characterisation, gas monitoring and groundwater quality assessment were carried out during 1998 and early 1999. The waste was found to be very variable in both its physical and chemical properties and to contain significant contamination, including oils, solvents, heavy metals and asbestos. Risk assessments concluded that without remediation there was a potential for contamination of the underlying groundwater and for migration of gases and volatile organic compounds to neighbouring homes. Following discussions with the regulatory authorities, it was concluded that the only practical remedial option was complete removal of the $200,000m^3$ of waste to an engineered landfill and its replacement with clean imported fill. With hindsight, it is considered probable that had this scheme not been implemented, the site would have been identified by the local authority under Part IIa of the contaminated land regulations as a contaminated site requiring remediation.

Because of the close proximity to the site of existing houses and the knowledge that there had been complaints during the waste deposition, it was recognised that there could be significant environmental impact from the remediation, therefore certain controls were demanded within the planning permission. During the development and agreement of the initial detailed project method statement care was taken to consult all interested parties. The final agreed statement required that one member of the consulting engineer's staff would act as a full-time Environmental Supervisor, whose sole responsibility was to monitor the waste excavation and its environmental effects.

COMMUNICATIONS

Procedures to ensure close liaison between the parties to the remediation contract, local residents and the regulating authorities were considered essential and were adopted from the planning stage. Project representatives attended public meetings arranged to discuss the proposed scheme. The parish council appointed one of their members to liaise with the residents of the parish and the project staff. The selected representative was provided with appropriate personal protective equipment, given a full safety induction and subsequently allowed free access to the remediation site. Complaints to the representative by members of the public generally resulted in his immediate visit to the site to investigate. The parish council representative prepared an independent monthly report on the site work, focussing particularly on issues raised by, or of interest to, the public. This report was first presented at the Parish Council meeting, then published in a popular local newsletter.

Prior to the commencement of the work, and at monthly intervals thereafter, site co-ordination meetings were held attended by the parties to the remediation contract including: the landowner, the environmental body assisting with funding through the landfill tax credit scheme, the environmental consultant, the remediation contractor, the operator of the receiving landfill sites, the Environment Agency, the County, District and Parish councils and the NHBC. All site operations were openly discussed during these meetings and any complaints were discussed in detail. Copies of the minutes of these meetings and project reports were provided to the parish council representative for his use in discussions with residents and, where appropriate, these were deposited in the local library for reference.

ENVIRONMENTAL IMPACTS
Highway & Pedestrian Traffic
The access to the site used by the transport yard was narrow and unsuited to large numbers of heavy vehicle movements, therefore the new vehicular access approved for the future residential development was constructed prior to the start of the remediation contract. To minimise the impact on traffic flow on the already busy local road network, the number of vehicle movements was restricted by the planning permission to 100 per day. The contract specification required each vehicle leaving site carrying waste to the landfill to return with clean backfill material, to minimise both environmental impact and haulage costs. To protect pedestrians it was specified that the point where the main site haul road crossed a public footpath would be manned continuously. The planning conditions also demanded adequate sheeting of loads and vehicle wheel wash facilities.

Air Quality
Regular checks of on-site air quality were carried out by the Environmental Supervisor using a flame ionisation detector (FID) to measure the concentration of volatile organic compounds (VOCs) in air. The highest VOC concentrations were expected in the base of the excavation and the drum storage compound. Measurements were taken at the start of each shift, before the workforce was allowed access to these areas, to confirm that VOC concentrations were within the relevant occupational exposure limits[1]. When elevated levels of VOCs were measured, or strong odours detected, control measures were implemented and if necessary the wearing of appropriate organic vapour masks was instructed. Control measures included restricting the area of excavation to a minimum, covering any open areas of waste giving off odour with inert material and the use of a water based spray odour control system.

In addition, regular checks were made around the perimeter of the site to assess the concentration of VOCs leaving the site. In general, concentrations were below detectable limits but on a number of occasions when strong odours were noticeable outside the site, measurements were taken using Dræger tubes to give lower detection limits. It was demonstrated that exposure limits relevant to the general public were not being breached. The limits used were generally those recommended by the Department of the Environment Transport and the Regions[2], which for most substances were much more stringent than the corresponding occupational limits applied for those working on the site.

Large quantities of asbestos-cement waste were known to exist within the landfill and significant quantities of fibrous asbestos were encountered during waste excavation. The method statement included procedures for the sampling and identification of any suspect materials and their subsequent safe handling and removal. During asbestos removal air sampling was carried out both immediately adjacent to the working area and on the downwind site perimeter to monitor any fibre release.

The stockpiling of waste was restricted as a condition of the planning permission, both as a dust control restriction and to limit the visual impact of the works. A series of static dust monitoring points was set up around the site perimeter prior to the commencement of waste removal. Measurements were taken to provide baseline data to which dust emissions during the work could be compared. In general little dust arose from the waste as the freshly excavated material was generally damp, however it was necessary to apply water sprays to the site haul roads as these regularly became a dust source in dry weather. Additional pumped samples of fine particulates were taken on several occasions but it was noted that regional air quality had more effect on the results of this work than did site operations.

Groundwater Quality

Within the waste were areas of low permeability clay fill, often supporting localised perched water tables. In some areas this perched water was heavily contaminated with oils and solvents, arising either from crushed and corroded drums deposited within the waste or perhaps from liquids which had in the past been allowed to soak away into the body of waste. A system was set up to collect and treat any contaminated water encountered during excavation to prevent its loss through the highly permeable sands and the possible adverse effects on underlying groundwater quality.

Vibration

During the initial stages of the backfilling operations a vibratory roller was used for compaction of the sand fill, leading to complaints from nearby residents that the vibration from site activities was causing damage to their properties. Reference to published information[3] predicted that such vibrations would not be expected to cause damage to structures at such distances. To reduce the residents' concerns, the compaction plant was changed to a heavier dead-weight roller but although the frequency of complaints was much reduced, they did continue. To allow comparison of actual vibration intensities with published standards, vibration measurements were made in the gardens of two properties adjoining the site. The measured vibration levels resulting from the site work were low, only comparable to vibrations resulting from heavy vehicles passing on the main road. Although structural surveys of selected properties showed no damage attributable to vibration from the site work, the perception of several residents remained that vibration was an issue.

Noise

Site working hours were restricted to minimise the impact on neighbouring residents. Regular noise monitoring was not considered necessary, since the noise generated by normal excavation and fill operations was within acceptable daytime limits. On particular occasions when noisy operations were in progress, such as during operation of the tyre shredding machine, or if any plant had to be left running overnight, noise measurements were taken. In these cases it occasionally proved necessary for the contractor to provide screening to reduce the noise measured at the site boundary to acceptable levels.

Radioactivity

Following recommendations from the Health and Safety Executive, a radiological survey of the site was carried out, although there was no evidence, either documentary or from hearsay, to suggest that any radioactive waste had ever been deposited on the site. No radioactivity above background levels was detected in the initial survey, nor during further monitoring.

THE ODOUR ISSUE

Although odour control measures were included in the specification for the remediation, it had been expected that odour was unlikely to be a significant problem because of the original restrictions on the waste deposited in the landfill and the low organic matter content identified in the investigations.

During mid-1999, as excavation progressed deeper, large volumes of waste heavily contaminated with oils and solvents were encountered which on occasion, particularly on hot summer days, gave rise to a strong oily odour. Off site measurements indicated very low or undetectable levels of VOCs, however the odour caused serious concern to nearby residents. Adverse press reports appeared referring to a stench and questioning the potential health effects of the remediation works. Despite the availability of reassuring test results provided through the local Environmental Health officers and the parish representative, the residents could not be satisfied of the safety of the work through the existing channels of communication and, through the press, called for the work to be halted.

PARTICIPATION BY RESIDENTS GROUP

A public meeting was called by the residents, at which not only was criticism levelled at the remediation contractor but the independence of the Borough Council officers and Parish Council representative in overseeing the works was questioned. Since nothing was likely to be resolved at this vociferous public meeting, the Borough Council proposed that elections be held to form a truly representative residents' committee, which would then participate in the Borough Council's decision making processes regarding the site. A committee of eight residents was duly elected by secret ballot. The residents group held regular formal meetings, initially weekly, then less frequently as outstanding issues were resolved. The first action of the committee was to select an independent firm of Environmental Consultants who were then appointed, and funded, by the Borough Council, to review the site work. The consultant's brief included checking that the project environmental management was adequate, and confirming that there were no off-site health risks to residents. The independent consultant's work included flux box testing immediately above the waste and air sampling both adjacent to the working area and at the site boundary, in order to identify the VOCs being released and measure the variation in their concentrations with distance. The field data was subsequently input into a numerical model to assess "worst case" scenarios. The consultants report concluded that under adverse conditions there could be an odour detectable as far as 1km from the site but that this did not present a risk to health, as in many cases the odour threshold[4] was considerably lower than the exposure limit. This report was presented orally at a committee meeting to ensure that any technicalities could be clarified immediately without leading to misapprehensions.

ROLE OF THE LOCAL HEALTH AUTHORITY

Because of the residents' concerns for their health, the Borough Council arranged that a Medical Officer of Health from the South Cheshire Health Authority (SCHA) should participate in the site co-ordination meetings. Since the Health Authority had little experience in such matters, they called in their retained consultant from the Guy's and St. Thomas' NHS Trust Chemical Incident Response Service (CIRS) to visit the site and assess the environmental controls and off-site health risks. The CIRS report was presented direct to the residents so that its independence from the project team was clear. The CIRS report also confirmed the adequacy of site procedures, and proved of considerable value in gaining the confidence of the residents. Additionally, the SCHA carried out a study comparing the health records held by local General Practitioners during the period of the works with records from the same practices in previous years, and with records from other Cheshire practices during the same period. The small sample made definite conclusions impossible, however it was reported that whilst there was some increase in the number of asthma consultations with GPs there was no corresponding increase in drug prescribing rates.

DISCUSSION

The appointment early in the project of the parish council representative was valuable, since he was often able to respond immediately to queries raised by residents, explaining what was being done and why. Alternatively, in the case of more technical queries, he was able to raise the query with the appropriate member of the project team, ensure a reply was obtained, and then explain the reply to the resident. Some residents, however, perceived him as part of the project team, rather than their representative.

Although the Borough Council had been instrumental in setting the standards for many of the site environmental controls within the planning permission, it was not perceived by residents as properly representing their interests: one local resident even instigated an Ombudsman investigation of the environmental health department. However, the Ombudsman found no maladministration. The subsequent appointment of an elected residents' committee and their adoption of structured formal meetings improved relations considerably and allowed reasoned dialogue between the parties. The residents' group took up the Borough Council's offer to allow them to select the consultant carrying out the review of the work on behalf of the Borough Council and therefore accepted that the review was independent. In the same way although existing technical data and reports were often apparently perceived by the residents as in some way biased towards the views of the project team, reports prepared in response to queries raised by the committee were accepted.

CONCLUSION
Despite extensive environmental monitoring and controls, and strenuous efforts to involve all interested parties in the development of the project from the outset, the perception grew among certain residents that there was a potential health problem related to the site and that their representatives were not managing this correctly on their behalf. Once such a perception is established in the minds of the residents it becomes reality for them[5]. Only by giving the residents more involvement in the project could this perception be changed and their confidence gained. Ultimately, the remediation work removed the potential hazard of a contaminated site from the locality and rendered a previously unusable site suitable for housing.

ACKNOWLEDGEMENTS
The authors would like to thank the representatives of the following organisations involved in the remediation works for their cooperation:

Landowner   P. E. Jones (Contractors) Limited
Registered Environmental Body   Cheshire Environmental Services
Main Contractor   Alfred McAlpine (Civil Engineering) Limited

REFERENCES
1. Health and Safety Executive, Occupational Exposure Limits, EH40/99, 1999.
2. Department of the Environment, Transport and the Regions, Expert Panel on Air Quality Standards.
3. Transport Research Laboratory, Groundborne Vibrations Caused by Mechanised Construction Works, TRL Report 429, 2000.
4. National Environmental Technology Centre, Odour Measurement and Control, 1994.
5. SNIFFER (Scotland and Northern Ireland Forum for Environmental Research), Communicating Understanding of Contaminated Land Risks, Environment Agency R & D Technical Report P142, 1999.

Natural Attenuation, Fate and Transport of Pollutants including Gas

Visualization of Unstable Fingered Flow and Measurement of Fluid Contents by Light Transmission in Transient-Three-Phase Oil-Water-Air Systems in Sand

Christophe J.G. Darnault[1], David A. DiCarlo[2], Tim W.J. Bauters[3], James A. Throop[3], Tammo S. Steenhuis[3], J.-Yves Parlange[3], and Carlo D. Montemagno[3]

[1]Malcolm Pirnie, Inc., Independent Environmental Engineers, Scientists & Consultants, 11832 Rock Landing Drive, Suite 400, Newport News, VA 23606, USA

[2]Department of Petroleum Engineering, Stanford University, Stanford, CA 94305, USA

[3]Department of Agricultural and Biological Engineering, Riley-Robb Hall, Cornell University, Ithaca, NY 14853, USA

ABSTRACT

Most three-phase flow models lack rigorous validation because very few methods exist that can measure transient fluid contents on the order of seconds of whole flow fields. This research presents the application of the light transmission method by which fluid content can be measured rapidly in three-phase systems. The method uses the hue and intensity of light transmitted through a slab chamber, to measure fluid contents. The water is colored blue with $CuSO_4$. The light transmitted by high frequency light bulbs is recorded with a color video camera in RGB (Red, Green and Blue) and then converted to HSI (Hue, Saturation and Intensity). Calibration of hue and intensity with water, oil and air is made using cells filled with different combinations of the three fluids. The hue and water content are uniquely related over a large range of fluid contents. Total liquid content is a function of both hue and light intensity. The air content is obtained by subtracting the liquid content from the porosity. In the transient experiments, unstable fingered flow were formed by dripping water on the surface in a two-dimensional slab chamber with either oil, air or oil-air-saturated sand. The light transmission method is able to capture the spatial resolution of the fluid contents and can provide new insights in rapidly changing, two-phase and three-phase flow systems.

INTRODUCTION

Non-aqueous phase liquids (NAPL's) enter the vadose zone as a result of spills, inadequate disposal practices, or leaking underground storage facilities, thus contaminating groundwater resources (Schwille, 1988). Measuring the contaminant concentrations is an important factor in assessing risk to human health and the environment and in developing effective remediation strategies. Multiphase flow and transport phenomena in the unsaturated and saturated zones of the subsurface environment are the focus of numerous research efforts (Parker, 1989). One of the less well understood transient flow phenomena, in three-phase systems, is unstable fingering (Kueper and Frind, 1988). Fingering decreases fluid retention time in the vadose zone, thus increasing groundwater contamination. Consequently, there is a need for fast, non-destructive, and accurate measurements of transient three-fluid phase flow in porous media.

Transient visualizations (but not direct measurement of fluid contents) have been made in Hele-Shaw cells with smooth walls (Saffman and Taylor, 1958) or with imprints of porous media on glass (Schwille, 1988). Currently, very few methods exist that allow rapid determination of fluid contents in three-phase, NAPL-air-water systems. Most of the methods that allow determination of fluid contents in three-phase, NAPL-air-water systems involve some form of radiation. These include dual energy gamma radiation (Ferrand et al., 1986), X-

Geoenvironmental impact management, Thomas Telford, London, 2001.

ray attenuation (McBride and Miller, 1994), and computerized tomography (Morton et al., 1999). One method that does not use x-rays is magnetic resonance imaging (Johns and Gladden, 1998). The disadvantage of these methods is that they cannot measure transient flow phenomena. Synchrotron X-rays allow accurate and fast measurements of fluid contents in transient flow fields, in any soil type, but can measure only a small section of the flow field at one time due to the small beam size of 1 mm by 8 mm, (DiCarlo et al.,1997). The light transmission method (LTM) is a non-destructive method that allows measurement of fluid contents in transient air-oil-water flow occurring in sandy porous media (Darnault et al., 2001). The LTM has been found to yield reliable fluid contents in steady-state, air-oil-water flow fields, as verified by comparing the results of a static experiment with synchrotron x-rays. The advantages of the LTM are that it does not involve radiation and that it is able to visualize fluid-content changes, over the whole flow field, with a time resolution of tenths of seconds.

This research presents the application of the light transmission method developed for three-phase flow systems to investigate unstable fingered flow in soil-air-oil-water system.

LIGHT TRANSMISSION METHOD

For fluid measurements in porous media, the light transmission method involved placing a two-dimensional chamber in front of a uniform light source and recording the transmitted light (Glass et al., 1989; Darnault et al., 1998; Darnault et al., 2001). Experiments were performed at a constant temperature of 20°C. A light source composed of a bank of 24 fluorescent, high-frequency light bulbs, located in front of a white background, was used. The transmitted light was recorded with a Sony Color Video Camera employing three ½ inch CCD (Charge Couple Device) images, each having a total of 250000 effective picture elements. The camera was located 1 m in front of the chamber with constant settings (zoom = 0.95 m and aperture = f 5.6). The images were stored on a Hi8 video cassette in RGB format. Recorded images were converted from RGB to HSI format and analyzed using a Power Macintosh equipped with a video digitizer (RasterOps 24 XLT from RoasterOps Corporation, 1992) and scientific image processing software (IPLab Spectrum V3.00 software from Signal Analytics Corporation, 1989-1995). The advantage of the HSI format is that it treats color roughly the same way that humans perceive and interpret color.

In three-phase systems consisting of air, water and oil, Darnault et al., (2001) developed a calibration method to measure the water, oil and air contents as a function of hue, intensity and porosity with LTM. This calibration technique involved a two-dimensional calibration chamber consisting of cells filled with a porous media and known quantities of fluid ratios - oil and water, air and water, oil and air, or air-oil-water - allowing to obtain relationship for water, oil and air contents as a function of hue, intensity and porosity.

The oil used was Soltrol 220, a transparent isoparaffine solvent from the Phillips 66 Co. The distilled water was dyed blue with $CuSO_4$ at a concentration of 28%. The calibration chamber was 62 cm high, 52 cm wide, and 1 cm thick. It was divided into 24 cells: 3 cm high by 23 cm wide. The cell walls were constructed from the same 1 cm thick Hyzod polycarbonate sheet (Sheffield Plastic, Inc) as the experimental chamber. The cells were packed to a porosity of 38% with 20/30 sieve size, industrial-quartz silica sand (Union Corporation). The sand packed cells were then filled with all possible fluid combinations. The fluid contents of each of the three fluids were varied by 0.076 cm^3/cm^3 increments (Figure 1).

Average hue and intensity values were plotted versus the corresponding water and liquid contents to obtain the calibration curves (Figures 2&3); fluids contents were then expressed by the following set of equations (Darnault et al., 2001). Different equations were used to relate H (hue) to θ_w (water content) above and below 0.076 cm^3/cm^3 (Darnault et al., 2001):

$$\theta_W = 0.076 + 0.304 * \left(\frac{H - H_K^\theta}{H_S^\theta - H_K^\theta} \right) \qquad \text{for } 0.076 < \theta_W \leq \theta_S \qquad [1]$$

$$\theta_W = 0.076 * \left(\frac{H - H_0^\theta}{H_K^\theta - H_0^\theta} \right) \qquad \text{for } 0 \leq \theta_W \leq 0.076 \qquad [2]$$

where θ_W is the volumetric water content, H is the hue of the transmitted light, H_K^θ is the hue at 0.076 cm^3/cm^3 water, H_S^θ is the hue at 0.38 cm^3/cm^3 water (saturation), H_0^θ is the initial hue at 0 cm^3/cm^3 water, and H_K^θ is the hue at 0.076 cm^3/cm^3 water from the calibration chamber.

Thus for constant water content, the liquid content θ_L can be found as (Darnault et al., 2001):

$$\theta_L = \theta_S * \left[\frac{(\zeta I - I_\omega)}{I_S(\theta_W)} \right] \qquad \text{for } 0.03 \leq \theta_L \leq \theta_S \qquad [3]$$

where θ_S is the saturated volumetric liquid content, equal to 0.38 cm^3/cm^3, I is the observed intensity, ζ is the correction factor, I_ω is the intensity value of the anchor point, and $I_S(\theta_W)$ is the saturated liquid intensity value with a water content θ_W.

To find the saturated liquid intensity value, $I_S(\theta_W)$, two sets of equations are used depending if the water content is above or below 0.076 cm^3/cm^3 (Darnault et al., 2001):

$$I_S(\theta_W) = (I_0^\theta - I_\omega) - \frac{\theta_w}{0.076}(I_0^\theta - I_K^\theta) \qquad \text{for } 0 \leq \theta_W \leq 0.076 \qquad [4]$$

$$I_S(\theta_W) = (I_K^\theta - I_\omega) - \frac{(\theta_w - 0.076)}{0.304}(I_K^\theta - I_S^\theta) \qquad \text{for } 0.076 < \theta_W \leq \theta_S \qquad [5]$$

where θ_W is the volumetric water content calculated by Equations [1] and [2] , I_0^θ is the intensity for 0.38 cm^3/cm^3 oil, I_K^θ is the intensity for 0.076 cm^3/cm^3 water with 0.304 cm^3/cm^3 oil, I_S^θ is the intensity for 0.38 cm^3/cm^3 water, and I_ω is the intensity of anchor point.

For liquid contents below 0.03 cm^3/cm^3 we used a linear relationship between the observed intensity and the intensity of air dry sand - intensity value (40).

The oil and air contents, θ_O and θ_A were obtained from the liquid content and porosity:

$$\theta_O = \theta_L - \theta_W \qquad [6]$$

$$\theta_A = \theta_S - \theta_L \qquad [7]$$

APPLICATION

Unstable fingered flow experiments were performed with a slab chamber filled with quartz silica 20/30 sand and saturated with either oil or a combination of oil-air-system to meet the initial experimental conditions. The experimental chamber had 1 cm thick polycarbonate walls and interior dimensions of 45 cm wide, 1 cm thick, and 55 cm tall. A manifold of five fluid ports was located at the bottom. To form a finger, water was either applied as a point source or as a rainfall simulation through a needle using a cam driver, 1 cm above the sand surface, at a rate of 3 ml/min. The oil level was kept constant via overflow tubing located 1 cm below the oil in the chamber. Vertical profiles were analyzed for fluid contents, at the center of the finger, as the finger progressed into different sand-oil-air saturated phases.

Fluid content profiles for the fully formed fingers in different soil-oil-air-water systems are presented in Figures 4a,b and Figure 5d. Figures 5a,b&c show the image of a fully form water in soil-air-oil systems in (a) RGB format, (b) hue image and (c) intensity image. As expected from previous experiments (Glass et al., 1989; and DiCarlo et al., 1999), and as clearly

Water (cm²/cm²)	Oil (cm²/cm²)	Air (cm²/cm²)	RGB	HUE	INTENSITY	Water (cm²/cm²)	Oil (cm²/cm²)	Air (cm²/cm²)	RGB	HUE	INTENSITY
0	0.380	0				0	0.380	0			
0	0.304	0.076				0.076	0.304	0			
0	0.228	0.152				0.152	0.228	0			
0	0.152	0.228				0.228	0.152	0			
0	0.076	0.304				0.304	0.076	0			
0	0	0.380				0.380	0	0			
0.076	0.076	0.228				0.380	0	0			
0.076	0.152	0.152				0.304	0	0.076			
0.076	0.228	0.076				0.228	0	0.152			
0.152	0.076	0.152				0.152	0	0.228			
0.152	0.152	0.076				0.076	0	0.304			
0.228	0.076	0.076				0	0	0.380			

Figure 1. Visualization of the different combinations of fluid contents (oil-water-air in 12/20 silica sand) from the calibration chamber under RGB, hue, and intensity image analysis.

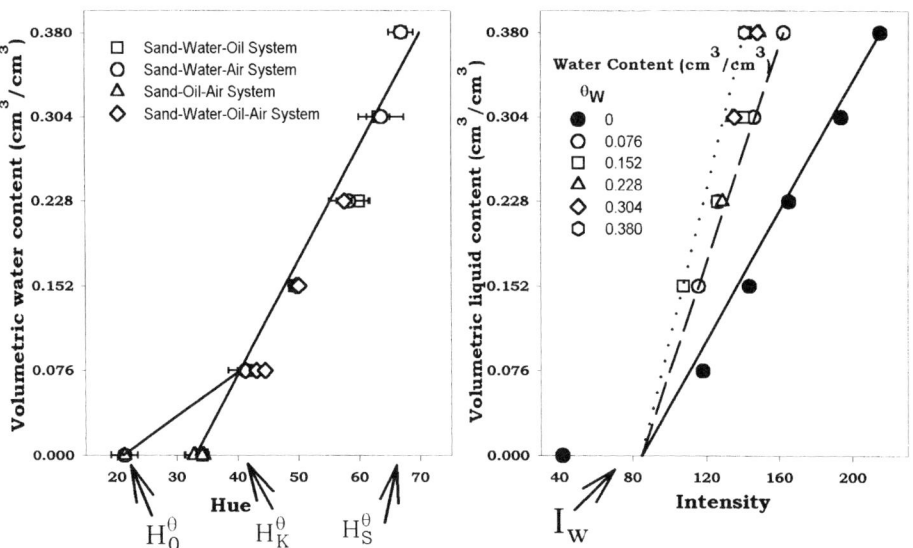

Figure 2. Water contents in soil-oil-air-water systems versus hue values.

Figure 3. Liquid contents in soil-oil-air-water systems versus intensity values.

observed in Figures 4a&b and Figure 5d, the tip of the finger is the wettest while the remainder of the finger is much dryer. In all fingering experiments, the finger does not expand once it is formed due to hysteresis in the soil constitutive relationships. The LTM measurements, in conjunction with simultaneous matrix potential measurements, can be used to determine the constitutive relationship for fingered flow in porous media.

The LTM, a non-destructive method, allows qualitative and quantitative measurements over the whole flow field of transient air-oil-water flow – unstable fingered flow - occurring in sandy porous media with high resolution. Qualitative data, such as visualization of the phenomena, and quantitative data, such as the dimensions, velocity, and fluid contents of the flow, are important in understanding flow patterns in soil-air-oil-water systems. Additionally, the LTM provides data to validate 1-D and 2-D computer codes for transient air, oil and water flow, to develop models of three-phase flow phenomena, to simulate groundwater pollution scenarios, and to simulate groundwater remediations like those using surfactants.

ACKNOWLEDGEMENTS
This research was performed at the Department of Agricultural and Biological Engineering, Cornell University, and was sponsored by the US Air Force Office of Scientific Research, under grant/contract number F49620-94-1-0291.

REFERENCES
Darnault, C.J.G., J.A. Throops, A. Rimmer, D.A. DiCarlo, T.S. Steenhuis and J.-Y. Parlange. 1998. Visualization by light transmission of oil and water contents in transient two-phase flow fields. Journal of Contaminant Hydrology. 31: 337-348.

Darnault, C.J.G., D.A. DiCarlo, T.W.J. Bauters, A.R. Jacobson, J.A. Throop, C.D. Montemagno, J.-Y. Parlange, and T.S. Steenhuis. 2001. Measurement of fluid contents by light transmission in transient three-phase oil-water-air systems in sand. Water Resources Research. In Press.

DiCarlo, D.A., T.W.J. Bauters, T.S. Steenhuis, J.-Y. Parlange and B.R. Bierck. 1997. High speed measurements of three-phase flow using synchrotron x rays. Water Resources Research. 33: 569-576.

DiCarlo, D.A., T.W.J. Bauters, C.J.G. Darnault, T.S. Steenhuis and J.-Y.Parlange. 1999. Rapid determination of constitutive relations with fingered flow. Proceedings International Workshop on Characterization and measurements of the hydraulic properties of unsaturated porous media. Riverside, CA. October 22-24. 1997. p: 433-440.

Ferrand, L.A., P.C.D. Milly and G.F. Pinder. 1986. Dual-gamma attenuation for the determination of porous medium saturation with respect to three fluids. Water Resources Research. 22: 1657-1663.

Glass, R.J., T.S. Steenhuis and J.-Y. Parlange. 1989. Mechanism for finger persistence in homogeneous, unsaturated, porous media: Theory and verification. Soil Sc. 148: 60-70.

Johns, M.L. and L.F. Gladden. 1998. MRI study of non-aqueous phase liquid extraction from porous media. Magn. Reson. Imaging. 16(5-6): 655-657.

Kueper, B.H. and E.O. Frind. 1988. An overview of immiscible fingering in porous media. Journal of Contaminant Hydrology. 2: 95-110.

McBride, J.F. and C.T. Miller.1994. Nondestructive measurements of phase fractions in multiphase porous-media experiments by using x-ray attenuation. Cent. Multiphase Res. News. 1: 10-13.

Morton, E.J., R.D. Luggar, M.J. Key, A. Kundu, L.M.N. Tavora and W.B. Gilboy. 1999. Development of a high speed x-ray tomography system for multiphase flow imaging. IEEE Transactions on Nuclear Science. 46(3): 380-384.

Parker, J.C. 1989. Multiphase flow transport in porous media. Review of Geophysics. 27: 311-328.

Saffman, P.G. and G. Taylor. 1958. The penetration of a fluid into a porous medium or Hele Shaw cell containing a more viscous liquid. Pro. Royal Soc. London A. 245: 312-331.

Schwille, F. 1988. Dense chlorinated solvents in porous and fractured media. Translated by J.F. Pankow, Lewis, Chelsea, MI, 146pp.

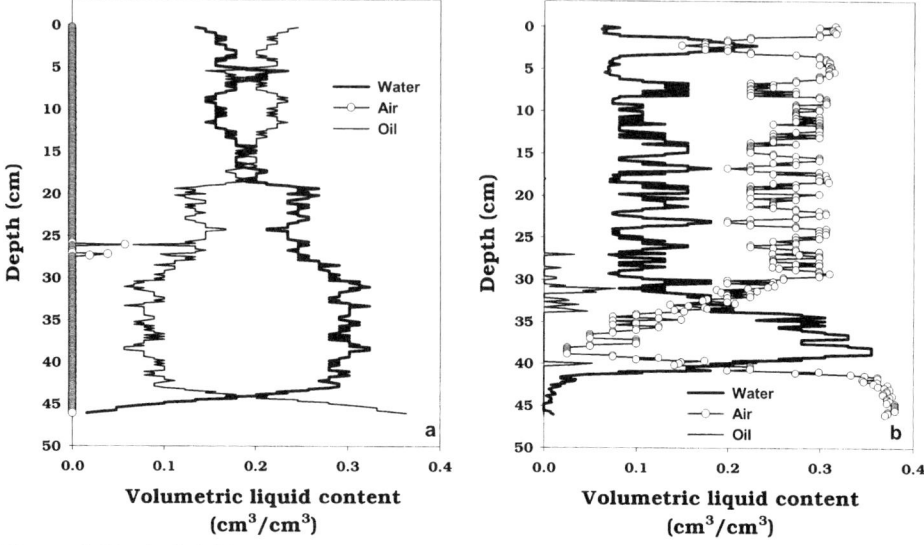

Figure 4. Vertical fluid content profile of unstable fingered flow in two-phases flow system: (a) water finger in soil-oil system, (b) water finger in soil-air system.

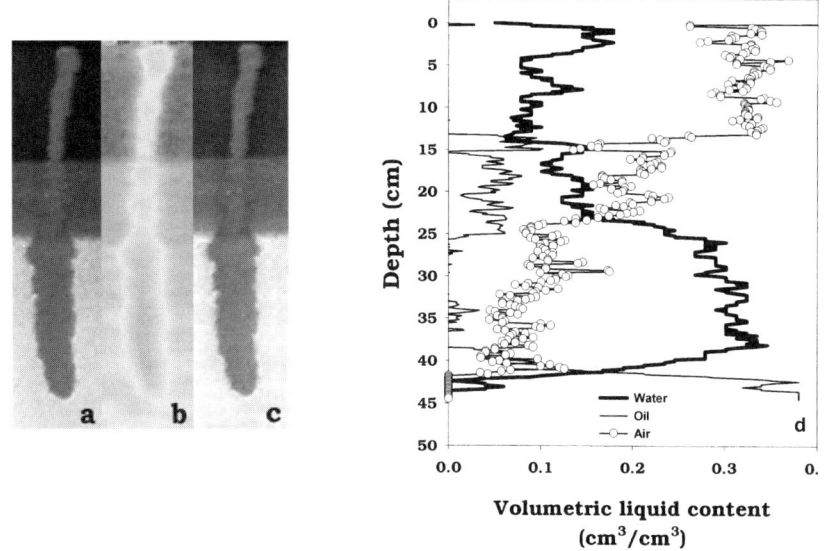

Figure 5. Visualization of water fingering phenomena in soil-air-oil system using (a) RGB system, (b) hue system, (c) intensity system. Vertical fluid content profile of unstable fingered in three-phases flow system: (d) water finger in soil-air-oil system.

Growth and Migration of Gas Bubbles for In-situ Bioremediation

Mark Dyer and Jennifer Faulconbridge, University of Durham, UK

ABSTRACT
The insitu bioremediation of contaminated land can often involve the injection or production of biogenic gas bubbles in the subsurface. The flow patterns for such gas bubbles and the effects on aqueous permeability are both complex and difficult to predict. A series of optical tests were carried out using an artificial sand to directly observe the growth and migration of gas bubbles for different injection rates, pore water pressure, injection point and soil heterogeneity. The results provide information on the relative size and stability of bubbles, the mode of upward movement in saturated sand and the obstructions presented by slight changes in soil grading. The results can be used to predict airflow patterns for soil column and field studies and potential effects on groundwater flow for in-situ bioremediation.

INTRODUCTION
There is a growing interest with the use of in-situ remediation technologies for the treatment of polluted soil and groundwater, particularly bioremediation. There are a number of factors that influence the suitability of in-situ bioremediation for a site, ranging from non-technical issues such as time, resources and finance to the biogeochemcial characteristic of the subsurface (Barr et al 2001). One of the key issues is the permeability of the ground and its influences on the delivery of carbon substrates, nutrients or oxygen to indigenous microbes. Consequently, in-situ bioremediation techniques are typically used in coarse-grained soils or fissured rock (e.g. gravel and sands) rather than finer grained soils (e.g. silts or clays) (Suthersan 1996, Anderson 1995). The higher permeability of coarse-grained soils enables liquids and gases to flow much more readily through the pore space than in fine-grained soils (e.g. silts and clays).

The insitu bioremediation of contaminated land can often involve the injection or production of biogenic gas bubbles in the subsurface. In the case of air sparging, air is typically injected below the lowest known level of pollution. Due to buoyancy effects, the injected air rises through the saturated zone towards the surface. Through a variety of mass transfer processes, the contamination is either stripped into the migrating air or oxygen as it is transferred into the subsurface, effectively assisting in the aerobic degradation of the pollutant. The air injection rates are typically 1000 to 2000 ml/min. A limited number of studies have been performed to investigated airflow in soils for in-situ bioremediation (Reddy and Adams 2001, Zhang and Burns 2000, Semer et al 1998, Fuchsberger and Semprich 1995 and Ji et al 1993). The work by Reddy and Adams was particulary useful in providing a more systematic investigation into the affects of soil heterogeneity on airflow patterns.

In comparison with the direct injection of air into the subsurface, the in-situ biodegradation of carbon substrates and pollutants such as chlorinated solvents can generate a wide range of biogenic gases at potentially slower rates. For example, butanediol fermentation pathway for glucose as a carbon substrate would produce short chain acids (formic, lactic, acetic), alcohol (ethanol), carbon dioxide and hydrogen (Singleton 1999). In contrast with air injection the rate of gas production can be several orders of magnitude less and the main concern is a decrease in soil permeability and obstruction to groundwater flow that could inhibit the release of a carbon substrate. For example, the laboratory anaerobic soil column tests by Dyer et al (2000) reported gas production at a rate of 0.1ml./min/gm of molasses. This rate of gas production was sufficiently to complete block the column within 48hrs and prevent groundwater flow. Although the column tests only modelled two dimensional groundwater flows at atmospheric pressure, the results highlighted potential problems with gassing in soils for insitu bioremediation that could be compounded with soil heterogeneity by variations in sedimentary facies. More recently alternative methods of supplying carbon substrates for in-situ bioremediation of chlorinated solvents have involved the injection of gas mixtures comprising nitrogen and methanol or methane to promote anaerobic reductive dehalogenation

TEST APPARATUS AND PROCEDURE

The laboratory tests were carried out in a modified 100mm diameter Perspex cylinder previously used for constant head permeability tests, as illustrated in Figure 1. The column was filled with particles of crushed borosilicate glass immersed in liquid paraffin (light grade). The similarity in refractive indices rendered the granular medium transparent (Dyer 1986). The borosilicate glass had been crushed using a mechanical crusher and sorted by wet sieving. The majority of tests were carried out using a particle size range equivalent to coarse sand (0.6-2.0mm).

Pore pressure was applied within the cylinder using a compressed air supply via a liquid/air interface in the volume-measuring device, which also measured the volume change in the pore liquid caused by the presence of the injected air. The pore pressure could be altered with the use of a pressure regulator, which operated between 2 – 22psi (13.8 – 151kPa). Air was injected into the base of the cell using a peristaltic pump (capable of working pressures of up to 170kPa and flow-rates between 0.23 – 20.77ml/min). Two pressure transducers were calibrated and the relationships between the mV reading and the actual pressure in kN/m^2 were determined. These were used to quantify both the pore-pressure and the entry pressure. During all tests the volume changes in the pore fluid were recorded and the behaviour of the air migration, including bubble diameters (growth and shrinkage), size distributions and appearance of channels, were all monitored and photographed with and without transparent sizing grids placed over the cell.

Three series of experiments were conducted. The first studied the pattern and behaviour of airflow injected into an aperture in the base of the test cell, positioned off centre to improve visibility. In the first series a total of three tests were conducted subject to the following conditions: (i) air was injected at a rate of 0.23ml/min under pore pressure due solely to the head of paraffin in the column; (2) pore-pressure was subsequently increased to50 kN/m^2 by compressed air via interface in the volume-measuring device; (3) air injection later halted and pore pressure progressively increased up to 200 kN/m^2. A second series of tests were carried out to study the effect of moving the injection point to mid-height within the sample. Mid-height injection was intended to investigate previous soil column tests by Dyer et al 2000, which used a plug of molasses as a carbon substrate for reductive dehalogenation of dichloroethane. In this second series the first tests were performed under a pore pressure

$50kN/m^2$ with the injection rate varied between 0.23 and 5.06ml/min. A second test was performed at an injection rate of 0.23 ml/min but with pore pressure increased to $100kN/m^2$. A third and final series of tests was carried out to investigate the effects of heterogeneity by layering the sample with bands of coarse, and medium grained sand size particles overlain by gravel size partilces. The layer of medium sand was positioned as a central band with areas of coarser sand towards the back of the cell (opposite the injection point) and at each side. It was positioned to investigate whether bubbles would travese laterally until reaching the zone of higher permeability. An injection rate of 1ml/min was used and a pore-pressure of $50kN/m^2$ was applied.

AIRFLOW PATTERNS
The photographs shown in Figure 2 are typical of airflow patterns observed for injection through a lance at mid-height of the sample. The patterns of airflow were created by two injection rates of 0.23ml/min and 5.05 ml/min respectively. The slower injection rate is similar to the rate of gas generated by fermentation of mollases in previous soil column experiments for bioremediation a chlorinated hydrocabron (Dyer et al 2000), whilst the higher rate models air sparging at a much reduced scale. The photographs show quasi-static columns of bubbles after 30 minutes of air injection. An examination of the photographs and accompanying sketches reveal certain features. Firstly, the migration of bubbles through saturated sand was characterised by clustering and collisions that trigger further upward movement. The clustering of bubbles appeared to increase the overall bouyancy force and so overcome resistance from the surface tension of the pore fluid between particles as well as the drag from fluid viscosity. In addition, clustering of bubbles left temporary gaps in the airflow pattern.

Increasing the injection rate to 5.05ml/min had the effect of significnatly broadening the airflow pattern. Although the injection rate was relatively low compared with the rates used for air sparging (typically 1000 to 2000 ml/min), the results demonstrate a beneficial increase in the air-water contact area, which in practice would increase the likelihood of volatilisation of the contaminants. Furthermore, the test showed that when the injection process was halted, existing bubble channels often collapsed and became saturated with the fluid. When injection was resumed new pathways were formed. The observation indicates that the pulsating the airflow (i.e. turning the system on and off at specified intervals) could provide a greater distribution and mixing of the air in the sub-surface as sometimes used for air sparging.

COEFFICIENT OF PERMEABILITY
There already exists a wealth of data about the effects of gas bubbles on the aqueous permeability of coarse-grained soils. Early work by Christiansen 1944, Corey 1957 and Wyckhoff and Botset 1936 provide valuable information on the relationship between degree of saturation and soil permeability as shown in Figure 3. More recently, researchers have used this information to explain transient flow through unsaturated (LeBihan JP and Leroueil S 1998). The data by Corey illustrates how the coefficient for permeability of water (K_{rw}) in sand can be dramatically reduced by a slight decrease in the degree of saturation below 100%. The permeability of water is shown to drop sharply when air invades the system and reaches a very small value while saturation is still considerably greater than zero. In comparison, the permeability of air (K_{ra}) approaches 100% at a saturation greater than zero.

However, there are doubts about the relevance of the data produced by Corey and others in the context of in-situ bioremediation. The results relates to a test condition where a uniform mixture of air and water was injected into a horizontal tube of sand. In the soil column tests on bioremediation gas was generated within the column. Likewise, the optical test results show a localised clustering of gas bubbles. The degree of saturation within the sample is clearly non-uniform, making the use of published data on relative changes in aqueous permeability difficult to apply in this context. Nevertheless, the results illustrate how fermentation gases would locally reduce aqueous permeability and possibly entrap bubbles for a loner term effect. In the field, the upward migration of gas bubbles could likewise significantly effect groundwater flow and mixing of the carbon substrate within an aquifer.

BUBBLE SIZE

With regard to the size and stability of individual bubbles, injection from the lower orifice (internal diameter 2mm) at the base of the column generally produced larger diameter bubbles. Whilst, smaller diameter bubbles were more frequently observed when air was injected through the smaller diameter lance (internal diameter 1mm) at sample mid-height. Figure 4 compares the distribution of bubble diameters produced by injection from the base and mid-height of the sample. These results were produced from observations made after 30 minutes of injection at 0.23ml/min with a pore-pressure of $50kN/m^2$. A greater number of bubbles are shown for injection from the base compared with the mid-height of the sample simply because the bubbles had a greater distance to travel and hence accumulate in the sample. Both distributions are skewed. The modal value of bubble diameter for both sets of data is similar to the injection orifice diameter. This would indicate that the orifice injector type and diameter have significant control on the bubbles size. In practice, the air sparging process would benefit from small diameter bubbles that transport as discrete elements through the soil. Decreasing the diameter of bubbles offers several advantages. Smaller diameter bubbles have larger surface area to volume ratio per volume of gas, which is favourable for mass transfer, and they have decreased buoyancy, which is accompanied by an increased residence time within the contaminated area.

REFERENCES

Barr D, Bardos P, J Weeks, Finnamore J and Nathanail P 2001. Biological treatment for contaminated land. CIRIA Funders Report RP625.

Corey AT 1957. Measurement of water and air permeability in unsaturated soil. Soil Science Society Proc.

Dyer MR, 1986 Observation of Stress Distribution in Crushed Glass with Application to Soil Reinforcement. DPHil Thesis, Oxford University

Dyer MR, Van Heiningen E and Gerritse J. 2000. In-situ bioremediation of 1,2-dichloroethane under anaerobic conditions. Geotechnical and Geological Engineering Vol 18(4) pp 3131-334

Fuchsberger M and Semprich S 1995. Air flow through partially saturated cohesionless soil. Proc. 1st Int. Conf Unsaturated Soils. Paris.

Ji W, Dahmani A, Ahlfeld DP, Lin JD and Hill E 1993. Laboratory study of air sparging: air flow visualization. Groundwater Monitoring Review. 13(4)

Reddy KR and Adams JA 2001. Effects of soil heterogeneity on airflow patterns and hydrocarbon removal during in-situ air sparging. Geotechnical and Geoenvironmental Engineering pp234-247.

Semer R, Adams JA and Reddy KR 1998 An experimental investigation of airflow patterns in saturated soils during air sparging. Geotechnical and Geological engineering Vol. 16(1).

Singleton P 1995. Bacteria in biology, biotechnology and medicine. Chap 5. Wiley

Wyckhoff RD and Botset HG 1936. The flow of gas-liquid mixtures through unconsolidated sands. Physics (7)

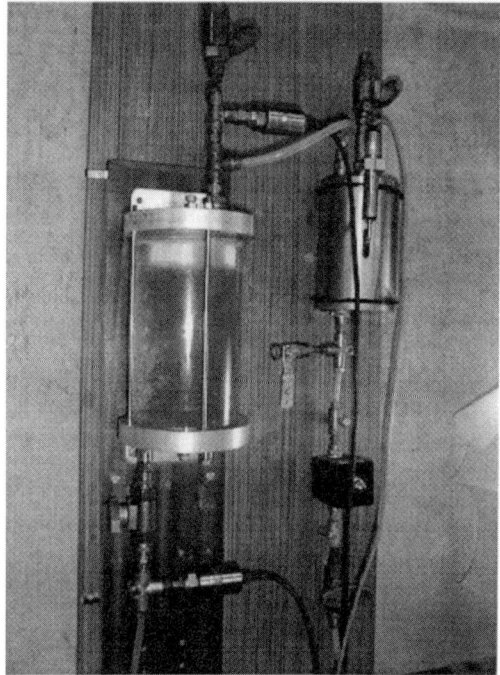

Figure 1 Test Apparatus

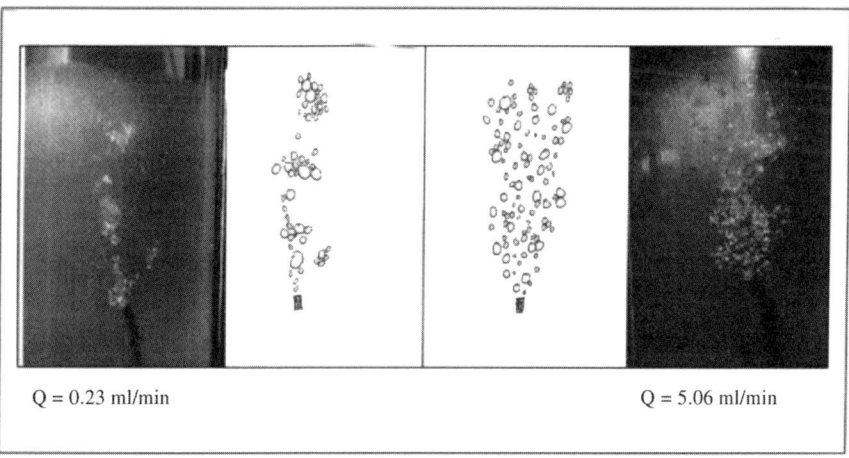

Q = 0.23 ml/min

Q = 5.06 ml/min

Figure 2 Preliminary tests results showing quasi-static column of air bubbles
for different flow rates (Q)

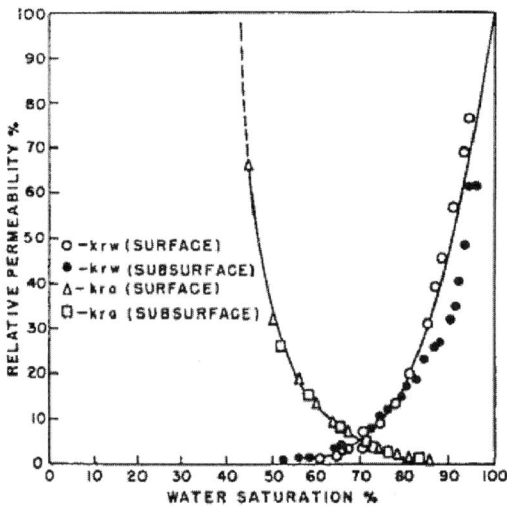

Figure 3 Relative permeability of air (K_{ra}) and water (K_{rw}) as a function of degree of saturation (after Corey 1957)

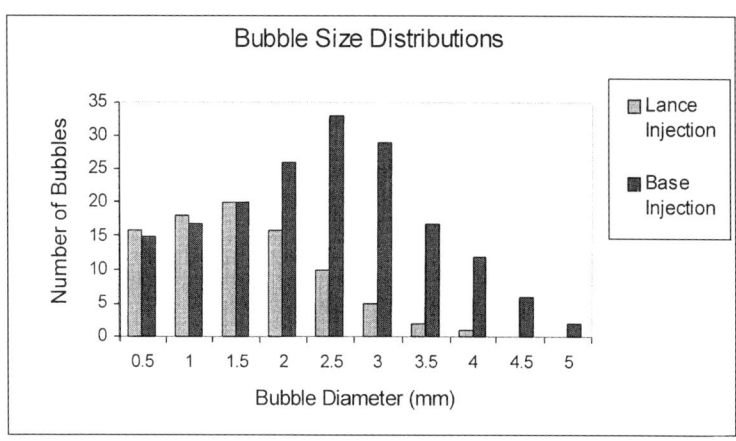

Figure 4 Distribution of bubble sizes for two injection points

OBSERVATIONS ON THE MIGRATION OF CHLORINATED SOLVENTS IN POROUS MEDIA

Mark Dyer and Claire Lambert, University of Durham, UK

ABSTRACT
Preliminary laboratory tests have been carried out to investigate the migration of chlorinated compounds in coarse-grained soils above and below the water table. The laboratory tests were carried out by staining trichloroethylene (TCE) with the red dye Sudan IV (Fisher Scientific) and monitoring the migration through a sample of glass ballotini. The results illustrate the downward migration TCE as a finger and eventually pooling at the base of the container. In the dry sample, TCE preferentially coated the ballotini as the wetting fluid. In a moist sample of ballotini, water displaced TCE as the wetting fluid and created discrete globules of TCE (termed residuals) within a broader band of pollution. Penetration of TCE below the water table (once sufficient head overcame capillary pressures between soil grains) was characterised by preferential wetted pathways.

INTRODUCTION
Organic pollutants are often the most troublesome types of pollutants found at contaminated sites. The chemicals can be environmentally significant at low aqueous and gaseous concentrations and difficult to remove. A particularly problematic group of organic contaminants are dense non-aqueous phase liquids (DNAPLs), which include chlorinated hydrocarbons (commonly used as degreasers and cleaners or bulk chemicals for production of plastics). The relatively low viscosity, high density and low aqueous solubility typically leads to deep-seated pollution, where the immiscible liquid penetrates the aquifer as a finger of non-aqueous liquid that can pool at obstructions. The obstructions are commonly due to changes in soil permeability, such as a layer of fine sand, silt or clay. Subsequently, the contaminant can pollute an aquifer by a slow rate of dissolution from residual globules or pools of pure product (Pankow and Cherry 1996). The authors have encountered this scenario at several sites, where subsequent clean-up strategy depending on an understanding about the likely distribution of the chlorinated compounds in the sub-surface (Dyer et al 2001).

There have been a limited number of previous studies into the mechanisms controlling the migration of chlorinated compounds into the sub-surface. Using the experimental technique developed by Schwille (1988), a series of preliminary tests were carried out to examine the migration of chlorinated solvent in a coarse-grained soil. In particular, a comparison was made between migration above and below the water table, as well as the effect of wettability on the difference of a hydrophobic chlorinated compound in moist soil.

TEST METHOD
The chlorinated compound trichloroethylene (TCE) was stained red using the dye Sudan IV (Fisher Scientific) at a concentration of 1g/l. Sudan IV is soluble in TCE and insoluble in

water. Glass ballotini was selected as the porous media. The ballotini were transparent around the sides of the sample, which meant that migration of the chemical could be easily observed. The glass spheres ranged in size from 1.5 to 2mm in diameter. Falling head tests carried out on samples of ballotini recorded a permeability of 2.55×10^{-2} m/s for a porosity of 0.418. The specific gravity is 2.94. All photographs were taken using a digital camera set on 'macro' function, thus enabling accurate recording of the results at sufficient magnification.

Controlled quantities (up to 40ml) of TCE were released onto samples of ballotini using a syringe as shown in Figure 1. The ballotini was contained in a 1 litre glass beaker. The level of water in the sample was controlled using a peristaltic pump as shown in figure 1b.

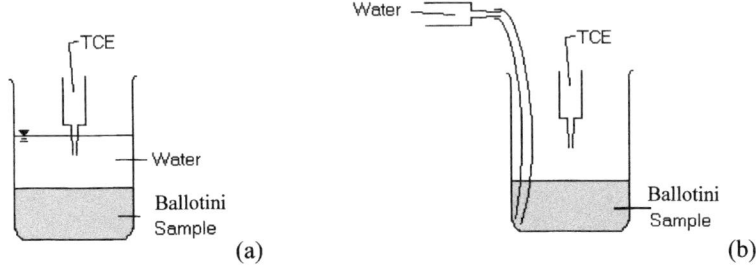

(a)   (b)

Figure 1 Apparatus for (a) saturated and (b) unsaturated tests

THEORY
The distribution of a chlorinated compound in saturated coarse-grained soil depends on several physical and chemical properties of the chemical, groundwater and soil. Interfacial tension is one of the key properties of the chlorinated compound that controls the hydraulic pressure needed to overcome capillary pressure between soil particles to migrate into the sub-surface as well as the eventual spreading of the pollutant as a pool. Interfacial tension occurs, as the name implies, at the interface between two immiscible fluids. It arises due to an imbalance of cohesion forces on molecules at the surface of a fluid. Consequently both surfaces of the fluids acts as a stretched membrane. If any surface molecule is raised or depressed slightly, the molecular bonds between it and the adjacent molecules are stretched and so a restoring force tends to pull the molecule back into its original position of equilibrium. As a result, surfaces contract to the smallest possible surface area, thus minimising any excess free energy present at the interface. Hence, small droplets of water in air are spherical is due to interfacial tension. Mechanical work must be done for there to be any increase in surface area. Interfacial tension is defined as the amount of work necessary to separate a unit area of one substance from another and is measured in dynes/cm (10^{-5} N/cm). It is the means by which two fluids can exist adjacent to one another while remaining at different pressures.

The forces of attraction between similar molecules within the same substance are known as cohesive forces. The attractive forces between molecules of one substance and molecules of another substance, such as another fluid or a solid are known as adhesive forces. They are often of a different magnitude to cohesive forces. This is the basis of interfacial forces. If the adhesive forces of a solid surface are greater than the cohesive forces between molecules in a liquid, the liquid molecules are attracted to the solid and so spreads of it. The surface of the fluid becomes curved as the liquid spreads over the solid surface, a process known as 'wetting'. When wetting occurs, the affinity of two fluids at an interface for the solid surface can be measured by looking at the contact angle (Tipler 1991). The contact angle can be calculated using the following expression, where σ represents interfacial tension between solid, gas or liquid.

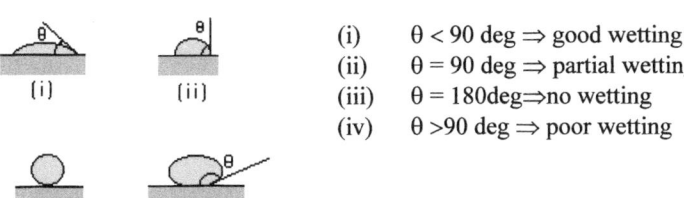

$$\cos\theta = \frac{\sigma_{sg} - \sigma_{sl}}{\sigma_{gl}} \qquad \text{(eqn 1)}$$

If the angle $\theta > 90$, the gas (g) will be the wetting fluid. If $\theta < 90$ the liquid (l) is the wetting fluid, as shown in Figure X. The extent of the interface and therefore, the spreading of the drop are manifested as the contact angle (Rao 1972).

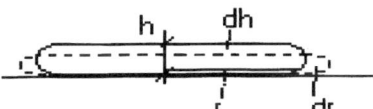

(i) $\theta < 90$ deg $\Rightarrow$ good wetting
(ii) $\theta = 90$ deg $\Rightarrow$ partial wetting
(iii) $\theta = 180$deg$\Rightarrow$no wetting
(iv) $\theta > 90$ deg $\Rightarrow$ poor wetting

Figure 2 Wettability defined by contact angles

For the case of a chlorinated compound pooling on an obstruction in the sub-surface, a relationship can be determined between the spreading coefficient and the height of a sessile drop by considering a circular drop of radius r, height of h and constant volume (V) as shown below.

If the drop is assumed to spreads by a small amount, the radius can be considered to increase by dr and the height of the centre of gravity decrease by a corresponding amount from h/2 to (h-dh)/2. To maintain equilibrium, the total change in energy must be equal to zero. Hence a direct relationship between the contact angle and thickness of a drop can be expressed as follows. A full derivation is given by (Davies and Rideal 1963).

$$\theta = \cos^{-1}\left(1 - \frac{1}{2\sigma_{la}}\Delta\rho_l gh^2\right) \qquad \text{(eqn 2)}$$

When the wetting angle is known, the head of pressure (H) needed to overcome capillary pressure between soil particles can be determined using the following expression. A full derivation can be found in Tipler (1991) and Lambert (2001)

$$H = \frac{2\sigma\cos\theta}{\rho rg} \qquad \text{(eqn 3)}$$

Since $\Delta P = \rho gH$, the above equation can be modified to apply to any opening with a pore throat of radius (r) (Pankow JF and Cherry JA, 1996)

$$\Delta P = \frac{2\sigma\cos\theta}{r} \qquad \text{(eqn 4)}$$

RESULTS

Migration and Pooling in Unsaturated Zone

When released onto the surface of a sample of dry ballotini, TCE was observed to quickly migrate downwards as shown in Figure 3. The chemical preferentially coated the particles as the wetting fluid. A pool formed at the base of the sample, which was approximately 5mm thick. The pool was slightly thicker below the point of injection of the TCE as the residual coating on the beads meets the pool here. At the base of the sample, the pool spread continuously with an even coating as shown in Figure 4. This was due to the wettability of the chemical relative to air. It could displace air from the pores, thus filling the pore space. Only when the whole base of the beaker was covered did the thickness of the pool begin to increase.

When the TCE was released onto moist ballotini, the chemical again migrated downwards through the sample as shown in Figure 5. However, the coating on the ballotini was less even due to water preferentially wetting the beads and hence residual globules of TCE were formed in the pore spaces. In addition, the downward migration of TCE resulted in wide front due to the displacement of TCE by water. A pool eventually formed at the base of the sample, which again was thicker below the point of injection. In contrast with the dry sample of ballotini, the pool was not continuous and did not completely envelop the beads. This gave the base of the pool a mottled appearance. The pool was on average approximately 6mm thick. However, the thickness was very uneven, due to the presence of water in some pores and not others.

Migration and Pooling in Saturated Zone

TCE was released onto a sample of ballotini partially submerged with water. The water had been injected from the bottom of the sample using a peristaltic pump. Hence, the beads above the water table were completely dry. The TCE was observed to migrate quickly downwards through the dry ballotini until reaching the water table again preferentially wetting the dry beads. Subsequently, TCE pooled on the surface of the water until a sufficient head of liquid overcame the capillary pressures between ballotini for penetration to continue. TCE penetrated the saturated sample as two separate fingers of pollution below the point of injection as shown in Figure 6. As additional TCE was added the pool on the water surface spread over the whole sample and the thickness of the pool increased. However as the surface pool increased in thickness, no additional fingers were created to penetrate into the water table. In fact, increasing the thickness only resulted in further penetration of the initial fingers that eventually reached the bottom of the sample directly below the point of initial penetration. Once the initial fingers had been formed, this provided a preferential 'wetted' path for the chemical so it would have been easier for these fingers to penetrate further than to create new fingers. The fingers penetrating the saturated ballotini were substantially thinner than the pollution penetrating the dry sample. This shows the funnelling effect at the capillary zone, noted by Pankow and Cherry (1996).

TCE was later released onto a saturated sample of ballotini with an inclined surface. The chemical was injected below the water surface at the top of the sloped sample, ran down and pooled at the lowest point of the surface. The TCE ran down the slope as a series of small globules, thus minimising any free surface energy. Some of these globules were left on the slope as residual pollution. When more TCE was injected, the larger drops collided with the residuals, which were transported down to the main pool, hence reducing the amount of residual present on the slope. The pool increased in thickness until penetration occurred at the deepest part of the pool. This thickness was measured as 12mm. In comparisons, equation (4) can be used to calculate the thickness needed to overcome capillary pressures between ballotini for penetration as shown below. A contact angle of 134 deg was used based on measurements made on single sessile drop of TCE in earlier tests. The measured and calculated thicknesses are in good agreement.

Ballotini radius = 2mm Opening - radius = 0.03mm	$\Delta P = \rho gh = \dfrac{2\sigma \cos\theta_w}{r}$ Assuming an opening of a radius 0.03mm, $\Delta P = \dfrac{2 \times 34 \times \cos 134}{0.03} = 1.47 \times 981 \times h$ $\Rightarrow h = 1.1cm$

DISCUSSION

Although only a limited number of tests were carried out, the results provided a useful insight into the migration of chlorinated solvents into coarse-grained soils such as fluvial or glacial sands and gravels. The tests did not model any effects from soil sorption, which are more relevant to soils containing clay particles and organic carbon. The results illustrated the importance of wettability when considering the distribution of a chlorinated solvent as residual globules in moist and saturated soils. Residuals in the vadose zone can lead to significant pollution of soil vapour as reported by Dyer et al (2001). Pooling of a chlorinated solvent was also observed both on the water table and at the base of the sample. The pooling of a dense non-aqueous phase liquid on water table is not often mentioned in textbooks on pollution from chlorinated solvent, since the liquid is denser than water and hence attention is normally given to deep seated sources of pollution. However as the tests showed until the pressure head exceeds capillary pressures, pooling would take place on the water table. Subsequently, penetration of TCE was characterised by preferential pathways, which did not increase in number with release of additional TCE.

REFERENCES

Davies JT, Rideal EK, 1963. Interfacial Phenomenon. Academic Press, London.

Dyer M, Zutphen M and Hetterschijt R, 2001. Improved site characterisation of contaminated land using pump and treat data. ICE Proc Geotechnical Engineering 11892

Lambert C 2001. Physical and numerical modelling of chlorinated solvents in groundwater. Final year MEng project, School of Engineering, University of Durham.

Pankow JF and Cherry JA, 1996. Dense chlorinated solvents and other DNAPLs in groundwater: history, behaviour and remediation. Waterloo press, Portland, Orgeon.

Rao SR, 1972. Surface Phenomena. Hutchinson Educational, London.

Schwille F, 1988. Dense chlorinated solvents in porous and fractured media – model experiments. Translated by JF Pankow. Lewis Publishers.

Tipler P 1991, Physics for Scientists and Engineers – Extended Version. 3rd Edition

Figure 5 Migration of TCE through moist ballotini

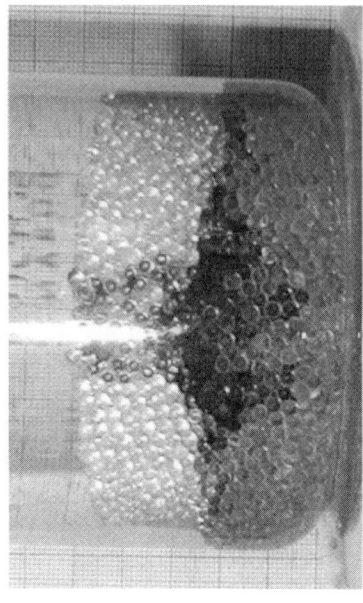

Figure 6 Migration of TCE through dry and saturated ballotini

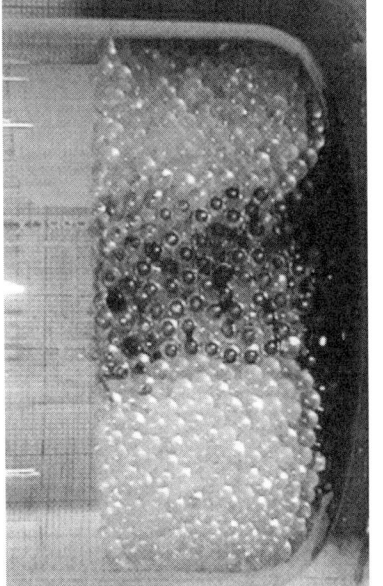

Figure 3 Migration of TCE through dry ballotini

Figure 4 Spreading of TCE at base of dry ballotini

Soil Acidification Effect on some Physico-Chemical Soil Properties of Clayey Materials

M. Elzahabi[a] & R.N. Yong[b]

[a] Department of Civil and Environmental Engineering, University of Carleton, Ottawa, Canada

[b] Geoenvironmental Engineering Research Center, Cardiff School of Engineering, University of Wales, Cardiff, U.K.

ABSTRACT

In naturally acidic soils, pollutant laden acid rain as well as acid leaching and flushing remediation approaches makes the transport of heavy metals in the vadose zone an important field of investigation. A series of geotechnical and geochemical laboratory tests before and after soil washing with nitric acid were used to investigate the effect of the soil pH and the presence of carbonate on certain physico-chemical illitic soil properties. The overall geotechnical properties of the treated soils revealed no significant changes after the washing procedures.

Meanwhile, reducing the soil pH has a considerable effect on the geochemical aspects of the treated soils. The loss of cation exchange capacity in the soils is associated with the presence of carbonate, low soil pH, and the exchangeable cations in the soil. The significance of these factors and how they are likely to affect the movement of metals in clayey soils are also discussed.

INTRODUCTION

One approach to treating contaminated sites is physical separation and removal of the contaminants from the soil. Physical separation can be achieved in situ by introducing a fluid into the soil that will flush out the contaminants, while leaving the soil matrix intact. Soil flushing uses water, a solution of chemicals in water, or organic extractant to recover contaminants from the in situ material. Solutions such as hydrochloric acid (HCl), ethylenediamine tetraacetic (EDTA), and calcium chloride ($CaCl_2$) can be used as flushing agents, (EPA, 1997). Soil flushing is an in situ extraction of contaminants from the soil. This technology is applicable for both organic and inorganic contaminants and for metals in particular. The use of soil flushing chemicals may involve adjusting the soil pH, chelating metal contaminants or displacing toxic cations with nontoxic cations. Several chemical and physical phenomena control the mobility of metals in soils. The finer soil sized fractions can bind metals electrostatically as well as chemically. Numerous soil factors affect the sorption of metals and their migration in the subsurface. Such factors include particle size, specific gravity, water content, Atterberg limit, hydraulic conductivity, soil type, soil density and pH, organic and amorphous content measurement, cation exchange capacity (CEC), specific surface area (SSA) and the carbonate content, (Elzahabi & Yong, 1999).

In order to investigate the effect of acid leaching on certain physico-chemical illitic soil properties, soil pH was reduced to the desired values by washing the soil several times with a 1:10

Geoenvironmental impact management, Thomas Telford, London, 2001.

ratio of distilled water to a nitric acid over a long period of time to insure complete soil pH equilibrium. The soil was then air-dried and ground to pass through a 2 mm sieve.

MATERIALS AND METHODS
A series of geotechnical and geochemical laboratory tests before and after soil washing with nitric acid were used to investigate the effect of acid leaching on certain physico-chemical illitic soil properties.

GeotechnicalCcharacterization
Measurements of grain size distribution were performed experimentally using the mechanical (sieve analysis) and the hydrometer method according to ASTM standards method D422. The specific gravity, water content and Atterberg limits were performed experimentally according to ASTM standard methods D854, D2216-80 and D4318-84, respectively. Determination of optimum moisture content and maximum dry density relationships for soils were carried out following ASTM standard D698-78, impact method, using a 2.49 kg rammer and 305 mm drops.

Geochemical Analysis
The cation exchange capacity (CEC) for the soil was determined using two different methods: the silver-thiourea method described by Chhabra et al. (1975) and the barium chloride method described by Hendershot and Duquette (1993). The specific surface area of the material was determined using the Ethylene Glycol-Monoethyl Ether method, following the procedures described by Eltantaway and Arnold (1973), and Carter et al. (1986). The carbonate content of the soil was determined in accordance with the titration method described by Hesse (1971) in the soil chemical analysis handbook. Soil pH was measured in a 1:2 ratio of soil to distilled water solution. 20 ml of distilled water was added to 10 g weight of 2 mm soil and shaken for 30 minutes. Then, the suspension was allowed to settle for 30 minutes. The pH of the soil was measured using a Beckman I^{TM} 12/pH/ISE meter, according to the method described in the analytical methods manual edited by Sheldrick (1984). The organic content of the soil was determined using the titration method described by Jackson (1956). The presence of amorphous materials such as: silicon dioxide (SiO_2), iron oxide (Fe_2O_3), and aluminum oxide (Al_2O_3), was determined using the method described by Segalen (1968).

RESULTS & DISCUSSION
Geotechnical Aspects
The predominant mineral used in this study is illite with varying amounts of chlorite, quartz, feldspar and calcite. The results of the soil properties and composition analysis are close to those found by Macdonald (1994). A similar study by Karczewska (1987), showed that the mineralogical analysis of soil samples after acid leaching revealed no significant changes in the soil composition. The reduction of the soil pH does not alter the mineralogical composition. The predominant minerals for both treated and untreated soil were unchanged and these minerals are: illite, chlorite and carbonates such as calcite, and dolomite.

The standard proctor compaction test was used according to ASTM D698 to determine the maximum dry density and the optimum moisture content of soil A at pH 9.5 and soil B at pH 6.9. Results show that at low soil pH, the maximum dry density and moisture content revealed no significant changes.

The carbonate contents in soil A and B were considered to be significant and varied between 15.18% and 11.73%. This may explain their high values of soil pH. The carbonate contents for soil C and D decreased with decreasing soil pH and ranged between 5.22% and 4.60%.

Particle Size Distribution

Particle size distribution affects the surface area for adsorption of heavy metals. The particle size distributions were obtained by using mechanical sieve analysis in combination with the hydrometer method. Sieves were used to obtain the percentages of the following soil components: a) Sieve No.4 (4.75 mm), gravel; b) Sieve No. 4 (4.75 mm) and No. 40 (0.420 mm), coarse to medium sand; c) Sieve {No. 40(0.42mm to No. 200 (0.075mm)}, fine sand, and d) silt passes through sieve No. 200. For clay, where particles are less than 0.002 mm, the hydrometer method was used to determine the fraction down to 0.001 mm.

Grain size analysis of the samples A, B, C, and D showed the following composition:

Soil A : clay 44%, silt 40%, sand 16%, gravel 0 %
Soil B: clay 41%, silt 40%, sand 19%, gravel 0%
Soil C : clay 38%, silt 38%, sand 24%, gravel 0%
Soil D: clay 35%, silt 41%, sand 24%, gravel 0%

It was found that during the washing stages as the carbonates were extracted, the soil lost some fine clay particles and the portion of the silt and the sand was slightly increased. The significance of this factor is how it is likely to affect the movement of H.M. (Heavy metals) along the soil column. Variations in adsorption between different size fractions are mostly a reflection of their carbon content (Yong et al., 1992a). The capacity for H.M. movement is increased not only by low pH and low carbonate content but also by large particle size.

Atterberg Limits

Liquid limits (LL) and plastic limits (PL) of soils were determined using ASTM D4318-84. The relation between Atterberg limits, soil pH and carbonate content for soil A (pH 9.5), soil B (pH 6.9), soil C (pH 4.0) and soil D (pH 3.5) are illustrated in Figure 1. Results show that when the soil is mixed with distilled water, the liquid and plastic limits of the soil increase slightly with a decrease in soil pH and carbonates content (soil D & soil C). This might occur due to an increase in the amount of water trapped between particles and due to the flocculated arrangement of acidic soil, where the edges of clay particles become positively charged after attracting extra hydrogen ions (Hoppe, 1986). Changes of the LL may be used as a first indicator of the effect of a liquid on a soil (Bowders and Daniel, 1987) whereas a drastic change in the LL indicates that the structure of the clay may be affected by contact with the specific leachate. Figure 2 compares the LL of soil A at pH 9.5 and soil B at pH 6.9, mixed with distilled water and lead solution. It is observed that the LL of soil A (pH 9.5) and soil B (pH 6.5) was not affected by the soil mixing with a high lead solution (5000mg/l). This indicates that the percolation of highly concentrated Pb solutions did not affect the structure of the illitic clay.

Hydraulic conductivity

This particular case shows that the slight variations in the overall hydraulic conductivity results cannot be considered important as they fall within the same order of magnitude. This indicates that reducing the soil pH by dissolving some of the carbonate bonds does not alter significantly the hydraulic conductivity of the illitic soil. The latter means that the mineralogical composition of the treated soils remain unchanged.

Geochemical Aspects

Cation Exchange Capacity

The Cation Exchange Capacity (CEC) of soil A (pH 9.5) was determined using the silver thiourea method while the barium chloride method was used for soil B (pH 6.9), soil C (pH 4.0) and soil D (pH 3.5). The Cation Exchange Capacity values for soil A, B, C, and D, as a function of soil pH and carbonate content are presented in Figure 3. The CEC value of soil A, rich in carbonates, was calculated to be 34.20 meq/100g. The results show that the dissolution of carbonates and the acidification of the soil cause a decrease in the CEC values obtained, ranging from 34.20 meq/100g (Soil A, illite at pH 9.5) to 7.94 meq/100g (soil B, pH 6.9), to 6.81 meq/100g (soil C, pH 4) and to 5.34 meq/100g (soil D, pH 3.5).

The CEC obtained for soil A, falls well within the range of reported values which are usually between 20 to 40 meq/100g (Grim,1968; Yong et al., 1992a; and Hausenbuiller, 1985; Ouhadi,1997). The CEC measured on the illitic untreated soil is much higher than the CEC of the soil with carbonates extracted. It is clear that the loss of the CEC is associated with the presence of carbonates in the soil. Some researchers have reported similar decreases in the CEC. MacDonald and Yong (1997) indicated that the removal of carbonates resulted in a drop in the CEC from 24 meq/100g to 8.2 meq/100g. This drop might be explained as a result of the dissolution of soil carbonates during CEC measurement causing an artificially high CEC when the carbonates are present. The carbonate dissolution process showed a great excess in Ca^{2+} and Mg^{2+} cations in the pore fluid. Karczewska (1987) studied the effect of acid leaching on illitic soil properties. He demonstrated that acid precipitation has a major effect on soil pH and on CEC. Clayey samples showed the greatest changes in the soil pH (up to 46%) and revealed a large decrease in the cation exchange capacity (up to 73%) especially when soil is leached with strong acid of pH 2.0. Furthermore, it is observed from Figure 3 that the CEC was found to be pH dependent. It increases with increasing soil pH and decreases with decreasing soil pH and increasing soil acidity (Robitaille, 1982; Yong et al., 1992a). The influence of pH on the CEC is well known by many researchers. The pH-dependent cation exchange capacity data measured by Pratt (1961) showed that the CEC for some soil samples changed continuously with pH and increased from pH 4.5 to 8.0. CEC measured in other soil samples with pH 3 to 4.5 gave a constant value. The present study agrees with this finding as the CEC measured between soil pH 4 and soil pH 3.5 is only slightly different.

Moreover, results showed that by lowering the soil pH, the soluble cations such as Ca^{2+}, Na^+ and K^+ were reduced and leached out during the soil preparation procedures and their amounts decreased with the soil pH adjustment. Only calcium and magnesium cations, released from the soil structure, significantly increased with lower soil pH. It is observed from Figures 4 that the amount of Ca^{2+} released from the soil structure to the soluble phase increases with decreasing soil pH, and reaches its maximum at soil pH 6.9, then decreases with decreasing soil pH and carbonate content. The amount of Mg^{2+} released from the soil structure shows an increase with decreasing soil pH, whereas, the amount of Na^+ and K^+ released to the soluble phase shows only a slight decrease with decreasing soil pH.

Furthermore, these research results showed that the accumulation of nitrate leads to an increase in H^+ ion concentration which replaces the basic cations such as Na^+, Ca^{2+} and Mg^{2+} on the clay surface and increases their leaching out of the soil. Leaching of cations leads to an increase in soil acidity and therefore, lowers the soil pH, as demonstrated by Robitaille (1982). In addition to leaching of nutrients, an increase in soil acidification leads to the mobility of other toxic elements such as Al, Mn, etc. (Dale and Turner, 1982). One might expect a significant decrease in the CEC at soil pH 3.5. However, instead, with the presence of exchangeable cations and the increase in calcium and magnesium exchangeable cations, the CEC of the soil only decreased slightly (Pratt, 1961) and was still able to almost completely retain the introduced metals under unsaturated conditions, (Elzahabi & Yong, 1997).

Specific Surface Area

The SSA results are presented in Figure 5 in terms of soil pH and carbonate content. It was observed that the SSA increases with decreasing the soil pH and the presence of the carbonate content in the soil. As shown, a SSA of 76.41 m^2/g (soil A, pH 9.5) was measured when soil carbonates were present. However, the decrease of soil pH and the removal of soil carbonates resulted in an increase in the soil SSA to 86.09 m^2/g (soil B, pH 6.9), to 100.41 m^2/g (soil C, pH 4) and to 101.91 m^2/g (soil D, pH 3.5). The extraction of carbonates resulted in a SSA which was greater than the SSA of the illitic soil A. The surface area of the acidic soil tends to increase as the carbonate content and the pH of the soil decrease. This increase in the surface area when the soil pH decreases and soil carbonates solubilize was expected, given that soil carbonates act as a strong bonding agent in soil.The removal of the carbonate bonding provided the opportunity for greater particle dispersion and thus provided a larger SSA as recorded by many researchers (Yong and Warkentin ,1975; Mitchell, 1993; Kersten and Förstner, 1989; MacDonald and Yong,1997). The resulting increase in SSA due to particle dispersion is likely to affect the sorption of heavy metals. In fact, it should be noted that retention of heavy metals is due to the availability of exposed clay particle surface. Bear in mind that the formation of clusters in the case of compacted materials will considerably decrease the effective SSA, resulting in less adsorption.

SUMMARY

The overall geotechnical properties of the treated soils revealed no significant changes after the washing procedures. The standard proctor compaction test results did not show any significant changes in the maximum dry density of the soil by lowering the soil pH. Only a slight decrease occurred in the optimum moisture content. The liquid limits of the soil show a slight increase with a decrease in the soil pH and carbonate content. In addition, it was observed that during the washing stages and as the carbonate content was extracted, the soil lost some fines clay particles and the portion of silt and sand increased slightly resulting in an increase in voids ratio. These changes might increase H.M. movement along the soil column. The slight variations in the overall hydraulic conductivity results cannot be considered important as they fall within the same order of magnitude. This indicates that reducing the soil pH by dissolving some of the carbonate bonds did not significantly alter the hydraulic conductivity of the illitic soil. Furthermore, the mineralogical composition of the treated soils remains unchanged. Meanwhile, reducing the soil pH has a considerable effect on the geochemical aspects of the treated soils. The loss of CEC in the acidic soils is associated with the presence of carbonate, low soil pH, and the exchangeable cations in the soil. The removal of carbonates decreases the soil pH, increases the cations leaching and results in a drop of the CEC. In addition, the removal of the carbonate bonding provided the opportunity for greater particle dispersion and thus a larger SSA.

The significance of these factors is how they are likely to affect the movement of H.M. along the soil. Variations in adsorption between different size fractions are mostly a reflection of their carbon content. The capacity for H.M. movement is increased by low pH, low carbonate content, and large particle size. Furthermore, the resulting increased exposure of the effective SSA and the CEC may also be involved when an increase in heavy metal sorption has been observed. These factors are considered when sorption of heavy metal in unsaturated illitic soil is investigated, (Elzahabi & Yong, 1999).

REFERENCES

Bowders, J.J., Jr. and Daniel, D.E. , 1987. Hydraulic conductivity of compacted clay to dilute organic chemicals. Journal of Geotechnical Engineering, ASCE 113-12:1432-1448.
Carter, D.L., Mortland, M.M. and Kemper, W.D. 1986. Specific surface area measurement in methods of soil analysis. Klute, A. et al., 2nd ed., Part I, American Society of Agronomy,

Madison, Wisconsin, pp. 413-423.

Chhabra, R., Pleysier, J. and Cremers, A., 1975. The measurement of the cation exchange capacity and exchangeable Cations in soil: a new method. Proceedings of the International Clay Conference. pp. 439-448. Applied Publishing Ltd., Illinois, U.S.A.

Dale J. And Turner J., 1982. The effects of acid rain on forest nutrient status. J. Water

Eltantaway, I.N. and Arnold, P.W., 1973. Reappraisal of ethylene glycol mono-ethylether (EGME) method for surface area estimations of clays. Soil Science 24:23-238

Elzahabi M. and Yong R.N.,(1999), pH Influence on Sorption Characteristics of Heavy Metal in the Vadose Zone, Proceedings of the 2nd Geoenvironmental Engineering Conference on Ground Contamination organized by the British Geotechnical Society and the Cardiff School of Engineering, and held in London at the Institution of Civil Engineers, September 1999, pp.255-263

Elzahabi M. and Yong R.N., 1997. Vadose zone transport of heavy metals. Proceedings of the First Geoenvironmental Engineering Conference organized by the British Geotechnical Society and the Cardiff School of Engineering, University of Wales, Cardiff, September 1997, pp. 173-180.

EPA, 1997. Recent Developments for In Situ Treatment of Metal Contaminated Soils. EPA-542-R-97-004

Grim, R.E., 1968. Clay Mineralogy. 2nd edition, McGraw-Hill, N.Y., 596.

Hare, F.K. and Hutchinson, T.C., 1989. Human environmental disturbances. Environmental Science and Engineering Handbook, edited by Henry J.G. and Heinke G.W., pp.114-142.

Harter, R.D., 1983. Effect of soil pH on adsorption of lead, copper, zinc and nickel. Soil Sci. Soc. Am. J. 47:47-51

Hausenbuiller, R.L., 1985. Soil Science, Principles and Practice, Third Edition, Wm.C.Brown Publishers, Dubuque, IA.

Hendershot, W.H., Lalande, H. and Duquette, M., 1993. Ion exchange and exchangeable cations. Chapter 19. Soil Sampling and Methods Analysis. Canadian Society of Soil Science Press. pp. 167-176.

Hesse, P.R., 1971. A Textbook of Soil Chemical Analysis. William Clowes and Sons, London, 519 pp.

Hoppe, E., 1986. The influence of acid rain on the engineering properties of a sensitive clay. Msc. Thesis, Department of Civil Engineering and Applied Mechanics, McGill University, Montreal.

Jackson, M.L., 1956. Soil chemical analysis- advance course. Published by the Author. University of Winsconsin, Madison.

Karczewska, H., 1987. The effect of acid leaching on some physico-chemical properties of Quebec soil. Master of Engineering Thesis, Department of Civil Engineering and Applied Mechanics, McGill University, Montreal.

Kersten, M. and Förstner, U. 1989. Speciation of trace elements in sediments. In Trace Element Speciation, Analytical Methods, and Problems (G.E. Batley, ed.) pp. 245-317.

MacDonald E., 1994. Aspects of competitive adsorption and precipitation of heavy metals by a clay soil. Msc. Thesis, Department of Civil Engineering and Applied Mechanics, McGill University, Montreal.

MacDonald E. and Yong R.N., 1997. On the retention of lead by illitic soil fractions. Proceedings of the First Geoenvironmental Engineering Conference organized by the British Geotechnical Society and the Cardiff School of Engineering, University of Wales, Cardiff, September 1997, pp.116-127.

Mitchell, J. K.,1993. Fundamentals of Soil Behaviour. John Wiley and Sons, Inc. Toronto, 437pp.

Ouhadi, V.R., 1997. The role of marl components and ettringite on the stability of stabilised marl. Ph.D. Thesis, McGill University, Montreal.

Robitaille, G, 1982. The effect of acid deposits on soil. Environment Canada- Quebec.

Segalen, P., 1968. Note sur une methode de determination des produits mineraux amorphes dans certains sols a hydroxides tropicaux. Cah, Orstom ser. Pedel 6:105-126.

Sheldrick, B.H, 1984. Analytical Methods Manual 1984. Land Resource Research Institute, Ottawa, Ontario. No:84-30. pp.2.

Yong R.N., Mohamed, A. M. O. and Warkentin, B. P., 1992a. Principles of contaminant transport in soils. Elsevier, Amsterdam, pp.211.

Yong , R.N., and Warkentin, B.P., 1975. Soil properties and behaviour. Elsevier, New York.

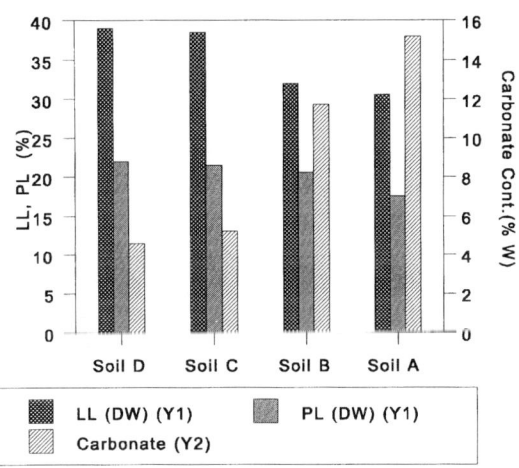

Figure 1   Results of Some Atterberg Limit Tests

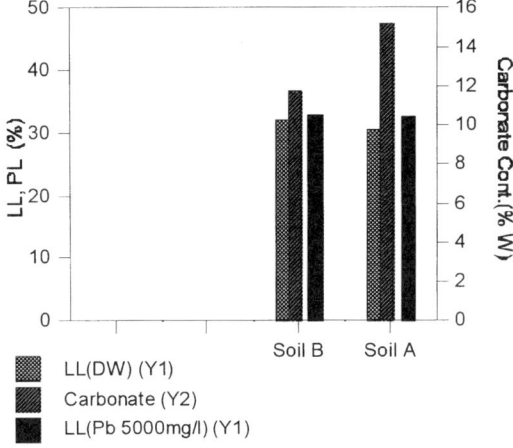

Figure 2 Liquid Limits Obtained with DW and 5000mg/l Pb-solution

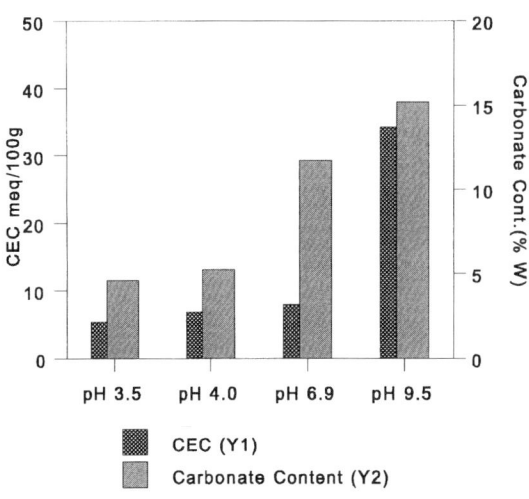

Figure 3   Cation Exchange Capacity

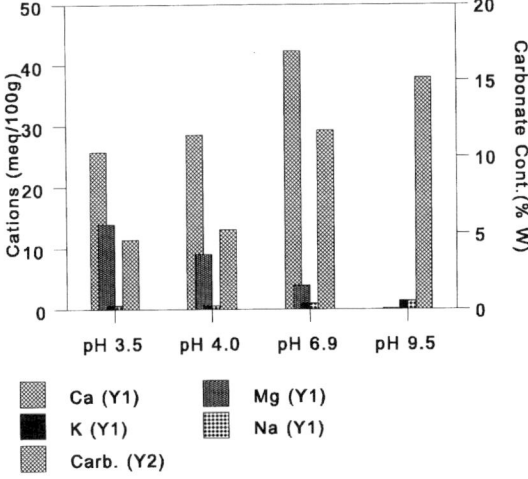

Figure 4   Soluble Cations

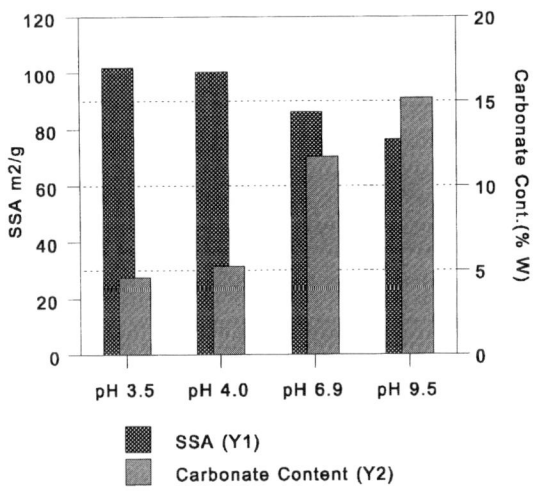

Figure 5   Specific Surface Area

Adsorption-Diffusion Problem in Bentonite Clay by a Method of Molecular Dynamics and Multiscale Homogenization Analysis

Y. ICHIKAWA
Nagoya University, Nagoya 464-8603, Japan
K. KAWAMURA
Tokyo Institute of Technology, Meguro-ku, Tokyo 152-8551, Japan
K. KITAYAMA
Nuclear Waste Management Organization of Japan, Chiyoda-ku, Tokyo 100-8118, Japan

ABSTRACT

We present a numerical scheme to calculate adsorption-diffusion behavior of chemical species in saturated bentonite. The scheme covers molecular characteristics and micro-/macro-continuum behavior. We show an example of diffusivity of tritium comapred with experimental data.

INTRODUCTION

Bentonite is a micro-inhomogeneous material, which consists of clay minerals (mainly montmorillonite), macro-grains such as quartz fragments, water, air and others. Properties of the saturated bentonite are characterized by the montmorillonite and water. We here treat montmorillonite hydrate with pure and solute water and analyze its molecular and macroscopic behavior. The local properties are calculated by a molecular dynamics (MD) simulation, and the diffusion behavior is analyzed by a multiscale homogenization analysis (HA). We use the local characteristics calculated by MD in HA (Kawamura *et al.* 1997; Ichikawa *et al.* 1998). An adsorption-diffusion problem of multi-component chemical species in saturated bentonite is treated.

MD SIMULATION OF MONTMORILLONITE HYDRATE

The key issue for the molecular simulation is how to determine the interatomic potentials. We introduce a new interatomic potential model. By using this model diffusive and viscous properties are calculated for some chemical species in the neighborhood of a montmorillonite crystal. We here analyze a beidellite hydrate $Na_{1/3}Al_2[Si_{11/3}Al_{1/3}]O_{10}(OH)_2 \cdot nH_2O$ which involves only Al^{3+} in the octahedral layer, and a montmorillonite crystal involves Al^{3+} and Mg^{2+}. Since both have similar properties, we call them 'montmorillonite hydrate' symbolically.

We calculated diffusivity and viscosity of water neighboring to a montmorillonite mineral

Geoenvironmental impact management, Thomas Telford, London, 2001.

by using the following newly developed interatomic potential function (2-body term)

$$U_{ij}(r_{ij}) = \frac{z_i z_j e^2}{4\pi\varepsilon_0 r_{ij}} + f_0(b_i + b_j)\exp[\frac{a_i + a_j - r_{ij}}{b_i + b_j}] - \frac{c_i c_j}{r_{ij}^6}$$

$$+ D_{1ij}\exp(-\beta_{1ij}r_{ij}) + D_{2ij}\exp(-\beta_{2ij}r_{ij}) + D_{3ij}\exp\{-\beta_{3ij}(r_{ij} - r_{3ij})^2\}.$$

A 3-body term is added to the H-O-H interaction because of its sp^3 hybrid orbital:

$$U_{ijk}(\theta_{ijk}, r_{ji}, r_{jk}) = f_k[1 - \cos\{2(\theta_{ijk} - \theta_0)\}](k_i k_j)^{1/2}; \quad k_i = \frac{1}{\exp\{g_r(r_{ij} - r_m)\} + 1}$$

Parameters $\{z, a, b, c\}$ and $\{D, \beta, r_1, r_2, r_3\}$ for the 2-body term, and $\{f_k, \theta_0, g_r, r_m\}$ for the 3-body term are specified so as to reproduce structural and physical properties of several oxide crystals such as quartz, corundum and feldspars (Kawamura 2001).

A snap shot of the montmorillonite and external water is shown in Figure 1(a). A montmorillonite mineral is illustrated at the left side. We divide the clay-water system into 50 slices with ca. 0.2 nm thickness in the z-direction (i.e., in the c-axis of the montmorillonite mineral), and we can calculate the mean square displacement (m.s.d.) and the diffusivity (i.e., the slope of m.s.d.) in each slice. Next by applying the Stokes-Einstein relationship, the viscosity of water at each sliced region is determined. Calculated diffusivity and viscosity are shown in Figure 1(b).

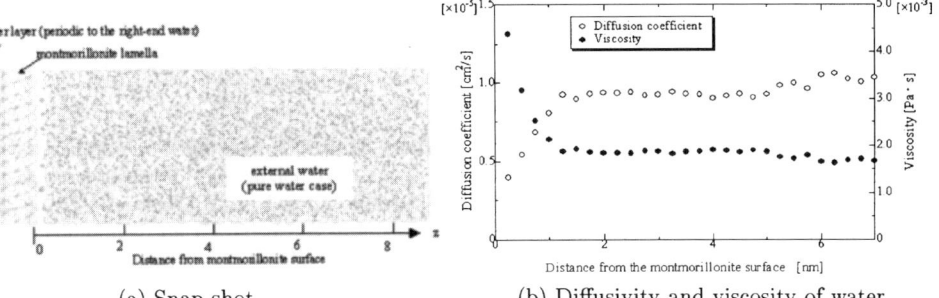

(a) Snap shot   (b) Diffusivity and viscosity of water

Figure 1. MD results for the montmorillonite hydrate with external water.

MULTISCALE HOMOGENIZATION THEOTY FOR DIFFUSION PROBLEM IN POROUS MEDIA

Adsorption phenomena of soil have been evaluated as a distribution factor K_d for solid phase together with the concept of adsorption isotherm. This involves essential difficulty in order to interpret the microscopic adsorption phenomenon. We present a new procedure for resolving the difficulty by extending the micro-characteristics to macro-behavior. That is, we show a coupled molecular dynamics (MD) and homogenization analysis (HA) method where the microscopic properties are calculated by MD, and HA is used to analyze the micro-/macro-continuum behavior.

Governing Equations of Diffusion Problem in Bentonite

We treat the diffusion problem of multicomponent solution in undeformable solid skeleton with n-chemical species. In the microscale problem we think only of the flow region, so the mass conservation law can be given as

$$\int_{\Omega_f} \left[\frac{\partial(\rho c_\alpha)}{\partial t} + \frac{\partial}{\partial x_i}(\rho c_\alpha v_i^\alpha) + \dot{\gamma}_\alpha \right] dv - \int_{\Gamma_{fs}} \zeta_i^\alpha n_i \, ds = 0 \tag{1}$$

where ρ is the average mass density of solution, c_α the mass-percent concentration of the α-th component, v^α the particle velocity, $\dot{\gamma}_\alpha$ the source term due to chemical reaction and so on, Ω_f the flow region, Γ_{fs} the internal fluid-solid interface, ζ_i^α the mass flux of the α-th component on Γ_{fs} adsorbed from the solution and n_i the unit outward normal.

The average velocity of solution, v, and the diffusing mass flux of the α-th component, j^α, are defined by

$$v = \frac{1}{\rho}\sum_{\alpha=1}^{n}\rho_\alpha v^\alpha = \sum_{\alpha=1}^{n} c_\alpha v^\alpha, \qquad j^\alpha = \rho_\alpha(v^\alpha - v) = \rho c_\alpha(v^\alpha - v). \tag{2}$$

Applying the Fick's law for diffusion to the second term of LHS, we have the following diffusion equation in Ω_f :

$$\frac{\partial(\rho c_\alpha)}{\partial t} + v_i \frac{\partial(\rho c_\alpha)}{\partial x_i} - \frac{\partial}{\partial x_i}\left(\sum_{\beta=1}^{n} \rho D_{ij}^{\alpha\beta} \frac{\partial c_\beta}{\partial x_j} \right) + \dot{\gamma}_\alpha = 0 \qquad \text{in } \Omega_f. \tag{3}$$

Let us consider the adsorption-diffusion behavior in the neighborhood of a clay mineral. Clay minerals adsorb cations on the edges Γ_e. Our MD analyses show that cations lie near to the clay surface Γ_i in the interlayer space and they diffuse along the surface Γ_i (surface diffusion). Then the adsorption condition of the α-the species can be written as

$$\zeta_i^\alpha n_i = \begin{cases} \sum_{\beta=1}^{n} s_{\alpha\beta}(c_\alpha^l - c_\alpha), & \text{if } c_\beta < c_\beta^l \\ 0, & \text{if } c_\beta \ge c_\beta^l \ (c_\beta = c_\beta^l) \end{cases} \qquad \text{on } \Gamma_e \tag{4}$$

where $s_{\alpha\beta}$ is the adsorption coefficient of the β-the species with respect to the α-the species which is a function of the hydrogen ion exponent, pH, $etc.$, and c_β^l is the marginal concentration of adsorption.

We guess there is no chemical adsorption on interlayer surfaces ($\zeta_i^\alpha n_i = 0$ on Γ_i), and surface diffusion takes place in the film water layer Ω_{sd} neighboring to the clay surface. So we introduce the following diffusion matrix $(D_{ij}^{\alpha\beta})_{\text{local}}$ for the local coordinate system $x^{2\prime}$ shown in Figure 2:

$$(D_{ij}^{\alpha\beta})_{\text{local}} = \begin{pmatrix} D_s & 0 & 0 \\ 0 & D_s & 0 \\ 0 & 0 & 0 \end{pmatrix} \quad \text{in } \Omega_{sd} \tag{5}$$

where D_s is the surface diffusion coefficient.

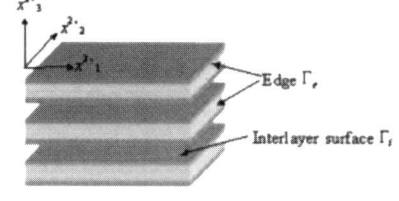

Figure 2. Edge and interlayer structure of clay minerals.

For simplicity we henceforth treat a diffusion problem of one chemical species ($\alpha = 1$), so we set $c^\varepsilon = c_\alpha$ and $\rho f^\varepsilon = \dot{\gamma}_\alpha - \partial \zeta_i^\alpha / \partial x_i$. Here the superscript $^\varepsilon$ gives a scale factor for a unit cell. Then under incompressible condition of the solution ($\rho = constant$, $\partial v_i^\varepsilon / \partial x_i = 0$) we have the following system of partial differential equations:

Governing equation;
$$\frac{\partial c^\varepsilon}{\partial t} + v_j^\varepsilon \frac{\partial c^\varepsilon}{\partial x_j} - \frac{\partial}{\partial x_i}\left(D_{ij}^\varepsilon \frac{\partial c^\varepsilon}{\partial x_j}\right) + f^\varepsilon = 0 \quad \text{in } \Omega_f, \tag{6}$$

Boundary conditions (BC);

(Dirichlet BC);
$$c^\varepsilon(\boldsymbol{x}, t) = \hat{c}^\varepsilon(t) \quad \text{on } \partial\Omega_c \tag{7}$$

(Neumann BC);
$$- D_{ij}^\varepsilon \frac{\partial c^\varepsilon}{\partial x_j} n_i = \hat{q}(t) \quad \text{on } \partial\Omega_q \tag{8}$$

(Internal adsorption BC);
$$\zeta_i n_i = \begin{cases} s(c^l - c^\varepsilon), & \text{if } c^\varepsilon < c^l \\ 0, & \text{if } c^\varepsilon \geq c^l \ (c^\varepsilon = c^l) \end{cases} \quad \text{on } \Gamma_e \tag{9}$$

Initial condition;
$$c^\varepsilon(\boldsymbol{x}, t) = c_0^\varepsilon(\boldsymbol{x}) \quad \text{at } t = t_0 \tag{10}$$

Micro-, Meso- and Macro-scale Equations

We introduce the following perturbation for the global concentration function $c^\varepsilon(\boldsymbol{x}; t)$ with respect to the macro-, meso- and micro-coordinate systems $\boldsymbol{x}^0$, $\boldsymbol{x}^1$ and $\boldsymbol{x}^2$, respectively:

$$c^\varepsilon(\boldsymbol{x}; t) = c^0(\boldsymbol{x}^0; t) + \varepsilon c^1(\boldsymbol{x}^0, \boldsymbol{x}^1, \boldsymbol{x}^2; t) + \cdots; \qquad \boldsymbol{x}^1 = \frac{\boldsymbol{x}^0}{\varepsilon}, \quad \boldsymbol{x}^2 = \frac{\boldsymbol{x}^1}{\varepsilon} \tag{11}$$

where c^α s ($\alpha = 0, 1, 2, \dots$) are X^1-/X^2-periodic functions:

$$c^\alpha(\boldsymbol{x}^0, \boldsymbol{x}^1, \boldsymbol{x}^2; t) = c^\alpha(\boldsymbol{x}^0, \boldsymbol{x}^1 + \boldsymbol{X}^1, \boldsymbol{x}^2; t), \quad c^\alpha(\boldsymbol{x}^0, \boldsymbol{x}^1, \boldsymbol{x}^2; t) = c^\alpha(\boldsymbol{x}^0, \boldsymbol{x}^1, \boldsymbol{x}^2 + \boldsymbol{X}^2; t).$$

The differentiation with respect to $\boldsymbol{x}$ can be written in the terms of $\boldsymbol{x}^0$, $\boldsymbol{x}^1$ and $\boldsymbol{x}^2$:

$$\frac{\partial}{\partial x_i} = \frac{\partial}{\partial x_i^0} + \frac{1}{\varepsilon}\frac{\partial}{\partial x_i^1} + \frac{1}{\varepsilon^2}\frac{\partial}{\partial x_i^2}. \tag{12}$$

We substitute (11) and (12) into the governing equation (6), then we get the following partial differential equations corresponding to each ε-term.

$O(\varepsilon^{-4})$-term :
$$\frac{\partial}{\partial x_i^2}\left(D_{ij}^\varepsilon \frac{\partial c^0}{\partial x_j^2}\right) = 0. \quad c^0 \text{ is a function only of } \boldsymbol{x}^0, \text{ so this is satisfied.}$$

$O(\varepsilon^{-3})$-term : First characteristic function

$$\frac{\partial}{\partial x_i^1}\left(D_{ij}^\varepsilon \frac{\partial c^0}{\partial x_j^2}\right) + \frac{\partial}{\partial x_i^2}\left\{D_{ij}^\varepsilon\left(\frac{\partial c^0}{\partial x_j^1} + \frac{\partial c^1}{\partial x_j^2}\right)\right\} = 0.$$

Terms $\partial c^0/\partial x_j^2$ and $\partial c^0/\partial x_j^1$ vanish, so c^1 is independent from $\boldsymbol{x}^2$, which suggests existence of the *first characteristic function* $N_1^k(\boldsymbol{x}^1)$ such as

$$c^1(\boldsymbol{x}^0, \boldsymbol{x}^1; t) = -N_1^k(\boldsymbol{x}^1)\frac{\partial c^0(\boldsymbol{x}^0; t)}{\partial x_k^0} \tag{13}$$

$O(\varepsilon^{-2})$-term : Second characteristic function and microscale equation [MiSE]

$$\frac{\partial}{\partial x_i^2} \left\{ D_{ij}^\varepsilon \left(\frac{\partial c^0}{\partial x_j^0} + \frac{\partial c^1}{\partial x_j^1} + \frac{\partial c^2}{\partial x_j^2} \right) \right\} = 0. \tag{14}$$

Since the term $\partial c^0 / \partial x_j^0 + \partial c^1 / \partial x_j^1$ is a function only of $\boldsymbol{x}^0$ and $\boldsymbol{x}^1$, we have the *second characteristic function* $N_2^k(\boldsymbol{x}^2)$ such as

$$c^2(\boldsymbol{x}^0, \boldsymbol{x}^1, \boldsymbol{x}^2; t) = -N_2^k \left(\delta_{kl} - \frac{\partial N_1^l}{\partial x_k^1} \right) \frac{\partial c^0}{\partial x_l^0} \tag{15}$$

where (13) is used. Substituting this into (14) yields the following *microscale equation* [MiSE] together with X^2-periodic boundary condition for the microscale domain Ω_2:

$$\frac{\partial}{\partial x_i^2} \left\{ D_{ij}^\varepsilon \left(\delta_{jk} - \frac{\partial N_2^k}{\partial x_j^2} \right) \right\} = 0 \quad \text{in } \Omega_2. \tag{16}$$

$O(\varepsilon^{-1})$-term : Mesoscale equation [MeSE]

$$\frac{\partial}{\partial x_i^1} \left\{ D_{ij}^\varepsilon \left(\frac{\partial c^0}{\partial x_j^0} + \frac{\partial c^1}{\partial x_j^1} + \frac{\partial c^2}{\partial x_j^2} \right) \right\} + \frac{\partial}{\partial x_i^2} \left\{ D_{ij}^\varepsilon \left(\frac{\partial c^1}{\partial x_j^0} + \frac{\partial c^2}{\partial x_j^1} + \frac{\partial c^3}{\partial x_j^2} \right) \right\} = 0. \tag{17}$$

We introduce a volume average $< \cdot >_2$ for Ω_2, then the last term vanishes because of X^2-periodicity. Substituting (13) and (15) into this we get the *mesoscale equation* [MeSE] together with X^1-periodic boundary condition for Ω_1:

$$\frac{\partial}{\partial x_i^1} \left\{ D_{ik}^{H_2} \left(\delta_{kl} - \frac{\partial N_1^l}{\partial x_k^1} \right) \right\} = 0 \text{ in } \Omega_1; \quad D_{ik}^{H_2} = \frac{1}{|\Omega_2|} \int_{\Omega_2} D_{ij}^\varepsilon \left(\delta_{jk} - \frac{\partial N_2^k}{\partial x_j^2} \right) d\boldsymbol{x}^2. \tag{18}$$

$O(\varepsilon^0)$-term : Macro-scale equation [MaSE]

$$\frac{\partial c^0}{\partial t} + v_j^\varepsilon \left(\frac{\partial c^0}{\partial x_j^0} + \frac{\partial c^1}{\partial x_j^1} + \frac{\partial c^2}{\partial x_j^2} \right) - \frac{\partial}{\partial x_i^0} \left\{ D_{ij}^\varepsilon \left(\frac{\partial c^0}{\partial x_j^0} + \frac{\partial c^1}{\partial x_j^1} + \frac{\partial c^2}{\partial x_j^2} \right) \right\}$$
$$- \frac{\partial}{\partial x_i^1} \left\{ D_{ij}^\varepsilon \left(\frac{\partial c^1}{\partial x_j^0} + \frac{\partial c^2}{\partial x_j^1} + \frac{\partial c^3}{\partial x_j^2} \right) \right\} - \frac{\partial}{\partial x_i^2} \left\{ D_{ij}^\varepsilon \left(\frac{\partial c^2}{\partial x_j^0} + \frac{\partial c^3}{\partial x_j^1} + \frac{\partial c^4}{\partial x_j^2} \right) \right\} + f^\varepsilon = 0.$$

We substitute (13) and (15) and make average for Ω_2 and Ω_1, then we finally get the *macroscale equation* [MaSE]:

$$\frac{\partial c^0}{\partial t} + v_l^H \frac{\partial c^0}{\partial x_l^0} - \frac{\partial}{\partial x_i^0} \left(D_{il}^H \frac{\partial c^0}{\partial x_l^0} \right) + f^H = 0 \quad \text{in } \Omega_0 \tag{19}$$

where

$$v_l^H = \frac{1}{|\Omega_1|} \int_{\Omega_1} v_k^{H_2} \left(\delta_{kl} - \frac{\partial N_1^l}{\partial x_k^1} \right) d\boldsymbol{x}^1, \quad D_{il}^H = \frac{1}{|\Omega_1|} \int_{\Omega_1} D_{ik}^{H_2} \left(\delta_{kl} - \frac{\partial N_1^l}{\partial x_k^1} \right) d\boldsymbol{x}^1,$$

$$v_k^{H_2} = \frac{1}{|\Omega_2|} \int_{\Omega_2} v_j^\varepsilon \left(\delta_{jk} - \frac{\partial N_2^k}{\partial x_j^2} \right) d\boldsymbol{x}^2, \quad f^H = \frac{1}{|\Omega_1|} \int_{\Omega_1} f^{H_2} d\boldsymbol{x}^1, \quad f^{H_2} = \frac{1}{|\Omega_2|} \int_{\Omega_2} f^\varepsilon d\boldsymbol{x}^2.$$

Diffusion of tritium water HTO in bentonite

Tritium water, denoted as HTO, is free from adsorption, so it is appropriate to evaluate the effective diffusivity that is equivalent to the homogenized diffusivity D^H. Diffusion coefficient of HTO in free water is $2.44 \times 10^{-5} cm^2/s = 769.48 cm/year$ (Klitzsche *et al.*, 1976). The target bentonite, Kunigel V1, consists of about 50% of montmorillonite in weight and the rest is macro-grains, mainly quartz. That is, one third of the mesoscale domain is an impermeable quartz region and the rest is montmorillonite hydrate. Based on our measurement we set the size of quartz grains as $15\mu m$ and the size of meso-domain is set as $45\mu m$. The size of a montmorillonite lamellar crystal is about $100 \times 100 \times 1 nm$, and a lamellae group is formed by several parallel platelets. The followings are also assumed: a lamellae group consists of 4, 6 and 8 platelets, the dry density is $2.0 g/cm^3$, the interlayer distance is 0.56nm, which corresponds to two-layers of hydrated water molecules, and in the neighbor of montmorillonite D^ε of HTO varies in the same profile as the normal water given in Figure 1(b). Under these geometrical and physical conditions we form a micro-scale domain, and another micro-scale domain is filled with quartz.

By using this model we calculate the homogenized diffusivity. Since the microscale model is pseudo-one-dimensional and we have huge number of lamellae in the mesoscale level, D^{H_2} is assumed to be isotropic and of a value of one third of the D_{11}-component of the above pseudo-one-dimensional microscale model. Calculated results are shown in Table 1 together with experimental data. We observe the 4-layer model is consistent with the experiment data. However since we do not account for a geometrical tortuosity effect, the 6-layer model is also conformable.

Table 1. Calculated and experimental effective diffusion coefficient of HTO in saturated bentonite with its dry density $2.0 g/cm^3$ (experiment, JNC 1999).

	Effective diffusivity of HTO (cm^2/year)	
	Calculated	Experiment
8-layers	28.384	
6-layers	20.892	16.083
4-layers	16.068	

CONCLUSIONS

We formulated an adsorption-diffusion problem in saturated bentonite based on the molecular and micro-/macro-characteristcs of clay-hydrate. The scheme used the molecular dynamics (MD) simulation and a mutiscale homogenization analysis (HA). We showed the calculated effective diffusivity of tritium consistent with experimental data.

REFERENCES

1. Ichikawa, Y., *et al.*, "Unified molecular dynamics and homogenization analysis for bentonite behavior", *Engineering Geology*, 54, 21-31 (1999).
2. JNC, *Second Progress Report on Research and Development for Geological Disposal of High-level Waste in Japan* (1999).
3. Kawamura, K., *et al.*, "New approach for predicting the long-term behavior of bentonite: the unified method of molecular simulation and homogenization analysis", *Proc. Sci. Basis for Nuclear Waste Management XXI*, Davos, Material Research Society, 359-366 (1997).
4. Kawamura, K., "Physical properties and chemistry of water and clay minerals", *Bull. Earthquake Research Institute, Univ. Tokyo* (2001) to be appeared.
5. Klitzsche, C.S., *et al.*, "Grundwasser der Zentralsahara: Fossile Vorrate", *Geol. Rundschau*, 65, 264 (1976).

MTBE - an overview of a new problem

C. JESUS-RYDIN, GEOL. ENG., and O. C. HANSEN, MSc (Biology), Danish
Technological Institute, Environmental and Waste Technology, Gregersensvej P.O.Box 141,
2630 Taastrup, Denmark

ABSTRACT
Methyl tertiary-butyl ether (MTBE) is a synthetic chemical used in unleaded gasoline as an
additive to enhance the octane number and to increase oxygen content in the gasoline. The
latter has the advantage of improving the combustion and, therefore, to convert carbon
monoxide emissions (product of incomplete auto engines combustion) to carbon dioxide.

Reformulated gasoline may provide economic and air quality benefits, as it complies with
legislation on air quality standards and helps to reduce the production of ozone. Nevertheless,
there are important concerns regarding potential harmful effects of MTBE on humans and
environment.

Several events have raised concern over the safety of MTBE, especially after being
considered a potential carcinogen to humans and also due to its physical and chemical
characteristics which makes it persistence in the environment. In 1996, the city of Santa
Monica, in the USA, closed some of its major drinking water wells after discovering MTBE
contamination. Furthermore, the U.S. Geological Survey reported that in investigations
carried out between 1994-1995, MTBE to be the second most common contaminant in
shallow urban aquifers.

In 1985, MTBE was first introduced in Denmark as an additive to unleaded gasoline in an
attempt to phase out the lead additives. Less than 2- decades after, in April 2000 MTBE was
included in the Danish EPA list of unwanted chemicals.

In January 2000 the EU agreed that the vapour pressure of gasoline should be reduced, but
since ethanol will increase the vapour pressure, it can no longer be considered as a possible
European replacement for MTBE. Furthermore, the EU policies aim to reduce the aromatic
compounds content (particularly benzene) by 2005, this might lead to a need for increasing
the MTBE content.

In Denmark 95% of the drinking water is originated by groundwater. Since 1998, the Danish
authorities have carried out a survey to evaluate the extent of MTBE contamination of
groundwater. Furthermore, Denmark has implemented an action-plan to phase-out MTBE
provisionally limited to 1 January 2005.

INTRODUCTION
Methyl tertiary-butyl ether (MTBE) is a synthetic chemical used in unleaded gasoline as an
additive to enhance the octane number and to increase oxygen content in the gasoline. The
latter has the advantage of improving the combustion and, therefore, convert carbon
monoxide emissions (product of incomplete auto engines combustion) to carbon dioxide.

These provide economic and air quality benefits, as it complies with legislation on air quality standards and helps to reduce the production of ozone.

MTBE was first used in USA gasoline in 1979, primarily in premium grades of gasoline at levels of 2-3% by volume, as an octane booster and replacement for lead in gasoline. In Denmark, MTBE was first introduced as an additive to unleaded gasoline in 1985, in most cases only to the 98-octane gasoline. Currently, MTBE constitutes up to 17% by volume of gasoline both in USA and Europe.

In the last decade several events have raised concern over the safety of MTBE, especially after being considered a potential carcinogen to humans and also due to its physical and chemical characteristics which makes it persistent in the environment and a potential contaminant of drinking water.

In 1996, the city of Santa Monica, in the USA, closed some of its major drinking water wells after discovering MTBE contamination. Furthermore, the U.S. Geological Survey reported that in investigations carried out between 1994-1995, MTBE to be the second most common contaminant in shallow urban aquifers.

Recently, the widespread use of this substance as an additive for motor vehicle gasoline has increasingly been discussed in EU member states and at the EU level. The debate has focused on its persistence and mobility in soil of the substance, the widespread contamination of drinking water wells by MTBE in USA and the findings up to a few years ago from those EU countries where MTBE has been monitored for. Also the future possibility for contamination of groundwater in EU has been debated, since an increase in the use of the substance is expected as a result of a new EU directive on control of air pollution from motor gasoline emissions.

Less than 2 decades after its introduction in Denmark, in April 2000 MTBE is included in the Danish EPA list of unwanted chemicals.

MTBE'S PHYSICAL AND CHEMICAL PROPERTIES
MTBE as other oxygenates is rather soluble in water, with significantly higher solubility than benzene, toluene, xylene and other petroleum hydrocarbons. This presents significant problems when considering the fate and transport of these pollutants in the environment, as well as when discussing treatment options.

Given its high solubility in water, 42 to 50 g/l, MTBE is quite mobile in the environment. It partitions weakly to the organic fraction in soils, sediments and suspended particles, preferentially remaining in the aqueous phase. This compound is expected to move essentially at the same rate as the groundwater flow, with nearly no retardation due to sorption. The adsorption coefficient related to organic carbon, log Koc, is 0.9 to 1.3.

Due to the weak bond to organic compounds, MTBE follows the movement of the water and, since MTBE is not easily naturally degraded in the environment, it acts like a conservative substance in nature.

MTBE is considered to be highly volatile with a vapour pressure of 330 hPa at 20°C. Nevertheless it is believed that evaporation of MTBE from groundwater is limited, while the evaporation from the petrol-phase is relatively high. This suggests that an early action, in the unsaturated zone, at the source of contamination is relatively suitable for stripping remediation techniques.

Recent studies have found that MTBE is very slowly degraded under aerobic and anaerobic conditions. Anaerobic studies have shown that MTBE is degraded under specific conditions: low organic content and pH around 5.5. MTBE degradation did not proceed in rich soils, possibly due to the abundant availability of other substrates for microbial activity.

RISK ASSESSMENT
EU is currently reviewing MTBE as part of the existing chemicals programme. The result is not yet finalised. A risk assessment performed according to the EU risk assessment directive includes a hazard and exposure assessment to human health and the environment

Environmental emissions of MTBE are closely related to gasoline, its storage, distribution and use in combustion engines which covers more than 98% of its use. Vehicle exhaust gasses are a major source of MTBE in ambient air. The leaking from gasoline storage tanks is the major source of MTBE to soil and groundwater. Because of a low adsorption capacity MTBE released to soil from subterranean tanks and pipes ends up in the groundwater. Using a Mackay fugacity level 1 model, the theoretical equilibrium distribution of MTBE based on physico-chemical properties would be 93.9% to air, 6.05% to water and 0.5% to soil. MTBE is not ready biodegradable. The degradation rates of MTBE are under debate but generally considered very low. Under natural environmental conditions, MTBE is considered persistent. MTBE is stable to hydrolysis and photolysis in water. Some indirect photochemical photolysis may occur in the atmosphere but with an estimated half-life of 3 to 6 days, MTBE may be considered stable to photolysis in air. As a hydrocarbon, MTBE is considered an ozone precursor. Due to the high vapour pressure (330 hPa at 25°C), MTBE is one of the major VOC components in oxygenated gasoline.

The acute toxicity to aquatic organisms is low with acute EC_{50} values above 100 mg/l: LC_{50} values for fish 672-1054 mg/l, Daphnia 340-681 mg/l and algae 184->800 mg/l. The effects to humans are still under discussion.

Whatever the result on toxicity, a prominent problem with MTBE appears to be the influence on water quality since MTBE has a strong effect on taste and odour of water. For instance, Young *et al.* (1996) observed that the geometric mean odour and taste thresholds for humans were 34 and 48 µg/l, respectively, with the lowest concentrations 15 and 40 µg/l. The US-EPA has set a Secondary Maximum Contaminant Limit (SMCL – an advisory guideline set for aesthetic, non-health effect parameters) at 15 µg/l for taste and odour in drinking water. The California EPA has adopted a SMCL of 5 µg/l (OEHHA 1999). In Denmark, the Danish EPA has set an preliminary water quality limit of 30 µg/l. This value is expected to be reduced. In a recent Danish study, odour detection threshold 7.4 µg/l and taste detection threshold 7.3 µg/l were determined. The lowest detections were at 3 µg/l (Danish EPA 2000).

The EU risk assessment has not been finalised yet but the ECB Newsletter (ECB 2000) states in a progress report that for several of the assessed topics, a conclusion III has been reached. Conclusion III indicates that there is a need for limiting the risk and risk reduction measures should be considered. The conclusion III was reached on releases to surface water from terminal site storage tank bottom waters. Based on the data gathered on groundwater concentrations of MTBE from several Member States there is a reason for concern. Furthermore an investigation of possible avoidance behaviour of wildlife related to MTBE contaminated water is necessary (ECB 2000).

The effect on wildlife of taste and odour such as for instance avoidance behaviour is unknown. The consequences of releasing a persistent substance to the environment may affect more than just the aesthetic value of drinking water. Chronic studies on aquatic and terrestrial organisms are currently not available.

EXPERIENCE RELATED TO MTBE CONTAMINATION
In connection with the polemic around MTBE, a groundwater survey has been carried out in Denmark, particularly in connection to petrol station's sites.

Data from the counties
As illustrated in Table 1, only 6 out of 16 counties had looked for MTBE by the summer of 1997. This data refers to surveys carried out by the counties on petrol stations groundwater.

Table 1. Results from the counties MTBE survey (summer 1997)

County	Survey for MTBE	No. samples	Samples with MTBE	Concentration range µg/l
Københavns Amt	-			
Frederiksborg	+	3	0	<DL
Roskilde	-			
Vestsjælland	-			
Stortstrøm	-			
Bornholm	-			
Fyn	+	12	8	3-6000
Sønderjylland	-			
Ribe	+	> 10	6	1000-3700
Vejle	+ [1]			
Ringkøbing	+ [2]			<0.1-400
Århus	+	6	2	22000-550000
Viborg	+ [2]			<DL
Nordjylland	-			
Københavns Kommune	+	12	2	1-42 [4]
Frederiksberg Kommune	+	3	0	0
- no survey for MTBE in 1997; + has included MTBE in the surveys; ? number not known; 1 detected in soil sample on one site; 2 only surveyed by OM[1]; 3 Copenhagen Water Supply is excluded; 4 in addition MTBE was detected in other 5 samples at concentration <1 µg/l				

Frederiksberg County has in addition to the samples taken from abstraction wells surveyed 16 observation wells in the county. MTBE was found in 5 wells in concentrations from 0.11 to 3.5 µg/l.

In beginning of November 2000, the *Funen County* (second largest island in Denmark) published results from a survey on MTBE in the petrol stations groundwater carried out in the last couple of years.

Results show that MTBE could be traced in the shallow aquifers at 62 out of 72 of the investigated sites. In nearly half the sites presented concentrations above 30µg/l [2]. In 51 sites the concentrations were greater than or equal to 2 µg/l [3]. In 15 cases, concentrations were above 1 mg/l and maximum was 53 mg/l.

---

[1] The Danish Oil Industries Association for Remediation of Retail Sites.
[2] Current provisional Danish limit value based on taste/odour, presently under revision.
[3] Limit for sensitive persons to taste MTBE in tap water.

In this county, the raw water at drinking waterworks was also surveyed. MTBE was analysed at 20 sites, of which 7 presented MTBE. In 2 of these 7 sites the concentrations of MTBE were above the current limit value of 30 µg/l. In all 7 sites MTBE has spread from contaminated soils to groundwater in deeper located main aquifers.

The drinking water[4] from 8 drinking waterworks was analysed, of which 2 presented MTBE at concentrations of 0.6 and 9.4 µg/l.

The status on registering Petrol Contamination with MTBE (OM May 2000),
OM is the Danish oil clean-up association, responsible for the remediation plan on closed petrol stations. This plan is being implemented since 1992 and until 1999 a number of 1377 sites had been investigated/remediated.

In connection with the action-plan for registering petrol contamination with MTBE, analyses were carried out on soil and groundwater. MTBE was added to petrol since 1985 and OM identified 479 disused petrol station sites in operation since 1985, which had been selling MTBE-containing petrol.

The results from the soil analyses of the 479 sites are divided into 3 groups:

- 52 sites showed no contamination;
- 41 sites had less than 30 tons of contaminated soil;
- 386 sites had more than 30 tons of contaminated soil;
- 427 sites, approx. 89% of the sites surveyed were contaminated to some extent.

Out of the 479 sites, only in 293 sites were taken groundwater samples, mainly due to no screened wells or because the wells were too shallow. The results from the groundwater analyses are divided into 2 groups:

- 126 sites showing no groundwater contamination;
- 167 sites showing BTEX and/or MTBE contamination.

Of the 167 sites with contamination, MTBE was found[5] in the groundwater of 102 sites. In 34 sites, the provisional Danish limit value was exceeded.

Examples of downstream groundwater contamination
This information is based on the casual available data, since no systematic survey was carried out yet. Therefore, it is not known to what degree this data represents the extension of MTBE downstream contamination on a national level.

The Table 2 shows examples from selected cases from Funen and Zealand where petrol spills had occurred and downstream contamination with MTBE was identified.

---

[4] Drinking water in Denmark is typically aerated and sand bed filtrated and distributed to the consumer without further treatment.
[5] detection limit from 0.1 to 1 µg/l

Table 2 – Examples on downstream groundwater contamination

Location	Source concentration (μg/l)	Downstream observation well		Abstraction well	
		Distance from source (m)	Concentration (μg/l)	Distance from source (m)	Concentration (μg/l)
Solhøj abstraction area	0.3 - 680	1900	0.25 – 0.4	3000	0.2 – 0.3
Taastrup/Valby abstraction area	7100	20	4.1	630	0.23
Bystævnet and Søndersø municipality, Funen	730 - 39	17	8.3	30	1.1
		20	2.8	75	0.37
Odensevej and Nyborg municipality, Funen	55 - 23000	18	2.9 - 55	-	-

Surface waters

In Copenhagen area, the 3 lakes being used for water abstraction have been surveyed for the MTBE content. The results are stated in Table 3.

Table 3 – MTBE survey in lakes used for drinking water production

Lake	Date	Concentration MTBE μg/l
Gyrstinge lake	15 December 1997	< 0.1
Haraldsted lake	15 December 1997	< 0.1 and 0.6 [1]
Gyrstinge lake	1 April 1998	< 0.1
Haraldsted lake	1 April 1998	< 0.1
Arre lake	1 April 1998	< 0.1
Arre lake	24 August 1998	< 0.1
Arre lake	16 August 1999	< 0.1
1 Two laboratories analysed the same water sample, but only one laboratory detected MTBE		

THE ACTION-PLAN ON MTBE

In Denmark, 95% of the drinking water is provided from groundwater sources and the remaining 5% from lakes, which just needs to be filtered and aerated to be usable. Therefore, the Danish authorities are extremely interested in protecting the groundwater against contamination. In 1998, the Danish EPA thus drew up an MTBE action plan aimed at preventing groundwater contamination with MTBE. What was previously believed to be an American problem had now become a national problem. The monitoring data revealed that leaking of MTBE to groundwater from refuelling stations had taken place to a considerable extent.

The main objective for this action plan was to determine the extension of MTBE contamination of the groundwater in order to determine whether it comprises a major problem. MTBE was therefore included in the national groundwater monitoring programme. The results of the first two years are presented in Table 4.

Table 4 - Groundwater monitoring in Denmark

County, Location		Samples analysed	Positive of analysed	% positive	Concentration μg/l	Reference
GRUMO*	1998	17	0			GEUS 1999
	1999	29	1	3.6%	1.4	GEUS 2000
Water work wells	1998	191 (186 wells)	22 (19 wells)	10.2%	Median: 0.33 Max.: 33	GEUS 1999
	1999	208 (164 wells)	54 (28 wells)	17.1%	Median: 0.29 Max. 45 μg/l	GEUS 2000

*GRUMO: Groundwater monitoring in primary aquifers, 10-20 metres below soil surface

The conclusion of the first analysis from groundwater at or near filling stations and the monitoring programme is that the point sources may result in concentrations up to hundreds of milligrams per litre close to the source (table 1). However, the diffuse emission may result in very low concentrations in surface water and groundwater of less than 1 to a few micrograms per litre (Table 2 to 4).

In June 2000, the Danish EPA started a campaign to convince the public to change from the 98 to 95 octane gasoline, since in most cases in Denmark MTBE is only added to the 98 octane gasoline (Table 5). In connection with this campaign a leaflet has been published.

Table 5 – Average content of MTBE in Gasoline sold in Denmark during 2000 (approx. %)

Octane specification	Winter specifications	Summer specifications
98 octane	5%	11%
95 octane (mixture of 98 and 92 octane)	2.5%	5.5%
95 octane	0.05%	3.0%
92 octane	0.04%	3.0%

The Danish EPA has held detailed discussions with the Danish Petroleum Association regarding a solution to the problem. In November 2000, the Association has subsequently declared that MTBE will gradually be removed from petrol with an octane rating lower than 98 during the course of 2001. Furthermore, the number of filling stations selling MTBE-containing 98-octane petrol will be gradually reduced to less than 10% of the present number, in order to reduce the number of potential sources of contamination.

The oil and petrol industry's phase-out is provisionally limited to 1 January 2005, when the EU policies aim to reduce the aromatic compounds content (particularly benzene) in petrol might necessitate the addition of MTBE to 95-octane petrol once again. In fact, in January 2000 the EU agreed that the vapour pressure of gasoline should be reduced. Because a potential alternative like ethanol would increase the vapour pressure, it can no longer be considered as a possible European replacement except in a few EU Member States where exemption was given on grounds of arctic climate (Finland, Sweden and United Kingdom).

In order to reduce the general risk of spillage of petrol from petrol and oil products from filling stations, new more strict regulations governing filling station safety are being drawn up in Denmark. The draft of the new regulations sets the end of 2004 as the deadline for upgrading all existing filling stations.

REFERENCES
Danish EPA 2000. MTBE in groundwater. Danish Environmental Protection Agency, Copenhagen.

Danish EPA 2000. The petrol additive MTBE comprises a threat to the drinking water supply. Danish Environmental Protection Agency, Copenhagen.

Danish EPA 2000. New monitoring data show increased number of MTBE polluted sites. Danish Environmental Protection Agency, Copenhagen.

Danish EPA 2000. Taste and odour study on MTBE performed by ISO method with measured concentrations. Danish Environmental Protection Agency, Copenhagen.

ECB 2000. European Chemicals Bureau Newsletter. Issue No. 2. European Chemicals Bureau, Ispra, Italy (http://ecb.ei.jrc.it)

GEUS. 1999. Grundvandsovervågning 1999. Danmarks og Grønlands Geologiske Undersøgelser (Groundwater monitoring 1999. Geological Survey of Denmark and Greenland). In Danish.

GEUS. 2000. Grundvandsovervågning 2000. Danmarks og Grønlands Geologiske Undersøgelser (Groundwater monitoring 2000. Geological Survey of Denmark and Greenland). In Danish.

Jesus-Rydin, C. 2000. Systematic Investigation of Petrol Stations in Denmark. Estratégias de Descontaminação de Solos e Águas Subterrâneas em Ambiente Urbano e Industrial. Lisboa (in port.)

Jesus-Rydin, C. 2000. MTBE a potential environmental problem. Seminar at Lund University.

OEHHA 1999. Secondary maximum contaminant level for methyl-tert-butyl ether and revisions to the unregulated chemical monitoring list. Title 22, California Code of Regulations. California EPA Office of Environmental Health Hazard Assessment.

Young WF, Horth H, Crane R, Ogden T, Arnott. 1996. Taste and odour threshold concentrations of potential potable water contaminants. Water Research 30: 331-340.

Contribution of Anaerobic Microbial Activity to Natural Attenuation of the "B" in BTEX

S. J. JOHNSON, K. J. WOOLHOUSE, H. PROMMER, D. A. BARRY & N. CHRISTOFI
School of Civil and Environmental Engineering/Contaminated Land Assessment and Remediation Research Centre, The University of Edinburgh, Edinburgh EH9 3JN United Kingdom.

ABSTRACT
Anaerobic biodegradation of hydrocarbons, using a variety of terminal electron acceptors, is increasingly being reported both in laboratory studies and in the field. Of all the petroleum hydrocarbons, benzene is considered the most problematical due to its high toxicity and relatively high aqueous solubility. These, combined with its peculiarly stable structure mean that it has long been considered recalcitrant in all but aerobic conditions. There is now a small, but growing, literature to suggest that this may not in fact be the case. We present a mini-review of the field, encompassing reviews up to 1997 and original papers published since then. It appears that benzene is indeed degraded anaerobically, but that organisms capable of doing so are not ubiquitous. In addition, benzene degradation may be competitively inhibited by the presence of more readily degraded compounds such as toluene. Certainly, the occurrence and rate of benzene attenuation under anaerobic conditions is far more site-specific than for other BTEX compounds. We discuss a mathematical method for modelling redox-dependent, differential degradation rates.

INTRODUCTION
Petroleum contains, in addition to many other hydrocarbon constituents, benzene, toluene, ethylbenzene, and xylenes (BTEX). These are the most significant components in terms of pollution potential, as they are the most soluble. Leaks of petroleum, leading to contamination of soil and groundwater by BTEX compounds, are widespread. Thus, dissolved BTEX compounds in the subsurface environment are candidates for removal via naturally occurring processes, whereby redox reactions mediated by autochthonous microorganisms result in the production of less harmful, even benign, products [1].

Benzene typically makes up less than 2% of petroleum [2], but is important since it is considered the most toxic and persistent of all petroleum components. It has limited solubility in water (1.78 g l⁻¹ [3]), yet is the most soluble of petroleum hydrocarbons [4]. Its structure and shape make it difficult to oxidise and degrade. Reasonable evidence exists showing that the TEX compounds all degrade naturally in groundwater systems [5], whereas for benzene the picture is mixed. Indeed, it is thought that benzene degradation may be inhibited in the presence of other hydrocarbons, such as toluene [6], though the mechanism for this is unclear.

Geoenvironmental impact management, Thomas Telford, London, 2001.

BTEX degradation occurs most rapidly under aerobic conditions. However, aquifers are often anoxic. In the absence of dissolved oxygen in groundwater, benzene degradation rates decrease or can stop altogether. The aerobic degradation of benzene, via catechol, is well established [5], and will not be further considered here. The application of oxygen to anoxic soil, sediments and groundwater is possible by a variety of means (biopiles, injection of O_2/air/aerated water/hydrogen peroxide or the injection of chlorite, which is degraded by perchlorate-reducing bacteria to yield oxygen *in situ* [7]. These are intrusive and, therefore, expensive measures. Hence, monitored natural attenuation (intrinsic bioremediation) is likely to remain the most widespread remediation technique for BTEX-contaminated aquifers

Natural attenuation encompasses a host of physical processes (e.g. dispersion, dilution, sorption and volatilisation) as well as chemical and biological degradation. Biodegradation depends on microbial activity that varies with hydrogeological site characteristics and aquifer geochemistry. Here, anaerobic benzene biodegradation is examined, considering evidence from both laboratory and the field. Literature surveys reveal conflicting evidence on conditions required for its degradation. Thus, it is difficult to predict *a priori* the occurrence/rate of benzene removal without a detailed understanding of aquifer conditions.

A 1997 review of likely mechanisms [8] suggested that benzene could be degraded via benzoate under a range of conditions. No single microbial species has been shown to completely degrade the compound under anaerobic conditions, although stable benzene-degrading enrichment cultures are known. Toluene-degrading organisms, however, have been identified, and include members of the nitrate-reducing genera *Azoarcus* and *Thauera,* and the iron-reducing *Geobacter metallireducens* as well as a variety of unnamed sulphate-reducers. The only organisms known to degrade BTEX compounds anaerobically are bacteria, but it has been suggested that the currently poorly studied anaerobic fungi might prove to be involved.

Aronson and Howard [5] reviewed a large number of laboratory and field investigations in 1997. The majority of published studies failed to demonstrate anaerobic benzene degradation. Those that did indicated that benzene was degraded under nitrate-, Fe(III)- and manganese-reducing, and sometimes under methanogenic conditions. Many authors attributed lack of degradation to insufficient residence time. Others suggested that since benzene degradation appears to be inhibited in the presence of other carbon sources, it might be that the degradation seen in the field was due to aerobic degradation at the plume periphery.

Since these reviews, a number of pertinent papers have been published. Space does not allow for the comprehensive review of papers published prior to 1997. We therefore suggest this paper should be read in conjunction with the earlier reviews [5;8;9].

BENZENE DEGRADATION UNDER DIFFERENT REDOX REGIMES
Cellular respiration comprises a chain of oxidation-reduction couples, whereby energy is extracted via a stepwise oxidation (i.e., removal of electrons) of organic and inorganic molecules. In order to proceed, there needs to be a relatively more oxidised chemical species available to prevent the accumulation of electrons that would hinder the reaction kinetics. These are known as terminal electron acceptors (TEAs). In aerobic respiration, the TEA is molecular oxygen, but in the absence of oxygen, a number of less highly oxidised TEAs may serve, assuming organisms capable of making use of them are present. Available TEAs are

generally used in the environment in decreasing order of oxidation-reduction potential. Possible TEAs include NO_3^-, $Fe(III)$, $Mn(IV)$, SO_4^{2-}, and CO_2.

Nitrate-reducing conditions

Benzene has long been considered recalcitrant in the field under nitrate-reducing conditions. Where it has been seen, it has been much slower than under aerobic conditions and it appears that O_2 is still required as a substrate for the oxygenases that mediate the oxidative cleavage of the aromatic ring, even if it is not used as a TEA. Benzoate is often considered to be a central metabolite in the degradation of monoaromatic hydrocarbons, and it has been shown to be degraded under denitrifying conditions [8] though some workers still point to the apparent inability of nitrate-reducers to degrade benzene [10]. Nales *et al.* [11] demonstrated benzene degradation under nitrate-reducing conditions, but found that TEX substrates competitively inhibited its degradation. They also demonstrated benzene degradation under sulphate- and Fe(III)-reducing conditions, but not with methanogenesis. More recently, benzene has again been shown to be degraded in microaerobic, nitrate-reducing conditions, with O_2 required for oxygenases [12]. Burland & Edwards [13] link benzene degradation to reduction of nitrate to nitrite (but not to conversion to gaseous nitrogen).

Iron–reducing conditions

Iron is considered to be especially significant in hydrocarbon degradation in marine sediments, with Fe(III) chelated to a variety of compounds shown to stimulate benzene oxidation in anaerobic sediment [14;15]. Kazumi *et al.* [16] showed that benzene was degraded in methanogenic, sulfate-reducing, and iron-reducing conditions. Benzene loss also occurred in the presence of Fe(III) in sediments from freshwater environments. Heider *et al.* [9] noted recently that benzene was degraded under iron-reducing conditions but that no single benzene-degrading organism had been isolated. A community including members of the genus *Geobacter* was implicated. Many primitive benzene-degrading bacteria (hyperthermophiles), previously thought to require SO_4^{2-}, have been shown to grow using Fe(III) as an electron acceptor [17;18]. Caldwell and Suflita [19] found evolution of phenol and benzoate under a range of conditions (Fe(III)- and sulphate-reduction, and with methanogenesis), supporting the theory that benzoate is a central metabolite in anaerobic degradation of aromatics.

Sulphate-reducing conditions

Several workers have demonstrated benzene degradation under sulphate-reducing conditions in soil collected from contaminated sites (e.g. Phelps *et al.* [20]). Chaudhuri & Wiesman [21] showed that degradation was via benzoate. Benzene degradation was comprehensively demonstrated in sulphate-reducing sediments from San Diego Bay [22]. Reinhard *et al.* [23] investigated BTEX degradation under a range of redox conditions, but benzene degradation was only associated with sulphate reduction. Enrichment of aquatic sediments with known benzene-degraders leads to degradation of benzene and growth of benzene-degrading organisms, suggesting that the lack of benzene degradation in some aquifers is due to failure of appropriate organisms to colonise the aquifer, rather than adverse environmental conditions [24]. A sulphate-reducing consortium was found to remain relatively complex despite the culture's long exposure to benzene as the only carbon and energy source (over 3 years) and repeated dilutions of the original enrichment [25]. Conversely, complete mineralization of benzene to CO_2 has been demonstrated apparently within single cells in microcosms [22], and more recently in a contaminated aquifer [26]

Methanogenesis
Where no electron acceptors other than CO_2 are present, it is suggested that benzene might be degraded to CO_2 and methane. In such a situation, it would not be possible to mineralise all the benzene. However, in the absence of any other TEA, this might play a role in limiting the extent of the contaminant plume. Benzene has been shown to be converted to CH_4 and CO_2 with no lag phase in the absence of other electron acceptors [16;27;28].

MODELLING DIFFERENTIAL DEGRADATION RATES
Modelling of the complex biogeochemical interactions is a valuable means of quantifying the varied and complex interactions between contaminants, the local hydrogeology (groundwater flow) and the local hydrogeochemistry and, ultimately, making predictions of (i) the viability of natural attenuation as a remediation technique for BTEX compounds in aquifers and (ii) the feasibility/efficiency of enhanced remediation schemes. For the quantification of hydro-carbon compounds, models of different levels of complexity exist. Probably the most common mathematical formulation is based on linking the removal of organic compounds to the microbial growth rate. For a single organic compound, this growth rate can be expressed by

$$\frac{\partial X_{growth}}{\partial t} = v_{max} \frac{C_{org}}{K_{org} + C_{org}} \frac{C_{ea}}{K_{ea} + C_{ea}} X , \tag{1}$$

where t is time, X is the local microbial concentration (subject to both $_{growth}$ and $_{decay}$), v_{max} is an asymptotic maximum specific uptake rate, C_{org} and C_{ea} are the aqueous concentrations of the organic compound (substrate) and TEA, respectively. K_{org} and K_{ea} are the half-saturation constants for the organic compound and the TEA, respectively. For the complete mass-balance of the bacterial group, microbial decay needs to be considered, leading to

$$\frac{\partial X}{\partial t} = \frac{\partial X_{growth}}{\partial t} + \frac{\partial X_{decay}}{\partial t} , \tag{2}$$

$$\text{with} \quad \frac{\partial X_{decay}}{\partial t} = -v_{dec} X , \tag{3}$$

where v_{dec} is a decay rate constant. During microbial growth, both organic substrate and TEA are consumed at rates that are proportional to v_{max}. Thus, for a known reaction stoichiometry, the degradation rate can be easily determined from.

$$\frac{\partial C_{org}}{\partial t} = Y \frac{\partial X_{growth}}{\partial t} , \tag{4}$$

where Y is a stoichiometric factor. As written, the above formulation treats multiple hydro-carbon compounds as one single compound with similar physico-chemical properties and, consequently cannot mimic the above-discussed differential degradation (or recalcitrance) of compounds under varying redox conditions. To simulate this, (1) needs to be modified to

$$\frac{\partial X}{\partial t} = \left[\left(\sum_{n=1,n_{org}} \frac{\partial X_n}{\partial t} \right) - v_{dec} \right] X , \tag{5}$$

where each of the growth terms $\partial X_n/\partial t$ is derived from a compound-specific term similar to (1) where the uptake rates v_{max} can then differ between different substrates. If more than one TEA is involved in the degradation process, this will typically require simulation of concentrations of multiple microbial groups, each associated with a particular TEA. To inhibit growth of bacteria in the presence of a thermodynamically more favourable TEA, one or more inhibition terms of the form

$$I_{inh,ea} = \frac{K_{inh,ea}}{K_{inh,ea} + C_{inh,ea}} \tag{6}$$

can be included as a factor in (1) and/or (5), where $C_{inh,ea}$ is the concentration of a more favourable TEA and $K_{inh,ea}$ is an inhibition constant that needs to be much smaller than typical concentrations of the more favourable TEA. The Monod-type inhibition term $I_{inh,ea}$ will then remain ≈ 0 as long as the more favourable TEA is present in significant amounts but reaches its maximum value ≈ 1 as soon as the more favourable TEA is depleted (no growth-inhibition).

With the above equations incorporated into a numerical model [29], it is then possible to simulate redox-dependent, differential degradation rates, e.g., benzene might degrade at the same or faster rate as toluene under aerobic conditions whereas under selected anaerobic conditions it might degrade more slowly or not at all.

CONCLUSIONS
Although there is generally a paucity of research on the anaerobic degradation of benzene, it is evident from a number of studies that the prevailing redox, physicochemical and biological conditions all play a part in determining the rate and extent of benzene degradation. The overall picture of anaerobic benzene degradation is that it can and does occur in a variety of conditions, but that organisms capable of utilising benzene anaerobically are by no means ubiquitous. There are also contrasting data concerning the use of TEAs in degradation, with studies by Nales *et al.* [11] showing benzene degradation under NO_3^--, SO_4^{2-}- and Fe(III)-reducing conditions and not under methanogenic conditions, while Kazumi *et al.* [16] report degradation under methanogenic, SO_4^{2-}- and Fe(III)-reducing conditions only. Where rapid benzene degradation is seen in the field it is often associated with shallow aquifers and it is suggested that most of this degradation is aerobic, along the margins of contaminant plumes, with a limited amount occurring anaerobically within the plume body [5]. Anaerobic degradation of benzene is clearly far more site-specific than for the remaining TEX compounds, its extremely stable structure making it less susceptible to microbial attack and thus highly dependant on both biotic and abiotic factors. A high organic fraction has been shown to inhibit anaerobic benzene degradation [11], as has the presence of alternative energy sources [30]. In the presence of high concentrations of BTEX compounds, it may be that removal of the more easily degraded TEX component and associated TEA may be responsible for the persistence of benzene in the environment. The use of enrichment cultures may favour faster-growing species at the expense of slower-growing microorganisms which are capable of degrading benzene over longer incubation periods/residence times [31]. It is likely that a combination of laboratory experiments and modelling of hydrogeology and hydrogeochemistry will aid in determining the potential for natural attenuation at a given site. While still poorly understood, anaerobic biodegradation is acknowledged as a factor in the natural attenuation of a variety of compounds, including some, such as benzene, once considered to be recalcitrant in all but aerobic conditions.

REFERENCES

1.   **Wiedemeier, T. H., M. A. Swanson, J. T. Wilson, D. H. Kampbell, R. N. Miller, and J. E. Hansen.** 1996. Approximation of biodegradation rate constants for monoaromatic hydrocarbons (BTEX) in ground water. Ground Water

Monitoring and Remediation **16**:186-194.

2. **Irwin, R. J.** 1997. Environmental Contaminants Encyclopedia. National Park Service, Fort Collins, CO, USA.

3. **Stephen, H. and T. Stephen**. 1963. Solubilities of Inorganic and Organic Compounds. Pergamon Press, Oxford.

4. **Alexander, M.** 1999. Biodegradation and Bioremediation. Academic Press, San Diego.

5. **Aronson, D. and P. H. Howard**. 1997. Anaerobic biodegradation of organic chemicals in groundwater: a summary of field and laboratory studies. Syracuse Research Corporation, North Syracuse, NY, USA.

6. **Krumholz, L. R., M. E. Caldwell, and J. M. Suflita**. 1996. Biodegradation of 'BTEX' hydrocarbons under anaerobic conditions, p. 61-99. *In* R. L. Crawford and D. L. Crawford (eds.), Bioremediation: Principles and Applications. Cambridge University Press, Cambridge.

7. **Coates, J. D., R. A. Bruce, and J. D. Haddock**. 1998. Anoxic bioremediation of hydrocarbons. Nature **396**:730.

8. **Harwood, C. S. and J. Gibson**. 1997. Shedding light on anaerobic benzene ring degradation: a process unique to prokaryotes? Journal of Bacteriology **179**:301-309.

9. **Heider, J., A. M. Spormann, H. R. Beller, and F. Widdel**. 1999. Anaerobic bacterial metabolism of hydrocarbons. FEMS Microbiology Reviews **22**:459-473.

10. **Kao, C. M. and R. C. Borden**. 1997. Site-specific variability in BTEX biodegradation under denitrifying conditions. Ground Water **35**:305-311.

11. **Nales, M., B. J. Butler, and E. A. Edwards**. 1998. Anaerobic benzene degradation: A microcosm survey. Bioremediation Journal **2**:125-144.

12. **Durant, L. P. W., P. C. D'Adamo, and E. J. Bouwer**. 1999. Aromatic hydrocarbon biodegradation with mixtures of O_2 and NO_3^- as electron acceptors. Environmental Engineering Science **16**:487-500.

13. **Burland, S. M. and E. A. Edwards**. 1999. Anaerobic benzene biodegradation linked to nitrate reduction. Applied and Environmental Microbiology **65**:529-533.

14. **Lovley, D. R., J. C. Woodward, and F. H. Chapelle**. 1996. Rapid anaerobic benzene oxidation with a variety of chelated Fe(III) forms. Applied and Environmental Microbiology **62**:288-291.

15. **Caldwell, M. E., R. S. Tanner, and J. M. Suflita**. 1999. Microbial metabolism of benzene and the oxidation of ferrous iron under anaerobic conditions: implications for bioremediation. Anaerobe **5**:595-603.

16. **Kazumi, J., M. E. Caldwell, J. M. Suflita, D. R. Lovley, and L. Y. Young**. 1997. Anaerobic degradation of benzene in diverse anoxic environments. Environmental Science and Technology **31**:813-818.

17. **Anderson, R. T., J. N. Rooney-Varga, C. V. Gaw, and D. R. Lovley**. 1998. Anaerobic benzene oxidation in the Fe(III) reduction zone of petroleum contaminated aquifers. Environmental Science and Technology **32**:1222-1229.

18. **Rooney-Varga, J. N., R. T. Anderson, J. L. Fraga, D. Ringelberg, and D. R. Lovley**. 1999. Microbial communities associated with anaerobic benzene degradation in a petroleum contaminated aquifer. Applied and Environmental Microbiology **65**:3056-3063.

19. **Caldwell, M. E. and J. M. Suflita**. 2000. Detection of phenol and benzoate as intermediates of anaerobic benzene biodegradation under different terminal electron-accepting conditions. Environmental Science and Technology **34**:1216-1220.

20. **Phelps, C. D., J. Kazumi, and L. Y. Young**. 1996. Anaerobic degradation of benzene in BTX mixtures dependent on sulfate reduction. FEMS Microbiology Letters **145**:433-437.

21. **Chaudhuri, B. K. and U. Wiesmann**. 1995. Enhanced anaerobic degradation of benzene by enrichment of mixed microbial culture and optimization of the culture-medium. Applied Microbiology and Biotechnology **43**:178-187.

22. **Lovley, D. R., J. D. Coates, J. C. Woodward, and E. J. P. Phillips**. 1995. Benzene oxidation coupled to sulfate reduction. Applied and Environmental Microbiology **61**:953-958.

23. **Reinhard, M., S. Shang, P. K. Kitanidis, E. Orwin, G. D. Hopkins, and C. A. Lebron**. 1997. In situ BTEX transformations under enhanced nitrate- and sulfate-reducing conditions. Environmental Science and Technology **31**:28-36.

24. **Weiner, J. M. and D. R. Lovley**. 1998. Anaerobic benzene degradation in petroleum-contaminated aquifer sediments after inoculation with a benzene-oxidising enrichment. Applied and Environmental Microbiology **64**:775-778.

25. **Phelps, C. D., L. J. Kerkhof, and L. Y. Young**. 1998. Molecular characterization of a sulfate-reducing consortium which mineralizes benzene. FEMS Microbiology Ecology **27**:269-279.

26. **Anderson, R. T. and D. R. Lovley**. 2000. Anaerobic bioremediation of benzene under sulfate-reducing conditions in a petroleum-contaminated aquifer. Environmental Science and Technology **34**:2261-2266.

27. **Grbic-Galic, D. and T. M. Vogel.** 1987. Transformation of toluene and benzene by mixed methanogenic cultures. Applied and Environmental Microbiology **53**:254-260.

28. **Weiner, J. M. and D. R. Lovley.** 1998. Rapid benzene degradation in methanogenic sediments from a petroleum-contaminated aquifer. Applied and Environmental Microbiology **64**:1937-1939.

29. **Prommer, H., D. A. Barry, and G.B. Davis.** 1999. A one-dimensional reactive multi-component transport model for biodegradation of petroleum hydrocarbons. Environmental Modelling and Software **14**:213-223.

30. **Corseuil, H. X., C. S. Hunt, R. C. F. Dos Santos, and P. J. J. Alvarez.** The influence of the gasoline oxygenate ethanol on aerobic and anaerobic BTX biodegradation. Water Research **32**:2065-2072.

31. **Rabus, R., H. Wilkes, A. Schramm, G. Harms, A. Behrends, R. Amann, and F. Widdel.** 1999. Anaerobic utilization of alkylbenzenes and n-alkanes from crude oil in an enrichment culture of denitrifying bacteria affiliating with the β−subclass of *Proteobacteria*. Environmental Microbiology **1**:145-157.

Experimental Investigation on the Properties of DNAPLs Migration

M. KAMON[1], K. ENDO[2], and T. KATSUMI[3]

[1]Professor, Disaster Prevention Research Institute, Kyoto University, Kyoto, Japan
[2]Graduate Student, Department of Civil Engineering, Kyoto University, Kyoto, Japan
[3]Associate Professor, Department of Civil Engineering, Ritsumeikan University, Shiga, Japan

INTRODUCTION

Attempts have been made to treat soil and groundwater which have been contaminated with dense non-aqueous phase liquids (DNAPLs), such as volatilized organic compounds (VOCs), by the pump-and-treat method, bio-remediation, and permeable reactive barriers, and by containing them with impermeable barrier walls. DNAPLs have little solubility in water; thus, they can exist and migrate in soil pores as independent phases. Many numerical models, called multiphase flow models, have been proposed in relation to immiscible phases (Collins, 1961; Helmig, 1997), but few works have been done on the experimental aspects of DNAPL migration (Garnier et al., 1998; Vogler et al., 1999) because an effective experimental method has not yet been developed.

In modeling a multiphase flow, unlike the porosity, the dry density, the viscosity, and the intrinsic permeability, which are usually assumed to be time independent parameters, the degree of saturation, the relative permeability, and the capillary pressure head all vary according to the elapsed time, and yield so called k-S-p relations. To model a simplified multiphase flow system, intrinsic permeability K and the relation among relative permeability k_r, capillary pressure head h_c, and the degree of saturation S should be clarified. The goals of this research are to obtain the k-S-p relations by developing an experimental system with a new type of probe, and then to estimate its validity.

EXPERIMENTAL APPROACH

The apparatus used to measure the migration properties of an immiscible two-phase flow is illustrated in Figure 1. Rotary tube pumps and vacuum pump are assembled in order to control the water content of a sandy medium in a column 3.5 cm in inside diameter and 50 cm in length. Toyoura sand with a soil particle density of 2.64 g/cm³ is used, and the sandy medium is prepared with a porosity of 0.38 according to the underwater falling method. The pore water, which is a sodium-chloride solution and has a concentration of 0.05 mol/L, is induced from the bottom of the column, while continuing to pump the corresponding volume into the sand from the top.

Three hydrophilic and three hydrophobic tensiometers as well as five electrical conductivity probes are inserted into the medium from the side of the column. As substitutes for DNAPLs, Hydrofluoroehter, HFE-7100, and Performance Fluid, PF-5080, (both liquids produced by 3M™) are used for the sake of safety. The properties of these fluids and the representative

Geoenvironmental impact management, Thomas Telford, London, 2001.

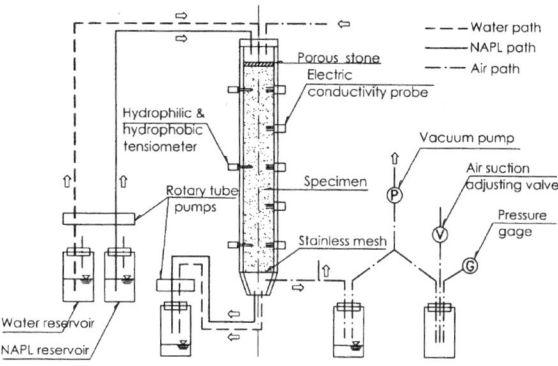

Figure 1. Apparatus for column tests on the multi phase flow

VOCs are summarized in Table 1.

Electrical Conductivity Probe to Measure the Degree of Water Saturation

A method to measure the water content in porous media was developed by Archie (1942) , in which changes in the electrical resistivity of soil systems caused by the water content and the porosity, were utilized. Since then, a method which uses four-electrode probes has been the standard for the electrical investigation of soil systems in both the field and laboratories (Garnier et al., 1998). In this study, three-electrode probes are developed which can design a multi channel system easier than the four-electrode probes, and an efficient estimation of the degree of water saturation, in terms of the measurement of the electrical conductivity of soil systems under an immiscible two-phase flow condition, is achieved. The principle of the measurements taken by the probes is described in the following.

An electric power source, of 1.0 Vpp output voltage and 0.1 Hz frequency, is selected. An alternating current is used for the probes because a direct current would produce errors due to the electrolysis around the electrodes. The impedance of the power source r is 50 Ω. A single-end type data logging system with over 1 MΩ input impedance is used to record the output signal from each probe. Details and the electrical circuit are shown in Figure 2. Among the three gilded electrodes, the two outer ones have equal voltage. When parts of the probes are put into the soil, the resistance R_1 of the probes will change, R_2 and R_3 are set in the experiment to be 10 kΩ and 1 MΩ, respectively. The composite resistance of the parallel RC circuit in

Table 1. Properties of the DNAPL substitutes and the representative VOCs (20 ℃)

Terms		HFE-7100	PF-5080	TCE	PCE	Water
Chemical formula	–	$C_4F_9OCH_3$	C_8F_{18}	C_2HCl_3	C_2Cl_4	H_2O
Specific gravity	–	1.52	1.76	1.464	1.623	0.998
Relative visosity	–	0.58	1.23	0.59	0.90	1.00
Surface tension	mN/m	13.6	15	29.30	31.30	72.75
Interfacial tension	mN/m	—	—	34.50	44.40	none
Vapor pressure	kPa	28	3.87	7.73	2.13	2.34
Solubility in water	ppm	12	11	110	15	none

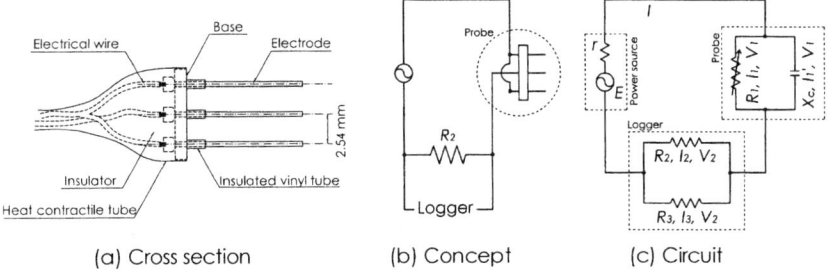

(a) Cross section   (b) Concept   (c) Circuit

Figure 2. Details and the electrical circuit of the developed three-electrode probe

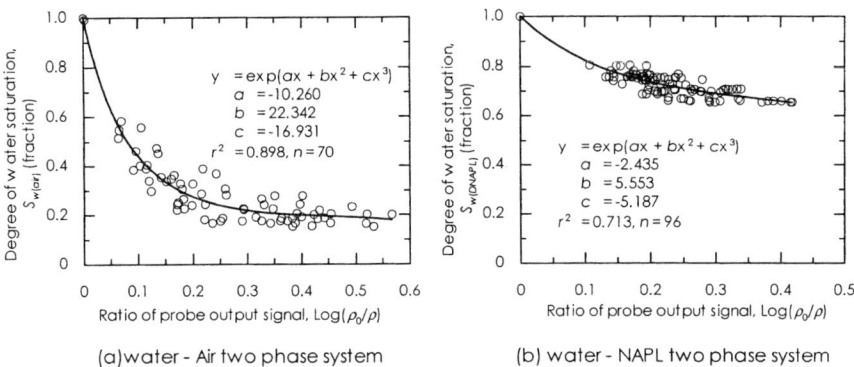

(a)water - Air two phase system   (b) water - NAPL two phase system

Figure 3. Calibrated results of the developed electrical conductivity probes

the soil medium is defined R_x. Since 0.05 mol/L of the sodium-chloride solution (electrical conductivity of 4.73 mS/cm) is used instead of the pore water, and an alternating current with low frequency is applied, the effect of the condenser component in the RC circuit is negligible. Thus, the total circuit forms a series circuit of two resistances, namely, R_x and R_2; consequently, changes in voltage difference V_1 in the medium can be measured via voltage V_2 in the logging system.

Calibrated Results for the Two Phase Flow of Water and Air
Let ρ_0 (mV) be the output voltage from the probes inserted into the medium saturated with the sodium-chloride solution. By setting this as the initial state, the pore water is pumped out, without inflow, until the outflow from the bottom stops. Then, the sand media located at the probes are extracted as a set of specimens and the water content is measured by the oven-dry method. This procedure is repeated by changing the rate of draining. Figure 3(a) shows the degree of water saturation $S_{w(air)}$ versus $\log(\rho_0/\rho)$, where ρ is the measured output voltage. From this figure, an approximate relation between $S_{w(air)}$ and $\log(\rho_0/\rho)$ ($\equiv x$) can be obtained as $S_{w(air)} = \exp(-10.260x + 22.342x^2 - 19.931x^3)$.

Calibrated Results for the Two Phase Flow of Water and DNAPL
The experimental method here is the same as that above, however, an equal amount of DNAPL is supplied to the drained pore water from the top. Therefore, no air saturation exists in this case.

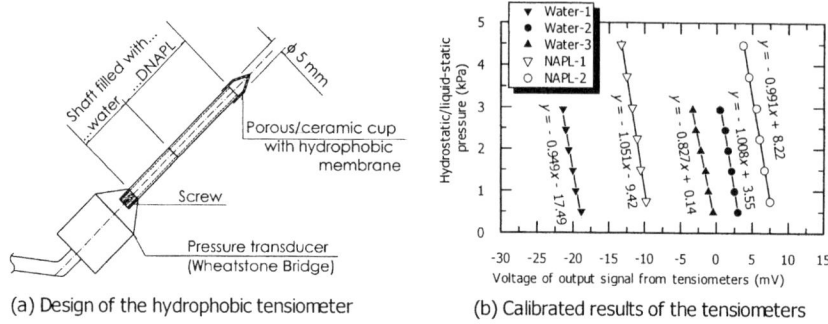

(a) Design of the hydrophobic tensiometer   (b) Calibrated results of the tensiometers

Figure 4. Calibration and design of hydrophilic and hydrophobic tensiometers

The calibrated results are shown in Figure 3(b), which lead to the approximated expression $S_{w(DNAPL)} = \exp(-2.435x + 5.553x^2 - 5.187x^3)$.

Tensiometer to Measure the Capillary Pressure Head of Pore Liquids

In general, a tensiometer consists of a porous cup saturated with deaired water, a shaft filled with deaired water, and a pressure transducer. Two types of tensiometers are used in this study. One is a standard hydrophilic tensiometer used to measure the pore water pressure, and the other is a hydrophobic tensiometer, shown in Figure 4(a), used to measure the pore DNAPL pressure. The hydrophobic tensiometer consists of a porous cup with a hydrophobic membrane, saturated with DNAPL, and a shaft filled with DNAPL and deaired water.

The sensitivity of a tensiometers response depends on whether or not the water in the shaft and the pore water in the soil are continuous throughout the porous cup. To confirm this, the hydrostatic/liquid-static pressure is measured by increasing the depth of the liquid step by step. The results are shown in Figure 4(b). The horizontal axis represents the output voltage (mV) from each tensiometer and the vertical axis represents the liquid-state pressure (kPa). The gradient of each line corresponds to the correction coefficient of the output voltage. The responses of both tensiometers are satisfactory, and their validity has been confirmed.

RELATION BETWEEN THE DEGREE OF WATER SATURATION AND THE CAPILLARY HEAD

The relation between the degree of water saturation, S_w, and the capillary pressure head, h_c, of the unsaturated porous media with immiscible fluid phases, pair i and j, is referred to as a water retention curve, where i and j stand for air (a), water (w), and non-wetting fluid (n) (in this case, DNAPL). The empirical parametric form, called the VG model, was given by van Genuchten (1980): $S_{ej} = [1 + (\alpha h_c)^n]^{-m}$ for $h_c > 0$, where α and n are the VG parameters, and $m = 1 - 1/n$. $S_{ej} = (S_j - S_{jr})/(1 - S_{jr})$ defines the effective saturation of the j phase with S_{jr} being the residual or the irreducible saturation. Another empirical form of the water retention curve, called the BC model, was given by Brooks and Corey (1964), namely, $S_{ej} = (h_d/h_c)^\lambda$ for $h_c \geq h_d$, where λ is the pore distribution index and h_d is the displacement head of the reference two-phase system.

Figure 5(a) shows the relation between the degree of water saturation and the capillary pressure head for the water-air, the water-HFE, and the water-PF systems. The capillary pressure head

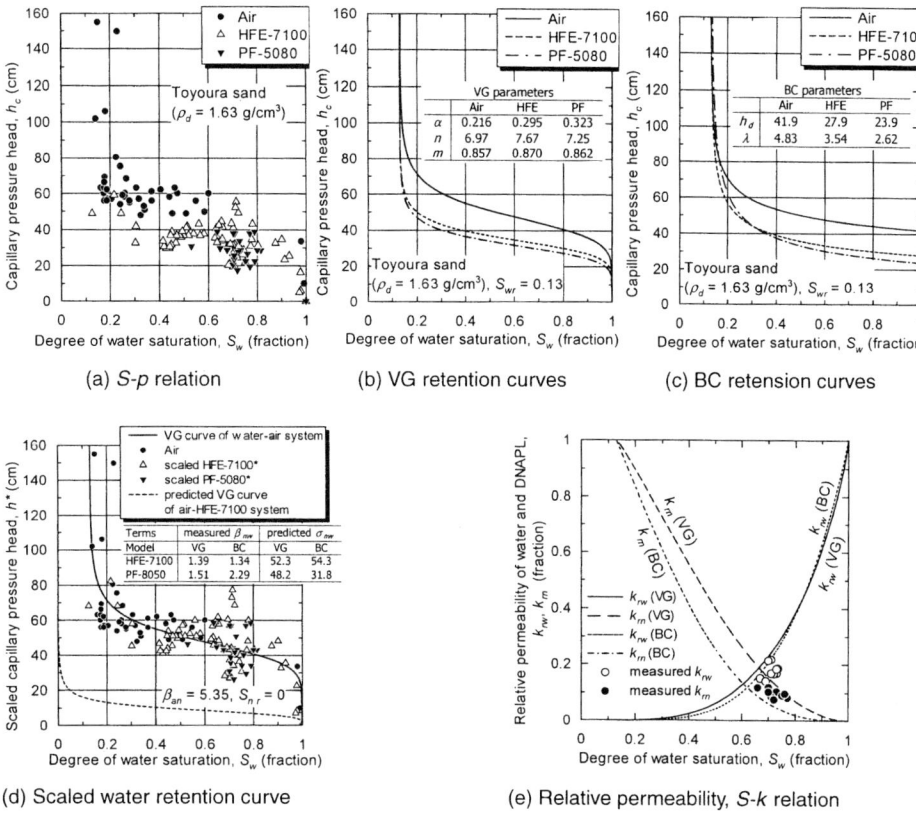

Figure 5. Fitted and scaled *S-p* and *S-k* relations

values for the water-HFE and the water-PF systems are plotted lower than those of the water-air system. This tendency is consistent with the results shown by Parker (1989). Figures 5(b) and (c) show the curves fitted by the VG and the BC models, respectively.

The difference between the water-air and the water-NAPL systems is scaled by introducing scaling factor β_{ij} (Lenhard and Parker, 1987). The scaling factor is defined as $\beta_{ij} = \sigma*/\sigma_{ij}$, in which $\sigma*$ is the interfacial tension of the reference fluid pair and σ_{ij} is the interfacial tension between fluids i and j. Choosing a reference system is arbitrary, but a natural choice would be the surface tension of the air-water fluid pair. Thus, the scaled VG model can be expressed as $S* = [1 + (\alpha h*)^n]^{-m}$, where $S* = S_{ej}$ and $h* = \beta_{ij} h_{ij}$. Applying the same scaling procedure to the BC model, $S* = (h_d/h*)^\lambda$ can be obtained. The scaled results between the degree of water saturation and the capillary head are plotted in Figures 5(d), and the predicted curve for the air-HFE system with $\beta_{an} = \sigma_{aw}/\sigma_{an} = 5.35$ is shown by the broken line. The measured scaling factor, β_{nw}, and the predicted interfacial tensions by the factor, $\sigma_{nw} = \sigma_{aw}/\sigma_{nw}$, are summarized in this figure.

RELATION BETWEEN THE DEGREE OF WATER SATURATION AND THE RELATIVE PERMEABILITY

The method for predicting relative permeability k_r is also proposed using the VG and the BC models, and the value of the relative permeability is estimated from the S-p relation obtained above. The relative permeability usually ranges from 0 to 1, and it reaches 1 when the porous media are saturated with the reference phase. The relative permeability of the water, k_{rw}, and the DNAPL phase, k_{rn}, are given by the following equations. By the VG model, $k_{rw} = S_e^{\varepsilon}[1 - (1 - S_e^{(1/m)})^m]^2$ and $k_{rn} = (1 - S_e)^{\gamma}[1 - S_e^{(1/m)}]^{2m}$, where ε and γ representing the connectivity of pore structure in usual take $\varepsilon = 1/2$ and $\gamma = 1/3$ (Helmig, 1997), while by the BC model, $k_{rw} = S_e^{((2+3\lambda)/\lambda)}$ and $k_{rn} = (1 - S_e)^2(1 - S_e^{((2+\lambda)/\lambda)})$. The predicted curves for the relative permeability and the measured plots are shown in Figure 5(e), where the open circles indicate the measured relative permeability of water and the solid circles indicate that of the DNAPL, which are obtained using the hydraulic gradient computed by the hydrophilic and hydrophobic tensiometers, respectively.

CONCLUSION

A three-electrode probes, available for the measurement of the electrical conductivity of pore water even in a small experimental system, has been developed, and satisfactory results for the electrical conductivity of both water-air and water-DNAPL two-phase flow systems have been obtained. In addition, a hydrophilic tensiometer has been found to be useful not only for the air-water system, but also for the water-DNAPL system. A hydrophobic tensiometer can detect DNAPL and measure its pressure, although the size and the state of the pore space cannot be inferred from the data. In conclusion, the experimental system developed in this paper is expected to be utilized for evaluating the migration properties of water-DNAPL two-phase flow systems in sandy media.

REFERENCES

Archie, G. E. (1942) : The electrical resistivity log as an aid in determining some reservoir characteristics, *Trans. A.I.M.E.*, Vol. 146, pp. 54–61.

Brooks, R. H. and Corey, A. T. (1964) : Hydraulic properties of porous media, in *Hydrology Paper no.3*, Colorado State University, Fort Collins, pp. 1–27.

Collins, R. E. (1961) : *Flow of Fluids through Porous Materials*, Reinhold Publishing Corporation.

Garnier, J., Thorel, L., and others, (1998) : NECER: Network of European centrifuges for environmental geotechnical research, in T. Kimura, O. Kusakabe, and J. Takemura eds., *Centrifuge 98*, Balkema, Rotterdam, pp. 957–982.

Helmig, R. (1997) : *Multiphase Flow and Transport Processes in the Subsurface*, Springer.

Lenhard, R. J. and Parker, J. C. (1987) : Measurement and prediction of saturation-pressure relationships in three-phase porous media systems, *J. Contam. Hydrol.*, Vol. 1, pp. 407–424.

Parker, J. C. (1989) : Multiphase flow and transport in porous media, *Reviews of Geophysics*, Vol. 27, No. 3, pp. 311–328.

van Genuchten, M. Th. (1980) : A closed-form equation for predicting the hydraulic conductivity of unsaturated soils, *Soil Sci. Soc. Am. J.*, Vol. 44, pp. 892–898.

Vogler, M., Arslan, U., and Katzenbach, R. (1999) : The influence of capillarity on multiphase flow within porous media. - a new model for interpreting fluid levels in groundwater monitoring wells in dynamic aquifers, in R. N. Yong and H. R. Thomas eds., *Geoenvironmental Engineering*, Thomas Telford, London, pp. 318–325.

The influence of fluctuations in the groundwater table on Monitored Natural Attenuation

PROF. DR.-ING. R. KATZENBACH, Institute and Laboratory of Geotechnics,
Darmstadt University of Technology, Germany,
DR.-ING. MATTHIAS VOGLER, Ingenieursozietät Professor Dr.-Ing. Katzenbach GmbH,
Frankfurt am Main, Germany,
DIPL.-ING. ANKE FEHSENFELD, Institute and Laboratory of Geotechnics,
Darmstadt University of Technology, Germany

INTRODUCTION

Spills and leaks of chlorinated solvents and of petroleum hydrocarbons such as gasoline, diesel, motor oils and similar materials cause widespread contamination in the environment. With the moment of infiltration into soil or groundwater, several natural processes begin to destroy or alter the chemical components of the contaminants. Theses processes explain what Monitored Natural Attenuation (MNA) means when the term is used to describe a potential strategy to remediate a contaminated site.

WHAT IS NATURAL ATTENUATION?

Natural attenuation, also known as passive bioremediation or intrinsic remediation, is a passive remedial approach that depends upon natural processes to degrade and dissipate organic constituents in soil and groundwater. Natural Attenuation processes include a variety of physical, chemical and biological processes that under favourable conditions act without human intervention to reduce mass, toxicity, mobility, volume or concentration of contaminants in soil and groundwater. These processes, mainly influenced by advection, include biodegradation, abiotic degradation, diffusion, dispersion, dilution, volatilisation, sorption and chemical or biological stabilization, transformation or destruction of contaminants (ASTM 1998, Wiedemeier et al. 1999). The influences of the several components are shown by example of figure 1.

Geoenvironmental impact management, Thomas Telford, London, 2001.

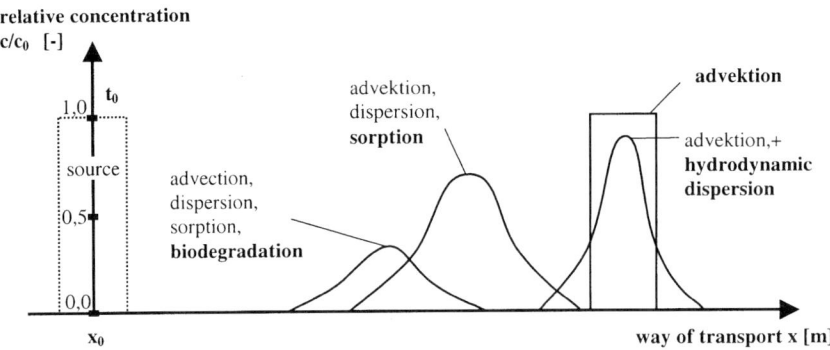

Figure 1. Processes of natural attenuation

Physical processes include mass transfer, such as sorption and ion exchange (Van Impe et al. 2001). Chemical processes involve chemical reactions, such as oxidative degradation of halogenated hydrocarbons in the presence of oxids. The aerobic and anaerobic biological processes involve biological degradation or transformation of contaminants, such as the bacteria mediated degradation e.g. of benzene (figure 1).

All these processes run in the multi-phase system soil-water-air-contaminant. The basis for modelling of large-scale contaminant transport in groundwater including the above mentioned processes is the realistic simulation of the contaminant characteristics (e.g. density, volatility, solubility, wettablility) in the porous medium soil. In general the contaminants like BTEX or chlorinated solvents are present in NAPL form (non-aqueous phase liquid) and also as dissolved contaminants in the groundwater. That means the basis for modelling is the knowledge of the processes which take place at the interfaces between the water-, the air- and the contaminant surface.

EXPERIMENTAL AND NUMERICAL INVESTIGATIONS
OF MULTIPHASE FLOW

Before starting any remediation the infiltrated oil volume and the concentration have to be calculated from fluid levels in monitoring wells. The multi-phase flow of contaminants in porous media is strongly influenced by the porous medias's capillary characteristics. The capillary properties are described by the capillary pressure-saturation relation and the relative permeability-saturation relation in the multi-phase system (Vogler et al. 1999). Due to the absence of capillary effects in monitoring wells it is impossible to evaluate the contaminant volume in the soil directly from nonaqueous phase levels in the monitoring wells. As a result, the volume of contaminants in the soil respectively in the groundwater will be considerably overestimated if only using the information given by monitoring wells. Confronted with this phenomenon a new Simulation Technique was developed. With this new approach including the effects of hysteresis due to the irregular pore geometry and to phase entrapment it is possible to calculate the NAPL volume on the basis of measurements made in monitoring wells considering the decrease and the increase of the groundwater table (Vogler 1999). If the capillary effects are not considered the NAPL volume will be overestimated. The validation of the numerical simulation (figure 3) is carried out by means of laboratory experiments in the Darmstadt Multiphase Simulation Model (figure 2).

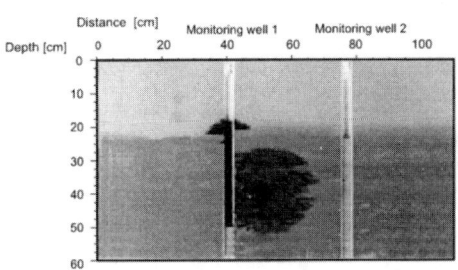

Figure 2: Laboratory experiment

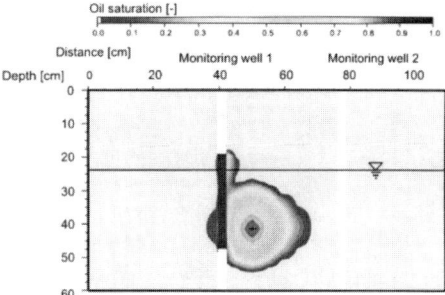

Figure 3: Numerical simulation

The realistic simulation of the multiphase flow of the water, air and contaminant phases in the granular porous medium, which define the initial and boundary conditions, is the basis for modelling of contaminant transport. In porous media the distribution of the residual oil changes depending on fluctuations in the groundwater table.

DISTRIBUTION OF NAPL IN THE PORES

The distribution of the residual saturation in the pores at or near the water table is subjected to leaching from the rise and fall of the water table (seasonal). Figures 4 and 5 indicate the general scenario of a release of NAPL into the soil which subsquently moves vertically under the forces of gravity and soil capillarity through the unsaturated zone to the water table. Contaminants in the porous media can be found up to four different states at a contaminated site. Some of the liquid may be trapped in the soil pores (residual saturation); some may evaporate (volatilisation); some become to be sorbed to the surface of the soil particles (sorption) and some may be dissolved in the groundwater (dissolved plume).

In figure 4 the distribution of NAPL phase without considering the capillary effects is shown. In this case the nonaqueous phase liquids will form a large, connected body of continuous NAPL (mobile NAPL). The volume of the continuous NAPL-phase is V_1 and the surface of the NAPL-phase is A_1. The saturation of the NAPL phase is equal to 1, consequently the relative permeability of water K_w has a value of zero. That means there is no flow of water through the NAPL-phase in the porous media thus there is no supply of natural bacteria, microorganism and fresh water. The result is the centre of the NAPL phase is not influenced by chemical and biological reactions degrading the mass and concentration of the NAPL phase.

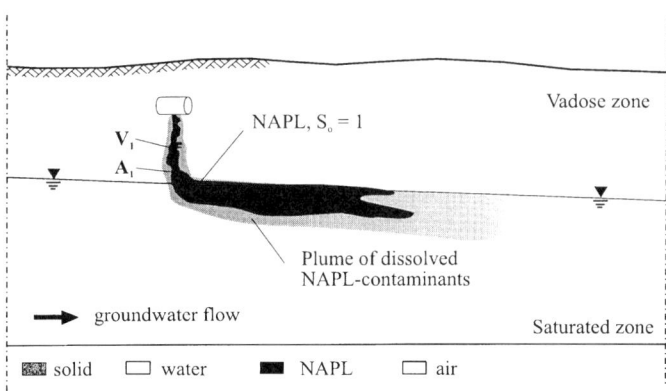

Figure 4. NAPL phase: $S_o = 1$

In natural porous media there is no continous organic phase in the soil (figure 5). The residual organic liquids are trapped as small, immobilised drops, so called "blobs". These individual blobs are not connected to each other. They are immobile for capillary forces. The size, shape and spatial distribution of these blobs affect the dissolution of organic liquids into the water and air phase.

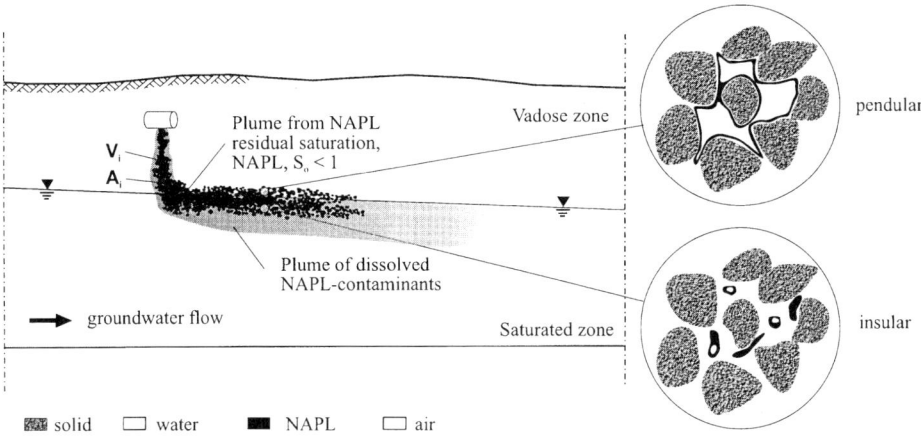

Figure 5. NAPL phase: $S_o < 1$; pendular - insular

In the extreme the total amount of NAPL volume V_i (figure 5) is equal to the NAPL volume V_1 in figure 4. But the sum of the surface areas of all NAPL blobs in figure 5 is much greater than the surface area in figure 4. The relative permeability of water K_w is greater than zero;

that means both water and NAPL could flow simultaneously. However, as saturation of either phase increases, the relative permeability of the other phase correspondingly decreases. The distribution of NAPL in the pores depends on the moisture conditions.

If there is a low water saturation in the porous medium (unsaturated zone), the NAPL residual saturation will wet the grains in a pendular state. After increasing the groundwater the wetting fluid, which is water, will occupy the pores adjacent to grains and the NAPL will be present as isolated drops in the open pores. This state is called insular. Following the pendular and insular distribution of NAPL is introduced.

In the vadose zone most of the pores may be filled with all fluid phases. Note that the NAPL phase will be the wetting phase at **pendular** distribution. In this case, the NAPL will be in direct contact with the solid and occupy the smallest pores. The NAPL forms rings around the grain contact points (figure 5). At this low NAPL saturation they do not form a continuous phase, except a very thin film of (adsorbed) NAPL onto the solid surface. The interface between NAPL and air is very big, that means the microbial activities have quite good conditions. At pendular state the residual NAPL is in contact with immobile residual water. Because of its long duration of contact a high concentration of soluble contaminants adjust to the water. If this water will be replaced to fresh water (e.g. by precipitation) a high concentration of contaminant is entered to the moving groundwater. The residual water could be mobilised after increasing the groundwater, as well, with the effect of a peak in the dissolved contamination.

After increasing groundwater the pendular rings expand until small continuous NAPL phase are formed. These **insular** blobs of NAPL in one or a few pores are shown in figure 5 for the saturated zone. The biggest pores are filled with the wetting fluid water including the NAPL blobs in the middle. The NAPL surface of each blob is smaller in comparison with the thin oil film in the pendular state. That means there are bad conditions for chemical reactions and destruction of contaminants. The rate of biodegradation will depend, in part, on the supply of molecular oxygen by fresh water to the contaminated area. At levels of dissolved oxygen below 1 to 2 mg/l in the groundwater, aerobic biodegradation rates are very slow (EPA 510-B-95-007). Because of a relatively high permeability to water at the insular state, the groundwater migrates fast and there is a large supply of these nutrients.

Summarising the residual NAPL may act as a source of soluble components which dissolve in the groundwater in the saturated zone forming a dissolved plume and to the air in the vadose zone. The amount of dissolution is depending on the surface of NAPL that is exposed to the migrating groundwater (thin oil film formed on the grains or oil blobs), the size of oil blobs and the chemical characteristics of oil (e.g. the solubility of oil compounds) (Fetter 1999). In some cases, residual blobs will be positioned out of the flow path of moving groundwater (e.g. into dead-end-pores) and the rate of dissolution will be reduced. In this case the NAPL blob will persist for a long time.

Dissolution of the organic phase will also have an effect on the permeability to water and the increased permeability to water will effect the rate of dissolution. As the organic phases dissolve, its saturation becomes reduced and the permeability of water increases. Increasing permeability results in increased water flow under constant boundary conditions.

Figure 7 illustrates the relative permeability graphs for a wetting phase and a non-wetting phase for various saturations in the two fluid phase system. Three regions are existing: insular, funicular and pendular.

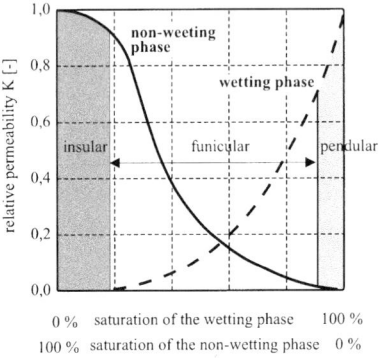

Figure 6. Relative permeability for a wetting and a non-wetting phase (van Dam 1967)

The insular region has a high saturation of the non-wetting phase and the permeability of the non-wetting phase is low. At funicular distribution the wetting and the non-wetting phases flow simultaneously and both are mobile phases. The pendular state exhibits a high saturation of the wetting phase. The flow of the wetting phase dominates and the non-wetting phase is immobile.

In the two-phase system NAPL-air, the NAPL-phase is the wetting fluid whereas in the two-phase system water-NAPL the water-phase is the wetting fluid.

CASE EXAMPLE

In the following example, the BIOSCREEN model is used to simulate natural attenuation of a benzene plume. The BIOSCREEN Natural Attenuation software is based on the Domenico (1987) three-dimensional analytical solute transport model. The Domenico solution accounts for the affects of advective transport, dispersion, sorption and biodegradation. In our examples we investigate the following model cases for our contaminated site:

case 1: no variation

case 2: transport with no sorption and first-order decay

case 3: transport with sorption and variation of the first-order decay

The input data set is summarized in table 1.

	v_a [m/s]	α_L [m]	R [-]	λ [pro Jahr]
case 1	$1,6 \cdot 10^{-6}$	25	1,7	0,17
case 2	$8,0 \cdot 10^{-7}$	25	1,7	0,074
case 3	$1,6 \cdot 10^{-6}$	25	1,7	0,085

Table 1. Data input set

v_a: apparent flow velocity   R: Retardation factor

α_L: longitudinal dispersivity λ: first-order decay coefficient

Solute transport with biodegradation modelled as a first-order decay was used to reproduced the plume length (about 500 m). The first-order decay model (ASTM, 1998) was used to determine the value of λ for the different cases. The source concentration is equal to 2,3 mg/l in the centre of the source. The results of the calculations are shown in sections .

In case 1, figure 7, no good match was reached with the calculated concentrations over time and the measured concentrations. In the next calculation was on the assumption that there is no sorption. The results of case 2, figure 8, show, that the processes of sorption are not the decisive factors for this contaminated site. In the last calculation the influence of the first-order decay factor was investigated. The first two runs overestimated the biodegradation, so the amount of λ was decreased in case 3. A good match of the plume length and the measured concentrations was reached with this input data set, shown in figure 9. The results show, the biodegration is a very sensitive factor to determine how effective natural attenuation is for attaining site remediation goals.

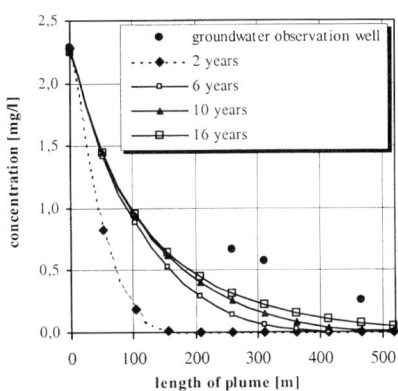

Figure 7: Case 1

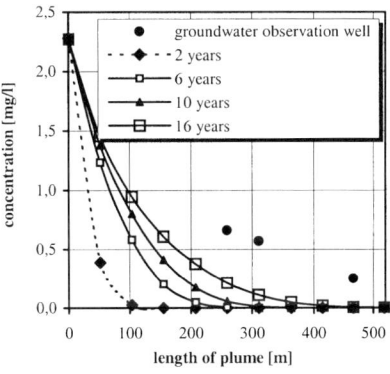

Figure 8. Case 2

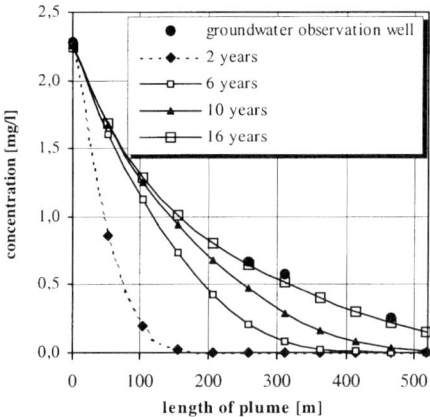

Figure 9: Case 3

CONCLUSION

Because of the interaction between the dissolution of components and the distribution of the residual NAPL-phase in the pores multiphase-fluid flow and the contaminant transport have to be investigated considering the effect of hysteresis. The natural processes are influenced by the interactions between the air-, water- und contaminant phase, e.g. by changing supply of molecular oxygen. The solubility of organic compounds varies due to time the groundwater stays inside the contaminated area, as well.

Summarising a remediation by Monitored Natural Attenuation requires the following site characterization information:

- Laboratory and field investigation of soil mechanical and chemical properties of the soil and of the hydrological, chemical and biological properties of the fluids e.g. locations of potential receptors, direction of ground water flow and hydraulic conductivity.
- Modelling of the processes running in the soil and groundwater.
- Validation of the numerical processes by small and large model tests in the laboratory and in-situ and evaluation remedial processes.
- Appropriate monitoring program, based on the principals of the observational method, which is a well known and successful in classic soil mechanical approach (deep excavations, slopes, tunnelling, high loaded foundations.
- Checking the estimated natural processes.

However the interaction between the multiphase-fluid flow and contaminant transport have to be considered to estimate the effectiveness of natural attenuation.

REFERENCES

ASTM (1998)

Standard Guide for Remediation of Ground Water by Natural Attenuation at Petroleum Release Sites (E 1943-98), West Conshohocken, Pennsylvania.

Domenico, P.A. (1987)

An analytical model for multidimensional transport of decaying contaminant species. Journal Hydrology 91, 49-58.

EPA (1995)

How to Evaluate Alternative Cleanup Technologies for Underground Storage Tank Sites: A Guide for Corrective Action Plan Reviewers. Office of Underground Storage Tanks, EPA 510-B-95-007

Fetter C.W. (1999)

Contaminant Hydrogeology. Second Edition. Prentic-Hall International, London

Van Dam, J. (1967)

Die Ausbreitung von Mineralöl in grundwasserführenden Schichten. Deutsche Gewässerkundliche Mitteilungen, Jg. 11, Heft 1, 15-26

Van Impe, W.F., Katzenbach, R., Ennigkeit, A. (2001)

TC5 – Summarized Activities. Mitteilungen des Institutes und der Versuchsanstalt für Geotechnik der Technischen Universität Darmstadt, Heft 55

Vogler, M. (1999)

Einfluss der Kapillarität auf die Mehrphasenströmung bei der Sanierung von Mineralölschadensfällen im Boden. Mitteilungen des Institutes und der Versuchsanstalt für Geotechnik der Technischen Universität Darmstadt, Heft 45

Vogler, M., Arslan, U., Katzenbach, R. (1999)

The infuence of capillarity on multiphase flow within porous media. A new model for interpreting fluid levels in groundwater monitoring wells in dynamic aquifers. 2nd BGS International Geoenvirionmental Engineering Conference, London, UK, 318-325

Wiedemeier, T., Rifai, H., Newell, C., Wilson, J. (1999)

Natural Attenuation of Fuels and Chlorinated Solvents in the Subsurface. John Wiley & Sons

Contaminant transport in shallow free surface aquifers

I. KAZDA, S. FERINOVÁ
REAT – Research and Educational Academy, Prague, Czech   Republic

ABSTRACT
A large number of sites in the Czech Republic have been contaminated due to non-ecological disposal of industrial waste during the last forty years. These so-called old economic loads are necessary to be gradually eliminated. As remediation measures to be applied at the respective sites demand considerable costs, for each particular case such a remediation technology has to be selected to ensure not only reaching the remediation targets, but also economic efficiency.

Numerical simulation of solute transport requires good approximation of aquifer geometric shape. Therefore, the exploitation of quasi-spatial numerical model need not be adequate while modelling solute transport in a shallow aquifer with a extensive variable thickness. Using a three-dimensional model in such case will allow reliable setting of the contaminant propagation within the aquifer.

The example of contamination of a waste repository and a shallow aquifer with a free surface illustrates the advantages of the solute transport numerical modelling application in the very first stage of remediation planning, when not all data is available in the satisfactory amount, yet it is necessary to get an idea of the current condition of the aquifer contamination and its further development. It demonstrates the effect of the contamination water yield and the promptness of corresponding remediation intervention. Compatible hardware and software will make it possible to assess, in a short time and with low costs, the relevant remediation alternatives and evaluate the effect of interaction of the solutes transported by groundwater with the percolated medium.

INTRODUCTION

From 1950 to 1989, many localities in the Czech Republic were contaminated by industrial waste. These, so-called, old environmental loads appeared due to extensive socialist economy, which did not take into account the consequences of non-ecological toxic waste disposal. Improperly set up landfills and repositories were often sources of contamination of groundwater, which represents an irreplaceable natural resource in the Czech Republic. That is why, after 1990, the Czech Government made a decision to implement gradual remediation of these old environmental loads. The presumed remediation costs have come up to over three billion USD. For this reason, it is of special importance to design an adequate remediation technology for each locality so as not only to fulfil remediation targets, but also reach economically acceptable costs for the remediation measures applied.

In preparing remediation of each contaminated locality, it is necessary to provide a correct assessment of its hydrogeological situation, evaluate the current pattern and extent of solute transport in groundwater and the impact of remediation methods available for the respective case. For these purposes the application of a numerical modelling method, based on using the finite-element method, has proved effective. Among great advantages of this method there is a possibility of accurate modelling of all characteristic properties of percolated rock continuum, simulating any time pattern of not only solute transport itself, but also of the effect of remediation methods on the contaminated aquifer and on the environment in its vicinity. For these purposes the application of a numerical modelling method, based on using the finite element method, has proved effective (Kazda 2000).

NUMERICAL MODELLING OF GROUNDWATER FLOW

A design of remediation measures for a contaminated locality is based on the assessment of a groundwater flow pattern. Numerical modelling of a groundwater flow pattern or solute transport within a given domain of the rock continuum must be based on concrete data, established by means of a hydrogeological survey. The governing equation of a steady groundwater flow in a heterogeneous anisotropic three-dimensional domain may be written using a summation convention of the Cartesian tensor analysis as

$$\frac{\partial}{\partial x_i}\left(k_{ij}\frac{\partial h}{\partial x_j}\right) + Q^* = 0 \quad i,j = 1, 2, 3 \tag{1}$$

where k_{ij} is the hydraulic conductivity component (a second order tensor), h is the hydraulic head and Q^* represents the sources and sinks in the given domain:

$$Q^* = \sum_{k=1}^{n} Q_k\, \delta(x_1 - x_{1k})(x_2 - x_{2k})(x_3 - x_{3k}) \tag{2}$$

where n is the number of sources and sinks and δ is the Dirac distribution. If a shallow aquifer allows neglecting the vertical seepage velocity component, approximating thus the groundwater flow as a horizontal plane flow with a free surface, the equation (1) may be integrated in the direction of the axis z (Kazda 1990):

$$\frac{\partial}{\partial x_i}\left(k_{ij}H\frac{\partial h}{\partial x_j}\right) + Q^* = 0 \quad i,j = 1, 2 \tag{3}$$

where H is the thickness of a percolating groundwater flow.

In numerical modelling of a groundwater flow in a shallow aquifer, a simpler conceptual model is frequently used, based on the equation (3), which is of a quasi-spatial type and, therefore, allows using two-dimensional finite elements for its solution. The preparation of a three-dimensional numerical model, on the contrary, is much more time consuming. Setting the three-dimensional mesh of finite elements requires detailed data, characterising individual geological layers and all prominent discontinuities within the modelled domain. The choice of correct boundary conditions on the three-dimensional domain's boundary and setting the initial condition is also a more difficult task.

In order to prepare a mesh of two-dimensional elements, various pre-processors may be used, which make its setting substantially easier and faster. The two-dimensional mesh can be, at the same time, easily checked by means of drawing it. With three-dimensional meshes, however, the pre-processors applied must be much more sophisticated, and checking by means of drawing the mesh is not so simple any more. A convenient means to be used in this case is a pre-processor that allows not only drawing the entire three-dimensional mesh, but also its set part only. In this way, searching for potential mistakes is substantially simplified.

The dimensionality effect of the numerical model applied may be shown by solving a ground-water flow pattern in a shallow aquifer, situated in Southern Bohemia and contaminated by waste water from an ore-dressing plant. Fig. 1 displays the shape of this aquifer, applied subdivision into elements and boundary conditions.

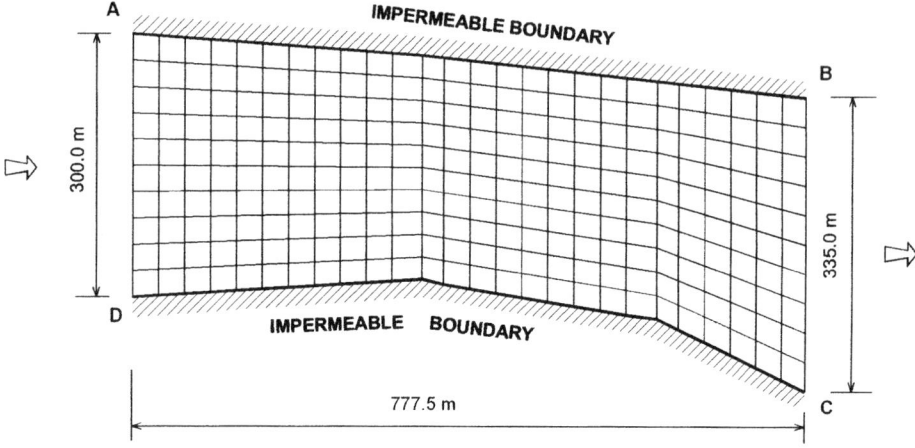

Figure 1 Plan of a modelled aquifer and set boundary conditions

The groundwater flow and solute transport in this aquifer were first solved by means of a quasi-spatial model (Kazda, Vaníček 1999). In order to create this model of a horizontal plane groundwater flow, two-dimensional isoparametric elements with eight nodes were used, characterised by an incomplete bi-quadratic approximation polynomial.

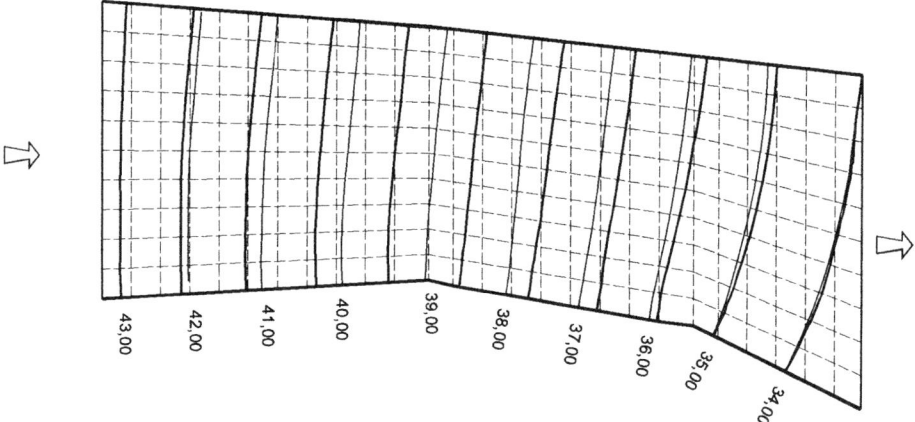

Figure 2 Free surface contour lines for quasi-spatial (thin lines) and 3-D (thick lines) solutions

The three-dimensional numerical model, created by means of isoparametric elements with eight nodes and a trilinear approximation polynomial, had such boundary conditions set on its boundary to correspond to a quasi-spatial model. The results of a groundwater flow solution within a modelled domain by means of a three-dimensional model are shown in Fig. 2 using thick line free surface contours. Thin lines in the same picture represent free surface contours, computed by means of a quasi-spatial model. By comparison it becomes evident that the free surface curve in both solutions differs namely in the medium part of the domain. As the average gradient of a free surface is roughly 0.1 %, slight changes in the computed piezometric heads will result in marked shifts of the free surface contours.

The above-mentioned comparison points out the importance of using a dimensionally correct numerical model. Within the given domain, the impermeable underground layer surface is irregular in shape, while in cross-sections the aquifer's thickness is markedly lowered towards its both edges. As a consequence, the quasi-spatial model does not depict the given domain with sufficient accuracy, and the respective shallow aquifer must be approximated using a three-dimensional numerical model.

Fig. 2 also shows the significant role of boundary conditions, set at the inflow part and the outflow part of the boundary. In the vicinity of these two parts, due to boundary conditions, the differences between the solutions using both models are very small. Another significant fact is that the free surface contours set by means of a three-dimensional model are equally smooth as the free surface contours for the quasi-spatial model, even though the three-dimensional elements applied here are of a lower level of approximation polynomial than the two-dimensional elements of the quasi-spatial model. The applied mesh of three-dimensional elements is, therefore, sufficiently detailed for solving the groundwater flow problems.

MODELLING OF SOLUTE TRANSPORT
Solute transport in groundwater must always be solved as an unsteady problem. The governing differential equation of solute transport contains both the advection member and the diffuse member:

$$n_e \frac{\partial c}{\partial t} = -\frac{\partial}{\partial x_i}\left(ncV_i\right) + \frac{\partial}{\partial x_i}\left(nD_{ij}\frac{\partial c}{\partial x_i}\right) + c_Q Q^* + W \qquad i, j = 1, 2, 3 \tag{4}$$

In this equation n_e is the effective porosity, c is the solute concentration, t is the time, V_i is the seepage velocity component in the direction of x_i - axis, D_{ij} is the coefficient of hydrodynamic dispersion and W represents the solute reactions during transport.

A numerical solution of the initial problem defined by this equation is much more demanding than that of the groundwater flow. Solute transport in the respective aquifer may be solved only providing the vector field of seepage velocities is known. This condition, in turn, can be easily fulfilled only if the groundwater flow is of a steady type. In such case seepage is solved for the given domain, while for solute transport identical seepage velocities are used for the entire simulated period. The transport computation algorithm is much more complicated in the case of an unsteady groundwater flow. Then, actual values of seepage velocities must be set for each particular time step of transport simulation. As a consequence, a numerical solution of solute transport is much more demanding for computer time.

The finite element mesh for simulating solute transport must be of greater detail than that used for solving a groundwater flow. The dimensions of finite elements must necessarily comply with the criteria given by the limit values of the Peclet and Courant numbers (Kazda 1990).Numerical modelling of solute transport much more frequently requires the usage of a three-dimensional model than is the standard procedure in solving a groundwater flow. This results from the fact that the source of contamination usually has a distinct three-dimensional character, which affects pollution propagation within the aquifer.

SIMULATION OF SOLUTE TRANSPORT IN A SHALLOW AQUIFER

As shown in Fig. 2, the seepage pattern in a shallow aquifer is affected by the exploitation of a

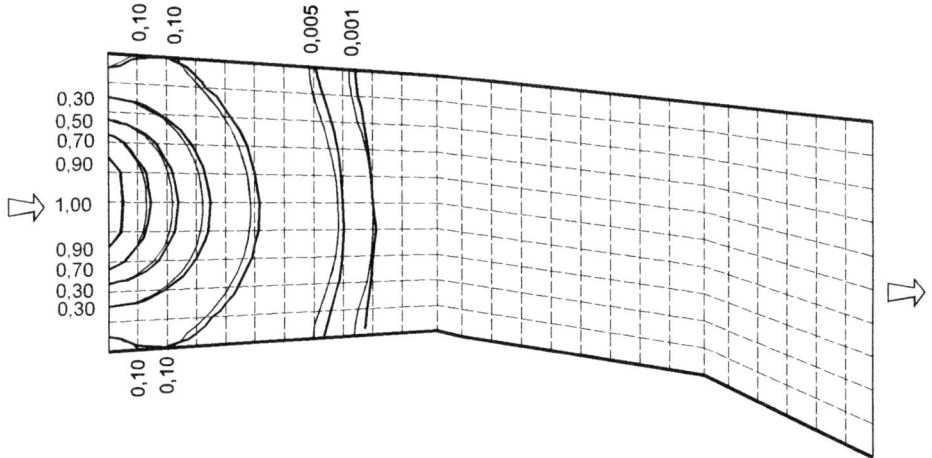

Figure 3 Aquifer contamination 15 years later

more detailed three - dimensional model of an aquifer. The results of solute transport, set by a steady distribution of contaminant concentrations in the inflow part of the aquifer, are compared for both applied models in Fig. 3, using concentration contour lines for the elapsed time of 15 years, while in Fig. 4 contour lines for 30 years after the beginning of contamination are used. Thick contour lines, again, correspond to a three-dimensional model, and thin lines to a quasi-spatial one.

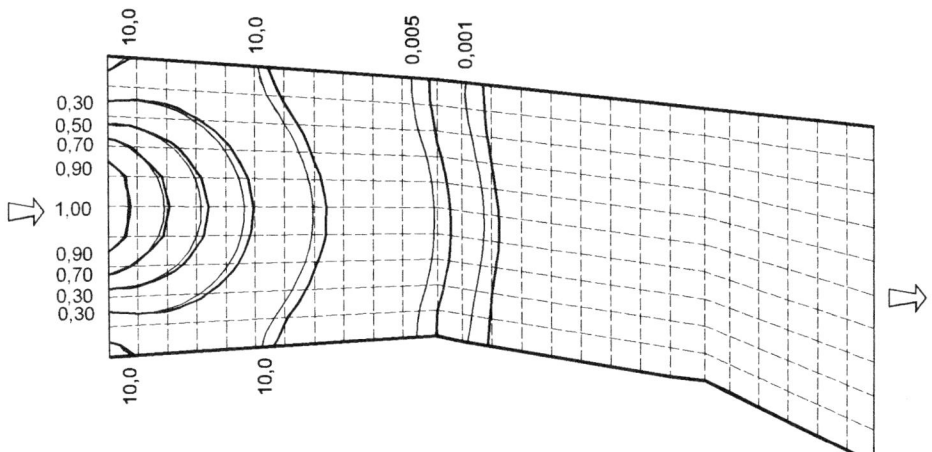

Figure 4 Aquifer contamination 30 years later

For the time of 15 years back, there are slight differences between the contour lines of both solutions. The effect of a more accurate modelling of the aquifer shape by means of a three-dimensional model is evident namely at places where the contour lines $c = 0.005$ and 0.001 approach the impermeable parts of the aquifer boundary. The differences between the both models will become more distinct only after the expiry of 30 years (Fig. 4), in particular in the case of the contour lines for small concentrations of contaminants.

CONCLUSIONS

The pollution of groundwater as a consequence of contaminants leaching from landfills and waste repositories is a frequent phenomenon in the Czech Republic. The advantages and possibilities of numerical modelling application used in such cases were illustrated by an example of a contaminated free surface shallow aquifer in Southern Bohemia.

Numerical modelling of solute transport by groundwater allows an easy and prompt evaluation of the actual state and future development of contamination of a polluted shallow aquifer, as well as prediction of the effect of the available remediation measures. The use of the finite element method at the same time enables to simulate the time pattern of all remediation measures whose effect should be compared.

Numerical solution of solute transport requires careful approximation of a geometric shape of the respective aquifer. Therefore, the exploitation of a quasi-spatial model need not be adequate while solving transport in a shallow aquifer with a variable thickness along its cross-section. Using a three-dimensional model in such case, on the contrary, will allow reliable setting of the velocity of contaminant propagation within the aquifer.

ACKNOWLEDGEMENTS

This paper was prepared as part of the research project *Numerical modelling of environmental problems in general multi-parameter rock systems* and it was supported from grant No. 103/99/0751 by the Grant Agency of the Czech Republic. The support is gratefully acknowledged.

REFERENCES

Kazda, I.: Finite Element Techniques in Groundwater Flow Studies with Applications in Hydraulic and Geotechnical Engineering. Developments in Geotechnical Engineering 61, Elsevier, Amsterdam 1990.

Kazda, I.: Modelling of the Waste Repository – Groundwater Interaction. 12th Regional Central European Conference IUAPPA and 4th International Conference on Environmental Impact Assessment – Prague 2000, Conference Proceedings, Section B, pp. 209 – 214.

Kazda, I. – Vanícek, I.: Finite Element Analysis of Transport and Fate of Pollutants in Complex Hydrogeological Systems. *In*: Yong, R. N. – Thomas, H. R. (eds.): Geoenvironmental Engineering - Ground Contamination: Pollutant Management and Remediation, pp. 272 – 279, Thomas Telford, London 1999.

Natural attenuation of mine drainage-borne zinc in shallow subsurface soils over permafrost, central Yukon, Canada

DYLAN MACGREGOR [1] AND LORETTA Y. LI [2]
(1) M.A.Sc. candidate, Department of Civil Engineering, The University of British Columbia, 2324 Main Mall, Vancouver, B.C. Canada, V6T 1Z4.
(2) Assistant Professor, Department of Civil Engineering, The University of British Columbia, 2324 Main Mall, Vancouver, B.C. Canada, V6T 1Z4.

ABSTRACT
Zinc-bearing mine effluent of pH ~6 discharges from an inactive underground silver mine on a hillside in central Yukon, Canada, near the northern limit of the discontinuous permafrost zone. The effluent percolates through shallow soils overlying permafrost and discharges to small stream in the valley below. Water sampling undertaken during summer 2000 shows progressive removal of zinc from drainage waters in the downslope direction; typical effluent zinc concentrations of 150 mg/l at the adit mouth are reduced to <2 mg/l where the drainage discharges to the creek. Investigation of the shallow subsurface revealed a poorly decomposed organic horizon overlying a poorly sorted fine grained mineral soil. These two soil types were sampled at numerous locations across the study area and subsequently subjected to batch adsorption tests as a measure of the natural attenuation capacity of the organic and mineral soils in contact with the mine drainage.

KEYWORDS
natural attenuation, mine drainage, zinc, permafrost

INTRODUCTION
Galkeno 300 mine, Keno Hill mining district, central Yukon Territory, Canada, was a silver-lead-zinc producer during the 1950's. Recent changes in mine hydrology have resulted in discharge of waters containing 150 mg/L zinc to the environment. This effluent percolates into the shallow subsurface downslope of the discharge point; shallow permafrost restricts downward movement of water, limiting flow to the upper 60 cm of soil. Flow traverses a path length of 1.7 km before discharging to a fish-bearing stream in the valley bottom. Concern over impacts of metal-bearing mine drainage on aquatic life led to an investigation of natural attenuation at the site. This investigation was undertaken from June through August, 2000 and involved flow monitoring, water sampling for chemical analysis and soil sampling for chemical analysis and laboratory studies. This paper presents the results of this field investigation and of preliminary lab testing.

SITE DESCRIPTION
Keno Hill mining district is centered at 63° 55' N, 135° 25' W in central Yukon, Canada. The Galkeno 300 adit is situated on an east-northeast facing hillside (Figure 1) at an elevation of 1160 m (3800 ft) above sea level (asl), approximately 300 m (1000) ft above the valley floor.

Geoenvironmental impact management, Thomas Telford, London, 2001.

Vegetation consists of open mature black spruce forest with willow and dwarf birch understory at higher elevations and transitions to a mixed forest dominated by black and white spruce, paper birch, balsam poplar and trembling aspen with willow and dwarf birch understory at lower elevations. Golden moss is the common ground cover across the site. Slope of the Galkeno 300 hillside generally varies between 10° and 20°, with a single local section at 25°. Near the valley bottom, the slope has two important areas of lower gradient, between 3° and 5°; these flats are characterised by significant accumulations of undecomposed and partially decomposed organic matter. The site configuration and location of flow paths is shown in Figure 2.

The site is largely undisturbed by human activity. However, the anthropogenic features that do exist exert a significant influence on downslope water movement. The site is cut by a power line right-of-way that parallels the slope, running from the base of the Galkeno 300 waste rock dump to the intersection of Christal Creek and the Silver Trail Highway. Approximately 2/3 of the distance down the slope, the site is cut by a 1950's-era road which parallels the 975 m (3200 ft) contour; older roads and wagon trails can still be recognised on the lower flats and are commonly coincident with the flow path. The site is underlain by permafrost which limits vertical infiltration of drainage water to the upper 0.6 m of soil; this unfrozen soil consists of an upper 10-30 cm layer of poorly decomposed organic material and a lower 30-50 cm layer of well graded sandy silt with minor gravel (glacial till).

SITE HISTORY

The Galkeno ore body was mined by underground methods during the 1950's and early 1960's. The Galkeno 300 adit is the highest of two portals into the workings; prior to 1997, there was no effluent reported from this level. The lower portal, the Galkeno 900 adit, has produced metal-rich effluent since the onset of mining. In 1995, a cement plug was emplaced in the lower adit in an attempt to prevent further effluent drainage. Beginning in 1997, two years after installation of the plug in the lower adit, metal-rich discharge from the upper adit has been continuous and flow volumes are reported to have increased from the onset of discharge to the present.

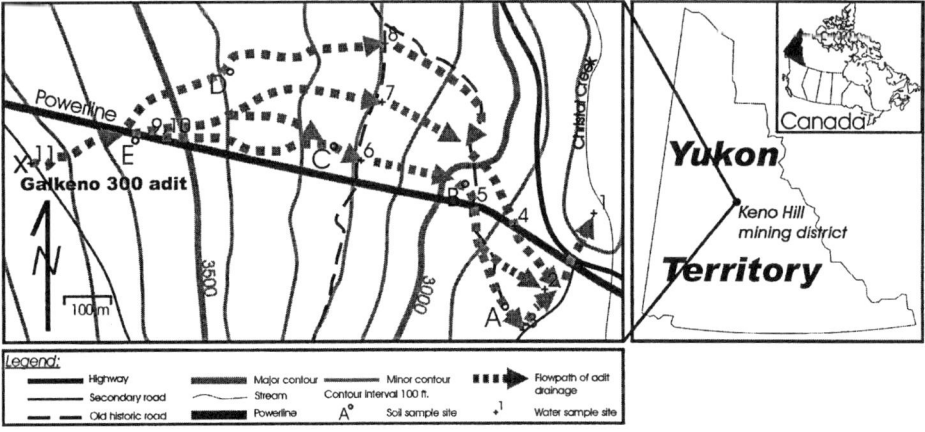

Figure 1. Regional maps showing location of field site; site map displaying physiography, site features and location of sampling sites.

Figure 2. Looking west at field site. Galkeno 300 adit is at top left; impacted receiving waters of Christal Creek are at bottom right. White lines mark the location of effluent flow paths.

SITE INVESTIGATION

Water Sampling
To assess changes in drainage chemistry from the adit to the creek, water samples were collected at a number of points along the flow path (Figure 1). Temperature, pH and conductivity were measured in the field; samples were analysed for total and dissolved metals via atomic absorption spectrophotometry. Sampling began in early June and ran through late August; the results provide a spatial and temporal assessment of drainage chemistry.

Flow Monitoring
Two rectangular weirs with end contractions (Grant and Dawson, 1995) were constructed, one in the adit mouth and one just upstream of the adit drainage-Christal Creek confluence, to monitor input and output flow volumes. Daily precipitation inputs were measured using a Rain Gauge Type B (Environment Canada, 1992) to monitor the potential for dilution. Inspection of the site indicated no water inputs other than precipitation and adit drainage. Sampling of all flows identified adjacent to the adit drainage 'catchment' indicates that all adit discharge reports to Christal Creek via the flow path under investigation. Flow data are used in conjunction with water quality data to estimate metal fluxes from the adit and to Christal Creek; the difference represents the attenuated mass of contaminant.

Soil Sampling
Samples of mineral and organic horizons were collected from test pits at five locations (Figure 1); duplicate samples were collected from within the flow path as well as from similar material immediately adjacent to, but isolated from, adit drainage. Samples adjacent to flow path are thought to represent uncontaminated examples of the material in contact with the mine drainage.

PRELIMINARY LABORATORY STUDIES
Batch Adsorption Testing
Samples of mineral and organic soils were air dried and subjected to batch adsorption tests (U.S.E.P.A., 1992) to determine the capacity of each soil type for removing aqueous zinc for solution. Dosing solutions were prepared using reagent grade zinc nitrate hexahydrate

(ZnNO$_3$·6H$_2$0) dissolved in 0.5 % nitric acid (HNO$_3$), diluted to desired concentration with distilled water and adjusted to pH 6.5. This initial pH is a representative value that approximates conditions across the site. Soil and solution were mixed at a 1:20 ratio (2.000 g oven-dry equivalent mass soil: 40 mL solution) in 50 mL polyethylene centrifuge tubes for 24 hours at 25°C in a mechanical rotator at 22 rpm. Following mixing, mineral samples were centrifuged, organic samples were filtered, and equilibrium zinc concentrations were measured via atomic absorption spectrophotometry.

RESULTS

Water Sampling

Aqueous metal concentrations show no temporal trends for the duration of the site investigation; results from each sampling period indicate that water quality at each sampling site varied little over the period of June to August, 2000. Spatially, aqueous metal concentrations decreased with increasing distance from the adit discharge; rate of decrease is greater after flow enters subsurface 200 m downslope of discharge. Figure 3 demonstrates the continuous removal of zinc along the subsurface length of the flowpath. A summary of important metal species concentrations at the adit discharge and at the sampling point immediately upstream of the mine drainage confluence with Christal Creek is presented in Table 1.

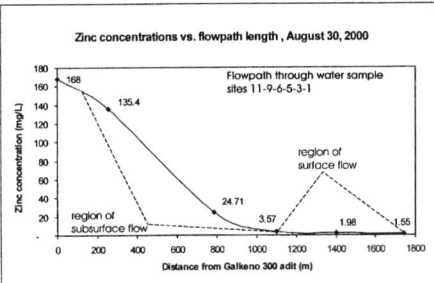

Figure 3. Removal of aqueous zinc from mine drainage with increasing distance along flowpath. Majority of mass is removed in region of subsurface flow.

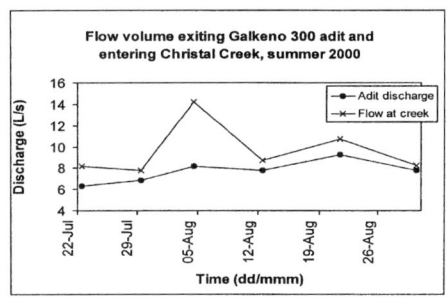

Figure 4. Flow volumes at adit discharge and discharge to Christal Creek, July and August, 2000.

Flow Monitoring

Flow monitoring data for the period of late July and August is shown in Figure 4. The peak in discharge to Christal Creek on August 4 reflects a preceding rainfall event; other monitoring periods indicate on average 16% dilution of adit discharge occurs during periods of base flow. This level of dilution is of little significance in light of the reductions in aqueous metal concentrations noted in Table 1 which span 2 orders of magnitude. Flow monitoring data coupled with water quality data also permit estimates of metal fluxes (Table 1) and of masses of metals removed over the length of the flow path.

Table 1. Water quality data at Galkeno 300 adit and immediately upstream of confluence with Christal Creek, August 30, 2000.

	Galkeno 300 adit	Adit drainage above confluence with Christal Creek	Metal reduction factor	Daily mass flux from adit (kg)	Daily mass flux to creek (kg)
discharge (L/s)	7.8	8.3	n/a	n/a	n/a
pH	6.34	7.26	n/a	n/a	n/a
Zn	168 (mg/L)	1.55 (mg/L)	108	113	1.11
Mn	225 (mg/L)	0.488 (mg/L)	461	152	0.350
Cd	0.592 (mg/L)	<0.006 (mg/L)	'complete' removal	0.40	below detection
As	0.20 (mg/L)	<0.06 (mg/L)	'complete' removal	0.01	below detection

n/a: criteria not applicable to specified parameter

Batch Adsorption Testing

Batch adsorption test results are displayed in Figures 5 and 6. The organic soil displays a greater capacity to adsorb aqueous zinc at the upper limit of concentrations tested than does the mineral soil. Both soils exhibit similar adsorption characteristics at the lower initial solution concentrations which are more representative of site conditions.

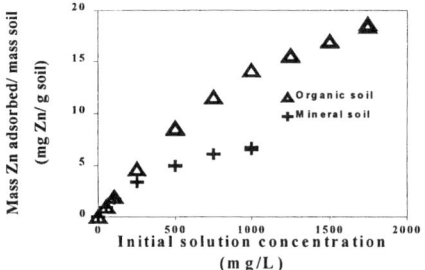

Figure 5. Batch adsorption test results illustrating mass of zinc adsorbed per mass of soil vs. initial solution Zn concentration.

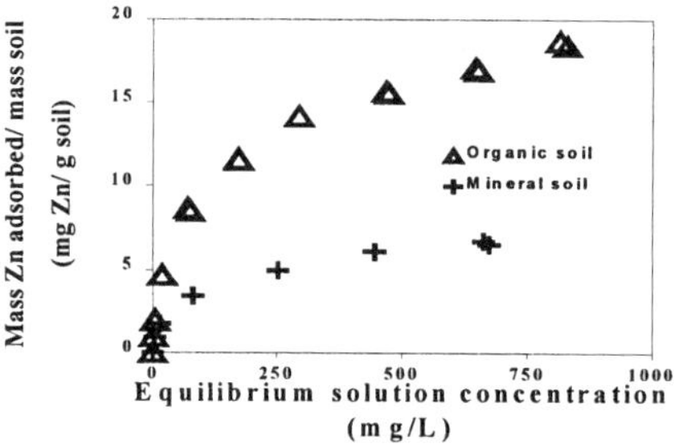

Figure 6. Batch adsorption test results illustrating mass of zinc adsorbed per mass of soil vs. equilibrium solution Zn concentration.

SUMMARY AND CONCLUSIONS

Aqueous zinc is removed from mine effluent as it flows through shallow soils; concentration of zinc in effluent decreases by 2 orders of magnitude over a path length of 1700 m. While precipitation inputs provide a degree of dilution, water balance data combined with the scale of reduction of aqueous zinc concentrations indicate that dilution does not play a major role in the observed natural attenuation. Batch adsorption tests on the soil materials in contact with the effluent indicate that the upper organic rich soil layer found at the site is capable of removing significant masses of aqueous zinc from solution. Further testing is necessary to ascertain the performance of this material when exposed to a continuous source of aqueous zinc at typical field concentrations.

FURTHER STUDIES

This work is part of a ongoing study aimed at determining the natural capacity of shallow soils to adsorb zinc from mine drainage waters at the Galkeno 300 site. Future work will include further batch testing including sequential washings at field zinc concentrations, followed by selective sequential extraction to determine the mechanism(s) of zinc adsorption. Following batch testing, a series of column leaching tests will be undertaken to evaluate the removal of zinc under continuous source conditions at field pH and zinc concentrations; following zinc breakthrough, pH of influent will be decreased to simulate the onset of acidic drainage. Desorption of zinc will be subsequently monitored to assess the zinc retention capacity of the soil under acidic drainage conditions.

ACKNOWLEDGEMENTS

This research would not have been possible without the support of Environment Canada, the Mining Environment Research Group, the Northern Research Institute and the Northern Scientific Training Program in the form of in-kind contributions, fellowships and grants to D. MacGregor. Additional support in the form of a NSERC grant to L. Li is greatly appreciated.

REFERENCES

Environment Canada, 1992. Manual of Climatological Observations (3rd edition). Weather Services Directorate. Ottawa, Canada.

Grant, D.M. and Dawson, B.D. 1995. Isco Open Channel Flow Measurement Handbook (4th ed.). Isco, Inc. Lincoln, USA.

U.S.E.P.A. 1992. Batch-type procedures for estimating soil adsorption of chemicals. EPA/530/SW-87/006F. U.S. Environmental Protection Agency. Washington, D.C.

Flow and Transport through Clay Membrane Barriers

MICHAEL A. MALUSIS[1], CHARLES D. SHACKELFORD[2], and HAROLD W. OLSEN[3]
[1] Project Engineer, GeoTrans Inc., 9101 Harlan St., Suite 210, Westminster, Co, USA
[2] Professor, Dept. of Civil Engineering, Colorado State University, Fort Collins, Co, USA
[3] Research Professor, Division of Engineering, Colorado School of Mines, Golden, Co, USA

ABSTRACT
The flux equations for liquid and solute migration through clay membrane barriers (CMBs) used in waste containment and remediation applications are discussed. A simplified analysis of flow through a geosynthetic clay liner (GCL) using measured values for the chemico-osmotic efficiency coefficient (ω) of the GCL shows that membrane behavior can result in a total liquid flux that counters the outward Darcy flux due to chemico-osmotic counter flow. Also, the effect of a CMB is shown to reduce the contaminant mass flux through the barrier relative to the contaminant mass flux that would occur in the absence of membrane behavior due to two mechanisms: chemico-osmotic counter flow, and contaminant (solute) restriction. An important consideration with respect to solute restriction is the implicit correlation between ω and the effective diffusion coefficient, D^*, of the contaminant.

INTRODUCTION
The ability of clay soils to act as membranes that restrict the passage of solutes (e.g., aqueous miscible chemicals) is well documented (e.g., Kemper and Rollins 1966, Olsen 1969, and Olsen et al. 1990). Restricted movement of charged solutes (ions) through the pores of a clay soil is attributed to electrostatic repulsion of the ions by electric fields associated with the diffuse double layers (DDLs) of adjacent clay particles (e.g., Hanshaw and Coplen 1973, Fritz 1986, Keijzer et al. 1997). Non-electrolyte solutes (uncharged species), such as aqueous miscible organic compounds, also may be restricted from migrating through clay soils when the size of the solute molecule is greater than the pore sizes of the clay soil (Grathwohl 1998). Membrane behavior also results in chemico-osmosis, or the movement of liquid in response to a solute concentration gradient (e.g., Katchalsky and Curran 1965).

The extent to which a clay soil acts as a membrane in the presence of a concentration gradient is termed chemico-osmotic efficiency, and often is expressed quantitatively in terms of a chemico-osmotic efficiency coefficient, ω, or reflection coefficient, σ (e.g., Kemper and Rollins 1966, Olsen et al. 1990). The theoretical chemico-osmotic efficiency coefficient of an "ideal" membrane that completely restricts the movement of solutes is unity (i.e., $\omega = 1$), whereas $\omega = 0$ for a material that exhibits no solute restriction (e.g., Mitchell 1993). In most cases, the pores in clay soils that exhibit chemico-osmotic membrane behavior vary over a range of sizes such that not all of the pores are restrictive. In such cases, ω typically falls within the range $0 < \omega < 1$, and the clay soils are referred to as "non-ideal" membranes (Kemper and Rollins 1966, Olsen 1969, Barbour and Fredlund 1989, Mitchell 1993, and

Keijzer et al. 1997). The term "semi-permeable membrane" also is frequently used when $0 < \omega \leq 1$, but this term refers to the relative ability of water (solvent) to pass through the membrane rather than the restriction of the solutes.

Factors Affecting Clay Membrane Behavior
The value of ω is affected by several factors, including the state of stress in the soil, the types and amounts of clay minerals in the soil, and the types (species) and concentrations of the solutes in the pore water (Kemper and Rollins 1966, Olsen et al. 1990, Mitchell 1993). In general, ω increases with increase in the effective stress in the soil (or decrease in void ratio or porosity), increase in the activity of the clay soil, and decrease in the solute charge and/or solute concentration (Kemper and Rollins 1966, Olsen 1969, Mitchell 1993).

In particular, the ability of sodium bentonite to exhibit membrane behavior in the presence of common electrolytes (e.g., NaCl) has been illustrated extensively (e.g., Kemper and Rollins 1966, Keijzer et al. 1997). These results suggest that membrane behavior is significant in clay soils containing an appreciable amount of sodium montmorillonite. Clay soils containing a significant amount of sodium montmorillonite, such as sodium bentonite, also are desirable for use in waste containment barriers (e.g., soil-bentonite cutoff walls, geosynthetic clay liners, compacted sand-bentonite liners) due to the low hydraulic conductivity (e.g. $\leq 10^{-9}$ m/s) typically required in these applications. Thus, the existence of membrane behavior resulting from the sodium montmorillonite content in clay soil barriers may have an effect on the migration of contaminants through such barriers.

The effects of solute concentration and charge on the chemico-osmotic efficiency coefficient, ω, of bentonite specimens are illustrated in Fig. 1. The data in Fig. 1 indicate that ω decreases as the average salt concentration increases for a given specimen porosity (n) and salt solution, whereas the data in Fig. 1a also indicate that ω decreases with an increase in charge (Ca^{2+} versus Na^+) for a given porosity and average salt concentration. Both of these trends are consistent with expected behavior based on DDL theory in that the thickness of the DDLs of adjacent clay particles and the resulting extent of influence of the ion-restricting electric fields inside the soil pores decreases as the ion concentration and cation valence in the pore water increases (e.g., Fritz 1986).

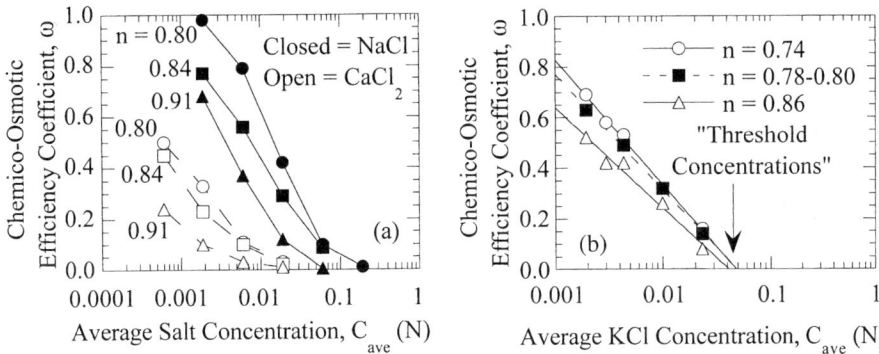

Figure 1. Chemico-osmotic efficiency coefficients as a function of average salt concentration across the specimen and specimen porosity (n) for (a) bentonite specimens (data from Kemper and Rollins 1966), and (b) a geosynthetic clay liner (data from Malusis 2001).

The data in Fig. 1 also indicate that the membrane effect of the bentonite specimens is essentially destroyed (i.e., $\omega = 0$) for average monovalent salt (NaCl or KCl) concentrations ranging from approximately 0.05 N to 0.1 N or average divalent salt (CaCl$_2$) concentrations of ~ 0.02 N. While these limiting or "threshold concentrations" (see Fig. 1b) may seem to be low (dilute), a 0.05 N KCl concentration, for example, corresponds to 1,773 mg/L of Cl⁻. By comparison, the current drinking water standard (DWS) for Cl⁻ in the U. S. is 250 mg/L, and the regulated maximum contaminant levels, or MCLs, for many of the toxic heavy metals, such as arsenic, cadmium, mercury, and selenium, are ≤ 0.05 mg/L. Thus, even though membrane behavior is primarily associated with dilute concentrations, these dilute concentrations still may exceed significantly regulated limits such that the range of concentrations for which clay membrane behavior typically is significant also may be of environmental concern. Therefore, determination of the "threshold concentrations" above which the clay soil no longer exhibits membrane behavior ($\omega = 0$) for a variety of chemical species and testing conditions represents an important practical component in terms of characterizing membrane behavior for clay soil barriers.

FLOW AND TRANSPORT THROUGH CLAY MEMBRANE BARRIERS

As illustrated in Fig. 2, clay barriers that exhibit membrane behavior, or clay membrane barriers (CMBs), can be used in both vertical and horizontal containment scenarios. Vertical containment scenarios typically involve the insertion of a vertical cutoff wall to prevent or minimize the spread of contamination in the subsurface, and usually are employed in applications involving the remediation of contaminated sites (e.g., Shackelford and Jefferis 2000). Horizontal containment applications typically involve the construction of a liner system used to prevent subsurface contamination in waste disposal applications (e.g., solid waste landfills). However, in both types of applications, the objective of the barrier is to maintain a contaminant concentration at the exit end of the barrier, C_L, lower than the source concentration of the same contaminant, C_0. Thus, by definition, $C_L < C_0$.

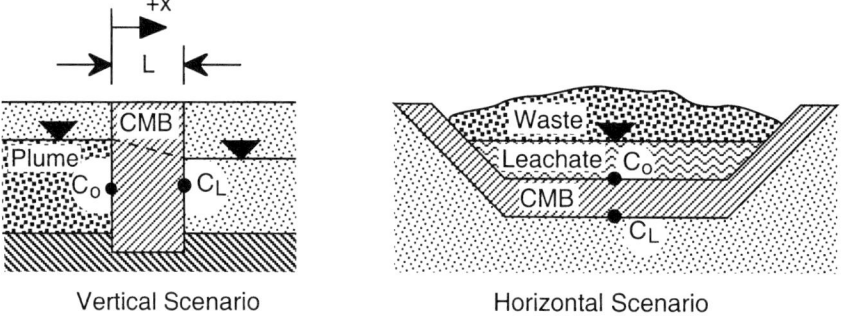

Figure 2. Vertical and horizontal containment scenarios ($C_0 > C_L$) for clay membrane barriers (CMBs) (from Shackelford et al. 2001).

Membrane Behavior and Chemico-Osmotic Flow

The total liquid flux through a clay membrane barrier (CMB), q, at steady state includes a hydraulic liquid flux, q_h, in response to the difference in hydraulic head, and a chemico-osmotic liquid flux, q_π, in response to a difference in solute concentration (e.g., Katchalsky and Curran 1965, Kemper and Rollins 1966, Olsen et al. 1990), or

$$q = q_h + q_\pi = -k\frac{\Delta h}{L} + \omega\frac{k}{\gamma_w}\frac{\Delta\pi}{L} \tag{1}$$

where k = the hydraulic conductivity of the CMB, Δh (< 0) = the head loss across the CMB, L = the thickness of the CMB, γ_w = the unit weight of the solution (i.e., essentially the same as water for dilute solutions), and $\Delta\pi$ (< 0) = the theoretical chemico-osmotic pressure resulting from the difference in solute concentration across the barrier.

The theoretical chemico-osmotic pressure difference, $\Delta\pi$, in Eq. 1 can be calculated based on the salt concentrations at the specimen boundaries in accordance with the van't Hoff expression (Katchalsky and Curran 1965). The van't Hoff expression is based on the limiting assumption that the electrolyte solutions are ideal and dilute and, therefore, provides only approximate values of the chemico-osmotic pressure difference. However, Fritz (1986) notes that the error associated with the van't Hoff expression is low (< 5 %) for 1:1 electrolytes (e.g., NaCl, KCl) and concentrations < 1.0 M. For a single-salt system, and considering the notation given in Fig. 1, the theoretical chemico-osmotic pressure difference according to the van't Hoff expression is given as follows:

$$\Delta\pi = \nu RT\Delta C = \nu RT\left(C_L - C_o\right) \tag{2}$$

where ν = the number of ions per molecule of the salt, R = the Universal gas constant [8.314 J mol^{-1}K^{-1}], T = the absolute temperature [$^\circ$K], and C = the salt concentration in molarity (M). For example, for KCl solutions ($\nu = 2$), Eq. 2 becomes:

$$\Delta\pi = 2RT\left(C_L - C_o\right) \tag{3}$$

and Eq. 1 becomes:

$$q = -k\frac{\Delta h}{L} - \omega k\frac{2RT}{\gamma_w}\frac{\left(C_o - C_L\right)}{L} \tag{4}$$

As indicated by Eq. 4, the chemico-osmotic flow through the CMB, q_π, occurs from lower solute concentration to higher solute concentration and, therefore, opposes the hydraulic liquid flux, q_h, in typical containment applications. Thus, if the effect of chemico-osmotic efficiency in the CMB is ignored (i.e., $\omega = 0$), all liquid flow occurs as a hydraulic liquid flux in accordance with Darcy's law (i.e., the first term in Eqs. 1 and 4).

The relative significance of chemico-osmotic counter flow through a CMB can be illustrated by considering the ratio, q/q_h, expressed for a simple salt solution (e.g., KCl) as follows:

$$\frac{q}{q_h} = \frac{q}{q_{\omega=0}} = 1 + \omega\frac{2RT}{\gamma_w\Delta h}\left(C_o - C_L\right) \tag{5}$$

As indicated by Eq. 5, $q/q_h = 1$ when either $C_o - C_L = 0$ or $\omega = 0$ since no chemico-osmotic flow occurs through the CMB. For $\omega > 0$ and $C_o - C_L > 0$, upward chemico-osmotic flow opposes the downward hydraulic liquid flux. Values of $q/q_h < 0$ indicate that the chemico-osmotic counter flow is sufficiently high such that the net liquid flux is directed into the containment facility, whereas values of $q/q_h > 0$ indicate that the net liquid flux is directed out of the containment facility.

The potential significance of membrane behavior on the movement of liquid through the CMB is illustrated herein by considering a geosynthetic clay liner (GCL) containing sodium bentonite as a liquid-containment barrier as shown schematically in Fig. 3. The depth of ponded liquid is assumed to be 305 mm (1.0 ft), in accordance with the maximum allowable leachate depth based on current U. S. regulatory standards. Also, the solute concentration in the leachate, C_o, is assumed to be greater than the solute concentration, C_L, at the exit boundary of the GCL, and the GCL is assumed to be saturated by prehydration.

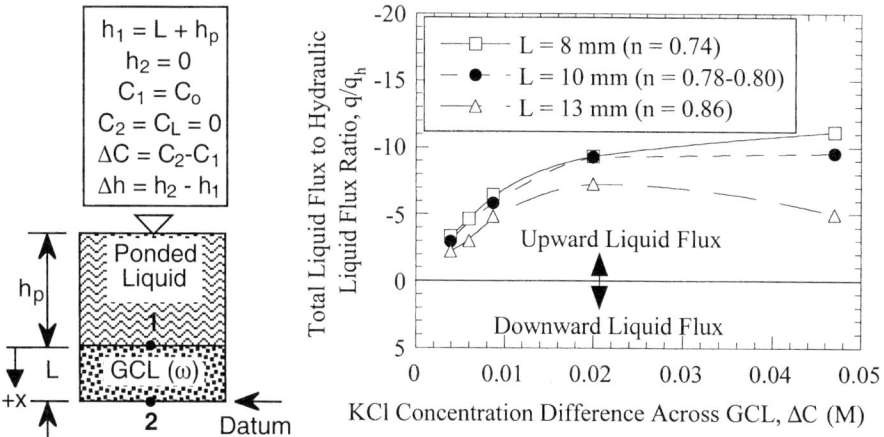

Figure 3. Results of simplified analysis of the potential effect of membrane behavior on the liquid flux through a GCL (n = porosity, q = total liquid flux, q_h = hydraulic liquid flux).

Values of q/q_h versus the difference in KCl concentration across the GCL (i.e., $\Delta C = C_o - C_L$) for the measured ω values previously shown in Fig. 1b also are shown in Fig. 3. The results based on the measured data indicate that upward liquid flux is likely to occur through the GCL for the range of ΔC considered in this study. However, the effect of the membrane behavior eventually will be destroyed and q/q_h will approach unity as values of ΔC (= C_o) approach the threshold concentrations (see Fig. 1b).

A net liquid flux into the containment facility results in negative advection (i.e., advection in the opposite direction of diffusion) and, thus, is potentially beneficial from the viewpoint of reducing the net rate of contaminant migration out of the containment facility. However, in addition to the limitations resulting from the simplifying assumptions previously noted, the results shown in Fig. 3 neglect the increase in solute concentration with depth due to diffusion and the potential for solute-clay interactions (i.e., compatibility). Both of these factors tend to reduce and eventually eliminate the beneficial contribution of chemico-osmotic counter flow due to compression of the DDLs and a corresponding decrease in ω as the concentration within the GCL increases.

Membrane Behavior and Solute Transport

The general expression for total contaminant (solute) mass flux, J (mass/area/time), in a fine-grained soil (i.e., neglecting mechanical dispersion) that exhibits membrane behavior can be written for one-dimensional transport as follows (e.g., Mitchell 1993, Malusis 2001, Shackelford et al. 2001):

$$J = \underbrace{(1-\omega)q_h C}_{J_{ha}} + \underbrace{q_\pi C}_{J_\pi} - \underbrace{nD^* i_c}_{J_d} \tag{6}$$

where D^* = the effective diffusion coefficient as defined by Shackelford and Daniel (1991), i_c = the concentration gradient, and the other parameters are as previously defined.

The hyperfiltrated advective mass flux, J_{ha}, in Eq. 6 represents the traditional advective transport term that is reduced by a factor of $(1 - \omega)$ due to the membrane behavior of the soil. In physical terms, the factor $(1 - \omega)$ represents the process of hyperfiltration whereby solutes are filtered out of solution as the solvent passes through the membrane under an applied hydraulic gradient. The second term, J_π, represents the counter advective transport of solutes due to chemico-osmotic counter flow, q_π. The third term, J_d, represents solute diffusion through soil in the form of Fick's first law as defined by Shackelford and Daniel (1991).

In the limit as $\omega \to 0$ and, thus, $q_\pi \to 0$, Eq. 6 reduces to the traditional advective-diffusive solute mass flux expression. However, as $\omega \to 1$ (i.e., an "ideal" CMB), $J \to 0$. Thus, the effect of a CMB is to reduce the contaminant mass flux through the barrier relative to the contaminant mass flux that would occur in the absence of membrane behavior. This reduction in the contaminant mass flux results from two mechanisms: viz., (1) the counter mass flux due to the chemico-osmotic counter flow (i.e., J_π), and (2) the solute restriction inherent in the J_{ha} and J_d terms. The solute restriction inherent in J_{ha} is an "explicit" effect in that $J_{ha} \to 0$ as $\omega \to 1$ (Eq. 6). However, the solute restriction inherent in J_d is not explicitly stated in Eq. 6 and, therefore, is an "implicit" effect. This implicit solute restriction results from the correlation between ω and D^*.

For example, consider the data shown in Fig. 4 based on D^* and ω values recently measured by Malusis (2001) using the apparatus and procedures described by Malusis et al. (2001) for the same GCL for which the data in Fig. 1b were measured. As indicated in Fig. 4, D^* must approach 0 as $\omega \to 1$ because, by definition, no contaminant can enter a perfect membrane. At the other extreme, the maximum value of D^*, or $D^*_{(max)}$, will occur when there is no membrane behavior ($\omega = 0$). Thus, D^* for CMBs must decrease from $D^*_{(max)}$ to 0 as ω increases from 0 to 1, respectively, as shown in Fig. 4. As a result, any correct simulation of the contaminant mass flux through a CMB must be based on values of D^* measured at the correct concentration for the application and, therefore, the appropriate ω value. The use of D^* values measured separately from ω at concentrations greater than the "threshold" concentration at which $\omega = 0$ (e.g., to decrease the testing time) in simulations performed to evaluate contaminant transport through CMBs will neglect this implicit correlation between D^* and ω and, therefore, will be fundamentally incorrect.

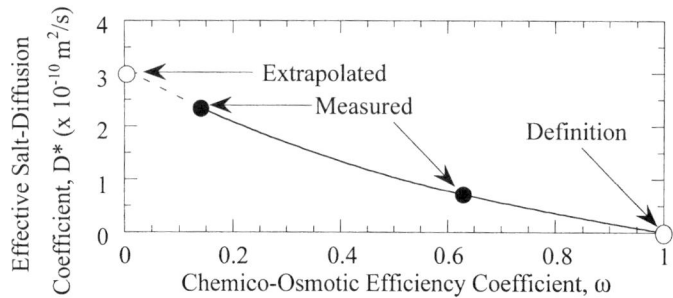

Figure 4. Effect of chemico-osmotic efficiency on the effective salt-diffusion coefficients for KCl diffusion through a GCL at steady state (measured data from Malusis 2001).

SUMMARY AND CONCLUSIONS

After a brief introduction defining membrane behavior, the major factors controlling membrane behavior in clay soils are discussed. In particular, clay soils containing a significant amount of sodium montmorillonite, such as sodium bentonite, have been shown to exhibit membrane behavior. Since such clay soils also are typically used as waste containment barriers (e.g., soil-bentonite cutoff walls, geosynthetic clay liners, compacted sand-bentonite liners), the existence of membrane behavior in such clay soil barriers may have an effect on the migration of solutes through such materials.

The flux equations for flow and transport through clay membrane barriers (CMBs) used in waste containment and remediation applications are presented and discussed. A simplified analysis of flow through a geosynthetic clay liner (GCL) using measured values for the chemico-osmotic efficiency coefficient (ω) of the GCL shows that membrane behavior can result in a total liquid flux that actually counters the outward Darcy flux due to chemico-osmotic counter flow.

Also, the effect of a CMB is shown to reduce the contaminant mass flux through the barrier relative to the contaminant mass flux that would occur in the absence of membrane behavior due to two mechanisms: chemico-osmotic counter flow, and contaminant (solute) restriction. An important consideration with respect to solute restriction is the implicit correlation between ω and the effective diffusion coefficient, D^*, of the contaminant.

ACKNOWLEDGEMENTS

This paper represents an extension of a recently completed joint research effort between Colorado State University (CSU) and the Colorado School of Mines (CSM), funded by the U. S. National Science Foundation (NSF), Arlington, VA, under Grant CMS-9634649 for CSU and Grant CMS-9616855 for CSM. The opinions expressed in this paper are solely those of the writers and are not necessarily consistent with the policies or opinions of the NSF.

REFERENCES

Barbour, S. L. and Fredlund, D. G. (1989). Mechanisms of osmotic flow and volume change in clay soils. *Canadian Geotechnical Journal*, 26, 551-562.

Fritz, S. J. (1986). Ideality of clay membranes in osmotic processes: A review. *Clays and Clay Minerals*, 34(2), 214-223.

Grathwohl, P. (1998). *Diffusion in natural porous media, Contaminant transport, sorption/desorption and dissolution kinetics*. Kluwer Academic Publ., Norwell, MA, USA.

Hanshaw, B. B. and Coplen, T. B. (1973). Ultrafiltration by a compacted clay membrane – II. Sodium ion exclusion at various ionic strengths. *Geochimica et Cosmochimica Acta*, 37(10), 2311-2327.

Katchalsky, A., and Curran, P. F. (1965). *Nonequilibrium thermodynamics in biophysics*. Harvard University Press, Cambridge, MA, USA.

Keijzer, Th. J. S., Kleingeld, P. J., and Loch, J. P .G. (1997). Chemical osmosis in Compacted clayey material and the prediction of water transport. *Geoenvironmental engineering, contaminated ground: Fate of pollutants and remediation*, R. N. Yong and H. R. Thomas, eds., Thomas Telford Publ., London, 199-204.

Kemper, W. D. and Rollins, J. B. (1966). Osmotic efficiency coefficients across compacted clays. *Soil Science Society of America, Proceedings*, 30, 529-534.

Malusis, M. A. (2001). *Membrane behavior and coupled solute transport through a geosynthetic clay liner*. Ph.D. Dissertation, Department of Civil Engineering, Colorado State University, Fort Collins, CO, USA.

Malusis, M. A., Shackelford, C. D., and Olsen, H. W. (2001). A laboratory apparatus to measure chemico-osmotic efficiency coefficients for clay soils. *Geotechnical Testing Journal*, 24(3), 229-242.

Mitchell, J. K. (1993). *Fundamentals of soil behavior*, 2nd Ed. John Wiley and Sons, New York, USA.

Olsen, H. W. (1969). Simultaneous fluxes of liquid and charge in saturated kaolinite. *Soil Science Society of America, Proceedings*, 33, 338-344.

Olsen, H. W., Yearsley, E. N., and Nelson, K. R. (1990). Chemico-osmosis versus diffusion-osmosis. *Transportation Research Record No. 1288*, Transportation Research Board, Washington D.C., 15-22.

Shackelford, C. D. and Daniel, D. E. (1991). Diffusion in saturated soil: I. Background. *Journal of Geotechnical Engineering*, 117(3), 467-484.

Shackelford, C. D. and Jefferis, S. A. (2000). Geoenvironmental engineering for *in situ* remediation. *International Conference on Geotechnical and Geoenvironmental Engineering (GeoEng2000)*, Melbourne, Australia, Nov. 19-24, Technomic Publ. Co., Inc., Lancaster, PA, USA, Vol. 1, 126-185.

Shackelford, C. D., Malusis, M. A., and Olsen, H. W. (2001). Clay membrane barriers for waste containment. *Geotechnical News*, BiTech Publ. Ltd., Richmond, BC, Canada, 19(2), 39-43.

GIS And Analytical Techniques for the Evaluation of the Occurrence, Transport And Fate Of Estrogenic Hormones in Waste and Surface Waters

A. MARINO*, A. RINALDINI*, S. BELLAGAMBA*, C. SIMEONI**, G. LUDOVISI*
and A. MOCCALDI***
* ISPESL/DIPIA; Roma, IT
** ISPESL/DIPIA, Monteporzio Catone (RM), IT
*** ISPESL, Roma, IT

ABSTRACT
This paper outlines the potential of the integration of GIS and laboratory techniques as a tool for the environmental management of areas exposed to water pollution phenomena. The presence in natural environment of compounds with estrogenic properties has become a major subject of world-wide growing concern, because these compounds may interfere with the reproduction of man, livestock and wild-living animals. As such, recently much research is directed towards the occurrence, effects and risks of these compounds. One of the groups of compounds under investigation are the natural estrogenic hormones, primarily synthetized in the female body and essential for female characteristics and reproduction, and closely related synthetic hormones. A problem in the routine analysis of hormones in surface and waste waters in the absence of a sufficiently sensitive analytical procedure. Thus the fist objective of this study was to develop an analytical procedure (using mass spectrometry and High Performance Liquid Chromatography) that enable analysis of four estrogenic hormones in concentration below 1ng/l in surface and waste water.
The second objective of this study was to obtain a general idea about the occurrence of the hormones in water environment near Rome and the possible patterns of surface water contamination from hormones. GIS techniques offered good advantages. In particular the possible patterns of surface waters pollution from estrogenic compounds have been considered. Analytical data have been inserted in a specific GIS database and integrated with further information like lithology, geological structure, stratigraphy, geomorphology, hydrogeological features, DEM, waste water treatment plants location and characters. The processing of data layers was performed, using a dedicated software, through typical GIS operators like indexing, recoding, matrix analysis, proximity analysis. In particular the possible patterns of surface water pollution have been considered.
According to our results this methodology can be a reliable, sensible and easy to update support to competent Authorities in environmental management.

INTRODUCTION
This paper outlines the potential of the integration of GIS and laboratory techniques as a tool for the environmental management of areas exposed to water pollution phenomena. The presence in natural environment of compounds with estrogenic properties has become a major subject of world-wide growing concern, because these compounds may interfere with the reproduction of man, livestock and wild-living animals. As such, recently much research

is directed towards the occurrence, effects and risks of these compounds. One of the groups of compounds under investigation are the natural estrogenic hormones, primarily synthetized in the female body and essential for female characteristics and reproduction, and closely related synthetic hormones. A problem in the routine analysis of hormones in surface and waste waters in the absence of a sufficiently sensitive analytical procedure. Thus the first objective of this study was to develop an analytical procedure (using mass spectrometry and High Performance Liquid Chromatography) that enable analysis of four estrogenic hormones in concentration below 1ng/l in surface and waste water.

The second objective of this study was to obtain a general idea about the occurrence of the hormones in water environment near Rome and the possible patterns of surface water contamination from hormones. GIS techniques offered good advantages. In particular the possible patterns of surface waters pollution from estrogenic compounds have been considered. Analytical data have been inserted in a specific GIS database and integrated with further information like lithology, geological structure, stratigraphy, geomorphology, hydrogeological features, DEM, waste water treatment plants location and characters. The processing of data layers was performed, using a dedicated software, through typical GIS operators like indexing, recoding, matrix analysis, proximity analysis. In particular the possible patterns of surface water pollution have been considered.

ANALYTICAL PROCEDURE
Chemicals and reagents
Estrone (1,3,5(10)-estratriene-3-ol-17-one), 17β-estradiol (1,3,5(10)-estratriene-3,17β-diol), estriol (1,3,5(10)-estratriene-3,16α,17β-triol), and ethinylestradiol (17α-etinyl-1,3,5(10)-estratriene-3,17β-diol) were from Sigma (St. Louis, MO, USA). The chemical structures of the estrogens used in this study are shown in Figure 1. To obtain HPLC-grade water, distilled water was purified by passing it through a Milli-Q RG apparatus (Millipore, Bedford, MA, USA). Acetonitrile of LC gradient grade, and methanol and dichloromethane, both of LC grade, were from Carlo Erba (Milano, Italy). Reagent grade hydrochloric acid and glacial acetic acid were purchased from Carlo Erba. The 6 mL cartridges packed with 500 mg of ENVI-CARB, a graphitized nonporous carbon with a surface area of 100 m^2/g and sieved at 120/400 mesh (Supelco, Bellefonte, PA, USA), were used for solid-phase extraction.
Preparation of standards
Estrogens were dissolved in an appropriate volume of methanol to yield a 1 μg/μL stock solution for each individual estrogen. A 500-μL aliquot of each estrogen stock solution was mixed and diluted with methanol to give a primary working solution containing each estrogen at 100 ng/μL. The diluted working solutions were daily prepared with methanol to give solutions at the appropriate concentrations. All the stock solutions and the primary working solution were stored at -18°C and brought to room temperature before use.
Sampling of sewage samples
All the samples were kept at 4°C and analysed within 48 h in order to keep microbiological degradation to a minimum. No preservative was added.
Extraction of sewage samples
Prior to the extraction, each sewage sample was vacuum filtered through a 55 mm Whatman (Maidstone, UK) GF/C filter with a pore size of 1.2 μm, and the retained particulate material was washed with 10 mL of methanol which was added to the aqueous sample. The estrogen chemicals were then extracted from samples by the ENVI-CARB cartridge. For recovery experiments, a portion of the free estrogens was spiked into the samples. The samples were stirred for 30 min. prior to SPE procedure.

The ENVI-CARB cartridge was fitted into a side-arm filtering flask and, before processing the sample, it was sequentially washed with 20 mL of water acidified with hydrochloric acid

(10 mM), 5 mL of methanol, 7 mL of CH_2Cl_2/CH_3OH (60:40, v/v), 3 mL of methanol, and finally 10 mL of water. For the clean-up 1 L of sample was forced through the extraction device under vacuum from a water pump. After the sample was passing through the ENVI-CARB cartridge, the vacuum was reduced and the cartridge was sequentially washed with 10 mL of water, 10 mL of CH_3OH acidified with CH_3COOH 100 mM and 3 mL of CH_3OH.
In order to achieve a back-flushing elution, a suitably drilled cylindrical teflon piston with one conical indented base and a Luer tip (device home-made, as reported by Di Corcia et al. [13]) was forced into the cartridge until it reached the upper frit. The cartridge was turned upside-down, a glass vial was placed below it, and the estrogen chemicals were back-eluted by 10 mL of a CH_2Cl_2/CH_3OH (60:40, v/v) solution. In a thermostatic bath (ASAL, Cernusco, Italy) set at 40°C the eluate was brought to dryness under a gentle nitrogen stream. The residue was reconstituted with 250 μL of a H_2O/CH_3CN (50:50, v/v) solution. For the LC/APCI-MS/MS analysis 100 μL of final extract were injected into LC/MS/MS apparatus.

Flow injection analysis-mass spectrometry

A PE Sciex (Concord, ON, Canada) API 365 triple quadrupole mass spectrometer, equipped with an atmospheric pressure chemical ion source interface, was used. Nitrogen was supplied as nebulizing, drying, curtain, and collision gas, and flow injection analysis was used for the recording of positive mass ion spectra. The settings for the nebulizer, curtain, and collision gases were 12, 10, and 3 (arbitrary) on the API-365. The HN probe was maintained at 500°C, and the gas-phase chemical ionization was effected by a corona discharge needle at 3 μA. Samples were introduced into the mass spectrometer through a Rheodyne Model 7125 injector which was connected to the heated nebulizer (HN) probe by PEEK tubing of 50 cm length and 0.005 in. I.D.. The solvent used was H_2O/CH_3CN (50:50, v/v) at a flow rate of 1 mL/min. MassChrom 1.0 software, from PE Sciex, was used for data acquisition and processing on a Power MacIntosh G3.

Analysis by LC/HN-MS/MS

Liquid chromatography was carried out using a Perkin-Elmer binary LC pump 250 (Perkin-Elmer) equipped with a Rheodyne 7125 injector with a 100 μl loop. The four estrogens were chromatographed on a 25 cm x 4.6 mm I.D. column filled with 5 μm (average particle size) Alltima LC-18 packing (Alltech, Deerfield, IL,USA), with a precolumn (Supelguard 2 cm x 4.6 mm I.D.) supplied by Supelco. Analysis was carried out using a solvent gradient. The initial composition of the mobile phase was 35% of acetonitrile and 65% of water. The gradient was programmed to linearly increase the amount of acetonitrile up to 45% in 10 min, then adjusted to 50% immediately, and then to increase up to 60% in 15 min. To clean the column the amount of acetonitrile was kept constant at 100 % for 5 min. The flow rate of the mobile phase was 1 mL/min, and the column effluent was introduced into the APCI source.
The triple quadrupole mass spectrometer (API 365) was operated in positive ion (PI) mode. The multiple reaction monitoring (MRM) mode was chosen for quantitation. Data acquisition was divided into three retention time periods, and in each period tuning parameters were optimized for different MRM transition pairs to enhance the sensitivity for ions that required different instrument state files.

Analysis by GC/MS

For purposes of comparison, the clean-up, derivatization, and GC/MS detection were also performed as described by Lee *et al.* [10]. A Hewlett-Packard 5890 Series II GC equipped, with a 5972 mass-selective detector(MSD), and a 30 m x 0.25 mm I.D. x 0.5 film thickness HP-5-MS column, was used for quantitation of estrogens. GC conditions were as follows: injection port, 250°C; interface, 280°C; initial oven temperature, 70°C with 1 min hold; programming rates: 30°C/min (from 70°C to 180°C); and 5°C/min (from 180°C to 290°C). Carrier gas (helium) linear velocity was held constant at 36.9 cm/s. Splitless injections (1 μL) were made by a HP7673 autosampler with a splitless time of 1 min. The electron energy and electron multiplier voltage were 70 eV, and 400 V above the autotune value, respectively. The MSD was tuned each day by using perfluorotributylamine (PFTBA) and the standard autotune program. Full-scan mass spectral data were collected from m/z 50 to 600. For quantitation of estrogens in a sample extract, selected ion monitoring (SIM) was used. Response factors for the estrogens were generated by injecting mixtures of their pentafluoropropionyl (PFP) derivatives at 2 levels (1000 and 200 pg/μL). The following quantitation and confirmation ions, respectively, were used in SIM work: m/z 372 and 416 (E_1), m/z 564 and 401 (E_2), m/z 563 and 399 (E_3), and m/z 359 and 306 (EE_2).

GIS

The analytical data have been inserted in a GIS database and integrated with further information like lithology, geological structure, hydrology, hydrogeological characteristics, that were imported in the GIS as layers. A Digital Elevation Model (DEM) has been realised from interpolation of topographic data (IGM maps). Furthermore the same methodology allowed to reconstruct the spatial patterns of ground water from isopiestic lines.

The information regarding the exact position and the characteristics of industrial plants were obtained from ISPESL database. The EASI-PACE (PCI) software allowed to consider and process information both in raster (like land-use map and DEM) and vectorial (like surface hydrology, isopiestic lines) format.

The processing of data layers was performed, using the same software through typical GIS operators like indexing, recoding, matrix analysis, proximity analysis. The matrix analysis assigns a different output value to each unique combination of input values. The proximity analysis creates an output layer showing successive zones of proximity (distances) to a specified entity or group of entities. The choice of recoding indexes, weights and matrices was based on references, personal experience and knowledge of the area. In particular the use of matrix analysis instead of simple weighted overlaying allowed a better control on parameters and variables. The recoding and indexing was applied in order to obtain a limited number of classes for each layer (map), for a better understanding of the information.

RESULTS AND CONCLUSIONS
Climatic conditions

The area is characterised by Mediterranean Temperate Climate: hot temperate climate with extension of summer and a mild winter. Minimum value of relative humidity during August (%59%), while maximum value occurs during winter (86%). Pluviometric regime is typically Maritime

Sampling Of Untreated Waste Water And Treated Effluents

In 1999 untreated waste water and final treated effluents of five WWTPs (Marino in September, Genzano in July, Roma N in February, Roma S in March), mainly domestic, were sampled and analysis were performed. The results of analytical procedures, regarding four estrogenic hormones in final treated effluents are reported in Fig. 1.

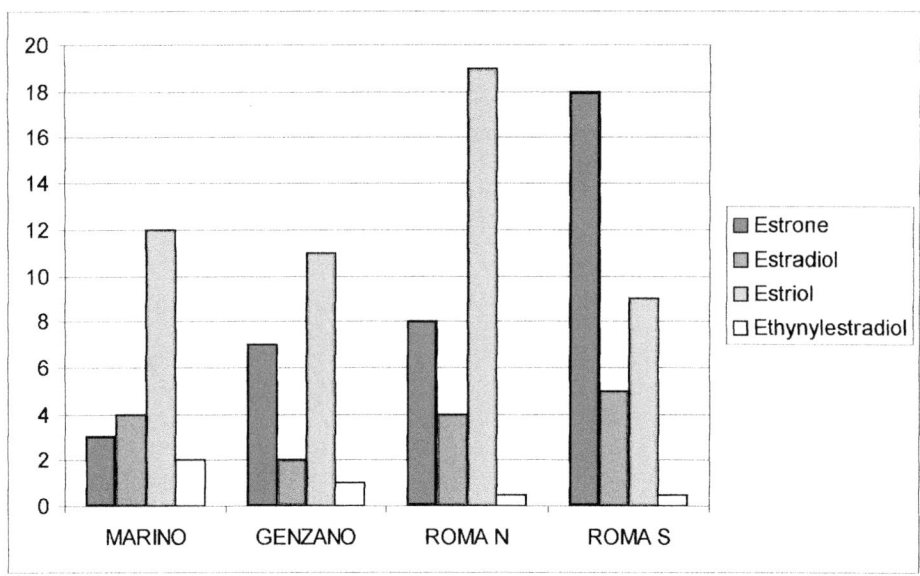

Fig.1 Results of analytical procedures on samples of treated effluents (in ng/L)

Castelgandolfo
The WWTP is located inside the crater that contains Albano lake (or Castelgandolfo lake), just a few meters above the lake water level. A pipe conveys treated water in an underground drain that crosses the crateric-wall just below Castelgandolfo, flowing inside a tunnel ending at "le Mole". The drain was dug by the Romans subsequently to the siege of Veii (396 B.C.) in order to avoid repeated floods due to the overflowing of waters from the rim of the depression. In 1882 the average flow of this drain was 105 l/sec. At the moment no water from the lake flows in the drain because of the fall in lake level due to a range of different reasons. The drain is about 1200 m long and emerges, at *Quarto Mole* or *Mole di Albano*, where inputs in Rio Albano channel. Flowing in channels having different names, the most important is Fosso di Vallerano, merges with Tiber River at Tor di Valle South of Rome.
The Fosso di Vallerano with its elongated shape stretches for about 23 km. It is about 23 km long, 7 km wide and its surface is about 99kmq. The higher part of the basin is constituted by an hilly region, while the lower part is characterised by a plain region.
The water streams flow in an area characterised by volcanic rocks of Colli Albani Complex and quaternary continental sediments in the Tiber Valley.
Inside the border of Fosso di Vallerano Basin, downstream of the drain, are located many urban settlements like EUR quarter (Roma), Le Frattocchie, S. Maria delle Mole, Cecchignola, Castel di Leva, Falcognana, Quarto Cesareto.
Genzano-Nemi
The WWTP is located south of Genzano. Treated waste water flows in a channel that merges with Fosso Grande a few km downstream.
The Fosso Grande Basin stretches for about 20 km NE-SW and is about 10 km wide. The Fosso Grande flows in the Tirrenian Sea SW of Ardea The higher part of the basin is constituted by an hilly region, while the middle and lower parts are characterised by a plain, but quite cut by creeks, region. Inside the borders of the basin are located many urban settlements, namely Pescarello, Checchina, S. Procula Maggiore, S. Palomba and Ardea.

The water streams flow in an area characterised by volcanic rocks of Colli Albani Complex in the upper part and by quaternary continental and coastal sediments in the middle and lower part.

Rome North

The WWTP Rome North treat domestic waste waters from north-eastern part of the town for a served population of about 600000 inhabitants and an average yield in dry weather of about 3 m³/s. The active sludge treatment technique is applied.

The Grottarossa WWTP is on the right bank of the river just north of Rome and conveys the treated water directly in Tiber River.

Fig. 2 WWTP Rome North

Rome South

The WWTP Rome North treat domestic waste waters from south-western part of the town for a served population of about 1100000 inhabitants and an average yield in dry weather of about 8.5 m³/s. The active sludge treatment technique is applied.

The Acilia WWTP is located on the left side of the Tiber River a few km downstream of the town. The WWTP conveys treated water directly in the Tiber river.

Fig.3 WWTP Rome South

The developed analytical method was validated by evaluating recovery, precision, calibration curve, and limit of quantitation. The sensitivity for each analyte was determined to 0.5 ng/L for estradiol and ethynylestradiol, and 1 ng/L for estrone and estriol. The lower level of quantitation was defined as the lowest concentration on the calibration curve with an accuracy within 20% and precision within 20%.

The application of GIS techniques allowed the reconstruction of occurrence, transport and fate of estrogenic hormones in waste and surface waters (rivers and sea). Furthermore the methodology allowed to outline the areas and thus the population and environment interested by pollution hazard deriving from this dangerous and persistent substances. According to our results this methodology can be a reliable, sensible and easy to update support to competent Authorities in environmental management.

REFERENCES

A.C. Belfroid, A. Van der Horst, A.D. Vethaak, A.J. Schafer, G.B.J. Rijs, J Wegener and W.P. Corfino, *Analysis and occurrence of estrogenic hormones and their glucuronides in surface water and waste water in the Netherlands*, The Science of Total Environment, 225, 101-108, 1999.

C. Boni, P. Bono and G. Capelli, *Carta idrogeologica del territorio della regione Lazio*, L. Salomone, Roma, 1988.

P. A. Brivio, G.M. Lechi and E. Zilioli, *Il telerilevamento da aereo e da satellite,* Edizioni Delfino, Milano, 1990.

P. A. Brivio and E. Zilioli, *Il telerilevamento da satellite per lo studio del rischio ambientale,* Edizioni dell'Ulisse, Roma, 1995.

A. Laganà, A. Bacaloni, G. Fago and A. Marino, *Trace analysis of estrogenic chemicals in sewage effluent using liquid chromatography combined with tandem mass spectrometry,* Rapid Commun. Mass Spectom. 14, 401-407, 2000.

D.G.J. Larsson, M. Adolfsson-Erici, J Parkkonen, M. Petterson, A.H. Berd, P.E. Olsson and L. Forlin, *Ethinyloestradiol and undesired fish contraceptive?,* Aquatic Toxicology, 45, 91-97, 1999.

U. Poli, M. Ippoliti, A. Marino, I. Alberico and P. Bragatto, "Studio del rischio di inquinamento delle acque superficiali e del danno conseguente nell'area metropolitana di Roma fino alla foce del Fiume Tevere, mediante tecniche di telerilevamento e GIS" *Seminars of Metropolitan areas and rivers,* pp. 107-115, Acea, Roma, 1996.

T.A. Ternes, M. Stumpf, J. Muller, K. Haberer, R.D. Wilken and M. Servos, *Behaviour and occurrence of estrogens in municipal sewage treatment plants – I. Investigations in Germany, Canada and Brazil,* The Science of Total Environment, 225, 81-90, 1999.

Metal Adsorption onto Gram-Negative Bacteria from an Oil-contaminated Site

BRYNE T. NGWENYA
Department of Geology & Geophysics, University of Edinburgh, Grant Institute, West Mains Road, Edinburgh EH9 3JW.

ABSTRACT

This paper reports thermodynamic parameters for metal adsorption onto a *gram-negative* bacterium, which would allow bacterial effects to be included in models of metal transport in the subsurface. Potentiometric titrations were used to determine the different types of sites present on bacterial cell walls. Stability constants for adsorption of a variety of metals to each site was determined from batch adsorption experiments at varying pH with constant metal concentration, and the number of sites involved in metal uptake was confirmed by sorption isotherm experiments at constant pH. Titrations revealed three distinct acidic surface sites on the bacterial, with pK values of 4.3±0.1, 6.9±0.5 and 8.9±0.5, corresponding to carboxyl, phosphate and hydroxyl/amine groups, with surface densities of 5.0±0.7x10^{-4}, 2.2±0.6 x10^{-4} and 5.5±2.2 x10^{-4} mol/g of dry bacteria. Only carboxyl and phosphate sites are involved in metal uptake, with surface affinities of the order Cu>Pb>Zn based on the following stability constants: *Log* $K_{carboxyl}$: Zn^{2+} = 3.3±0.1, Pb^{2+} = 4.3±0.2, and Cu^{2+} = 4.4±0.2, *Log* $K_{phosphate}$: Zn^{2+} = 5.1±0.1 and Cu^{2+} = 6.0±0.5. Some of these stability constants are similar to those for common organic compounds found in natural soil solutions, suggesting that bacteria may play an important role in metal attenuation in the subsurface.

1. INTRODUCTION

Adsorption of metals to organic matter, clay minerals and iron oxides is critical for natural attenuation of metals in the subsurface (Davis & Kent, 1990; Wen et al., 1998). The adsorption process occurs by metals binding to active sites present on these particles. Thus a thorough understanding of the way that metals interact with the different sites is important in predicting metal transport in the environment. Recent studies suggest that bacteria also possess a variety of sites whose affinity for metal binding may be similar to those of the soil constituents above (Fein et al., 1997; Seki et al., 1998). Their small size and ubiquitous presence in near-surface environments presents a large surface area for metal adsorption.

These factors have led to a renewed interest in how bacteria mediate geo-environmental processes so that their effects can be accounted for in computer models of contaminant transport (Fein, 2000). In order to realise this, however, it is necessary to develop a quantitative model of metal adsorption onto bacterial surfaces. Such a model requires the determination of thermodynamic constants for different sites on the bacterial surface, and the metal-site stability constants for those sites involved in metal uptake under natural and contaminated conditions. This paper discusses recent experiments designed to evaluate these thermodynamic parameters for metal adsorption onto a *gram-negative* bacterium isolated from an oil-contaminated site.

Geoenvironmental impact management, Thomas Telford, London, 2001.

2. THEORY

Bioassays (e.g. Beveridge & Murray, 1980) and/or spectroscopic analysis (e.g. Pagnanelli et al., 2000) suggest that up to five different sites may occur on most bacterial surfaces, including carboxyl, phosphoryl, hydroxyl, amine and phosphodiester functional groups (Cox et al., 1999). These sites are generally protonated below pH 2.2, but each site dissociates at a different pH as the pH increases (Fein et al., 1997). A generic acid-base equilibrium can be written for each of these dissociation reactions:

$$R - S - H^0 \Leftrightarrow R - S^- + H^+ \tag{1}$$

where R is the cell wall and $-S-H$ stands for a protonated surface functional group. The corresponding mass action equation for reaction (1) is:

$$K_1 = \frac{[R - S^-]a_{H^+}}{[R - S - H^0]} \tag{2}$$

where K_1 is the deprotonation constant for reaction (1) etc., a_i is the activity of species i in solution and the square brackets represent concentration of surface species. Assuming that deprotonation of the surface site occurs prior to adsorption of the metal ion (M^{m+}) to the site (Fein et al., 1997; Seki et al, 1998) leads to the following generic equilibrium reaction for metal adsorption:

$$R - S^- + M^{m+} \Leftrightarrow R - SM^{(m-1)} \tag{3}$$

This reaction corresponds to the following mass action equation:

$$K_3 = \frac{[R - SM^{(m-1)}]}{[R - S^-]a_{M^{m+}}} \tag{4}$$

The constants K_1, K_3 etc. are necessarily conditional and cannot be used to model metal adsorption behaviour under all natural conditions. This is because the charge on the bacterial surface, which depends on external conditions, imposes an electric field around the bacterial cell, which can affect metal adsorption equilibria. These electrostatic interactions can be corrected for using the Boltzmann factor (Stumm & Morgan, 1996), yielding an intrinsic thermodynamic constant ($K_{intrinsic}$) thus:

$$K_{instrinsic} = K_n \exp\left(\frac{\Delta Z \psi F}{RT}\right) \tag{5}$$

where F is the Faraday constant, ΔZ is the change in the charge of the surface species, ψ is the surface potential, R is the gas constant, T is the temperature and $n = 1, 2\ 3$ etc. Several theories exist for relating the surface potential to surface charge (Stumm & Morgan, 1996), all of which can model adsorption data equally well. The constant capacitance model is used here to relate surface charge (σ) to potential *via* a constant capacitance (C):

$$C = \frac{\sigma}{\psi} \tag{6}$$

Equations (1) to (6) constitute a quantitative thermodynamic model for metal adsorption to bacterial surfaces which is applicable under most conditions. Acid-base titrations have been conducted on bacterial suspensions to determine the number and concentration of different sites and their intrinsic dissociation constants. Metal adsorption experiments were conducted to establish which of these sites are involved in metal adsorption, and to determine their metal stability constants.

3. EXPERIMENTAL METHODS

A copper-resistant strain of gram-negative bacteria was isolated from soil collected from a diesel contaminated site by incubating the soil for three weeks in a 2mM copper medium containing 30g/L tryptone soya broth of which 0.5% was yeast extract. Experimental batches were grown and treated according to a method described in Ngwenya & Sutherland (*in press*). Deprotonation constants and surface site concentrations were determined from acid-base titrations on suspensions of bacteria in a background electrolyte of 0.01M $NaNO_3$ under a positive oxygen-free N_2 pressure. Titrations were carried out using an automated Mettler Toledo DL53 burette assembly on 50-ml suspensions of varying bacterial concentration. The 0.5M NaOH titrant was prepared fresh every day and standardised in triplicate against potassium hydrogen pthalate (KHP). The titrator was programmed in dynamic mode with successive titrant additions after a stability of 0.1mV/s had been attained.

Sorption experiments were conducted using 25ml suspensions of the bacteria, both at constant *p*H with variable starting metal concentrations, and as a function of *p*H with constant initial metal concentration. For constant *p*H experiments, metal concentrations ranged from 4 ppm to 60 ppm, while for the *p*H variable experiments, initial concentrations were 10±0.5 ppm. The metal-bacterial suspensions were equilibrated for 30 minutes, based on kinetic experiments showing that equilibrium adsorption was attained within 15 minutes. Sampling involved pelleting the cells and pipetting 10ml of the supernatant into an acid-cleaned bottle. The sample was acidified to 1% v/v HNO_3 and analysed by Flame Atomic Absorption Spectroscopy using matrix-matched standards.

4. RESULTS AND DISCUSSION

4.1 Acid-base titrations

Figure 1 shows a typical acid-base titration of a suspension of the bacteria as well as the electrolyte alone. We note that the presence of bacteria imparts a significant buffering capacity to the electrolyte, due to deprotonation of active functional groups on the bacterial surface. However, the bacterial suspension shows relatively weak inflection points, apparently due to an overlap in the pH over which the different sites deprotonate (Seki et al., 1998). In order to calculate the number of surface sites and their surface density, data from each titration was fitted with a model in which the number of sites on the bacterial surface was varied successively from one to five. This was carried out using FITEQL 4, a speciation program for calculating equilibrium constants from experimental (Herbelin & Westall, 1999). Convergence of the model, coupled with the variance was used to judge goodness of fit of each model. Input to the models included the concentration of the bacteria in the suspension, thermodynamic constants for the dissociation of water and the acid-base behaviour of the electrolyte, taken from Smith & Martell (1976). Activity coefficients were calculated using the Davies equation (Stumm & Morgan, 1996). A surface area of 140 m^2g^{-1} of dry bacteria and a surface capacitance of 8 Fm^{-2} were found to provide good fits to the data.

The best fit to the experimental data required a three-site model, consistent with results from previous studies on both gram-positive (Plette et al., 1995; Fein et al., 1997) and gram-negative (Seki et al., 1998) bacteria. The mean pK_{int} values (n = 6) for the three sites for this strain were 4.3±0.1, 6.9±0.5 and 8.9±0.5. The first two values were assigned by comparison with deprotonation constants for well-known functional groups to carboxyl (4.3) and phosphoryl (6.9) groups. The third value is on the lower end of those reported by other workers, but is within the range for hydroxyl or amine groups (Cox et al., 1999). The site concentrations were $5.0\pm0.7\times10^{-4}$, $2.2\pm0.6\times10^{-4}$ and $5.5\pm2.2\times10^{-4}$ mol/g dry bacteria for the carboxyl, phosphoryl and hydroxyl/amine groups respectively. It is unclear as to origin of the large error associated with the hydroxyl/amine groups, although this may be indirect evidence that more than one functional group dissociates in this region. The site concentrations are generally lower than those calculated for gram-positive bacteria, which may be due to the lower peptidoglycan content of gram-negative bacteria (Cox et al., 1999).

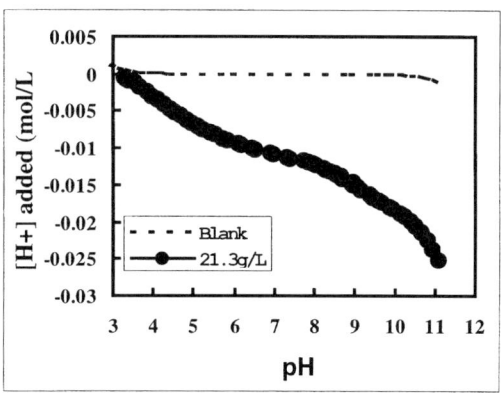

Figure 1. Acid-base titrations of a bacterial suspension in 0.01M NaNO3 electrolyte and the electrolyte blank, showing that the presence of bacteria buffers the electrolyte pH.

4.2 Metal adsorption experiments

In general, the number of sites involved in metal uptake can be determined from adsorption isotherm experiments at constant pH. In most cases, the total amount of metal adsorbed per unit mass of bacteria (M, mol/g) is related to the amount of metal in solution at equilibrium (C, mol/L) by a generalised Langmuir equation:

$$M = \frac{M_1^{max} K_1 C}{1 + K_1 C} + \frac{M_2^{max} K_2 C}{1 + K_2 C} + ... \tag{7}$$

where M^{max} (mol/g) is the maximum amount of metal that can be adsorbed, K (L/mol) is the conditional Langmuir equilibrium constant for each site, and the subscripts refer to the different sites (1, 2 3 etc) on the particle surface. Figure 2 shows representative sorption isotherms of the three metals Pb, Cu and Zn at pH 5.5 (symbols). The curves are best-fit lines generated from non-linear regression of the adsorption data to equation (7), using the Marquardt-Levenberg algorithm (Marquardt, 1963). Only fits with two sites converged, indicating that two sites were required to fit the data at this pH. The data is consistent with the following Langmuir parameters: for Cu: $K_1 = 5.9\times10^3$, $M_1^{max} = 1.3\times10^{-4}$, $K_2 = 9.9\times10^4$, $M_2^{max} = 1.5\times10^{-4}$; for Zn, $K_1 = 4.2\times10^3$, $M_1^{max} = 1.6\times10^{-4}$, $K_2 = 1.0\times10^4$, $M_2^{max} = 7.3\times10^{-5}$; and for Pb, $K_1 = 7.5\times10^3$, $M_1^{max} = 1.3\times10^{-4}$, $K_2 = 1.0\times10^5$, $M_2^{max} = 1.1\times10^{-4}$.

The equilibrium constants given above are conditional and only hold at the particular pH of the experiment. To determine intrinsic stability constants, data from pH variable experiments was fitted to equations (3) to (5) using FITEQL 4. The deprotonation constants and site densities determined from acid-base titrations were used as input in metal adsorption data, plus metal hydrolysis reactions whose stability constants were taken from Smith & Martell (1976).

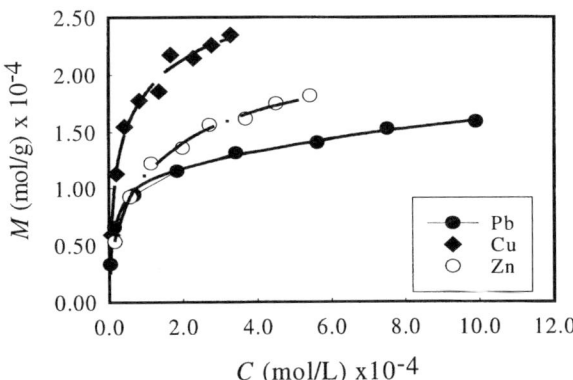

Figure 2. Cu, Zn and Pb isotherms at pH 5.5 (symbols), showing Langmuir adsorption behaviour. The curves depict fits to equation (7) with two surface sites, indicating that adsorption occurs to two sites.

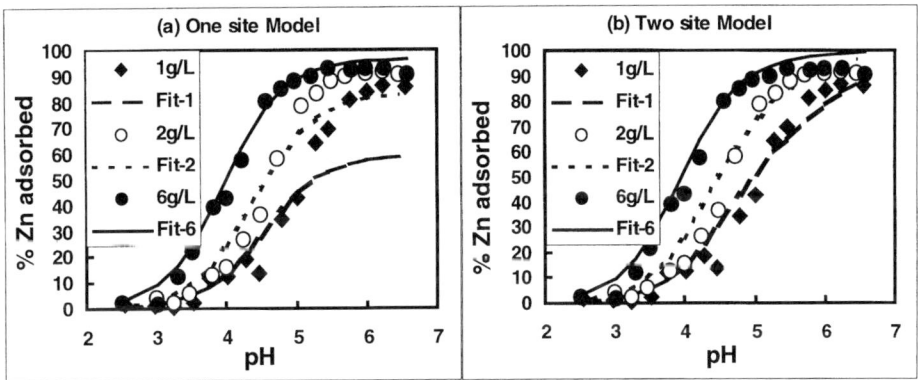

Figure 3. Adsorption edges (symbols) and FITEQL modelling (curves) for Zn with varying bacterial suspension concentrations. FITEQL modelling shows that two sites are required to fit all the data at low bacterial concentrations, consistent with isotherm data above.

The lines shown in Figure 3 represent FITEQL-generated curves assuming that adsorption occurs to one site (a) or two sites (b). As with the titration modelling, the variance was used to select the best-fitting model, in conjunction with visual adherence of the model curves to the experimental data. For all three metals, the one site model was adequate for high bacterial suspension concentrations, but low bacterial suspensions required two sites to fully fit the data. These observations confirm the inference above, that at least two sites are required to fit adsorption data under most natural pH conditions. Furthermore, the dependence on bacterial concentration suggests that site concentration is a primary control on adsorption. The model calculations help to constrain the type of sites involved in metal uptake, and are consistent with adsorption at low pH occurring mostly to carboxyl sites, with the following stability constants for the three metals, given as $Log\ K_{carboxyl}$: Zn^{2+} = 3.3±0.1, Pb^{2+} = 4.3±0.2, and Cu^{2+} = 4.4±0.2. Meantime $Log\ K_{phosphoryl}$, calculated from the low bacterial concentration data are Zn^{2+} = 5.1±0.1 and Cu^{2+} = 6.0±0.5. The two-site model did not converge for Pb data, and so the Pb-phosphoryl stability constant could not be evaluated. The stability constants calculated from this study differ from those found by other workers on gram-negative bacteria (e.g. Seki et al., 1998), possibly reflecting inherent differences between species. However, they are also of similar magnitude to metal-stability constants for common organic ligands (e.g. oxalate), so bacteria are likely exert important control on metal attenuation in the subsurface.

5. SUMMARY
Deprotonation constants for surface sites and stability constants for metal adsorption to sites on a gram-negative bacterial surface have been determined. The surface contains three different sites, and at least two these are involved in metal uptake under natural pH conditions, as confirmed by two independent types of experiments. The metal-stability constants are similar to those for common organic compounds found in natural soil solutions, suggesting that bacteria may play an important part in metal attenuation in the subsurface.

6. ACKNOWLEDGEMENTS
This work was supported by a NERC research grant (GR8/3676), which is gratefully acknowledged. I thank Ian Sutherland and Lynn Kennedy for help with microbial cultures, Ann Mennim for the AAS analysis and Ben Finney for conducting some of the experiments.

7. REFERENCES
Beveridge, T.J. & Murray, R.G.E. (1980). *J. Bacteriol.* 141, 876-887.

Cox, J.S.; Smith, D.S.; Warren, L.A. & Ferris, F.G. (1999). *Environ. Sci. Technol.* 33, 4514-4521.

Davis, J.A. & Kent, D.B (1990). In: *Mineral-Water Interface Geochemistry*, Hochella, M.F.; White, A.F. Eds.; Mineralogical Society of America, Washington, D.C. pp. 177-260.

Fein, J.B. (2000). *Chem. Geol.*, 169, 265-280.

Fein, J.B.; Daughney, C.J.; Yee, N. & Davis, T.A. (1997). *Geochim. Cosmichim. Acta.* 61, 3319-3328.

Herbelin, A.L. & Westall, J.C. (1999). *FITEQL 4.0: a computer program for determination of chemical equilibrium constants from experimental data*; Report 99-01, Department of Chemistry Oregon State University, Corvallis.

Marquardt, D.W. (1963). *J. Soc. Industr. Applied. Maths.* 11, 431-441.

Ngwenya, B.T. & Sutherland, I.W. (*in press*). In: *Proc. 10th Symposium on Water-Rock Interaction*, Fanfani et al (Eds.), Balkema, Rotterdam.

Pagnanelli, F., Panini, M.P., Toro, L., Trifoni, M & Veglio, F. (2000). *Environ. Sci. Technol.*, 34, 2773-2778.

Plette, A.C.C., Van Reimsdijk, W.H., Benedetti, M.F. & Van der Wal, A. (1995). *J. Colloid Interface. Sci.*, 173, 354-363.

Seki, H., Suzuki, A. & Mitsueda, S-I. (1998). *J. Colloid Interface. Sci.*, 197, 185-190.

Smith, R.M. & Martell, A.E. (1976). *Critical Stability constants. IV. Inorganic complexes*, Plenum Press.

Stumm, W. & Morgan, J.J. (1996). *Aquatic Chemistry*, Wiley, Chichester, 1022pp

Wen, X.; Du, Q.; & Tang, H. (1998). *Environ. Sci. Technol.*, 32, 870-875.

A Finite Element Analysis of Migration of Landfill Gas in Unsaturated Soil

B. PALANANTHAKUMAR and W.J.FERGUSON
Department of Civil and Offshore Engineering,
Heriot-Watt University, Edinburgh EH14 4AS, United Kingdom.

Abstract

A two-dimensional, two-phase flow numerical model for the migration of gas in unsaturated soil was developed based on the finite element method employing the Galerkin discretisation technique. The governing system of non-linear partial deferential equations of the model was derived from a mechanistic approach. The model treated the migration of liquid water, heat, dry air, carbon dioxide and methane gases one by one with five-independent system variables of pore water pressure, temperature, pore air pressure and molar concentration of carbon dioxide and methane. The numerical results are compared against previously published experimental data.

1   Introduction

Landfills have been widely used as a method for the disposal of solid waste throughout the world. Degradation of waste in landfills under anaerobic conditions produces landfill gas and leachate. The migration of these hazardous products from landfills can cause environmental damage and can have a detrimental effect on human health and the local environment. As a result, the movement of contaminants under the ground surface is becoming an increasingly important area of research The control of these potential contaminants in modern landfills is achieved by the construction of basal, perimeter and capping liners. However, some uncertainties remain about the effectiveness of compacted clay barriers in preventing lateral gas migration. Uncontrolled migration of landfill gases beyond the confines of a waste disposal facility can represent a significant risk to public safety due to the potentially explosive nature of landfill gas [1]. In order to avert a disaster, an effective means of controlling the landfill gas is required. This, in turn requires an accurate investigation of the gas migration mechanisms in the subsurface of landfills.

Many studies [2, 3] have investigated the migration of landfill gas numerically in the past. However, they all considered the migration of either only methane or a mixture of gas containing methane and carbon dioxide. The objective of the study is to develop a model that treats the transport of methane and carbon dioxide individually.

2   Theoretical Formulations

An unsaturated porous medium can be represented as a three-phase (gas, liquid and solid) system. In this study, the liquid phase is considered to be pure water containing a small quantity of dissolved air and two contaminant gases (CO_2 and CH_4), the gas phase assumed to

Geoenvironmental impact management, Thomas Telford, London, 2001.

be a multi mixture of water vapour, dry air and two contaminant gases. Both the liquid and gas phase are assumed to flow through a rigid porous matrix. Hence, the model reduces to two-phase system. A mechanistic approach is adopted which treats the flow of liquid water, heat, dry air and two-contaminant gases independently.

The mechanistic approach employs Darcy's law and Fick's law to describes the convective gas flow and diffusive gas flow respectively. By applying the principle of mass conservation, these two laws are combined to produce the governing equation of flow through the porous media. These equations are expressed in terms of five primary variables; pore water pressure (P_w), temperature (T), pore air pressure (P_g), molar concentration of CO_2 (C_{gi}) and molar concentration of CH_4 (C_{gj}).

(I) Moisture Transfer: Moisture in unsaturated soil exists in two phases namely liquid water and vapour. Vapour flow is assumed to flow due to two effects i.e. under the influence of a vapour pressure gradient and as part of the bulk flow of air. Therefore, applying the conservation of mass into the moisture transfer gives;

$$\frac{\partial(\phi\rho_l S_l)}{\partial t} + \frac{\partial(\rho_l\theta_v)}{\partial t} = -\nabla(\rho_l V_l) - \nabla(\rho_l V_v) - \nabla(\rho_v V_g)$$ (1)

where ϕ, ρ_l, ρ_v, S_l, θ_v, V_l, V_v and V_g are porosity, liquid density, density of water vapour, liquid saturation, the volumetric vapour content and the velocity of liquid, water vapour and pore air velocity respectively.

By expanding the equation (1) and substituting density and velocity expression in term of gas pressure, pore-water pressure and temperature, it could be written as follows [3];

$$C_{11}\frac{\partial P_w}{\partial t} + C_{12}\frac{\partial T}{\partial t} + C_{13}\frac{\partial P_g}{\partial t} = \nabla.(K_{11}\nabla P_w) + \nabla.(K_{12}\nabla T) + \nabla.(K_{13}\nabla P_g) + \nabla.(K_{16}\nabla Z)$$ (2)

(II) Heat Transfer: Applying the conservation of energy law for heat flow through soil media indicates that the time derivative of the heat content, $\overline{H}$, is equal to the spatial derivative of the heat flux Q. i.e. $$\frac{\partial \overline{H}}{\partial t} = -\nabla Q$$ (3)

Equation (3) could be converted into following form[3];

$$C_{21}\frac{\partial P_w}{\partial t} + C_{22}\frac{\partial T}{\partial t} + C_{23}\frac{\partial P_g}{\partial t} + C_{24}\frac{\partial C_{gi}}{\partial t} + C_{25}\frac{\partial C_{gj}}{\partial t} = \nabla.(K_{21}\nabla P_w)$$
$$+ \nabla.(K_{22}\nabla T) + \nabla.(K_{23}\nabla P_g) + \nabla.(K_{26}\nabla Z)$$ (4)

(III) Dry Air Transfer: Dry air in unsaturated soils can be considered to exist in two forms, bulk air and dissolved air. The bulk air transfer is driven by a gradient of air pressure, whilst the dissolved air transfer is coupled to the flow of pore liquid. Henry's law is used to define the proportion of dry air contained in the pore liquid. Applying the law of mass conservation for flow of dry air in pore of the soil states that;

$$\frac{\partial\{\phi C_{ga}(S_g + H_a S_l)\}}{\partial t} = \nabla\{(D_{ga} + H_a D_{la})\nabla C_{ga}\} - \nabla\{(V_g + H_a V_l)C_{ga}\}$$ (5)

where H_a is Henry's volumetric coefficient of solubility, D_{ga} is the diffusivity coefficient of dry air and D_{la} is the hydrodynamic dispersion coefficient of dry air

Expanding the equation (5) and substituting C_{ga} in terms of C_{gi} and C_{gj} gives [3];

$$C_{31}\frac{\partial P_w}{\partial t} + C_{32}\frac{\partial T}{\partial t} + C_{33}\frac{\partial P_g}{\partial t} + C_{34}\frac{\partial C_{gi}}{\partial t} + C_{35}\frac{\partial C_{gj}}{\partial t} = \nabla.(K_{31}\nabla P_w)$$
$$+ \nabla.(K_{32}\nabla T) + \nabla.(K_{33}\nabla P_g) + \nabla.(K_{34}\nabla C_{gi}) + \nabla.(K_{35}\nabla C_{gj}) + +\nabla.(K_{36}\nabla Z)$$

(6)

(IV) Carbon Dioxide (CO₂): Applying the principle of mass conservation for the carbon dioxide component of the gas phase and assuming that gas and liquid velocities of the contaminant gas equal those of the gas mixture gives,

$$\frac{\partial\{\phi C_{gi}(S_g + H_iS_l)\}}{\partial t} = \nabla\{(D_{gi} + H_iD_{li})\nabla C_{gi}\} - \nabla\{(V_g + H_aV_l)C_{gi}\}$$

(7)

where D_{gi}, D_{li}, and H_i are the effective diffusion coefficient, the hydrodynamic dispersion coefficient and Henry's law coefficient for the carbon dioxide respectively.

This could be rewritten in following form,

$$C_{41}\frac{\partial P_w}{\partial t} + C_{42}\frac{\partial)}{\partial t} + C_{43}\frac{\partial P_g}{\partial t} + C_{44}\frac{\partial C_{gi}}{\partial t}$$
$$= \nabla.(K_{41}\nabla P_w) + \nabla.(K_{43}\nabla P_g) + \nabla.(K_{44}\nabla C_{gi}) + \nabla.(K_{46}\nabla Z)$$

(8)

(V) Methane (CH₄): Similarly in the above section, the methane-transfer gives following governing equation,

$$C_{51}\frac{\partial P_w}{\partial t} + C_{52}\frac{\partial)}{\partial t} + C_{53}\frac{\partial P_g}{\partial t} + C_{54}\frac{\partial C_{gi}}{\partial t}$$
$$= \nabla.(K_{51}\nabla P_w) + \nabla.(K_{53}\nabla P_g) + \nabla.(K_{55}\nabla C_{gi}) + \nabla.(K_{56}\nabla Z)$$

(9)

3   Finite Element Formulation

The system variables (P_w, T, P_g, C_{gi} and C_{gj}) are approximated in terms of nodal values and submitted into the equations (2), (4), (6), (8) and (9). A residual error is obtained, which is then minimised using the Galerkin method. Applying Greens theorem to the dispersive term involving second order derivative yields a relation which may be expressed in matrix form as;

$$[K][\phi] + [C]\frac{d[\phi]}{dt} + [J] = 0$$

(10)

The matrix equation (10) generates a system of linear differential equations of first order. When the time derivative is replaced by a finite difference approximation (backward level time stepping scheme) the equation becomes;

$$\left\{[K]^{n+1} + \frac{[C]^{n+1}}{\Delta t}\right\}[\phi]^{n+1} = \left\{\frac{[C]^{n+1}[\phi]^n}{\Delta t} + [J]^{n+1}\right\}$$

(11)

The superscript n refers to the time level and Δt is time step. It can be seen that the solution for ϕ at time level n+1 can be obtained directly from the stiffness (K) and capacitance (C) matrices, and ϕ at time level n, which is known

4   Result and Discussion

The model was verified against an experiment data published by Visvanathan *et al.*, 1999 [4]. In this experiment, a soil column which contents 70% sand, 15% clay and 15% silt was used to find the methane and carbon dioxide profile along the column. Fig 1 presents a schematic diagram of the column under investigation. A gas mixture of 60% methane and 40% carbon dioxide was purged at the rate of 6.5 mL/min through the bottom of the column under pressure of 101421.32 Pa, and released through the top of the column at atmospheric pressure. Throughout the experiment, the average moisture content of the column soil was 11.7% and ambient temperature was 32°C. The material properties employed in the numerical simulation were taken directly from the literature [2,3,4] and are summarised in Table 1.

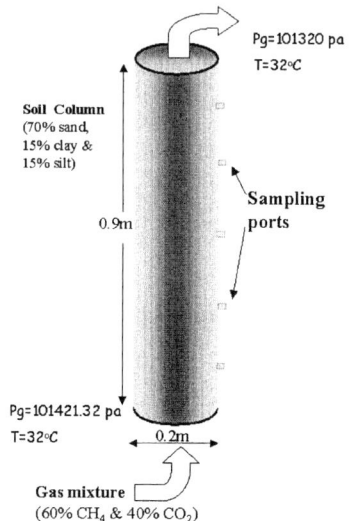

Table 1	Material Properties	
Property		**Value**
ϕ		0.4
K_w	m s^{-1}	0.000145
k	m^2	10^{-13}
C_{ps}	J kg^{-1} K^{-1}	837
ρ_s	kg m^{-3}	2650
λ_t	W m^{-1} K^{-1}	$0.5 + 1.5 S_l$
L	J kg^{-1}	2.4×10^6
C_{pl}	J kg^{-1} K^{-1}	4184
C_{pv}	J kg^{-1} K^{-1}	1870
C_{pgi}	J kg^{-1} K^{-1}	35.44
C_{pgi}	J kg^{-1} K^{-1}	29.22

Fig 1   Schematic Diagram of Soil Column

The finite element mesh constructed to represent the soil column is comprised of 93 nodes and 18 eight-node elements in rectangular shape. The initial conditions throughout the domain under investigation are uniform and are pore water pressure of 59210.5 Pa which is equivalent to a liquid saturation of 46.5%, temperature 32°C, gas pressure of 101320 Pa and contaminant gases 0.0 mol/m^3. The top surface nodes of the column were assumed to be at atmospheric pressure of 101320 Pa and 0.0 mol/m^3 contaminant gas whilst the bottom surface nodes of the column were fixed at gas pressure 101421.32 pa with 1.088×10^{-4} mol/m^2 s gas flux boundary condition. Methane oxidation capacity of soil taken from the literature [4] was adopted as source terms. A variable time step was used in the numerical solution with initial, minimum and maximum time steps of 1 sec , 0.001 sec and 1 hr respectively.

The model was simulated for 50 days, and results of the system variables at different time intervals were obtained and compared. It is noticed that all variables became stable within 50 days. Therefore, time-dependant behaviour was not observed afterwards. Fig 2 shows profiles of pore-water and gas pressure on 50th day. The stable pore-water pressure varies linearly along the height of column. Gas pressure stabilised in a parabolic shape (concave upward). It matches with the gas profile investigated in the past [5].

Comparison of model results against the experimental data at 50th day is presented in Fig 3. The pattern of predicted curves and experimental data points behave in same manner. The estimated carbon dioxide is slightly higher than the experiment whilst the methane is less. It reflects inaccuracy of methane oxidation capacity of the soil, which was indeed measured in batch experiments where experiment conditions was deferent from the column experiment

[4]. A drastic decrease in the concentration of methane is evident that methane oxidisation played a major role in consuming methane during its migration. Incorporation of precise methane oxidation into this model could yield better results.

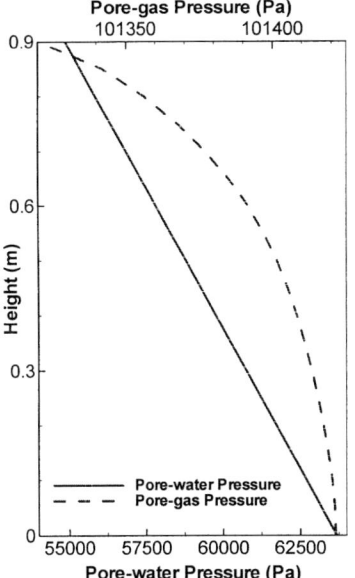

Fig 2   Pore-water and Gas Profile

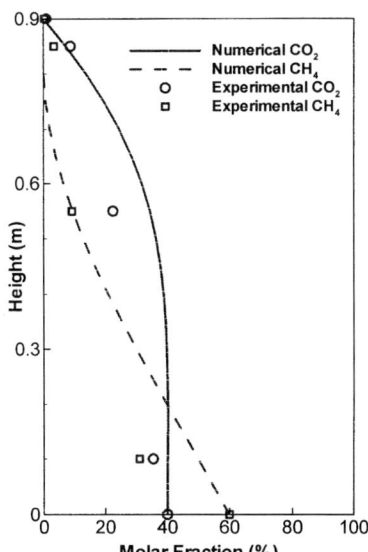

Fig 3   Comparison of Numerical Results against Experimental Data

5   Conclusion

This paper presents a finite elements analysis of landfill-gas migration in unsaturated soil. The governing equations of the model are derived from a mechanistic approach where the mass and energy conservation laws are defined for a particular phase into which Darcy's law and Fick's law are substituted. The model treats the movement of liquid, heat, air, methane and carbon dioxide independently with five-independent system variables of pore water pressure, temperature, pore air pressure and, molar concentration of carbon dioxide and methane. The comparison of model results against experimental data is reasonable. It is yet to verify against the movement of carbon dioxide (CO_2) and methane (CH_4) in real landfill sites. Though the model is intended to use for migration of landfill gas it could be used to assess the migration of any hazardous vapour in soil.

Acknowledgments: We are thankful to Dr.C.Visvanathan and D. Pokhrel (AIT, Thailand) for providing the experimental data which were used to validate the numerical results.

Appendix. Equation coefficients
Capacity Coefficients:

$$C_{11} = \phi(\rho_1 - \rho_v)\frac{\partial S_1}{\partial P_w} + \rho_o\phi S_g\frac{\partial h}{\partial P_w}, C_{12} = \phi(\rho_1 - \rho_v)\frac{\partial S_1}{\partial T} + \rho_o\phi S_g\frac{\partial h}{\partial T}, C_{13} = \phi(\rho_1 - \rho_v)\frac{\partial S_1}{\partial P_g} + \rho_o\phi S_g\frac{\partial h}{\partial P_g}$$

$$C_{21} = (T - T_r)\left\{A_{Tl}\frac{\partial S_1}{\partial P_w} + \phi S_g C_{pga}\frac{\partial C_{ga}}{\partial P_w} + \phi S_g C_{pv}\rho_o\frac{\partial h}{\partial P_w} + \phi S_l C_{pl}\frac{\partial P_l}{\partial P_w}\right\} - \phi\rho_v L\frac{\partial S_1}{\partial P_w} + \phi S_g L\rho_o\frac{\partial h}{\partial P_w}$$

$$C_{22} = H + (T-T_r)\left\{A_{Tl}\frac{\partial S_l}{\partial T} + \phi S_g C_{pga}\frac{\partial C_{ga}}{\partial T} + \phi S_g C_{pv}(h\beta + \rho_o\frac{\partial h}{\partial P_w}) + \phi S_l C_{pl}\frac{\partial P_l}{\partial T}\right\} - \phi\rho_v L\frac{\partial S_l}{\partial T} + \phi S_g L(h\beta + \rho_o\frac{\partial h}{\partial T})$$

$$C_{23} = (T-T_r)\left\{A_{Tl}\frac{\partial S_l}{\partial P_g} + \phi S_g C_{pga}\frac{\partial C_{ga}}{\partial P_g} + \phi S_g C_{pv}\rho_o\frac{\partial h}{\partial P_g} + \phi S_l C_{pl}\frac{\partial P_l}{\partial P_g}\right\} - \phi\rho_v L\frac{\partial S_l}{\partial P_g} + \phi S_g L\rho_o\frac{\partial h}{\partial P_g}$$

$$C_{24} = (T-T_r)\left\{\phi S_g C_{pgi} + \phi S_g C_{pga}\frac{\partial C_{ga}}{\partial C_{gi}}\right\} \qquad C_{25} = (T-T_r)\left\{\phi S_g C_{pgj} + \phi S_g C_{pga}\frac{\partial C_{ga}}{\partial C_{gj}}\right\}$$

$$A_{Tl} = \phi\rho_L C_{pl} - \phi C_{pv}\rho_v - \phi C_{pga}C_{ga} - \phi C_{pgi}C_{gi} - \phi C_{pgj}C_{gj}$$

$$C_{31} = \phi(S_g + H_a S_l)\frac{\partial C_{ga}}{\partial P_w} + \phi C_{ga}(H_a-1)\frac{\partial S_l}{\partial P_w} \qquad C_{32} = \phi(S_g + H_a S_l)\frac{\partial C_{ga}}{\partial T} + \phi C_{ga}(H_a-1)\frac{\partial S_l}{\partial T}$$

$$C_{33} = \phi(S_g + H_a S_l)\frac{\partial C_{ga}}{\partial P_g} + \phi C_{ga}(H_a-1)\frac{\partial S_l}{\partial P_g} \qquad C_{34} = \phi(S_g + H_a S_l)\frac{\partial C_{ga}}{\partial C_{gi}} \qquad C_{35} = \phi(S_g + H_a S_l)\frac{\partial C_{ga}}{\partial C_{gj}}$$

$$C_{41} = \phi C_{gi}(H_i-1)\frac{\partial S_l}{\partial P_w} \qquad C_{42} = \phi C_{gi}(H_i-1)\frac{\partial S_l}{\partial T} \qquad C_{43} = \phi C_{gi}(H_i-1)\frac{\partial S_l}{\partial P_g} \qquad C_{44} = \phi S_g + H_i\phi S_l$$

$$C_{51} = \phi C_{gj}(H_j-1)\frac{\partial S_l}{\partial P_w} \qquad C_{52} = \phi C_{gj}(H_j-1)\frac{\partial S_l}{\partial T} \qquad C_{53} = \phi C_{gj}(H_j-1)\frac{\partial S_l}{\partial P_g} \qquad C_{55} = \phi S_g + H_j\phi S_l$$

Kinetic Coefficients:

$$K_{11} = \frac{K_w}{g} + \rho_l K_{vl}, K_{12} = \rho_l K_{v2}, K_{13} = \rho_v K_g + \rho_l K_{v3}, K_{16} = \rho_l K_w + \rho_v \rho_g g K_g$$

$$K_{21} = \rho_l L K_{vl} + C_{pv}\rho_l K_{vl}(T-T_r) + \frac{C_{pl}\rho_l K_w}{\gamma_l}(T-T_r) \qquad K_{22} = \lambda + \rho_l L K_{v2} + C_{pv}\rho_l K_{v2}(T-T_r)$$

$$K_{23} = \rho_l L K_{v3} + \rho_v L K_g + C_{pv}\rho_l K_{v3}(T-T_r) + (T-T_r)K_g(C_{pv}\rho_v + C_{pga}C_{ga} + C_{pgi}C_{gi} + C_{pgj}C_{gj})$$

$$K_{26} = L\rho_v \rho_g g K_g + C_{pl}\rho_l K_w(T-T_r) + (T-T_r)\rho_g g K_g(C_{pv}\rho_v + C_{pga}C_{ga} + C_{pgi}C_{gi} + C_{pgj}C_{gj})$$

$$K_{31} = (D_{ga} + H_a D_{la})\frac{\partial C_{ga}}{\partial P_w} + \frac{C_{ga}H_a K_w}{\gamma_l} \qquad K_{32} = (D_{ga} + H_a D_{la})\frac{\partial C_{ga}}{\partial T} \qquad K_{33} = (D_{ga} + H_a D_{la})\frac{\partial C_{ga}}{\partial P_w} + C_{ga}K_g$$

$$K_{34} = (D_{ga} + H_a D_{la})\frac{\partial C_{ga}}{\partial C_{gi}} \qquad K_{35} = (D_{ga} + H_a D_{la})\frac{\partial C_{ga}}{\partial C_{gj}} \qquad K_{36} = K_w H_a C_{ga} + K_g C_{ga}\rho_g g$$

$$K_{41} = \frac{K_w C_{gi}H_i}{\gamma_l} \qquad K_{43} = K_g C_{gi} \qquad K_{44} = D_{gi} + H_i D_{li} \qquad K_{46} = K_w H_i C_{gi} + C_{gi}\rho_g g K_g$$

$$K_{51} = \frac{K_w C_{gj}H_j}{\gamma_l} \qquad K_{53} = K_g C_{gj} \qquad K_{55} = D_{gj} + H_j D_{lj} \qquad K_{56} = K_w H_j C_{gj} + C_{gj}\rho_g g K_g$$

[The abbreviations used in Thomas and Ferguson (1999) were also used in this study]

Reference

[1]. Williams, G.M., and Aitkenhead, N., 1991. Lessons from Loscoe-the uncontrolled Migration of Landfill Gas. *Quarterly Journal of Engineering Geology,* 24: 191-207.

[2]. Metcaff, D.E. and Farquhar, G.J., 1987. Modeling Gas Migration through Unsaturated Soils from Waste Disposal Sites. *Water, Air and Soil Pollution,* 32 : 247-259.

[3]. Thomas, H.R., and Ferguson, W.J., 1999. A Fully Coupled Heat and Mass Transfer Model incorporating Contaminant gas Transfer in an Unsaturated Porous Medium. *Computers and Geotechnics,* SA2: 307-317.

[4]. Visvanathan, C., Pokhrel, D., Cheimchaisri, W., Hettiaratchi, J.P.A. and Wu, J.S., 1999. Methanotrophic Activities in Tropical Landfill Cover Soils: Effects of Temperature, Moisture and Methane Concentration. *Waste Management & Research,* 17: 313-323.

[5]. Alzaydi, A.A., More, C.A. and Rai, I.R., 1978. Combined Pressure and Diffusional Transition Region Flow of Gases in Porous Media. *AIChE Journal,* 24(1): 35-43.

Theoretical comparison of contaminant transport parameters derived from batch adsorption, diffusion cell and column tests.

S.A.RICHARDS and A.BOUAZZA
Department of Civil Engineering, Monash University, Melbourne, Australia.

INTRODUCTION

With the increased focus on "risk assessment" and "natural attenuation", in contaminated site and waste management, reliable contaminant transport modelling has become increasingly important. Many analytical and numerical models exist that are all based on the same governing equation for contaminant transport. Selecting the right model for the situation at hand is vital, as each model is prepared with a different set of boundary conditions and generalisations about material behaviour. However, whatever the model, the reliability of the output is ultimately limited by the appropriateness of the input. Adsorption is a characteristic of geological deposits that can significantly affect contaminant transport. The adsorptive behaviour of soils and rocks can have many external influences. Therefore, empirically derived parameters are necessary for accurate modelling. There are many methods available for acquiring these parameters. Methods such as batch adsorption provide an indication of adsorption parameters in bulk solution, whereas diffusion cell or column tests simulate contaminant transport through the material under near in-situ conditions. It has long been noted that adsorption parameters derived from batch adsorption tests varied from those extrapolated from column tests. Several researchers have sought to explain the variation and/or modify models to better suit the source of parameters. Ideally, a material would be assessed, through column testing, with the solute that is to be encountered on site, under the hydraulic gradient and confining pressure that is expected. However, column tests are time consuming, and hence expensive, and results can be misleading if the test conditions vary significantly from the site conditions. If a suitable relationship can be derived to link extended batch adsorption results and actual contaminant transport, then reliable contaminant transport modelling may be performed within the time and cost limitations generally faced by site assessment consultants and waste containment facility designers.

CONTAMINANT TRANSPORT THEORY

The general governing equation for contaminant transport in porous media, can be written as follows:

$$\frac{\delta C}{\delta t} = D \frac{\delta^2 C}{\delta x^2} - v \frac{\delta C}{\delta x} - \frac{\rho_s}{n}\left(\frac{\delta S}{\delta t}\right) \tag{1}$$

where, C is the resident pore fluid concentration, S is the concentration in the solid phase, D is a lumped diffusion/dispersion parameter, v is solute velocity, ρ_s material bulk density and n is material porosity.

Geoenvironmental impact management, Thomas Telford, London, 2001.

The right hand side of the equation can be broken down into three parts. The first term relates to diffusion and dispersion, which are generally considered to be governed by the pore size and connectivity in the material and the size and activity of the contaminant molecule. The second term relates to advection and is governed by the hydraulic gradient across the material and the permeability of the material. The third term relates to adsorption. The amount of adsorption is governed by the ratio of mass of adsorbed material to mass of pore fluid, represented by the ρ_s/n term, and the rate of uptake of contaminant on to the solid, represented by $\delta S/\delta t$.

Since the adsorption term contains a derivative with respect to time, it is often placed on the left hand side of the equation. The most common form of the governing equation (Freeze and Cherry, 1979) replaces the adsorption term with a lumped parameter called the retardation coefficient. Physically, this coefficient represents the reduction in contaminant velocity due to adsorption.

Thus, the governing equation becomes:

$$R_d \frac{\delta C}{\delta t} = D \frac{\delta^2 C}{\delta x^2} - v \frac{\delta C}{\delta x},$$ (2)

where R_d is the retardation coefficient.

The retardation coefficient is expressed as:

$$R_d = 1 + \frac{\rho_s}{n} \frac{\delta S}{\delta C},$$ (3)

where $\delta S/\delta C$ is the ratio of the concentration of contaminant in the solid phase to concentration in the liquid phase. This ratio can be graphically represented by an adsorption isotherm. If the material does not adsorb the contaminant of interest, then $\delta S/\delta C$ is zero and R_d equals 1. If the adsorption isotherm is linear, then $\delta S/\delta C$ equals the slope of the line, which is known as the partition coefficent, K_d.

TEST METHODS

Batch Adsorption Test
The adsorption isotherm can be obtained in a number of ways. The most direct way of obtaining a relationship between the concentrations in the solid and liquid phases is to perform batch adsorption tests (Roy et al., 1992). These tests are performed by mixing solutions of known concentration and volume with material of known weight, until equilibrium is reached. The phases are then separated and the concentrations in solution are determined. The difference between the original concentrations and the final concentrations is considered to have been adsorbed. By this method, a plot of equilibrium solution concentration vs adsorbed concentration, can be obtained quickly and directly.

Column Test (Advective/diffusive)
Column tests are an attempt to duplicate in-situ conditions in a laboratory environment. The test is similar to a permeability test, in that a solute is passed through the material under a hydraulic gradient. The influent and effluent (and often several intermediate locations) concentrations are monitored over several pore volumes of flow. Analytical or numerical solutions, to the governing equation of contaminant transport, can be used to fit contaminant breakthrough predictions to the effluent concentrations obtained.

Diffusion Test

Diffusion tests are similar to column tests in that contaminant transport, through a bulk sample of the material of interest, is directly observed. The difference is that the advective component of transport is removed in diffusion tests. There are several ways to conduct a diffusion test. Each requires a concentration gradient across the sample. Figure 1 shows diagrams of several methods set-up, the initial conditions and the actions required. Similar to the column test, an appropriate model can then be used to fit a curve to the concentration changes observed.

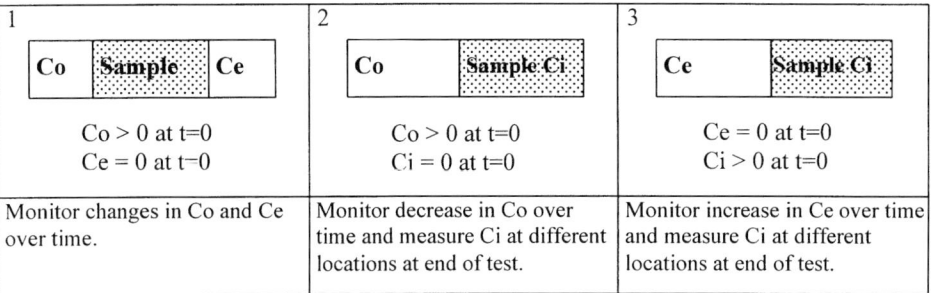

1	2	3
Co [Sample] Ce	Co [Sample Ci]	Ce [Sample Ci]
Co > 0 at t=0 Ce = 0 at t=0	Co > 0 at t=0 Ci = 0 at t=0	Ce = 0 at t=0 Ci > 0 at t=0
Monitor changes in Co and Ce over time.	Monitor decrease in Co over time and measure Ci at different locations at end of test.	Monitor increase in Ce over time and measure Ci at different locations at end of test.

Figure 1. Schematics of several diffusion test set-ups.

LIMITATIONS OF TEST METHODS AND MODELS

Each of the test methods described above, provides information about the contaminant transport in the material under the conditions of the test. Transport parameters can vary under different conditions. Therefore, it is important to consider the conditions of the test and the conditions of the situation that is to be modelled carefully.

Batch Adsorption Test

The batch adsorption test reflects pure adsorption at equilibrium, where no concentration profiles exist in the solute or the solid phase. A limitation of the batch adsorption test is that it only reflects the partitioning at equilibrium. Separate kinetics tests need to be performed to observe the rate of adsorption. It also provides no information about diffusivity through the bulk material.

Column Test

The column test reflects advective, dispersive, diffusive and adsorptive behaviour in a bulk material. The benefits of column testing are that the material is assessed in a bulk condition, similar to that encountered on site, and the effects of several transport mechanisms can be observed at the same time. However, these benefits are also related to the disadvantages of this type of test. If the material has a low permeability, the hydraulic gradient is often increased to expedite the test. This alters the relative effect of each of the transport mechanisms. Unless several tests are run under various conditions (similar to sensitivity analysis), the effect of changes to each mechanism, can not be separated. This can lead to inaccurate selection of empirical parameters. When these parameters are input to a model to simulate a different set of conditions, the error can become significant.

Concentration profiles would exist in a layer around each aggregate and through the adsorbent aggregates (see Figure 2.). The layer thickness, around each aggregate of adsorbent particles, would depend on the velocity of the pore fluid. The faster the fluid is moving, due to advection, the greater the mechanical dispersion, which would result in a thinner layer around the aggregates (Taylor and Krishna, 1993). However, since advection would be greater the magnitude of the concentration gradient (the difference between the pore fluid concentration and the concentration at the aggregate surface) is likely to be larger, as the adsorption and diffusion into the solid phase is limited by other factors. The concentration profile through the aggregate depends on the rate of diffusion into the aggregate and the rate of adsorption to the particles. None of these elements is considered in the standard models used to analyse the test data.

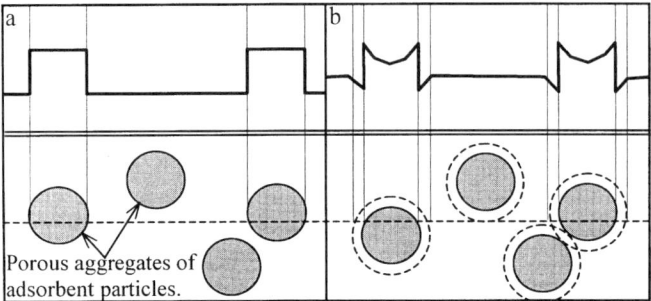

Figure 2. Schematic of concentration profiles; a) as modelled and, b) actual.

Diffusion Test
The diffusion test reflects diffusion and adsorption in the bulk material. The diffusion test can provide good information on adsorption (desorption, in the third case shown in Figure 1.) and diffusion depending on the appropriateness of the model used to assess the data. However, diffusion through low permeability materials can be very slow and tests may take months. The mechanisms at work are similar to those described above for the column test. However, since contaminant transport through the pore fluid is solely due to diffusion, and not advection and diffusion, then concentration gradients in the film and through the aggregate are likely to be less pronounced than in the column test.

Since there is no pore fluid velocity through the sample, the only conditions that can be varied in a diffusion test are the bulk material density and the concentration gradient. A change in bulk material density is likely to change the size or density of the aggregates within the material, which would affect rate of diffusion into the aggregates. Applying a concentration range that is significantly different from the range of exposure expected in the field, may lead to an incorrect prediction of adsorption behaviour if the adsorption isotherm is non-linear.

Modelling
The models currently utilised, to analyse test data and model contaminant transport situations, make a few notable simplifications.

When the derivatives with respect to time are lumped together and the retardation coefficient created, the time dependent adsorption term is removed.
i.e.

$$\frac{\rho_S}{n}\left(\frac{\delta S}{\delta t}\right) \xrightarrow{becomes} \frac{\rho_S}{n}\left(\frac{\delta S}{\delta C}\right)\frac{\delta C}{\delta t} \tag{4}$$

This means that the rate of adsorption is simply described by the rate of change of pore fluid concentration and the partition ratio, between equilibrium concentration in liquid to solid phases. This assumes instantaneous equilibrium between the liquid and solid phases.

In some cases this model may be appropriate. For instance, where there is low flow and fast equilibration or no adsorption. If the column or diffusion test conditions are very similar to the conditions expected in-situ, then the inaccuracies of the existing models are inconsequential. It only becomes relevant if the rate terms are likely to change. For instance, if the flow velocity, matrix structure or concentration gradient differs between the test and the modelled situation.

The dependency on consistent test conditions could be reduced if the model used to analyse test data and predict contaminant transport in-situ reflected the micro mechanisms at work. This improvement could mean that test conditions could be changed to reduce test duration, without sacrificing the quality and appropriateness of results.

BASIS FOR PROPOSED CHANGES TO TRANSPORT MODEL
Many other disciplines study phenomena analogous to contaminant transport through porous media. Valuable insight can be gained by studying the approaches taken by other disciplines and applying the components relevant to the application of interest. A majority of the advances in mass transfer modelling, utilised in chemical engineering, has come from studies in heat transfer. Conduction is analogous to diffusion, convection to advection and enthalpy of reaction to adsorption/desorption (there is no mass transfer equivalent to radiation).

In process engineering, the rate of adsorption is most simply described by a linear driving force model. This model can be written as follows:

$$\frac{\delta S}{\delta t} = K\left(S^{eq} - S\right),$$

(5)

where K is the overall mass transfer coefficient, S^{eq} is the concentration adsorbed on the solid phase at equilibrium and S is the current solid phase concentration.

The mass transfer coefficient is, in effect, a lumped parameter that can take into account diffusion through the film layer, diffusion into an aggregate and adsorption kinetics. (5) can be direct substituted into (1) and K can be determined empirically. However, to better reflect the mechanisms involved the mass transfer coefficient can be described as follows:

$$\frac{1}{K} = \frac{V}{A}\frac{1}{k_f} + \frac{1}{k_p}$$

(6)

where, V is the volume of the average aggregate, A is the surface area of the average aggregate and k_f is the mass transfer resistance due to the film and k_p is the intra-aggregate mass transfer resistance. For spherical aggregates, V/A becomes $r_p/3$, where r_p is the aggregate radius. The adsorption of the contaminant molecule to the adsorbent particle, once they are in intimate contact, is assumed to be instant.

The mass transfer resistance due to diffusion through the film layer is related to the fluid dynamics of the system. As mentioned previously, the velocity of the pore fluid affects the thickness of the film. Empirical relationships have been obtained to link the Reynolds and Schmidt numbers to the Sherwood number. The most popular relationship is the Ranz and Marshall equation, which was developed for heat transfer. Their equation was prepared from data obtained from single spheres and has not been found to correlate well to packed beds.

The following equation was prepared using solid-liquid phase mass transfer data from packed beds (Wakao and Kaguei, 1982).

$$Sh = 2 + 1.1Sc^{1/3} \, Re^{0.6} \tag{7}$$

where, Sh is the Sherwood number $\left(\dfrac{k_f . 2 . r_p}{D_m}\right)$, Sc is the Schmidt number $\left(\dfrac{\mu_w}{\rho_w D_m}\right)$ and Re is the

Reynolds number $\left(\dfrac{2r_p \rho_w v}{\mu_w (1 - p)}\right)$. Where, r_p is the aggregate radius, D_m is the molecular diffusivity, v is the pore fluid velocity, ρ_w is the density of water, μ_w is the viscosity of water and p is the porosity of the bulk material.

The mass transfer resistance due to intra-aggregate diffusion can be written as (Tien, 1994):

$$k_p = \frac{15 D_e}{r_p^2} \tag{8}$$

where, D_e is the intra-aggregate diffusivity and r_p is the radius of the aggregate. This is commonly called Glueckauf's approximation and is an approximation of the theoretical solution for diffusion into a sphere.

The structure of the model has been developed and tested over time and is widely used in other fields. These equations have been developed for beds packed with uniform, spherical, adsorbents. However, unlike process engineering, the materials encountered in environmental contaminant transport are not manufactured and therefore do not have consistent characteristics. The difficulty that lies in applying these equations is in characterising the voids in geologic materials and the rate of diffusion through them. Some consideration needs to be made of the void and aggregate geometry within the bulk materials being tested and a method of approximating spherical aggregates.

CONCLUSION
The expressions presented above, for mass transfer resistance, provide a means of modifying the contaminant transport model to reflect the micro mechanisms taking place. Empirical coefficients are still utilised and in the early stages of use the model may still be subject to misinterpretation. However, as pore size distribution testing develops and becomes more widely conducted, more and more rigorous theoretical components can be added to the model.

It is envisaged that this modified governing equation can be used to include the effects of diffusion into aggregates of adsorbent particles and may be able to explain the observed differences between adsorption parameters obtained by batch testing and the behaviour observed in column tests.

REFERENCES
Freeze, R. A. and J. A. Cherry (1979). Groundwater. Englewood Cliffs, NJ, Prentice-Hill Inc.
Roy, W. R., I. G. Krapac, S. F. J. Chou and R. A. Griffin (1992). Technical Resource Document - Batch-type procedures for estimating soil adsorption of chemicals. Washington, United States Environmental Protection Agency.
Taylor, R. and R. Krishna (1993). Film Theory. Multi Component Mass Transfer. New York, John Wiley and Sons, Inc.: 152-219.
Tien, C. (1994) Adsorption calculations and modeling. Newton, Butterworth-Heinemann.
Wakao, N. and S. Kaguei (1982). Heat and mass transfer in packed beds. New York, Gordon and Breach Science Publishers.

Migration of LNAPL in an Unsaturated/Saturated Sand

R S SHARMA and M H A MOHAMED

Department of Civil and Environmental Engineering
University of Bradford, U.K.

ABSTRACT

Accidental spills of hydrocarbons, such as Light Non-aqueous phase liquids (LNAPLs), is one of the most common sources of subsurface contamination. Migration of LNAPL in a porous medium is influenced by various factors such as the number of fluids present in the unsaturated/saturated zones and the proportion of pores occupied by each fluid. The results for relationship between matric suction and degree of saturation are presented in this paper, with a brief discussion, for water-air, water-LNAPL and LNAPL-air systems in a sand. Furthermore, results from column tests, simulating the influence of fluctuating groundwater table on the distribution of LNAPL, are compared with the results correlating matric suction and degree of saturation.

INTRODUCTION

Leaks from underground storage tanks and accidental spills of hydrocarbons, such as Light Non-aqueous phase liquids (LNAPLs), pose a serious risk of subsurface contamination. A LNAPL migrates downwards under the influence of gravity and capillary forces, contaminating the subsurface region around the spill as well as deteriorating the quality of groundwater. Migration of LNAPL in a subsurface is influenced by various factors such as the number of fluids present in the unsaturated/saturated zones and the proportion of pores occupied by each fluid. Furthermore, fluctuations of groundwater table cause LNAPL to redistribute itself in the subsurface. Since fluctuations of groundwater table commonly occur in any geographical region, it is important to understand the implications of such fluctuations on the distribution of LNAPLs in subsurface. One way of analysing and quantifying the effects of fluctuations of groundwater table on LNAPL distribution is to examine the nature of the relationship between matric suction and degree of saturation. Detailed investigation, aimed at developing a new understanding of the relationships between matric suction and degree of saturation, is the focus of this paper.

Migration of a liquid in a saturated porous medium is expressed by Darcy's law, which can also be extended to describe a multi-phase flow (see, for example, Bear, 1979 and Sharma, 2000). In a multi-phase flow system, flux of each fluid is a function of the effective degree of saturation of the fluid (Brooks and Corey, 1964). For accurate simulation of LNAPL distribution in unsaturated porous medium, the unsaturated hydraulic conductivity, which is linked with the degree of saturation and matric suction, should be known. However, the relationships between matric suction, degree of saturation and unsaturated hydraulic conductivity are different during drying and wetting processes. This difference is attributed to a phenomenon called "hydraulic hysteresis".

Geoenvironmental impact management, Thomas Telford, London, 2001.

Recently, researchers have tried to investigate the influence of groundwater fluctuations on the distribution and trapping of LNAPL in unsaturated porous media either by using numerical models or performing laboratory tests using 1-D models. Noticeable improvements in the prediction have been reported when the effects of hysteresis are considered (see, for example, Ostendorf et. al., 1993 and Steffy et. al., 1998). However, some studies suggest that either there is a minor effect of hysteresis on contaminant transport (Mitchell and Mayer, 1998) or non-hysteretic model gives close prediction to 1-D column experiments (Oostrom and Lenhard, 1998). This means that the influence of groundwater fluctuations on LNAPL distribution is not settled yet and further research is needed.

A programme of research is in progress at the University of Bradford, involving small-scale tests, 1-D column tests and tests on 2-D and 3-D laboratory models. In this paper, results for the relationship between degree of saturation and matric suction are presented with a brief discussion, for water-air, water-LNAPL and LNAPL-air systems in a sand. Furthermore, results from column tests, simulating the influence of fluctuating groundwater table on the distribution of LNAPL, are compared with the results correlating matric suction and degree of saturation.

MATERIALS AND TESTING PROGRAM
Buchner funnel was used for initial investigation of the relationship between degree of saturation and matric suction for different fluid systems. Samples were prepared by pouring sand from a fixed height into the water-filled funnel by keeping the water level always above the sand level. The samples were thus initially fully saturated, which were then subjected to increasing value of matric suction, by lowering the burette to a given height, to obtain the main drying curve. Likewise, wetting curves were obtained by reducing the suction. Similarly inner drying and wetting curves (scanning curves) were obtained. The degree of saturation was calculated in each step. Further details of sample preparation and measurement of various parameters are given in Mohamed and Sharma, 2000.

Mineral oil was used as a light non-aqueous phase liquid. Mineral oil was selected because of its negligible solubility in water, its very low volatility at room temperature and it is safe to work with. Measured specific gravity of mineral oil was 0.828. The values of mineral oil interfacial tension with water and air were 49.0 and 30.1 mN/m respectively (Scroth et. al., 1995). In general, all sand samples were classified as poorly graded coarse sand ($C_u = 1.35$ and $C_c = 0.92$) with D_{50} range between 0.80 and 0.85 mm. The porosity of all the samples was found to vary in a narrow band, 43 (+/- 1)%.

Seven experiments were performed to investigate the relationship between matic suction and degree of saturation for three different fluid systems (water-air, water-LNAPL and LNAPL-air). Additionally, column tests were carried out to simulate the influence of fluctuations of groundwater table on LNAPL distribution. The sand was poured into the column, like in Buchner funnel, and then the water table was lowered to 3.4cm. It was followed by introducing 50cm^3 of LNAPL on the sand surface after equilibrium condition of water along the column height was established.

RESULTS
The results from seven tests using different fluid systems (water-air, water-LNAPL and LNAPL-air) in coarse sand samples are presented here in order to examine the nature of the relationship between degree of saturation and matric suction. The results for water – air

system are shown in Figure 1. It includes results from 5 tests, which are presented (1 to 5), in Figure 1. Separate samples prepared by the same procedure were used in each test. The sample was subjected to increasing followed by decreasing suction to obtain drying and wetting curves. The highest values of suction head for tests 1 to 5 are shown in Figure 1 by a, b, c, d and e respectively. It is clear from these results that there is hysteresis in the relation between matric suction and degree of saturation. The amount of hysteresis is described here by the difference between the values of matric suction head in drying and wetting respectively at a given value of degree of saturation. In water – air fluid system, the average amount of hysteresis is 5.5cm measured in the straight-line part of the curve. During the drying path, water starts to drain out of the sample after reaching 8.5cm head. This head is called the bubbling pressure head. Eventually, the water saturation reached minimum value at which no further reduction in degree of saturation is obtained even with further increase in matric suction head. This degree of saturation is 13.0% and called residual saturation. Furthermore, during the wetting path, the degree of saturation is 92.0% although the matric suction head had dropped to zero. This suggested that 8.0% of air was trapped inside the sample. The amount of trapped air is 0, 3.0, 6.0, and 8.0% when the sample was wetted from degree of saturation of 91.6, 75.5, 43.1 and 27.1 % respectively (see 5, 4, 3, 2 respectively in Figure 1).

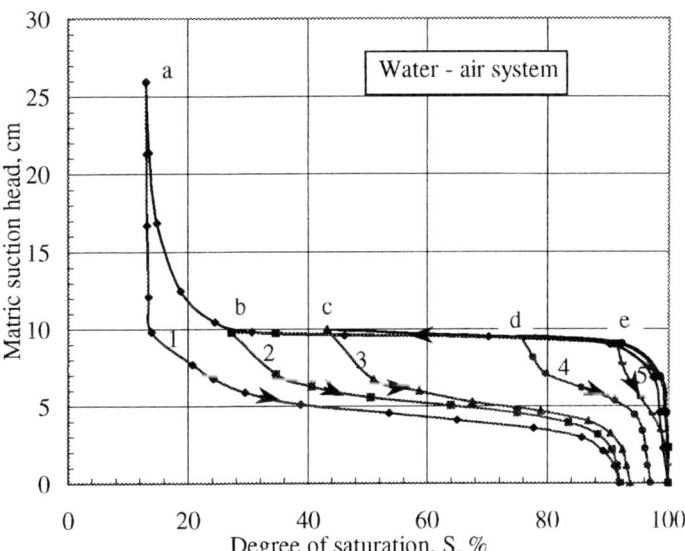

Figure 1: Matric suction versus degree of saturation

In case of water – LNAPL system, hysteresis in the relationship between matric suction and degree of saturation is again clear, as shown in Figure 2. The amount of hysteresis is, however, only 3.0cm, which is less than that for water – air system (5.5cm, see Figure 1). Furthermore, the amount of hysteresis appears to decrease with subsequent drying and wetting cycles. Inspection of Figure 2 shows that the residual saturation for water is 17.0% and the amount of trapped LNAPL in the sample is 11.0%. The bubbling pressure is estimated to be 4.0cm.

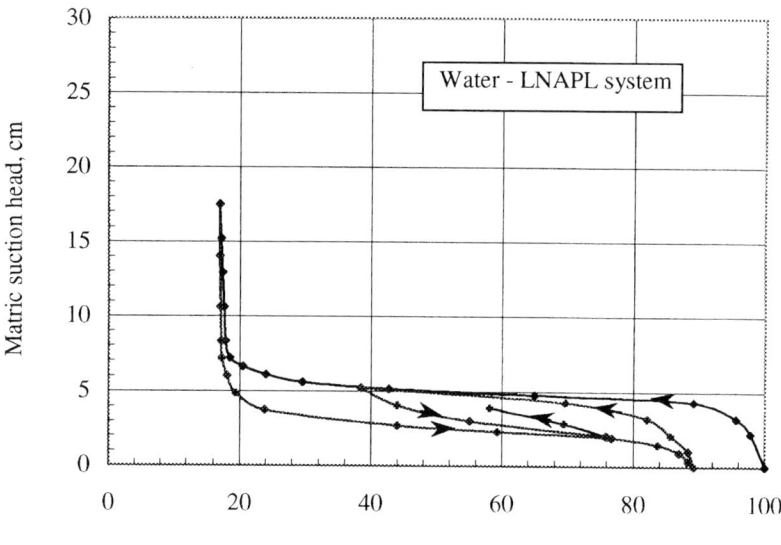

Figure 2: Matric suction head versus degree of saturation

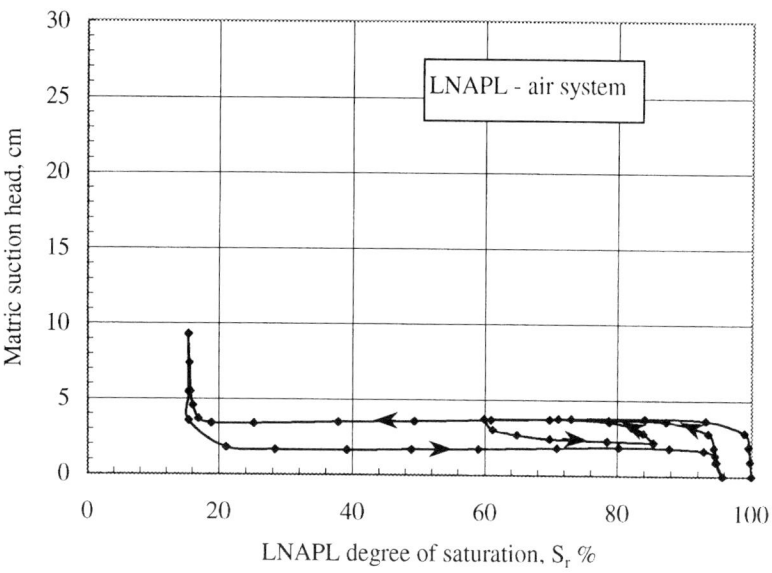

Figure 3: Matric suction head versus degree of saturation

Figure 3 shows the results for LNAPL – air system. In this case, the LNAPL suction head was converted to equivalent value of water head for the purpose of comparison. The hysteresis in the relationship of matric suction head and degree of saturation is again clear, but it is only 2.0 cm, implying that the hysteresis is the minimum for LNAPL – air system. The residual saturation of LNAPL and trapped air are 15.4 and 4.0% respectively. The bubbling pressure is 3.5 cm.

The experimental results suggest that there is no single relationship between matric suction and degree of saturation that can be implemented in a numerical model. Extensive research at the University of Bradford is in progress to analysis and quantify the effect of hysteresis on LNAPL migration using 2 and 3-D laboratory models.

After 3days   After 5 days   After 6 days

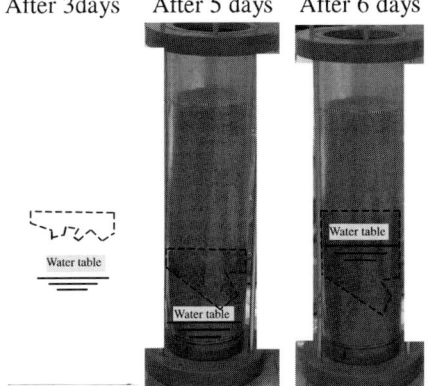

| Period | Water table level | LNAPL thickness |
days	cm	cm
3	8.5	1.5
5	3.4	4
6	11	9

Table (1) results of column experiments

Figure (4) Images of LNAPL distribution

Figure (4) shows the distribution of LNAPL under fluctuations of groundwater table. The LNAPL zones are delineated by the dashed lines. The thickness of LNAPL zone was measured after equilibrium condition was established between water and LNAPL distribution. Once LNAPL is introduced on the sand surface, it starts migrating downwards under the influence of gravity. Eventually, it distributes itself in 1.5cm thick zone above the top of the capillary fringe. Lowering water table from 8.5cm to 3.4cm causes LNAPL to redistribute itself in a relatively larger area than the initial zone. Finally, with rising water table, LNAPL redistribute itself under and above the water table. LNAPL zone became 9.0cm thick after 6 days (see figure 4). Table 1 summarises the results from column experiments.

DISCUSSION
The experimental results show that the relationships between matric suction and degree of saturation are different along the main drying and wetting paths. This difference suggested that there is hysteresis in the relationship, which is consistent with published results (see, for example, Abdul, 1988, Lenhard, 1992 and Sharma, 1998). Furthermore, it is clear that the relationship between matric suction and degree of saturation cannot be described as a linear relationship, but it takes the form of S-Shape curve. However, for the same porous medium, the amount of hysteresis varies depending mainly upon the interfacial tension of fluids present and the contact angle between each fluid and the soil grains during wetting and drying paths (see, for example, Mohamed and Sharma, 2000).

In this experimental programme three different fluid systems (water-air, water-LNAPL and LNAPL-air) were used and the amount of hysteresis varied for each system. For example, at 50% degree of saturation (see Figures 1, 2 and 3), the difference in the value of matric suction head between main drying and wetting paths for water-air, water-LNAPL and LNAPL-air are 5.5, 3.0 and 2.0 cm respectively for coarse sand samples. The amount of hysteresis appeared to reduce with further cycles of drying and wetting (see Figures 2 and 3).

The amount of trapped air increases with increasing value of the initial degree of saturation, when wetting commences from drying stage, as shown in Figure 1. For water-LNAPL system, trapped LNAPL was 11.0% at zero matric suction, whereas trapped air was 8% only for water-air system, implying that the amount of trapped non-wetting fluid varied for different fluid systems. The variation could be due to difference between the viscosity of a given non-wetting fluid and a wetting fluid.

In the column experiment, LNAPL first floats on the top of the capillary fringe with irregular interface. This suggested that LNAPL would flow in the sand column through preferential pathways, which are linked with the pore size distribution of the medium. During lowering of the water table from 8.5cm to 3.4cm, LNAPL distributed within 4.0cm. The migration of LNAPL, in this case, occurred due to gravity. However, raising the water table from 3.4cm to 11.0cm causes LNAPL to redistribute itself in 9.0cm thick zone. During the process of raising the water table, water pushes LNAPL upwards, trapping it either in single ganglia or large pockets. Finally, LNAPL redistributed itself in large zone. Results from column experiments were consistent with the results of matric suction and degree of saturation. Over a cycle of wetting and drying, there is a net increase in degree of saturation (see, for example, Sharma, 1998). This implies that with fluctuations of groundwater table, there would be a net increase in the degree of saturation, which in turn would enhance the migration of LNAPL. Results from column tests show that fluctuations of groundwater table could significantly increase the spread of LNAPL in subsurface. The overall spread would depend upon the speed of rising groundwater table, properties of fluids and porous medium.

CONCLUSIONS
Contamination of groundwater aquifers by Light Non-aqueous Phase Liquids (LNAPLs) is a serious problem in many areas around the world. A LNAPL migrates downwards under the influence of gravity and capillary forces. However, fluctuations of groundwater table significantly influence the distribution of LNAPL in unsaturated/saturated zones. An experimental investigation was carried out to obtain the relationships between matric suction and degree of saturation for water – air, water – LNAPL and LNAPL – air system. Additionally, column tests were performed to investigate the implications of groundwater fluctuations.

Following are the main conclusions from the experimental investigation:
- The relationship between matric suction and degree of saturation is hysteretic for all the three fluid systems. The relationship is non-linear and approximates to S-shape curve depending upon the saturation history.
- The amount of hysteresis varied for all the three fluid systems with minimum hysteresis for LNAPL – air system. The amount of hysteresis reduced with further drying and wetting cycles, but it needs further investigation.
- The bubbling pressure head decreased with decreasing interfacial tension between the fluids.
- The amount of trapped air depends upon the initial degree of saturation from drying stage.
- The column results appeared to have good correlation with the results of matric suction and degree of saturation. Fluctuations of groundwater caused LNAPL to spread out in large areas in sandy soils.

REFERENCES

Abdul, S.A. 1988. Migration of petroleum products through sandy hydrogeologic systems. Ground Water Monitoring Review 8(4):73-81.

Bear, J. 1979. Hydraulics of groundwater. McGraw Hill Int.

Brooks, R.H. and Corey, A.T. 1964. Hydraulic properties of porous media. Colorado State University. Hydrology papers (3): 27.

Lenhard, R.J. 1992. Measurement and modelling of three-phase saturation-pressure hysteresis. Journal of Contaminant Hydrology 9:243-269.

Mitchell, R.J. and Mayer, A.S. 1998. The significance of hysteresis in modelling solute transport in unsaturated porous media. Soil Sci. Soc. Am. J. 62:1506-1512.

Mohamed, M.H.A. and Sharma, R.S. 2000. Influence of hydraulic hysteresis on the migration of contaminants in subsurface. International Conference on Geo-engineering in Arid lands, AUE :395-402.

Oostrom, M. and Lenhard, R.J. 1998. Comparison of relative permeability-saturation-pressure parametric models for infiltration and redistribution of a light nonaqueous-phase liquid in sandy porous media. Advances in Water Resources 21(2):145-157.

Schroth, M.H., Istok, J.D., Ahearn S.J. and Selker, J.S. 1995. Geometry and position of light nonaqueous-phase liquid lenses in water-wetted porous media. Journal of Contaminant Hydrology 19:269-287.

Sharma, R.S. 1998. Mechanical behaviour of unsaturated highly expansive clays. DPhil thesis, University of Oxford, U.K.

Sharma, R.S. 2000. Keynote paper: Migration of non-aqueous phase liquids in subsurface. International Conference on Geoenvironment, Muscat, Sultanate of Oman. 2:571-583.

Steffy, D.A., Johnston, C.D. and Barry, D.A. 1998. Numerical simulations and long-column tests of LNAPL displacement and trapping by a fluctuating water table. Journal of Soil Contamination 7(3):325-356.

An assessment of the thermal and hydraulic interaction between a clay buffer and host rock.

H.R. THOMAS, P.J. CLEALL and H.P. MITCHELL
Geoenvironmental Research Centre, Cardiff School of Engineering, Cardiff University, P.O. Box 925, Newport Road, Cardiff, CF24 0YH, Wales, U.K.

N. A. CHANDLER and D.A. DIXON
Whiteshell Laboratories, AECL-Research, Pinawa, Manitoba, Canada, R0E 1L0.

ABSTRACT

A numerical analysis of the interaction between a clay buffer and the host rock of a deep geological nuclear waste repository is presented. A coupled thermo-hydro (TH) axisymmetrical finite element modelling approach has been adopted. In order to verify this approach, results from the numerical model have been compared with experimental values taken from a large in ground experiment performed by Atomic Energy of Canada Ltd (AECL). The theoretical formulation adopted and the preliminary results from the numerical modelling are presented. A good correlation between simulated and measured results was found for the temperature and moisture fields.

INTRODUCTION

Several decades of research and development have been dedicated to high level nuclear waste disposal. Currently the preferred option of a number of disposal authorities is a well-sited and engineered geological repository, (Backblom and Martin, 1999). The main purpose of this method of disposal is to ensure that no harmful material can escape to the biosphere in concentrations that will impose risk to human health or the local ecology. Although the host rock is generally highly impervious to water flow, construction of the caverns, tunnels and boreholes may create excavation damage zones, and consequently increased hydraulic conductivities. For these reasons the rock alone cannot be relied upon as a barrier. Therefore the interaction between the clay barrier, and the host rock is a significant focus for research

For many proposed repository designs a clay-based buffer will be placed in direct contact with the host rock. In this paper a bentonite buffer material is investigated. Such materials have a tendency to swell upon wetting, and exhibit large suctions when unsaturated. These and other desirable properties have resulted in the selection of bentonite as a candidate buffer material by a number of National waste disposal agencies, for example; AECL, Canada (Graham et. al. 1997); SKB, Sweden (SKB, 1992); and Nagra, Switzerland (Kristallin, 1994). The buffer is initially unsaturated when placed. However, it is thought that during the lifetime of the disposal vault the water table would return to the pre-construction level, and the buffer would become saturated. The time taken from placement to saturation is strongly influenced by the interaction between the buffer and the host rock. This paper presents a first assessment of the initial stages of the problem. The numerical approach adopted and the preliminary results achieved are presented here.

Geoenvironmental impact management, Thomas Telford, London, 2001.

THEORY

Unsaturated soil may be considered as a three-phase system comprising liquid, gas and solid particles. For this application, the liquid phase is considered to be water containing dissolved air, and the gas phase is considered to be a mixture of water vapour and dry air. A theoretical formulation for the flow of moisture, dry air and heat is given below. The primary variables of the model are pore water pressure, u_l, pore air pressure, u_a, and temperature, T.

Heat transfer

The law of conservation of energy states that the time derivative of heat content, Ω_H, is equal to the spatial derivation of heat flux, $\mathbf{Q}$, expressed as

$$\frac{\partial \Omega_H}{\partial t} = -\nabla \mathbf{Q} \tag{1}$$

where the heat content of a moist soil, includes latent heat of vapourisation and the classical components of heat capacity. Following the approach presented by Ewen and Thomas (1989) H_c may be expressed as

$$H_c = (1-n)C_{ps}\rho_s + n\left(C_{pl}S_l\rho_l + C_{pv}S_a\rho_v + C_{pda}S_a\rho_{da}\right) \tag{2}$$

where C_{ps}, C_{pl}, C_{pv} and C_{pda} are the specific heat capacities of solid particles, liquid, vapour and dry air respectively, and ρ_s, ρ_l, ρ_v, ρ_{da} are the densities of solid particles, pore-liquid, water vapour and dry air respectively. Following the approach by Thomas & King (1991), heat flux may be defined as

$$\mathbf{Q} = -\lambda_T \nabla T + \left(\mathbf{v}_v \rho_v + \mathbf{v}_a \rho_v\right)L + \left(C_{pl}\mathbf{v}_l\rho_l + C_{pv}\mathbf{v}_v\rho_l + C_{pv}\mathbf{v}_a\rho_v + C_{pda}\mathbf{v}_a\rho_{da}\right)(T - T_r) \tag{3}$$

where λ_T is the coefficient of thermal conductivity of unsaturated soil, and $\mathbf{v}_v$, $\mathbf{v}_a$, and $\mathbf{v}_l$ are the velocities of water vapour, air and liquid respectively.

Moisture transfer

The governing equation for moisture transfer includes the following fluxes; liquid flux, vapour flux, and vapour flux arising from the bulk flow of vapour due to the movement of pore air. It may be expressed as

$$\frac{\partial(\rho_l nS_l)}{\partial t} + \frac{\partial(\rho_v n(S_l - 1))}{\partial t} = -\rho_l \nabla.\mathbf{v}_l - \rho_l \nabla.\mathbf{v}_v \quad \nabla.\rho_v \mathbf{v}_a \tag{4}$$

Darcy's Law is used to define the velocities of the pore liquid and the pore air. The definition of vapour velocity follows the approach presented by Thomas & King (1991).

Numerical formulation

The governing differential equations are solved spatially using a finite element technique and temporally using a finite difference technique. In this paper the Galerkin weighted residual approach is adopted and the domain is discretised using eight noded isoparametric axisymmetric elements. The temporal discretisation utilises a fully implicit mid-interval backward difference time-stepping algorithm. The coupled processes are solved simultaneously within a time step and a converged solution is assumed to have been achieved when the difference between successive iterations falls below a certain tolerance.

COMPASS Code

The numerical modelling was performed using COMPASS (Code for modelling partially saturated soil) which has the capability to model fully coupled thermal, hydraulic, mechanical and chemical problems.

BUFFER/CONTAINER EXPERIMENT

Atomic Energy of Canada Ltd (AECL) performed a large in-ground experiment between 1991 and 1994. The aim of the experiment was to examine how heat affects the performance of the dense sand-bentonite buffer. The experiment involved the placement of a full-size heater (representing a nuclear fuel waste canister) in a 1.24m diameter, 5m deep borehole back-filled with 50:50 mix of bentonite and sand buffer material. The buffer was compacted into the borehole at a water content of 18%. The experiment may be divided into three distinct phases; the construction of the 240m deep underground facility, a 'dwell' period to allow stabilisation of pore water pressures between the buffer, and the rock, and a heating period of 896 days.

Thermal results

The initial rock temperature of 11°C was affected by an average air temperature of 15°C within the tunnels and caverns. At the start of the heating phase a constant power of 1000W was applied to the heater. After 26 days the power to the heater was increased to 1200W in order to reach the required heater temperature of 85°C. Rapid heating occurred within the experiment, although it was felt that steady state conditions were not in place at the end of the experiment, (Graham et. al., 1997).

Hydrogeological results

The construction of the underground caverns and drilling of the emplacement borehole resulted in a drawing down of water pressures from an initial value of 21MPa, affecting a few metres of rock close to the excavation. Following emplacement of the unsaturated buffer there was a 'dwell' period of 170 days to allow the water pressure in the buffer and rock to stabilise prior to heating. During this period there was a desaturation of the rock close the granite/buffer interface, and wetting of the buffer reducing the average suction from 3.98MPa to 3.3MPa. Following activation of the heater the pore water pressure in the rock increased rapidly, and then gradually dissipated. These early peaks may be due to thermal expansion of water in the rock. During the heating phase there was also considerable drying in the buffer close to the heater, and wetting of the outer edges.

SIMULATION OF THE EXPERIMENT

Material parameters for the sand and buffer materials were specified by Graham, (1997). However the unsaturated material parameters for the granite have not, to the author's knowledge, been directly measured. A review of the literature was therefore undertaken. Freig and Vomvoris (1994) had carried out a laboratory experiment to determine the hydraulic material properties of granite from the Central Area region. Gens et. al. (1998) fitted an equation to this data. Adopting this approach and using a threshold pressure of 5.5MPa based on the intrinsic permeability for the granite (Davies, 1991), the following relationships for degree of saturation and hydraulic conductivity were established,

$$S_l = \left[1 + \left(\frac{u_a - u_l}{P_0} \right)^{1/(1-\beta)} \right]^{-\beta} \quad \text{when } (u_a - u_l) \geq 0 \tag{5}$$

where P_0 is the air entry value, and β is a material parameter, taken as 0.33 for granite.

$$k_l = k_s S_l^{1/2} \left(1 - \left(1 - S_l^{1/\beta} \right)^\beta \right)^2 \tag{6}$$

where k_s is the saturated hydraulic conductivity.

Numerical modelling

The 15m by 16m axisymmetric domain shown in Figure 1 was adopted, and discretised into 1165, 8-noded, isoparametric, finite elements. The numerical modelling followed the phases of construction, and the influence from the construction of the underground caverns was considered first. Boreholes within the rock measured the initial drawdown of the pore water pressures following drilling of the borehole, and a good correlation was achieved between experimental and numerical results. Following placement of the buffer there was a dwell period of 170 days to allow the unsatured buffer to stabilise with the saturated rock at the interface prior to heating. The numerical modelling of this period was used to assess the selected material parameters for the granite.

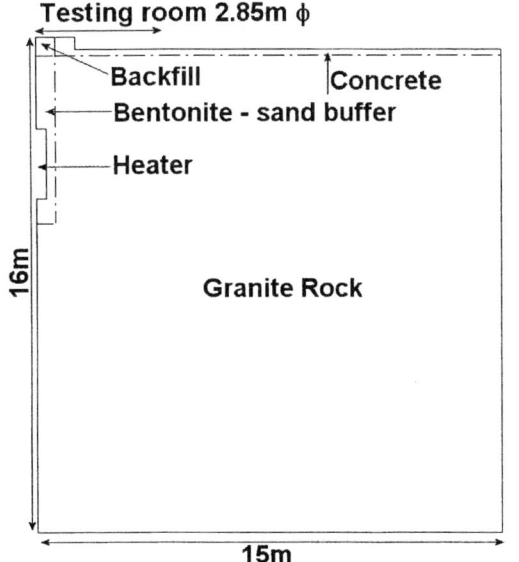

Figure 1: Axisymmetric domain

Although not shown, a good correlation was found between results for the dwell period. Desaturation occured within the rock for a distance of approximately 1.5m from the centre of the heater. It was noted within the experiment that the average suction within the buffer dropped from 3.98MPa to 3.3MPa during the dwell period. The numerical results show that the average suction within the buffer was 3.34MPa at the end of this period. The adopted relationship for granite was deemed reasonabale.

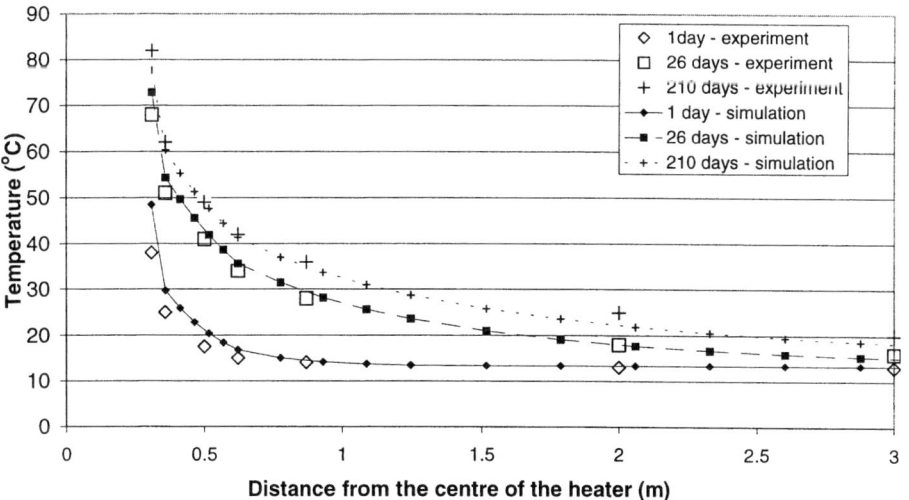

Figure 2: Temperature distributions along the mid-height of the heater

A coupled temperature and mass analysis for the heating phase was then undertaken. The results for the first 210 days of heating are shown in Figure 2. It can be observed that there is a good correlation between the numerical results and the experimental data throughout the sand, buffer and rock layers. In particular the effect of the varying value of thermal conductivity, due both to the different materials and the effect of drying near to the heater, can be clearly seen with distinct changes of gradient in the temperature profile.

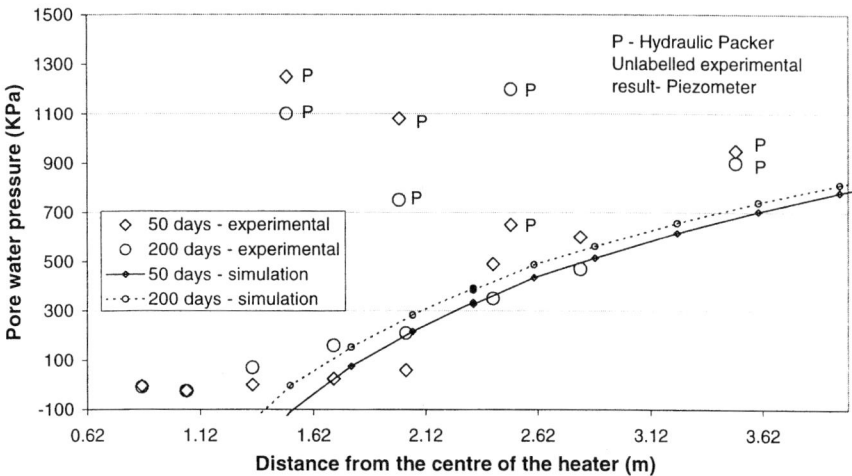

Figure 3: Pore water pressure profiles with time during the heating period

Results for the pore water pressures during the heating period are shown in Figure 3. Two sets of instrumentation were used to measure the pore water pressures within the rock; hydraulic packers, and hydraulic and pneumatic piezometers. It can be seen that these instruments gave what appear to be conflicting results. In Figure 3 it should be noted that all experimental values above 650KPa are from the hydraulic packers, whilst those below 650KPa are from the piezometers. The results at 200 days closely match the data from the piezometers. It was felt, by the experimentors, that the high values in the early packer readings may be due to localised increases in water pressure in the pores of the rock, due to the heating front, (Graham et.al., 1997). It can be seen that in the numerical results the pore water pressure is still increasing after 200 days and the results have not yet peaked. The hydraulic packers reached peak values relatively quickly. However the piezometers closer to the buffer took longer to reach their maximum values.

It is claimed that these results give a reasonable prediction of the behavior in the rock. Further analysis is now being carried out, and in particular an investigation of thermal expansion of the pore water is recognised to be required. A comparison of the suctions within the buffer is not made at this stage as it was felt that the psychrometers may have been affected by vapour pressure during the course of the experiment, resulting in lower readings than may otherwise have been expected, (Graham et. al., 1997).

CONCLUSIONS

A set of results have been presented for a simulation of the temperature and moisture phases of a clay buffer/host rock interaction zone. For validation purposes these results have been compared against the experimentally measured results from AECL's buffer/container experiment

Due to the uncertainty in the material parameters for the granite, a literature review was carried out and a semi-empirical equation based on a threshold pressure concept was adopted. During the dwell period the rock desaturates close to the rock/buffer interface, whilst the buffer becomes wetter. A good correlation between results was achieved for the pore water pressures within the rock and the average suction within the buffer.

During the experiment, hydraulic packers and hydraulic and pneumatic piezometers were used to measure the pore water pressures within the rock. The numerical results after 200 days of heating show a good correlation with the results from the piezometers. It is felt that the packers may be showing higher readings reflecting thermal expansion of water within the pores of the rock.

A good correlation between simulated and measured results was found for the temperature field.

It is felt that the buffer/rock interaction requires further analysis with an investigation of the thermal expansion of the pore water and of the stress strain behaviour of the system being required.

REFERENCES

Backblom, G., and Martin, C.D. (1999) "Recent experiments in hard rocks to study the excavation response: Implications for the performance of a nuclear waste geological repository" *Tunnelling and underground space technology* **14** 377-394.

Davies, P.B., (1991) "Evaluation of the role of threshold pressure in controlling flow of waste-generated gas into bedded salt at the waste isolation pilot plant", *Technical report* SAND-90-3246, Sandia National Laboratories, Albuquerque, New Mexico.

Ewen J., and Thomas H.R., (1989) "Heating unsaturated medium sand" *Geotechnique* **39**, No. 3, 455-470.

Frieg, B., and Vomvoris, S., (1994) "Investigation of hydraulic parameters in the saturated and unsaturated zone of the ventilation drift", *Technical report* 93-10, Nagra, Baden, Switzerland.

Gens, A., Garcia-Molina, A.J., Olivella S., Alonso, E.E., and Huertas, F., (1998) "Analysis of a full scale *in situ* test simulating repository conditions", *International Journal for Numerical and Analytical Methods in Geomechanics.*, **22**, 515-548.

Graham, J., Chandler, N.A., Dixon, P.J., Roach, To T., and Wan, A.W.L. (1997) "The Buffer/Container experiment: results, synthesis, issues", *Technical report* AECL-11746, COG-97-46-I.

Kristallin, I., (1994) "Safety assessment report", *Technical report* 93-22, Nagra, Switzerland.

SKB (1992) "SKB 91. Final disposal of spent nuclear fuel. Importance of the bedrock for safety" *SKB Technical report* 92-20. SKB, Stockholm.

Thomas H.R., and King S.D., (1991) "Coupled temperature/capillary potential variations in unsaturated soil" *ASCE Journal of Eng. Mech.* **117** No. 11, 2475-2491.

Coupled multispecies transport in geochemical problems

H. R. THOMAS, R. N. YONG, P. J. CLEALL and S.C. SEETHARAM
Geoenvironmental Research Centre, Cardiff School of Engineering, Cardiff University,
Newport Road, Cardiff, CF2 1XH, Wales.

ABSTRACT

This paper deals with the development of a theoretical model for the transport of multichemical species in unsaturated soil, including the effect of complex geochemical reactions. The numerical solution for the governing differential equation and the chemical interaction equation is briefly discussed, including the coupling between the transport and the geochemical model. An isothermal advection-diffusion and geochemical interaction problem is solved to demonstrate the capabilities of the model. The results show the ability of the model to predict the transport of chemicals, including the effect of sorption.

INTRODUCTION

Predicting the flow and fate of pollutants in the subsoil is crucial to enable engineering solutions to prevent damage to the environment (Yong, 2000). Pollutants' interactions with soil material can be varied and complex depending on the local environment and the type of soil present (Yong, 2000). Chemicals in transit may undergo dissolution or precipitation or get sorbed on to the surface of soil particles. Hence the movement of chemicals no longer becomes a transport problem alone, but one that includes the complex geochemistry.

Previous work (Thomas, et.al., 1997 ; Thomas, et.al., 2001a) presented the development of a theoretical and numerical formulation for coupled moisture and chemical transport, including geochemical interactions. The work presented in this paper extends the previous approach to include a more general multichemical species formulation, coupled with geochemical interaction models, which forms a part of a larger research program involving Thermo-Hydraulic-Chemical-Mechanical behaviour of soil (Thomas, et.al, 2001b).

In order to solve the transport problem, an existing finite element package COMPASS (COde for Modelling PArtly Saturated Soil) is further developed to include multichemical species transport. This is then coupled to a geochemical model, MINTEQA2 (Allison et.al., 1991). An isothermal problem is considered here involving both advection-diffusion of 2 chemical species, including geochemical reactions with the soil particles.

The coupled flow and geochemical model is compared with the transport model alone, to demonstrate the effect of the coupled model on the attenuation of chemicals due to the geochemical interactions with the soil particles

Geoenvironmental impact management, Thomas Telford, London, 2001.

THEORY

The soil is considered a three-phase porous medium containing, moisture, air and solid phase. The flow laws viz., Darcy's law and Fick's law are used to develop the governing equations for moisture, air and chemical solute transport in terms of the primary variables porewater pressure, pore air pressure, temperature and dissolved chemical concentration. The temperature formulation is presented elsewhere (Thomas and King, 1991).

Moisture Transport

Based on mass conservation the governing equation for the moisture phase can be stated as:

$$\frac{\partial \rho_l n S_l}{\partial t} + \frac{\partial \rho_v n S_a}{\partial t} = -\rho_l \nabla . \mathbf{v}_1 - \nabla .(\rho_l \mathbf{v}_v) - \nabla .(\rho_v \mathbf{v}_a) \tag{1}$$

Where, ρ_l is the density of water, ρ_v is the density of vapour, n is the porosity, S_l is the degree of saturation, S_a is the degree of saturation of pore air, $\mathbf{v}_1$, $\mathbf{v}_v$, $\mathbf{v}_a$ are the velocity vector for the water, vapour and dry air phase respectively.

The liquid flow law can be defined via a Darcy's law approach. In order to include the effect of osmotic gradient due to the presence of multichemical species, an additional osmotic term is included.

$$\mathbf{v}_1 = \mathbf{v}_1^1 + \mathbf{v}_1^{c_d} = -K_l \left[\nabla \left(\frac{u_l}{\gamma_l} \right) + \nabla z \right] + \sum_{j=1}^{n} K_l^{c_d}{}^j \left(\nabla c_d{}^j \right) \tag{2}$$

Where, j=1,2,.....,n number of species. $K_l^{c_d}{}^j$ is the hydraulic conductivity due to concentration gradient and K_l being the hydraulic conductivity of water.

Introducing the liquid flow law and rearranging in terms of the primary variables, the governing equation (1) can be rewritten as,

$$C_{ll} \frac{\partial u_l}{\partial t} + C_{lT} \frac{\partial T}{\partial t} + C_{la} \frac{\partial u_a}{\partial t} = \nabla .[K_{ll} \nabla u_l] + \nabla . \sum_{i=1}^{n} \left[K_{lc_d}{}^i \nabla c_d{}^i \right] +$$
$$\nabla .[K_{lT} \nabla T] + \nabla .[K_{la} \nabla u_a] + J_l \tag{3}$$

Where, i=1,2,.....,n number of chemical species, u_a, u_l, T, and c_d represents pore air, pore water, temperature and dissolved chemical concentration respectively and J_l the gravity term.

Multichemical species transport

The governing equation for chemical solute transport for a single chemical species situation, can be written as (Bear and Verruijt,1987)

$$\frac{\partial (n S_l c_d)}{\partial t} + \frac{\partial (n S_l s_s)}{\partial t} = -\nabla .(\mathbf{q}_{c_d}) \tag{4}$$

Where, $\frac{\partial s_s}{\partial t}$ is the sink term, which represents the quantity of chemical sorbed on to the soil particles and $\mathbf{q}_{c_d}$ is the chemical flux.

The above general chemical solute transport equation is solved individually for each different chemical species. It should be noted that the advective velocity of the ith chemical species, due to the chemical gradient of other chemical species in the solution (i.e., the chemical gradient due to (n-i) number of species) is required.

Hence the equation (4) can be rewritten in a generalised form for the multichemical species as,

$$
C_{c_d l}^i \frac{\partial u_l}{\partial t} + C_{c_d a}^i \frac{\partial u_a}{\partial t} + C_{c_d c_d}^i \frac{\partial c_d^i}{\partial t} + C_{c_d b}^i \frac{\partial s_s^i}{\partial t} = \nabla.\left[K_{c_d l}^i \nabla u_l\right] + \nabla.\left[K_{c_d a}^i \nabla u_a\right]
$$
$$
+ \nabla.\left[K_{c_d c_d}^i \nabla c_d^i\right] + \nabla.\left[\sum_{j=1}^{n,(i \in n)} K_{c_d c_d}^j \nabla c_d^j\right] + \left[v_l \nabla c_d^i\right] + J_{c_d}^i
$$

(5)

For $i = 1, 2, \ldots, n$ number of chemical species, and $j = 1, 2, \ldots, n$ number of chemical species, excluding the ith species. $J_{c_d}^i$ represents the advective gravity term for the species i.

Geochemical model
MINTEQA2 is used as the geochemical model in this study. MINTEQA2 is based on an equilibrium constant method and offers both electrostatic and non-electrostatic models to represent sorption, which forms the fourth term of the LHS of the governing equation for chemical transport (equation (5)).

The mass action expression for a system of n independent components for formation of m species can be represented as:

$$K_i = \{S_i\} \Pi X_j^{-a}{}_{ij}$$

(6)

where K_i is the equilibrium constant for the formation of species i, $\{S_i\}$ the activity of species i, X_j the activity of component j, a_{ij} the stoichiometric coefficient of component j in species i and Π indicates the product over all components in species i.

In addition to the mass action expression, the set of n independent components is governed by n mass balance equations of the form

$$Y_j = \Sigma\, a_{ij}\, C_i - T_j$$

(7)

where T_j is the total concentration of component j and Y_j is the difference between the calculated total concentration of component j and the known analytical total concentration of component j.

For the electrostatic models, charge balance equations are defined as

$$Y_\sigma = \Sigma\, a_{i\sigma} C_i - T_\sigma$$

(8)

where $a_{i\sigma}$ is the stoichiometry of the electrostatic component pertaining to σ in species i where σ is the surface charge. The solution is arrived by solving for equations 6 and 7 or 6 and 8 for the non-electrostatic and electrostatic models respectively.

NUMERICAL SOLUTION
The numerical solution of the governing equation is achieved by using a finite element method for the spatial discretisation and a finite difference time stepping scheme for temporal discretisation. The Galerkin weighted residual method is employed to formulate the finite element discretisation.

For the two governing equations described above, this method yields, in a concise notation,

$$A\phi + B\frac{\partial \phi}{\partial t} + C = \{0\}$$

(9)

where ϕ represents the variable vector $\{u_l, c_d\}^t$ and $\mathbf{A}$, $\mathbf{B}$ and $\mathbf{C}$ are matrices and vectors representing the discretised components of the governing equation.

To solve equation (9) an implicit mid-interval backward difference algorithm is implemented, since it is found to provide a stable solution for highly non-linear problems (Thomas and King, 1991). This approach yields

$$\phi^{n+1} = \left[\mathbf{A}^{n+\frac{1}{2}} + \frac{\mathbf{B}^{n+\frac{1}{2}}}{\Delta t} \right]^{-1} \left[\frac{\mathbf{B}^{n+\frac{1}{2}} \phi^n}{\Delta t} - \mathbf{C}^{n+\frac{1}{2}} \right] \tag{10}$$

where the superscript represents the time level. A solution can be found to equation (10) via a predictor-corrector iteration scheme (Thomas and King, 1991).

A solution to the chemical interaction equations is achieved by Newton-Raphson approximation method (Allison et.al., 1991). The coupling of the transport model with the geochemical model is achieved by sequentially solving the transport and then the chemical equations for each of the integration points for each iteration of the predictor corrector scheme mentioned above. With appropriate initial and boundary conditions the set of non-linear coupled governing differential equations can be solved via the numerical solution presented above.

APPLICATION

The objective of this example is to demonstrate the simulation of an isothermal 1D advection-diffusion problem including the modelling of geochemical interaction. Two chemical components are considered in this problem, namely, Copper and Zinc. The Langmuir model is used to describe the geochemical reactions, as this method allows a finite concentration of sorbent sites to be defined and hence can predict competitive sorption between the chemicals. The geometry of the problem a 1mm by 100mm column is defined with 100 elements each of 1x1 mm. The initial and boundary conditions for the problem is defined as follows:

Initial conditions:
For both Cu and Zn, $c_d = 1$ mol/m^3, and Porewater pressure = -5000 Pa for $0 \leq x \leq 0.001$ at t=0
Boundary condition:
For both Cu and Zn, $c_d = 1$ mol/m^3 at x = 0, t $\geq$ 0, Porewater pressure gradient = 9.81 Pa/mm
Material properties (for a medium clay):
Porosity=0.4, Hydraulic conductivity = 4.568 x 10^{-12} m/s, Dispersion coefficient= 6.342x10^{-10} m^2/s and 6.342x10^{-11} m^2/s, Initial Sorbent concentration = 10^{-3} moles/Litre, LogK value for Zn interaction = 1.4186, LogK value for Cu interaction = 2.6885, Enthalpy of the reaction is assumed to be zero

Three sets of problems were solved based on the above boundary and initial conditions.

A graph of dissolved concentration of the two chemicals against the distance from the source for transport coupled to a geochemical problem, is shown in figure 1, after 25000 seconds. This shows different gradients in each curve indicating different rates of transport and hence different rates of sorption by the soil particles depending on the values defined for the initial sorbent sites and the logK (log of the equilibrium constant) value of the chemical species. In this example it can be seen that Copper is preferentially adsorbed in comparison to Zinc as would be expected considering the higher logK value for Copper-Sorbent reaction. A lower value of logK for Zn interaction only suggests that the equilibrium concentration of Zn component in the solution is more than Cu leading to a higher mobility of Zn in the porewater.

The same problem was simulated with no geochemical model so as to solve an advection-diffusion transport problem, with the same diffusion coefficients. The profile of chemical concentration for this scenario is also shown in figure 1. This shows a less steep gradient in the curve compared to a coupled curve indicating that there is no attentuation of chemicals in the uncoupled problem.

The transport only problem was repeated with different diffusion coefficients. The model showed different rates of movements of chemicals as seen in the figure 2 due to different diffusion coefficients. Higher rates of movement were observed with higher the diffusion coefficient.

CONCLUSIONS
Three sets of application problems were considered, covering transport coupled to geochemical problem and transport alone. The model was able to demonstrate the effect of sorption on the chemical movement. In addition the model was able to predict different rates of movements at different diffusion coefficients. Overall, this model demonstrated that it is capable of representing a combination of advection-diffusion-sorption processes involved in the chemical solute transport problem. Further work needs to be carried out to validate the model against experimental results, for various scenarios involving multichemical speices.

REFERENCES
Bear, J and Verruijt, A (1987) *Modelling groundwater flow and pollution*. Dordrecht, D. Reidel Publishing Company.

Thomas, H.R. and King S.D. (1991). *Coupled temperature/capillary potential variations in unsaturated soil.* ASCE, Journal of Eng. Mech. 117(11): 2475-2491.

Allison, J.D., Brown, D.S. and Kevin J. Novo-Gradac (1991), *MINTEQA2 user manual version 3.0*, Environmental Research Laboratory, US EPA

Thomas, H.R., Yong, R.N. and Hashm A.A. (1997), *Numerical modelling of contaminant transport in Unsaturated soil,* Proc. Int. Conf on Geoenviromental Engineering, Cardiff. Thomas Telford, London, pp 278-283

Yong R.N. (2000). *Geoenvironmental Engineering, Contaminated Soils, Pollutant Fate, and Mitigation, CRC Press*

Thomas, H.R., Yong, R.N. and Hashm, A.A. (2001a), *Modelling the transport of lead in partly saturated soil,* Proc. Int. Symposium on Suction, Swelling, Permeability and Structure of clays – Clay science for Engineering, Adachi and Fuke (eds) Balkema, Rotterdam, pp341-346

Thomas, H.R., Cleall, P.J. and Hashm, A.A. (2001b). *Thermal / hydraulic /chemical / mechanical (THCM) behaviour of partly saturated soil.*Proceedings of the 10th International Conference on Computer Methods andAdvances in Geomechanics, IACMAG 2001, Tuscon, Arizona, 7-12 January 2001

ACKNOWLEDGEMENTS
The work presented has been carried out as a part of a research programme supported by the European Regional Development Fund. This support is gratefully acknowledged.

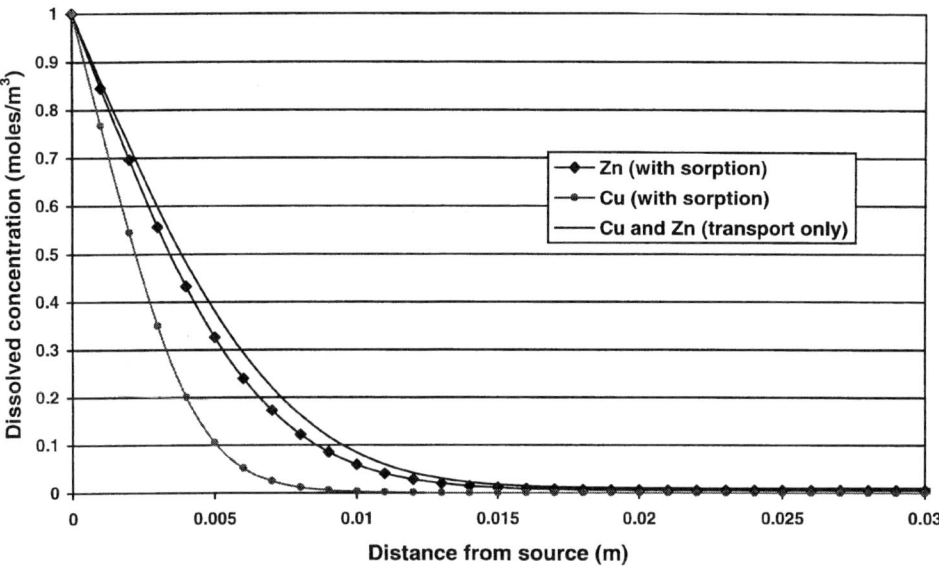

Figure 1 : Concentration profile of Cu and Zn with and without geochemical interaction (with same diffusion coefficients

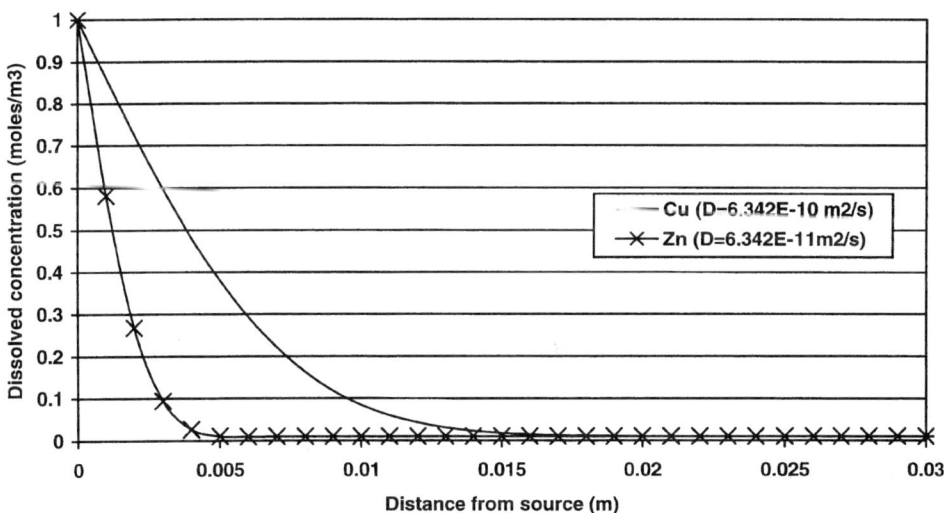

Figure 2 : Concentration profiles of Cu and Zn without geochemical interaction (with different diffusion coefficients

Non-isothermal migration of multiple gas species in partly saturated soil

Z. ZHOU, P. J. CLEALL and H.R. THOMAS
Geoenvironmental Research Centre, Cardiff School of Engineering, Cardiff University, PO Box 925, Newport Road, Cardiff, CF24 0YF, Wales, UK.

ABSTRACT

A theoretical formulation of multi-gas flow coupled with heat and moisture transfer in partly saturated soil is presented in this paper. The theory is developed for the simulation of generic gas transfer problems in porous materials so that a wide range of problems may be simulated. Suitable numerical methods (FEM) are used to solve the theoretical formulation. A simple boundary value problem with raising temperature and raising one gas pressure as two driving forces is analysed. Results are presented showing the interactions among the water pressure, gas pressures and temperature. Particularly, in this case, the effects of temperature and pressure of gas one on pore water pressure and pressure of gas two are observed from the numerical results

INTRODUCTION

Contaminant gas transfer is an important issue of environmental problems. A number of model have been developed by researchers (Moore, Rai and Alzaydi, 1979; Findikakis and Leckie 1979; Metcalfe and Farquhar, 1987, Ferguson and Thomas, 1997; Thomas and Ferguson, 1999 to simulate simple gas transfer problems. In many environmental problems gas migration involves the movement of a number of gases, whether it happens inside a waste repository across a clay liner or within ground adjacent to waste. Previously developed models have no been general or flexible enough to simulate the problems with any number of gas species. A theory able to consider the migration of multiple gas species is thus required for the investigation of such problems.

This paper presents a general theory for multiple gas species migration in unsaturated soil base upon a mechanistic approach. The diffusion of each gas species is described via Fick's law. The bulk flow of the gas mixture is described via Darcy's law approach. In order to include the effect of temperature and pore water of the soil, standard theory of heat and moisture transfer are also included. This allows the couplings between the multiple gas species, hydraulic and temperature fields to be fully incorporated. A numerical solution of the resulting governing differential equations is achieved via the finite element and finite difference methods. A computer program has been developed to implement the numerical solution. A simple problem is then presented with a fixed temperature and a fixed gas species pressure boundary condition applied to give two driving forces for flow. Any variation of pore water pressure or of the second gas species are caused by coupling effects. As such, the validity of the theory and the correctness of the computer program may be tested from the preliminary results of this simulation.

Geoenvironmental impact management, Thomas Telford, London, 2001.

THEORETICAL FORMULATION

Unsaturated soil is considered here as a three-phase system comprised of liquid, gas and solid grains. The liquid is assumed to be water and may contain certain amounts of dissolved gases. The gas phase may have any number of independent gas species plus a non-independent gas specie, namely, vapour. Each individual gas component is considered as an ideal gas. Although the focus of this paper is on gas transfer, moisture and heat flow are also considered.

Gas transfer

Although vapour is one of the gas species, its transfer will be described in section of moisture transfer. In a multi-component gaseous system, pressure driven flow of the gas mixture, $\mathbf{v}^b$, is described here via a Darcy's law approach, diffusive flow of each of the individual components is described by Fick's law, transport of dissolved gases are described by Henry's law. The total gas flux is thus

$$\mathbf{v}_i = \mathbf{v}_i^b + \mathbf{v}_i^d + \mathbf{v}_i^s$$
$$= \chi_i (k_a \nabla u_a) + k_i \nabla u_i + H_i v_l$$

$$(1)$$

where χ_i is the concentration of gas species i, k_a is the gas permeability and u_a is the total gas pressure, k_i is the gas diffusivity of gas i and u_i is the partial gas pressure of gas species i, H_i is the Henry's constant and v_l is the liquid velocity.

The law of conservation of mass leads to a governing equation for each gas species which can be expressed as

$$\frac{\partial\left[(\theta_i + H_i\theta_l)\rho_i\right]}{\partial t} = -\nabla.\left[\rho_i(v_i)\right] + R_i$$

$$(2)$$

where θ_i is the volumetric content of the gas i, θ_l is the volumetric water content, ρ_i is the density of the gas i and R_i is the source term.

Moisture transfer

Adopting the approach proposed by Darcy (1856), the velocity of liquid water can be expressed for a multiphase flow in unsaturated soil as

$$v_l = -k_l\left[\nabla\left(\frac{u_l}{\gamma_l}\right) + \nabla z\right]$$

$$(3)$$

where v_l is the liquid velocity due to pressure and elevation heads, k_l is the unsaturated hydraulic conductivity, γ_l is the unit weight of liquid and z is the elevation.

The contribution to vapour transport by diffusive flow is dealt, with in this study, by a flow law proposed by Philip and de Vries (1957) and extended by Ewen and Thomas (1989),

$$v_v^d = \frac{D_{atms}v_v n}{\rho_l}\left(\rho_0\frac{\partial h}{\partial s}\right)\nabla u_l - \frac{D_{atms}v_v n(\nabla T)_a}{\rho_l \nabla T}\left(h\frac{\partial\rho_0}{\partial T} + \rho_0\frac{\partial h}{\partial T}\right)\nabla T - \frac{D_{atms}v_v n}{\rho_l}\left(\rho_0\frac{\partial h}{\partial s}\right)\nabla u_a$$

$$(4)$$

where D_{atms} is the molecular diffusivity of vapour through air, n is the porosity, v_v is a mass flow factor, $(\nabla T)_a/(\nabla T)$ is the microscopic pore temperature gradient factor, ρ_0 is the density of saturated water vapour, h is the relative humidity, g is the gravitational constant and T is the temperature.

Applying the law of conservation of mass yields

$$\rho_1\frac{\partial(\theta_1)}{\partial t}+\frac{\partial(\rho_v\theta_a)}{\partial t}+\rho_1\nabla.\mathbf{v}_1+\rho_1\nabla.\mathbf{v}_v^d+\nabla.(\rho_v\chi_v\nabla u_a)=0$$

(5)

where θ_1 is the volumetric water content and θ_a is the volumetric content of the gas mixture.

Heat transfer

The law of conservation of energy dictates that the time derivative of the heat content, Ω, is equal to the spatial derivative of the heat flux, $\mathbf{Q}$. Mathematically this can be expressed as

$$\frac{\partial(\Omega)}{\partial t}=-\nabla.\mathbf{Q}$$

(6)

The heat content of moist soil per unit volume can be defined as (Thomas and He, 1995)

$$\Omega=H_c(T-T_r)+LnS_a\rho_v$$

(7)

where L is the latent heat of vaporisation. Following the approach presented by Ewen and Thomas (1989) the heat capacity of unsaturated soil H_c at reference temperature, T_r, can be expressed to include the contribution of each individual species as

$$H_c=(1-n)C_{ps}\rho_s+n\left(C_{pl}S_l\rho_l+C_{pv}S_v\rho_v+\sum_{i=1,i\neq v}^{v}C_{pi}S_i\rho_i\right)$$

(8)

where C_{ps}, C_{pl}, C_{pv} and C_{pi} are the specific heat capacities of solid particles, liquid, vapour and gas components air respectively, ρ_s is the density of solid particles, v is the total number of gases and v represents vapour.

The heat flux per unit area, $\mathbf{Q}$, is defined to include the contribution of each individual species as

$$\mathbf{Q}=-\lambda_T\nabla T+\left(\mathbf{v}_v\rho_v+\mathbf{v}^b\rho_v\right)L+\left(C_{pl}\mathbf{v}_l\rho_l+C_{pv}\mathbf{v}_v\rho_v+\sum_{i=1,i\neq v}^{v}C_{pi}\mathbf{v}_i\rho_i\right)(T-T_r)$$

(9)

where λ_T is the coefficient of thermal conductivity of unsaturated soil.

NUMERICAL SOLUTION

The set of partial differential equations (2), (5) and (6) is highly non-linear and it is therefore very difficult to derive an analytical solution for general cases. The finite element method is used to obtain a numerical solution. Spatial discretisation is achieved with the Galerkin weighted residual method. Temporal discretisation is achieved via a backward mid-interval time difference scheme. The resulting set of non-linear equations are solved via a predictor corrector iteration scheme. The numerical solution has been implemented in the computer code COMPASS. The code COMPASS and the algorithm of the numerical solution have been extensively verified during the last decade.

NUMERICAL ANALYSIS OF AN ASSUMED PROBLEM

Application

The problem considered here is a one-meter deep soil column with boundary conditions set for the top and the base of the column. The soil has been discretised into 100 equally sized iso-parametric elements. Two primary gas species are assumed in this case to demonstrate the capability of the approach. The species do not however relate to any specific gas. Temperature of 293 K is fixed at the top boundary and 287 K at the bottom boundary. Pore water pressure, gas pressures for two gases are only fixed at the bottom boundary with the value of -1.0×10^5 Pa, 7.0×10^4 Pa and 5.0×10^4 Pa respectively. The initial values for temperature, pore water pressures and two gas pressures are 287 K, -1.0×10^5 Pa, 0.5×10^5 Pa and 0.5×10^5 Pa respectively. As such, the problem set up has two main drive forces, one is the heating at the top boundary, the other is pressure raising for gas one at the bottom boundary.

Results

The temperature profiles up to 6000 seconds are shown in figure 1. The temperature near the heating boundary has increased significantly in the initial period of the simulation. The steep temperature front then progresses further down into the column. No palpable temperature increase can be seen at depth below 0.5m over the time period considered.

Pore water pressure profiles are illustrated in figure 2. The pore water pressure near the heating boundary decreases dramatically due to drying induced by heating. This heating process causes a certain amount of water near the heating boundary to become vapour. This vapour migrates to a nearby region and condenses there. This can be seen in the occurrence of a peak in the pressure profile. In the majority of the sample, an increase of pore water pressure can also be observed because of the increase of gas pressures. The influence of the fixed pore water pressure at the bottom of the sample can be seen with the pore water pressure in the bottom half of the sample being drawn down to the fixed boundary value.

The pressure profiles of gas one is shown in figure 3. The gas pressure increases near the top boundary is mainly caused by temperature increase, while the gas pressure at the base is mainly influenced by gas pressure applied at the boundary. Gas pressure in all of the sample increases significantly, mainly because the gas is relatively mobile. There is a small localised variation in the gas pressure near the top boundary because of an increase of vapour pressure leading to a movement of gas one away from this area. This gas pressure drop is most pronounced in the early stages of loading and gradually disappears as the vapour pressure dissipates.

The pressure profile of gas two is shown in figure 4. Again, increase of pressure for gas one is observed from the pressure profiles. A raised fixed pressure at the bottom boundary increases the total pressure of gas within the simulated domain. The strong coupling between gas one and gas two resulted in pronounced variation of pressure profile for gas two. The fixed gas pressure for this gas at the bottom of the soil column is kept the same as the initial gas pressure. This tends to reduce the pressure throughout the sample. Again, the influence of vapour can be recognised from the results near the heating end but is only significant in the very early stage of the simulation.

CONCLUSION

A theory for multiple gas transfer in unsaturated media under non-isothermal condition based on a mechanistic approach has been presented. A numerical solution has been developed using the finite element method. A simple problem has then been analysed and the results presented indicate both the significance of interactions between temperature, pore water pressure and gas pressures and the capability of the model.

REFERENCES

Edlefsen, N. E. and Andersen, A. B., (1943), The thermodynamics of soil moisture, Hilgardia, 16, 31-299.

Ewen, J. and Thomas, H.R. (1989) "Heating unsaturated medium sand", *Geotechnique* **39**, No. 3, 455-470.

Ferguson, W J and Thomas, H R, 1997, "A numerical study of the effect of thermal gradients on landfill gas migration". Proc Int Conf on Geoenvironmental Engineering, Cardiff, 181-186.

Findikakis, A.N. and Leckie,J.O., Numerical simulation of gas flow in sanitary landfills, ASCE Journal of the Environmental Engineering Division, 105(EE5), 927-945, 1979.

Metcalfe, D.E. and Farquhar, G.J., Modelling gas migration through unsaturated soils from waste disposal sites, Water, Air and Soil Pollution, 32, 247_259, 1987.

Moore, C.A., Rai, I.S. and Alzaydi, A.A., Methane migration around sanitary landfills, ASCE Journal of the Geotechnical Engineering Division, 105(GT2), 131-144, 1979.

Philip, J.R. and de Vries, D.A. (1957) "Moisture movement in porous materials under temperature gradients", *Trans. Amer. Geophys. Union*, 38, No. 2., 222-232.

Thomas, H R and Ferguson, W J, 1999, "A fully coupled heat and mass transfer model incorporating contaminant gas transfer in an unsaturated porous medium". Computers and Geotechnics, 24, 65-87.

Thomas, H.R & He, Y. (1995) "An analysis of coupled heat, moisture and air transfer in unsaturated soil", *Geotechnique*, **36**.

Thomas, H.R. & Samson, M.R. (1995) "A fully coupled analysis of heat, moisture and air transfer in unsaturated soil", *Journal of Engineering Mechanics, Am. Soc. Civ. Eng.* **12**, No.3, 392-405

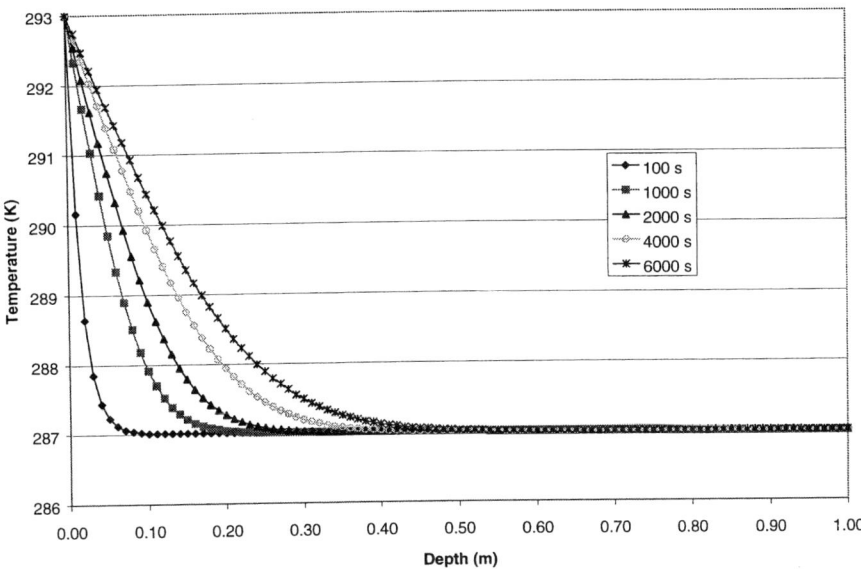

Figure 1 Temperature profiles

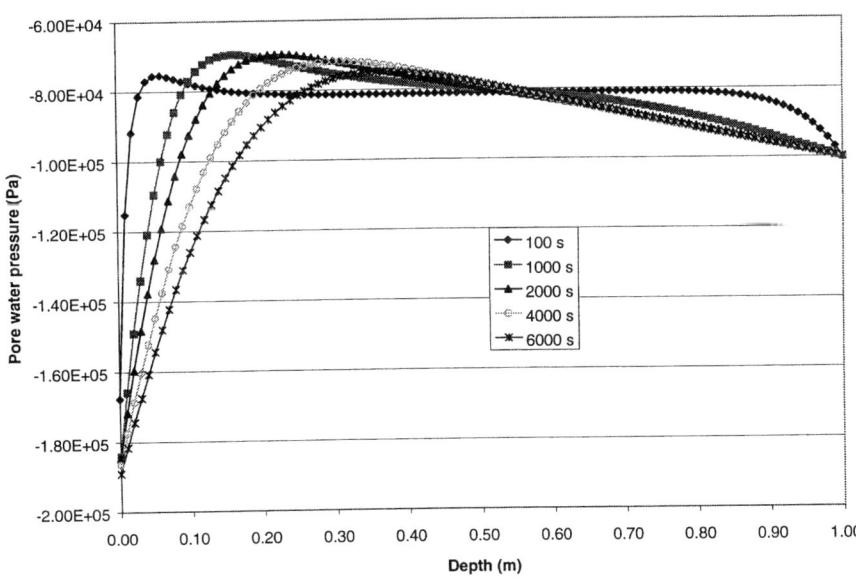

Figure 2 Pore water pressure profiles

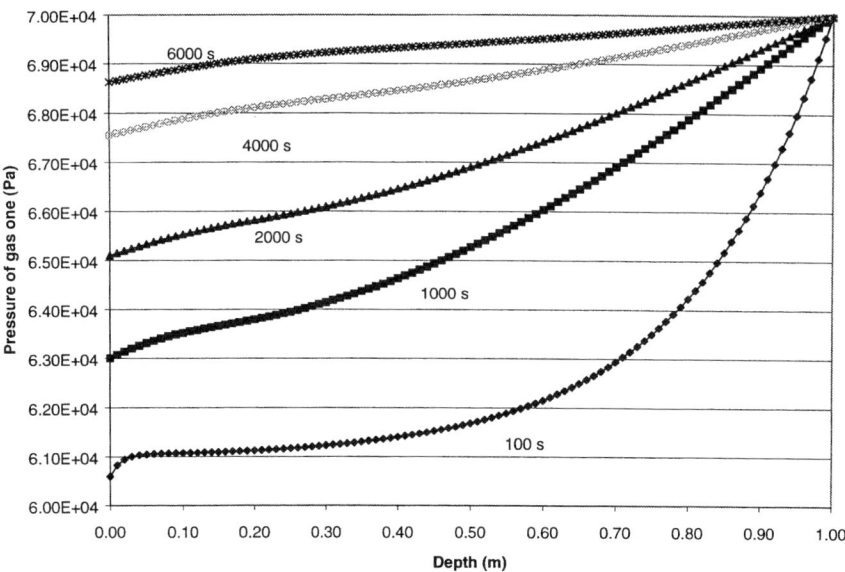

Figure 3 Pressure profiles for gas one

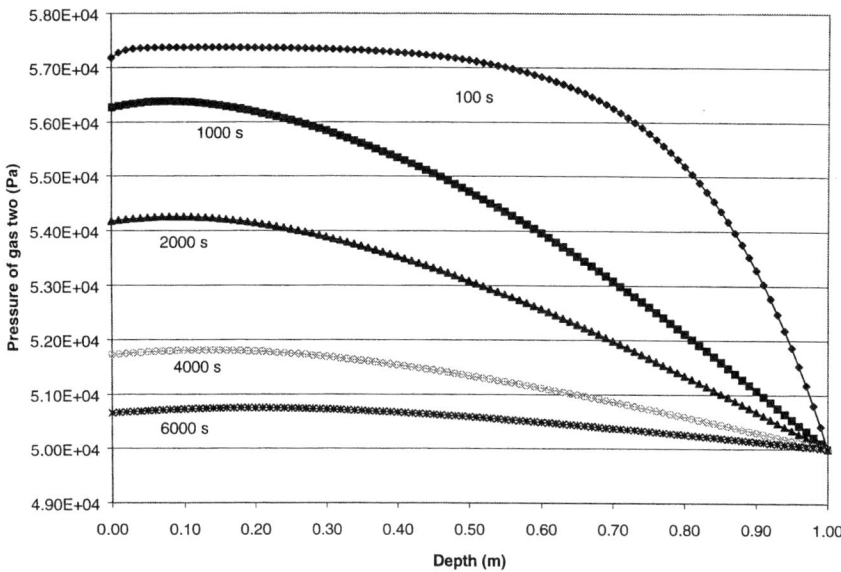

Figure 4 Pressure profiles for gas two

Design, Construction and
Operation of Landfill Sites

Low-frequency conductivity measurements to monitor ionic migration through landfill liners

J. BLEWETT, W. J. McCARTER, T. M. CHRISP, G. STARRS

Heriot-Watt University, Department of Civil and Offshore Engineering, Edinburgh, EH14 4AS, Scotland, U.K.

ABSTRACT
This paper considers the technical aspects associated with low-frequency conductivity measurements within liner materials. Experimental work is presented to show the direct application of electrical techniques to track ionic movement through clay. The potential for field monitoring of liner systems is discussed.

INTRODUCTION
Although many of the mechanisms governing contaminant flow through landfill liners are understood theoretically, the large number of variables involved make rigorous performance assessments difficult. In order to collect relevant performance data to produce empirical liner specifications, it is recognised that the long term durability of in-situ liner systems must be monitored. To make this operation cost-effective it would be preferable to develop a non-destructive approach to allow real-time monitoring of the liner. The alternatives are to excavate sections of the liners (which must be replaced) or to construct purpose built test sites. Indeed, the most widely used methods for assessing diffusion rates, both in the laboratory and in-situ, generally require a destructive approach. [1, 2].

The current work presents initial laboratory studies in the application of electrical conductivity techniques to monitor and model ionic migration through clay, in preference to the use of dielectric methods [3, 4]. The advantage of low-frequency conductivity measurements is that they are relatively straightforward to make in terms of accuracy and instrumentation required [5, 6]. Ultimately, a simple approach is preferred such that the technology can be transferred to field monitoring systems.

For one dimensional diffusive driven flow of non-reactive solutes in soil, Fick's second law gives,

$$\frac{\partial C}{\partial t} = D^* \frac{\partial^2 C}{\partial x^2} \tag{1}$$

where, D^* is the diffusion coefficient, C is the concentration of the ionic species under study, t is time and x is the depth.

The solution to equation (1) is,

$$C(x,t) = C_t + (C_s - C_t)erfc\frac{x}{\sqrt{4D^*t}} \qquad (2)$$

where, $C(x,t)$ is the concentration at depth x and time t, C_i is the background concentration (i.e. initial ionic concentration of the species in the soil prior to diffusion), C_s is the surface concentration of the diffusing species at time, t, *erfc* is the error function complement, and D^* is the diffusion coefficient of the ionic species.

The ratio D*/D₀ is used to characterize the tortuosity, τ, of the saturated soil [7, 8] where D_o is the limiting free solution diffusion coefficient of the species. Of interest in the current work, the Nernst-Einstein equation can be used to relate diffusion and ionic conductivity through,

$$\frac{D^*}{D_o} = \frac{\sigma_{bulk}}{\sigma_{pore}} = \tau \qquad (3)$$

where, σ_{bulk} is the bulk ionic conductivity of the saturated soil, σ_{pore} is the conductivity of the interstitial pore-fluid. Tortuosity is used here as a combined measure of the effect of the increased transport path and the reduced cross-sectional area available for transport and therefore incorporates the porosity of the material. This equation ignores the surface conductivity of the particles which is deemed to be significantly smaller than that of the free ions.

EXPERIMENTAL

Kaolin powder, supplied from ECC International, was used as an idealised soil sample (plastic limit 28%, liquid limit 54% and particle specific gravity 2.59). Analytical reagent grade sodium chloride and double-distilled water was used to prepare a solution of concentration 0.1 moles/l (0.1M). Kaolin was mixed with the appropriate volume of solution to produce a slurry of three times the liquid limit. Mixing took place in a rotary mixer for approximately 1 hour to ensure the uniformity. The slurry was poured into a 50x50x600mm (height) acrylic column. After pouring, the sample was further agitated to prevent any preferential layers being formed during placement. The sample was allowed to settle until no further vertical movement of the clay/water interface occurred. The conductivity of the free pool of water, which settled out at the top of the sample was measured. This water was assumed to represent the composition and conductivity of the pore fluid within the stabilised sample. The pool of water was then carefully decanted off the top of the clay and replaced with a 2.0M sodium chloride solution and diffusion allowed to take place. The top of the column was sealed to prevent evaporation of liquid and the solution was replaced daily to ensure a constant driving gradient.

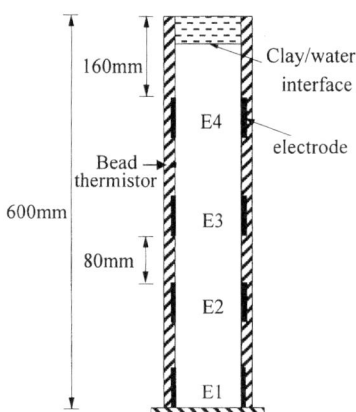

Fig. 1 Diffusion column used in the current work.

To facilitate electrical measurements, the column had pairs of 50mm stainless steel square plate electrodes positioned at discrete heights within the column (Fig. 1). The electrodes were positioned flush with the sides if the column. Electrical measurements were thus taken normal to the direction of diffusion. A bead-thermistor was mounted in the side-wall of the column thereby allowing monitoring of the clay temperature (although testing was carried out within the confines of a temperature controlled laboratory at 22°C±1°C). Electrical impedance measurements were taken using a Solartron 1260 frequency response analyser (FRA) which was connected through an Hewlett Packard HP3488A switch unit to provide five reading channels: one for each of the four pairs of electrodes and one connected to the thermistor. The FRA operated in a voltage-drive mode with a signal amplitude of 100mV. Sample impedance measurements were taken at 100 spot frequencies within the range 1Hz-1MHz. This allowed deconvolution of electrode polarisation effects from the sample response [9], hence accurate identification of the bulk resistance of the sample between each electrode pair. Lead inductive effects were nulled from the measured data, with measurements recorded on a 5 hour cycle.

Technical issues associated with bulk conductivity measurements
The following aspects were considered during the initial stages of the laboratory programme:

Calibration of electrodes
The bulk conductivity (σ_{bulk}, in Siemens/m) of a prismatic sample of clay slurry can be written,

$$\sigma_{bulk} = \frac{d}{RA} \text{ S/m} \qquad (4)$$

where, A is the electrode surface area (m^2) and d the electrode separation (m). R is the bulk resistance of the sample (in ohms) obtained from impedance measurements. For the diffusion column, the geometrical constant, A/d, for each pair of electrodes cannot be calculated directly as fringing effects occur in the electrical field. This constant was obtained through calibration of the column with liquids of known conductivity. Using this procedure, the A/d factor for electrode positions 2-4 was obtained as 6.92×10^{-2} m (±1%), whereas electrode position 1 had a value 5.96×10^{-2} m (±1%). Regarding the latter, the difference in these two factors results from the non-conductive boundary (i.e. the base of the column) in close proximity to electrode 1 position.

Electrode Polarisation (EP)
The problem of electrode polarisation in two-point electrical measurements on soil is normally eliminated by the use of alternating current. EP manifests itself as an impedance at the electrode/soil interface shunted in series with the sample impedance [4, 5]. As the frequency of the applied electrical field increases, EP effects reduce; however, only by monitoring the impedance over a sufficiently wide frequency range can accurate bulk resistance measurements be obtained. This is an aspect not normally considered in two-point measurements on soils and is particularly important for materials where conduction is primarily ionic in nature [5, 10]

Lead inductance
Lead inductive effects, which progressively degrade data with increasing frequency, must be nulled from the measured data. As the frequency of the applied field increases, lead inductive effects can dominate the measured data. In the current work, a refined calibration algorithm [11] was employed in preference to the on-board lead nulling facility of the FRA.

Temperature effects on conductivity

Electrical conduction is a thermally activated process and, for an ionically conducting material such as clay, the Arrhenius relationship is applicable:

$$\sigma_{bulk} = Ae^{-\left[\frac{E_a}{R_gT}\right]}$$

(5)

Where, σ_{bulk} is the bulk conductivity (S/m) of the material; T is the absolute temperature (K); A is the pre-exponential conductivity constant (S/m); E_a is activation energy for the conduction process (kJ/mol) and R_g is the gas constant (8.314kJ/mol/K). There is a dearth of data on the activation energy for electrical conduction in soils.

Although the laboratory used during this study is temperature controlled, small fluctuations in the ambient temperature will serve to distort the final pattern of results. In the field, however, temperature effects can be much more significant and conductivity values must be *corrected* to a predefined reference temperature. In order to undertake this correction process, an appropriate value of the activation energy is required. A column of kaolin was placed in a cooling cabinet for 48-hours and the sample temperature reduced to 5°C (278K). The column-thermistor and electrode pair 2 were connected to the data acquisition system and simultaneous measurements of temperature and resistance were obtained on a two minute cycle as the temperature of the sample rose to room temperature. The increase in measured conductivity with temperature is shown in Fig. 2 and plotted in an Arrhenius format. The activation energy was obtained as, $E_a = 17.8$kJ/mol.

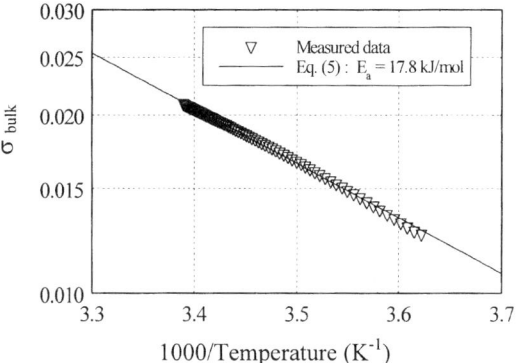

Fig. 2 Arrhenius relationship for kaolin.

From equation (6), the measured bulk conductivity, σ_{bulk}, recorded at a temperature T (in K) can now be corrected to and equivalent conductivity (σ_{ref}) at a reference temperature (T_{ref}) through a knowledge of the activation energy, E_a, for the conduction process viz,

$$\sigma_{ref} = \sigma_{bulk}e^{-\frac{E_a}{R_g}\left[\frac{1}{T_{ref}}-\frac{1}{T}\right]}$$

(6)

All data were corrected to 20°C (293K) using an activation energy of 17.8kJ/mol.

RESULTS AND DISCUSSION

Fig. 3 presents $\sigma_{bulk}/\sigma_{pore}$ (denoted normalised conductivity) for each electrode level just prior to the start of the diffusion test and obtained from bulk and pore fluid conductivity. From equation (3) above, this ratio must also give a semi-quantitative assessment of the tortuosity of the pore network within the clay. The normalised conductivity results indicate that the sample is in a relatively uniform state after sedimentation with an average tortuosity value, $\tau = 0.42$. The porosity of the clay in the column was 0.49. Since the free solution diffusion coefficient, D_0, of NaCl is $1.57 \times 10^{-9} m^2/s$ [8] then, applying equation (3) would yield D* for NaCl in the clay as $6.6 \times 10^{-10} m^2/s$.

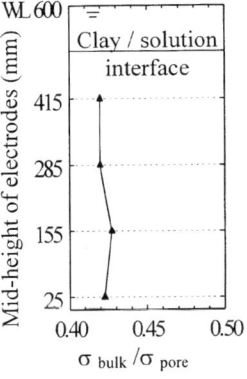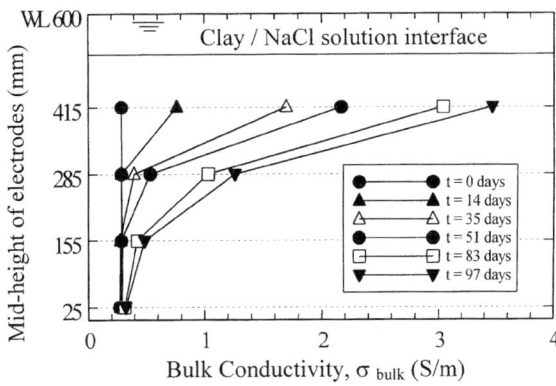

Fig. 3 Normalised conductivity prior to diffusion.

Fig. 4 Bulk conductivity profiles during diffusion.

Fig. 4 displays the development of the bulk conductivity profile through the column and presented at selected times during diffusion testing. Preliminary experiments established a relationship between NaCl concentration and solution conductivity, which was then used, in conjunction with σ_{bulk}, to relate conductivity changes measured during diffusion, to NaCl concentration in the interstitial pore fluid. Fig. 5 presents the values thus obtained as a function of time at each electrode level. The theoretical concentration versus time profiles at each electrode level, $C(x,t)$, predicted using equation (2) are presented on this Figure using a diffusion coefficient $D* = 6.6 \times 10^{-10} m^2/s$ (see above).

It must be emphasized that no allowance has been made for the possibility of structural changes occurring within the kaolin during diffusion. At present, the current predictions are taken to represent a first approximation. However, any structural effect on conductivity is likely to be small in comparison to the change in bulk concentration due to increasing NaCl concentration in the pore fluid.

Practical significance

The simplicity of the testing system offers the potential for adapting as a field monitoring technique. Currently, electrode configurations are being studied which can be inserted into into landfill liner material to obtain multiple, localised in-situ measurements. Further work by the authors will address these and other

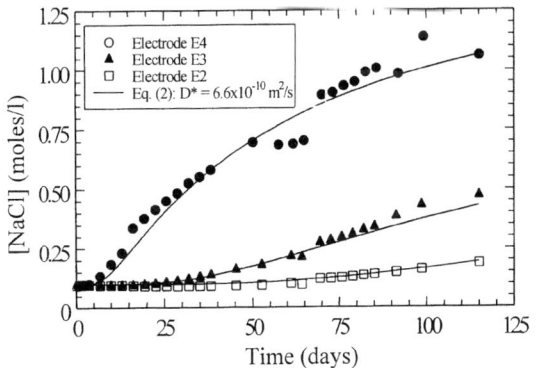

Fig. 5 Pore-fluid concentration versus time response from conductivity measurements and predicted response.

issues e.g. the development of a portable data acquisition system.

Bulk conductivity and pore-fluid conductivity measurements could also be exploited / developed as a rapid means of estimating the diffusion coefficient and tortuosity of saturated soil-water systems. Work is continuing in this respect.

CONCLUSIONS

In addition to highlighting a number of technical aspects relating to electrical measurements, it was shown that conductivity measurements could be used to track ionic movement through clay during diffusion. Conductivity values were converted to an equivalent concentration and used to develop the concentration versus time response for each electrode pair. A diffusion coefficient, determined from normalised conductivity measurements prior to the start of the test, was used to predict a theoretical concentration versus time curve. This agreed with the measured response.

ACKNOWLEDGEMENTS

The Authors wish to acknowledge the support of the Engineering and Physical Sciences Research Council (Grant GR/M47959).

REFERENCES

1. Rowe, R. K., Caers, C. J. and Barone, F. 'Laboratory determination of diffusion and distribution coefficients of contaminants using undisturbed clayey soil', *Can. Geotech. J,* 1988, 25, 108-118.
2. Crookes, V. and Quigley, R. M., 'Saline leachate through clay: a comparative laboratory and field investigation', *Can. Geotech. J,* 1984, **21**, 349-362.
3. Santamarina, J. C. and Fam, M., 'Changes in dielectric permittivity and shear wave velocity during concentration diffusion', *Can Geotech. J,* 1995, *32*, 647-659.
4. Carrier, M. and Soga, K., 'Dielectric measurements of clay as a potential method of contaminant detection', *Proc. 3rd Int. Congress on Environmental Geotechnics, Portugal, 7-11 September,* 1998, 491-496.
5. McCarter, W. J. and Desmazes, P., 'Soil characterisation using electrical measurements', *Geotechnique,* 1997, 47, 1, 179-183.
6. Blewett, J, McCarter, W. J., Chrisp, T. M. & Starrs, G., 'Monitoring sedimentation of a clay slurry', Accepted for pblication in *Geotechnique.*
7. Shackelford, C. D., and Daniel, D. E., 'Diffusion in saturated soil II: Results for compacted clay', ASCE, J Geotech. Div., 117, 3, 485-506, 1991.
8. Rowe, R. K., and Badv, K., 'Chloride migration through clayey silt underlain by fine sand or silt', ASCE, J Geotech. Div., 122, 1, 60-68, 1996.
9. McCarter W. J. and Brousseau, R., 'The A.C. response of hardened cement paste', *Cement and Concrete Research,* 1990, 20, 6, 891-900.
10. Carrier, M. and Soga, K., 'A four terminal measurement system for measuring the dielectric properties of clay at low frequencies', *Engineering Geology,* 1999, **53**, 115-123.
11. Starrs, G. and McCarter, W. J., 'Immittance response of cementitious binders during early hydration', *Advances in Cement Research* 1998, 10, 4, Oct, 179-186.

The improvement of cartographic documents for landfill site selection

K. S. DUARTE, Researcher, University of Brasilia, Brazil, and A. C. S. CORRÊA, Researcher, University of Brasilia, Brazil, and N. M. SOUZA, Professor, University of Brasilia, Brazil.

INTRODUCTION

In 1992 a MSc. thesis presented a research about the Environmental Protection Area of the São Bartolomeu River, Federal District, Brazil, and included 13 cartographic documents at final scale of 1:50,000, consisting of environmental, geographical and geotechnical aspects. One of those maps focused on possible areas for waste disposal. The main reason for that research was the illegal occupation of the area, which demanded a general risk prediction. Nowadays, with the legalization and increase of some urban occupation areas, the search for landfill potential sites in the surroundings has emerged.

In order to select a specific landfill site, recent remote sensing products, field investigation data, update land use data and the adoption of a geographic information system, SPRING, was used.

As a result, the area classified as adequate has been reduced, determining a potential site for waste disposal, Figure 1. Another consequence of this research was to show the importance of evaluating and updating frequently the cartographic documents, mainly where the land use changes quite fast, in order to produce practical and useful documents for local community.

METHODOLOGY

The basic cartographic document, (Duarte, 1992), was not very specific because the criteria adopted to define the location, as adequate, reasonable or inadequate, were limited to some technical attributes such as thickness and permeability of the unconsolidated superficial material, declivity and geological conditions of the underlying bedrock. The determination of a specific landfill site should consider the changes in the land use, the distances of new urban occupations and environmentally sensitive areas, main roads, as well as the minimum size accepted to a landfill implantation.

To compare the changes in the region, aerial photographs of 1991 and remote sensing images of 1997 (processed by a sampling classification method) were used. This processing manipulation allowed a fast classification of areas as urban occupation, terrain with vegetal cover and with exposed soil.

The legalization of occupied areas and the urban expansion in the region, required the definition of an area for waste disposal. The available map of 1992 indicated a very large area considered as adequate, which suggested the necessity of updating.

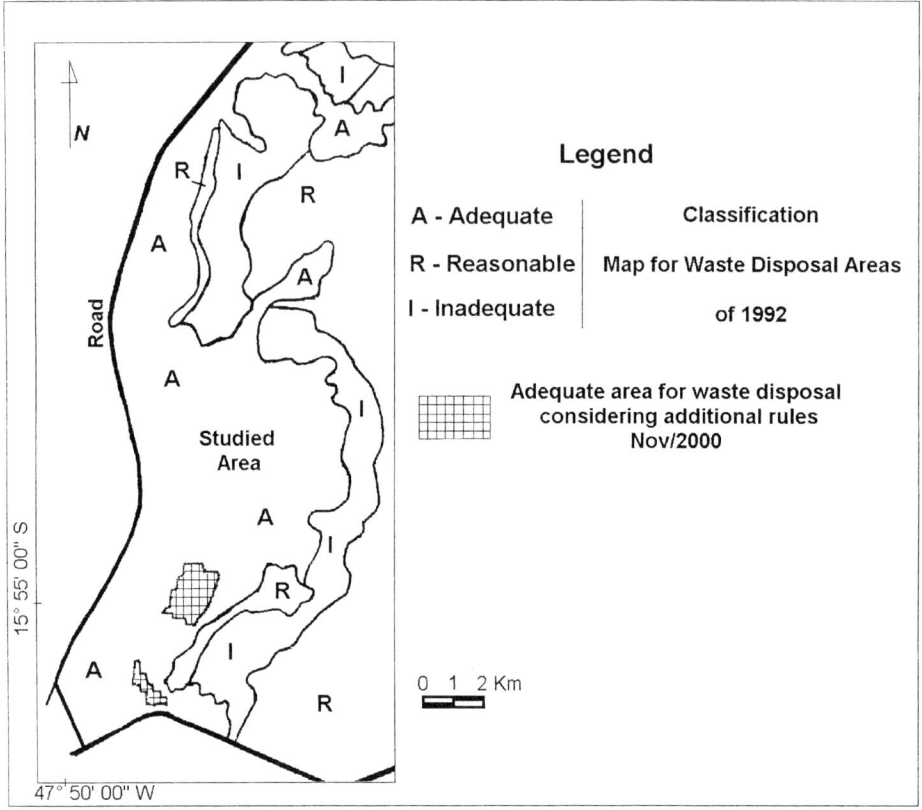

Figure 1. Comparison of areas considered adequate by mapping of 1992 and by additional rules.

Bonham-Carter, 1994, suggested some rules for the selection of an area for waste disposal that was not considered on previous map. One of these rules is the distance of main roads. The landfill construction either very close or very distant to the roads is not convenient. With the use of SPRING-GIS, that managed the improvement of previous cartographic documents, the map of distances of main roads was created after the creation of a net of distances.

The distances of urban areas were not considered in the map of 1992. The adequate distance from landfill to urban areas must not be very close or very distant, so the same previous process was conducted by SPRING-GIS.

Another aspect of extreme importance is the secure distance of environmentally sensitive areas. Some researches made in the region pointed the spring of a river and the waterside vegetation as an important area to protect. The SPRING-GIS was used to create a distance net in relation to the spring and to the section of the river. The map of distance was produced and a reasonable distance of these sensitive areas was considered as inadequate.

The declivity of selected area was less than 5% in conformity with the map of 1992. The soil map (EMBRAPA, 1978), had been reclassified to permit a correlation among the soil units, the photointerpretation data and the laboratory data. The latest selected area is compatible with the reclassified map of soil.

The geological conditions were taken into consideration in the map of 1992. The area selected by the additional rules is lying on convenient thick lateritic deposit of "porous clay", which is a typical kind of tropical residual soil formed by clays, silts and sand, in different percentages depending on the geological source. The voids ratio observed were between 1.2 and 2.2. This kind of soil presents a great metal attenuation capacity.

In the area considered to be adequate for waste disposal, the remote sensing image analysis pointed an evident difference of tonality not observed by the photointerpretation analysis and not considered in the previous map. The difference of tonality at the sensing image of exposed soil suggested a significant change of material that demanded a field investigation. Figure 2.

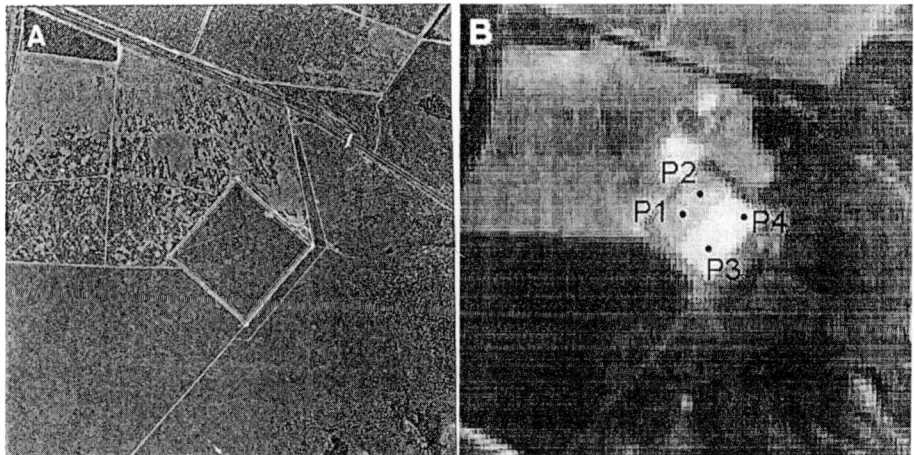

Figure 2. Comparison between (A) aerial photographic product (1991) and (B) remote sensing product (LANDSAT 1999 Band 5 - monochromic). The use of bands 5,4 and 3 in the RGB canals, present a better contrast with different colours to different kinds of soil. The points from P1 to P4 were points of soil sampling.

The field investigation confirmed the difference of material observed in the image. In points P1 and P2 the soil is more humid and less dense than in points P3 and P4. In these points, samplings to soil characterization tests were made. The results can be observed on Table 1.

Field investigation coupled with some laboratory tests found that the differences pointed out by the satellite image were in fact, more related with texture and humidity than with a difference of soil geology. Indeed, the site suffered an intense soil movement. The excavation of the higher side modified the natural declivity. In its turns, the soil removed was compacted over the lower half of site. The difference observed in the image and verified in the field investigation confirms the importance of sensing image analysis as an important tool of investigation.

Table 1. Results of soil characterization tests

Sample	Natural Moisture (%)	Bulk Density KN/m³	WL (%)	WP (%)	IP (%)	Particle Size Distribution (%)			
						Gravel	Sand	Silt	Clay
P1	31	11.41	54	40	15	0	7.4	11.1	81.5
P2	33	11.48	46	35	11	0	10.8	10.2	79.0
P3	11	16.47	30	25	5	7.0	41.8	10.5	39.8
P4	10	18.30	31	26	5	13.9	38.6	10.9	36.6

CONCLUSIONS

Taking into consideration some rules suggested by Bonham-Carter, *op. cit.,* some restrictions observed by analyses of remote sensing and field works, the area considered as acceptable for waste disposal was reduced to a smaller area, but still suitable. On the other side, the Boolean rules can be evaluated with the intention of modifying the area considered as acceptable, in conformity with the local necessities and the adoption of an acceptable level of risk.

REFERENCES

BONHAM-CARTER, G.F. 1994. Geographic Information Systems for Geoscientists - Modeling with GIS. Pergamon. Ottawa, Canada. 398p.

DUARTE, K.S. 1992. Mapeamento da Margem Direita do Rio São Bartolomeu, Distrito Federal. MSc. thesis, ENC/UnB. Brasilia/DF. 13 maps, 130p.

Empresa Brasileira de Pesquisa Agropecuária (EMBRAPA) 1978. Levantamento de Reconhecimento dos solos do Distrito Federal. Technical Bol. No. 53. Rio de Janeiro.

One-dimensional Testing of Partly Saturated Soils at Elevated Temperatures: Preliminary Results

J. P. FOLLY and S. W. REES, Geoenvironmental Research Centre, Cardiff University, UK

ABSTRACT

Experimental investigation of the behaviour of partly saturated soils is a key component in the analysis and design of a wide range of geoenvironmental problems. Furthermore, in some circumstances the influence of temperature variation needs to be considered in addition to hydraulic and mechanical behaviour. To this end, temperature and suction controlled oedometers have recently been developed at Cardiff University. Two one-dimensional cells have been built that permit parallel testing of partly saturated soils over a temperature range of room temperature to 90°C. Suction control is achieved using the axis translation technique and is currently limited to a maximum of 1.5 MPa. Vertical stress is applied via an air pressure controlled loading ram, operating up to maximum stress of 2.5 MPa. This paper briefly describes the design and development of the equipment and presents some preliminary experimental results.

INTRODUCTION

The behaviour of partly saturated soils at elevated temperatures is recognised to be important in a number of geoenvironmental problems. Practical examples include; optimisation of geothermal energy utilisation (Rees *et al.*, 2000), improved performance of buried services (Anders and Radhakrishna, 1988), and thermally enhanced clean-up of contaminated land (Lee *et al.*, 1999). A further important application area is the design of engineered clay barriers for use within the context of the deep geological disposal of high level nuclear waste. In some scenarios, clay/sand mixtures will be placed, around the waste canisters, in an initially partly saturated state. The swelling properties of the barrier will be utilised to provide a low permeability seal around the waste canisters. High temperatures associated with the decaying waste and severe pore pressure gradients between the barrier and the host material add to the complexity of the problem. In these circumstances an understanding of the thermo-hydro-mechanical (THM) behaviour of the barrier is vital. A considerable effort has been directed towards the development of suitable theoretical and numerical frameworks for the analysis of the type of problem (e.g. Thomas and Cleall, 1999). However, notwithstanding the significant developments already made, experimental determination of related soil properties remains a challenge.

In order to address some of the above problems, two temperature and suction controlled oedometer (TSO) cells have been developed at Cardiff. Preliminary experimental investigations have been undertaken to calibrate the cells and demonstrate the type of information that can be obtained. In the first instance, speswhite kaolin has been chosen for the initial programme of work. The moderate swelling properties of kaolin and an abundance of related experimental data make it ideal for this phase of the work.

Geoenvironmental impact management, Thomas Telford, London, 2001.

EQUIPMENT DESIGN AND DEVELOPMENT

The approach adopted at Cardiff broadly follows the work of Romero *et al.* (1995). At this stage two cells have been designed and built in-house. Figure 1 shows a photograph of the cell components. Figure 2 provides a schematic of the auxiliary equipment required for operating the cells.

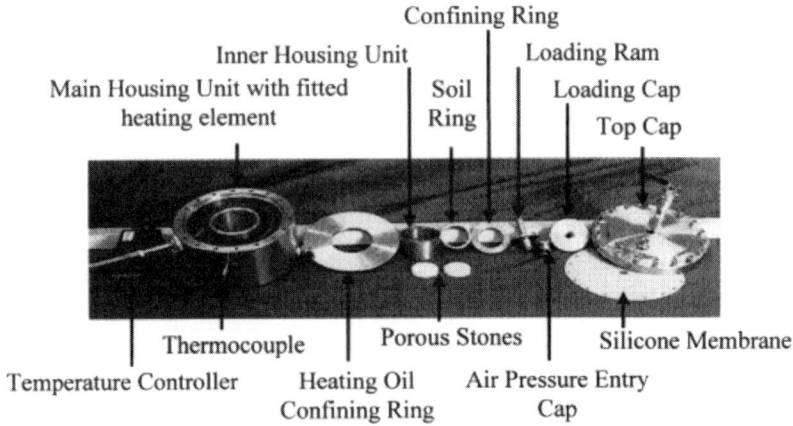

Figure 1. Photograph of TSO cell components

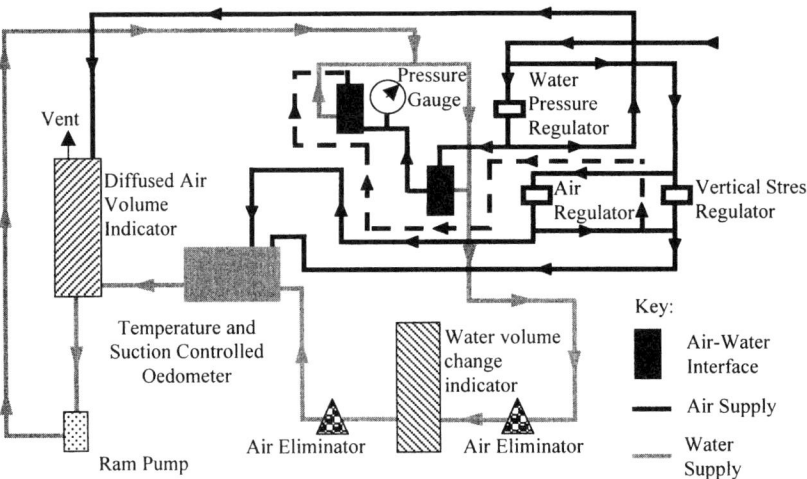

Figure 2. Schematic of the auxiliary equipment

Temperature control is achieved using a stainless steel sheath wire heater that is submerged in heating oil and coiled in a sealed chamber surrounding the soil sample. A PID (Proportional, Integral and Derivative) temperature controller with type-T thermocouples controls the heater, and an insulation jacket surrounds the cell to minimise heat loss and protect users. The temperature testing range of the cells is from room temperature to 90°C. However, Romero *et al.* (1995) suggested that compaction of soil may only be carried out in the cell at temperatures below 60°C. Above this, a separate temperature controlled preparation mould is recommended in order to avoid desaturation of the porous stones. To avoid this requirement, the initial experimental programme has therefore been restricted to a maximum temperature of 60°C.

Temperature calibration has been carried out for the new cells. Figure 3 shows that thermal equilibrium within the soil compartment is established within four hours.

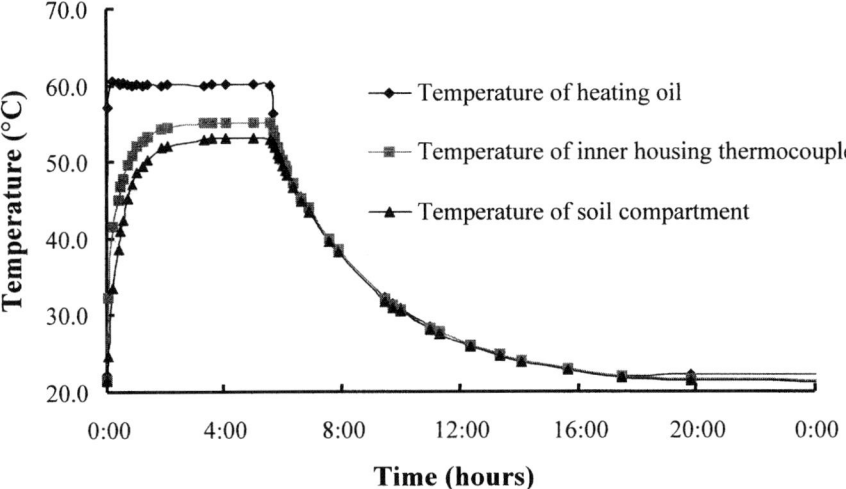

Figure 3. Temperature measurements during heating and cooling phases

Suction control is achieved via the axis translation technique utilising a 1.5 MPa high air entry ceramic disc. Water pressure control is delivered to the base of the sample through inlet points located beneath the ceramic disc. The air pressure is transmitted to the top of the sample via a standard porous stone. The maximum suction is limited to less than 1.5 MPa due in part to the fact that a back pore-water pressure must be applied to the ceramic disc during testing to prevent it from drying. Soil-water content changes are calculated by measuring the amount of water crossing the ceramic disc using a water volume change indicator. The results are compensated for evaporation losses and diffused air. The volume of diffused air is quantified using a diffused air volume indicator (Fredlund, 1975). Suction control has been checked by comparisons with pressure plate extractor (PPE) results (Fig. 4). Reasonable agreement has been achieved.

The vertical load is applied to the soil sample by pressurising a chamber between an impermeable membrane and the top cap using regulated nitrogen gas. The pressure is then transferred to the soil via a loading ram. Net vertical stress is determined as the difference between the applied vertical load and the sample pore-air pressure applied to generate a required suction. Therefore, the generation of suction and net stress are interdependent and self-limiting. At present the maximum net vertical stress is limited to less than 2.5 MPa. Figure 5 shows that cell deformation decreases with an increase in temperature due to thermal expansion. Furthermore, calibration exercises have established that frictional resistance generated between the loading ram and guidance hole in the top cap is negligible.

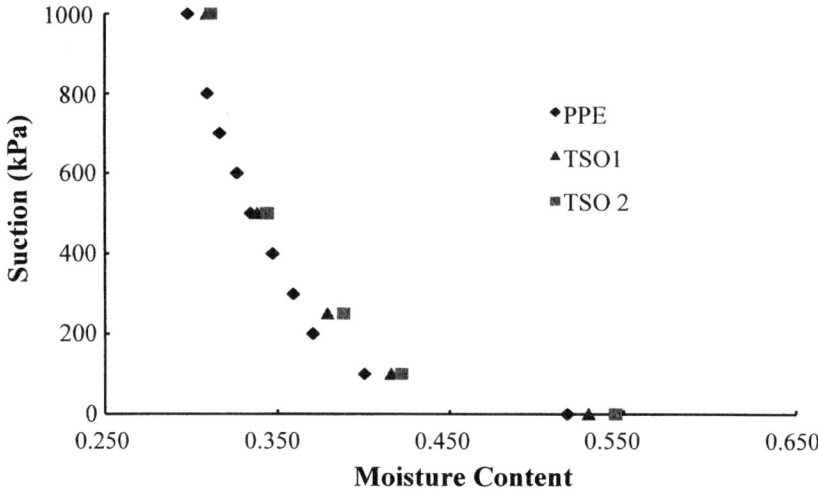

Figure 4. Suction control checks

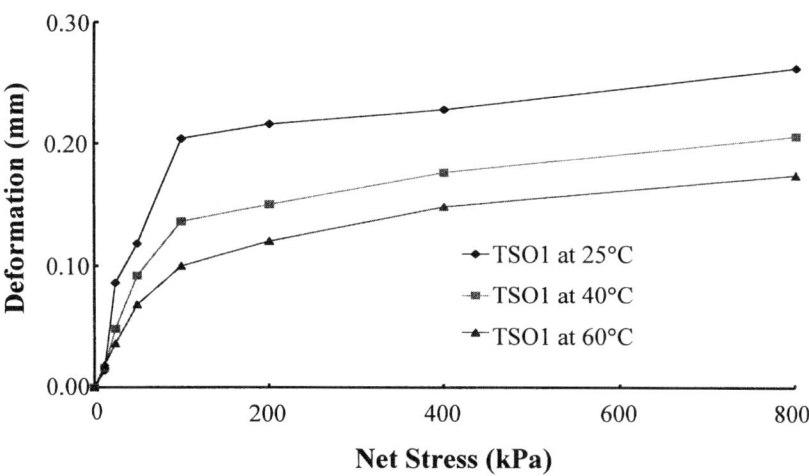

Figure 5. Elasticity of cell components at elevated temperatures

PRELIMINARY EXPERIMENTAL RESULTS

For the reasons outlined above, preliminary tests have been carried out on speswhite kaolin. Related, isothermal, data can be found in the literature, for example; Josa (1988), Sivakumar (1993), and Wheeler and Sivakumar (1995). More recently, Sivakumar and Wheeler (2000) and Wheeler and Sivakumar (2000) have reported on the influence of compaction procedures on the mechanical behaviour of speswhite kaolin.

In order to obtain confidence in the results obtained from the new cells a series of saturated tests have been undertaken. Samples were statically compacted at 25°C and 60°C using the procedure outlined by Sivakumar (1993). The mechanical behaviour of the samples was then investigated, maintaining these temperatures, by applying net stress in accordance with the standard ASTM D 2435 time and loading scheme up to 0.8 MPa.

Figure 6 shows that prior to application of the initial net stress the void ratio of both samples was similar. However, after increasing the net vertical stress to 12 kPa, a lower void ratio is evident in the sample prepared at the higher temperature (Fig. 6). Correspondingly, the sample compacted at 60°C can be seen to experience greater compressive strain (Fig. 7). The results indicate that initial compaction process was not greatly influenced by the elevation of the cell temperature. However, a thermal softening is evident as the loading sequence is applied. Figure 7 indicates that the temperature rise has had a quite marked effect on the compressive strength of the kaolin during the initial phase of the test. The results presented are very much a first attempt at this type of testing. However, it is encouraging to note that both Finn (1951) and Campanella and Mitchell (1968) have reported similar observations.

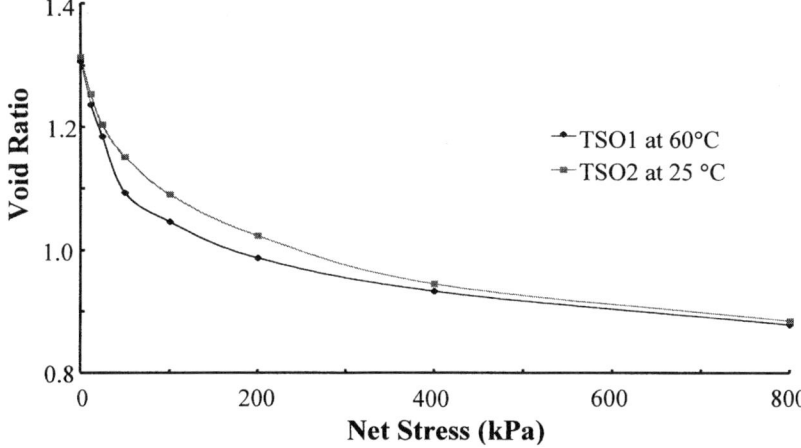

Figure 6. Effect of temperature on isotropic consolidation behaviour of speswhite kaolin

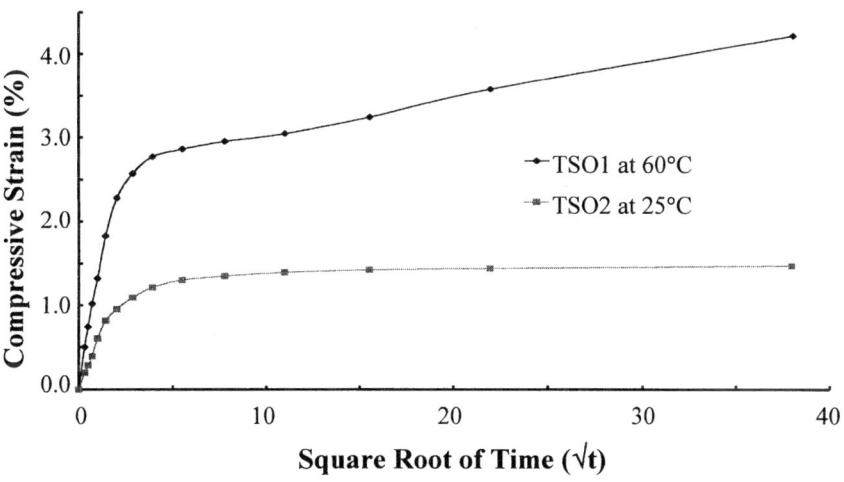

Figure 7. Effect of temperature on the compressive strain for speswhite kaolin.

CONCLUSIONS
The design and calibration of new temperature and suction controlled oedometer cells has been described. The equipment can operate to a maximum temperature of 90°C. Suction control is limited to less than 1.5 MPa and net vertical stress is limited to less than 2.5 MPa. The results of preliminary experimental work performed on speswhite kaolin have been presented. These show that the new equipment is able to achieve suction control with reasonable accuracy in comparison with pressure plate tests. Encouraging results have also been obtained from saturated compression tests performed at various temperatures. Further work is now necessary to investigate the influence of a range of temperatures, suction and net stress combinations. The work will then be extended to investigate clays more commonly used in practical geoenvironmental problems.

ACKNOWLEDGEMENTS
The work presented has been partly supported by the Engineering and Physical Sciences Research Council. This support is acknowledged.

REFERENCES
Anders, G.J. and Radhakrishna, H.S. (1988). "Computation of temperature, field and moisture content in the vicinity of current carrying underground power cables." *IEE Proc.*, 135, 51-62.
ASTM (1998). "Annual book of ASTM standards." *American Society for Testing and Materials, 100 Barr Harbor Drive, West Conshohocken, PA 19428-2959.*
Campanella, R.G. and Mitchell, J.K. (1968). "Influence of temperature variations on soil behaviour." *J. Soil Mech. Found. Div. ASCE,* 94, 709-734.
Finn, F. N. (1951). "The effect of temperature on the consolidation of soils." *Special Technical Publication No. 126, ASTM,* 65-72.
Fredlund, D.G., (1975). "A diffused air volume indicator for unsaturated soils." *Canadian Geotechnical Journal,* 12: 533-539.
Romero, E, Lloret, A. and Gens, A. (1995). "Development of a new suction and temperature controlled oedometer cell." *Proc. 1ˢᵗ Int. Conf. Unsaturated Soils, Paris,* 2, 553–559.
Josa, A. (1988). "Un moledo elasto-plastico para suelos no saturados." *Tesis Doctoral, Universitat Politecnica de Catalunya, Barcelona.*
Lee G., Glascoe, N. M., Wright S. J. and Abriola, L.M. (1999). "Modeling the Influence of Heat/Moisture Exchange During Bioventing." *J Environ Eng.* ASCE, 125(12), 1093-1103.
Rees S. W., Adjali M. H., Zhou Z, Davies M., and Thomas H. R. (2000) "Ground Heat Transfer Effects on the Thermal Performance of Earth-Contact Structures." *J. Renewable and Sustainable Energy Reviews,* 4(2000) 213-265.
Sivakumar, V. (1993). "A critical state framework for unsaturated soil." *Ph.D. thesis, University of Sheffield.*
Sivakumar, V. and Wheeler, S.J. (2000). "Influence of compaction procedure on the mechanical behaviour of an unsaturated compacted clay. Part 1: Wetting and isotropic compression." *Geotechnique,* 50, 359-368.
Thomas, H. R. and Cleall, P.J. (1999) "Inclusion of expansive clay behaviour in coupled thermo hydraulic mechanical models." *Engineering Geology,* Volume 54, Issues 1-2, September 1999, 93-108.
Wheeler, S.J. and Sivakumar, V. (1995). "An elasto-plastic critical state framework for unsaturated soil." *Geotechnique,* 45, 35-53.
Wheeler, S.J. and Sivakumar, V. (2000). "Influence of compaction procedure on the mechanical behaviour of an unsaturated compacted clay. Part 2: Shearing and constitutive modelling." *Geotechnique,* 50, 369-376.

Comparison of leachate attenuation characteristics of colliery spoils in field, column and batch tests

R.J. FREEWOOD, J.C.CRIPPS, AND C.C. SMITH.
Department of Civil and Structural Engineering, University of Sheffield.

INTRODUCTION

Waste Management paper 26B (DoE, 1995) recognises the validity of site specific designs to remove the leachate hazard to surface and ground water posed by landfills, and thus decrease the problem of "over engineering" of low-risk landfill sites. This includes the consideration of materials in landfill systems to attenuate the hazardous components of leachate before they reach vulnerable ground or surface waters. The process of leachate attenuation includes; physical (e.g. dilution), physicochemical (e.g. sorption, ion exchange), chemical (e.g., precipitation and complexation), and microbial (e.g. degradation). However, before leachate attenuation is considered in the risk assessment procedure for landfill design a thorough understanding of the processes governing the attenuation of leachate are required.

Coal mining and associated coal utilising industries have led to waste material often referred to as colliery spoil. Colliery spoils have been used in construction of landfill liners due to the large redundant volumes of little economic use. However, previous investigations into the attenuation capacity of spoil are limited, and due to its inherent variability a large number of samples may need to be tested during material selection to ensure adequate performance. Rapidity of the test becomes significant when comparing a large number of samples, which would be the case during material selection in the quality assurance phase of landfill liner construction. The two main types of laboratory test used in leachate attenuation investigations are batch equilibrium and column flow tests. Both tests are used to obtain contaminant retention parameters for use in risk assessment studies of contaminant transport from landfill sites. Laboratory tests are more frequently performed than field based tests due to better control on the test conditions and relative ease of test monitoring. Batch tests are routinely used in the laboratory in preference to column tests because of the relatively short test time, reduced need for leachant sample analysis, and better test precision. However, column tests are considered to be more realistic than batch tests in determining sorption parameters since they simulate the migration of contaminants in the field more accurately. This paper aims to discuss the use of batch and column tests for assessing colliery spoil as a leachate attenuating material within landfill liner systems. The results of a large-scale field test are also presented for comparison with laboratory test results.

MATERIALS AND METHODS

The colliery spoil used in the present study was from Selby Colliery obtained from the Welbeck Land Reclamation Scheme, Nr Wakefield. The mineralogy comprised on average 17% quartz, 38% illite, 13% mixed layer clays, 9% kaolinite, 3% siderite, 0.6% pyrite, and 14% total organic matter. The cation exchange capacity of this material was on average 7.2 meq/100g determined by the methylene blue method (Taylor, 1985). The spoil was sieved below 2mm for the laboratory tests and coarse unsieved spoil was used in the field test. In addition, acid washed silica sand was used to function as a blank for comparison with colliery spoils. The sand was also used in the construction of the field test fluid sampling devices.

Geoenvironmental impact management, Thomas Telford, London, 2001.

Both natural and synthetic landfill leachates were used in the batch tests, and synthetic leachates were used in the column and field tests. A range of single-solute (i.e. only one solute present) and multi-solute (i.e. various solutes present) synthetic leachates were used that contained key contaminants of environmental concern within the approximate concentration range observed by other researchers in landfill leachates (e.g. Christensen et al., 1994). The natural methanogenic landfill leachate was obtained from ALcontrol Laboratories Ltd, and was used in both unspiked and spiked form in which solutes used to produce the synthetic leachates were added. Sodium azide (NaN_3) was also added to a number of the leachates to inhibit biodegradation of organic solutes. The average leachate compositions of the multisolute tests are shown in Table 1. Solute concentrations for metals were determined by ICP-AES, anions and ammonium by flow injection ion chromatography, and organic solutes by HPLC.

Component	NL (mg/l)	NL spiked-NaN$_3$ (mg/l)	NL spiked-NaCl (mg/l)	MSC (mg/l)	MSI-Field (mg/l)
pH	7.87	7.35	7.31	6.06	6.64
EC(mS/cm)	3.84	4.95	5.48	3.77	1.23
Eh (mV)	193	162	161	205	233
Alkalinity	835.9	1480.8	1220.4	339.8	18.9
Chloride	489.4	493.7	792.46	578.3	474.1
Sulfate	158.3	610.7	668	392.6	194.5
NH$_4$	165.2	162.6	144.2	156.1	67.8
Na	1428	4327	8457	221.5	67.0
Mg	50.2	44.52	45.4	137.9	46.2
K	265.9	290.6	342.4	133.9	56.4
Ca	126.4	98.7	93.1	73.5	56.2
Mn	0.0462	21.8	21.43	176.4	70.2
Fe	0.637	0.024	0.012	0.0	0.0
Ni	0.232	18.32	17.61	25.3	5.7
Cu	0.09	0.647	0.442	0.2	1.6
Zn	0.03	0.325	0.348	5.1	9.2
Cr	0.016	0.396	0.231	0.3	0.3
phenol	bdl	23.90	15.59	31.2	-
o-cresol	bdl	23.75	11.02	44.4	-
p-cresol	bdl	26.51	24.22	35.7	-

bdl-below detection limit, NL-natural leachate, MSC-combined inorganic and organic leachate MSI-inorganic leachate

Table 1. Natural leachate, natural spiked leachate and synthetic leachate compositions

The majority of the batch tests used a 1:10 solid to solution ratio and were constantly agitated at room temperature for 7 days. A number of tests were performed using various solid:solution ratios and time periods to investigate the effect of solid concentration and equilibration time on solute sorption respectively. On test completion the samples were centrifuged, a fluid samples taken for analysis and the pH determined. Blank tests without sorbent present were also performed to investigate changes in leachate composition.

Column flow tests were performed to simulate the flow of contaminants through colliery spoils. Both saturated (leachate flow in upward direction) and unsaturated (downward leachate infiltration) column tests were performed to investigate the influence of the presence of oxygen in the pore spaces on attenuation and hydraulic processes in the sample. The column cells were 80mm long by 50mm diameter and made from borosilicate glass. The flow rates were maintained using a peristaltic pump and fluid samples were obtained at the outflow end of the column. The tests involved an initial water leaching phase followed by a synthetic leachate flow phase with a water desorption phase at the end of the test. The test physical and hydraulic parameters are shown in Table 2.

The field test used pore fluid sampling devices placed in rows of three at 3 levels within a 7m diameter by 2.5m high conical colliery spoil test heap. The leachate, which was supplied by a gravity feed leachate reservoir, infiltrated into the spoil at the top of the test heap. Levels 1, 2 and 3 were 0.25m, 1m, and 1.75m vertically below the leachate input respectively. The test involved an initial water leaching phase for 160 days followed by a synthetic leachate flow phase using leachate shown in Table 1 for 255 days with a water desorption phase for 207 days at the end of the test. On completion the test heap was excavated and bulk samples taken for analysis. The test physical and hydraulic parameters are shown in Table 2.

Test Material/Leachate	Bd Mg/m^3	n	ne	pore vol ml	DV m/s	ALV m/s	D$_L$	Pe	RT hrs
Column (unsaturated)									
Sand MSI	1.24	0.16	0.14	145	2.4E-06	1.7E-05	3.9E-07	6.42	2.3
Spoil MSI	1.41	0.09	0.55	86	2.4E-06	4.4E-06	4.7E-08	13.54	9.1
Column (saturated)									
Sand MSC	2.13	0.38	0.38	113	7.9E-06	2.1E-05	6.8E-07	4.60	2.0
Spoil Ni-SS	2.03	0.54	0.53	160	7.9E-06	1.5E-05	4.8E-07	4.70	2.8
Spoil NH$_4$-SS	1.91	0.50	0.50	160	7.9E-06	1.6E-05	3.9E-07	6.09	2.6
Spoil MSI	2.00	0.56	0.56	165	7.9E-06	1.4E-05	1.9E-07	11.25	2.9
Spoil MSC	2.06	0.55	0.45	161	7.9E-06	1.8E-05	2.2E-07	12.15	2.4
Spoil MSC pH4	1.91	0.43	0.43	126	7.9E-06	1.8E-05	2.0E-07	13.73	2.3
Spoil MSC slow flow	1.94	0.44	0.44	119	7.9E-07	1.8E-06	1.1E-07	2.42	23.1
Field Test									
Level 1	1.63	0.16	0.16	7E+04	8.9E-08	5.6E-07	5.1E-08	2.75	125
Level 2	1.71	0.16	0.16	3E+05	8.9E-08	5.6E-07	2.4E-07	2.28	500
Level 3	1.73	0.16	0.15	5E+05	8.9E-08	5.9E-07	2.8E-07	3.73	820

SS-single solute, MSC pH4-acidified leachate, Bd- bulk density, n- porosity, ne- effective porosity, DV-Darcy velocity, ALV-average linear velocity, D$_L$-hydrodynamic dispersion coefficient, Pe-Peclet Number, RT-unretarded residence time

Table 2. Physical and hydraulic parameters for column and field tests.

RESULTS

The results of the batch tests are plotted as isotherms for each solute with the equilibrium concentration in solution (mg/l) on the abscissa against the amount removed from solution in the solid phase (mg/g). The results for the combined inorganic and organic (MSC) synthetic leachate sorption tests with the spoil are shown in Figure 1.

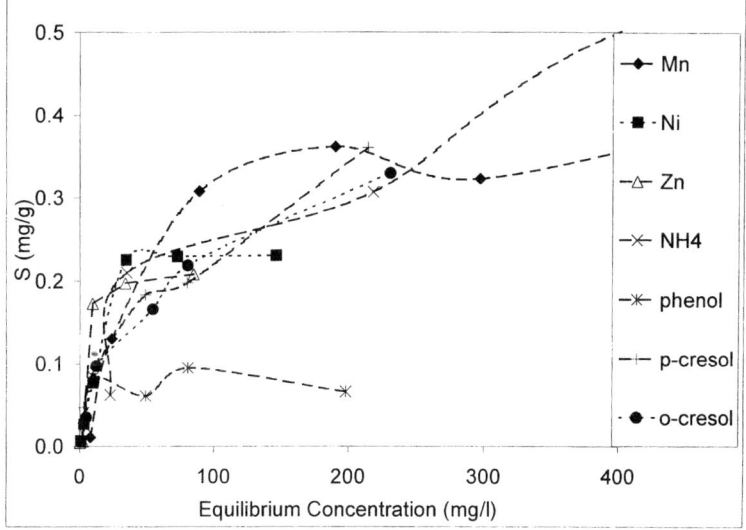

Figure 1. Batch test isotherms for synthetic MSC leachate with colliery spoil

All isotherms show non-linear forms over the solute concentration range studied with an approximate linear range over the first portion of the curve up to approximately 100 mg/l. The linear portions of the isotherms were used to determine solute partition coefficients (K_d values) for sorption of solutes to allow comparisons between the column and field test results. The K_d values obtained in the batch tests using colliery spoil and sand as a blank are summarised in Table 3. The K_d values for the natural leachate tests were similar to those in the synthetic leachate tests. This indicates that the effect of the natural leachate matrix (e.g. presence of humic substances etc.) has not reduced the sorption of organic or inorganic solutes onto the spoil. Organic solutes in samples without azide additions all showed increased K_d values that may be attributed to microbial degradation of the organic compounds. Batch tests carried out for 6 hr, 24 hr, 48 hr, 4 days, 7 days, 15 days, 28 days, and 60 days duration showed no significant changes in solute concentration occurring after approximately 5 days indicating that chemical equilibrium had been reached by this time. Concentrations of Ca, Na, and K in solution varied the greatest, attributed to desorption of these cations from the spoils due to cation exchange processes with the leachate components.

The solute most influenced by the solution:solid ratio was Ni, showing an increasing K_d with increasing solution:solid ratio with a factor of 5 variation between the 2.5:1 and 500:1 solution:solid ratios. Similar results were observed in the single solute Ni test with increased sorption values for higher ratios. The distribution coefficients for phenol and ammonium increased significantly above a ratio of 100:1. However, Mn sorption was relatively unaffected by the solution:solid ratio. The dissolved silicon content decreased with increasing solution:solid ratio due to the lower proportion of solid present in both multisolute and single solute tests.

Column tests
Both saturated and unsaturated test results are plotted as column test breakthrough curves with C/Co against pore volumes eluted from the column, with sorption and desorption phases included. Retardation (R_d) values were obtained by assuming breakthrough occurred at C/Co =0.5, since the input solute concentration used was within the linear range observed in the batch tests, and were converted to K_d values where it was assumed that sorption processes were linear. The contaminant breakthrough curve for unsaturated contaminant flow through spoil is plotted in Figure 2, with C/Co against volumetric water content eluted. The results are summarised in Table 3.

The breakthrough curve for Na shows negligible retardation since it emerges from the column at the same time as the unreactive chloride. K, Mg, and Mn show some retardation, as does NH_4. Zn and Ni are more retarded within the spoils than the other solutes. Extrapolation of the curves suggests breakthrough of Ni and Zn occur at approximately 300 pore volumes. The pH of the effluent increases from pH 4 at the start of the input phase to pH 4.5 at the end of the phase. Solute concentrations rapidly decrease when deionised water was passed through the column. Ni and Zn decrease most rapidly indicating the least amount of desorption, whereas K shows high concentrations in the leachate indicating desorption was occurring. The pH of the effluent increased during the deionised water leaching phase from pH 4.5 to pH 5.0. The spoils retained significant quantities of Zn, Ni, Mn, and to a lesser extent NH_4 after desorption with deionised water. This indicates that the majority of the retention mechanisms for these solutes are irreversible with water. Ammonium concentration was not observed to decrease and nitrate levels did not increase in the column test suggesting that ammonium nitrification was not occurring.

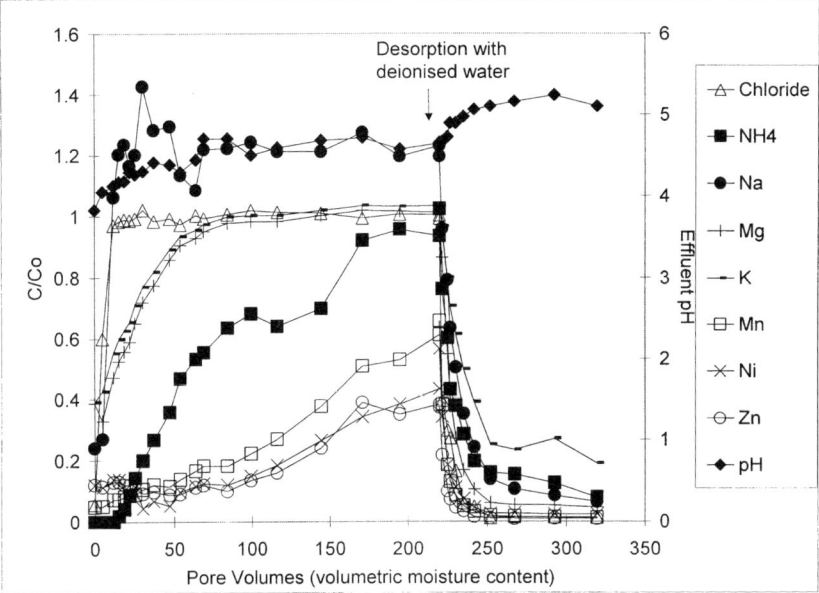

Figure 2. Multisolute breakthrough curve during unsaturated flow test with spoil.

The solutes were not significantly retarded by the quartz sand column since all of the breakthrough curves emerge rapidly from the column with solute K_d values factors of approximately 20 less than in the tests with spoils. Some solute retardation occurred however, which is attributed to precipitation onto quartz particles that function as nucleation sites and/or weakly bonded pH dependent sorption onto quartz particles (Yong et al, 1992). When the column was flushed with deionised water the solute concentration decreased rapidly with no significant tailing indicating that sorption processes were predominantly irreversible.

Comparing reduced flow rate and normal flow rate saturated test results indicated increased sorption if the water velocity within the column was reduced. This is attributed to the increased residence time for the solutes within the column allowing more time for sorption to occur. This was most significant for samples containing more organic compounds since the increased sorption time apparently results in greater diffusion of solutes into the organic matter.

Field test
During the field test only the centre three sampling devices positioned vertically below the leachate input yielded sufficient pore fluids for analysis to be performed. The breakthrough curves for solutes in Level 1 are presented in Figure 3. The solute concentrations were corrected for the leached concentration from the spoil before calculating C/Co for the breakthrough curves. This assumes that the leached concentration from the spoil was constant. The K_d values obtained in the field test are summarised in Table 3.

Similar values of R_d were obtained for Levels 1 and 2 in the field test. The long tailing on the desorption curves indicate that sorption is slowly reversible. Ammonium breakthrough was not observed in Level 2 and no cation breakthrough was observed in Level 3 after 8.5 pore volumes of leaching.

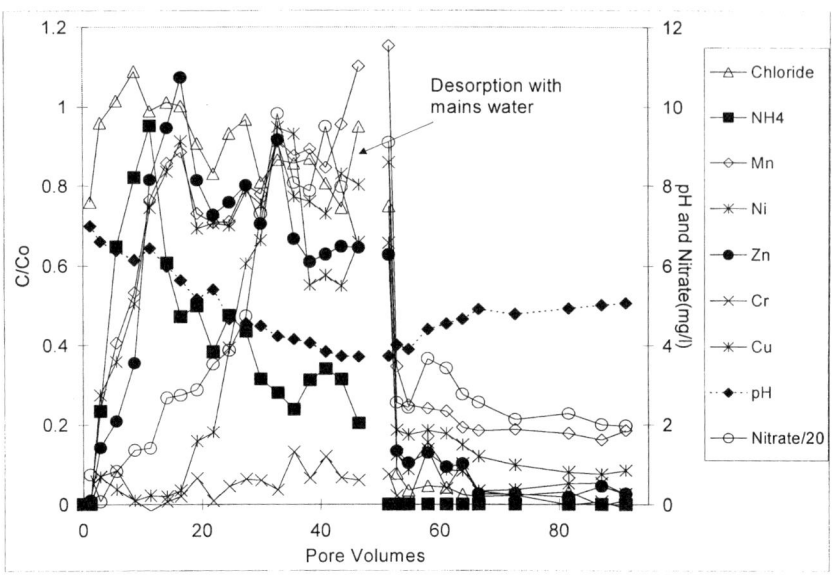

Figure 3. Field test Level 1 breakthrough curves (0.25m below input)

The ammonium concentration decreased steadily in Level 1 after the first 10 pore volumes and the nitrate concentration increased significantly. The leachate pH decreased from pH 7 to pH 3.7 during the leachate flow phase, and the effluent alkalinity decreased to zero shortly after leachate flow commenced. After the multi-solute leaching phase ceased the pH of the pore fluids slowly increased during the water desorption phase. However, pH did not recover to the values obtained during the initial water leaching phase. Nitrate, Ni and Mn levels decreased slowly showing significant tailing on the breakthrough curves.

TEST		Material	NH$_4$	Mn	Ni	Zn	Cu	Cr	phenol	o-cresol	p-cresol
Batch											
Single-solute		Spoil	3.3	-	14.0	-	-	-	8.2	-	-
Multisolute(MSI)		Spoil	3.8	3.2	4.9	7.8	-	-	2.5	4.0	4.8
Multisolute(MSI)		Sand	0.4	0.7	1	0.9	-	-	-	-	-
Multisolute(MSC)		Spoil	1.2	1.9	6.5	17.8	-	-	1.2	2.7	3.1
Natural Leachate		Spoil	1.6	-	10.5	11.6	6.6	5.1	bdl	bdl	bdl
Nat. leachate(spiked NaN3)		Spoil	1.7	-	12	3.1	7.3	85	3.4	4.8	5.9
Nat. leachate (spiked NaCl)		Spoil	1.9	-	10.4	4.5	5.3	114	12.1	6.9	27
solution:solid	2.5 (l/kg)	Spoil	1.2	2.2	6.8	0.3	-	-	6.9	7.1	10.8
"	10 (l/kg)	Spoil	1.9	3.9	13.9	3.6	-	-	5.0	9.5	10.4
"	100 (l/kg)	Spoil	4.0	3.4	21.4	16.9	-	-	7.6	8.3	11.5
"	500 (l/kg)	Spoil	19.3	2.7	29.9	45.7	-	-	22.8	15.2	14.6
Column											
Unsaturated-MSI		Sand	0.2	0.9	0.7	0.6	-	-	-	-	-
Unsaturated-MSI		Spoil	4.8	14.2	16.5	16.5	-	-	-	-	-
Saturated-MSC		Sand	0.2	0.1	0.2	0.5	-	-	0.1	0.2	0.1
Nickel single solute (SS)		Spoil	-	-	27.7	-	-	-	-	-	-
NH$_4$ single solute (SS)		Spoil	-	-	2.5	-	-	-	-	-	-
Multisolute-MSI		Spoil	1.4	2.1	2.8	3.8	-	-	-	-	-
Multisolute-MSC		Spoil	1.5	2.6	3	4.5	-	-	1.9	3	3
MSC pH4		Spoil	1	0.4	2	1.4	8.3	12.7	1.4	1.6	1.8
MSC slow flow		Spoil	2.3	2.3	3.7	nbt	-	-	3	5.6	6
Field Test											
Level 1 (0.25m)		Spoil	0.6	1.1	1.3	1.4	4.0	-	-	-	-
Level 2 (1m from input)		Spoil	-	1.4	1.4	1.6	-	-	-	-	-

Table 3. Summary table of K$_d$ (l/kg) values obtained from attenuation tests.

DISCUSSION

The order of solute sorption based on K_d values for spoil was similar in multisolute batch and column tests. The quantity of cations sorbed in terms of charge equivalents from the pH4 leachate column test (MSC pH4), where significant precipitation was not thought to be occurring, in decreasing order of charge sorbed was (in meq/kg); $NH_4^+(11.7)>Mn^{2+}$ $(3.8)>Ni^{2+}(3.0)>Zn^{2+}$ (1.9). The total cations sorbed in this test was 60.4 meq/kg, which was slightly less than the cation exchange capacity of the spoil indicating cation exchange processes were prevailing at low pH.   This sorption order closely follows the unhydrated cation radii similar to that observed by Yong et al, (1992) as (in nanometers); $NH_4^+(0.143)>$ $Mn^{2+}(0.091)> >Ni^{2+}(0.078)>Zn^{2+}(0.074)$.   However, ammonium sorption was expected to be higher than for the other solutes studied due to the higher concentration used in the leachates creating a greater mass action for ammonium sorption. The order of organic solute sorption onto the spoil in terms of K_d was the same in all tests. In decreasing order, which inversely follows the order of aqueous solubility of these compounds, the order was (in g/l at 25°C); p-cresol(23)>o-cresol(31)>phenol(87).   This indicates that hydrophobic sorption onto spoil organic matter is the predominant sorption process for the organic solutes.

In general, the flow test (column and field) metal K_d values were lower than those obtained in the batch test, which is most significant for the pH4 MSC leachate. Furthermore, field test K_d values were approximately factors of 2 less than those obtained in the column tests. Column test organic K_d values were slightly higher than those for batch tests, with the exception of the acidified leachate tests that were significantly lower than batch test values. Ammonium K_d batch values were not significantly different to the flow test, with the exception of the field test K_d that showed a factor of 6 reduction in K_d compared with the batch values. These results suggest that using batch tests to determine K_d values may overestimate the retention of contaminants within the landfill liner. This is considered to be due to a number of factors including larger spoil surface available for sorption in the batch tests, and the failure to attain local chemical equilibrium within the column tests. This is highlighted by the observed equilibration time of up to 5 days required in the batch test method with the multisolute leachate. Therefore, chemical equilibrium may not be reached within the column since the average retention time for retarded solutes is rarely in excess of 1 day. However, longer solute residence times should occur within an attenuating medium in a field situation giving sufficient time for local equilibrium conditions to be attained.

The batch tests results indicated that the solution:solid ratio may also affect the sorption processes with higher K_d values obtained at higher solution:solid ratios. Similar results were observed by Voice et al. (1983) who attributed the cause of the particle concentration effect to dissolved particles (e.g. inorganic colloids or dissolved organic matter) originating from the solids that bind with solutes and hold them in solution producing decreased distribution coefficients.   The high Si content is indicative of inorganic colloids originating from clay minerals. Because column and field tests have a significantly lower effective solution:solid ratio than batch tests, sorption parameters may also be higher in batch tests compared with column tests due to the decreased colloidal concentration.

The decrease in pH observed in the field test Level 1 was not observed in any of the batch or column tests in the present study. The oxidation of pyrite within the colliery spoils was not thought to contribute significantly to the decrease in leachate pH since this process was not observed during the initial water leaching phase. This significant decrease in pH observed in Level 1 after the input of the synthetic leachate is considered to be caused by the oxidation of ammonium (NH_4) in the plume.   The oxidants are probably free oxygen in the partially saturated zone. Torstensson et al. (1998) suggested that NH_4 oxidation occurs by a two stage microbially mediated process in which ammonium is converted to nitrite followed by the conversion of nitrite to nitrate with the overall denitrifying reaction being;

$$NH_4 + 2O_2 \rightarrow NO_3 + 2H^+ + H_2O$$

The field test Level 1 test indicates an initial lag phase of approximately 60 days before significant oxidation of ammonium occurs. This may be attributed to the time required for development of a sufficiently vibrant nitrifying bacterial community. This may explain why no ammonium oxidation was observed in the column tests since the residence time in the unsaturated columns (no biocide used), was only 10 hours which would not be sufficient time for a nitrifying bacteria population to develop. Torstensson et al. (1998) suggested that this oxidation process would be restricted to the edges of a natural landfill leachate plume where oxygenated and contaminated groundwater mix. Therefore, ammonium oxidation may not be an effective attenuation mechanism in landfill liners due to anaerobic conditions prevailing.

CONCLUDING REMARKS

The results show that there is a potential for both inorganic and organic components of landfill leachate to be attenuated by colliery spoils. However, solute retention is dependent on many factors including test method and test conditions. The relatively short test time and the significantly lower number of samples required for determination of attenuation parameters for the batch test compared with the column test renders it most useful in assessing the attenuation characteristics of colliery spoils, particularly within a construction quality assurance framework. However, batch tests may overestimate K_d values compared with those apparently occurring in the field, and column test values may also overestimate K_d if the columns are not large enough to include field particle size distributions and opportunities for macroscopic channelling within the samples. Therefore, the test method used should be carefully chosen and standardised with the variability of laboratory obtained parameters to those occurring in the field being incorporated into risk assessment models. Further comparison of batch and column test results to data obtained from natural leachate migration through landfill liners is required since the results obtained from the field study, although useful, may not truly replicate attenuation processes occurring in -situ at the base of a landfill.

Acknowledgements

The research was funded by the Centre for Environmental and Engineering Geosciences, University of Sheffield. The authors would like to thank those at the Welbeck Land Reclamation Scheme for their co-operation during the research.

References

Christensen T.H., Kjeldsen P., Albrechestsen H, Heron G., Nielsen P.H., Bjerg P.L., and Holm P.E. (1994). Attenuation of landfill leachate pollutants in aquifers. Crit. Rev. Env. Sci. and Technol. Vol 24, No 2., 119-202.

Taylor, R.K. (1985). Cation exchange in clays and mudrocks by methylene blue. J. of Chem Technol and Biotech, vol 35A, 195-207

Torstensson, D., Thornton, S.F., Broholm, M.M., and Lerner, D.N. (1988). Hydrogeochemistry of pollutant attenuation in groundwater contaminated by coal tar wastes. In Contaminated Land and Groundwater: Future Directions. Geol. Soc. Lond. Eng. Geol. Spec. Pub., 14, 149-157

Voice, T.C., Rice, P., Weber, W.J. (1983). Effect of solids concentration on the sorptive partitioning of hydrophobic pollutants in aquatic systems. Env. Sci. and Tech., Vol. 17, No.9 pp 513-518.

Waste Management Paper 26B, (1995). Landfill design, construction and operational practice, Department of the Environment Publication. U.K.

Yong, R.N., Mohamed, A.M.O., and Warkentin, B.P. (1992). Principles of Contaminant Transport in Soils. Devel. in Geotech. Eng., 73. Elsiever Press, Amsterdam.

Chemical Compatibility of Modified Bentonite Permeated with Inorganic Chemical Solutions

TAKESHI KATSUMI[1], MASANOBU ONIKATA[2], SHINYA HASEGAWA[3], LING-CHU LIN[4], MITSUJI KONDO[5], and MASASHI KAMON[6]
[1] Department of Civil Engineering, Ritsumeikan University, Kusatsu, Shiga, Japan
[2] Hojun Co., Ltd., Annaka, Gunma, Japan
[3] Sansui Consultant Co., Ltd., Kyoto, Japan (Formerly, Ritsumeikan University)
[4] Formerly, Disaster Prevention Research Institute, Kyoto University, Uji, Kyoto, Japan
[5] Hojun Co., Ltd., Tokyo, Japan
[6] Disaster Prevention Research Institute, Kyoto University, Uji, Kyoto, Japan

ABSTRACT: The chemical compatibility of multiswellable bentonite (MSB), permeated with sodium chloride (NaCl) and calcium chloride (CaCl$_2$) solutions, were evaluated to discuss its applicability as landfill barrier material. From the results of free swell, liquid limit, and hydraulic conductivity tests, MSB was found to exhibit higher swelling power and lower hydraulic conductivity than natural bentonite (NB) for all levels of concentration of 0-2 mol/L NaCl and 0-0.5 mol/L CaCl$_2$ solutions. In particular, the hydraulic conductivity values of MSB were one to two orders of magnitude lower than those of NB for all concentration levels tested. There is a possibility that the liquid limit, rather than the swelling power, may provide appropriate predictions for the hydraulic conductivity as an index. It is concluded that MSB is expected to be an excellent alternative barrier material for landfill liner systems.

INTRODUCTION

Bentonite-soil mixtures and/or geosynthetic clay liners (GCLs) have recently been proposed as mineral barriers for landfill liners and cover systems, since bentonite exhibits high swelling power which results in extremely low hydraulic conductivity. However, there is a significant concern as to their performance against chemical solutions and waste leachates, because these liquids might adversely affect the osmotic swelling of the bentonite. One of the most important results from several recent research works on the chemical compatibility of GCLs is that the first wetting liquid, rather than the following permeant, may dominate the hydraulic conductivity, and it is likely that the prehydrated GCLs can maintain low hydraulic conductivity against the permeation of chemical solutions even though the direct permeation of these solutions would have a significant effect on the non-prehydrated GCLs (e.g., Daniel et al. 1993, Ruhl and Daniel 1997, Petrov and Rowe 1997, Jo et al. 2001). Shackelford et al. (2000) pointed out, however, that the hydraulic conductivity tests reported in most of these research works were conducted for only a short duration such that a chemical equilibrium had not yet been achieved. Furthermore, the results of the hydraulic conductivity tests on prehydrated GCLs, permeated with CaCl$_2$ solutions, showed that prehydration does not always result in low hydraulic conductivity (Vasco et al. 2001).

The use of chemical resistant bentonite is considered to be a greatly anticipated measure against chemical attack. Thus, several types of modified bentonites have recently been developed to improve the chemical incompatibility (e.g., Onikata et al. 1996 and 1999, Lo et al. 1997). Prior to the practical application of these modified bentonites, the performance of these bentonites must be evaluated. The hydraulic conductivity and swelling

Geoenvironmental impact management, Thomas Telford, London, 2001.

characteristics of multiswellable bentonite (MSB), developed by Onikata et al. (1996 and 1999), as well as natural bentonite, were evaluated in this study in order to discuss their applicability as landfill barrier materials.

MULTISWELLABLE BENTONITE (MSB)

Bentonites have been considered excellent materials for waste containment barriers because of their low hydraulic conductivity and high adsorption capacity. The remarkable swelling of natural bentonite, which results in extremely low hydraulic conductivity, is attributed to the osmotic swelling of smectites. The interlayer cations of smectites play an important role in osmotic swelling. Na- and Li-rich bentonites infinitely swell with water (H_2O) due to osmotic swelling. However, when polyvalent cations exist at the exchangeable sites of smectite, limited swelling is observed because only crystalline swelling, not osmotic swelling, occurs. The chemical compositions of permeants also have a significant effect on the hydraulic conductivity of GCLs. Divalent and trivalent cations in permeants result in a significant increase in the hydraulic conductivity of GCLs. In addition, a high concentration of monovalent cations limits the osmotic swelling of bentonite.

Onikata et al. (1996) found that propylene carbonate (PC) forms complexes with smectite and activates osmotic swelling, even in aqueous electrolyte solutions such as sea water. Multiswellable bentonite (MSB) is bentonite which has been treated with propylene carbonate (PC). It activates the osmotic swelling of bentonite even in chemical solutions, where natural bentonite (NB) does not swell. MSB, mixed with 25% PC, was used for the experiments in this study. The NB used in this study was produced by Hojun Kogyo Co., Ltd. The MSB was processed from the same NB, also produced by Hojun Kogyo Co., Ltd.

EXPERIMENTAL RESULTS AND DISCUSSION
Index Properties - Free Swell and Liquid Limit

Free swell tests were conducted according to ASTM D 5890. The swelling power (swell volume in mL/2 g-solid) of NB and MSB for both NaCl and $CaCl_2$ solutions is plotted in Figure 1. In the $CaCl_2$ solutions, increases in the concentration of $CaCl_2$ reduced the swelling power, and concentrations higher than 0.3 mol/L resulted in a significant effect on the swelling. In the NaCl solutions, concentrations lower than 0.5 mol/L provided higher swelling power for both NB and MSB than distilled water (0 mol/L concentration). This is probably attributed to the effects of the flocculation and the dispersion of bentonite particles, which occurred in the low concentration monovalent cations. The low concentration of Na^+ is considered to contribute to the dispersement of more bentonite particles, and to result in a larger swelling volume.

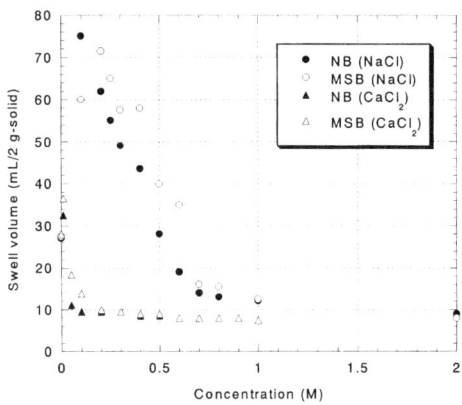

Fig. 1.   Swelling power of NB and MSB in NaCl and $CaCl_2$ solutions

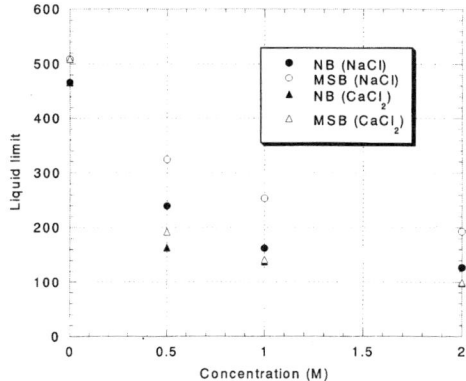

Fig. 2.   Liquid limit of NB and MSB with NaCl and CaCl$_2$ solutions

In NaCl solutions, the swelling power of MSB is clearly greater than that of NB for concentrations lower than 0.6 mol/L, while the difference in swelling power is negligible for concentrations higher than 0.7 mol/L. A larger swelling power for MSB than NB was observed for CaCl$_2$ concentrations lower than 0.5 mol/L. The swelling power of 10 mL/2 g-solid, which is considered the minimum value for the swelling power of this bentonite, was achieved for NB with CaCl$_2$ concentrations higher than only 0.1 mol/L, while the same value was not achieved for MSB when the CaCl$_2$ concentrations were lower than 0.3 mol/L. It can be concluded, therefore, that MSB exhibits higher swelling power than NB for electrolyte solutions which have concentrations lower than a certain level.

The liquid limit values, which were measured according to JIS A 1205, are plotted in Figure 2. The CaCl$_2$ solutions provided lower liquid limit values than the NaCl solutions for all concentration levels, because fewer water molecules are contained in the interlayer of smectite. MSB exhibits higher liquid limit values than NB, and this tendency is clearer for NaCl solutions than for CaCl$_2$ solutions. NaCl solutions with concentrations lower than 0.5 mol/L resulted in lower liquid limit values than distilled water. This is not consistent with the results from the swelling power indicated in Figure 1. The swelling power of the NaCl solutions for the same concentration levels was clearly higher than that for distilled water. As previously indicated, not only osmotic/crystalline swelling, but also the dispersion and the flocculation of smectite particles, are considered to reflect the free swell test results. For the purpose of this research, that is, to evaluate the hydraulic containment performance, evaluating only the swelling is considered more appropriate for index tests. Thus, the liquid limit is considered in order to provide a better index for hydraulic containment barrier materials.

Hydraulic Conductivity

Hydraulic conductivity tests were performed using flexible-wall permeameters, according to ASTM D 5084. Due to the limited number of permeameters, specimens with a diameter of 6 cm were prepared for NaCl solutions as permeants, while specimens with a diameter of 10 cm were used for the CaCl$_2$ solutions. NB or MSB granules were placed between the top and the bottom pedestals to achieve a thickness of approximately 1 cm. This condition is supposed to simulate the performance in cases where these bentonites are used as GCLs, as well as simply to evaluate the chemical compatibility of the bentonites. Each specimen had a dry density of 0.79 g/cm^3. Under a cell pressure of 20-30 kPa, the bentonite specimens were exposed to the permeant liquid for longer than 24 hours. Then, a hydraulic gradient of 80-90 was applied to the specimens. The hydraulic conductivity was calculated from the measured influent and effluent volumes. In addition, pH and electrical conductivity were determined for effluent to check the chemical equilibrium, as suggested by Shackelford et al. (2000), although these data are not reported in this paper.

The changes in hydraulic conductivity with flow volume are shown in Figure 3 for the permeation of NaCl solutions and in Figure 4 for $CaCl_2$ solutions. As indicated in these figures, bentonites which have low hydraulic conductivity values, permeated with low concentration solutions, were not tested for a long enough duration to achieve a chemical equilibrium. These tests are still ongoing, but only the recently obtained results are discussed in this paper. The latest values for hydraulic conductivity are plotted in Figure 5.

The hydraulic conductivity values of NB, permeated with NaCl solutions, were lower than 10^{-8} cm/s for concentrations lower than 0.5 mol/L, but were higher than 10^{-7} cm/s for concentrations higher than 1.0 mol/L. Petrov and Rowe (1997) conducted hydraulic conductivity tests on needle-punched GCLs permeated with 0.6 and 2.0 mol/L NaCl solutions, and reported that the hydraulic conductivity values with a 0.6 mol/L NaCl solution was 10^{-7} cm/s. The results of Jo et al. (2001) indicated that the permeation of a 0.1 mol/L NaCl solution on similar needle-punched GCLs exhibits a hydraulic conductivity value of 10^{-9} cm/s, which is almost equal to the values permeated with de-ionized water, while the permeation of a 1.0 mol/L NaCl solution increased the hydraulic conductivity significantly. The results reported in this paper are consistent with the previous findings.

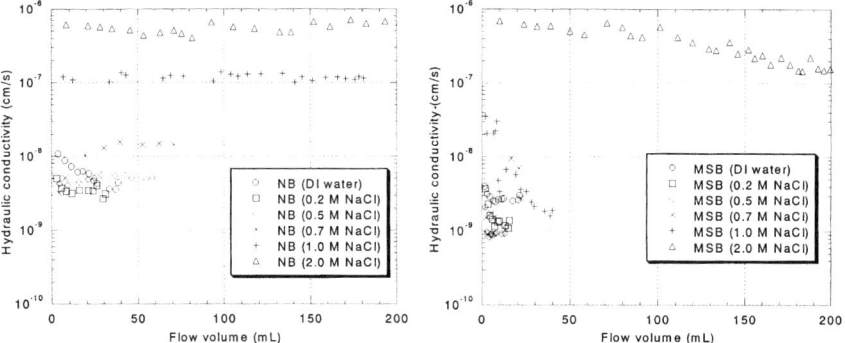

Fig. 3.   Change in hydraulic conductivity of NB and MSB permeated with NaCl solutions

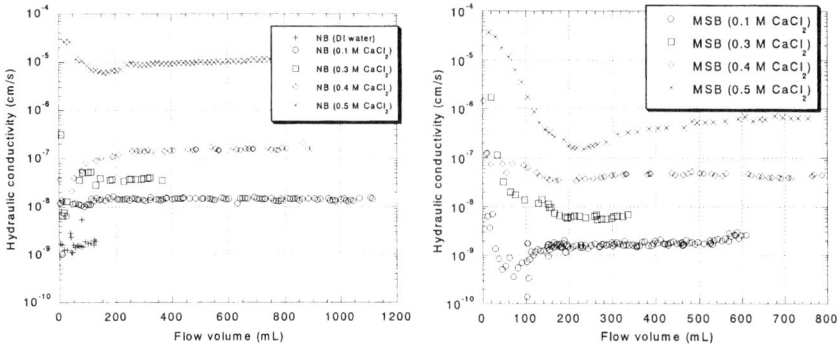

Fig. 4.   Change in hydraulic conductivity of NB and MSB permeated with $CaCl_2$ solutions

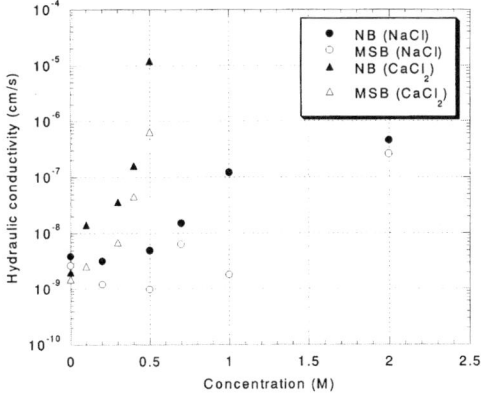

Fig. 5.   Hydraulic conductivity values of NB and MSB with the concentration of NaCl and CaCl$_2$ solutions

MSB exhibited a hydraulic conductivity value lower than 10^{-8} cm/s against NaCl solutions lower than 1.0 mol/L, and only the case of a permeation of a 2.0 mol/L NaCl solution increased the hydraulic conductivity value to higher than 10^{-7} cm/s. As indicated in Figure 5, MSB exhibited lower hydraulic conductivity values than NB for all concentration levels tested. Thus, it can be concluded that MSB is more compatible against NaCl than NB. It is recommended, however, that the tests be continued to achieve a chemical equilibrium until the final conclusion is derived.

Hydraulic conductivity tests with CaCl$_2$ solutions were conducted for 0.1, 0.3, 0.4, and 0.5 mol/L. Even the 0.1 mol/L solution increased the hydraulic conductivity value to 10^{-8} cm/s, and the hydraulic conductivity value for the 0.5 mol/L solution was in the order of 10^{-5} cm/s. Jo et al. (2001) reported that significant increases in the hydraulic conductivity of needle-punched GCLs were observed when the concentration of divalent cations increased from 0.01 to 0.1 mol/L, and that the hydraulic conductivity value was higher than 10^{-5} cm/s for concentrations higher than 0.1 mol/L. Thus, the bentonite used in this study is considered more compatible against CaCl$_2$ than the bentonite tested by Jo et al. (2001). As indicated in Figure 5, MSB always resulted in approximately one order of magnitude of hydraulic conductivity values lower than NB for all concentration levels. In particular, CaCl$_2$ concentrations lower than 0.4 mol/L contributed to a hydraulic conductivity value lower than 10^{-7} cm/s.

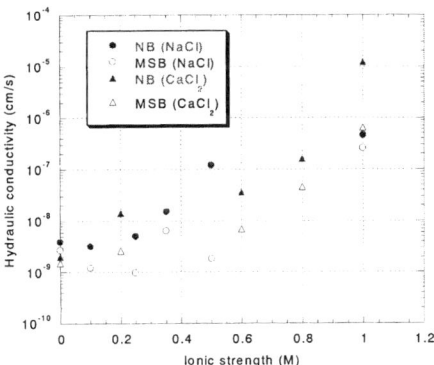

Fig. 6.   Hydraulic conductivity values of NB and MSB with the ionic strength of cations of NaCl and CaCl$_2$ solutions

Figure 6 illustrates the hydraulic conductivity values with the ionic strength of cations. Ionic strength is defined as $I = \Sigma \ c_i \ z_i^2$ (where c_i = concentration in mol/L and z_i = valence), but only the cations are considered herein. Ionic strength is considered to be an index used to explain the behavior of clay-chemical interactions. For example, the thickness of diffuse double layers on clay particles is in inverse proportion to the square route of the ionic strength of the cations (Mitchell 1993). As shown in Figure 6, the hydraulic conductivity can be determined from the ionic strength for both NaCl and CaCl₂ solutions. For NB, ionic strength values higher than 0.2 mol/L resulted in hydraulic conductivity values higher than 10^{-8} cm/s, while the hydraulic conductivity of MSB did not increase, but stayed at almost the same value for distilled water when the ionic strength value was lower than 0.5 mol/L. Even though the ionic strength value is higher than 0.5 mol/L, MSB exhibited one to two orders of magnitude of hydraulic conductivity lower than NB.

CONCLUSIONS

Free swell, liquid limit, and hydraulic conductivity tests were conducted on MSB and NB with NaCl and CaCl₂ solutions. The experimental results show that MSB exhibits higher swelling and lower hydraulic conductivity values for NaCl at all levels of concentration of 0-2 mol/L, and for 0-0.5 mol/L CaCl₂ solutions. In particular, the hydraulic conductivity values of MSB are always one to two orders of magnitude lower than those of NB. There is a possibility that the liquid limit may provide predictions for the hydraulic conductivity. In conclusion, MSB is expected to be an excellent alternative barrier material for landfill liner systems.

REFERENCES

Daniel, D.E., Shan, H.-Y., and Anderson, J.D. (1993): Effects of partial wetting on the performance of the bentonite component of a geosynthetic clay liner, *Geosynthetics'93*, IFAI, pp.1482-1496.

Jo, H.Y., Katsumi, T., Benson, C.H., and Edil, T.B. (2001): Hydraulic conductivity and swelling of non-prehydrated GCLs permeated with single species salt solutions, *Journal of Geotechnical and Geoenvironmental Engineering*, ASCE (accepted for publication).

Lo, I.M.C., Mak, R.K.M., and Lee, S.C.H. (1997): Modified clays for waste containment and pollutant attenuation, *Journal of Environmental Engineering*, ASCE, Vol.123, No.1, pp.25-32.

Mitchell, J.K. (1993): *Fundamentals of Soil Behavior, Second Edition*, Jon Wiley & Sons.

Onikata, M., Kondo, M., and Kamon, M. (1996): Development and characterization of a multiswellable bentonite, *Environmental Geotechnics*, M. Kamon (ed.), Balkema, pp.587-590.

Onikata, M., Kondo, M., Hayashi, N., and Yamanaka, S. (1999): Complex formation of cation-exchanged montmorillonites with propylene carbonate: Osmotic swelling in aqueous electrolyte solutions, *Clays and Clay Minerals*, Vol.47, No.5, pp.672-677.

Petrov, R.J. and Rowe, R.K. (1997): Geosynthetic clay liner (GCL) - chemical compatibility by hydraulic conductivity testing and factors impacting its performance, *Canadian Geotechnical Journal*, Vol.34, pp.863-885.

Ruhl, J.L. and Daniel, D.E. (1997): Geosynthetic clay liners permeated with chemical solutions and leachates, *Journal of Geotechnical and Geoenvironmental Engineering*, ASCE, Vol.123, No.4, pp.369-381.

Shackelford, C.D., Benson, C.H., Katsumi, T., Edil, T.B., and Lin, L. (2000): Evaluating the hydraulic conductivity of GCLs permeated with non-standard liquids, *Geotextiles and Geomembranes*, Elsevier, Vol.18, Nos.2-3, pp.133-161.

Vasko, S.M., Jo, H.Y., Benson, C.H., Edil, T.B., and Katsumi, T. (2001): Hydraulic conductivity of partially prehydrated geosynthetic clay liners permeated with aqueous calcium chloride solutions, *Geosynthetics Conference 2001*, IFAI, pp.685-699.

Evaluation of Influence of Transport Mechanisms on Contaminant Penetration through Landfill Lining System

JIRI KOSTAL
Department of Geotechnics, Faculty of Civil Engineering, The Czech Technical University
Thákurova 7, 166 29 Prague 6, Czech Republic
E-mail: kostalj@email.cz

ABSTRACT

This paper focuses on the possibility how to design clay liners of waste disposals more efficiently in comparison to the requirements of the relating norms. The approach to more effective design comes from one dimensional analytical solution of the contaminant transport equation for simplified assumptions. The Author graphically shows the influences of each transport process (advection, diffusion and sorption) on the contaminant movement through the clay liner. At the end the influence of the values of coefficients (k, D*, Rd) of each transport mechanism is evaluated.

INTRODUCTION

Clay lining takes important role for protection of the environment in the case of contaminant spreading from the different types of the landfills. This is due to the low permeability and high sorption capacity of the clay. Significant is also the high swelling potential of the clay, which can seal some defects in the lining (Vanicek (1998)). For the mineral protection barrier its thickness is more significant in comparison with membrane sealing mainly from the time of contaminant breakthrough.

For the evaluation of the influence of the clay lining two main principles can be used:
- Usage of the codes of practice (norms), which define required thickness of the clay barrier for fulfilling of the value of the permeability coefficient for the different types of waste deposits. The norms expect that when fulfilling theirs requirements the protection of the environment is sufficient.
- Usage of the individual design of the sealing barrier by calculation of the contaminant spreading through this sealing. Afterwards it is necessary to check the outflow of the contaminant concentration and amount at the bottom of the barrier and compare it to the defined limits.

Here we assume the second principle for the determination of the influence of different parameters on the contaminant spreading through the clay barrier. Because not only coefficient of permeability is important, but also the other transport mechanisms like diffusion, dispersion and sorption are significant.

ANALYTICAL SOLUTION OF THE CONTAMINANT TRANSPORT EQUATION

The basic equation for the contaminant transport in one dimensional problem can be by Shackelford (1996) written as:

$$R_d \frac{\partial c}{\partial t} = D\frac{\partial^2 c}{\partial x^2} - v\frac{\partial c}{\partial x} - \lambda_w c - (R_d - 1)\lambda_s c \qquad (1)$$

Using the equation it is possible to define the contaminant concentration c in specified point x and time t, if the hereafter declared properties of the clay soil are known.

- mean filtration velocity through the lining is defined by $v = v_D / n$, where n is total porosity of soil and v_D is flow velocity from Darcy's law, in which $v_D = k \cdot i$, where k is coefficient of permeability and i is hydraulic gradient;
- retardation factor R_D;
- coefficient of hydrodynamic dispersion D;
- first-order reaction rate constants λ_w a λ_s;

If we presume that the reaction constants are equal ($\lambda_w = \lambda_s = \lambda$), than the contaminant transport equation becomes:

$$R_d \frac{\partial c}{\partial t} = D\frac{\partial^2 c}{\partial x^2} - v\frac{\partial c}{\partial x} - R_d \lambda c \qquad (2)$$

Typical reaction constant for radioactive materials is theirs half-time decay. When there is not decaying contaminant in the solution, the reaction rate constant becomes zero.

Analytical solution of the above mentioned partial differential equation without the influence of decaying can be written for the following border and initial conditions as:

$$\frac{c(x = L, t)}{c_0} = 0,5 \cdot (erfc(\xi_1) + \exp(\xi_2) \cdot erfc(\xi_3)) \qquad (3)$$

border conditions: $c(0,t) = c_0$

$c(\infty,t) = 0$

initial condition: $c(x,0) = 0$

In the equation (3) the variables ξ_1, ξ_2 and ξ_3 are defined as:

$$\xi_1 = \frac{R_d L - v t}{2\sqrt{R_d D t}} \quad ; \quad \xi_2 = \frac{v L}{D} \quad ; \quad \xi_3 = \frac{R_d L + v t}{2\sqrt{R_d D t}} \qquad (4)$$

The above mentioned analytical solution was transformed to the computer model using spreadsheet. The description of the computer model is presented in Vanicek, Kostal and Vanicek (1998).

TESTED MODEL

Typical example, which exists most often, is the landfill of solid communal waste. For the protection barrier made from clay liner it is mostly defined by the requirement that the thickness should be at least 0,6 m with coefficient of permeability lower than $1 \cdot 10^{-9}$ m/s. Schematic cross-section of the landfill during the filling can be seen in figure 1. together with the detail of the bottom sealing. In this case the clay lining is in direct contact with the subsoil, which has significantly higher permeability. Above the sealing layer there is a drainage layer, which is below deposited waste material.

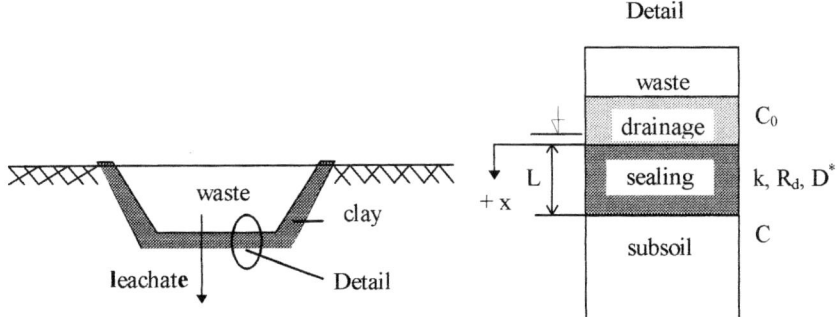

Figure 1.: Schematic cross-section of landfill

For the calculation of the penetration of the contaminant through the sealing layer the following assumptions have been used.

Filtration velocity is constant and is defined by the coefficient of permeability k and hydraulic gradient i. In this case we assumed, that the maximum height of the leachate in the drainage layer should be 0,05 m. This assumption is conservative. The next assumption is higher permeability of the subsoil, so the hydraulic gradient is calculated from the hydraulic head of 0,65 m and the flowing distance of 0,60 m.

Due to the fact that the calculation is general, we used constant contaminant concentration in the leachate c_0 equal to 1. Concentration of the contaminant determined in the profile of the clay sealing layer is described as c and in our case it is in fact a ratio from the initial concentration. Most important for us is the ratio of concentration in which the contaminant is leaving the sealing layer and entering the subsoil. Constant concentration of contaminant in the leachate is big simplification. In fact the concentration will decrease with time because of the chemical processes in the waste material and due to the fact that the landfill will be covered and only an minimum of water can get in.

Dispersion and diffusion processes are increasing the contaminant spreading. Both of these processes can be described by one coefficient of hydrodynamic dispersion $D = D_M + D_{ef}$, where D_M is coefficient of mechanical dispersion and D_{ef} is effective coefficient of diffusion. Due to the fact, that the coefficient of the mechanical dispersion is related to the flow velocity, its influence is in this case negligible. So the hydrodynamic dispersion D is given only by the coefficient of the effective diffusion D_{ef}, which is also written as D^* ($D = D_{ef} = D^*$).
Approximate values in the literature of the required coefficients varied based on the author and theirs definition. Comparison of different definitions can be seen in Shackelford and Daniel (1991). The basic equation for the coefficient of effective diffusion D^* can be written as:

$$D^* = D_0 \cdot \tau \qquad (3)$$

where D_0 is the diffusion coefficient of the contaminant in the solution and τ is tortuosity (influence of the non straight contaminant movement in soil). Some authors takes into account also the viscosity of the solution in soil, influence of the ion exchange or volumetric water content. For the basic definition of the coefficient D^* its value ranges in literature (Mitchell (1993)) from $2 \cdot 10^{-10}$ to $2 \cdot 10^{-9}$ m^2/s.

The next important process in contaminant spreading is sorption, which can be defined by the retardation factor R_d, and which decreases the contaminant movement. The value of the retardation factor is by Sutherson (1997) in the range between 1,0 and 100,0. The value of 1,0 represents the soil without the sorption capacity for the contaminant. The R_d value very much depends on the composition not only of the soil, but also of the contaminant leachate. For the simplest case the retardation factor R_d can be expressed by the equation:

$$R_d = 1 + K_d \frac{\rho_d}{n} \qquad (4)$$

where K_d is distribution coefficient, which is defined as the ratio between the sorbed concentration and concentration in solution, n is porosity and ρ_d is dry bulk density of the soil. For our modelling purposes we concentrated on the influence of separate parameters on the contaminant spreading in time t from 1 to 50 years.

INFLUENCE OF THE PERMEABILITY COEFFICIENT

In modelled case we assumed constant values for the D^* and R_d ($D^* = 2 \cdot 10^{-10}$ m^2/s, $R_d = 1,7$) and we changed the value of the permeability coefficient k in the order of one magnitude ($1 \cdot 10^{-9}$ m/s or $1 \cdot 10^{-10}$ m/s). The results of the ratio of contaminant concentration of this comparison for the whole thickness of the liner are presented in figure 2. for different times (1, 5, 10, 20, 30, 40 and 50 years). The differences are not so big, while the decrease of penetration is viewable. It can be seen from the results, that when permeability is low, diffusion is more important that advection. But the positive influence of the lower value of permeability coefficient is also in lower amount of the water flowing through the liner.

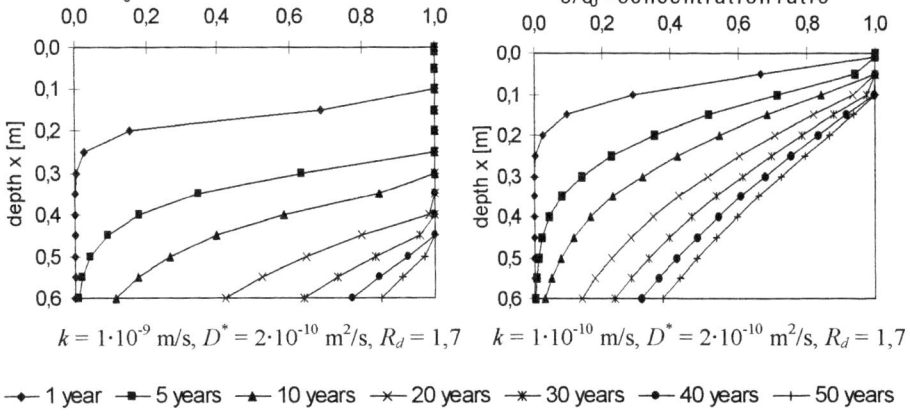

$k = 1 \cdot 10^{-9}$ m/s, $D^* = 2 \cdot 10^{-10}$ m^2/s, $R_d = 1,7$ $k = 1 \cdot 10^{-10}$ m/s, $D^* = 2 \cdot 10^{-10}$ m^2/s, $R_d = 1,7$

—♦— 1 year —■— 5 years —▲— 10 years —✕— 20 years —✳— 30 years —●— 40 years —┼— 50 years

Figure 2.: Contaminant penetration through the clay barrier for different permeability coefficients

INFLUENCE OF THE DIFFUSION COEFFICIENT

For the comparison of the influence of the diffusion coefficient D^* we kept the constant values of the permeability coefficient $k = 1 \cdot 10^{-9}$ m/s and of the retardation factor $R_d = 4,5$. The used range of diffusion coefficient in the modelling describes the maximum and minimum values ($D^* = 2 \cdot 10^{-9}$ m^2/s or. $2 \cdot 10^{-10}$ m^2/s). The results from this comparison are presented in figure 3. From the picture it can be seen the positive influence of lower value of the diffusion coefficient on the contaminant penetration through the sealing system. Roughly the slowdown of the contaminant movement is about half.

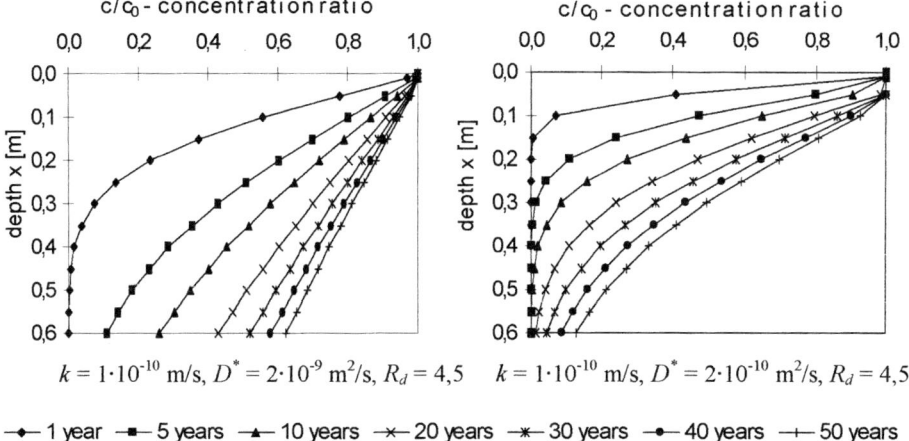

Figure 3.: Contaminant penetration through the clay barrier for different diffusion coefficients

INFLUENCE OF THE RETARDATION FACTOR

Like in the above examples the influence of the retardation factor has been studied for constant values of the others parameters, the coefficient of permeability $k = 1 \cdot 10^{-9}$ m/s and the diffusion coefficient $D^* = 2 \cdot 10^{-10}$ m^2/s - in its lower value. For retardation factor the values of 1,5 and 14,5 were investigated. The value of 1,5 represents materials with high mobility - phenols, alcohol and acetone. The value of 14,5 is typical for pollutants with intermediate mobility - naphthalene. From the result presented in figure 4. it can be seen obvious influence of pollutant's mobility on penetration through the clay barrier. When the higher value of retardation factor is used, the speed of penetration is roughly halved.

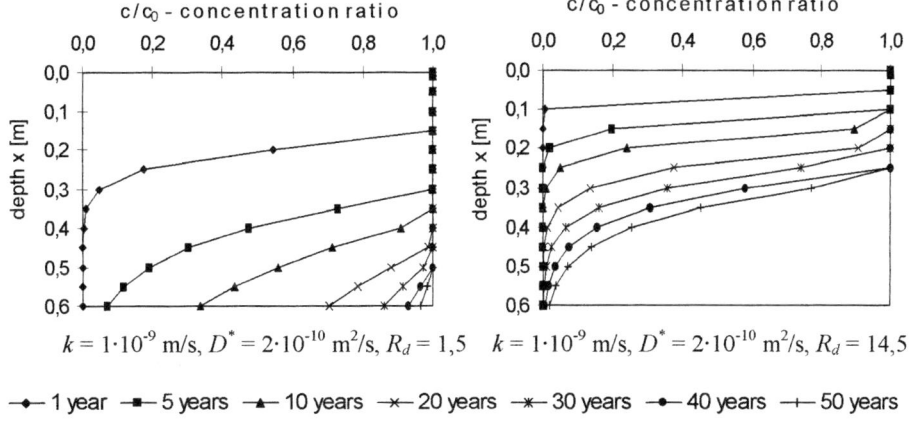

Figure 4.: Contaminant penetration through the clay barrier for different retardation factors

CONCLUSION

The real evaluation of the contaminant penetration through the clay sealing layer using the analytical one-dimensional solution of transport process shown itself as more realistic and gives us more understandable visualisation of the contaminant spreading then only fulfilment of requirement on the clay sealing liner from codes of practice. The advantage of the modelling is especially for the cases where one sort of waste is deposited in the landfill. On the other hand the modelling give us the possibility to assess the influence of different parameters on contaminant spreading. The coefficient of permeability has not such a big influence on the penetration but it decrease the amount of leakage through the sealing system. The influence of the values of the diffusion coefficient and of the retardation factor is quite high. The presented results give us new point of view on the design of the landfills and on the former practise in the landfill design.

REFERENCES

Mitchell, J.K., *Fundamentals of soil behaviour,* John Wiley & Sons, New York, second ed. 1993.

Shackelford, C. D., Modelling and analysis in environmental geotechnics, State of the Art Report, In: Proc. 2nd Int. Congress on Environmental Geotechnics, 1996, Osaka. pg. 141-171.

Shackelford, C.D., Daniel, D.E., Diffusion in Saturated Soil I - background, *Journal of Geotechnical Engineering, ASCE, Vol. 117, No 3,* March 1991.

Sutherson, S.S., *Remediation engineering. Design Concepts,* CRC Lewis Publishers, New York, 1997, pg. 3 - 25.

Vanicek, I.., Clay protection barriers and theirs geotechnics models, *Stavební obzor,* 1998, In Czech, pg.168 - 174.

Vanicek, M., Kostal, J., Vanicek, I., Analytical modelling of impact clay barriers on leakage of contaminants, In: Proc. New requirements on materials and constructions, 1998, Prague, In Czech., pg. 47 - 52.

Modelling biodegradation in landfill: a parametric study of enzymatic hydrolysis

J.R. McDOUGALL & I.C. PYRAH
School of Built Environment, Napier University, Edinburgh, UK

INTRODUCTION
The EU Landfill Directive has set a decreasing limit on the amount of degradable waste that nation states may landfill. In the meantime, and until organic matter is completely removed from the landfill waste stream, an opportunity for innovative landfilling techniques exists. The 'bioreactor' landfill is just such a technique. With optimum conditions, the bioreactor provides a near complete breakdown of the biodegradable fraction of waste refuse. From an engineering point of view, accelerated breakdown through control of the environmental conditions leads to earlier stabilisation, settlement and eventual re-use of landfill sites.

However, degradation processes at landfill scale are not well understood. Small scale testing reveals process fundamentals but at field scale the interpretation of test data is complicated by a mixture of biochemical and physical effects. To improve understanding of the landfill environment, the Authors have devised a model of landfill behaviour that incorporates physical effects and some of the key factors controlling biodegradation.

A modelling framework has been developed to investigate the hydraulic, biodegradation and settlement behaviour of landfilled refuse. The techniques employed provide a means of simulating these three aspects of landfill behaviour during the filling phase and for a given site geometry, waste type and climate. This paper, however, deals solely with the parametric performance of the biodegradation component of the model, in particular on the parameters controlling the enzymatic hydrolysis of organic material.

BIODEGRADATION MODEL
The main aim of the biodegradation model is to determine the loss of mass due to conversion of the solid organic fraction of waste. A simplified biodegradation process is defined by two distinct but interdependent systems: (a) the concentration of volatile fatty acids (VFA) in the aqueous phase and (b) a methanogenic biomass (MB) which is also concentrated in the aqueous phase. Equations describing the transport, growth and decay of each system provide a modelling framework but, due to interdependencies between the VFA and MB systems, must be solved jointly. Without modification, the biodegradation model resembles the type of two-stage anaerobic digestion model used by public health engineers. However, unlike more diffuse, aqueous substrates such as sewage sludge, waste refuse is predominantly a solid structured material where the rate and progress of decomposition are constrained by physical factors such as accessibility of the solid substrate to enzyme attack and low moisture content. Modifications to the two stage model have therefore been made and are considered briefly below; a more detailed description can be found in McDougall & Pyrah (1999).

Geoenvironmental impact management, Thomas Telford, London, 2001.

Enzymatic hydrolysis of solid waste
Following experiments on insoluble cellulose, Lee & Fan (1982; 1983) concluded that the kinetics of enzymatic hydrolysis are controlled by the structure of the cellulose substrate and the interaction activity between enzyme and substrate. Since the main degradable constituent of waste refuse is cellulose, their findings provide a useful starting point for a functional description of enzymatic hydrolysis in waste refuse. The parameters identified by Lee & Fan (1982) to describe the conversion of organic material in the solid phase to VFA in the liquid phase include a maximum hydrolysis rate, a product inhibition factor and a relative digestibility term. To capture the effect of moisture content on organic depletion in solid refuse, the Authors have added an effective moisture content term.

Maximum hydrolysis rate
This is the maximum or initial rate of hydrolysis of solid organic matter occurring under the most favourable substrate structure and interaction conditions. Cecchi et al (1988) and Wang & Banks (2000) indicate maximum volatile solid (VS) reduction rates in the range 4000g to 5000g totalVS/m^3 solution/day. Estimates of the maximum hydrolysis rate can also be made from VFA growth vs. time plots. From Barlaz et al (1989), a VFA growth rate of about 1800 mg VFA/L solution/day can be found. For the baseline simulation, a value of 2500 mg VFA/L solution/day has been selected.

Relative digestibility and structural transformation parameter
The presence of highly degradable organic matter and/or the initial colonisation and enzymatic attack of exposed waste surfaces means that initial hydrolysis rates are rapid. Remaining organic matter, having become less accessible to, or shielded from, enzymatic attack, or with an increased crystallinity, becomes less digestible and is hydrolysed at slower rates. Lee & Fan (1982) suggested that a lumped parameter, referred to as the relative digestibility, be used to reflect the combined effects of changes in accessible surface area and crystallinity. They found relative digestibility, ϕ, to be related to the extent of substrate conversion by a single parameter, n, the structural transformation parameter, i.e.

$$1] \qquad \phi = 1 - \left[\frac{S_o - S}{S_o} \right]^n$$

where S is the solid organic fraction remaining and S_o is the initial solid organic fraction.

Lee & Fan (1983) reported a value of 0.36 for the structural transformation parameter but indicated that it is probably strongly dependent on the structural features of the cellulose. In their tests, "Solka-Floc", a commercially available delignified cellulose was used but lignin, a substance which is resistant to enzymatic hydrolysis and can shield cellulose, comprises up to 15% by dry weight of the organic fraction of waste refuse (Bookter & Ham, 1982). Calculations performed on data presented by Wald et al. (1984) for rice straw, a lignified cellulose, reveal a higher value (0.7) for the structural transformation parameter.

Product inhibition
To simulate reductions in enzyme-substrate activity due to product inhibition, a function P, based on a form defined by Lee & Fan (1982) has been adopted, i.e.

$$2] \qquad P = \exp\left(- k_{VFA}(c)\right)$$

where k_{VFA} is a product inhibition factor and (c) is VFA concentration. Values for k_{VFA} are chosen to ensure that maximum VFA concentrations correspond to those reported in sites or other installations which are known to have 'soured' or 'stuck'. For the baseline simulations a value of 2×10^{-4} L/mg results in peak values of VFA of about 16,000 mg/L.

Moisture content
There is widespread agreement that one of the most important factors, and one that can be controlled most easily during the life of a landfill, is moisture content (e.g. Rees, 1980; Anon, 1995). Whilst the decay of VFA and the growth/decay of methanogenic biomass are assumed to occur within the bulk aqueous phase, hydrolysis of solid organics is a surface phenomenon occurring at the interface between the aqueous and solid phases. To control the influence of moisture on these surface processes, an effective moisture content term, which acts directly on the VFA growth function, has been introduced. It is defined as,

3]
$$\theta_E = \frac{\theta - \theta_R}{\theta_S - \theta_R}$$

where θ is volumetric moisture content and subscripts E, S & R refer to effective, saturated and residual respectively.

Combined functional description of enzymatic hydrolysis
Combining the effective moisture content term, relative digestibility, product inhibition factor and a maximum hydrolysis rate leads to,

4]
$$r_g = \theta_E b \phi P = \theta_E b \left[1 - \left[\frac{S_0 - S}{S_0} \right]^n \right] \exp\left(-k_{VFA}(c) \right)$$

where r_g is the enzymatic hydrolysis or VFA growth function

PARAMETRIC SENSITIVITY OF BIODEGRADATION MODEL TO FACTORS CONTROLLING ENZYMATIC HYDROLYSIS
A series of numerical simulations were run to evaluate the performance of the biodegradation model algorithm in relation to the enzymatic hydrolysis parameters. The impact of these factors is followed through both stages of the biodegradation process and measured using VFA and MB concentrations, and the depletion of the solid organic fraction (SOF). The simulations depict a waste volume with moisture flow effects set to zero so that growth and decay effects can be observed without interference from transport processes. A baseline simulation was established using the parameter values enumerated above. These parameters, including methanogenic stage parameters which were fixed for all the tests described herein, are given in Table 1. Results of the baseline simulation are shown in Fig.1 and compare favourably with recognised landfill behaviour.

Maximum hydrolysis rate and effective moisture content.
Since maximum hydrolysis and effective moisture content act identically on the enzymatic hydrolysis function, they could be combined into a single moisture-dependent hydrolysis rate however, it is the opinion of the Authors that the separation of these two terms offers clarity and aids parameter identification. For the purposes of these tests, the effective moisture content is fixed (at 1.0) while the maximum hydrolysis rate is varied; Figs. 2a-c show VFA, MB and SOF responses for rates of 2×, 1×, 0.5× and 0.25× baseline value (i.e.

Table 1. Details of biodegradation model parameters and initial conditions with selected values.

Parameter Type	Parameter	Notation	Value
Hydrolysis	Max. hydrolysis rate	b	2500 mg/L/day
	Structural transformation	n	0.7
	Product inhibition	k_{VFA}	2E-4 L/mg
VFA decay/MB growth	Specific growth rate	k_0	0.02 day^{-1}
	Half-saturation constant	k_{MC}	4000 mg/L/day
MB decay	Methanogen death rate	k_2	0.002 day^{-1}
	Yield coefficient	Y	0.2
Initial conditions	VFA	c	300 mg/L
	MB	m	250 mg/L
	SOF	S	310 kg/m^3

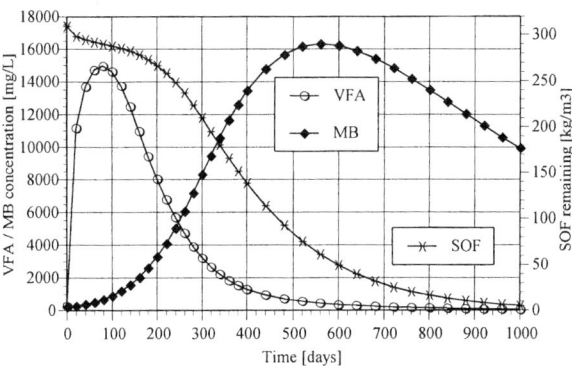

Figure 1 Variation of VFA, MB & SOF with time in baseline simulation.

2500mg/L/day). Both peak VFA and MB concentrations vary directly and significantly with the maximum hydrolysis rate. The timing of peak VFA concentrations are insensitive to the maximum hydrolysis rate whereas peak MB concentrations are noticeably delayed at lower rates. Rates of SOF depletion are also dependent on maximum hydrolysis rates. Values of 0.25× baseline clearly have a marked effect and similar results have been obtained by manipulation of the effective moisture content. In fact, at an effective moisture content of 0.01 (not shown), decomposition is virtually halted. It is generally recognised that once a methanogenic biomass is established, hydrolysis becomes the rate limiting factor in the biodegradation process.

Product inhibition
Simulations using production inhibition factors (PIF) of 2×, 1× and 0.5× baseline (i.e. 2×10^{-4} L/mg) are shown in Figs. 3a-c. Whilst peak VFA concentrations are inversely related to the PIF value, the effect on MB concentrations and their peak value positions, and on SOF depletion, is less pronounced. This behaviour is consistent with the postulated biodegradation pathway. Initial VFA accumulation/organic depletion is very sensitive to product inhibition, the impact of which can be traced through to the early stages of the SOF depletion plots. In the longer term, there is a more muted albeit persistent effect; a vigorous methanogenic population readily consumes hydrolysis products so the inhibition mechanism only plays a minor role.

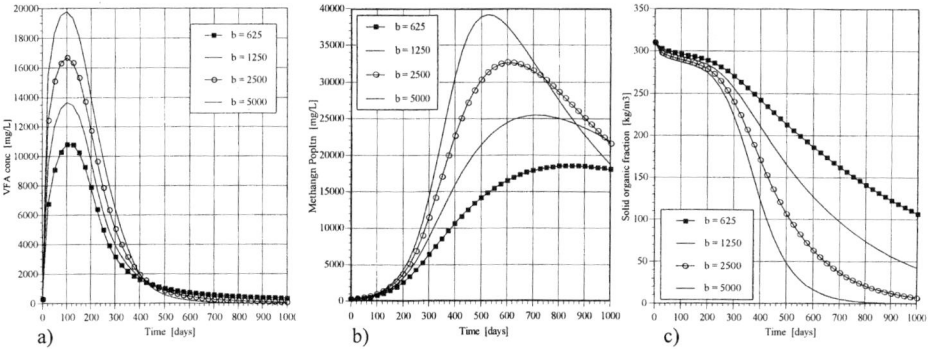

Figure 2. Variation of VFA (a), MB (b) & SOF (c) with time for selected maximum hydrolysis rates

Structural transformation

The simulations presented in Figs. 4a-c show that the structural transformation parameter has little effect on VFA concentrations; it does, however, have a longer-term effect on MB and SOF depletion, i.e. lower values produce a slower accumulation of methanogens with smaller peak concentrations and a deceleration in the long-term rate of organic depletion. This behaviour can be explained in terms of the physical process of hydrolysis. The structural transformation parameter reflects the proportion of relatively digestible to recalcitrant organic matter; a value of unity would result in a linearly decreasing level of digestibility, values of less than unity reflect an increasing proportion of recalcitrant material so the progress of decomposition decelerates more rapidly.

CONCLUSIONS

The test simulations have shown the qualitative performance and parametric sensitivity of the modified biodegradation model. Effective moisture content and maximum hydrolysis rate have an identical and significant effect on the model variables. Product inhibition is used in the biodegradation model to constrain the peak VFA concentrations occurring in the early stages of landfill decomposition and to simulate 'souring'; the product inhibition factor is set in conjunction with the maximum hydrolysis rate. Structural transformation reflects a real physical phenomenon and offers additional control over the conversion process but its practical benefit and identifiability have yet to be established.

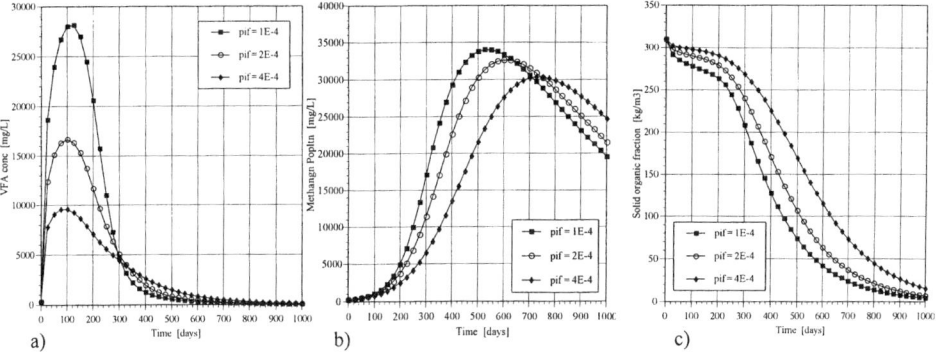

Figure 3. Variation of VFA (a), MB (b) & SOF (c) with time for selected product inhibition factors

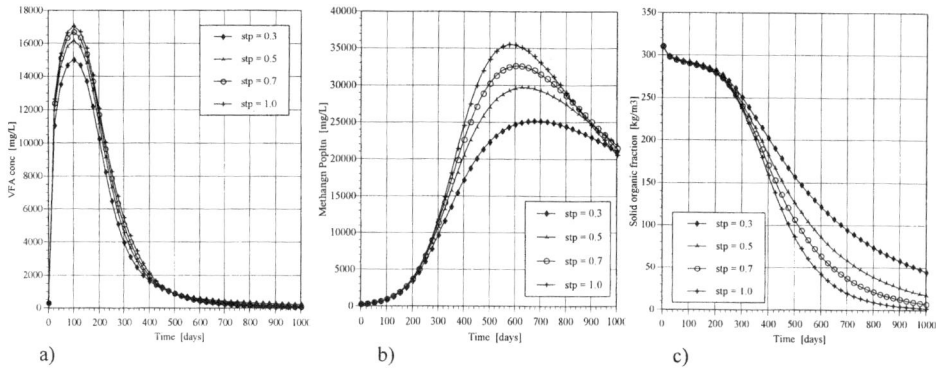

Figure 4. Variation of VFA (a), MB (b), & SOF (c) over time for selected transformation parameters

REFERENCES

Anon (1995) Landfill design, construction and operational practice. *Waste Management Paper 26B*, Dept of Environment, H.M.S.O., London.

Barlaz M.A., Ham R.K. & Schaefer D.M. (1989) Mass balance analysis of anaerobically decomposed refuse. *A.S.C.E., J. Env. Eng. Div,* 115, 6, 1088-1102

Bookter T.J. & Ham R.K. (1982) Stabilization of solid waste in landfills. *A.S.C.E., J. Env. Eng. Div,* 108, EE6, 1089-1100.

Cecchi F., Traverso, P.G., Claney, J. & Zaror, C. (1988) State of the art of R&D in the anaerobic digestion process of municipal solid waste in Europe. *Biomass,* 16, 257-284.

Jones K.L. & Grainger J.M. (1983) Application of enzyme activity measurement to a study of factors ... in domestic refuse. *Appl. Microb. & Biotechn.,* 18, Springer-Verlag, 181-185.

Lee Y-H. & Fan L.T. (1982) Kinetic studies of enzymatic hydrolysis of insoluble cellulose: (I) Analysis of the initial rates. *Biotechn. & Bioeng.,* 24, 2383-2406.

Lee Y-H. & Fan L.T. (1983) Kinetic studies of enzymatic hydrolysis of insoluble cellulose: (II) Analysis of extended hydrolysis times. *Biotechn. & Bioeng.,* 25, 939-966.

McDougall J.R. & Pyrah I.C. (1999) Moisture effects in a biodegradation model for waste refuse. *Proc. Sardinia '99 7th Waste Management and Landfill Symposium*, CISA, Cagliari, Italy. Vol 1, 59-66.

Rees J.F. (1980) The fate of carbon compounds in the landfill disposal of organic matter. *J. Chem.Techn. & Biotechn.* Vol.30, 161-175.

Wald S., Wilke C.R. & Blanch H.W. (1984) Kinetics of the enzymatic hydrolysis of cellulose. *Biotechn. & Bioeng.,* 26, 221-230.

Wang Z. & Banks C.J. (2000) Accelerated hydrolysis and acidification of municipal solid waste in a flushing anaerobic bio-reactor using treated leachate recirculation. *Waste Man. & Res.,* 18, 215-223.

ACKNOWLEDGEMENTS

This research has been funded by Hanson Waste Management through the Landfill Tax Credit Scheme and administered by ESART. The views expressed herein are those of the Authors.

Air and Water Permeability of a Compacted Soil Used in a Solid Waste Landfill in Recife, Brazil

MARINHO, F.A.M[1]., ANDRADE, M. C. J[2]., JUCÁ, J.F.T[3].
[1]*University of São Paulo, Brazil*
[2]*State University of Pernambuco, Brazil - Former MSc Student at USP*
[3]*Federal University of Pernambuco, Brazil*

INTRODUCTION

The final cover layer permeability of a solid waste landfill (SWL) is of extreme importance to reduce the water flow into the SWL and the flow of gas to the atmosphere. The determination of the parameters related to the flow of water and gas is an important aspect to adequately design the cover system.

When flowing a fluid (liquid or gas) may change its shape and also its flow characteristics, according to the physical and chemical peculiarity of the media. In this paper only the physical aspect will be considered. In this way any flow property of a fluid in porous media is affect by one of the following factors, *inter alia:* porosity of the media, maximum pore size, pore-size distribution and degree of saturation (or suction). The effect of the degree of saturation on the water or air permeability is probably the most difficult aspect to be investigated.

The focus of this paper is the flow properties of the soil used as cover layer at the Municipal solid waste landfill of Muribeca located at Pernambuco State, Brazil. The objective of this study are twofold: one is to provide additional geotechnical data for the soil; and the other is to infer the use of the Brooks and Corey model's in predicting the permeability air function. The soil water characteristic curve (SWCC) was obtained from samples compacted using the standard Proctor energy. The experimental permeability of the soil in its saturated state and the air permeability of the soil at different initial degree of saturation are presented.

SOIL CHARACTERIZATION

The soil tested presented the following Atterberg limits: w_L = 52% and I_P = 17%. The specific gravit is 2.67 and the percentage lower than 2µm is 28. Three sets of samples were prepared for three different types of tests, i.e., soil water characteristic curves (SWCC), water permeability and air permeability. Each set of samples was labelled according to the test to be performed. Table 1 presents the nomenclature adopted for the samples according with the test performed.

The initial characteristic of the samples of each set is shown in Figure 1. The samples were compacted using the standard Proctor energy. As can be seen in Figure 1 the samples of the sets B and C are shifted to the left. This was due to unintentional drying after compaction. In terms of SWCC it is considered that the difference in the initial conditions has a small reflection on the SWCC (e.g. Marinho & Stuermer, 2000).

Table 1. Samples Nomenclature

Set	Samples	Test
A	S1,S2,S3,S4,S5	SWCC
B	S1',S2',S3',S4',S5'	Water permeability
C	S1",S2",S3",S4",S5"	Air permeability

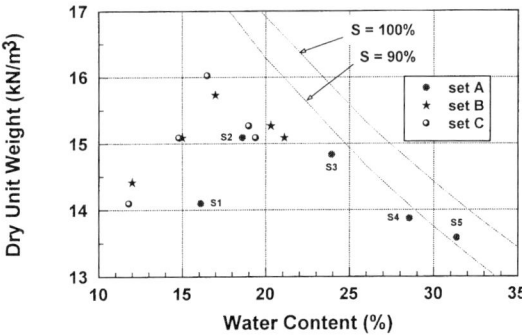

Figure 1 - Initial characteristic of the samples

SOIL WATER CHARACTERISTIC CURVES

The drying path of the SWCC was obtained for each sample. All of them started at its compaction water content. The suction was measured using the filter paper technique. The filter papers were placed in direct contact with the soil sample and left for at least seven days, attempting to measure matrix suction.

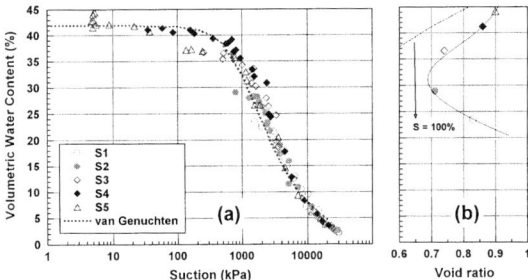

Figure 2. (a) SWCC in terms of volumetric water content and suction (b) Initial volumetric state of the samples.

Figure 2a shows the relation between the volumetric water content and suction for the samples (set A). Figure 2b presents the initial conditions in terms of volumetric water content and void ratio. After the suction measurement the volume of the samples was obtained. A vernier calliper was used to measure the dimension of the samples. The van Genutchen equation for the soil water characteristic curve was fitted into the experimental data shown in figure 2. Manually fitting the van Genuchten model to the experimental data one can obtain the following parameters: $\theta_r = 0$; $\theta_s = 42\%$; $\alpha = 0.0009 kPa^{-1}$ and $n = 1.75$. The parameters of that equation will be used to predict the unsaturated permeability function (van Genuchten, 1980).

WATER PERMEABILITY OF SATURATED SAMPLES

The water permeability tests using saturated samples were carried out in samples of the set B. Figure 3a presents the initial condition for the samples tested. The tests were performed using a flexible membrane apparatus. Each sample was allowed to absorb water under a confining pressure slightly higher than 200kPa and a backpressure of 200kPa, giving a low effective confining pressure. Figure 3b presents the saturated permeability (k_w) for the five samples tested. A reduction in the permeability with the increase of the compaction water content was obtained.

AIR PERMEABILITY OF SAMPLES IN ITS ORIGINAL COMPACTED STATE

The air permeability (k_a) was measured using the same flexible membrane apparatus used for the water permeability. The system to perform the air permeability test was adapted following the set up suggested by Maciel and Jucá (2000). Figure 3a gives the initial conditions of the samples before starting the air permeability tests. Each test was performed with the sample at its original state, after the compaction. No suction control system was used. Figure 3b presents the results of the air permeability tests, showing a rapid decrease in the permeability with increase of the initial water content.

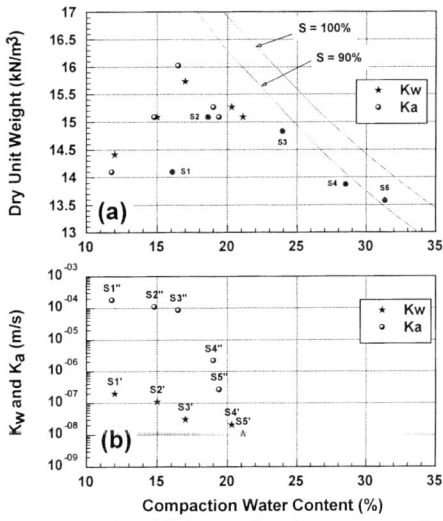

Figure 3. (a) Initial state for the compacted samples used (b) Water and air permeability as a function of their initial water content.

BROOKS AND COREY'S MODEL

The relation between the effective degree of saturation (S_e) and suction plotted on a straight line, if data for saturation higher than approximately 0.85 are omitted and a suitable value for the residual degree of saturation (S_r) is chosen (Corey, 1994). The effective degree of saturation is defined as:

$$S_e = \frac{S - S_r}{1 - S_r}$$

As pointed out by Corey (1994) the significance of S_e is related to flow, and that is the reason for using the residual degree of saturation as a limit. The wet phase state below the residual degree of saturation play a small role on the convective flow. The Brooks and Corey's model for representing the relation between effective degree of saturation and suction can be expressed as:

$$S_e \approx \left(\frac{(u_a - u_w)_b}{(u_a - u_w)} \right)^{\lambda}$$

The value $(u_a - u_w)_b$ is the suction at the air entry point. This expression is valid for $(u_a - u_w) >$ $(u_a - u_w)_b$ and $S > S_r$. Normally this expression is used for the drying path of the SWCC.

Figure 4a presents the data obtained using samples from set A, plotted in terms of effective degree of saturation versus suction. In order to obtain the parameters for the Brooks and Corey model's some points were selected. The air entry value was taken as 600kPa. Only points with suction higher than the assumed air entry value were considered. Figure 4b presents the points used for obtaining the parameters for the model. The value of λ was found to be 0.76 with a $r^2 = 0.93$. The result is shown in Figures 4a and 4b as a dotted line.

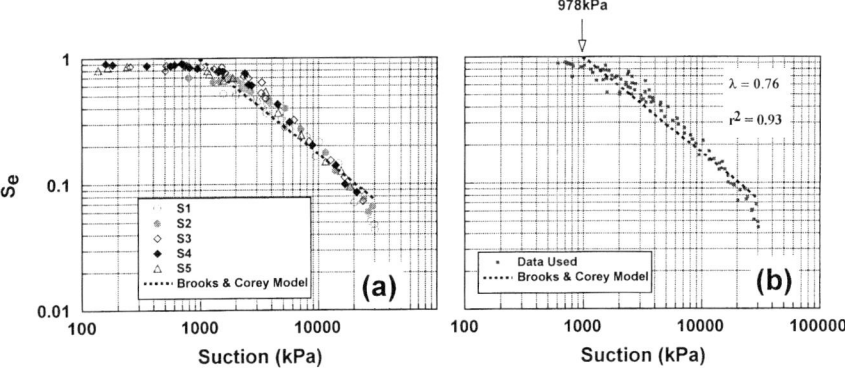

Figure 4. (a) Relation between the effective degree of saturation and suction (b) The use of the Brooks and Corey's method.

Applied to Water Permeability Function

A semi-empirical relation for the water permeability function $(k(\theta))$ was presented by Brooks and Corey (1964), as follow:

$$k(\theta) = k_w \left(\frac{(u_a - u_w)_b}{(u_a - u_w)} \right)^{(2+3\lambda)}$$

$k(\theta)$ is the water permeability at a specific value of suction and k_w is the water permeability for a degree of saturation of 100%. The value of λ is the same one obtained as shown previously.

Figure 5 presents two unsaturated permeability functions $(k(\theta))$ for the soil tested. They were obtained from Brooks and Corey model and from van Genutchen model. Both models require the permeability of the soil in the saturated state. The values of the permeability of the saturated soil are also shown in figure 5 and associated to a suction of 10kPa for reference purpose only. The average value of permeability was used in the models. The van Genutchen equation is presented as a comparison and its determination was based on the parameters presented previously. For the moment there is no experimental data to compare with the prediction. It can be seen that the Brooks and Corey model and the van Genutchen one are reasonably comparable for values of suction above the air entry value.

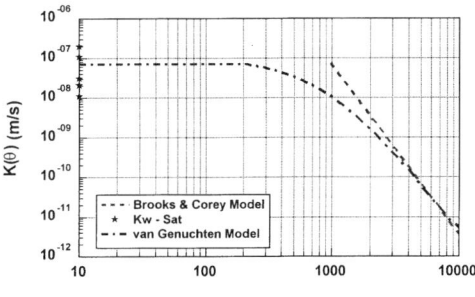

Figure 5 - Unsaturated soil permeability functions and the water permeabilities of the saturated samples.

Applied to Air Permeability

Brooks and Corey (1964) also propose another semi-empirical relation for the air permeability function. The equation, in terms of suction is:

$$k_a = k_{dry}\left[1 - \left(\frac{(u_a - u_w)_b}{(u_a - u_w)}\right)^\lambda\right]^2\left[1 - \left(\frac{(u_a - u_w)_b}{(u_a - u_w)}\right)^{2+\lambda}\right]$$

The values for λ and $(u_a - u_w)_b$ are the same determined previously.

Figure 6 presents the experimental data obtained for the air permeability performed with samples from set C. The value of suction associated with k_a was obtained via SWCC. It can be observed that there is a rapid increase in the permeability as the initial suction increases or the degree of saturation decreases. The increase in the pemeability is reduced when the air entry value is approached. The air entry value obtained form Brooks and Corey(B &C) model was 978kPa. The dotted line shown in figure 6 is the Brooks and Corey model applied to the data shown in figure 4b. The results follow reasonably well the general behaviour, with a difference been observed for low value of suction. By changing the value of the air entry value from 978kPa to 1100kPa, and keeping $\lambda = 0.75$, a better result is obtained. The variability of the results may occur in both tests, i.e., SWCC and air permeability. In figure 6a the B & C model is adjusted (using a manipulated $(u_a - u_w)_b$) to fit the data (dashed line).

CONCLUSIONS

In terms of SWCC the results are consistent with literature data and show some influence of the initial water content on the initial portion of the curve. It was assumed that these differences do not change significantly with a small change in the compaction energy. This assumption is important for the use of the SWCC in predicting the water permeability function ($k(\theta)$) and the air permeability function ($ka(\theta)$).

The water permeability of the soil in its saturated state decreased with the increase in the compaction water content. For the range of water content used, from 12% to 21%, k_w varied approximately one order of magnitude.

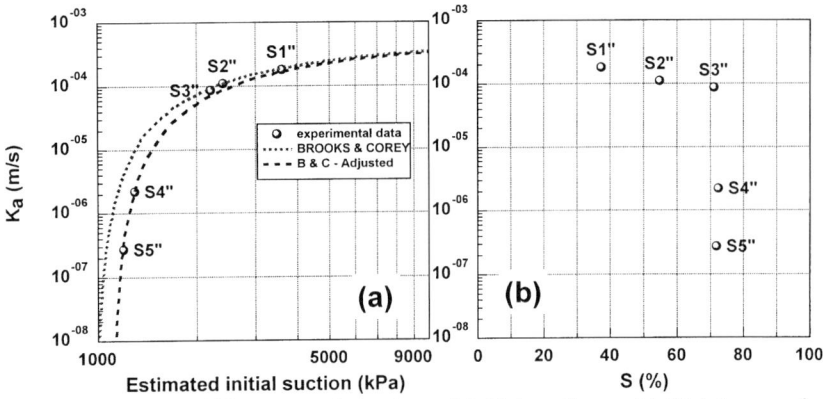

Figure 6. Air permeability versus the estimated initial suction and initial degree of saturation, showing the predictions using Brooks and Corey's model.

The air permeability(k_a) measured on the samples showed that when the compaction water content is lower than the optimum, k_a presents only a small change. For values of w/c higher than the optimum there is a reduction in the air permeability of approximately three orders of magnitude. The inflection point is directly associated with the air entry value, that seems to varies from 600kPa to 1200kPa. The Brooks and Corey model was evaluated for the air permeability function. Although there is some variability on the SWCC, affecting the prediction using the model, the result suggested that the B & C model could be a useful tool for predicting the air permeability function. Considering that the experimental determination of the air permeability is much easier than the determination of the unsaturated permeability, the B & C model could be used to predict the unsaturated permeability function.

REFERENCES
Brooks R.H. and Corey, A.T. (1964) - "Hydraulic properties of porous media. Colorado State University Hydrology Paper No. 3 March 27p.

Brooks R.H. and Corey, A.T. (1966) - "Properties of porous media affecting fluid flow" Journal of Irrigation and Drainage Div. Proc. ASCE, vol 92, IR 2, pp-61-88.

Corey, A.T. (1994) - "Mechanics of Immiscible Fluids in Porous Media" - Ed. Water Resources Publications - 252p.

Maciel, F.J and Jucá, J.F.T (2000) - "Laboratory and filed tests for studying gas flow through MSW landfill cover soil"- Geotechnical Special Publication - 99. Edited by Charles D. Shackelford, Sandra L. Houston and Nien-Yin Chang. - pp. 569-585.

Marinho, F.A.M. and Stuermer, M.M. (2000) - "The influence of the compaction energy on the SWCC of a residual soil"- Geotechnical Special Publication - 99. Edited by Charles D. Shackelford, Sandra L. Houston and Nien-Yin Chang. - pp. 125-141.

van Genuchten (1980) - "A closed-form equation for predicting the hydraulic conductivity of unsaturated soils" -Soil Science Society of America Journal , 44 - pp. 892-898.

ACKNOWLEDGEMENTS

The authors are grateful to Mr. A. R. de Brito from the Soil Mechanics and Instrumentation Laboratory of the Federal University of Pernambuco, Brazil, where the tested were performed. The Authors would like to thanks Dr. Scandar Ignatius for his important contributions to the analyses.

A Case Study of Landfill Construction as a Means of Land Reclamation

STUART MOLLARD
Babtie Group, Cardiff, UK

ABSTRACT
It was proposed to construct a new landfill site at the location of a backfilled former opencast coal site in South Lanarkshire, Scotland. No site restoration had been implemented on completion of opencasting operations and the site represented an area of dereliction, comprising largely unvegetated irregular mounds of opencast backfill material. A comprehensive ground investigation and programme of sampling/monitoring was required to assess the geotechnical and hydrogeological setting of the site. In order to provide a sustainable and cost effective approach to landfill construction, it was decided to review and assess the potential suitability of the clay-rich opencast backfill materials as a fill source for a basal mineral liner. This paper will discuss the investigations and assessments that were undertaken, with particular reference to the material suitability assessments, which facilitated a sustainable approach to the overall site design and provided sufficient information to satisfy the requirements of the Scottish Environment Protection Agency.

INTRODUCTION
The proposed landfill site is located south-east of East Kilbride in a sparsely populated area, currently derelict and used only for informal recreation. The site area was formerly subject to opencast mining, with coal seams excavated to a maximum depth of 18m. Opencast backfill materials, comprising re-worked natural superficial deposits and solid strata, which were loosely placed during opencast works, cover large parts of the site area. The site is bounded to the north-west by a stream, to the north-east by a closed former dilute and disperse landfill site and to the south-west and south-east by areas of rough grazing. The site slopes gently towards the stream, although the land surface is broken by irregular heaps of colliery spoil and water-filled depressions. Sparse vegetation and scrub exists in the area and no agricultural use is possible for the land in its present form (Figure 1). Environmental assessments had identified no significant ecological or landscape interest for the site in its current condition and landfill represented the only commercially realistic means of site reclamation and long-term improvement.

The area proposed for landfill development covers approximately 150,000m^2 and the proposed total void space for waste within the landfill area is approximately 1,000,000m^3. Restoration proposals for the site involve a landform generally in keeping with the surrounding environment, with areas of trees and shrubs of local provenance and the recreation of areas of marsh and scrapes and large boulders to provide habitats for flora and fauna currently resident on the site.

Geoenvironmental impact management, Thomas Telford, London, 2001.

Figure 1. Existing Site Condition

GROUND INVESTIGATION
In order to provide information on the geological, geotechnical and hydrogeological conditions at the site, a ground investigation was carried over an initial intensive six week period. The investigation comprised a total of 11 boreholes, and 35 trial pits. The nature, disposition and thickness of the materials encountered during the ground investigation are summarised below

Opencast Backfill: The backfill is variable in character, comprising mainly sandy clays and fine to coarse gravels, with local areas of predominantly gravel/cobbles. Field permeability tests carried out in boreholes within the backfill material confirmed the loosely compacted and free draining nature of the material in its current condition, with typical values of 10^{-4} to 10^{-6} m/s.

Natural Superficial Deposits: The ground investigation identified alluvium across much of the undisturbed parts of the site, underlain by Glacial Till, both of which are cohesive in character.

LANDFILL DESIGN
The basic principles used in the development of the preliminary site design are, in summary

- Minimum void space of 1 million m^3
- Base of site to be at least 1m above maximum groundwater level (DoE, 1995)
- Permeability of landfill basal liner to be a minimum of 1×10^{-9} m/s (DoE, 1995)
- Small landfill cells, progressively capped and restored to minimise the area of waste exposed at any time, thereby minimising leachate production volumes

- Overall site containment to meet pollution risk modelling requirements ('Landsim') to the satisfaction of SEPA, in relation to potential impacts on the adjacent stream, taking account of existing baseline monitoring data (DoE, 1994; NRA, 1994)

Following collation of the ground investigation results, assessment of the above parameters was undertaken to produce an outline design for review by the site developer. A key engineering aspect of the design at this stage was the suggestion that the backfill materials could provide a suitable source of cohesive material to form an appropriate low permeability liner for the site. This had obvious benefits for the development of the site in terms of utilising on-site materials, thereby reducing transport movements to and from the site, providing a more sustainable approach to the site reclamation and a considerably improved cost-effective construction methodology. Preliminary earthworks calculations suggested a total excavation volume of 650,000m^3, with a total of 250,000m^3 available for re-use within peripheral bunds, capping layers and the mineral liner. The balance of 400,000m^3 was required to recontour the southern edge of the site in the formation of a 'complimentary landscape area', which comprised an integral part of the planning application for the site reclamation proposals.

The ground investigation had revealed that the backfill materials present at the site were largely derived from reworked glacial till and cohesive alluvial deposits, together with broken down and highly weathered mudstone. Laboratory compaction and permeability tests were carried out on selected cohesive opencast backfill samples to provide a preliminary assessment of possible mineral liner applications for the material. The results of these tests suggested permeability in the range 1×10^{-10} to 8×10^{-11}m/s. However, the appropriate specification (BSI, 1990) for sample preparation for the tests required that the coarse fraction of the samples had to be removed and the test therefore was only recording the permeability of the fine fraction of the backfill materials. In order to demonstrate that the in-situ backfill material was capable of achieving the required in-situ permeabilities, it was necessary to carry out further investigation and assessment.

SUPPLEMENTARY INVESTIGATIONS
The first element of additional field assessment was a large-scale trial pit investigation to allow visual assessments to be made of material consistency and to facilitate additional, larger bulk samples to be taken for classification purposes. Following the excavation of a series of 'big pits' across the proposed landfill area, it was demonstrated that sandstone boulders within the backfill were relatively isolated and could be easily separated using excavation plant. Classification tests further demonstrated the consistently cohesive properties of the backfill, with plasticity limits varying little between samples (plasticity indices in the range 13 to 17%).

Given the positive results of the additional trial pits, it was recommended that a field trial be carried out to demonstrate compaction/permeability relationships and to determine the size of particle that should be excluded from the opencast backfill for use in the mineral liner, together with assessment of the optimum layer thickness for compaction. By using a field trial involving variation of moisture content and layer thickness, it would also be possible to assess the compaction characteristics of the opencast backfill, and combined with further permeability testing after compaction of the material, this would assist in development of the specification of the mineral liner. The following site works were undertaken to meet these general objectives

- Excavation of material from borrow pit locations within the proposed landfill area
- Preparation of trial area, by stripping vegetation and topsoil, removing boulders and rolling to provide a firm foundation
- Controlled placement and compaction of material in varying layer thicknesses, in four adjacent panels approximately 15m x 15m in area (Figure 2), generally in accordance with the Highways Specification (DoT, 1991)
- Development of 1m total liner thickness for each panel
- Use of water bowser and spray on site to adjust the moisture content of compacted material
- In-situ testing of compacted layers in each panel, comprising nuclear density tests, with validatory sand-replacement tests (BSI, 1990)
- Trench excavation through completed panels for logging of compacted materials
- Undisturbed sampling from each panel for laboratory permeability testing in a triaxial cell
- Installation of infiltrometers (2.4m x 1.2m x 0.6m tanks) on two completed panels, including automatic water level data loggers, to measure in-situ vertical permeability

Figure 2 Trial Panel Construction

Supplementary Investigation Results
The results of constant head permeability tests, carried out in triaxial cells, on samples taken from each of the trial panels are summarised in Table 1.

Panel	Initial Moisture Content	Initial Dry Density	Permeability Coefficient
1	13%	1.95 Mg/m^3	1.2 x 10^{-10} m/s
2	13%	1.98 Mg/m^3	5.6 x 10^{-11} m/s
3	12%	1.93 Mg/m^3	7.3 x 10^{-11} m/
4	16%	1.83 Mg/m^3	7.4 x 10^{-11} m/s

Table 1: Laboratory Permeability Test Results From Undisturbed Samples

In addition, four bulk samples were taken from excavated backfill materials prior to compaction, with subsequent recompaction to a bulk density of 2.20 Mg/m^3 (4.5kg rammer at natural moisture content) and permeability testing in a triaxial cell, as above. The results of these tests ranged from 1.0 x 10^{-10} to 5.9 x 10^{-11} m/s. Similar permeability tests on recompacted bulk samples carried out as part of the earlier ground investigation for the site gave results in the range 2.7 x 10^{-10} to 8.3 x 10^{-11} m/s. The in-situ density tests results are summarised in Table 2.

Panel	Description	mc(%)	DD(Mg/m³)	%Compaction
1	4 x 250m layers, 1 test per layer, at natural moisture content	11.6	1.98	97
		6.8	2.04	100
		16.8	1.85	90
		10.7	1.84	90
2	4 x 250mm layers, 1 test per layer, moisture content adjusted	11.1	1.97	97
		10.5	2.00	96
		11.7	1.95	96
		10.1	2.03	99
3	3 x 350mm layers, 1 test per layer, moisture content adjusted	13.0	2.04	100
		12.6	1.98	97
		12.2	2.02	99
4	Layers 400mm, 400mm, 200mm, 1 test per layer, moisture content adjusted	9.5	1.88	92
		12.0	1.77	87
		10.5	2.02	99

Table 2: Results of In-situ Density Tests

Both infiltrometers showed a gradual drop in water level over the monitoring period (60 to 70mm in total), but with regular fluctuations in recorded water levels. It was apparent from the data that a complete seal around the tanks was not achieved and that water losses from leakage at the tank margins and evaporation, together with possible input from precipitation, occurred. During the latter stages of the monitoring period, a better seal on the tanks was achieved and the water level changes more accurately reflect infiltration rates; during the final week of monitoring, an average rate of 0.6mm/day was recorded in tank 2, as illustrated in Figure 3.

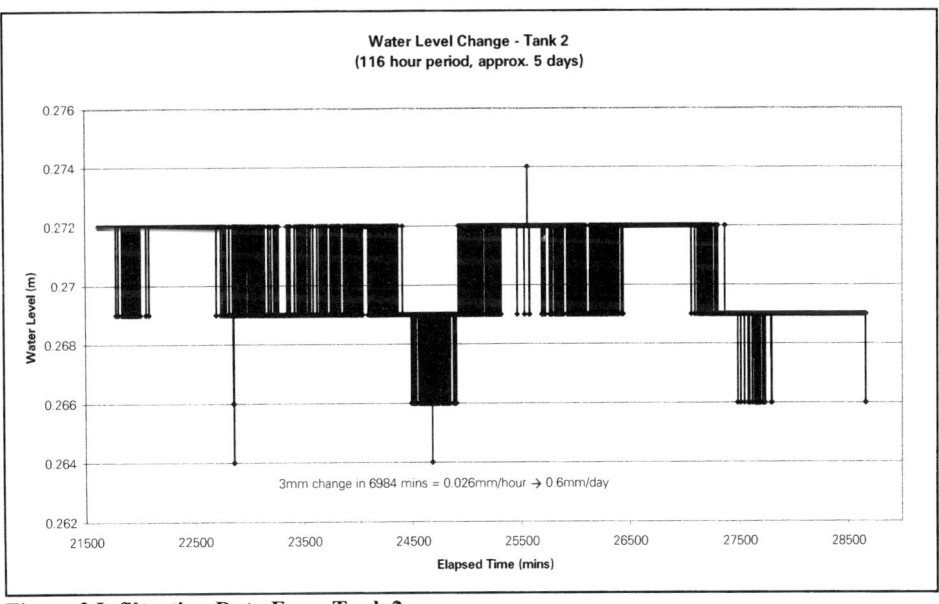

Figure 3 Infiltration Data From Tank 2

In summary, the results of the site liner trial provided the following conclusions applicable to the site design:

- Excavation and selection of material could be achieved to provide a suitably graded fill;
- The natural moisture content of the backfill is slightly drier than its optimum moisture content; wetting of the material in-situ therefore produces a greater degree of compaction;
- Data from infiltrometers installed following construction of trial panels demonstrated that bulk infiltration was not influenced by large-scale effects in the compacted materials. Therefore laboratory permeability testing of samples in a triaxial cell provides a representative estimation of in-situ permeability values.
- Given that all laboratory tests carried out produced permeability coefficients lower than 1×10^{-10} m/s, it can be demonstrated that the compacted material is capable of meeting mineral liner permeability requirements.

CONCLUSIONS

Construction of a landfill site and associated landscaping areas facilitated reclamation of an otherwise derelict and unutilised area of land. An initial geotechnical review of the materials present allowed the development of site earthworks designs based on a sustainable use of the site, with re-use of all excavated materials as landscaping or engineering fill, including a basal mineral liner.

Following a comprehensive ground investigation, a series of large scale site trials were set up to provide a practical assessment of in-situ properties and characteristics of the materials present within the proposed landfill area. A combination of laboratory test data, practical field test results and hydrogeological modelling was then used to demonstrate that the engineering proposals provided a suitable barrier against pollution from the landfill. The resulting design is a simple re-engineering of the site and re-use of existing site deposits to produce a secure waste disposal facility in the short term and a new restored landform complimentary with the natural surrounding environment in the long term.

REFERENCES

Department of the Environment, Waste Management Paper 26B, Landfill Design, Construction and Operational Practice, HMSO 1995

Department of the Environment, Waste Management Paper No 4, Licensing of Waste Management Facilities, Third Edition, HMSO 1994

National Rivers Authority, Landfill Liners, Internal Guidance Note No 7, 1994

BSI, BS 1377, Methods of Test for Soils for Civil Engineering Purposes, 1990

Department of Transport, Specification for Road and Bridgeworks, 1991

ACKNOWLEDGEMENTS

This paper has been published with the permission of Viridor Waste Management Ltd. The views expressed in the paper are those of the author and do not necessarily reflect those of Babtie Group or Viridor Waste Management.

Factors Affecting Groundwater Flows in Worked Coal Measures

PAUL MORRISON
Babtie Group, Glasgow, Scotland.

INTRODUCTION

The legacy of both deep and opencast mining activities throughout the United Kingdom has resulted in vast areas of disused and derelict terrain in various states of restoration. Many of these areas have now been developed as landfill facilities, disposing of vast amounts of domestic and industrial wastes generated by the surrounding town and city areas. Derogation of both groundwater and surface water quality may, potentially, occur through pollution by either mine or landfill derived contaminants. This can result in the loss of current or future water resources. Regulation 15 of the Waste Management Licensing Regulations (SI 1994/1056), is the key regulation from the point of view of groundwater protection, with regard to both operational and proposed landfill sites. As a result, a heavy reliance is currently placed upon the ability of predictive groundwater models to simulate flow behaviour, irrespective of hydrogeological setting. The applicability of these models to landfills situated above mined strata is questionable. The hydrogeological characteristics of these areas are undoubtedly complex, and a number of separate flow regimes with very different levels of magnitude may exist. Many researchers place little emphasis upon the development of an accurate conceptual model. This is indicative of overconfidence in the ability to characterise most groundwater systems, by the direct input of hydrogeological data into modelling packages deemed *acceptable* by the Environment Agencies.

This paper describes, in outline terms, factors affecting groundwater flows beneath a landfill site in the Central Coalfield of Scotland. The complexity of the groundwater system at the site, which is not in itself uncommon for many areas subject to extensive deep mining operations, necessitated the formation of a conceptual model representative of the processes, and their boundaries, which control groundwater flow behaviour. A sufficient level of information is presented to throw doubt upon investigative methodologies currently adopted at such sites and, in particular, the applicability of computational modelling packages which fail to represent, to a sufficient level of accuracy, these complex groundwater systems.

SITE DESCRIPTION AND OUTLINE GEOLOGY

The site lies within the Midland Valley of Scotland, on the western side of the Central Coalfield, in an area that has been worked extensively for coal, ironstone and fireclay for over 150 years. As a result the natural aquifer systems have been altered by the widespread underground mine workings and more recent opencast operations. A large flooded cut containing 1000 Ml of water constitutes the eastern extremity of the proposed landfill development, and a prominent watercourse, whose original configuration has been disturbed by opencast activities, meanders through the northern site area. Opencast voids of various

Geoenvironmental impact management, Thomas Telford, London, 2001.

sizes characterise the site and the surface pattern of 'crown holes' caused by the collapse of shallow stoop and room workings are evident at some locations.

The sedimentary rocks within the site date from the Westphalian succession of the Carboniferous period and incorporate the Middle and Lower Coal Measures. These comprise cyclic sequences of fine to medium grained white or buff sandstones, grey siltstones and mudstones interspersed with seams of fireclay, clayband and blackband ironstones and coal. The majority of the coal seams within the site are of the common bituminous type.

The area is traversed by numerous faults, which break up the strata into a large number of fault blocks and wedges. To the north of the site a quartz-dolerite dyke, associated with the Midland Valley Sill-Complex, has intruded along the line of an east-west fault. South of this feature a thinner dyke, not exposed at the surface, crosses a northwesterly watercourse and terminates at its intersection with a north-south trending fault. Evidence exists of burnt coal seams along a narrow margin either side of this smaller intrusion.

In areas of the site undisturbed by opencast operations the bedrock is overlain by a sandy boulder-clay, up to 18 m in thickness, arranged in smooth gently rounded ridges, generally trending east-west. Lying in the hollows between the boulder-clay ridges are deposits of peat, which in some cases are up to 6 metres in depth.

PHYSICAL FACTORS AFFECTING GROUNDWATER FLOWS
Basis for Interpretation
The desk study provided the information necessary to compile a series of thematic maps, detailing deep and opencast mining activities, fault positions and shaft and adit locations. Supplemented by data from a targeted ground investigation, a number of geological cross sections were compiled detailing lithology and groundwater observations. Information from mine abandonment plans enabled the accurate positioning of faults, and permitted voids and bands of packed waste encountered during drilling operations to be assigned, in most cases, to a particular horizon of mine workings in a specific coal seam. Site works also provided data describing the fracture characteristics and permeability values, relative to each element of the underlying aquifer system. Geological mapping of rock exposures was also undertaken for joint, bedding and shear zone characteristics.

Ground investigation works
Former opencast excavations and improperly sealed boreholes may create vertical pathways between hydrogeologically separated bands of strata. This was a major influence upon groundwater behaviour across the site, due to the extensive drilling and opencast operations that had taken place. Although measures are taken to plug boreholes above old mine workings, achieving an effective seal is extremely difficult. This is further compounded by localised effects surrounding the drilled hole in the soft rocks of the Coal Measures.

Mine workings
By far the greatest influence on groundwater behaviour at the site will be the subsurface characteristics either directly, or indirectly, linked to the extensive mining operations. Studies of groundwater flow through, and into, abandoned mine workings in Coal Measures rocks, have been made by numerous researchers notably; ALDOUS, SMART and BLACK, 1986., and XIAO, IRVIN and FARMER, 1991. Tracer tests, conducted in flooded mine workings, have shown that flows are concentrated into major flowpaths such as abandoned road

driveways, many of which are still accessible after 125 years. In addition, collapsed panels comprise layers of relatively high permeability material, which facilitate preferential groundwater flows across an area in specific mined horizons.

A series of thematic plans relative to each worked coal seam detailing position, thickness and base level, provided a graphical history of deep mining operations in the area. Although many older and relatively shallow mines remain unrecorded, these plans provide the most efficient means of assessing the disturbance, and potential pathways for groundwater flows, afforded by the mining activities, at a variety of stratigraphic levels. By cross referencing these plans with borehole logs, recorded waste horizons, voids, closed workings, and highly fractured 'roof' strata could be assigned a position in the mining sequence. The location and thickness of open voids encountered during drilling works, associated with mine workings, were tabulated. Open mine voids of up to 2.4 m were encountered beneath the site area.

Stoop and room mining methods were adopted at the northwest of the site, due to the shallow depth of the Ladygrange coal, and the variable condition of the overlying strata. However, as the majority of worked seams at the site were thin (generally less than 1 m), the predominant method of working was longwall. Abandoned, but relatively intact, stoop and room workings act as subsurface reservoirs of a considerable size. If hydraulic conditions are suitable, then these areas provide a zone of 'detention' for groundwaters. A change in these conditions may then facilitate a 'release' of these waters, locally disturbing the regional flow patterns.

The site area was littered with abandoned shafts, blind pits and adits, which connected a variety of worked coals at different stratigraphic levels.   Information regarding the composition and size of abandoned shafts in the area was minimal. The consolidation and/or capping of a number of these features may have taken place upon closure of the various collieries, or at a later date during restoration works. It was common practise in this area to erect timber staging at a position within the shaft, above which infilling took place. This led in many cases to the eventual decay of the timber, resulting in a collapse of infill material comprising colliery spoil, demolition rubble, wood and soil. These features may, in a similar manner to investigative bores sunk across the site, comprise vertical conduits for groundwater flow.   However, the main difference between a badly sealed borehole, and a partly consolidated shaft position, is one of size. Two abandoned shafts beneath the flooded cut feature, linked to Virtuewell coal workings 25-30 m below its base level, were thought to facilitate the interaction of surface waters with the underlying groundwater regime.

Although a large number of mine adits were recorded, predominantly at the northwest of the site, opencast operations had destroyed many of these features. However, some adits still existed, in a partly or fully collapsed form, to facilitate a direct link to zones of shallow mineworkings. These features behave in a manner analogous to a french drain structure, and during a period of prolonged and heavy rainfall, a localised recharge of the groundwater regime may occur. This leads to the formation of infiltration mounds, afforded by the relative extent of the mineworkings, with a life cycle proportional to the duration and intensity of the storm event, and the permeability of the surrounding strata.

Localised changes to permeability relative to collapsed and highly fractured zones of strata, will affect groundwater movements.   Opencast activities may further 'aggravate' these processes and create hydraulic links to fractured zones directly above a mined horizon, by the generation of an additional zone of fracturing, emanating from the pavement level of the surface mine. This was a very important hypothesis with regard to this study, as the flooded cut feature comprises a large opencast void known to have been extensively undermined. Fracturing processes, at depth and from the surface, may facilitate through a highly fractured and permeable zone of rock, a direct interaction between surface and subsurface waters.

Faults

Faults are often interpreted as being impermeable barriers to lateral groundwater flows, but they often act as pathways, which allow the ascent of deeper groundwater. Flows across fault regions may be impeded, but enhancement of flows parallel to the line of a fault may be induced. A heavy inflow of water, associated with the breaching of a 12.7 m fault at the site, led to the flooding of an opencast excavation. The subsequent cessation of mining operations created, in time, an extensive flooded cut feature. Abandonment plans also record flooding and water troubles in close proximity to this fault zone.

The throw amount, persistence, and strike direction details of 159 fault positions, recorded in a variety of coal seams, were assessed. The 'pattern' of faulting evident at the site is indicative of two separate fault systems. The predominant strike direction is northwest-southeast, but the limited number of east-west faults, effect a 'control' upon the position of the northwest striking faults. The termination, and in some cases deflection, of the northwest faulting by east-west faults, infers the northwest series to be the 'later' of the two systems.

The complexities of the hydrogeological characteristics afforded by faults, are compounded in this instance by extensive mine workings. A hydraulic barrier provided by a fault (or indeed any geological structure), may be entirely negated, conceptually, by the presence of intact driveroads driven through the fault region. These 'conduits' may confine groundwater through preferential flowpaths influencing groundwater potential variations across the site. This will be further compounded by the relative permeability of the surrounding rocks.

Dykes

Dykes almost invariably constitute impermeable structures to groundwater flows, afforded primarily by their low intrinsic permeability. The line of the two quartz-dolerite dykes at the north of the site, may comprise a 'no flow' boundary condition in this area. Evidence of the lateral distribution of these intrusions within the strata, is provided, in part, by records of 'burnt' coal shown on abandonment plans. Both dykes appear to persist, along a predominantly vertical attitude, through all of the worked coal seams in the area and there is no evidence to suggest the presence of small, localised, concordant intrusions associated with either dyke. In addition, there is no evidence to suggest the 'breaching' of either dykes at depth which would facilitate a link between workings. The boundaries of these coal workings are 'controlled' by the horizons of altered strata and the margins of 'spoilt' coal. Although the smaller intrusion terminates at the junction of a fault region, the larger dyke is of a considerable persistence in 'regional' terms (i.e. over 18 kms). If this feature forms a lateral barrier to flow, then groundwater will be diverted either to the east, or to the west, of the site. Depending on the facility for these diverted waters to flow along fractured zones, joints, packed waste horizons, brecciated margins of faults, and mine driveroads in a southerly direction through the site, the dyke may result in elevated potentiometric levels immediately to the south of its boundary. This 'model' would be further compounded by the influence of the extensive fault regions in the area, and in particular their potential facility to restrict flows across their boundaries.

OVERVIEW OF GROUNDWATER BEHAVIOUR

The frequency, orientation, connectivity and aperture size of fractures, will heavily influence the mechanism for groundwater flows within a rock mass. However, in worked Coal Measures strata, groundwater movements may occur as a combination of the following;

a) Major flows along abandoned, but still intact, mine driveroads, rooms and shafts.
b) Flows through loosely or tightly packed zones of mine waste.
c) Flows along the brecciated margins of fault zones.
d) Flows through highly fractured bands of sandstone, siltstone, mudstone and coal.
e) Seepage through 'intact' bands of sandstone, siltstone, and mudstone.

The primary mechanism for groundwater movements in the area, will relate to the artificial features associated with deep mining activities. The properties and characteristics of the fractures, inherent within the 'overall' system, will facilitate a secondary level of pathways for groundwater flow. The ratio of artificial, subsurface features, beneficial to fluid movements, to the fracture dominated flowpaths within unmined strata, will determine the 'weighting' that can be applied to the influence of each mechanism for flow.

The composite form of aquifer present beneath the site comprises, as its framework, cyclic sequences of sandstone, siltstone and mudstone, interspersed with seams of fireclay and coal. Sub-artesian groundwater conditions predominate, with various levels of confinement afforded by mudstone layers within the strata, and at rockhead by clay soils that characterise the superficial deposits. These leaky confined aquifers are locally disturbed, and in hydraulic terms 'short circuited', by highly permeable channels afforded by packed waste horizons and open, partly collapsed or fully collapsed mine driveroads.

A number of heavily faulted zones were identified bounded by prominent 'primary' fault positions. These zones will affect groundwater movements and, in particular, may accelerate the dissipation of stored waters impounded within abandoned mineworkings. The dyke features to the north of the site area were designated as 'no flow' boundaries, based on the classical role of these features as impermeable structures to groundwater flows.

Shafts and adits across the site comprise significant conduits for groundwater and surface water movements. These features facilitate localised zones of recharge or discharge, depending upon the potentiometric surface in the area, and the level of connectivity afforded. Although opencast operations have destroyed many adit features some still exist, in a partly or fully collapsed form, and facilitate a direct link to zones of shallow mineworkings.

The flooded cut feature is interacting, intermittently, with the underlying groundwater regime, providing a partial recharge of the composite aquifer system. The mechanism for this migratory process is provided, primarily, by mineshafts in the base of the feature, linked to abandoned mineworkings, inclusive of intact mine driveroads. Highly fractured zones of rock emanating downwards from the base of the cut, attributable to the mechanical disturbance associated with opencast operations, form a link with the fractured zones above workings in the Virtuewell coal seam and effect a secondary mechanism for water loss.

CONCLUSIONS

Abandonment plans provide a wealth of information essential to realistically assess the hydrogeological characteristics of areas subjected to extensive mining operations. Many plans detail the location and geometry of the faults encountered within the workings, and provide a subsurface map describing mine driveroad, shaft and adit positions. Notes of difficulties encountered within the workings are often included, detailing areas of water ingress and rock instability, which is of great assistance in the investigative process. The collation and assessment of these details permits the compilation of subsurface connectivity maps, which can form the basis for both conceptual and computational modelling studies.

The complexity of this form of groundwater system necessitates the formation of an accurate conceptual model, which draws together hydrological, geological and geochemical data in a realistic manner. The combination of relationships corresponding to each aspect of such a study, permits a conceptual model to be developed representative of the processes, and their boundaries, which control groundwater flow behaviour at a site. The conceptual modelling phase of a hydrogeological investigation is an essential prerequisite to numerical modelling studies, and should not in any circumstances be omitted.

Although there are innumerable forms of groundwater modelling packages available, the complexities of groundwater movements in extensively mined rock strata, limit the applicability of virtually every type of modelling approach to this hydrogeological setting. As demonstrated 'channels', afforded by deep mining activities, constitute the primary pathways for groundwater movement in extensively mined areas. It follows therefore that the full inclusion of these features is essential within any 'valid' form of computer model. Neither Darcian nor Discrete Fracture Network modelling approaches, permit the level of interaction afforded by these features, to be incorporated within a groundwater model.

Unless the modelling of groundwater flows in extensively mined strata is to become yet another feature of stochastic modelling techniques, further research is required to increase the accuracy of investigative methods applied in these areas. These studies would assist in the development of a model which would represent not only the movement of groundwaters through 'channels' afforded by mining activities, but in addition, groundwater flows associated with fracture patterns throughout the rock mass. It would also be necessary to incorporate fault patterns as hydrogeological characteristics within the model, affording barrier and/or pathway features throughout the network. Ultimately, the accurate simulation of groundwater flows in these areas will require a multi-layered approach to the modelling process, representative of the complexities of the aquifer systems. The alternative approach, where the complexities of the system are represented in an oversimplified manner, will never result in a model which can truly adopt the 'predictive' role necessary to promote a meaningful risk assessment for either landfill or contaminated land sites.

REFERENCES

Aldous, P.J., Smart, P.L. and Black, J.A. 1986. *Groundwater management problems in abandoned coal-mined aquifers: a case study of the Forest of Dean, England.* Quart. Journal of Eng. Geol, 19. 375-388.

Xiao, G.C., Irvin, R.A. and Farmer, I.W. 1991. *Water Inflows into Longwall Workings in the Proximity of Aquifer Rocks.* Mining Engineer, 151, July. 9-13.

Laboratory permeability measurements with Mercia Mudstone

E.J. MURRAY[1], J. DAVIS[2], P. KEETON[3] AND R.G. HILL[1]
[1] Murray Rix Limited, 13 Willow Park, Stoke Golding, Warwickshire U.K.
[2] Environment Agency, 15-17 Lower Queen Street, Sutton Coldfield, W.Midlands, U.K.
[3] Soil Mechanics, Askern Road, Carcroft, Doncaster, S.Yorkshire, U.K.

INTRODUCTION

As the environmental regulator, the Environment Agency insists on a strict standard of construction quality for lining and capping environmental protection systems at landfill sites. Current requirements for mineral liners dictate permeability testing in accordance with the British Standard constant head triaxial test (BS 1377 Part 6, Method 6) (BS test). In this test method the saturation, consolidation and permeability stages are carried out as distinct and sequential operations. However, there have been proposals from the waste industry to adopt an Accelerated Permeability Test (AP test) as an alternative means of assessing permeabilities. The Agency commissioned Murray Rix Ltd to carry out research to investigate the possible use of the AP test, where the consolidation, saturation and permeability stages are carried out as one, which might speed up the availability of results.

The brief was to investigate the AP test methodology and compare the results from controlled laboratory tests using the two methods. The paper presents results of tests carried out on one of the materials examined, Mercia Mudstone (MM). The test data presented includes the results of measurements of volumetric (density) and moisture content changes throughout the permeability tests. Permeability test results are presented for samples prepared at water contents and densities likely to be used in actual construction. Interesting conclusions may be drawn with respect to the significance of effective stress and hydraulic gradient on the permeability values and soil behaviour.

MATERIAL PROPERTIES

At source the MM material comprised firm to stiff friable red brown clay and very weak fragmented mudstone (gravel size). This was screened to remove/break down material >10mm in size. X-Ray Defraction indicated the clay mineralogy to comprise predominately chlorite and muscovite. The properties of the material following screening are outlined in Table 1.

TABLE 1 - Material Properties

Material	LL	PL	PI	Classification	Particle Size Distribution				Activity
					Clay (%)	Silt (%)	Sand (%)	Gravel (%)	
MM	34	20	14	CL	25	26	43	6	0.56
					34	23	41	2	0.41

The initial gradation is based on careful analyses with every effort taken to avoid breaking down of the coarser fractions and gives a picture of the gradation prior to specimen preparation and testing. The second grading distribution is based on normal BS testing techniques where the weaker mudstone fractions are broken down during the drying and sieving process. Careful grading analyses after specimen preparation and permeability testing

Geoenvironmental impact management, Thomas Telford, London, 2001.

indicated breaking down of the coarser fractions to an overall gradation intermediate between the ranges shown in Table 1.

PERMEABILITY TEST PROCEDURES

The test procedures were designed to provide information on the specimen condition throughout the permeability tests. The volume and moisture content changes necessary to determine the foregoing were based on the initial condition of the specimens prior to insertion in the triaxial cell. The test methodologies suggested that there were likely to be greater errors resulting from calculations based on end of test conditions as a result of uptake of water by the specimens, and the release of air from solution, on reduction in cell pressure. All tests were on recompacted laboratory prepared specimens of nominal 100mm diameter and 100mm length to minimise scale effects (Mitchell and Younger, 1967). As illustrated in Fig.1, volume change indicators A, B and C were included on the top and bottom back-pressure lines, and the cell pressure line, respectively. Indicators A and B allowed for measurement of the water entering and leaving the specimen while C allowed measurements of volume change of the specimen during each stage, after allowing for calibrated cell expansion. A pore water pressure transducer was incorporated in the pore pressure line at the base of the specimens.

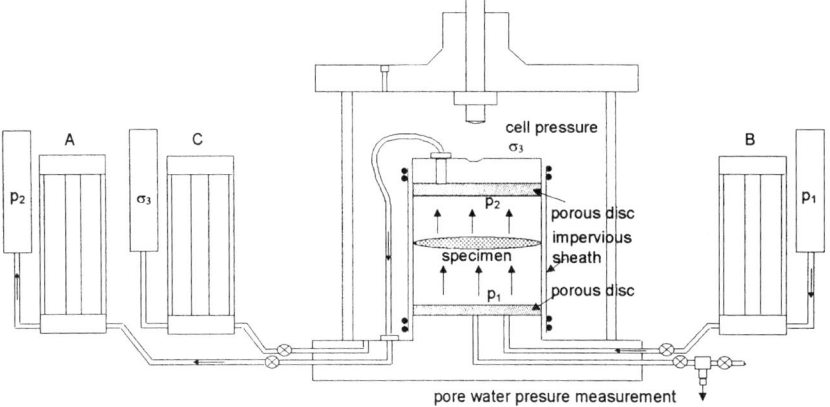

Fig. 1   Triaxial Cell Set-up

The BS Test

BS1377:1990 allows for alternative methods of saturation and consolidation prior to the permeability stage in the constant head test. In addition, discussions with the various laboratories consulted indicated different interpretations of the test procedure primarily to reduce on testing times. The following procedures were adopted in the permeability tests:

Saturation Stage – Alternating increments of cell pressure and back-pressure were applied while maintaining a small positive effective stress. Back-pressure was applied to the top of the specimen only and pore water pressures recorded at the base. At each level of total stress, B values were determined. A value of $B \geq 0.95$ was taken as indicative of adequate saturation.

Consolidation Stage – The procedure adopted was to allow drainage upwards while recording the pore water pressure at the base. Under these conditions the requirement for at least 95% dissipation of excess pore water pressure was ensured. Throughout the BS tests reported herein, an average effective stress of 187.5kPa was used as in the AP tests.

Permeability Stage – BS1377:1990 allows for permeability measurements under vertical upwards or downwards flow. In the tests undertaken, the hydraulic gradient was applied to achieve upward flow as adopted in most commercial laboratories. Though the average effective stress was maintained constant, the BS tests comprised two stages corresponding to an increase in the hydraulic gradient. A hydraulic gradient of 30 was applied in Stage 1 of the BS procedure but subsequently increased in Stage 2 to that of the AP test of 125[1]. An example of cumulative flow against time in a BS test is given in Fig.2.

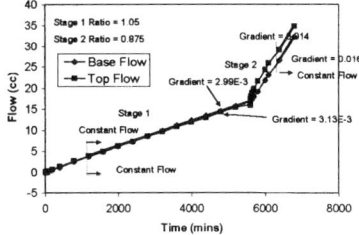
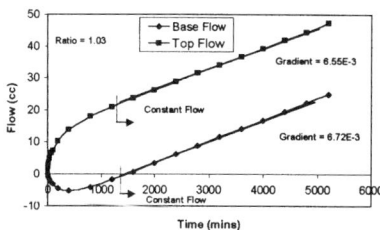

Fig.2 Cumulative Flow against Time – Test Series 1, MM(BS1)

Fig.3 Cumulative Flow against Time - Test Series 1, MM(AP1)

The AP Test
Following consultations with a large number of commercial laboratories, a consensus was reached to adopt imposed cell and back-pressures as detailed in Table 2.

Combined Saturation, Consolidation and Permeability Stages – The test uses hydraulic gradients and effective stresses in excess of those normally employed in the BS test. As no separate saturation or consolidation stages are employed, concerns have been raised on the influence of the test methodology on the measured permeability values and the degree of saturation of the specimens at the end of the tests. This is addressed in the following. An example of cumulative flow against time in an AP test is given in Fig.3.

TEST SERIES AND SPECIMEN IDENTIFICATION
Permeability tests were carried out on paired BS and AP test specimens prepared to initially similar conditions. As an example of the full specimen designation used in Table 2: MM(AP3) defines the material as Mercia Mudstone (MM) and the test as Accelerated Permeability Test number 3 (AP3 in Fig.4). Though the research programme included for three test series, only Test Series 1 and 3 were carried out on MM material and are outlined below:
Test Series 1 - Tests at 2 different target moisture contents and target air voids contents of 5% and 10%. The lower moisture content was the mean optimum from BS (2.5kg) Light Hammer Compaction tests and the higher moisture contents corresponded to an undrained shear strength of approximately 50kPa (BS1 to 4 and AP1 to 4 of Fig.4).
Test Series 3 - Tests were carried out at one target moisture content and two target air voids contents of 5% and 10%. The moisture content was the optimum from BS (4.5kg) Heavy Hammer compaction tests (BS5 and 6 and AP5 and 6 of Fig.4).

---

[1] In the BS test the average effective stress is specified and the hydraulic gradient is adjusted to achieve measurable flow. In the AP test both the cell pressure and top and bottom back-pressures, thus the hydraulic gradient and average effective stress, are specified.

Though the preparation moisture contents were in part determined from BS 2.5kg and 4.5kg compaction tests, the specimens were prepared by static compaction to the required dry densities as this was felt to provide a more uniform test specimen.

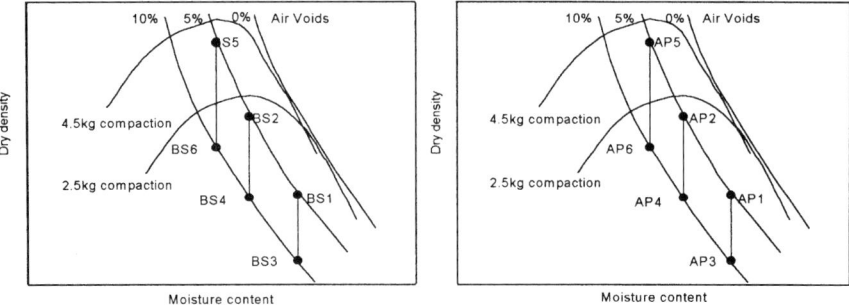

Fig. 4 Specimen Identification

DISCUSSION OF RESULTS

Examples of the plots of variation in moisture content against dry density in the paired BS and AP tests are presented in Figs 5 to 7.

TABLE 2 - Test Details and Results for MM Material

Test Series	Specimen Designation	Cell Press. (kPa)	Bottom Back Press. (kPa)	Top Back Press. (kPa)	Av. Consol. Press.[2] (kPa)	Hydraulic Gradient	Permeability (m/s)
1	MM(BS1) Stage 1	602.5	430	400	187.5	30	2.00E-10
	MM(BS1) Stage 2	650	525	400	187.5	125	2.10E-10
	MM(AP1)	550	425	300	187.5	125	1.05E-10
	MM(BS2) Stage 1	502.5	330	300	187.5	30	8.80E-10
	MM(BS2) Stage 2	550	425	300	187.5	125	1.30E-9
	MM(AP2)	550	425	300	187.5	125	4.00E-10
	MM(BS3) Stage 1	602.5	430	400	187.5	30	2.90E-10
	MM(BS3) Stage 2	650	525	400	187.5	125	2.60E-10
	MM(AP3)	550	425	300	187.5	125	1.40E-10
	MM(BS4) Stage 1	502.5	330	300	187.5	30	8.30E-10
	MM(BS4) Stage 2	550	425	300	187.5	125	1.20E-9
	MM(AP4)	550	425	300	187.5	125	1.30E-8
	MM(AP4)r[1]	550	425	300	187.5	125	8.80E-9
3	MM(BS5) Stage 1	552.5	380	350	187.5	30	4.20E-10
	MM(BS5) Stage 2	552.5	427.5	302.5	187.5	125	7.40E-10
	MM(AP5)	550	425	300	187.5	125	1.40E-8
	MM(BS6) Stage 1	752.5	580	550	187.5	30	1.20E-9
	MM(BS6) Stage 2	752.5	627.5	502.5	187.5	125	3.80E-9
	MM(AP6)	550	425	300	187.5	125	1.10E-7

[1]Repeat test
[2]Average consolidation pressure (average effective stress) is defined as the cell pressure less the mean of the top and bottom back-pressures.

Specimen Preparation

Those specimens which were prepared to the initially drier and more densely compact conditions, were notably less compact than the target values on extrusion from the compaction moulds. This applied particularly to the paired specimens MM(BS2) and MM(AP2), and MM(BS5) and MM(AP5), as shown in Figs 6 and 7. The density reduction may be attributed to an 'elastic' expansion of the low plasticity clay, which contain mudstone 'peds', on reduction in confining stress.   Repeat compaction operations confirmed that this was a material response to the release of confining stress.   On reduction in cell pressure and removal from the triaxial cell after permeability testing, specimens of MM material again experienced notable expansion.

Density and Moisture Content Variations during Testing

The dry density against moisture content paths in the BS tests are marked as A-B-C-D-E on Figs 5 to 7.   In these tests, there was a very low effective stress (circa 10kPa) applied until the end of the saturation stage at C; A-B being the initial specimen response to opening the top back-pressure tap.   All specimens are shown to increase in moisture content between A and C. The less compact, wetter specimens (e.g. Fig.5) exhibited an increase in dry density but the more heavily compact and drier specimens, (e.g. Fig.7), exhibited a rapid and large decrease in density.   Specimen MM(BS2), (Fig.6), exhibited little volume change and presents an intermediate response between those of the other specimens.   During the subsequent consolidation stage C-D, there was an increase in dry density of the specimens with the plots running parallel to the zero air-voids line.   Following this, during the permeability stage D-E, only very small changes in moisture content and dry density occurred.

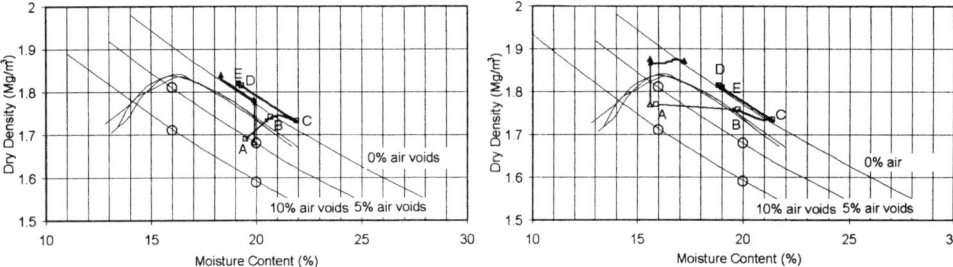

Fig.5   Dry Density/Water Content Paths– MM(BS1) and MM(AP1)

Fig.6   Dry Density/Water Content Paths– MM(BS2) and MM(AP2)

The behaviour during the AP tests contrasted with that during the BS tests.   The wetter and less compact AP tests (e.g. Fig.5) exhibited a rapid compression to near-zero air voids under the initial application of the cell pressures of 550kPa.   This was not the case for the drier and more compact specimens (e.g. Fig.7) where there was a gradual increase in moisture content at near-constant specimen volume as the tests proceeded.   The swelling exhibited during the initial stages of the MM(BS5) and MM(BS6) tests did not occur in the paired AP tests where the effective stress (187.5kPa) was well above that of the BS tests initially.

The BS and AP tests followed distinctly different stress paths.   At the end of permeability stage, the specimens were more compact in the AP tests even though the same average effective stress was adopted in the consolidation and permeability stages of the BS tests. Allowing the combined permeation and consolidation in the AP tests is considered to have facilitated migration of the finer particles into the voids between mudstone peds and a denser end condition for the specimens.   In the staged BS tests, permeation took place only after consolidation and very little subsequent volume change occurred (movement of particles during permeation is likely to have been more restricted).

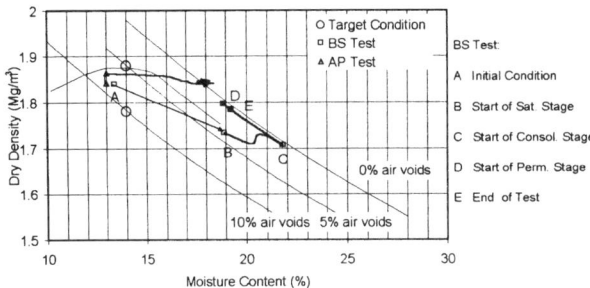

Fig.7 Dry Density against Moisture Content Paths – MM(BS5) and MM(AP5)

Of note is that at the end of both the BS and AP tests all specimens indicated near-saturated conditions based on determinations of air-voids contents. However, the B values for the AP test specimens were often around 0.9 after permeability testing. The relatively low B values are considered a consequence of the consolidated condition of the specimens.

Flow and Permeability Measurements
As illustrated in Table 2 and Fig.8, the permeabilities in tests MM(AP1), MM(AP2) and MM(AP3) were approximately half those of the paired BS tests. The plots are presented for Stage 1 of the BS tests where a hydraulic gradient of 30 was employed compared to a hydraulic gradient of 125 in the AP tests. As all specimens were shown to have been near-saturated, the lower permeabilities in these AP tests are considered primarily a consequence of the more compact state of the AP specimens.

The results of test MM(AP4), MM(AP5) and MM(AP6) indicate permeabilities one to two orders of magnitude (10 to 100 times) greater than in the paired BS tests. The AP test permeabilities were significantly greater than 1×10^{-9} m/s, the criterion often adopted as an upper limit to permeability for landfill sites, while the BS test permeabilities were generally less than this criterion. Establishing a reason for this and the significance to permeability testing of such soils is of the utmost importance in the design of landfills. The initial moisture content and density of the specimens is obviously important and influences the soil structure. The effective stress is considered the other major factor.

Influence of Soil Macro Structure
In the drier specimens, there is likely to have been a high degree of fissure continuity. This is consistent with the notable expansion of the more densely compact specimens on removal from the compaction mould. Under these conditions, discontinuities will have opened and pore water pressures will have reduced. It is estimated that suctions generally within the range 100 to 250kPa are likely to have been present on removal of samples from the compaction mould within the range of moisture contents of the tests. Subsequently, the BS test specimens MM(BS5) and MM(BS6) experienced a rapid and large increase in volume (reduction in dry

density) with a concomitant increase in moisture content when the specimens were exposed to water in the triaxial cell. The low moisture content and fissured and desiccated structure would be conducive to a relatively rapid uptake of water particularly along discontinuities. At points of contact between saturated 'packets' the stabilising influence of suction will have been reduced resulting in propagation of fissures.

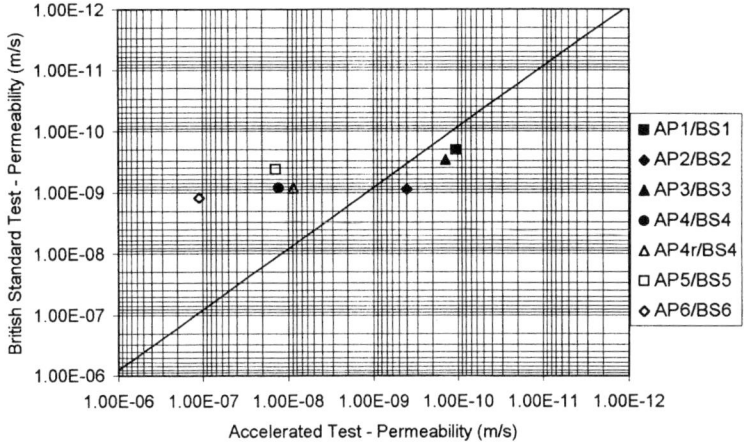

Fig.8 Comparison of AP and BS Permeability Measurements

Following the completion of the saturation stage, the subsequent consolidation stage will have resulted in closure of the fissures and healing (Fernandez and Quigley, 1990) of specimens. However, fissures will not have fully healed resulting in greater permeabilities in tests MM(BS4), MM(BS5) and MM(BS6) than obtained for the initially less compact specimens MM(BS1) and MM(BS3).

In the drier AP test specimens, the same process of expansion and fissure development on removal from the compaction mould may be inferred. However, there was not the low effective stress imposed as in the saturation stage of the BS tests. The plots of cumulative flow indicated uptake of water and positive pore water pressures of the order of 300kPa (roughly equivalent to the top back-pressure) as present on application of the elevated cell pressure. As there was no separate consolidation stage in the AP tests there is not thought to be the same degree of healing of fissures as in the paired BS tests resulting in greater permeabilities.

Influence of Effective Stress (and Hydraulic Gradient)
Boynton (1983) performed tests to illustrate the influence of effective stress on compacted clay. The average effective stresses varied between 14 and 103kPa. Where desiccation cracks were present the tests exhibited significantly greater permeability under low effective stress (well in excess of an order of magnitude greater) than under the higher effective stress. The influence on permeability of closing fissures by increasing effective stress is also evident from the results of Garcia-Bengochea et al (1979), Juang and Holtz (1986) and Murray et al (1996).

The influence of effective stress is not restricted to heavily fissured soils. Silver (1995) and Silver and Joseph (1999) concluded that the main impact on clay permeability is the structural changes in fabric due to effective stress induced deformations. This is compatible with the findings of other tests carried out as part of the research programme.

The influence of the hydraulic gradient, which is a measure of the rate of change of effective stress through a specimen, is less well appreciated. A considerable amount of evidence has been advanced to indicate that there is deviation from Darcy's law in fine-grained soils at low hydraulic gradients of less than about 6 (Mitchell and Younger, 1967). The deviation is likely to be influenced by the adsorbed double layer. There is also concern over elevated hydraulic gradients. It has been found experimentally that elevated hydraulic gradients can give rise to both increase (Schwartzendruber, 1969) and decrease (Mitchell and Younger, 1967) in permeability. Increase in permeability may be a result of mechanisms such as piping or hydraulic fracturing. Decrease in permeability appears to be a result of particle migration, causing clogging. Hird et al (1997) suggest from tests on colliery spoil that particle migration may give rise to both increase and decrease in permeability and that it is influenced by the hydraulic gradient. The direction of flow in permeability tests is likely to influence particle migration and may thus influence permeability measurements.

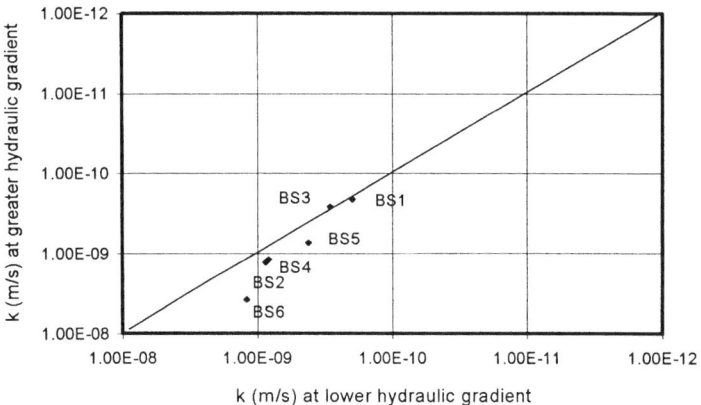

Fig.9 BS Permeability Values (m/s) for Changing Hydraulic Gradient from 30 to 125

In most of the tests carried out as part of the research there was no measurable influence due to change in hydraulic gradient within the range 10 to 125. This has been shown not to be the case for the drier MM specimens. At the end of the first stage of the BS tests the pressures were adjusted to examine the influence of increasing hydraulic gradient to that in the AP tests (i increased from 30 to 125). Though the average effective stress was maintained constant the increase in hydraulic gradient will have led to swelling at the inflow end of the specimen and further compression at the outflow. The results for Stage 2 shown on Fig.9 indicate that there was little influence on the wetter specimens MM(BS1) and MM(BS3) but the increase in hydraulic gradient resulted in an increase in permeability of up to 3 times for the drier more densely compact specimens.

CONCLUSIONS

(a) For the low plasticity MM material, depending on initial conditions, AP permeability results were often significantly greater (2 or 3 orders of magnitude greater) than the results for BS tests carried out with the same average effective stress.

(b) The stress paths followed in the BS and AP tests are significantly different, as are the end conditions. The AP specimens were saturated or near-saturated at the end of an AP tests but B values determined at the end of the test were often around 0.90.

(c) The average effective stress has a marked influence on permeability values. The hydraulic gradient also has an influence on the MM materials.

(d) Depending on initial conditions, swelling, elastic behaviour, yielding, irrecoverable plastic deformations, and collapse compression may be identified within the permeability tests.

(e) The Environment Agency accepts the BS test as an appropriate means of assessing the permeability of mineral liners and cappings for landfill sites. However, the sequential saturation, consolidation and permeability stages in a BS test are unlikely to be realised in practice. The potential variation in permeability results using different testing techniques is emphasised by the large variation in permeability recorded for the MM material using the BS and AP methods.

REFERENCES

Boynton, S.S. (1983). An investigation of selected factors affecting the hydraulic permeability of compacted clay. M.S. Thesis, University of Texas, Austin, TX, USA.

BS1377: 1990. *Methods of Tests for Soils for Civil Engineering Purposes, Code of Practice.* British Standard Institution, HMSO, London

Carpenter, G. W. and Stephenson, R. W. (1986). Permeability testing in the triaxial cell. *Geotechnical Testing Journal.* GTJODJ. Vol. 9, No. 1, pp 3-9.

Fernandez, F. and Quigley, R. M. (1991). Controlling the destructive effects of clay-organic liquid interactions by application of effective stresses. *Canadian Geotechnical Journal,* 28, pp 388-398

Garcia-Bengochea, I, Lovell, C.W. and Altschaeffl, A.G. (1979). Pore pressure distribution and permeability of silty clays. *Journal of the Geotechnical Engineering Division, ASCE,* 105, (GT7), p839-856.

Hird, C. C., Norton, F. and Joseph, J. B. (1997). Particle migration effects on colliery spoil liner permeability. *Sardinia '97, Sixth International Landfill Symposium, ed, S. Margherita di Pula, Cagliari, Italy,* pp 131-139

Juang, C.H. and Holtz, R.D. (1986). Fabric, pore size distribution, and permeability of sandy soils. *Journal of the Geotechnical Engineering Division, ASCE,* 112 (9), p855-868.

Mitchell, J. K. and Younger, J. S. (1967). Abnormalities in hydraulic flow through fine-grained soils. *Permeability and Capillarity of Soils, STP 417, ASTM,* pp 106-139.

Murray , E.J., Rix, D.W. and Humphrey, R.D. (1996). Evaluation of clays as linings to landfill. *Engineering Geology of Waste Disposal,* Geological Society Special Publication No.11, Ed. Bentley, S.P., pp251-258.

Schwartzendruber, D. (1968). *Soil Science Society of America Proceedings, Vol.32,* No.1, pp11-18.

Silver, R.K. (1995). The compaction and permeability performance of mineral landfill liners. *PhD Thesis, University of Paisley, Depart. Civil, Structural and Environmental Engineering, Paisley,* U.K.

Silver, R.K. and Joseph, J.B. (1999). Laboratory permeability testing of recompacted soils and landfills. *Proc. Sardinia 99, Seventh International Waste Management and Landfill Symposium,* pp 37-46.

Study of the water content balance between GCL and soil

Mohammad AL NASSAR

Gérard DIDIER et David CAZAUX

INSA LYON, URGC Géotechnique, Bat J.C.A. COULOMB, 34, Avenue des Arts, 69621 VILLEURBANNE CEDEX

INTRODUCTION

Geosynthetic Clay Liners (GCLs) are used in landfill liner systems as bottom, slope or cover lining application. In order to fill the sealing function, GCLs that have a very low hydraulic conductivity must be confined under normal stress and hydrated with water. Initial hydration is very important to confer the hydraulic performances of the GCLs subjected to a prolonged contact with leachate (von Maubeuge, 1995, Ruhl and Daniel 1997, Didier and Comeaga 1997). Indeed, the most pessimistic situation for a GCLs is a direct contact with the leachate when used in a landfill bottom liner system. The above researches showed that is necessary to hydrate the GCL before any contact with the leachate. To carry out this initial hydration, several way are available. An active process can be ensured by the rain or a controlled moistening of the GCL before or after the implementation of the confining layer. The bentonite can also be hydrated by vapor transfer of the water present in the support or confining soil.

We present in this paper the results of several series of tests performed to determine the water content balance of the GCL in contact with a sand (support or confining) with various initial water contents. Influence of the soil water content, the normal stress, the grain size distribution and the nature of the soil are also considered. The GCLs tested in this paper is a needle-punched GCL with a granular natural sodium bentonite.

TESTING PROCEDURE

All the tests were carried out in a laboratory at $20°\pm2°C$ on 9.5 cm diameter samples. Samples weight and thickness were measured before introduction into the test cell. The testing cell is a PVC cylinder with an internal diameter of 9.5 cm and 30 cm high.

Contact between GCL and confining soil

The sealing between the base of the column and the cylinder was ensured by a silicone joint. The sample was introduced in the cell and placed at the bottom of the column. A layer of 3 cm of the sand at a given water content is placed above the GCL sample. A FIBERGLASS probe is installed in a vertical manner into the sand layer in order to control its water content. The column is then filled with the same sand by successive compacted layers. A lid then insulated the column. In order to avoid any loss of water during the test, sealing of the probe wire to the lid was ensured by a silicone joint (figure 1.a).

The electrical resistance is measured against time. The moisture equilibrium was considered to be reached when the variation of the electrical resistivity against time was almost null.

Geoenvironmental impact management, Thomas Telford, London, 2001.

Contact between GCL and support soil

The column is filled by successive 3 cm thick compacted sand (Hostun RF sand) with a given water content. The GCL sample is then installed on the top of the sand. A piston allows the application of a normal stress on the GCL (7 kPa). Lateral sealing between the wall of the column and the piston is ensured by a silicone joint. A dial gauge allows to follow the swelling of the GCLs (figure 1.b).

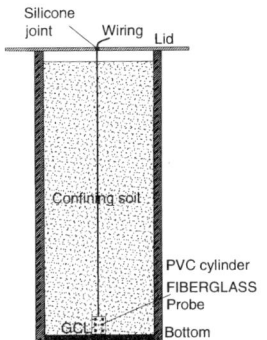
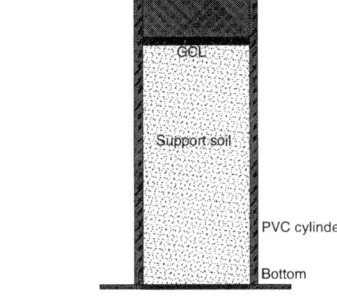

a. Contact GCL-Confining soil   b. Contact GCL-Support soil

Figure 1.

WATER CONTENT BALANCE BETWEEN GCL AND CONFINING SOIL

The study of the water content balance is carried out on a GCL with 3 types of soils with various controlled water contents:

HS : HOSTUN sand 0.16/0.63 mm, at five water contents: 5, 7, 10, 12 and 15%;
SS : River Saône sand, 0.5/2.5 mm, at a water content of 7%;
G : gravel 2/7 mm at a water content of 7%.

The GCL reveals an initial water content of 10,6 % and a dry mass per unit area of 5.92 kg/m^2. A normal stress of 5 kPa was applied to all the tested samples.
Figure 2 presents the variation of the electrical resistivity against time.
At the end of the second week of testing, all curves reached an asymptote which can be considered to the water content balance between the bentonite and the confining soil.

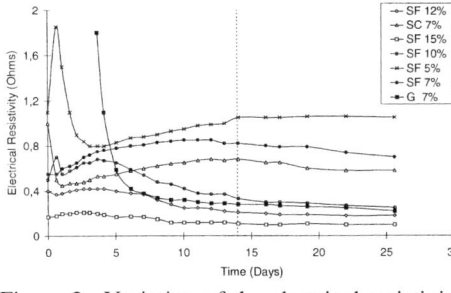
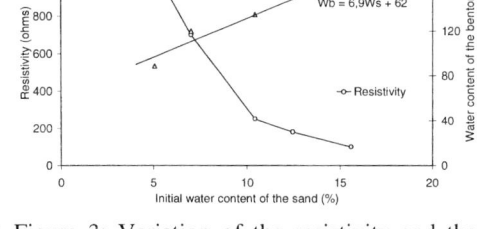

Figure 2 : Variation of the electrical resistivity against time.

Figure 3: Variation of the resistivity and the water content of bentonite versus initial water content in sand.

Table 1 gives the results obtained for the water content balance of bentonite:

Table 1: water content balance of the bentonite.

Type of sand	SH					SS	G
w% of the sand	5	7	10	12	15	7	7
w% of the bentonite	88.8	120.1	134.8	150.9	166.1	127.6	136.1

For the same initial water content of the confining soil (7%), the coarser is the sand, the higher is the water content of the bentonite.

For the Hostun sand, we note that the balance water content of bentonite increases proportionally with the initial water content of the sand from 89% to 166% (figure 3). The evolution of the water content is inversely proportional to the electrical resistivity of the sand close to the GCL.

Figure 4 presents the evolution of the water content of the sand at various levels in the column. We notice that the water content increases with the depth. The water in the sandy material is influenced by gravity especially for the higher initial water contents. This influence much depends on the sand grain size distribution.

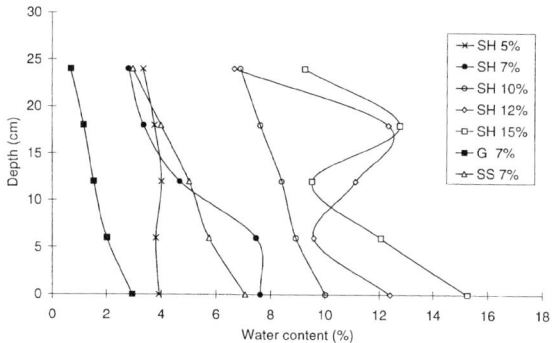

Figure 4: Variation of the water content of sand in the column.

Variation of the water content of bentonite against time

We followed the evolution of the water content of bentonite according to time when the GCL is in contact with Hostun sand at 12 % water content and under a normal stress of 5 kPa. We reproduced on 6 samples the same tests of balance GCL-confining soil.

The initial water content of bentonite is 17,33 %. The average mass per unit area of the GCL is 5.6 kg/m^2. Table 2 presents the water contents of the sand (in contact with the GCL) and the bentonite at the end of the test :

Table 2: Water content and swelling of GCL against time.

Duration of test (days)	Bentonite water content (%)	Swelling of GCL (mm)
1	43.2	2.62
2	70.8	3.42
4	148.6	3.84
8	181.8	3.88
12	179.4	3.89
23	189.0	4.13

Figure 5 presents the evolution of the water content of bentonite against time. For an initial water content of the sand of 12 % and a normal stress of 5 kPa, we note that the equilibrium water content and swelling are reached after about one week.

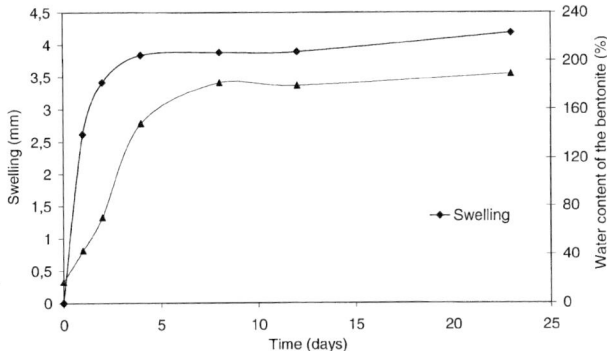

Figure 5: Variation of the water content and swelling against time.

WATER CONTENT BALANCE BETWEEN GCL AND SUPPORT SOIL

Influence of the initial water content of sand

The initial water content of the GCL was 10.9 % and the mass per unit area 5.90 kg/m2. The HOSTUN sand was used at various water content : 3, 5, 7, 10, 12,15 and 17%. The normal stress applied by the piston was 7 kPa for all the tests.

Swelling being almost stabilized after 24 days, we measured the water content of bentonite in the GCL and the water content of the sand at 5 regularly spaced levels in the column.
Table 3 gives the results obtained after 24 days.

Table 3: Final characteristics of the GCL

Initial water content of sand (%)	Bentonite water content after 24 days (%)	Swelling after 24 days Δh (mm)
3.0	39.2	0.36
5.0	66.38	0.87
7.0	90.66	2.34
10.0	100.7	2.58
12.0	118.7	2.62
15.0	148.4	2.90
17.0	160.6	3.46

Figure 6 shows the linear relationship between the initial water content of sand and the final water content of the GCL and between the GCL swelling and the water content of bentonite. The coefficients of these straight lines are undoubtedly a function of the type of GCL, the nature of the support or confining material and the applied normal stress.

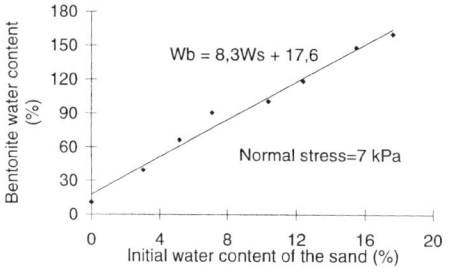
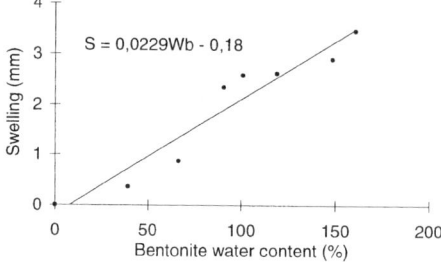

Figure 6 : Final water content of the bentonite versus initial water content of the sand.

Figure 7 : Swelling versus water content of the bentonite.

Influence of the normal stress

The objective was to determine the influence of the normal stress on the water content balance of a GCL in contact with Hostun sand. We used the procedure of previously described tests but we added weights on the pistons in order to increase the applied normal stress.

The initial water content of the GCL was 9.0 % and the mass per unit area 4.80 kg/m². Soil used was HOSTUN sand, its initial water content was 7%. The samples were confined under the following normal stresses : 7.0, 9.9, 14.1, 21.2, 28.2 kPa.

At the end of each test, we measured the water content of bentonite and the water content of the sand at 5 levels in the column (figure 7). Table 4 gives the results:

Table 4: Water content and swelling of the GCLs at the end of the tests

Normal stress (kPa)	Bentonite water content (%)	Swelling Δh (mm)
7.0	106.9	1.28
9.88	105.8	1.25
14.1	105.22	1.32
21.16	99.0	0.74
28.2	95.0	0.94

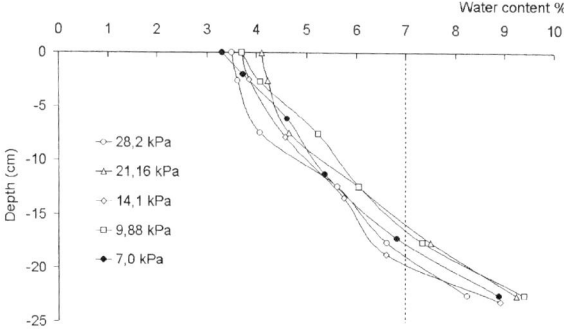

Figure 7 : Variation of the water content of the sand in the column.

For all the tests, the water content of the sand at the bottom of the column is higher than the initial water content (Ws=7%) which showing the influence of gravity on water content balance.

By making the mass balance of water into the sand, the quantity of water adsorbed by bentonite can be estimated. The water content balance of the bentonite and the swelling decrease when the normal stress increases (figure 8). Thus, the water content balance decreases of about 12.5 when the normal stress increases from 7 to 28.2 kPa.

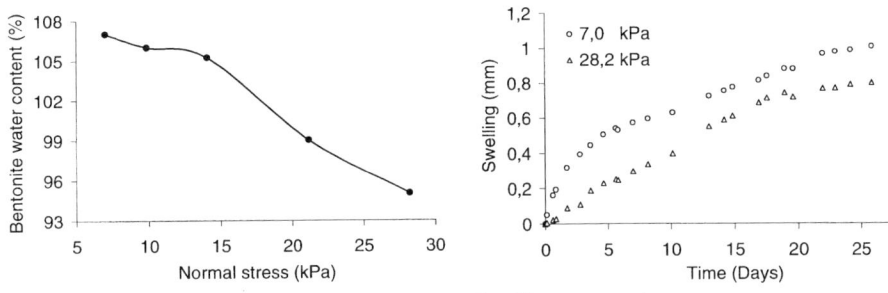

Bentonite water content versus normal stress   Swelling versus time

Figure 8.

We note that the water content balance and the swelling are reached at the end of three weeks. This means that the time of the balance is slower in the case GCL in contact with a wet soil support than with confining soil.

For the same conditions of soil water content and normal stress, the water content balance and swelling for the case GCL-confining soil are much higher than in the GCL-support case.

CONCLUSION

The series of tests GCL-confining soil shows that the bentonite is quickly hydrated in contact with a sand. Indeed, for a confining material with a water content of 12%, the tests showed that the water content of bentonite increased from 17% to 180% in 8 days.

The series of test GCL-support soil shows that the time of balance is longer (three weeks) than with confining soil. Indeed, gravity shows a strong influence in the case GCL-confining soil, where the balance water content is higher and the time of the balance is shorter. This observation is reversed in the case GCL-support soil. The balance water content of bentonite decreases according to the normal stress.

Figure 9 shows the evolution of the final water content of bentonite versus initial water content of the sand for both confining and support configurations. This evolution suggests the following comments:

- The final water content of bentonite increases proportionally with the initial water content of the sand for confining or support configurations
- The final water content of bentonite in contact with a confining soil is higher than with the same soil in support case (for the same conditions of water content of the soil).

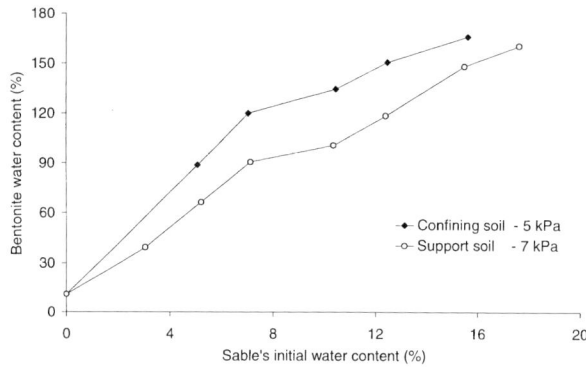

Figure 9 : Finale water content of the bentonite versus initial
water content of the sand.

A few recommendations can be proposed. GCL should be pre-hydrated before any contact
with leachate (Didier and Comeaga 1997). The series of tests showed that the pre-hydration of
the bentonite can be obtained by water absorption by suction in contact with a soil (sand) even
at low water contents (5%). It will be useful to study the water content balance in contact of
others types of soils (clay and silt for example).

REFERENCES

COMEAGA, L., Dispositifs d'étanchéité par géosynthétiques bentonitiques dans les centres de
stockage de déchets, Doctorat thesis in the Institut National des Sciences Appliquées de Lyon,
1997, 297 p., (in french)

DIDIER, G., COMEAGA, L. Influence of initial hydration conditions on Geosynthetic Clay
Liner leachate permeability. In: Testing and acceptance criteria for GCLs. ASTM STP 1308,
ed. Larry Well, USA, 1997. p. 181-195.

von MAUBEUGE, K.P. Performances of GCLs. In: Compte-rendu du Séminaire Les
Géocomposites Bentonitiques, Canada, October 1995. Publication SAGEOS 9525a. 1995. p.1-
16.

RUHL, J.L., DANIEL, D.E. Geosynthetic Clay Liner permeated with chemical solutions and
leachates. In: Journal of Geotechnical and Geoenvironmental Engineering, ASCE, 1997, Vol.
123, N° 4, p. 369-381

Effect of Waste Leachate on Geomechanical Properties of Boom Clay Used as Clay Barrier

OURTH ANNE-SOPHIE, Faculté Agronomique de Gembloux
Gembloux (Belgique) ; ourth.as@fsagx.ac.be
VERBRUGGE JEAN-CLAUDE, Université Libre de Bruxelles
Bruxelles (Belgique) ; jverbrug@ulb.ac.be

ABSTRACT

The influence of a leachate on geomechanical properties of a clay is examined. Various tests were conducted with and without leachate to show how clay reacts when in contact with pollutant.
Leachates vary from one landfill to another and also evolve with time according to the age of deposits. For this reason, a composed leachate as representative as possible of landfills in general is used, as well as one 'representative' clay.

First tests concern the changes in intrinsic properties of the clay : structure, moisture content, suction, Atterberg limits.

Mechanical properties of clay are studied too. Triaxial tests and oedometer tests are performed in laboratory for four combinations : clay prepared with water and water used during testing (reference) ; clay prepared with water and leachate used during testing (real case in landfill) ; clay prepared with leachate and leachate used during testing (particularly unfavourable case) ; and, clay prepared with leachate and water used during testing (unreal case to show influence of osmotic phenomenon). The influence of temperature is also taken into account.

As leachate naturally contains organic matters and micro-organisms, it evolves when it is in contact with the air. Furthermore, as the equipments (triaxial apparatus, GDS) could be damaged by the leachate, they have been adapted to perform the tests in acceptable conditions. The paper will overview the first results gained as well as the adaptations of the experimental devices.

TEST MATERIAL : CLAY AND LEACHATE

The clay used in the investigation is Boom clay, coming from a well-known deposit in Belgium. The leachate was taken from a municipal solid waste landfill site. It is a mixture of liquid from three cells of different ages. The mixture is then like a "mean" leachate.
As leachates evolve naturally, due to organic matter and micro-organisms, the conservation method to preserve its composition was to freeze the whole required quantity (about 1 m^3) without separation of components.

To study the possible modification in clay behaviour, the program involved tests carried out on clay contaminated for several months with the leachate. This 'soaking' and the materials identification have already been described by the authors (Ourth and Verbrugge, 1999 a & b). The most relevant characteristics of them are summarized hereafter

TEST PROGRAM AND FIRST RESULTS
Several kinds of tests are used in this investigation. The proposed test program is large but at the moment incomplete, and the results given here are those from the first investigations.

Material identification

Leachate
• The osmotic pressure of leachate, using a Wescor thermocouple psychrometer was found close to -450 kPa (Ourth and Verbrugge, 1999).

• The results of chemical analysis of leachate (before freezing) are given in table 1:

Table 1 : Chemical analysis of leachate (Ourth and Verbrugge, 1999)

N_{NH4}^+	(mg N/L)	248	TOC	(mg C/L)	316	Pb	(µg/L)	90
N_{NO2}^-	(mg N/L)	<0.02	Ca	(mg/L)	208	Cd	(µg/L)	1.1
N_{NO3}^-	(mg N/L)	2.98	Mg	(mg/L)	131	Zn	(µg/L)	258
N_{org}	(mg N/L)	10	K	(mg/L)	653	Hg	(µg/L)	0.13
SO_4^-	(mg/L)	217	Na	(mg/L)	2000	Cu	(µg/L)	48
Cl^-	(mg/L)	273	Fe	(mg/L)	4.73	Ni	(µg/L)	140
P_2O_5	(mg/L)	8.13	Mn	(mg/L)	0.459	Cr	(µg/L)	48

Clay
• Atterberg limits of the Boom Clay are :
    Liquid limit: 67 %
    Plastic limit: 28 %
    Shrinkage limit: 19 %

• The Proctor test using the CBR mould and the low compaction energy (LCPC, 1966) presented an optimum value for clay remoulded with water at $\gamma_d = 14.2$ kN/m^3 and $w_{opt} = 28\%$.

Effects of soaking and time on the mineral composition
• A X-Ray diffraction analysis was conducted on pure clay and on clay 'soaked' 9 month in leachate. It seems that soaking does not alter the qualitative composition of clay. Anyway, more tests and a more accurate interpretation will be done.

Geomechanical properties
The program involves triaxial tests on saturated and unsaturated samples and oedometer tests. The principal problem encountered for these tests is the evolution of the leachate when in contact with air and its corrosive effect on the testing apparatus. Some modifications have been done to prevent this corrosion and the deterioration of leachate.

For the tests, there are four possible combinations depending on the pore fluid used to prepare and to permeate the clay during test. Table 2 summarises the different combinations (Ourth and Verbrugge, 1999).

Table 2 : Combination of pore fluid for preparation and testing

Test series	Code	Preparation Fluid	Testing Fluid
1	W - W	Water	Water
2	W - L	Water	Leachate
3	L - W	Leachate	Water
4	L - L	Leachate	Leachate

Moreover, the symbol L* in the text below means that the test has been carried out on clay material prepared with the leachate after a short term period of two days and not on the 'soaked' one.

At present time different oedometer tests have been performed and some triaxial test.

Oedometer test
• Testing device :
First oedometer tests for W - W and L* - W (clay prepared with water and leachate : for notation, see table 2) were conducted in a standard 50 mm diameter oedometer. Tests with leachate as a permeant need an apparatus where the liquid is out of contact with the atmospheric air. So, cells combining the principles of the triaxial cell with regard to watertightness / airtightness and the oedometer were constructed. These cells are now used for all tests, even with water.

• Results :
Oedometer test data from the series W - W and L* - W were nearly the same for samples prepared either with water or with leachate, without 'soaking'.

Other tests have been made with soaked clay, at ambient temperature and at a temperature of about 60° C, which is the temperature observed in disposal sites.

The evolution of void ratio with applied stress for the various oedometers is shown on Figure 1. The thin lines correspond to ambient temperature (20° C) and the bold one to 60° C. The first stress (50 kPa) is always applied at ambient temperature. For high temperature tests, oedometer were warmed only when first settlement was stabilised. The next stresses (150, 350, 750 kPa) were applied at a temperature of about 60°. That is the reason why bold curve begins only after the first charge. The L - L and L - W samples used for high temperature tests have been soaking for 9 month in the leachate, those used at ambient temperature had only three months soaking.

Notwithstanding a scatter of curves, that seems mainly due to the initial conditions of samples, no general trend depending on temperature or water-leachate conditions can be put forward. Moreover, the C_c and λ values for the different tests are very similar and close to 0.5 showing a good accordance with the correlation's propounded by Biarez and Hicher (1994).

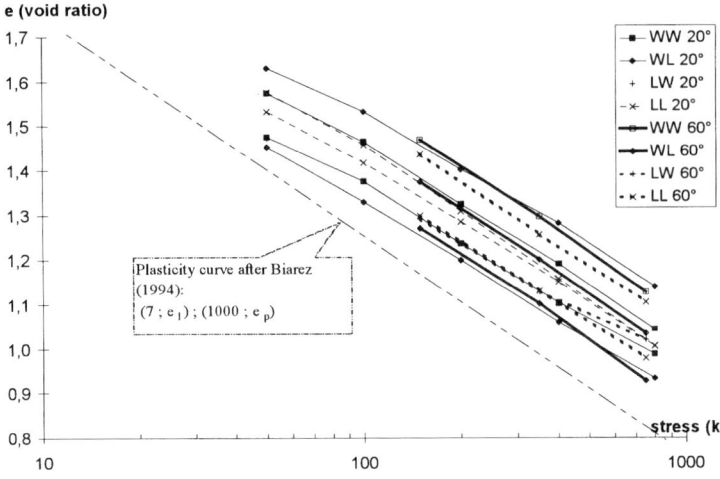

Figure 1. Oedometer test

Triaxial test
• Testing device :
To avoid damaging the apparatus and particularly the pressure volume controllers, a device is used to push the leachate inside triaxial cells. It is presented at Figure 2.

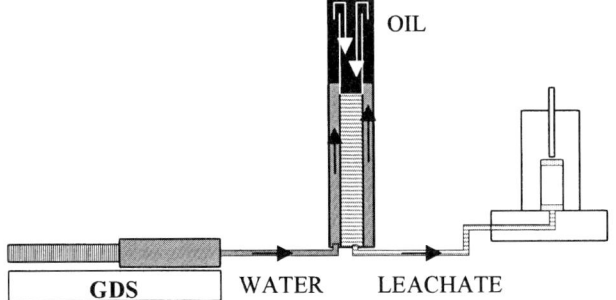

Figure 2. Triaxial testing device.

• Results :
At the moment some triaxial test has been lead at ambient temperature, and other are now performed at 60°C. For the second series results are not yet available. At this stage, the first results of test at about 20°C on clay prepared with water and prepared with leachate can anyway be compared. As for the oedometer tests, any behaviour difference, if anyway it exists is lesser than the scatter of the result. These observations are summarises by the curves given on Figure 3. There are two series of 3 samples W-W and two series L-L (for notation, see Table 2).

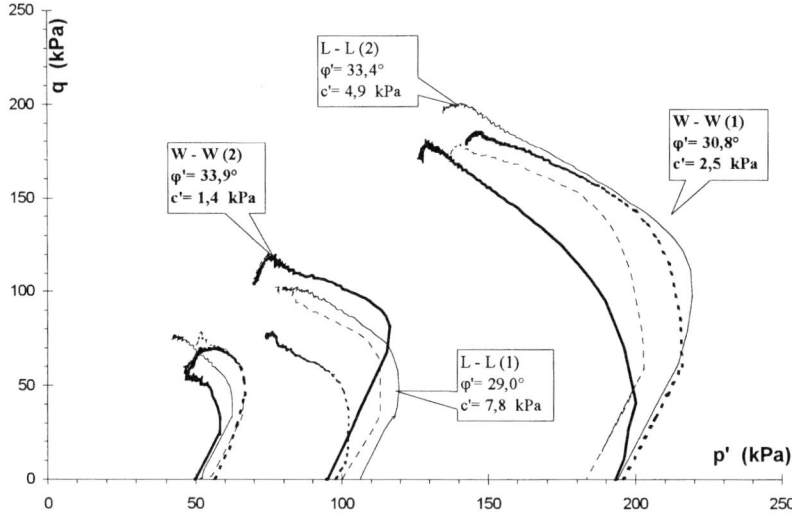

Figure 3. Triaxial test.

CONCLUSIONS

Different laboratory tests were performed with clay prepared with leachate or with water, with or without soaking. Water or leachate were used during tests performed at ambient temperature or at about 60° C. At present time, no significant differences between the results have been observed and if any, they are lesser than the scatter of the results.

The soaking realised in laboratory tries to imitate real landfill condition. However, it would be more "realistic" if the temperature is about 60° C. Some new soaking will be done at this temperature.

Finally, new tests are beginning with a second clay which seems more reactive in presence of leachate.

ACKNOWLEDGEMENTS

The research work described in this paper is supported by a grant from Belgian National Fund for Scientific Research (FNRS). Tests at 60° C were performed at the Ecole Centrale de Paris, thanks to all the staff for his collaboration. Both supports are gratefully acknowledged.

REFERENCES

Biarez, J and Hicher P-Y. (1994). "Elementary Mechanics of Soil Behaviour". A.A. Balkema, Rotterdam, Brookfield.

LCPC, Laboratoire Central des Ponts et Chaussées (1966). "Essais Proctor". Mode opératoire S.C., Dunod, Paris, pp 34.

Ourth, A.-S. (1998) Influence des lixiviats sur les propriétés géomécaniques, en conditions saturée ou non, de l'argile constituant des barrières d'étanchéité pour centre d'enfouissement technique et sites contaminés, Mémoire de fin d'études, Diplôme d'études approfondies en sciences agronomiques et ingénierie génétique, Faculté Universitaire des Sciences Agronomiques de Gembloux, Belgium, 74p.

Ourth, A-S and Verbrugge, J-C (1999-a). "Influence of Leachates on Geomechanical properties of Clay Used as Confining Barriers". Proceedings XIth Panamerican Conference on Soil Mechanics and Geotechnical Engineering, Iguassu, August, AA Balkema, Vol 1, pp. 497-500.

Ourth, A-S and Verbrugge, J-C (1999-b). "First results about the influence of leachate on the properties of Boom clay". Proceedings IInd Geoenvironmental Engineering Conference, London, September, Thomas Telford, pp. 47-53.

Seepage from landfill sites and from animal waste storage tanks

DR. P. PURCELL, DR. M. LONG and MS. HEATHER SCULLY Civil Engineering Department, University College, Earlsfort Terrace, Dublin 2, and MR. T. GLEESON, Teagasc, Kinsealy, Co. Dublin.

ABSTRACT

The paper describes a laboratory study of the suitability of Irish soils for use in the lining of landfill sites and animal waste storage tanks. The soils were taken from a number of locations and are typical of those found throughout Ireland. In the paper, laboratory investigations of the flow characteristics of the soils are presented. Basic soil tests, such as, for example, particle size distribution, were found to give a good indication of the suitability of the soils for lining municipal leachate or agricultural slurry containment facilities. In addition, agricultural slurries were found to cause a very significant reduction in the effective soil permeability due to the deposition of solids on the soil surface and within the pores of the soil.

INTRODUCTION

As part of a study into the suitability of Irish soils for use in the lining of landfill sites and animal waste storage tanks, four typical soils were selected and subjected to a series of laboratory tests. The soils were taken from four separate locations and are typical of those found throughout Ireland. In the paper, laboratory studies relating to (a) the flow of water and (b) the flow of slurry through soils are presented.

SOIL-WATER EXPERIMENTS

Some of the salient details of the soils and the sites studied are given on Table 1.

Soil	Glacial origin	Natural water content (%)	Liquid limit (%)	Plasticity index (%)
Carlow	Alluvial	20	34	16
Moorepark	Fluvio-glacial	25	24	11
Grange	Fluvio-glacial	14	17	15
Dublin till	Lodgement till	10	26	13

Table 1   Details of soils and sites studied

Particle size distribution curves for the materials are shown on Figure 1. All are reasonably well graded (a typical characteristic of Irish glacial soils), with the Carlow alluvial material having the highest fines content. The Moorepark and Grange fluvio-glacial deposits are somewhat gap-graded in the sand range with the Moorepark material having the greatest coarse content.

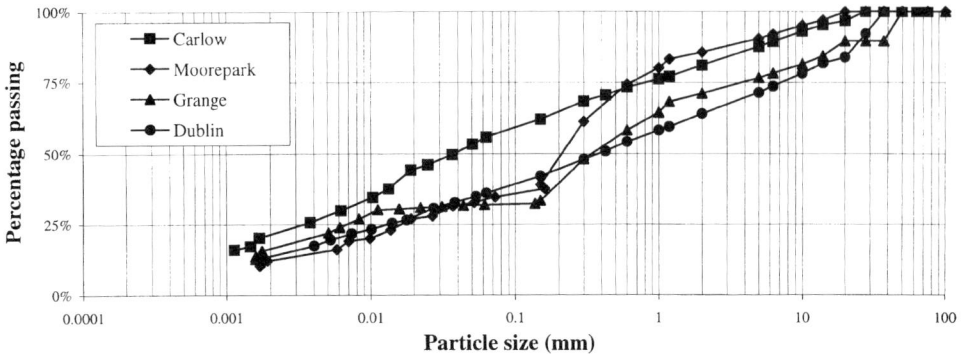

Figure 1   Particle size distribution chart

No specific standard exists for the assessment of soils suitable for use in animal waste storage tanks. However reference can be made to a standard for landfill sites. Typical suitable ranges of parameters for use as landfill clay liners are shown on Table 2 (EPA, 2000, McCullen and Long, 1999).

Number	Property	Units	Acceptable range
1	Permeability	m/s	$< 1 \times 10^{-9}$
2	Clay content ($< 2 \ \mu m$)	%	≥ 10
3	Fines content ($< 63 \ \mu m$)	%	20 - 30
4	Gravel content (> 2 mm)	%	≤ 30
5	Maximum particle size	mm	25 - 50
6	Plasticity index	%	10 - 30
7	Liquid limit	%	≤ 90

Table 2   Range of acceptable soil properties for use as clay liner in landfill sites

All of the soils pass criteria numbers 2 to 4, albeit with the Moorepark and Grange soils being borderline on one case and the Dublin till being borderline in two cases. All the soils pass criterion 5 as these particles were removed prior to testing. This obviously also could be achieved during pre-treatment on site. All the soils pass criteria 6 and 7, with again the Moorepark and Dublin till being borderline in one case.

A series of laboratory triaxial permeability (BS1377) tests were then carried out on the soils. The specimens were prepared by standard Proctor compaction to approximate in situ density with 2.5 kg ram. The objective of the tests was to check whether the method of placement influenced the permeability. In order to do this either the bulk density or moisture content values were systematically varied while keeping the other parameter constant.

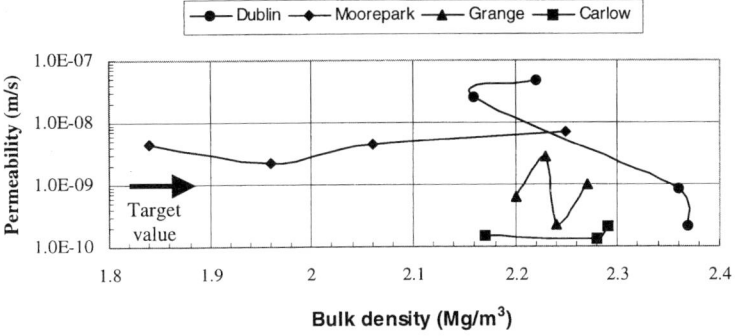

Figure 2   Variation of soil permeability with bulk density

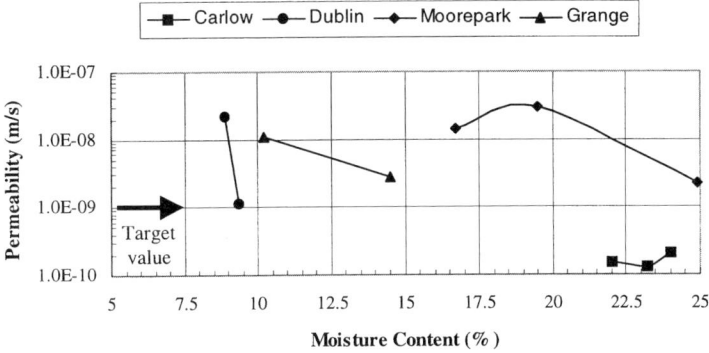

Figure 3   Variation of soil permeability with moisture content

The variations in permeability values with bulk density are shown on Figure 2. The soil specimens were compacted at their natural moisture contents, increasing bulk density being achieved for each soil specimen by increasing the number of layers compacted. Regardless of the bulk density, the fine Carlow material, not surprisingly, is acceptable in all cases. The coarser Grange material is also insensitive to bulk density but has a permeability value closer to the target value. All tests on the coarse Moorepark material fail the criterion. Only the Dublin till material appears to have a permeability which is sensitive to bulk density (this is for the reason that the material is well graded). Beyond a bulk density of about 2.3 Mg/m^3 the material has an acceptable permeability value. However, in order to achieve density values such as this in the field, very heavy compaction plant would be required.

In the case of the permeability/moisture content experiments shown in Figure 3, the natural moisture content of each soil specimen was altered by adding varying amounts of water and the soil was then compacted in three layers. As only relatively few data points are available, it is difficult to make any definitive conclusions regarding the relationship between moisture content and permeability. For the range of data available, again, only the Carlow material passes the acceptability criterion. The values obtained for the other material show little sensitivity of permeability to moisture content and generally fail the criterion.

SLURRY-SOIL EXPERIMENTS

Researchers (Culley et al.,1982) have established, in full-scale field trials, that animal slurries in contact with soil do cause a progressive sealing of the soil with time, thereby significantly reducing the effective permeability of the soil. The laboratory measurement of the permeability of porous media at very low values presents significant difficulties for conventional laboratory geotechnical equipment, such as the permeameter or manually operated triaxial apparatus, and requires the use of sophisticated computer-controlled apparatus such as that used in the soil-water experiments described above. However, animal slurries are unsuitable for use in such sophisticated apparatus due to problems with clogging of lines and potential damage to sensors. To overcome these difficulties, the specific resistance to filtration (srf) apparatus (Coackley and Jones, 1956) used in the laboratory assessment of the dewaterability of water and wastewater treatment sludges was adapted to study the flow of animal slurry through porous media (flow through a sludge cake and the filter medium is analagous to flow through a porous medium). In the srf test (see Fig. 4(a)), an aliquot of the sludge is placed onto a filter paper (Whatman no. 1,) in a Buchner funnel (70 mm diameter) and drawn through by vacuum pump; the cumulative filtrate volume is recorded as a function of time. In adapting the srf test to the case of animal slurry flowing through soil, it was considered to be inappropriate to draw the slurry through the soil by suction, because of the potential problem of air binding within the porous medium at sub-atmospheric pressures. Instead, the srf test apparatus was modified to enable a positive pressure to be applied to the upper surface of soil specimen, while maintaining the lower surface at approximate atmospheric pressure (Figure 4 (b)).

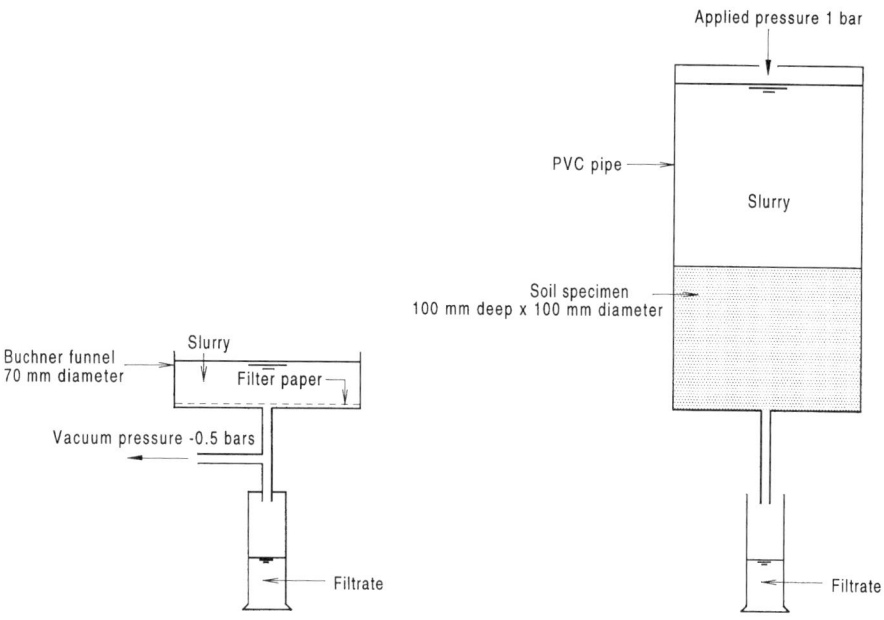

Figure 4   (a) SRF apparatus   (b) Modified SRF apparatus

A typical SRF result in the case of a cattle slurry with a solids concentration of 1.5% is presented in Figure 5.

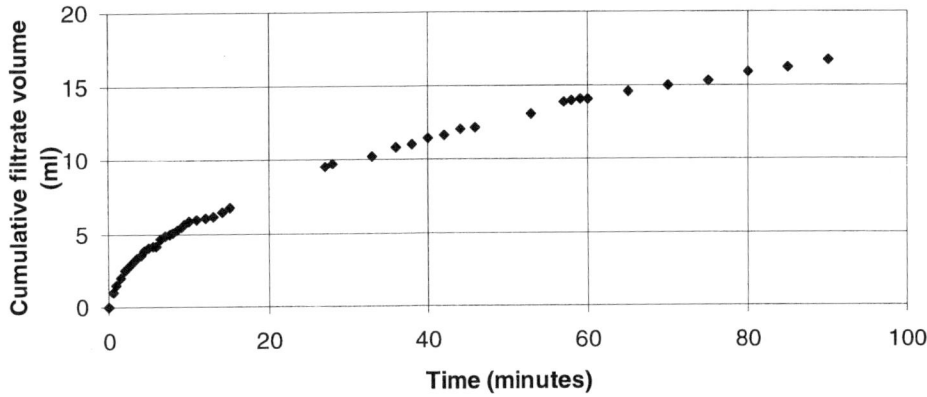

Figure 5   Slurry flow through laboratory filter paper

A typical result for the flow of the same cattle slurry through 100 mm deep by 100 mm diameter specimen of Leighton-Buzzard sand in the modified SRF apparatus is presented in Figure 6 .

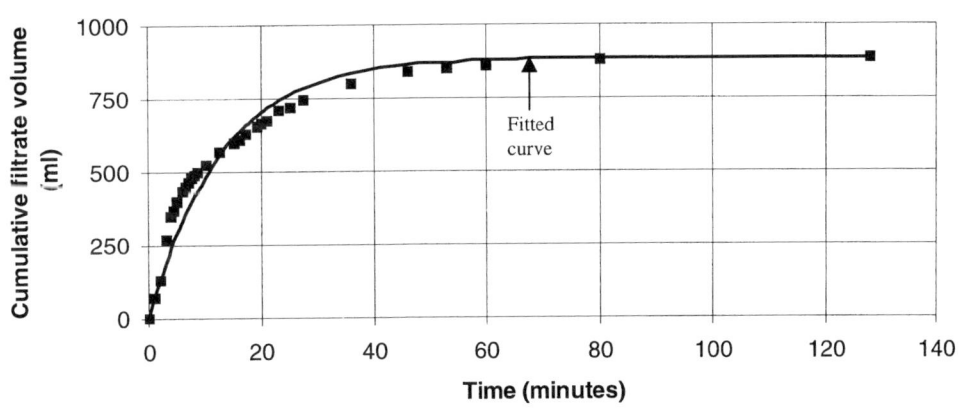

Figure 6   Slurry through Leighton Buzzard sand

Examination of Figures 5 and 6 shows that, although there is only a difference of a factor of two in plan areas of the respective test specimens, there is a considerably larger volume of filtrate in the case of the slurry flow through the sand specimen. Clearly, the filter paper and the slurry cake formed on the surface of the filter offers much greater resistance to the flow than does the much more porous sand. In the case of the filter paper, all the suspended solids in the slurry were retained on the surface of the filter, whereas in the case of the sand, some penetration of the slurry solids into the pores of the sand occurred and resistance to flow was a combination of both the surface cake and sand bed. The sand specimen was housed in a clear walled pvc pipe and visual inspection corroborated this observation (i.e. deposition of solids on sand surface and within the sand bed). Examination of Figure 6 would appear to suggest an exponential flow decay through the sand bed, and a best-fit curve of this form is indicated on the figure, which is described by the following equation:

$$V = 885(1 - e^{-0.08t})$$

where: V = cumulative filtrate volume (ml)

t = time from start of filter run (minutes).

The volumetric flow rate at any time t ($\dfrac{dV}{dt}$) is therefore: $70.8e^{-0.08t}$ (ml/minute)

The corresponding computed effective permeability (k) of the sand specimen at any time t is presented in Figure 7.

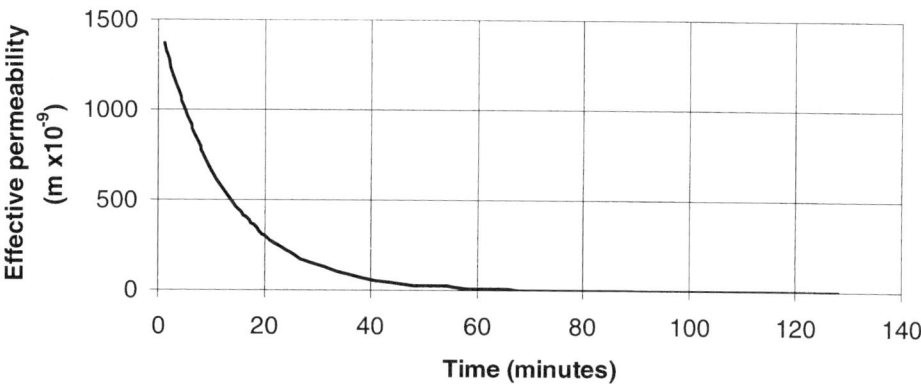

Figure 7 Effective permeability of Leighton-Buzzard sand due to progressive sealing by slurry

CONCLUSIONS

Basic soil tests such as, for example, particle size distribution give a good indication of the suitability of the soils for lining municipal leachate or agricultural slurry containment facilities. However, if a soil is only marginally within the specification in respect of such basic soil parameters it may not be acceptable from a permeability point of view, unless very large compaction plant is available. Agricultural slurries cause a very significant reduction in the effective soil permeability due to the deposition of solids on the soil surface and within the pores of the soil.

BIBLIOGRAPHY

Coackley, P. and Jones, B.R.S (1956) Vacuum Sludge Filtration, Sewage and Industrial Wastes, 28, 8, pp.963-976.

Culley, J.L.B. and Phillips, P.A. (1983) Sealing of soils by liquid cattle manure. Canadian Journal of Agricultural Engineering, 24: 87 – 90.

EPA Ireland (2000), Landfill site design, ISBN 1 84095 026 9.

McCullen, P and Long, M (1999) "Arthurstown landfill facility - geotechnical site characterisation". Proc Seminar. on Arthurstown Landfill held at the Institution of Engineers of Ireland 9/3/99. Also published in Trans. IEI 1998 / 99.

Parker, D.B., Schulte, D.D., Eisenhauer, D.E. (1999) Seepage from earthen animal waste ponds and lagoons – an overview of research results and state regulations.

Vesilind, P.A. (1975) Treatment and disposal of wastewater sludges, Ann Arbor Science, Ann Arbor, Michigan.

Stabilisation of Landfill Sites – Data Collection and analysis of Influencing Factors

H. TAMBLYN, A. ALANI, C.M. SANGHA & P.J. WALDEN, University of Portsmouth, Department of Civil Engineering, Burnaby Building, Burnaby Road, Portsmouth PO1 3QL

ABSTRACT

Sustainable waste management is one of the largest environmental challenges facing big cities. Despite increasing requirements to recycle waste and recover materials and energy, landfill is still a major method of waste disposal. Landfills are complex systems where an understanding of geotechnics, microbiology and chemistry is required to appreciate the processes occurring. Stabilisation, which can take up to 100 years, involves settlement of the ground surface in response to the degradation of materials in the waste. The microbial activity in the landfill breaks down materials to soluble then gaseous end products. Understanding the processes and the factors that affect them is important for managing the rate of biodegradation and the time to stabilisation.

This paper will report on data collected from five landfill sites in Hampshire, South of England, discussing behavior of "known" influencing factors in stabilisation process. Factors discussed in this paper are divided in two main categories a) site specific data including, geological, hydrogeological, topographical, waste type, clay liner type and engineering specifications and b) monitoring data including, COD, BOD, CH_4, Cl^-, NH_4, Rain Fall levels and etc.

Comparative study carried out between these landfill site aims to identify distinctive behaviors and similarities in the stabilisation process.

INTRODUCTION

The introduction of the 1995 Environment Act regulated that all landfilled waste should be enclosed in fully contained cells. During the years since the introduction of this act speculation has arisen as to the sustainability of fully contained cells, as if left un-assisted they could prove to show stabilisation times up to thousands of years. Over 3500 million tonnes of waste are disposed of each year on earth (D.O.E.) 2000, and the majority of this is simply left to degrade.

It is well documented that as waste ages changes occur in its composition affecting factors such as pH, ammoniacal nitrogen, chloride and biochemical oxygen demand (BOD). These changes occur as the waste breaks down, and the moisture content and temperature of the waste alters. Factors such as waste type, cell construction, waste density and climate will also affect the rate of change within the waste. Many investigations have been carried out studying the rate of stabilisation of waste with regards to key factors such as moisture content, pH and density, and considering the possibility of identifying the age or level of stabilisation of the cell from the condition of its constituents. Waste Management Paper 26 (WMP 26) clearly investigated the time scale for microbial decomposition, and produced the well-published graphs showing the degradation of landfill constituents. This paper details an investigation conducted to further investigations previously carried out, by considering some of the key factors detailed in waste management paper 26, and investigating them in local landfill sites.

Geoenvironmental impact management, Thomas Telford, London, 2001.

Some of the key factors considered within this investigation are:

pH. Which can rise from 6.2 when the waste is fresh, to approximately 7.2 during degradation. The pH level affects the levels of many other constituents such as chloride. pH must be considered when investigating seepage into local water courses as the solubility of heavy metals and many organic pollutants is higher at a pH <4 (Palmissano & Barlaz) 1996.
COD. Chemical oxygen demand measures the remaining oxidisable material within the waste, this therefore decreases as the site ages. The high levels of COD in young landfills results in the presence of fatty acids (D.O.E.) 1992.
TOC. Total organic carbon is an arbitrary measure of the oxygen status within a landfill, and decreases with age (D.O.E.) 1992.
Ammoniacal Nitrogen. The ammoniacal nitrogen level within a landfill, and therefore the ammonia content is important as an ammonia concentration in excess of 2500mg/l will hinder the rate of biological treatment (D.O.E.) 1992. Seepage of ammonia at any concentration above 0.5mg/l is of concern as levels above this can be toxic towards aquatic life.
Chloride. The chloride level affects the processes that occur in a landfill. Aerobic processes can tolerate chloride up to 2000mg/l whereas anaerobic processes appear more sensitive and can only tolerate concentrations up to 1000mg/l. the chloride level will rise as the landfill ages.

This paper reports on data collected from three landfill sites in Hampshire discussing the behavior of these influencing factors. Concentrating on three sites not previously detailed and following up the work previously carried out and documented in the paper titled Ageing processes and stabilisation of landfill sites, by the same authors, published in the proceedings of the Sixteenth international conference on solid waste technology and management, held in Philadelphia, U.S.A. December 2000.

Factors discussed within this paper are divided into two main categories, a) site specific data including geological, hydrogeological, topographical, waste type and liner type, and b) monitoring data including COD, methane, chloride, ammoniacal nitrogen, oxygen and carbon dioxide levels

Site specific data

(1) Bramshill	Bramshill landfill site is located on the east side of Bramshill, approximately 1.5km south waste of the village of Eversley, in Hampshire. The nearest large towns are Reading, which is 12km to the north, and Camberley, which is 13km to the east. The site is located in an area dominated by forestry plantations.
	Until the 1970s the area which is now filled with waste was worked for sand and gravel. When the voids were empty filling commenced. The site was initially based on the dilute and disperse method but in the past five years all cells were fully contained using a 2.5mm HDPE membrane, a total of five containment cells were filled. The site was licensed to receive primarily household waste, and filling on site ceased, and the site closed in 2000.
	The geology underlying the site is bagshot beds, consisting of sandy gravel overlying London clay, overlying chalk. The fissures in the chalk permit groundwater flow in a northerly direction, towards a small river about 2km away.
(2) Efford	Efford landfill site is located 3km south of Lymington in Hampshire, and adjoins the Keyhaven and Pennington marshes, which then border the solent. Landfilling has occurred at Efford since the late eighties and is filling in the

	voids left from mineral workings. Waste was filled in dilute and disperse cells until 1994 when fully contained cells were used, constructed from compacted clay, there are currently 12 fully contained cells constructed and used or waiting to be used at Efford . Efford landfill site is still receiving waste and is expected to remain open until 2006, if it continues to receive waste at a rate of approximately 200000 m³/year. Efford landfill site is licensed to receive household, commercial and industrial wastes. The geology underneath the site shows a deep band of clay overlain with sand and gravel. The Avon water runs along the western perimeter of the site.
(3) Hook Lane	Hook Land landfill site is located 4km waste of Fareham in Hampshire, and is surrounded by farming land situated between the coast, and the expanding urban development in the north. The site was used for agricultural land until 1941, when gravel extractions commenced. The site was initially filled as a dilute and disperse tip in the 1950s and the most recent five cells were all fully contained using a 2.5mm HDPE membrane. The site closed in 1997. The site was licensed to receive household, commercial and industrial wastes. The site lies approximately 1.25km from the coast and groundwater flows in a southerly direction out towards the sea. The site also lies halfway between the rivers Hamble and Meon, with a small tributary of the river Hamble running along the eastern side of the site. The geology of the site shows terrace gravels overlaying formations of the bracklesham beds above layers of clay and sand. The gravels would have a high permeability
(4) Paulsgrove	Paulsgrove landfill site is located 4km north of Portsmouth in Hampshire, Southern England, on reclaimed mud flats adjacent to Portsmouth harbor site of special scientific interest. The site is located close to the urban development of Paulsgrove. Adjacent to the current site is the Port Solent retail, marina and residential development which is situated on completed fill. Filling at the site commenced in the 1970s and was dilute and disperse until 1995, when fully contained cells came into use, constructed from compacted clay, six were constructed. The landfill is still being used, with landrises currently being constructed, based on a method of preferential drainage, and it is expected that the site will have sufficient space to continue receiving waste until 2005 if it continues to receive waste at a rate of 275000m³/year. The site is bordered on two sides by Portsmouth harbor and wetlands, with tidally influenced groundwater running underneath the site. The permeability of the geology underneath the site can be high, as Portsmouth is located in the Hampshire basin that is formed from Chalk overlying silty clay and alluvium. The site is licensed to receive mainly household and commercial waste with a small percentage of industrial wastes; it is not permitted to receive any kind of special wastes.
(5) Somerley	Somerley landfill site is located 3km north west of Ringwood in Hampshire, surrounded by forestry land. The area is undeveloped except for a residential lodge located adjacent to the southeastern corner. The site was a former sand and gravel extraction works, where landfilling commenced in 1965 on a very small scale until 1981 when the works expended and the rate of landfilling increased. The site was initially a dilute and disperse site, with five fully contained cells being used in the past five years. The liners are of composite construction combining HDPE membranes with compacted clay The site closed in 2000.

The site lies on the interfluve between the rivers Avon (1.5km form the site) and Crane (2km from the site). There are no tributaries that enter these rivers from adjacent to the site. The geology beneath the site consists of made fill above bracklesham beds consisting of sands & clays, above bagshot beds consisting of silt and fine sand.

The site is licensed to receive domestic, commercial, non-toxic industrial wastes and incinerator residues, all in both solid and liquid forms.

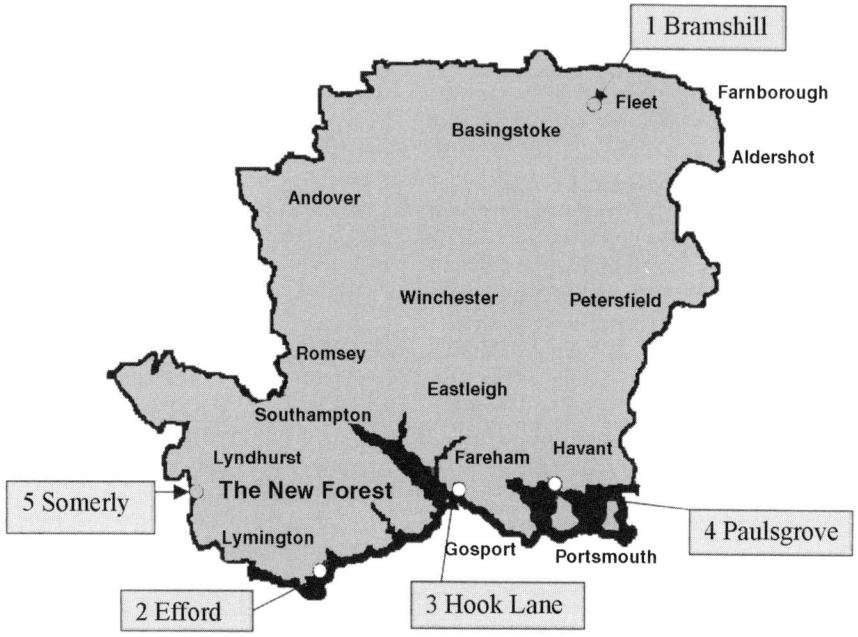

Figure 1 - Hampshire, showing Landfill Sites investigated

RESULTS AND ANALYSIS

Results of two or three boreholes per site from the three sites investigated are presented here, followed by a general conclusion drawn in conjunction with two further sites which have been previously investigated with results published in (Tamblyn, Alani *et al.* 2000)

1) Bramshill: from this site boreholes 9.7 and 9.25 are detailed from an investigation of approximately 20 boreholes.

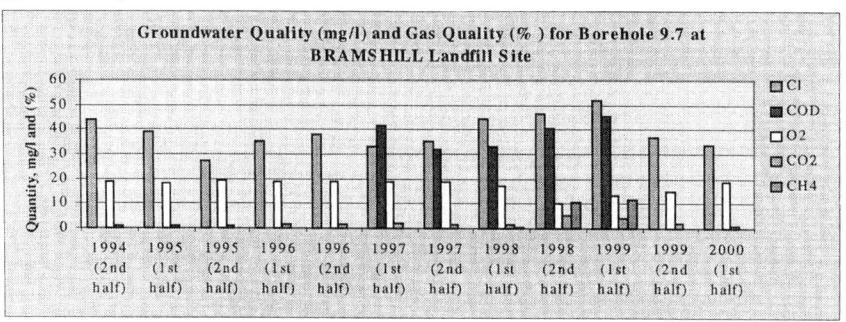

Figure 2 – Bramshill Landfill Site, Borehole 9.7

1.1). Borehole 9.7 is not located adjacent to any areas of fill, and is part of an area under a tree preservation area. It can be observed that in this borehole all levels are low, and not of concern, chloride tends to fluctuate around 40 mg/l, and chemical oxygen demand levels are very low, ammonia recorded in this borehole was below 1 mg/l. The gas exhibited in this borehole is also similar to air. The levels indicated in this borehole seem to indicate that this borehole is not being affected by landfill activities.

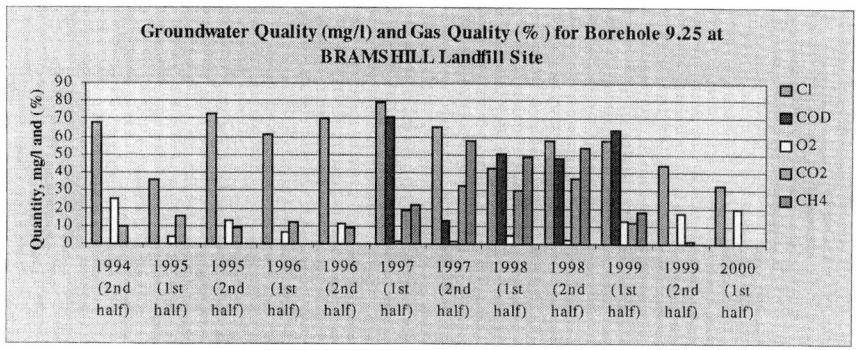

Figure 3 – Bramshill Landfill Site, Borehole 9.25

1.2). Borehole 9.25 is located adjacent to cell 1, which was filled between 1994 and 1996, it is located adjacent to land owned by the forestry commission. The chloride, chemical oxygen demand and ammonia levels in this borehole are also low during the monitoring period investigated and of no concern, but the oxygen levels are slightly lower, and a rise in the methane levels can be observed in 1997, approximately one year after filling commenced in this area. Little similarities were observed between these two boreholes, or any of the other boreholes on site.

2) Hook lane: from this site boreholes 4.24 and 4.26 were chosen for discussion purposes from an investigation of approximately 20 boreholes.

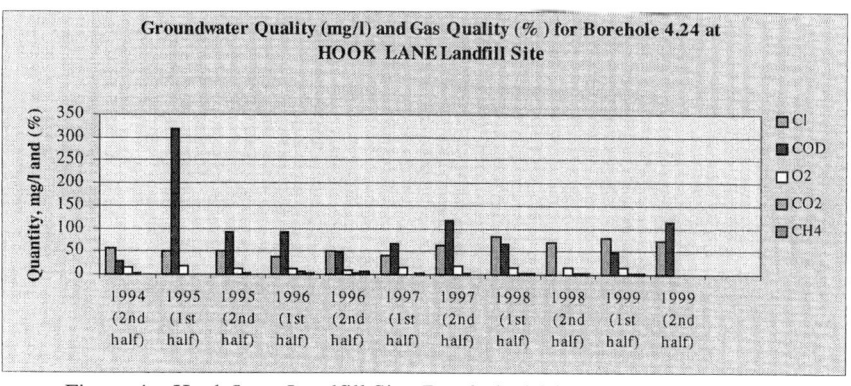

Figure 4 – Hook Lane Landfill Site, Borehole 4.24

2.2). Borehole 4.24 is located in close proximity to cells 4 and 5 which were both filled between 1996 and 1997, and is adjacent to farming land. The chloride levels in this borehole are low, but appear to have slightly risen over the past four years, no obvious trend can be observed for the chemical oxygen demand level, and no apparent reason has been determined for the sharp increase observed in early 1995. The gases are showing no signs for concern. The trends observed in this graphs appear to correlate to the graphs for the stabilisation of landfill organic and inorganic compounds detailed by the department of the Environment in 1994, possibly indicating that the fill is in stage 1 or 2. This has not been observed for any of the Bramshill graphs.

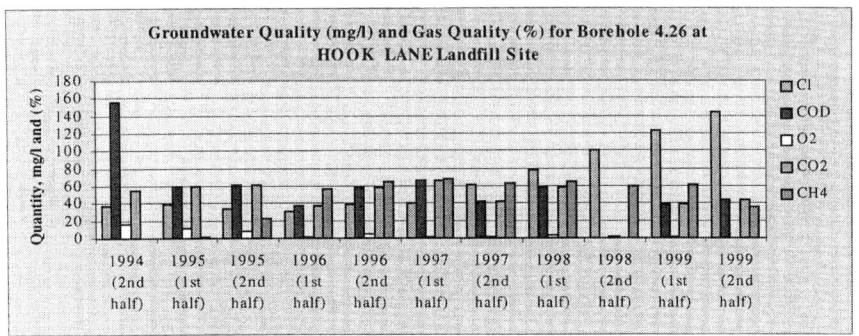

Figure 5 – Hook Lane Landfill Site, Borehole 4.26

2.2). Borehole 4.26 is located adjacent to cells 1 and 4, which were filled between 1994 and 1995, it is also in close proximity to previous dilute and disperse fillings adjacent to cell 1. This graph shows that the chloride in this borehole has been steadily rising and had reached values of around 140 mg/l at the end of 1999. Chemical oxygen demand values appear to be steadily decreasing, reaching a value of approximately 40 mg/l. Ammonia values were also negligible in this borehole. Oxygen levels are low, with carbon dioxide and methane levels being increased.

3) Somerley: from this site boreholes 6.11, 6.16 and 6.20 are displayed below, from an investigation totaling approximately 18 boreholes.

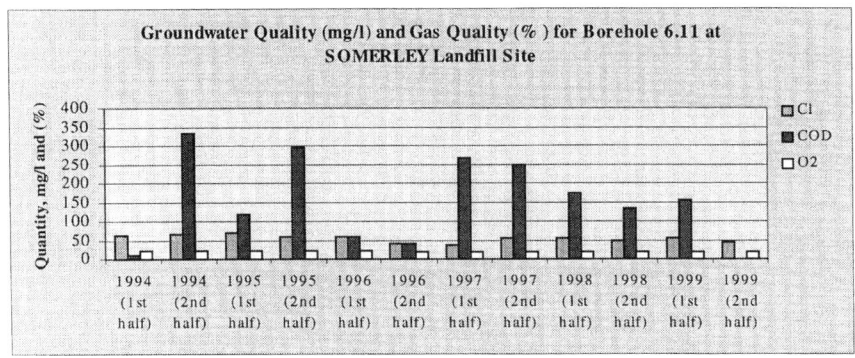

Figure 6 – Somerley Landfill Site, Borehole 6.11

3.1). Borehole 6.11 is located adjacent to previous dilute and disperse fillings and for this investigation chloride, chemical oxygen demand and oxygen are the only values detailed, as all other values were negligible. It can be observed that although the chemical oxygen demand levels have been quite high, they are now steadily decreasing.

3.2). Borehole 6.16 is located adjacent to an area filled between 1997 and 2000, located on land next to a new landfill. this borehole once again shows high unstable levels of chemical oxygen demand, with relatively low levels of chloride and oxygen.

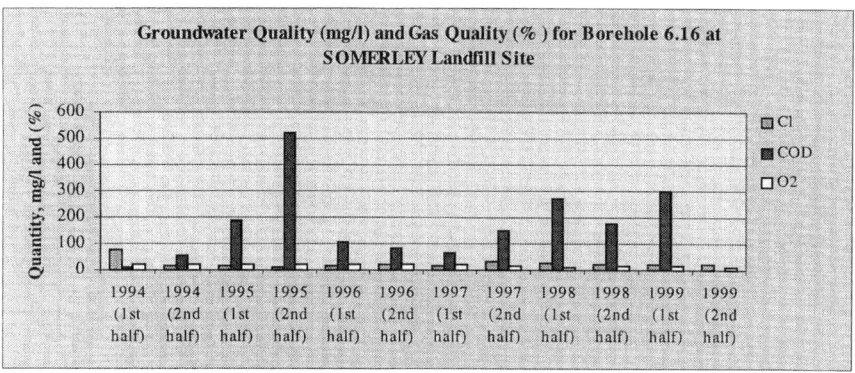

Figure 7 – Somerley Landfill Site, Borehole 6.16

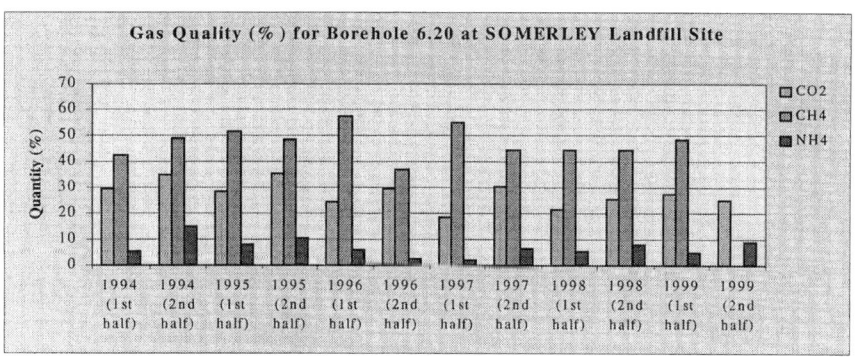

Figure 8 – Somerley Landfill Site, Borehole 6.20

3.3). Borehole 6.20 is located close to previous dilute and disperse fillings, away from the recent enclosed waste. This borehole exhibits high levels of methane, with fluctuating levels of carbon dioxide.

SUMMARY AND CONCLUSIONS
From the investigations carried out here and in previous work, a number of conclusions can be drawn;

- Although limited results are presented within this paper, the conclusions drawn are based on the entire amount of borehole information available from the five sites in Hampshire.
- On some occasions boreholes do exhibit the trends defined for landfill stabilisation by the department of the Environment in 1994, but in these cases the boreholes monitored may be close to dilute and disperse fillings.

- A large amount of scrutiny is needed when investigating stabilisation rates in and around landfill sites, and factors such as waste type, site geology and engineering information must be considered.
- Perimeter boreholes alone do not give a good indication of level of stabilisation, and leachate wells must also be investigated.
- In a parametric study of all boreholes in all five sites, no extraordinary excess of influencing factors ("acceptable" level) was observed, and although some similarities have been observed it was found rather difficult to draw a definitive conclusion with respect to the sites investigated.

REFERENCES
1. Department of the Environment, 2000, waste strategy 2000 for England and Wales. [online] Available from http://www.detr.gov.uk/environment/waste/strategy/cm4693/03.htm

2. Department of the Environment, 1986, fifth edition 1992. Waste Management paper no, 26 – Landfilling Waste. HMSO.

3. Department of the Environment, 1993, second edition 1994. Waste Management paper no 26a – Landfill completion. HMSO

4. Palmisano, A. C. & Barlaz, M. A., 1996. Introduction to solid waste decomposition – Microbiology of solid waste. Lewis.

5. Tamblyn, H., Alani, A.M., Sangha, C.M. & Walden, P. 2000. Ageing processes stabilisation of landfill sites. Proceedings of the sixteenth international conference on solid waste technology and management.

ACKNOWLEDGEMENTS
The authors wish to acknowledge the time and assistance given by Hampshire Waste Services Limited, in providing the raw data used, and financially supporting the project.

The initial behaviour of domestic refuse in a large-scale test

K.S.WATTS, P.TEDD, Building Research Establishment Ltd., UK.
R. LEWICKI, Shanks Waste Solutions Ltd., UK.

INTRODUCTION

The nature of refuse landfill, particularly its very poor engineering properties and the major health and safety issues associated with gas production, has so far largely prevented the reuse of recent refuse landfill sites for purposes other than agriculture or occasionally amenity use. Rates of settlement and waste stabilisation are extremely important for restoration schemes, modelling existing and future landfill engineering, aftercare management and monitoring, reuse for other purposes (sports fields, amenity, biodiversity schemes). As waste stabilisation processes may extend up to many decades, it is vital to develop improved methods of dealing with waste.

Inhibited gas production from saturated waste or excessively dry waste will delay the stabilisation of a landfill and may severely limit its potential for reuse. Re-circulating leachate is a means of ensuring the waste mass remains in a partially saturated state, promoting an increased and uniform rate of biodegradation and gas production over the full depth of the waste mass, as well as resulting in higher effective stresses within the waste. These factors play an important role in determining the rate and magnitude of volume reduction and hence waste settlement. It is essential that reliable data on fill behaviour is gathered in order that biodegradation/settlement models can be developed to aid the effective management of landfills and facilitate their early and safe reuse for future building or amenity purposes.

This paper describes a large scale test which is currently being carried out to identify the effect and benefits of leachate re-circulation and which forms part of a major project to study the long-term behaviour of refuse landfill. The recirculation test is being carried out by BRE in a sealed test pit containing some 96 cubic metres of domestic refuse. Surface and sub-surface settlement, *in situ* stresses, temperature, leachate levels and leachate chemistry are being monitored. Gas production is a critical element of the degradation process, and forms an important part of the experiment. A system has been installed to monitor and control gas extraction. The test represents a unique and detailed study of the behaviour of a representative volume of refuse which is considerably larger than is possible in large-scale laboratory apparatus but with all the advantages of complete control of the test environment. The test began in November 2000. This paper presents the early findings and the proposed future work.

REVIEW OF RECIRCULATION STUDIES

Leachate re-circulation is not a new concept. More than 25 years ago, studies began to show that leachate, if collected and re-circulated through a landfill, could enhance and accelerate the anaerobic decomposition of refuse leading to earlier stabilisation of the landfill and improving its ultimate usefulness. It has been recognised for a number of years that conventional landfilling of waste at its as-received moisture content would not prevent, but only postpone, the eventual pollution of groundwater (Lee and Jones-Lee, 1994). Because of the relatively slow rate of biodegradation this approach also implies a potential liability for some decades. However, by regarding a landfill as a treatment facility rather than simply a means of storage, it is possible to control the waste decomposition, encourage rapid waste stabilisation and minimise the long term risk to human health and the environment with consequent lower future liabilities. The advantages of leachate re-circulation include the distribution of nutrients and enzymes, pH buffering, the dilution of inhibiting compounds, redistribution of methanogens, liquid storage and evaporation. Re-circulation may reduce the stabilisation period from a few decades to a few years. However, barriers to the adoption of the concept include poorly defined design criteria and regulator reluctance, as well as a number of physical impediments.

Since the mid 1970s, laboratory scale studies have been carried out in a number of countries to investigate the effects of leachate re-circulation on leachate quality, waste stabilisation, waste settlement, gas production, the attenuation of heavy metals and other factors. The studies all showed that leachate re-circulation had the potential to accelerate the waste degradation process. Pilot-scale field studies have been carried out in several countries with broadly similar results and show that the process could be effective at full scale in the field. The principal objective of the closely controlled large-scale test described in this paper is to quantify the effect of leachate re-circulation on the geotechnical and biodegradation mechanisms causing settlement.

LARGE SCALE REFUSE TEST
A large test pit which was built specifically to study the behaviour of fill materials at a scale representative of field conditions has been filled with domestic refuse. The pit is 6m long by 4m wide and 4m deep and is constructed of 0.4m thick reinforced concrete. The pit is watertight and has a sump in one corner from which water can been pumped to control the level of leachate within the refuse. The pit has previously been used been used to study the mechanisms of collapse compression of inert fill such as colliery spoil (Skinner et al 1999). The pit has been lined with a 2mm thick HDPE geomembrane to prevent gas escape and leachate contamination of the concrete.

Figure 1 shows a schematic section of the pit filled with refuse, capped and sealed. It also shows the location of the instrumentation installed within the refuse, the gas extraction system and leachate re-circulation system. A 70mm thick granular drainage layer was placed at the bottom of the pit to allow percolating leachate to flow to the sump. A geogrid was placed on the drainage layer to separate it from the refuse. The well pipe extends below the sample into the pit sump to allow the leachate to be pumped to the re-circulation pipe network at the top of the sample.

Placement and encapsulation of the refuse
The refuse comprised typical domestic refuse supplied from North London. It contained plastic, paper, putrescible household waste, glass, etc. The refuse was spread on a concrete hard standing and crushed with several passes of the tracked excavator before being placed in nominally 0.5m thick layers in the pit. Each layer was compacted into place using the downward force of the excavator bucket. Waste was placed by hand around the instruments, and gas and water wells. A total depth of 3.6m of waste was placed in the pit. After placing the 0.8m thick surcharge layer of clay, equivalent to a loading of 19 kPa, the depth of the refuse decreased to 3.1m. The bulk density of the refuse was 0.48 Mg/m^3 before the surcharge was placed and 0.56 Mg/m^3 immediately after placing it.

The as delivered moisture content of the refuse was estimated to be 25% by weight. An additional 9.6 m^3 of water was added to increase the water content to an estimated total absorptive capacity based on work by Ryan (1988). The additional water also modelled the precipitation and infiltration over a period typical of uncapped exposure at an active landfill.

A geogrid separator and 100mm deep gravel layer was placed on top of the refuse. Perforated pipework for the leachate recirculation system was placed within this gravel to obtain an even distribution over the total surface area. A geotextile was placed between the gravel and the clay surcharge above. An HDPE geomembrane cap was then placed over the clay and welded to the HDPE pit lining to produce a gas tight seal. Specially designed gaiters were used to bring instrument connections, and the leachate and gas wells through the HDPE liner.

Instrumentation
Settlement
A magnet settlement gauge was installed through the full depth of the refuse with magnets at approximately 1m intervals, as shown in Fig. 1. This allows settlement, and hence the compression of the refuse, to be measured at different levels. At each magnet location a thermocouple was installed to measure temperature. Using a system of rods through the clay surcharge and HDPE membrane, the settlement of the top of the refuse can be measured at eight locations using precise levelling.

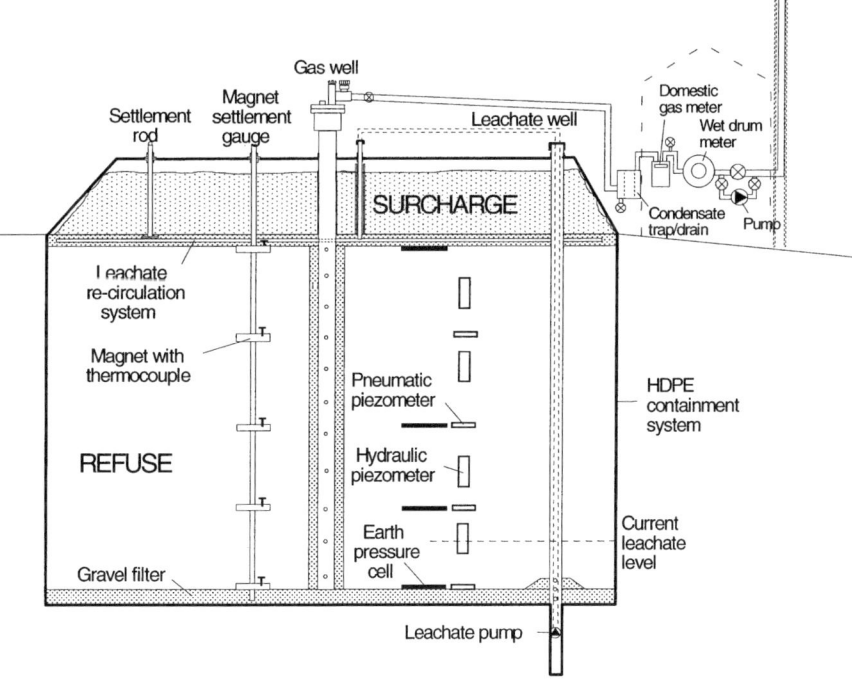

Fig. 1. The BRE test pit, showing monitoring facilities

Total and pore pressure measurements, permeability
Fluid filled total earth pressure cells were installed to measure total pressure at the top and bottom of the refuse and at two positions within the refuse. Pneumatic piezometers were installed at similar locations to determine pore pressures. From the readings taken on the total earth pressure cells and the piezometers as the experiment progresses, variations in effective stress with depth and time can be calculated. Hydraulic piezometers in gravel cells were installed within the refuse so that gas and fluid permeability measurements can be made. The gravel cells which remain above the leachate level have been used for discrete gas sampling.

Gas monitoring
A central gas collection well comprising a 130mm ID rigid plastic pipe with regular spaced 12mm diameter holes and encased in a coarse gravel annulus was built into the sample. A standard vertical *Accu-Flo* wellhead was placed on the gas well, allowing any pressure developed to be measured, the gas to be sampled and the constituents to be identified and quantified using a portable gas analyser, the *Gem 500*. Gas samples can also be taken at this location for external analysis. The gas extraction system from the top of the well head consists of a reduced diameter pipe leading to a condensate trap and gas flow meters. A pump has been installed to allow forced extraction of the gas which is vented to a stack.

Leachate monitoring and re-circulation
Leachate samples can be taken from the monitoring well. A basal drainage system has been provided to allow recirculation of leachate via a vertical well and horizontal pipe network laid within gravel below the clay surcharge.

INITIAL OBSERVATIONS
Settlement and pressures
Figure 2 shows the initial settlement including the effects of the surcharge. Initially, the refuse was settling at a rate equivalent to 0.6% compression per day. The surcharge loading of 19 kPa produced a settlement of 372mm or 12.9% vertical strain over the first 24 hour period giving a constrained modulus of the refuse of approximately 137 kPa. This is significantly less than field measurements which range from 0.7 to 6 MPa (Watts and Charles, 1999). A further 88mm settlement took place during the next nine days after which the rate of settlement decreased significantly. During the period since surcharging the rate of settlement reduced by an order of magnitude from 3mm per day to 0.3mm per day.

The total vertical earth pressure measured at the base of the sample prior to placing the surcharge was consistent with an estimated bulk density of 0.48 Mg/m^3. The measured increase in earth pressures due to placing the surcharge was consistent with the surcharge pressure at the top of the sample and within the sample but at the base of the sample it was approximately 20% lower than calculated. The leachate level was 0.5m above the base of the sample.

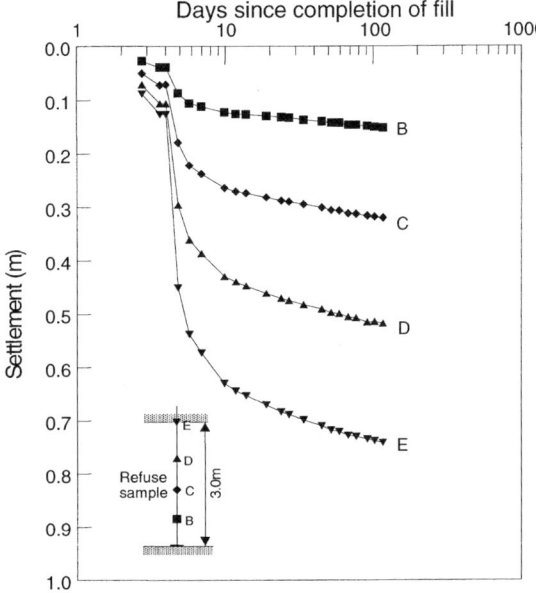

Fig. 2. Settlement measured with depth within the refuse against log time

Gas

Since sealing the pit, the gas has not been vented except when samples have been taken for analysis. No pressure has built up at the well head during the 90 days of the test, but small positive pressures have been measured within the fill at approximately 0.4 m below the top of the fill. Analysis of the main constituents of the gas indicated that within 21 days from placing the sample, the oxygen had been largely consumed and that CO_2 concentrations had reached between 75 and 80% with approximately 0.3% CH_4. The balance is assumed to be largely nitrogen. Hydrogen concentrations are less than 0.2%. The maximum temperature within the fill was measured at approximately 21 days. Wellhead concentration of CH_4 gradually increased to a maximum of 0.9% at 90 days followed by a subsequent decrease to 0.5%, with CO_2 decreasing to 60% and oxygen levels increasing to 5%. Average readings at points within the sample have remained stable and are shown for comparison in Table 1.

Table 1. Summary of gas analyses

Time after placing sample	Species	Concentration	
		at wellhead	internal (160 days)
10 days	Oxygen	4%	
	Carbon dioxide	65%	
	Methane	0	
	Balance	31%	
90 days	Oxygen	0	0
	Carbon dioxide	81.3%	82.0
	Methane	0.9 %	1.0
	Balance	17.8 %	17.0

Trace gases have been measured on a number of occasions. Volatile organic compounds were measured by pumping gas from the wellhead through a sampler with an absorbing material. A Perkin Elmer tube containing Tenax TA was used. The tubes are analysed by gas chromatography with a flame ionisation detector to measure amounts of VOCs and a mass spectrometer was used to identify the individual VOCs. A wide range of organic compounds were detected including aliphatic hydrocarbons, aromatic hydrocarbons, siloxanes, chlorinated compounds and oxygenated species. The total VOCs in Table 2 are summations of individual VOCs in the range C_{16}–C_{18}.

Table 2. Summary of trace gases

Total VOCs	90.9 mg/m^3
Tetrachloroethylene	17.0 mg/m^3
Toluene	5.3 - 24 mg/m^3
Organic fluorine	10 mg/m^3
Organic chlorine	100 mg/m^3
Organic silicon	1 mg/m^3

Leachate
A brief analysis of the leachate is shown in Table 3 and is compared with mean values of acetogenic leachates reported in Waste Management Paper 26B (DoE, 1995). Most values at the test site are within the range reported in Waste Management Paper 26B. The exception is the volatile fatty acids, which are very low in the samples from the test site.

Table 3 Analysis of leachate

Determinand	BRE Test Pit	From DoE, 1995	
		Mean	Range
pH	6.2	6.7	5.12 – 7.8
COD	32,000	36,817	2640 – 152,000
BOD	3100	18,632	2,000 – 68,000
Ammonia as N	280	922	194 – 3,610
Chloride	2800	1805	659 – 4,670
TOC	960	12,217	1,010 – 29,000
Volatile fatty Acids	150	8197	963 – 22,414
alkalinity	3100	7251	2,720 – 15,870
conductivity (• S/cm)	6200	16,921	5,800 – 52,000
Sulphate	590	674	<5 – 1,560
Iron	140	653	48.3 – 2,300
Sodium	1700	1371	474 – 2,400
Magnesium	180	384	25 – 820
Calcium	3000	2241	270 – 6,240

Notes: Results in mg/l except pH-value and conductivity

Refuse temperature

Measured temperatures within the refuse immediately following placing varied between 10 and 23°C and all showed a rise. The addition of 6.9 cubic metres of tap water at approximately 10°C resulted in substantial reductions in temperature particularly in the upper part of the sample. Following the addition of the water, the refuse temperature rose in the middle of the sample reaching a maximum of 20° C after about 15 days and has since shown a steady fall.

CONCLUSIONS AND FUTURE WORK

The large compression of the sample that has taken place during the first period of the experiment is probably largely due to mechanical effects, including the weight of the surcharge and due to creep. Settlements are nearly linear with log time. The amount of compression due to biological degradation is probably small since there has been no significant generation of gas and hence build up of excess pressure, although all the oxygen was initially used up. The leachate level has remained virtually constant. Temperatures within the refuse rose to a maximum of 20°C and have since started to gradually decrease.

The lack of gas generation and the decreasing temperature indicates that biodegradation appears to be taking place at a very slow rate or has ceased. It has therefore been decided to pump gas from the well and continue monitoring settlement, gas constituents, pressures and temperatures. At a later stage when steady state conditions have been reached, leachate will be re-circulated.

ACKNOWLEDGEMENTS

The work described in this paper is part of a wider project being carried out by BRE under the Landfill Tax Credit Scheme regulated by ENTRUST. E B Nationwide Limited are supporting the project with funding from the waste management company Shanks. The work is partly funded by DETR.

REFERENCES

DoE (1995). Waste Management Paper 26B. Landfill Design, Construction and Operation Practice. HMSO

Knox K (1996). Leachate re-circulation and its role in sustainable development. Proc. Institute of Waste Management, , pp.10-15.

Lee G F and Jones-Lee A (1994). Advantages and limitations of leachate recycle in MSW landfills. World Waste 73 (8):16, 19 August.

Ryan A M (1988). Water control of landfill sites. Proc of British Hydrological Society Symposium on Hydrology and Landfill Sites, Glasgow Paper E2.

Skinner H D, Charles J A and Watts K S (1999). Ground deformations and stress redistribution due to reduction volume of soil at depth. Geotechnique, vol49, no 1, Feb, pp 111-126.

Watts K S and Charles J A (1999) Settlement characteristics of landfill wastes. Proc. Instn Civ Engineers Geotech 137, pp225-233.

Watts K S and Charles J A (1990) Settlement of recently placed domestic refuse landfill. Proc. Instn Civ Engineers , Part 1 Dec., pp971-993.

Case Histories

Heat and ammonia associated with jet-grouting in marine clay

A. BRACEGIRDLE, Geotechnical Consulting Group, UK, and
S. A. JEFFERIS, University of Surrey and Geotechnical Consulting Group, UK.

Jet-grouting and other ground treatment can occasionally give rise to unexpected consequences. This paper describes a case study in which an annulus of jet-grout was formed in advance of tunnelling works in soft marine clay. When the tunnels were excavated inside the annulus, some six to twelve months after grouting, it was found that the ground temperature was 15°C higher than the ambient temperature, and ammonia gas pervaded the tunnel. Compressed air working was mandatory and working conditions within the tunnel became very unpleasant. The mechanisms of heat generation and dissipation and the generation of ammonia are discussed.

JET-GROUTING AND TUNNEL CONSTRUCTION

In 1984 and 1985, tunnels forming part of the Singapore Mass Rapid Transit were constructed through Marine Clay, a member of the Kallang Formation. The clay is very soft, contains relatively minor quantities of sulphate and chloride ions, but has a significant, albeit variable, organic content. Because of concern over the possible settlement and damage to buildings along the route of the tunnel, compressed air working was mandatory; the use of versatile earth pressure balance tunnelling machines capable of dealing with the variety of conditions along the tunnel route had not become widespread at the time of construction of the tunnels. The provision of a jet-grout annulus formed a last-minute addition to the contract, with the intention of providing additional security and the opportunity to reduce compressed air pressures.

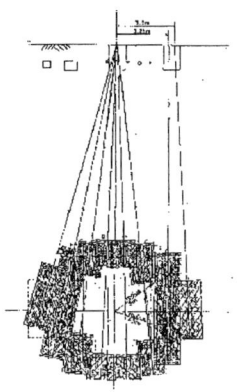

The process of jet-grouting involves the use of high-pressure jets from a specially adapted rotating drill stem, to form a cavity in which soil and cement grout are mixed. A number of different systems are available, all of which form a column of grout as the drill stem is withdrawn. Jet-grouting was pioneered in Japan in the mid-1960s, the early development of which is described by Miki (1985).

The annulus, shown in Fig. 1, was on average 2.0m thick, and was formed over a 500m length of the route of the twin running tunnels. The grout injected was

Figure 1: Jet-grout annulus

typically a 1:1 water/cement mixture with a lignosulphonate plasticiser.   The tunnels, constructed using an excavator mounted in a conventional shield, passed through the annuli between 6 and 12 months after jet-grouting. Soil temperatures of up to 45°C were recorded, which is 15°C in excess of the ambient soil temperature. In addition, ammonia gas at concentrations of up to 36ppm were recorded in the tunnel. Heavy demands were placed on the supply of compressed air to meet cooling and ventilation requirements.

HEAT GENERATION AND DISSIPATION

The quantity of heat released by the hydration of ordinary Portland cement is generally in the range 0.335 to 0.5 MJ/kg. The temperature rise of a grout mixture, assuming no dissipation of heat, Vo, may be calculated from the specific heat of the grout mixture. As may be seen from Fig. 2, a potential temperature rise of about 80°C can be expected for a 1:1 w/c ratio mix, as used on the project. Under ambient conditions, therefore, the temperature of a well-insulated grout mixture could potentially rise above 100°C as the cement hydrates. In practice, the temperature reached is influenced by the proportion of soil mixed in with the grout, and dissipation of heat into the soil. Fig. 3 shows the variation in temperature rise, Vo,

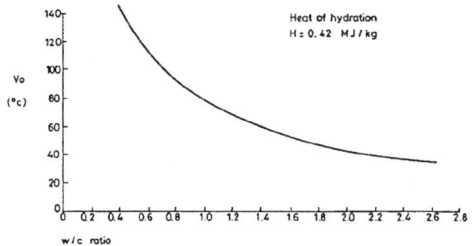

with soil content of the jet-grout mixture in situ.   The relationships shown in Figs 2 & 3 are based on the following values of specific heat:

cement:	1.26 kJ/kg °K
water:	4.18 kJ/kg °K
Marine Clay:	2.13 kJ/kg °K

Figure 2: Temperature (Vo) versus water/ cement ratio

Four analyses of the grout content of the grout/ soil mixture were made using samples taken from the tunnel face at different locations. The analyses, which were based on the calcium content of an acid extract and on the acid insoluble residue of treated soil, gave grout contents of between 39% and 55% by volume of the gout/ soil mixture. The average, 45%, corresponds to a grout/soil ratio of 0.8:1, implying a potential temperature rise, Vo, of between 35°C and 50°C.

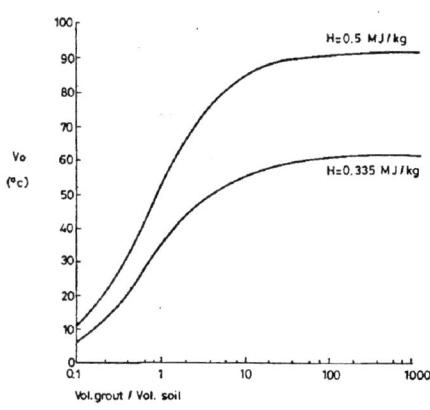

Figure 3: Temperature,Vo, versus proportion of grout

The rate of heat dissipation in the ground is a function of its thermal conductivity, K and diffusivity. In saturated soils, the K decreases with increasing moisture content, while the reverse is true in unsaturated soils. The mineral composition of soil is also important; the thermal conductivities of some minerals, K_g, commonly found in soils are given in Table 1.

Table 1: Thermal conductivity of minerals, W/m°K

mica:	0.67	silica:	1.13	quartz:	10.2
smectite:	0.8	calcite:	3.4		
illite:	1.1	quartz-feldspar:	4.18	water (K_w):	0.6

Johansen (1975) suggests that K_g for soils containing a mixture of minerals can be estimated from:

$$K_g = K^q_{quartz} \cdot K^{(1-q)}_{other}$$

where q is the quartz content of the solid soil grains. The following equation, adapted from Bloomer (1981) can then be used to estimate the thermal conductivity of a saturated soil.

$$K = \left[K_w^{\left(G - \frac{\gamma}{\gamma_w}\right)} \cdot K_g^{\left(\frac{\gamma}{\gamma_w} - 1\right)} \right]^{1/(G-1)}$$

The diffusivity is a function of thermal conductivity, specific heat and soil density. Heat flow is determined using the equation of conduction of heat (for example, see Carslaw and Jaeger (1986)), which can be solved numerically using a finite difference or finite element techniques. In this case, the problem can be simplified to one of radial flow only, and can be solved by hand calculation. The results of the simple analyses carried out are summarised in Figs. 4 and 5 in terms of non-dimensional temperature rise, $V_{t,r}/V_o$, and non-dimensional radius, $\bar{r}$.

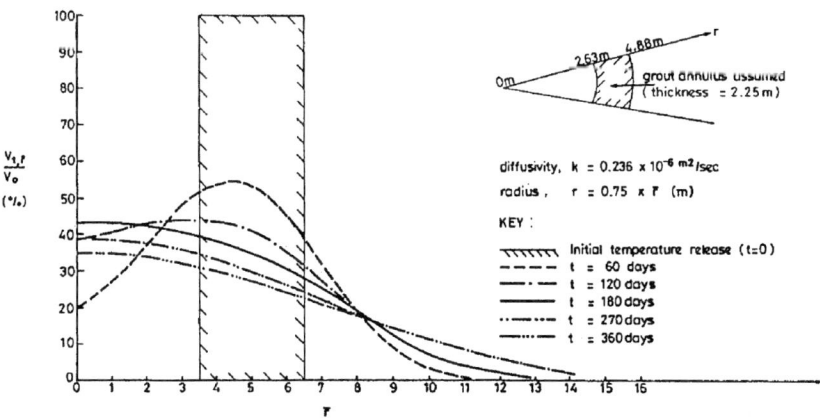

Figure 4: The variation of heat distribution with time and non-dimensional radius, $\bar{r}$

As may be seen from Fig. 4, redistribution of the initial temperatures occurs fairly rapidly, within the first 180 days, following which the material surrounded by the annulus cools relatively slowly. The sensitivity of the analysis to the thickness of the annulus is shown on Fig.5.

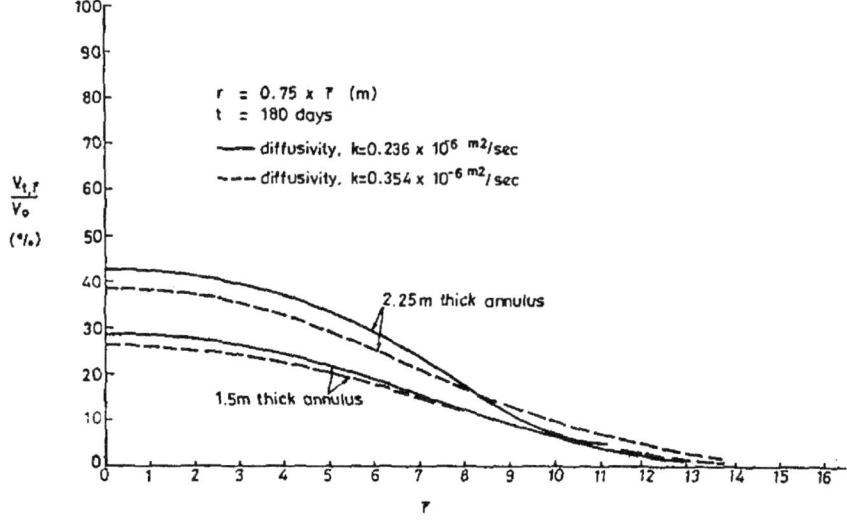

Figure 5: **Temperature variation with thickness of annulus, with non-dimensional radius,** $\bar{r}$

It follows from Fig. 2, that a potential for heating by about 40°C exists for an average grout/soil mix in the Marine Clay; Figs. 4 and 5 show that excess heat, amounting to about 40% of the initial peak jet-grout mix temperature, will be present some 6 to 12 months after the jet-grouting work. The analyses are therefore consistent with the soil temperatures observed during the tunnelling works, which were up to 15°C above ambient.

AMMONIA GENERATION

The tunnel records suggested that the ammonia levels in the tunnel, which reached a maximum of 36ppm by volume of free air, were associated with both the jet-grout and the Marine Clay. Estimates of the volume of ammonia lost into the tunnel atmosphere vary between about 1 and 7 kg/day/tunnel-heading, according to the rate of compressed air consumption and the air pressure operating in the tunnel. The quantity of soil excavated was of the order of 85m³ per typical working day, during which time the tunnel would pass through a volume of about 150m³ of treated ground.

Free ammonia comprises one molecule of nitrogen linked to three molecules of hydrogen. The fundamental building block of ammonium compounds, nitrogen, is available in soils in a number of different forms. Under anaerobic conditions, nitrogen is most likely to be present in reduced forms, such as aqueous ammonium ions (NH_4^+), dissolved or free ammonia gas

(NH₃), and organic nitrogen in varying stages of decomposition. Simple tests carried out by mixing small quantities of grout and soil in the laboratory gave no discernable ammonia smell, even when heated to 100°C. More sophisticated tests confirmed that the total nitrogen content of the Marine Clay was very small (0.07 to 0.08 % by weight), and the ammonium ion concentration about 0.01 to 0.02%. The equilibrium between ammonium ions and ammonia gas in solution is as follows:

$$NH_4^+ + OH^- \leftrightharpoons NH_3 + H_2O$$

The relative levels of ammonium ions and ammonia are controlled by pH and temperature. The higher the pH, the higher the concentration of hydroxyl ions (OH⁻), and the more the reaction will be driven to the right. At a pH of 12.5, and temperature of 0°C, over 99% of ammonium ions will convert to ammonia. The pH requirement is reduced at elevated temperatures; at 40°C for example, a pH of only 10.9 is required for the same result. Although the above data apply to simple aqueous solutions, the same general trends are applicable to the more complex chemistries likely to be present in soils.

The pH present in the soil prior to grouting was mildly acidic to mildly alkaline, and thus with minimal potential to release ammonia. The jet grouting procedure produces an intimate mix of soil, water and cement, resulting in an increase in pH to more than 11. This, combined with the elevated temperatures that occur as a result of hydration of the cement will cause any ammonium ions to be converted into ammonia. Under tunnel conditions, all of the available ammonia could be expected to escape to the tunnel atmosphere wherever the soil moisture is exposed; because of the quantities of soil involved and the confined working spaces, ammonia release, which was undetectable by smell in the simple laboratory tests, became highly significant.

Even at the very low concentrations of ammonium ions found in the soil, there could potentially be 0.31 kg of ammonia per cubic metre of soil. Thus, within the jet-grout/ soil mix the ammonia content could be 0.17 kg/m³. As discussed previously, the tunnel passed through approximately 150 m³ of soil/grout mix per day, giving access to a potential of about 25 kg of ammonia per day. The estimated peak rate of release corresponds to about 30% of the available ammonia gas present in the soil/grout mix. Despite the seemingly insignificant nitrogen content of the soil, the problems created were severe. The high pH of ammonia gas in solution in the eyes and nostrils of the tunnel workers placed severe physiological demands on them, especially when coupled with the high air temperatures within the tunnel.

HEAT AND AMMONIA: AN UNUSUAL PROBLEM?
While the mechanisms of heat generation during the hydration of cement and the equation of equilibrium of aqueous ammonia are well known, their combining to pose serious physiological and operational problems during ground treatment and tunnelling operations appears rare.

The problems to which this paper relates occurred over 15 years ago. The level of ammonium ions in the soil was small, and had this been known in advance of the works it is most unlikely that it would have caused concern. Since the conditions that caused the problem were not especially significant, it is not surprising that the condition emerges from time to time. Indeed, Pellegrino and Adams (1996) reported ammonia in a tunnel in similar ground conditions. Ammonia is only one class of chemicals in the ground that could cause severe disruption to construction works, even when present at levels that are generally considered environmentally benign. It is therefore appropriate to identify the principal risk factors for ammonia, which are believed to be as follows:

- Organic or sewage-contaminated soils.
- High pH environment and/or high temperature – cement injection can give both, while some chemical grouts can give high pH.
- Reducing conditions, as are typical of deep soils – shallow soils under oxidising conditions are expected to show a lower risk, as nitrogen will tend to be present as nitrate rather ammoniacal species.

It should be noted that a combination of both high temperature and high pH is not necessary for ammonia release. For example, ammonia release has occurred during slurry tunnelling work in organic soils using a bentonite slurry, which when fresh had a pH of 10.5. The pH of such slurries will normally drop in use, but can be raised by contamination with cement from contact grouting operations.

Finally, it should be noted that other problems associated with the placement of grout or concrete in the ground have been reported. Many practitioners will have noticed irregularities during the monitoring of pore water pressures adjacent to large diameter cast in-situ piles or diaphragm walls. Very high suctions can be imposed on concrete/soil interfaces during hydration, the affect of which is partially counteracted as warming of the surrounding soils occurs. The variation of water pressure with time will depend on the thermal and hydraulic conductivity and diffusivity of the soil. In a recent case involving the casting of a diaphragm wall in the Old Alluvium in Singapore, a suction wave was observed moving away from the wall, closely followed by temperature-induced positive pore pressures before ambient pore water conditions were re-established (Norrish, 2000).

REFERENCES

Bloomer, J.R. (1981) The thermal conductivities of mudrocks in the UK. Q.Jnl. Eng. Geol., London, Vol. 14, pp 357-362.

Carslaw, H. S. and Jaeger, J. C. Conduction of Heat in Solids. 2nd Edn.,Carendon Press.

Johansen, O. (1975) Thermal Conductivities of Soils. PhD Thesis, Trondheim, *(avail. in English: Cold Regions Research and Engineering Laboratory Report 637, US Army Corps of Engineers, Hanover, New Hampshire)*

Miki, G. (1985) Soil improvement by get-grouting. Proc. 3rd Int. Geotechnical Seminar, Soil Improvement Methods, Singapore, pp 45-52

Norrish, A. (2000) Pers. comm., LTA Changi Extension.

Pelligrino, G. and Adams, D. N. (1996) The use of jet-grouting to improve soft clays for open face tunnelling. Proc. Geotech. Aspects of Underground Constr. in Soft Ground, London, publ. Balkema, pp 423-428

Field Investigation into the Biodegradation of TCE and BTEX at a Redeveloped Light Industrial Site

Mark Dyer
School of Engineering, University of Durham, UK

ABSTRACT

The paper is based on a recent programme of groundwater monitoring at an industrial site in west London. Redevelopment of the site in 1997 revealed high levels of soil and groundwater pollution by hydrocarbon fuels, trichloroethylene (TCE) and soluble metal salts (e.g. free cyanide, chromium VI and nickel). The pollution originated from a previous metal plating and galvanising works at the site. As part of the redevelopment works, the owners undertook limited excavation works and groundwater extraction to remove the pollutant. However groundwater sampling has continued to show high levels of pollution. Following discussion with the Environment Agency in late 1998, a ground water monitoring programme was agreed to investigate the potential for co-degradation of the petroleum fuel and TCE. Groundwater samples have been taken from six boreholes (1C to 6C). The location of the monitoring boreholes relates to past pollution spillages and the layout of the new factory building.

INTRODUCTION

A programme of groundwater monitoring was carried out at an industrial trading estate in west London, where concerns exists over the potential gaseous concentrations of vinyl chloride in the vadose zone. Redevelopment of the site in 1997 by the owner occupier revealed high levels of soil and groundwater pollution by hydrocarbon fuels, trichloroethylene (TCE) and soluble metal salts (e.g. free cyanide, chromium VI and nickel). In agreement with the Environment Agency, groundwater quality was monitored at six boreholes (1C to 6C) shown in Figure 1. The first year of groundwater monitoring has just been completed. The location of the monitoring boreholes relates to past pollution spillages and the layout of the new factory building. Borehole 3C is positioned in the vicinity of the past TCE tank. Borehole 5C is located adjacent to a previous petroleum fuel tank. Borehole 1C is located in the vicinity of the original chromium plating works in the 1950's.

The site is adjacent to the River Longford concrete lined channel. In the 1920's the site was pastureland underlain by River Terrace Gravel (3rd Thames Terrace) overlying London Clay. By 1930's the site lay within a flooded gravel workings which extended up to the River Longford. Between 1940-50 the pits were backfilled, which lead to the construction of the Industrial Estate. Data from rotary boreholes show the centre of the site to be underlain by Made Ground to a depth of 6m comprising a lower unit of silty organic alluvium overlain by up to 2.5 metres of sand and gravel with brick rubble from the 1950's. The boundary of the site is underlain by River Terrace Gravel to a depth of 6m. The groundwater table is

Geoenvironmental impact management, Thomas Telford, London, 2001.

approximately 1.4 metres below ground surface with negligible seasonal variation. The site is one of several similar industrial units at the trading estate.

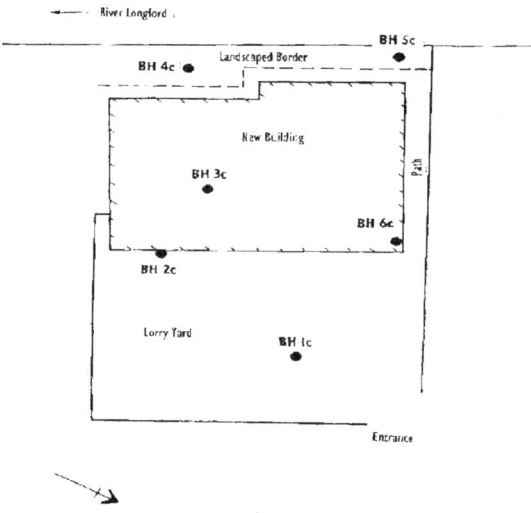

Figure 1 Site layout and location of boreholes

GROUNDWATER SAMPLING AND ANALYSES

The groundwater samples were retrieved using 'Waterra Inertial' pumps permanently installed within each borehole. Prior to sampling, the boreholes were purged of at least 30 litres of groundwater and then allowed to recover to the former standing level. Inflow rates were rapid and the groundwater stood at typically 1.1 to 1.4m below ground levels. Groundwater samples were analysed using a combination of field and laboratory techniques. Field measurements were made using a Burmarc Multiline P4 probe. The probe provided on-site measurements of dissolved oxygen content (expressed as % saturation), redox potential, conductivity and pH.

Laboratory analyses were carried out at Robertson Laboratories, North Wales and Scienco Ltd, Solihull. Laboratory analyses targeted two main groups of pollutants. The first group was organic contaminants comprising petroleum hydrocarbons and chlorinated solvents. The range of determinands was selected to investigate the potential for natural attenuation of chlorinated solvents as electron acceptors by reductive dechlorination (Weidemeier et al 1999). The potential pathway for reductive dehalogenation of perchloroethylene (PCE) and trichloroethylene (TCE) is shown in Figure 2 with the degradative by-products dichloroethylene (cis, trans and 1,1 isomers), vinyl chloride, ethene and ethane. In addition to chlorinated aliphatic compounds, the analyses included different sources of carbon as electron donors (i.e. total organic carbon, total petroleum hydrocarbons and BTEX) plus alternative electron acceptors (i.e. nitrate, sulphate). The second group of pollutants comprised inorganic contaminants, principally heavy metals as follows: cadmium, chromium, copper, nickel, zinc and cyanide (free and total). Although the groundwater analysed targeted these particular groups of pollutants, the groundwater samples were analysed for a much wider range of volatile organic compounds as part of standard suite of analyses carried out by Robertson Laboratories.

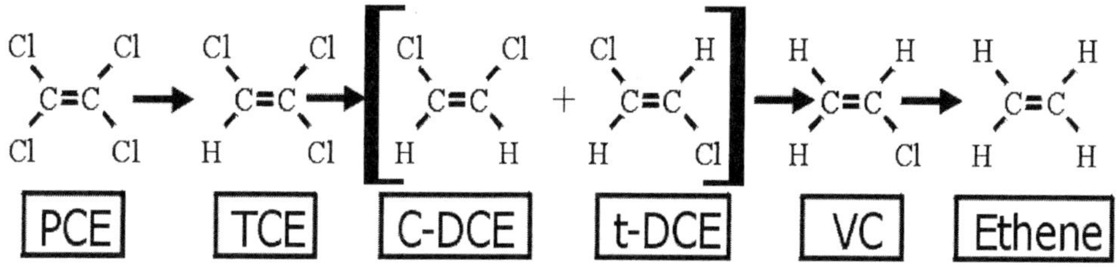

Figure 2 Reductive dehalogenation pathway for PCE and TCE
(Weidemeier et al 1999)

The volatile organics were analysed by Robertson using a Headspace GC-MS and expressed as μg/l. The concentrations for the remaining determinands were expressed as mg/l. A second laboratory Scienco Ltd analysed the groundwater samples for methanol, ethene and ethane, expressed as mg/l.

RESULTS

The chemical results have been collated in Figures 3 and 4. The former shows that dissolved oxygen levels in the majority of boreholes is low with correspondingly high sulphate concentrations and pH neutral conditions, which indicates a highly reducing aqueous environment. In comparison, groundwater analyses for two selected boreholes shown in Figure 4 indicate elevated concentrations of vinyl chloride and dichloroethene followed by trichloroethene. A similar trend was identified for borehole 1C. The aqueous concentrations of chlorinated hydrocarbons detected in the three boreholes (2C, 4C and 6C) were considerably lower apart from a few isolated readings.

Although the highest aqueous concentrations have been measured for vinyl chloride, the relative aqueous concentrations of other chlorinated aliphatics would suggest that the original pollutant was trichloroethylene (TCE) or possibly perchloroethylene (PCE). It would appear that TCE was being reduced to vinyl chloride. This interpretation is supported by the relatively high concentration of the cis-isomer for dichloroethene, which would normally be exceeded by the trans isomer in a purely chemical reaction due to steric effects (Weidemeier et al 1999, Bouwer 1994). It would appear that TCE is acting as an electron acceptor for the metabolism of another carbon substrate. However, different rates of reduction appear to result in an accumulation of vinyl chloride of up to 130mg/l in borehole 5C (i.e. 5% of saturated solubility at 20C). It is worth noting that in terms of molarity, the differences in aqueous concentrations between the lighter VC and cis-DCE widens further, where the molecular weights are 32g and 48g respectively. The accumulation of vinyl chloride is a cause for concern; partly because Henry' Law constant is relatively high (i.e. 3.58E+00 unitless at 20C, or 8.60E-02 atm-cum mol). Consequently, the gaseous concentration for vinyl chloride in the vadose zone could be up to 465 mg/l (or 465μg/m^3). If that had been the case, the gaseous concentration would be well in excess of guideline values published by regulatory authorities in the USA. For example, the inhalation unit risks for VC is given as 84μg/m^3 by the USEPA (OAQPS 1999).

The reduction of TCE by indigenous bacteria (such as sulphate reducing or denitrifying bacteria) is further supported by data on groundwater chemistry. Although not conclusive, the monitoring data generally indicates a highly reduced anaerobic environment with predominantly near neutral pH conditions. The carbon sources supplying electrons for reduction of chlorinated solvents are not clear from the monitoring data. As an initial assessment the past spillage of hydrocarbon fuels in the vicinity of borehole 5C would be a primary candidate as an electron donor. However a comparison of total organic carbon TOC with total petroleum hydrocarbon TPH in Figure 4 shows the organic source is well in excess of the petroleum hydrocarbons and furthermore TPH was only detected in significant quantities in borehole 5C. The reduction of TCE to vinyl chloride and ethene at other boreholes (3C and 1C) would suggest that organic carbon is acting as an electron donor for the reduction of TCE by halorespiration.

CONCLUSIONS
In summary, chemical analyses of groundwater samples from these three boreholes show elevated aqueous concentrations of chloroethenes with a classical reduction pathway for trichloroethylene (TCE) leading to an accumulation of vinyl chloride as shown in Figure 4. In addition, the results indicate anaerobic and pH neutral groundwater with a relatively high sulphate content (300mg/l). Elevated concentrations of total petroleum hydrocarbons (TPH) and single aromatic compounds (BTEX) of up to 5mg/l have been recorded for borehole 5C, whereas all other boreholes had a total organic carbon content of typically 20mg/l.

The results raise several questions about the potential for biodegradation of TCE at similar industrial sites where a dual spillage of TCE and petroleum hydrocarbons exists. Firstly, the reduction pathway results in an accumulation of vinyl chloride in aqueous solution of up to 130 mg/l. The high aqueous concentration for vinyl chloride could give rise to an even greater gaseous concentration within the vadose zone both outside and beneath buildings. Secondly, if the accumulation of vinyl chloride is problematic could the compound be oxidised by injection of an oxidant at low concentrations without harming obligate anaerobes, particularly SRB's. Thirdly, it is unclear which source of carbon is acting as the principal electron donor. Total hydrocarbon petroleum (TPH) is present at a maximum aqueous concentration of 5 mg/l (TPH) and coincides with the greatest reduction of TCE to VC as shown in Figures 3 and 4. However, the total organic content is 20mg/l. Fourthly, if TPH are the principal electron donors should further mixing of groundwater be encouraged to promote the reduction of TCE. Alternatively, the risk posed by accumulation of gaseous concentration of the chlorinated solvent trichloroethylene may be more acceptable.

Although the ground conditions at the site are relatively complex, the type and extent of pollution could be replicated at similar trading estates operating between the 1950's to 1980's when less control was exercised on the storage and handling of chemicals. The pollution presents both the site owner and the Environment Agency with an awkward problem to deal with in a rational and cost effective way.

REFERENCES
Bouwer EJ 1994   Bioremediation of subsurface contaminants in Environmental Microbiology. Wiley Lewis

Suthersan S 1996   Remediation engineering: Design concepts. CRC Lewls.

Weidemeier et al 1999   Natural attenuation of fuels and chlorinated solvents in the subsurface. Wiley

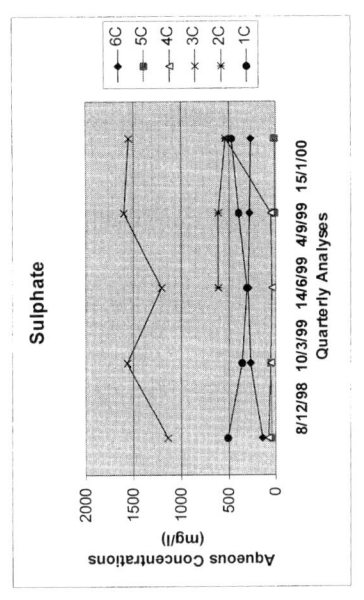

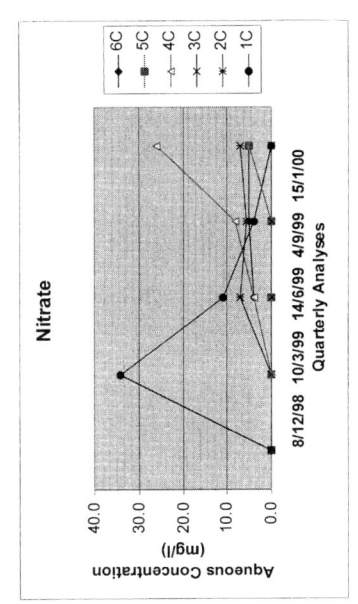

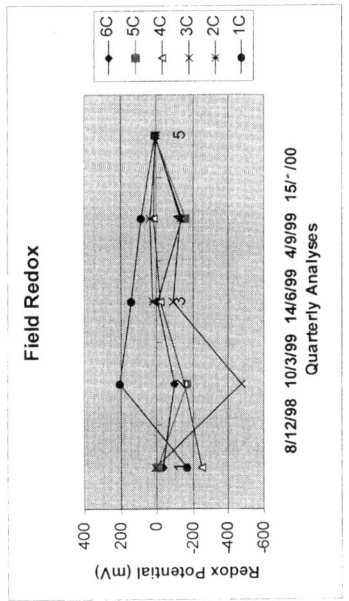

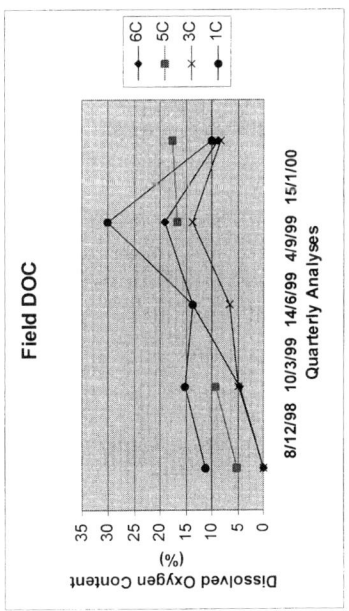

Figure 3 Groundwater chemistry

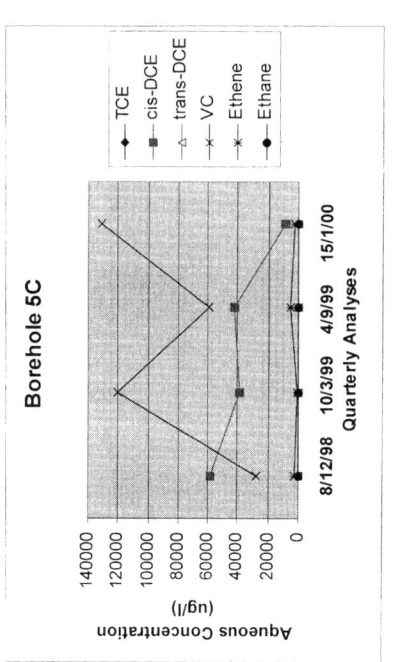

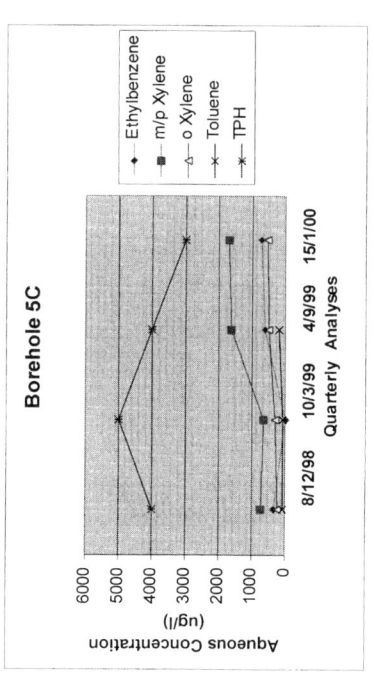

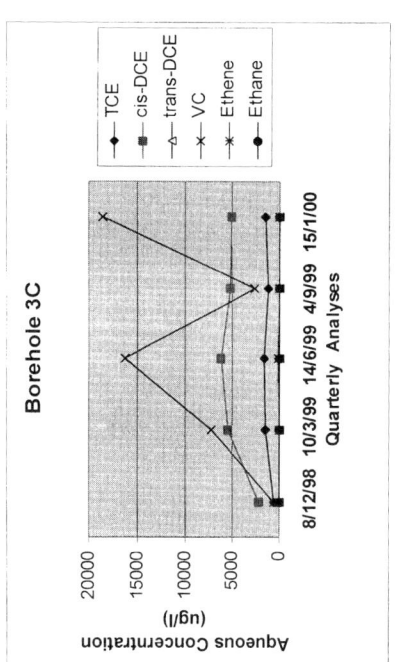

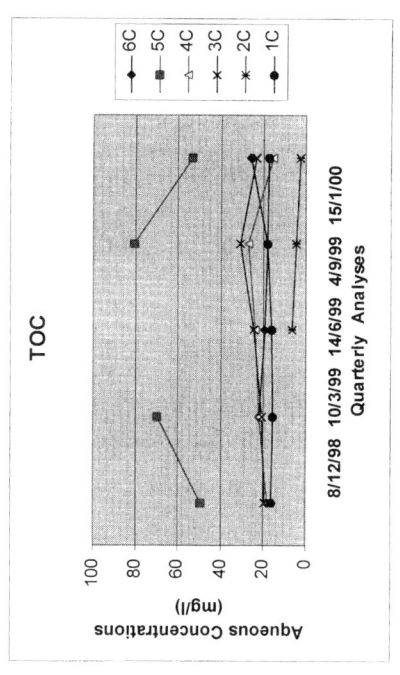

Figure 4 Aqueous concentrations of chlorinated hydrocarbons, BTEX, Total Petroleum Hydrocarbons (TPH) and Total Organic Carbon (TOC)

A Field Trial for In-Situ Bioremediation of 1-2, DCA

Mark Dyer, University of Durham, UK
Erwin van Heiningen and Jan Gerritse, TNO MEP, The Netherlands

ABSTRACT
Historic spillages of chlorinated hydrocarbons at a vinyl chloride plant (Rotterdam Botlek area in The Netherlands) have lead to contamination of the underlying aquifer. The principal contaminant is 1,2-dichloroethane (1,2-DCA). The contamination is temporarily contained by a pump-and-treat system. A field trial was carried out to investigate the feasibility of treating the dissolved phase of 1,2-DCA via reductive dechlorination by injection of an aqueous solution of methanol, ammonium chloride and sodium chloride into the confined aquifer using an array of 8 boreholes. Biodegradation of 1,2-DCA was localised. This was attributed to limited mixing of the carbon substrate within the test zone. In addition, clogging of recharge wells complicated groundwater circulation.

INTRODUCTION
The in-situ bioremediation of organic pollutants is critically dependent on the application of complex microbiological processes in a heterogeneous geological environment. In many cases, the main challenge is the engineering of environmental biotechnology using traditional geotechnical techniques. In the case of chlorinated solvents and other dense non-aqueous phase liquids (DNAPLs), the situation may be further complicated by the deep and intrusive nature of the pollution due to the chemical's high density, low viscosity and low aqueous solubility (Pankow and Cherry 1996). Consequently, the selection and delivery of nutrients and substrates for the bioremediation of chlorinated solvents can pose a major problem.

As part of an ongoing study into the selection and delivery of carbon substrates for the bioremediation of DNAPLs, the paper describes the results from field tests into the bioremediation of 1,2-dichloroethane (1,2-DCA. The site is located at a vinyl chloride plant in the Botlek area of Rotterdam harbour. It is reclaimed land overlying an extensive thickness of Quartenary deposits greater than 35 metres depth. The Quartenary deposits principally comprise silty clay between +1 to −21m NAP (Normaal Amsterdams Peil) with interbedded strata of silty clayey sand. A confined aquifer of sand is located between −21 to −29 metres NAP, which is further underlain by extensive deposits of silty CLAY. Overlying the whole site is a 4-metre thickness of made ground (sand) from earlier land reclamation. Sand columns extending down to the confined sand aquifer from previous ground improvements further complicated the ground conditions. Accidental leaks of chlorinated hydrocarbons have polluted the site. In particular, seepage of 1,2-DCA through the sand columns has polluted the underlying confined sand aquifer between 23 to 29m depth NAP.

GROUNDWATER CHEMISTRY
Data on groundwater chemistry from pumping wells at the site is shown in Table 1. The groundwater samples were taken using a peristaltic pump at the ground surface with a Viton pump tubing and polyethylene tubing extending down to the base of the wells at between 26

Geoenvironmental impact management, Thomas Telford, London, 2001.

to 29m NAP. Samples were collected in 40ml glass vials sealed with Teflon lined butylrubber septa. Samples were stored at 4 deg C until analyses. The data with indicate largely pH neutral conditions with a relatively low redox potential. In addition the results indicate elevated concentrations of chloride and bicarbonate ions. The chloride ions could have originated from saltwater intrusion and reduction of chlorinated solvents.

Item	Borehole/ Units.	GWMS-2	GWMS -4	MF11	MF12	P2507A	P 2507B	P507C
Cl	mg/l	780	1200	2000	950	2000	1600	2100
HCO3	mg/l	1100	1800	1300	1500	1100	1100	1200
SO4	mg/l	<0.2	<0.2	<0.2	7.8	<0.2	<0.2	<0.2
NH4+	mg/l	30	36	32	39	36	34	31
Fe	mg/l	3	3.8	0.93	0.74	5.5	13	11
Eh field	mV			-153	-143	-195	-183	-208
pH field	-			7.0	7.0	6.9	6.9	7.0
EC field	mS/m			680	390	630	570	680
EC lab	mS/m	350	500	700	400	650	580	690

Table 1 Data on groundwater chemistry from pumping wells

FIELD TESTS
An array of 8 boreholes were sunk at the site in 1996 using cable percussion techniques. The layout of the wells is shown in Figure 1. The location is downstream of the vinyl chloride production plant and abstraction well 2507B, which is used as part of a temporary pump-and –treat operation. The wells were arranged in a 12m x 14m plan area with four wells (E, F, G and H) used for water abstraction as shown in figure 2. Two other wells (A and B) were used as recharge wells for injection of water. The two remaining wells (C and D) were used to intercept flow through the bioactive zone and so monitor groundwater quality.

A pilot plant was assembled at the site to abstract groundwater from the wells, mix the water with a carbon substrate and nutrients and store in a dosing tank before injecting the aqueous solution into the confined aquifer via the recharge wells. Groundwater was abstracted at a total rate of 300l/hr from all four wells combined (i.e. 25l/hr from individual well filters). At the same time groundwater abstraction from well 2507B was maintained at 55 cum/day. Field data on pumping rates is shown in Figure 3. Early on problems were encountered with clogging of the recharge wells due to salt precipitation and biofouling. Recharge wells are extremely prone to biofouling where oxygenated groundwater can stimulate aerobes such as *Gallionella* to oxidise soluble iron and produce a soft red-brown gelatinous slime (biomass) even for iron concentrations below mg/l (Preene et al 2001). To overcome the problem, distilled water from the chemical works was used instead of groundwater to prepare an aqueous solution of methanol (4000mg/l) for injection via the recharge wells. The dissolved oxygen content for the distilled water was below 0.1mg/l.. Methanol was selected as the carbon substrate based on laboratory test results. In addition, ammonium chloride (11.5mg/l) was added to the aqueous solution as a nitrogen source, along with sodium chloride to match groundwater density and a lithium tracer (1200mg/l).

RESULTS
In-situ bioremediation was monitored within the test zone by sampling from borehole D from filter screens at three different levels. The results of laboratory analyses from groundwater samples are presented in Figures 4, 5 and 6. The graphs show aqueous concentrations of 1,2-DCA and degradative by-products by reductive dechlorination. The results are noticeable different for the three depths. There appears to be a localised biodegradation of 1,2-DCA within the lowest depth of the confined aquifer (31-32m NAP). At this depth the reduction in aqueous concentration of 1,2-DCA is matched by the production of methane as a degradative by-product. For other sampling depths, in-situ biodegradation was either absent (i.e. 27-28m depth) or unclear (i.e. 23-24m depth). This disparity was attributed to a limited mixing of the carbon substrate within the test zone, which restricted the bioavailability of the electron donor for reductive dechlorination of the dissolved phase of DCA.

GROUNDWATER MODELLING
In support of the field tests, Delft Geotechnics were commissioned to carry out a limited groundwater modelling exercise to investigate the transportation of the methanol solution within the test zone (Taat 2000). The 3-D groundwater model was developed for the site using the two finite difference programmes MODFLOW and MODPATH (McDonald & Harbaugh 1988, Pollock 1989). Longitudinal dispersivity was taken as 0.25m and transverse dispersivity as 0.0125m. The result indicated a very limited mixing of the methanol solution within the confined aquifer at a depth of 27-28m NAP. The result appear to explain the localised biodegradation observed in the field test due to localised distribution of the carbon substrate and hence its limited bioavailability as an electron donor for reductive dechlorination of 1,2-DCA to take place. Both results reinforce the need to deliver the carbon substrate through the aquifer for in-situ bioremediation to take place effectively.

8.0 CONCLUSIONS
The case study was a rare opportunity to investigate insitu bioremediation of a chlorinated hydrocarbon by reductive dehalogenation. The study had mixed success. Initially clogging of the recharged wells hampered groundwater circulation by salt precipitation and biofouling of the filter screens. These difficulties were overcome by using distilled water (with a low oxygen content of <0.1mg/l) from the chemical works for preparation of an aqueous solution of methanol. The resulting injection of methanol into the underlying confined aquifer resulted in a localised biodegradation of 1,2-DCA as detected by the lower filter screen for borehole D 31-32m NAP. At this level, the aqueous concentration of 1,2-DCA declined and was matched by similar increase in concentration of methane, which indicated that the chlorinated hydrocarbon was being reductively dechlorinated. For other sampling depths, in-situ biodegradation was either absent (i.e. 27-28m depth) or unclear (i.e. 23-24m depth). This localised biodegradation was understood to be due to a limited distribution of the carbon substrate within the test zone (12 x 14m area by 10m thickness), that restricted its bioavailability as an electron donor for reductive dechlorination of 1,2-DCA.

REFERENCES
McDonald and Harbaugh 1998. A modular three-dimensional finite difference groundwater flow model, US Geological Survey
Pankow JF and Cherry JA 1996. Dense chlorinated solvents and other DNAPLs in groundwater. Waterloo Press.

Pollock, W 1989. Documentation of computer programs to compute and display pathlines using results from the US Geological Survey modular three-dimensional finite difference groundwater flow model. US Geological Survey

Preene M, Roberts TOL, Powrie W and Dyer MR 2000. Groundwater control: design and practice. CIRIA Publication C515.

Taat J, van Ree D, Vos D, van den broek P, van Bijnen J.F., Pruijn M.F., van Winden S.C., Bosma T.N.P., Gerritse J., van Heiningen W.N.M., Meulenberg R.E. 2000. Final report: development of biotechnological in situ measures. Nobis Report 95-1-06

ACKNOWLEDGEMENTS
The field tests formed part of a project funded by the Dutch NOBIS programme for research into the application of in-situ bioremediation of contaminated land. The project consortium comprised Arcadis, Akzo Nobel, Delft Geotechnics, and TNO Institute of Environmental Biotechnology. Dr Dyer was supported by an Engineering Foresight Award from the Royal Academy of Engineering in the UK.

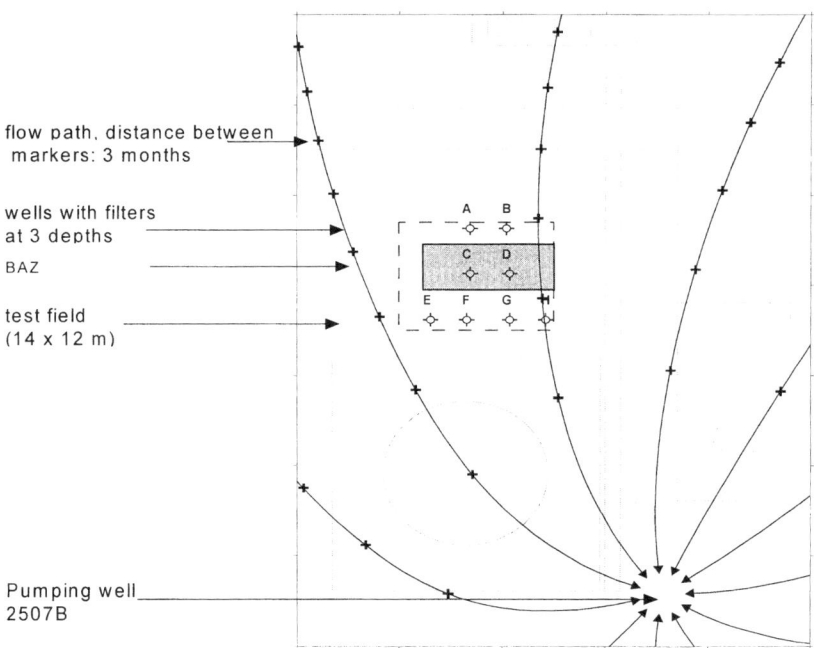

Figure 1 Location of field test

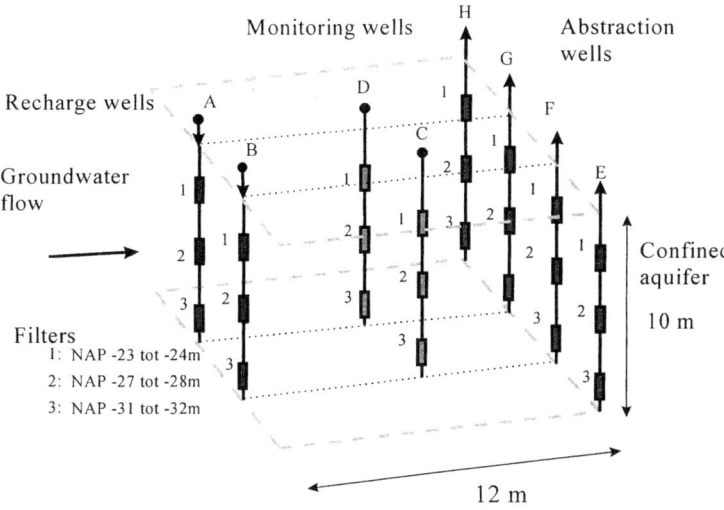

Figure 2 Layout of abstraction, recharge and monitoring wells

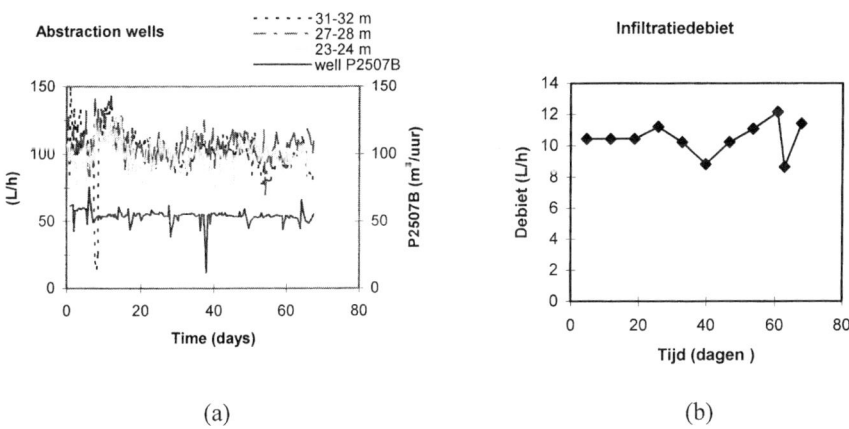

(a)   (b)

Figure 3 Field data on (a) combined abstraction rates for wells E,F,G,H at different filter screen levels and pumping well 2507B and (b) injection rates for individual filters for wells A and B

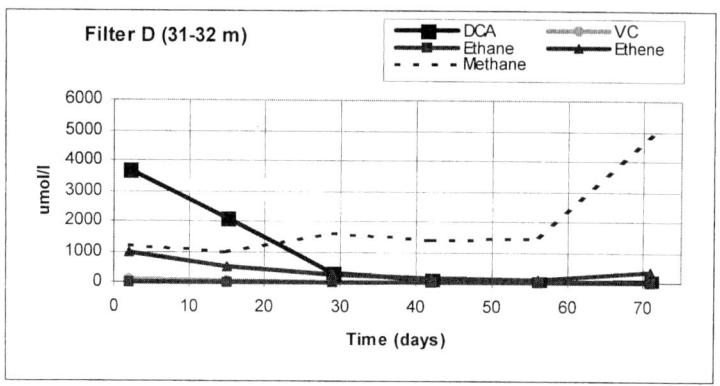

Figure 4 Analyses of groundwater samples for well D filter depth 31-32m NAP

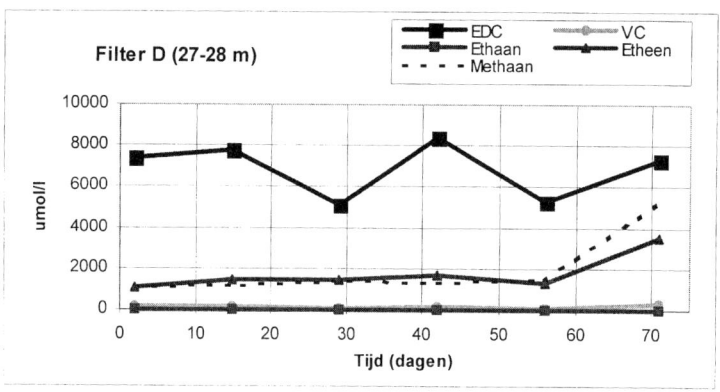

Figure 5 Analyses of groundwater samples for well D filter depth 27-28m NAP

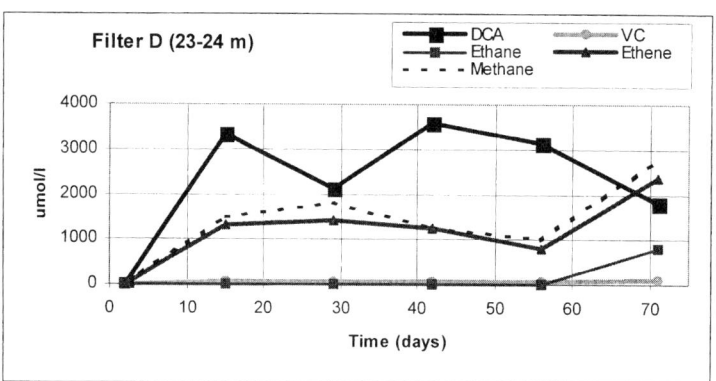

Figure 6 Analyses of groundwater samples for well D filter depth 23-24m NAP

The Echline road construction project: Environmental Assessment

Dr N Ghazireh and Dr H L Robinson[(*)]
Tarmac Ltd, Technical Services, Wolverhampton, UK
(*) Visiting Professor at Liverpool John Moores University

ABSTRACT
The Scottish Executive Development Department, Trunk Roads - Design & Construction Division have funded the construction and testing of a trial section of experimental road pavement at Echline, South Queensferry, Edinburgh. The trial, completed in April 2000, had a clear objective to maximise the use of sustainable, in-situ, recycled and cold lay materials and minimise the overall construction energy requirements.

This paper outlines the energy consumption data for each material installed in the road trial including aggregate and binder processing and final product mixing. A comparison is made between the construction energy requirements for the 'alternative recycled ' pavement and that for a conventional pavement deemed to have similar design life. The environmental benefits of using recycled and industrial by-product materials in road construction are highlighted.

1.   INTRODUCTION
The current DETR planning policy is to reduce the proportion of construction aggregate demand met from primary resources and to make up the shortfall with recycled and secondary aggregates. Present annual targets are 40 mt by 2001 and 55mt by 2006. DETR consider that the initial target has already been reached and have strongly suggested that the pending revised MPG 6 will include a new target of 25% secondary and recycled aggregates, which will equate to around an additional 25 mt. Government instruments currently in place to conserve primary minerals include the introduction of primary aggregates tax, increases in landfill tax and the Local Authority Agenda 21.

There remain barriers however which will need to be overcome for government recycling targets to be met. The construction industry must be sure it can manage the risk of performance failure through appropriate material selection and the operation of robust quality control systems. Innovative pavement design must be encouraged and finding ways to overcome the additional costs often associated with sourcing secondary arisings needs attention. As a consequence of current perceptions premium materials often find there way into low specification end uses where secondary aggregates could have been used. Thus opportunities to conserve primary mineral reserves and possibly gain other benefits such as energy savings are not taken.

To address some of these issues the Scottish Executive Development Department, Trunk Roads – Design and Construction Division contracted Tarmac to design, construct and assess the long term performance of an experimental road at Echline, Edinburgh. The overall objective was to maximise the use of recycled/secondary materials and to demonstrate savings in terms of reduced cost and construction energy. This paper summarises the environmental benefits delivered as a consequence of optimising the use of various secondary materials.

2. DESIGN PROPOSAL

The research trial consisted of a total area 150metres long by 6metres wide with a 100metre long experimental section sandwiched between two turning areas constructed by conventional methods. The technical design proposal submitted at tender stage addressed the following main criteria:

- Maximum use of recycled and secondary aggregates / binders
- Maximum energy savings in terms of overall road construction
- Maximum conservation of primary materials
- Establish the availability and sustainability of the materials used
- Quantify the environmental benefit overall
- Measure road performance over first year and assess durability aspects

The experimental pavement design was expected to meet the requirements outlined above and have a 40year design life (100msa traffic loading), although it is not expected to receive any traffic in the short to medium term. A flexible composite pavement design using by-product and recycled materials was proposed.

At design stage both the short-term and long-term requirements were considered. In the short-term each layer after construction must provide a platform for the construction of the next layer. The long-term consideration is to ensure the durability and performance requirements of the road are achieved over its expected service life. Taking all of these aspects into account, a design was proposed which optimised sustainable design and construction techniques. The Transport Research Laboratory (TRL) in Scotland, independently justified the proposal. The general layout of the design is illustrated in Figure 1.

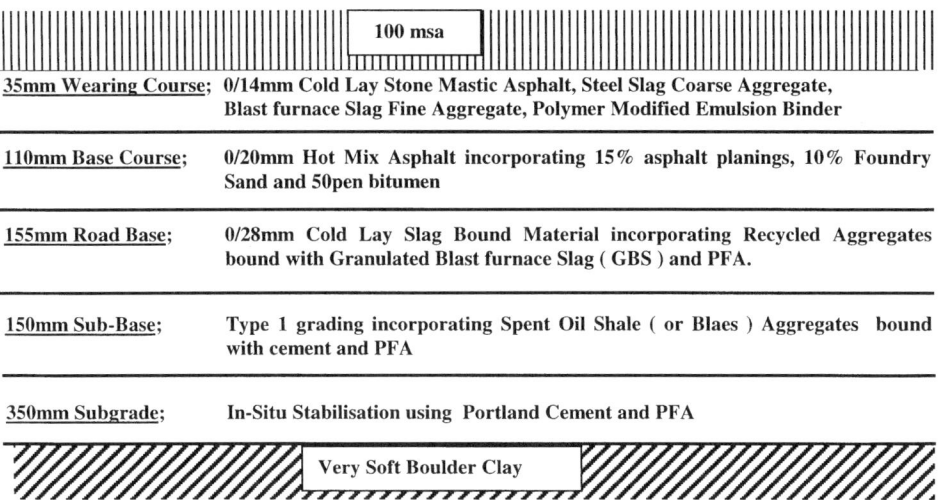

100 msa	
35mm Wearing Course;	**0/14mm Cold Lay Stone Mastic Asphalt, Steel Slag Coarse Aggregate, Blast furnace Slag Fine Aggregate, Polymer Modified Emulsion Binder**
110mm Base Course;	**0/20mm Hot Mix Asphalt incorporating 15% asphalt planings, 10% Foundry Sand and 50pen bitumen**
155mm Road Base;	**0/28mm Cold Lay Slag Bound Material incorporating Recycled Aggregates bound with Granulated Blast furnace Slag (GBS) and PFA.**
150mm Sub-Base;	**Type 1 grading incorporating Spent Oil Shale (or Blaes) Aggregates bound with cement and PFA**
350mm Subgrade;	**In-Situ Stabilisation using Portland Cement and PFA**
Very Soft Boulder Clay	

note ; the overall pavement thickness (450mm) plus in-situ stabilisation of the sub-grade is considered to be adequate to protect the soil from frost damage, allowing for frost penetration of 500mm at worst in Edinburgh.

Figure 1. Pavement Design Layout

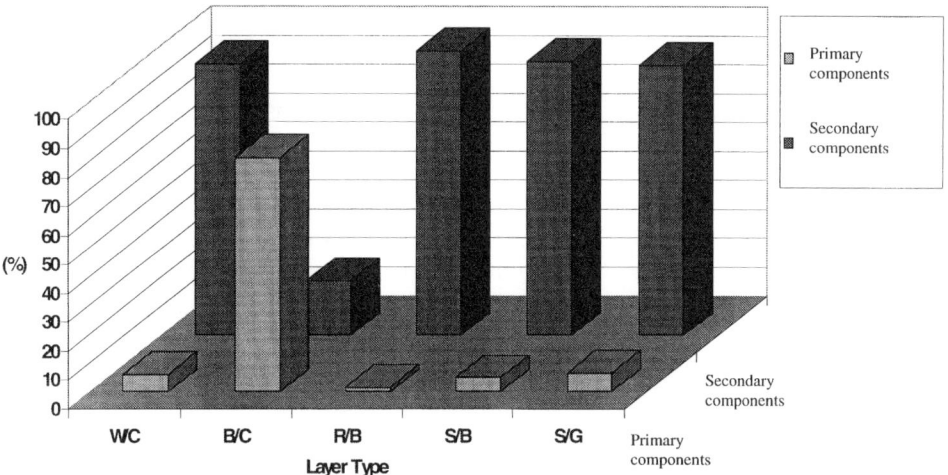

Figure 2: Comparison Between the Primary and Recycled/Secondary Materials
Used in the Construction

3.   ENERGY CONSUMPTION ASSESSMENT

3.1   Introduction
This section presents an evaluation of the energy consumed in terms of aggregate/binder and final product mixture production. Although the transport energy associated with hauling the recycled/secondary materials to the designated mixing plant plus delivery of finished product undoubtedly has a significant impact on the overall construction energy consumption, this factor is discounted in our calculations. This is justified by the experimental nature of the trial ie the small quantities involved could not justify commissioning local production plant thus some materials had to be hauled unusually long distances from production unit to the site. For example, the cold lay Stone Mastic Asphalt was produced at Tarmac's cold mix asphalt plant in Derbyshire , taking around seven hours to haul to site.

3.2   Material Rationale
Figure 2 compares the percentage by mass of the various materials used in each layer.
This figure shows that 81.2% by volume of the materials used are of secondary or industrial by-product origin and only 18.8% by volume are from primary source. For the conventional design construction similar calculations indicate that 18% by volume of the materials used are from non-primary source by taking account of the in-situ clay stabilisation.

3.3   Typical Energy Consumption For Material Production
Table 1 lists typical values for the energy consumption for operations relating to the production of binders and aggregates used in the research trial. These values are extracted from the general information report (GIR 49) published by ETSU [1].

The energy consumption of mixing processes, transport and site operations are summarised in Table 2 [1].

Table 1: Typical Energy Consumption For Material Production

Material (Aggregates and Binders)	Energy Consumption (MJ/tonne)
Portland Cement (PC)	5000
Pulverised Fuel Ash	25
Quicklime	5000
Granulated Blast Furnace Slag	40
Hydrated Lime	5000
Gypsum	5000
Lothian Shale	25 *(Assumed as Recycled Aggregates)*
Recycled Aggregates	25
Recycled Asphalt Planings	25
Foundry Sand	25 *(Assumed as Recycled Aggregates)*
Primary Crushed Rock	50
Steel Slag Aggregates	40 *(Assumed between Crushed Rock and Recycled Aggregates)*
Bitumen Emulsion	700
Bitumen	630

Table 2: Typical Energy Consumption For Production / Transport of mixed materials and the Pavement Construction Processes.

Process	Energy Consumption (MJ/tonne)
Hot-mix Asphalt (Production only)	300
Cold-mix Ex-situ (Production only)	70
Cold-mix In-Situ	70
Road transport per Kilometre	1.5 (range 1.35 – 3)
Rail transport per Kilometre	0.5 (range 0.4 – 0.7)
Bitumen transport per Kilometre	3
Excavate Untreated Material	25
Excavate Bound Material	50
Place and Compact Unbound Material	25
Place and Compact CBM and Black-top	50
Place and Compact Concrete	50 - 150

3.4   Energy Consumption of Mixtures

The data presented in Table 2 is used to compute the energy consumption for plant and in-situ produced materials.

Table 3 shows that compared to the fully flexible 100msa pavement design, the flexible composite alternative provides significant energy savings. With the exception of the capping layer, these figures do not account for energy consumption in transporting and laying materials ie typically ; 2MJ/t per Km road transport and 25 - 50MJ/t place & compact depending on whether unbound or bound material is being laid. The energy consumption required to transport the capping material 50km and then place and compact was included to provide comparison to the in-situ stabilisation process. The energy requirements for the transport and placement of the layers above capping are considered to be similar for both designs. Figure 3 shows the energy consumed in the production of mixtures

Table 3: Comparative Energy Consumption for 100msa Pavement Design

Pavement Type	Layer Thickness mm	tonnes / m^2	MJ / tonne Components/ production	MJ / m^2
Flexible	Combined Asphalt (HDM) layers 330mm			
	+ type1 GSB 150mm	0.792	378	299.4
	+capping (prodn) 350mm	0.360	50	18.0
	+capping haulage (50km)	0.840	50	42.0
	+ capping placement	0.840	2	84.0
		0.840	25	21.0
				total **464.4**
Alternative Flexible/composite	Coldlay W/C 35mm	0.0945	148	14.0
	Hot B/C 110mm	0.264	378	99.8
	SBM R/B 155mm	0.360	168	60.5
	Stab. S/B 150mm	0.240	322	77.0
	In-Situ Stab. 350mm	0.665	270	179.5
				total **430.8**

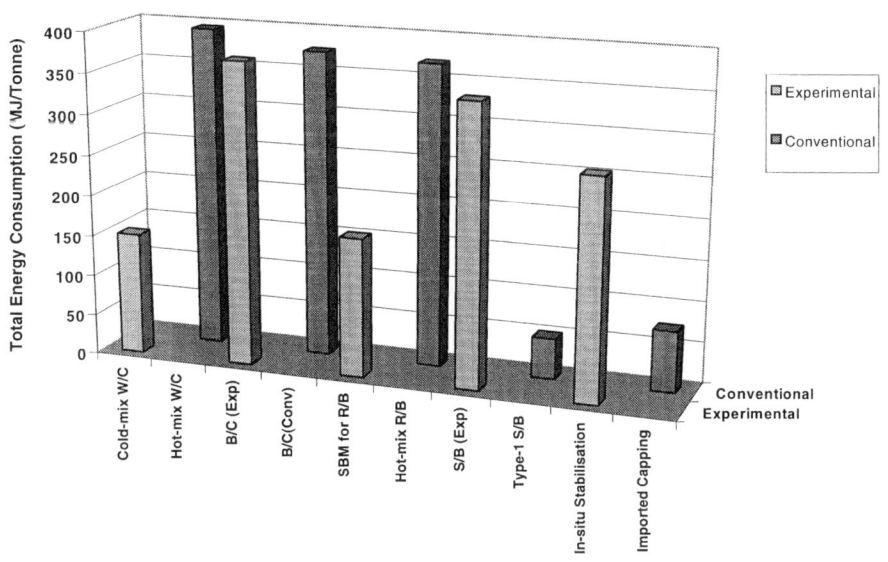

Figure 3: Comparative Energy Consumption of Mixtures Used in The Construction

4. CONCLUSIONS

Based on the energy values presented above the alternative design shows an overall energy saving of 11% compared to the conventional construction (excluding transport energy). However, the experimental section consumed 45% more transport energy than the conventional. This was expected because the Cold Lay SMA had to be hauled to site from Derbyshire, which represented the main contributing factor.

The trial has demonstrated that when alternative materials are coupled with good design and quality controlled production and construction practices, good pavement performance and environmental benefits can be delivered.

5. REFERENCE

1. ETSU (1997). ' Energy Minimisation in Road Construction and Maintenance'. General Information Report No.49, Department of the Environment, Transport and the Regions (DETR)' Energy Efficiency Best Practice Programme, Didcot, Oxfordshire, UK

Hydrological and land use controls on phosphorus distribution and movement in an upland landscape

C. KLIEN, BSc, DR M.T. WALTER, DR T.S. STEENHUIS, and S.W. LYON
Cornell University, Ithaca, NY USA

DEFINITIONS

The role of phosphorus (P) in eutrophication is well documented. In agricultural regions, nonpoint sources may be responsible for a large portion of the P found in surface waters. This paper focuses on larger-scale patterns of P at the landscape level. Soil samples were collected along two transects. Gravimetric soil moisture and soluble reactive P contents were determined for each sample. Land use and hydrology were dominant factors in determining P distribution on the landscape. Unlike traditional wisdom that P strongly adsorbs to the soil, P followed the flow paths of water and accumulated where the moisture does. Interflow, in particular, appears to play a significant role in P distribution on the landscape. The study suggests dissolved P losses need to be taken into account in predicting P concentrations in streams.

INTRODUCTION

The role of phosphorus (P) in eutrophication is well documented (Vollenweider, 1971; Sharpley and Smith, 1992). In fact, P enrichment is identified as one primary cause of impaired lakes and reservoirs according to the U.S. Environmental Protection Agency's Water Quality 305(b) reporting. In agricultural regions, nonpoint sources may be responsible for a significant portion of the P found in surface waters (Newman, 1995).

Historically, nonpoint water quality protection efforts in the USA have relied (and continue to rely) largely on the soil conservation infrastructure. This is based on the belief that by controlling soil loss all other nonpoint source pollution will also decrease. However, research during the last twenty or more years (e.g., Walter et. al.,1979; Gburek and Sharpley, 1998) showed that sediment transport processes are not applicable to all pollutants and, therefore, management practices designed for controlling sediment transport are often inappropriate for many other constituents. For example, soil conservation practices are often ineffective in controlling the transport of soluble chemicals, especially if they have a low soil partition coefficient. Furthermore, much of the early soil conservation research in the U.S. (1930's and '40's) was focused on the topographically flat U.S. Midwest in organic matter poor soils where Hortonian overland flow dominates. This is in sharp contrast to the well structured and high organic matter soils in humid climates that have infiltration capacities greater than rainfall intensities and runoff is generated from areas that saturate during the rainstorm. These runoff producing areas are called hydrologically sensitive because the rain that falls on saturated areas becomes overland flow and carries sediment and pollutants to the stream by saturated excess overland flow. Unlike Hortonian flow, the propensity of an area to produce runoff depends on its position in the landscape and is largely independent of rainfall intensity. The extent of the saturated areas differs with antecedent moisture conditions and storm duration. Thus, sensitivities of areas to contribute to storm flow and pollutant loads differ. Figure 1 exemplifies this well.

To simulate the loss of P from well vegetated watersheds, both the location of the saturated areas and the P content in these areas need to be known. Although the temporal and spatial distribution of these hydrological sensitive areas is now well understood (Frankenberger et al., 1999; Kuo et al., 1999), the interaction between P content and hydrological sensitivity has not been studied.

Historically, subsurface P movement has been largely marginalized, considered an unimpor-tant transport mechanism (Haygarth and Jarvis, 1999) relative to transport in surface runoff and in conjunction with erosion. Recently, however, research has shown signif-icant potential for P movement through soil especially via preferential flow paths (Jensen et al., 1998). Given this multiplicity and probable complexity of potential P flow paths, this paper takes a fresh look at the phenomena of P transport through the landscape by focusing on larger-scale patterns of P distribution to induce probable transport mechanisms rather than focusing on any one specific process.

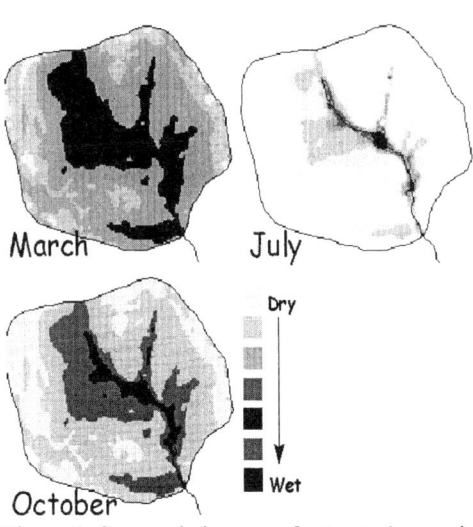

Figure 1. Seasonal changes of saturated areas in a watershed. (From Walter et al., 2000.)

SITE DESCRIPTION AND METHODS

Dairy farming is ubiquitous throughout New York State's Catskill Mountains and has historically been a primary economic activity in the region. The sites for this study were on one dairy farm located in a sub-basin of the Cannonsville Reservoir Watershed which is part of the NYC water supply system. Soils in the collection area are mostly shallow on relatively steep hillslopes. The climate in the area is humid continental with average annual temperatures of 8°C and an average precipitation of 112 cm yr^{-1}. The hydrology is dominated by shallow interflow and saturation excess runoff (Frankenberger et al., 1999). Results are equally valid for other dairy farms with soils with high organic matter.

Soil samples were collected along two transects, each identified as a probable flow path based on the local topography. Transect 1 was 300 m long and traversed a cropped upland hillside and crossed hay and corn fields. Transect 1 was selected to investigate P distribution patterns influenced by manure spreading, a practice applied to the entire hillside at one time or another over the course of a year. Transect 2 was 200 m long, lies downhill of the primary barnyard, and was selected to evaluate P distribution patterns with respect to a barnyard manure source. Shallow (0-5 cm) and deep (~15 cm) soil samples were taken every 10 m along each transect. The deep sample was typically taken right above the shallow fragipan common in these soils (Frankenberger et al., 1999). Occasionally the presence of rocks inhibited sampling.

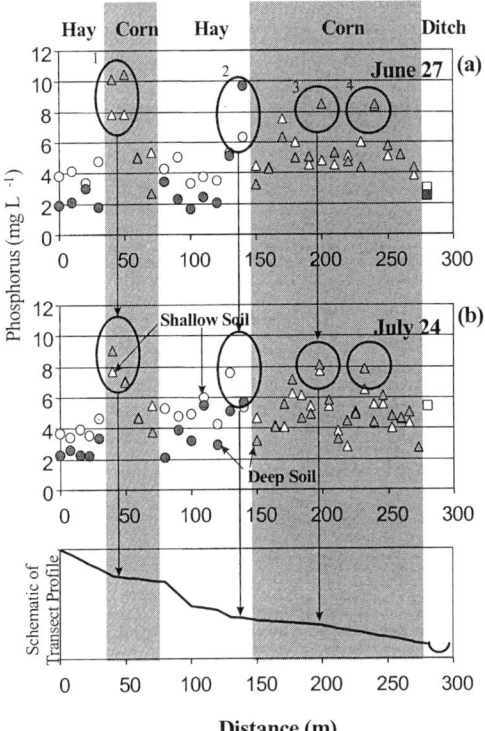

Figure 2. Phosphorus distribution along transect 1 on June 24 and July 27. (circles = hay, triangles = corn, squares = ditch, solid symbols = deep soil, and open symbols = shallow soil. The large circles indicate peaks in the data.)

Transect 1 was sampled twice: once on June 27, 2000 and once on July 24, 2000. The June sampling was during a relatively wet, hydrologically active period (3.8 cm rain in the previous week) and the July sampling during a prolonged dry period (no rain in the previous week). Vegetation in the hay fields was about 30 cm high. In the corn fields, the corn was approximately 60 cm high in June and 150 cm in July. The hay fields were generally steeper than the corn fields (Fig. 2). The corn fields received manure prior to spring planting and the hay fields last received manure spreading during the fall (1999) and winter (1999-2000). Manure spreading utilized a rear discharge, flail spreader.

Transect 2 was sampled on July 11, 2000 (0.3 cm rain in the previous week). Transect 2's location largely avoids potentially manure-spread land so that the effects of the barnyard-source could be isolated. A "level-lip spreader-board" is directly below the barnyard to evenly distribute barnyard runoff across the downslope area. The upper end of Transect 2 lies directly above the "spreader-board" where barnyard outwash is clearly visible. The top 10 m of the transect lies in a filter strip for the barnyard outwash. Below the buffer, the transect crosses 100 m of natural grassland or meadow. Below the meadow, the transect intercepts a drainage ditch and follows the ditch 70 m to its outlet at another ditch that drains a buffer area for agricultural runoff. Above the drainage ditch, the slope is relatively constant, 25-30%, and flattens near the ditch.

Gravimetric soil moisture and soluble reactive P (SRP) contents were determined for each sample. The SRP content was determined by Morgan's extraction (10% sodium acetate in 3% acetic acid buffered to pH 4.8, using a 1:5 soil to solution ratio (Morgan, 1941) and the resulting solution was colorimetrically analyzed (at 400 nm) using the ascorbic acid method with a phenolphthalein indicator (Standard Methods, 1985).

RESULTS AND DISCUSSION
Not surprisingly, land use is a dominant factor influencing P distribution on the landscape. The average Morgan P content for the meadow was 4.5 mg/l for the bottom layer and 7 mg/l for top soil layers. For corn these were 5 and 11 mg/l for the bottom and top layers, respectively, significantly higher for the surface layer at the 5% level.

Figure 2 shows the distribution of Morgan P along Transect 1 during the June and July samplings. Four peaks in P concentration appear along the transect and are identified with large circles in Fig. 2. At circles 1 and 2, the accumulation in P corresponds to points along the hillslope where the slope breaks from steep upslope to flat downslope. These areas are also known as hydrologically sensitive because interflow, traveling across the fragipan from a steep upslope, accumulates at this flat point with less hydraulic gradient. The moisture contents measurement taken during the June 27 sampling (with 3.5 cm of rain in the previous week) confirm this: The moisture content for the high P points in circles 1 and 2 were close to saturation (0.35 - 0.40 g/g) and was much higher than the surrounding soils (average moisture content 0.27 g/g). At locations 3 and 4, indicated by the large circles, high P contents (Fig. 2) were also associated with near saturated soils for the June 27 sampling. These areas were, therefore, hydrologically active. The reason for the high moisture content at these points was less clear and might be very well related to a shallower depth of the hard pan. For the July 27 sampling after a week without precipitation the soils were significantly drier (P = 0.05) with an average moisture content of 0.20 g/g and the wetness of the hydrologically active areas was less pronounced.

The relationship between moisture and Morgan P contents is further exemplified for Transect 1 for both land uses in Fig. 3. Although the regression coefficients are small due to the scatter in the data, there is a distinct and significant trend that the P increases with increasing moisture content for each of the two land uses for the June 27 sampling when the watershed was wet and the hydrological sensitive areas were active. The relationship is much weaker as expected for the July 24 sampling. On this date, evapotranspiration was dominant and erased the differences in moisture content between hydrologically sensitive areas and other parts of the watershed. There were no statistically significant changes in the concentrations of P found between the June 24 and July 27 samplings in either the hay or corn fields (Fig. 2). Thus the footprint of the high P areas in the hydrologically active area during a wet period remained in place even after the soil dried.

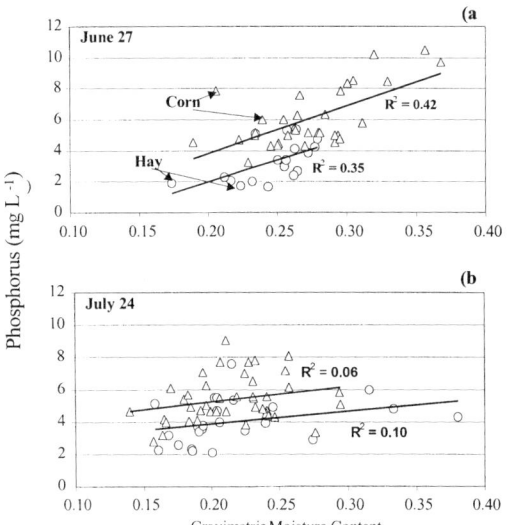

There are two reasons that the P contents in the hydrologically active areas were high. Phosphorus moves in a soluble form down the slope and accumulates in the same area where the water collects (Hergert et al., 1981). Secondly, previous studies show that P in surface soils from

Figure 3. Correlations between soil moisture and P for Transect 1 on June 24 and July 27. (circles = hay, triangles = corn.)

manure applications and excess fertilizers becomes more soluble and mobile under wet conditions (Baker and Laflen, 1982; Gaynor and Bissonette, 1992; Sharpley and Smith, 1992). Interestingly, though generally the shallow P concentration is higher than the deep concentrations, at the four peaks the relationship switches (in all but one instance) such that the deeper concentrations are higher than the shallow (Fig. 2). This further links P movement to interflow, i.e., the interflow persists deeper in the soil along the fragipan, thus accumulation of P in association with interflow would be expected at depth rather than at the surface

Figure 4 shows the distribution of Morgan P concentrations and gravimetric soil moisture along Transect 2. One of the striking features in Fig. 4a is the pronounced decrease in P content downslope, i.e., as one moves away from the barnyard. Although Fig. 4b suggests that trends in P concentration above the ditch are independent of soil moisture, Fig. 5 shows that the correlation is, in fact, very strong if the data from the ditch are removed. This strong correlation is expected since the barnyard is likely the source of both P and a substantial fraction of the water. This in fact is similar to the P transport phenomena for Transect 1. The barnyard serves the same role as the water from the upslope. Both have a high soluble P content which is then deposited in areas where the water accumulates/infiltrates. At the slope break in Transect 2 (~110 m downslope), the P content increases (Fig. 4a) similarly as in Transect 1's slope breaks (around 50 m and 140 m along the slope in Fig. 3).

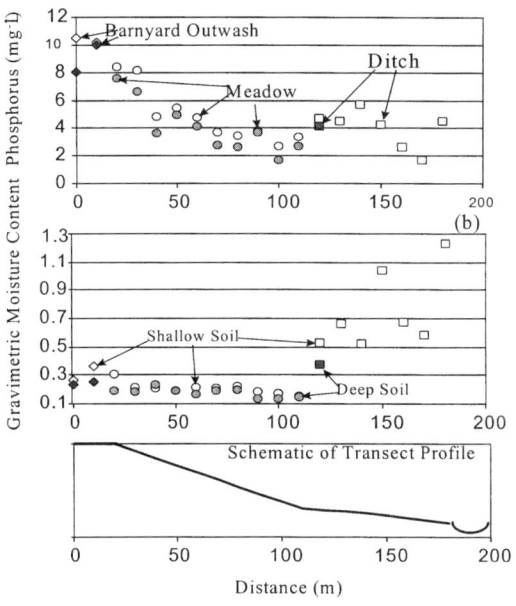

Figure 4. P concentration (a) and soil moisture (b) distribution along Transect 2 in relation to the transect profile. (diamonds = outwash, circles = meadow, squares = ditch, solid symbols = deep soil, and open symbols = shallow soil.)

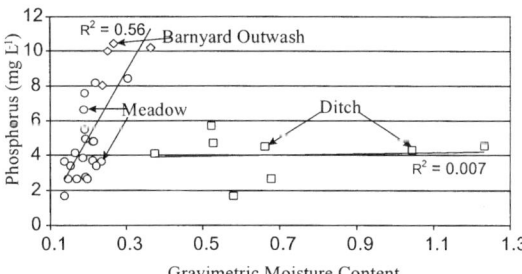

Figure 5. Correlations between soil moisture and P for Transect 2. (diamonds = outwash, circles = meadow, squares = ditch.)

Though discerning any obvious differences in vegetation at the time was difficult, the area between ~100 m and ~120 m on Transect 2 is managed as a "buffer area." This also appears to be the area of most significant P concentration. On Transect 2, the decrease in P content around 140 m is associated with a slight increase in downhill slope (Fig. 4a) and mechanisms involved are probably similar to those associated with P peak number 3 in Fig. 2a. Overall there are no other obvious trends of P concentration throughout the ditch. The ditch is wet and nearly saturated through the year and water travels through the ditch rather than infiltrating. In fact, Fig. 5 shows a very flat correlation between soil moisture and P content in the ditch.

CONCLUSIONS

The hydrology of a system and land use both influenced the distribution of phosphorus (P). The interaction between high P and hydrological active areas found in this study were remarkable. Water flow provides the mechanism for P transport. Phosphorus follows the flow paths of water and accumulates where the moisture does. Interflow, in particular, appears to play a significant role in P distribution on the landscape. This suggests that the loss of P from agricultural lands to nearby stream systems or water bodies can be reduced by binding the P in extremely active hydrological flow paths. When the hydrology of a system becomes inactive, the distribution of P is preserved, leaving a record of previous hydrological activity. More research is needed in other regions to examine if the interaction of hydrology and P movement holds.

REFERENCES

Baker, J.L. and J.M. Laflen. 1982. Effect of crop residue and fertilizer management on soluble nutrient runoff losses. Trans. ASAE 25:344-348.

Frankenberger, J.R., E.S. Brooks, M.T. Walter, M.F. Walter, T.S. Steenhuis. 1999. A GIS-based variable source area model. Hydro. Proc. 13:805-822.

Gaynor, J. and D. Bissonnette. 1992. The effect of conservation tillage practices on the losses of phosphorous and herbicides in surface and subsurface drainage waters. Final Report No. 60 to Southwestern Ontario Agr. Res. Corp., Agr. Canada Res. Sta., Harrow, Ontario. 134 pp.

Gburek, W.J. and A.N. Sharpley. 1998. Hydrological controls on phosphorus loss from upland agricultural watersheds. J. Env. Qual. 27:267-277.

Haygarth, P.M. and S.C. Jarvis. 1999. Transfer of phosphorus from agricultural soils. Adv. Agronomy 66:195-249.

Hergert, G.W., D.R. Bouldin, S.D. Klausner, and P.J. Zwerman. 1981. Phosphorus concentration water interactions in tile effluent from manured land. J. Env. Qual. 10:338-344.

Jensen, M., P.R. Jorgensen, H.C.B. Hansen, and N.E. Nielsen. 1998. Biopore mediated subsurface transport of dissolved orthophosphate. J. Env. Qual. 27:1130-1137.

Kuo, W.L., T.S. Steenhuis, C.E. McCulloch, C.L .Mohler, D.A. Weinstein, S.D. DeGloria, and D.P. Swaney. 1999. Effect of grid size on runoff and soil moisture for a variable-source-area hydrology model. Water Resour. Res. 35:3419-3428.

Morgan, M.F. 1941. Chemical soil diagnosis by the universal soil testing system. Conn. Agr. Exp. Sta. Bulletin 450.

Newman, A. 1995. Water pollution point sources still significant in urban areas. Env. Sci. Tech. 29(3):114A.

Sharpley, A.N. and S.J. Smith. 1992. Prediction of bioavailable phosphorus loss in agricultural runoff. J. Env. Qual. 21:32-37.

Standard Methods for the Examination of Water and Wastewater. 16th ed. 1985. Method 424F. AWWA/American Public Health Assoc., Washington, DC.

Walter, M.F., T.S. Steenhuis and D.A. Haith. 1979. Nonpoint source pollution control by soil and water conservation practices. Trans. ASAE 22:834-840.

Walter, M.T., E.S. Brooks, M.F. Walter, T.S. Steenhuis, J. Boll, K.R. Weiler. 2000. Hydrologically Sensitive Areas: Variable Source Area Hydrology Implications for Water Quality Risk Assessment. J. Soil Water Conserv. 3:277-284.

Vollenweider, R.A. 1971. Scientific fundamentals of the eutrophication of lakes and flowing waters, with particular references to nitrogen and phosphorus as factors in eutrophication. Organisation for Economic Co-operation and Development, Paris.

Use of Steel Sheet Piling: Todhole Burn a Case History

ANGUS MacARTHUR, PETER YOUNG, and JOHN THEOS
Enviros Aspinwall, Cambridge, UK and Corus Construction Centre, Scunthorpe, UK

INTRODUCTION
This paper is set at the former Ravenscraig Steelworks in North Lanarkshire. The site has seen heavy industrial activity for over 200 years, initially with coal mining and culminating with British Steel's Ravenscraig Steelworks which closed in 1992.This has left a significant environmental legacy and the use of steel sheet piling is presented as one of the methods used in the overall environmental strategy commissioned by Corus (formerly British Steel) for the remediation and ongoing environmental management of the site.

BACKGROUND
The site is underlain by sandstone and coal measures with multiple faulting across the area and is predominantly overlain with made-ground and some original drift material from the quaternary period. The coal measures have been significantly worked including the Ell, Pyotshaw, Main and Splint seams. Mining activity on the site ceased in the early part of twentieth century. There is still evidence of mining on the site including capped shafts, adits and pumping wells.

The South Calder Water flows from east to west across the northern half of site discharging into Strathclyde Loch, a major public amenity, 3km to the west of the site. Todhole Burn is a natural stream that has been artificially trained throughout the site since the 19th Century. At the time of the steelworks closure, only the final 200m remained exposed prior to discharge into the South Calder Water in the north-west of the site. The area around this discharge point was named the Todhole Burn site during the re-mediation works. At this point, the South Calder was also extensively trained, including a weir, to form a lagoon and bypass stream in order to control silting and scouring by the river.

THE PROBLEM
During the early days of steel making on site a consented discharge was established in the by-products plant by sinking a series of boreholes down to abandoned and disused mine workings below, comprising various worked seams between 30-130m below ground level. Wastewater, including wash oils, was disposed in the boreholes for over twenty years.
The hydrogeology of the area suggests that the groundwater flows tend towards the South Calder Water and, in particular, the area at the Todhole Burn site (see figure 1 below).

In the late 1970s, seepages of free phase wash oil and associated contamination required British Steel to protect the South Calder Water. It was subsequently established that this was linked to deep seated groundwater level recovery once pumping from the coal mines had ceased, allowing migration through the mine workings via faulting and a redundant pumping well to the Todhole Burn and adjacent area.

Geoenvironmental impact management, Thomas Telford, London, 2001.

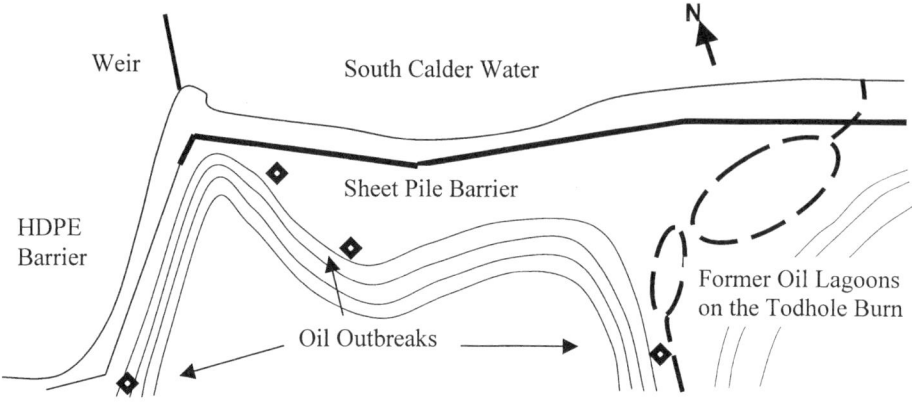

Figure 1 – Plan of Todhole Burn (not to scale)

During the operational life of the steelworks, environmental management using a series of open lagoons, which were farmed for oil, prevented contamination of the South Calder Water. However, on closure it was clear that a more rigorous and discrete method of long-term control of the outbreaks of contaminated groundwater was required to meet the expectations for the ultimate redevelopment of the site.

THE SOLUTION

Corus commissioned Enviros Aspinwall to act as their environmental consultants for the Ravenscraig site and a key issue was producing a long-term solution for the Todhole Burn area.

Extensive site investigations confirmed that a quick and permanent solution for removing or cleaning the groundwater was impractical as the contamination in the groundwater had spread below much of the former operational areas, and also over considerable depths in the various mine workings and faulted blocks of Coal Measures strata. However, when pathways were modelled, the only significant impact was predicted along the south bank of the South Calder Water, downstream of the culverted section. It was this area which became the focus of remedial measures to deal with the impacts of the historical groundwater contamination. The objectives of these measures were to:

- prevent further potential episodic outbreaks into the South Calder Water;

- reduce the risk of outbreaks from possible future deep seated ground movements;

- remove the wash oil lagoons and facilitate direct discharge of the Todhole Burn to the South Calder Water; and

- enhance the collection and disposal of recurring wash oil.

A design process and risk assessment approach identified that the following works would present the most effective solution for achieving the objectives.

- a physical barrier to prevent further outbreaks;

- a means of controlling shallow groundwater levels adjacent to the river along the south bank;

- an automated groundwater collection system;

- a pre-treatment plant for the contaminated groundwater; and

- culverting the remaining length of the Todhole Burn through the residual contaminated ground.

DESIGN

The design identified that a cut-off wall, with a back of wall drainage scheme, was required in the upper made ground and drift material down to the underlying rock head (see figure 2 below). The depth of the wall was set to 4-8m below ground level, running parallel with the south bank of the South Calder Water, which would allow effective interception of the free phase wash oil. This depth was also set to prevent the barrier penetrating the highly permeable, underlying sandstone, which contain greater flows of groundwater, which is not a threat to the quality of the river. A quantitative assessment of the river hydrology was also carried out that indicated that the reduction in groundwater flow to the river hydrology was well under 1% for the overall catchment area

During the early stages of the cut-off wall design a number of constraints were identified that had to be addressed when selecting an appropriate and effective containment method. Principally, these comprised the following:

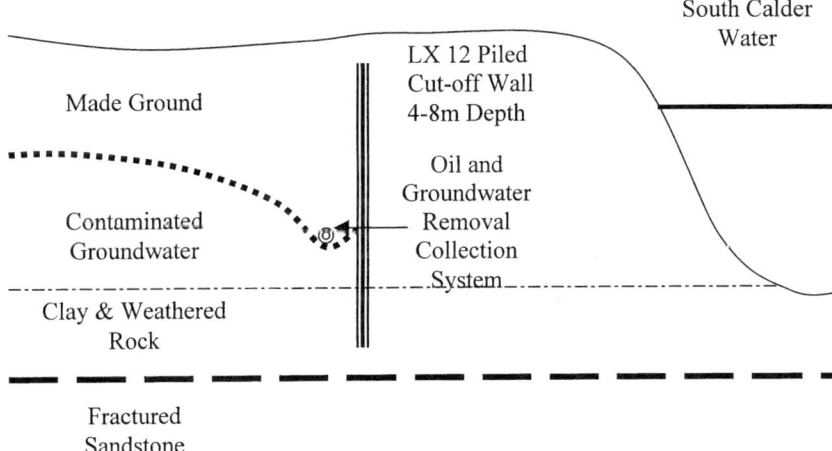

Figure 2 – Basic Barrier Construction (not to scale)

Existing Ground Conditions – The made ground had been placed at various periods using blaes (spent oil shale) and slag material, both of which had a high permeability, $\sim 10^{-5}$ m/s. The main fill areas included the old course of the Todhole Burn and the south bank of the lagoon formed in the South Calder Water. Any excavation work for a cut-off wall would give rise to the risk during construction of either opening new pathways into the river, or the river inundating the excavation when below the bed level.

Working Area – Space was limited and the cut-off wall followed the same route as the access track beside the river. Access was to be maintained at all times for the environmental management team in order to manage the existing pollution control infrastructure. This required the removal of wash oil arising from an open, abandoned pit no more than ten metres from the river on a regular basis. One section had no previous access and a narrow track was established on the steep river bank.

Secondary Environmental Risks – When adopting an active management system, secondary environmental risks arise and in this case the major issue was the disposal of the collected groundwaters. A treatment plant had been constructed at Todhole Burn to deal with the leachate arising from a secure containment facility containing previously excavated contaminated soils. The processes within the plant included separation, flocculation, chemical and biological treatments that would treat the leachate to a quality that would allow for ultimate disposal to surface waters. This plant would also be capable of treating the contaminated groundwaters provided that minor modifications were carried out to allow for the increase in flows through the plant.

Temporary Works – The design included the construction of pump chambers and back-of-wall drainage to draw off the contaminated groundwater and any free phase wash oil encountered. The cut-off wall required to act as part of the temporary works for this installation in preventing the inundation of the river water and minimising over-dig anticipated during the trench construction.

Construction Options - Construction techniques available for the construction of a suitable wall included cut-and-fill barriers, such as slurry trench cut-offs; mix-in-place barriers, including deep soil mixing and jet grouting; injection barriers using permeation grouting via boreholes; and displacement barriers comprising sheet steel piles or HDPE membrane.

Cut-and-fill barriers, whilst effective, were not suitable because of the existing ground conditions and the available working area. It was envisaged, that there was a significant potential for the river or high groundwater flows to inundate the excavation during the construction. Consideration was given to the excavation arisings that would include an elevated pH (pH~11) from the slag and oil contamination from the contaminated groundwater that would result in disposal under the Special Waste Regulations. Although HDPE membranes are classified as displacement barriers, the construction technique involves a similar open excavation methods and risks to that of cut-and-fill. Importantly, operatives are potentially at risk to a number of hazards when placing and jointing the barrier.

Jet and permeation grouted barriers: it was considered that it would either be difficult to check their integrity or that they would not be capable of forming low permeability barriers. In this case, ground conditions were known to be heterogeneous with buried watercourses and underground streams with the potential for excessive grout loss. More importantly, with the close proximity of the river there was a real potential for a 'blow-out' of grout into the river which would be an unacceptable risk to the river.

A steel sheet piling barrier provides a number of operational advantages over the other techniques during the construction phase of the works. These included, the role of sheet piling as temporary works for the back-of-wall drainage, minimisation of disruption to other activities during installation and a reduction in the construction programme for the works. There were concerns over the short-term permeability of the pile joints, especially during the installation of the back-of-wall drainage. However, Corus was able to offer an innovative, sealed joint based on their LX section piles marketed as the Halt Lock system.

The Halt Lock system selected for the barrier comprised a pair of standard Corus LX 12 sheet piles fully welded together in the production plant. On one side of the pair and enclosing the clutch, a special steel angle section was again factory-fitted using a continuous fillet weld. This angle piece, in conjunction with the pile clutch, creates a void that was filled at the plant with a sealing compound (see figure 3 below). When the pairs of piles are interlocked, the driving action displaces the angle section and effectively 'bites' into the incoming pair whilst at the same time, causing the sealant to diffuse throughout the joint.

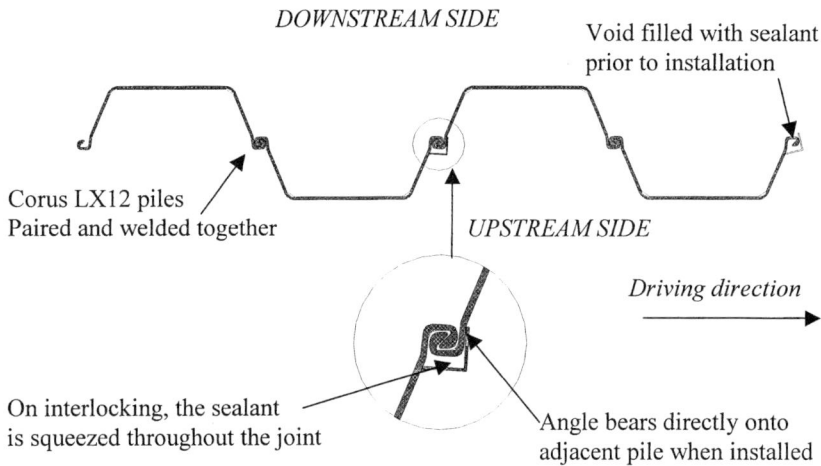

Figure 3 – Halt Lock Details (not to scale)

With any barrier system in contaminated or made ground, the durability of the barrier is an important issue. In this case, the initial design recognised that the pile sections may have needed corrosion protection using an appropriate coating applied to the piles to a level 1m above the anticipated groundwater level. Following consultation with Corus, the groundwater and soil conditions were considered not to be detrimental to the steel and a protective coating was deemed unnecessary due to the elevated pH of the soils. In addition, incorporation of an active wash oil removal system and design of the Halt Lock, such that the protective angle was installed on the side of the incoming wash oil, minimised exposure of the sealant.

Having reviewed the advantages and disadvantages of the various barrier technologies, sheet piling using the Halt Lock system was selected on the basis that it provided advantages in construction and operation over the other systems. However, there was a section where the depth to rock head was shallow (2-3 metres), so the temporary works properties of sheet piling were not required and an HDPE barrier was selected in preference to sheet piles.

BACK-OF-WALL DRAINAGE DESIGN
Studies of the shallow groundwater regime had shown that the groundwater levels varied seasonally. The original design set out that free phase wash oil was to be actively removed

and it was also considered, at that time, that the cut-off wall would promote the surface breakouts of wash oil during the winter periods upstream of the barrier.

Therefore, the design of the back-of-wall drainage promoted the continuous removal of wash oil with the minimum removal of groundwater. This was achieved by setting the invert level of the drainage to the seasonal low groundwater level coupled with shallow land drains to the areas associated with previous outbreaks for enhanced oil collection. The back-of-wall drainage is connected to four sumps incorporating pumps for onward disposal of contaminated groundwater to the plant for treatment and ultimate disposal.

CONSTRUCTION
Construction of the steel sheet pile work commenced in November 1997. Initially, a shallow proving trench was dug along the line of the barrier prior to the piling operation to locate any potential services and, in particular the new culvert for the Todhole Burn and the treatment plant effluent discharge. The piles were driven using a "Movax" vibrating hammer mounted on the boom of a 360^0 tracked excavator. It has been found that vibrating hammers are effective to use with the Halt Lock system as their mode of operation encourages the dispersion of the sealing compound into the complete clutch assembly.

The length of the sheet pile barrier was over 300 linear metres and the contractor's chosen method of working, along with good driving conditions, meant that the barrier was constructed at a rate of 100 piles or 60 linear metres per day. This enabled the wall to be completed well within the allotted time of three weeks. The design of the barrier called for the piles to be buried and any excess pile lengths were cut-off to 500mm below finished ground level. On completion of the piling, the back-of-wall drainage was installed with the invert levels set to the lowest point of the seasonal shallow groundwater regime with falls to discreet sumps containing oil skimmers.

CONCLUSION
Since fully commissioning the treatment plant and oil removal system in February 1999, there have been no further outbreaks of oil into the South Calder Water from the sheet pile barrier frontage. Similarly, where the new Todhole Burn Culvert and the effluent discharge pipework penetrate the sheet pile wall and into the river, there is no evidence of leaking. The original oil lagoons have since been removed and are now covered with a lined, engineered reed bed as part of the final treatment process for the contaminated groundwater. Booms previously placed across the South Calder Water have now been removed and the Todhole Burn is free flowing with no ingress of oils.

In general, the area has been landscaped and planted providing a habitat for a wide variety of flora and fauna. There are now no areas of exposed contamination, providing safe conditions for informal public access. Corus and Enviros, working in partnership, have provided an efficient and cost effective system for the long-term, sustainable management of the Todhole Burn Area.

REFERENCES
BRITISH STEEL plc (1997).The Piling Handbook, www.corusconstruction.com

CIRIA (1996). Barriers, liners and cover systems for containment and control of land contamination.Special Publication 124, www.ciria.org.uk

Characterisation of landfill waste and subsurface using electrical imaging

Magdeline POKAR [1], Meng Heng, LOKE [2] and Chong Yan, LEE [2]
[1] School of Earth Sciences, University of Leeds, Leeds, UK.
[2] School of Physics, Universiti Sains Malaysia, Penang, Malaysia.

This paper examines the usage of electrical resistivity on a sanitary landfill in Malaysia. It demonstrates the successful use of the electrical imaging method to delineate and characterise waste disposal sites. A Campus multi-electrode resistivity meter system was used to carry out the 2-D electrical imaging surveys and the RES2DINV software was used to invert the data to provide a geological model of the subsurface. As most contaminated sites tend to be of limited size, a modified Wenner-Schlumberger array was designed which enabled more measurement points to be made over a limited area. This survey was able to accurately differentiate wastes of different ages based on their resistivity values. It was also able to delineate the waste boundaries by the different resistivity values detected from the bund walls and clay liners. Possible salt-water intrusion was also detected underneath the landfill. The site was imaged from a depth of 1.3 m to 80 m.

INTRODUCTION

Landfill management is a very important aspect in sustainable development. Leachate from a badly managed landfill can contaminate local groundwater supply while gas pockets have the potential to ignite. Most subsurface sampling methods are highly intrusive (e.g. boreholes) and could act as a pathway for contaminant migration. Electrical imaging is non-intrusive and maps the subsurface by providing a resistivity distribution of the area. This distribution is then converted into an "image" which gives much needed information about the landfill. For example, monitoring the perimeter of a landfill enables the detection of leachate or gas migration outside the confines of the landfill. (Pokar & Loke, 1998).

A subsurface region to be studied is "sampled" by transmitting energy through it along many paths of known orientations, and from the properties of these transmissions of data, a cross-sectional image of the region of interest is constructed (Ramirez & Daily, 1995). Resistivity values are influenced by porosity, the amount of pore water and concentration of dissolved solids. Among the controlling factors that affect the rate and extent of bio-chemical decomposition in a landfill are the amount of moisture available, temperature and soil cover permeability (Ramly, 1989). Therefore different parts of a landfill may be at different decomposition stages at the same time. This results in different ion contents in the pore fluid and thus enabling resistivity to detect the differences. The results presented are based on work conducted on a landfill in Penang, Malaysia (Ampang Jajar Sanitary Landfill) which was constructed in 1989.

The landfill is located on a Quaternary coastal alluvial plain. Courtier (1974) estimated the alluvial plain to vary between 9-19 km wide with a general thickening of the alluvium towards the coast and free of any post depositional deformation. Seismic reflection surveys indicate the bedrock to be well over 200 m deep while a bore hole at Kampung Permatang Batu, 6 km South East of the landfill indicate clay layers to a depth of 8.5 m. (Kamaluddin, 1990). The thick clay layer gives the soil a low permeability, thus providing a natural liner to confine leachate migration.

METHODOLOGY
A multi-electrode resistivity meter system (Griffiths et. al. 1990) with 25 nodes was used to map the distribution of subsurface resistivity in a cross section below a profile (Figure 1).

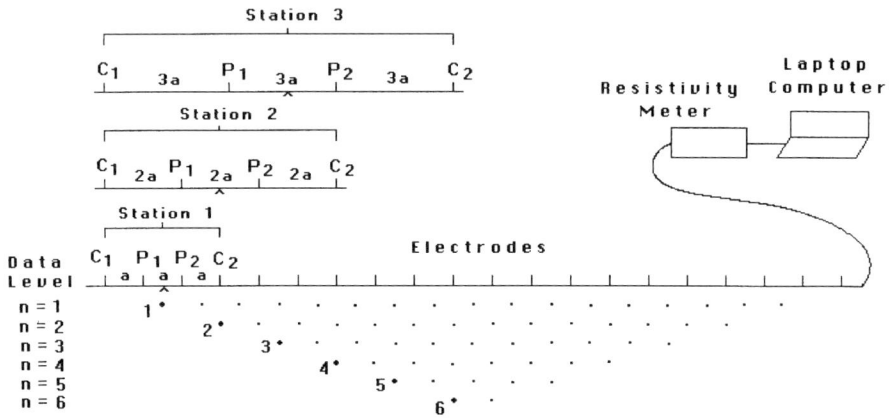

Figure 1 – Multi electrode data collection system

Three survey techniques were routinely utilised in the survey lines. The first being the traverse mode which measures all points within the first level (n=1) then continues with the second level of measurements followed by the roll on survey with consists of the forward roll and backward roll. Here measurements for the first station are taken for all levels before continuing with the second station. This survey mode is used when the survey line needs to be extended forward beyond the first 25 electrodes or extended backward beyond the first electrode.

A parameter file known as the modified Wenner Schlumberger array was created for the purpose of this survey. It is a combination of Wenner and Schlumberger array for the electrode spacings of "a" to "8a", for 5 levels of measurements (n = 1 to 5) and "a" is the unit electrode spacing (Figure 2). This array was designed to gather denser measurement points within the same length of traverse. This is important considering the general lack of space for most survey sites affected by subsurface contamination. The parameter file automatically selects 4 electrodes from the 25 electrodes available and measures the resistance values for each station. The survey design consists of 215 stations or measurements for each traverse and 72 measurements for each 4 electrode roll on.

The data is converted to apparent resistivity values and an inversion is then carried out on the data gathered. The purpose of a resistivity inversion is to generate a suitable geological model to suit the data that has been measured. A pseudosection consisting of apparent resistivity values measured in the field survey gives a distorted image of the true subsurface resistivity due to the geometry of the array used. An inversion of the results removes the distortions to give a realistic resistivity model by removing the geometric effects. (Loke, 1994). A resistivity inversion software, Res2Dinv (Loke, 1997) was used to carry out the inversion of measured data and generate geological models based on the field data.

A finite difference (discretisation by area) method is used to calculate apparent resistivity values in the forward modelling process (Dey and Morrison, 1979) while the inversion routine used by the program, Res2Dinv is based on the smoothness constrained least squares method.

RESULTS AND DISCUSSION
The results of 2 lines are presented here as they best delineate and characterise the wastes. These lines were conducted at Phases 4A and 4B, which were about 3 to 4 years old with Phase 4B being several months older than the other. They have had two successive bund walls built above one another. These bunds contain numerous compacted cells of wastes and act as barriers to the movement of wastes and leachate. Each bund is about 30 feet high and 15 to 20 feet wide. When a waste cell has been filled to capacity, a final soil cover is laid across the compacted cells of wastes and levelled to form a small flat-topped hill. If necessary a second or third bund will be built upon this layer, resulting in a "terraced hill structure".

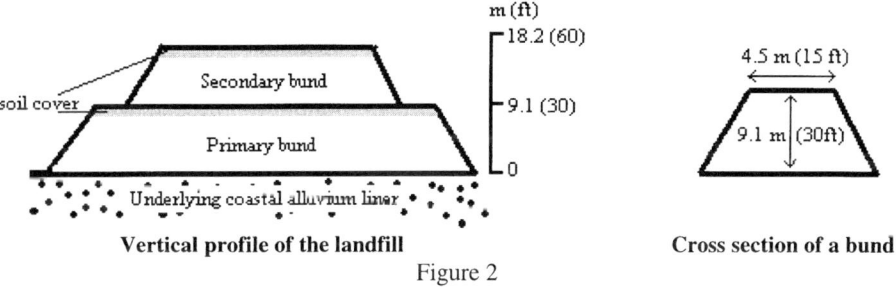

Vertical profile of the landfill   **Cross section of a bund**

Figure 2

Line 1 consisted of a survey 490 m long encompassing both Phase 4B and 4A. The survey consisted of a 5m spacing traverse and 18 four electrode roll along, thus imaging from a depth of 1.3 m to 21.7 m. With the effect of settling taken into account, the amount of wastes anticipated at the survey site was less than 18 m (60 feet)

The main resistivities can be divided into three groups. The first group ranging from 0.5 to 2 Ω.m, followed by the second group whose resistivities range from 4 to 8 Ω.m in the mid section of the inverse model section and the top part of the section range from 16 to 90.5 Ω.m. These three groups can each be interpreted as the underlying clay liner (0.5 to 2 Ω.m) and waste deposits (4 to 8 Ω.m and16 to 90.5 Ω.m). Resistivities here decreased with depth as deeper wastes were older and were in a more advanced decomposition state.

The resistivity inversion model confirms that the wastes in each phase have been deposited in two main successions with the older wastes located within the primary bund. The wastes in

4B(I) and 4A(I) are several months older than each other but they are both almost a year older than Phase 4B(II) and Phase 4A(II). The decaying stage of wastes in 4B(I) and 4A(I) should not be too different from each other but the wastes in layers 4B(I) and 4B(II), should be in very different decaying stages, exuding different by products and composition. This should result in a marked difference in resistivity values which enables the waste boundaries to be delineated.

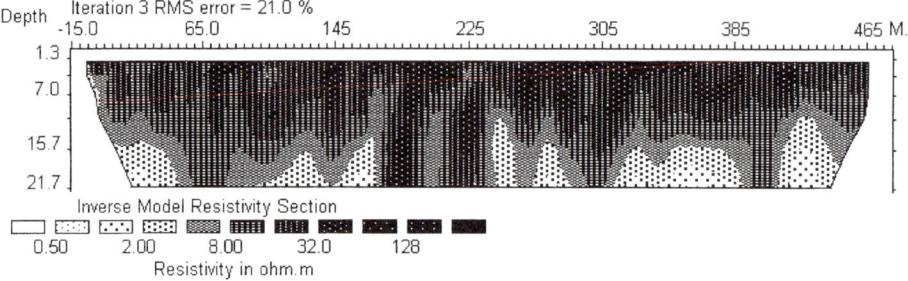

Figure 3 – Line 1 across Phase 4A and 4B which shows the waste boundaries.

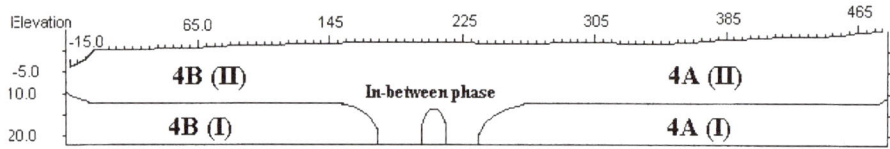

Figure 4 – Outline of waste boundaries based on resistivity data and information from landfill authorities.

Figure 3 shows the layering sequence with the top layer having resistivities ranging from 16 to 64 Ω.m followed by the second layer of 4 to 8 Ω.m. This represents 4B(II) and 4A(II) (16 Ω m -64 Ω m) while the resistivities of 4 to 8 Ω.m represent the primary bunds, 4B(I) and 4A(I). The underlying clay liners are represented by resistivity values of about 2 Ω.m. Figure 3 also shows the extent of the settling process where the newer wastes settle into the older wastes in the earlier bund. An area in between the two phases has also been detected where wastes have also been deposited there. This is a common practice in the landfill where the spaces are utilised to its limit. Based on the imaging results of Figure 3, Phase 4B is 190 meters long and phase 4A is 240 meters long while the additional in-between phase is only about 60 meters long. The outlines of these boundaries are shown in Figure 4.

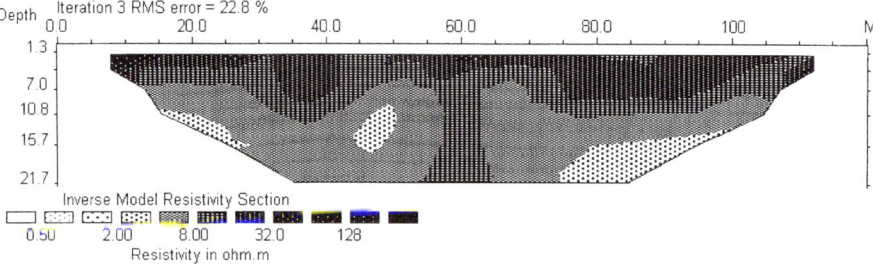

Figure 5 – Line 2 shows the boundaries between phase 4A and 4B of the landfill.

A second survey line was conducted several meters from the north-east boundary of phase 4B. The line was 120 metres long with 5 metres spacing. This line was especially successful in delineating the bund wall and shows the boundary between Phase 4A and 4B (Figure 5). Most of the wastes in 4B(II) range between 16 and 32 Ω m, indicating that the decaying process here is more advanced compared to the wastes in phase 4A (II). The wastes in the primary bund, 4B(I) have resistivities ranging from 4 to 8 Ω.m.

In order to investigate the conditions beneath the landfill, the survey was extended to a depth of 80 m. This was done with a 20 m spacing for each of the 25 electrodes. The data was then merged with the earlier 5 m spacing data (Line 1, Figure 3) to provide a complete section from 1.3 m depth. As expected, the wastes were confined to a depth of less than 20 m (Figure 6). The resistivity of the underlying alluvium (marine origin) beneath the landfill was about 2 Ω m. However a large zone of very low resistivity was detected. This zone begins at a depth of about 30 m and is about 90 m wide. Smaller zones of these very low resistivities are also located at a depth of 20 to 40 m underneath phase 4A of the landfill.

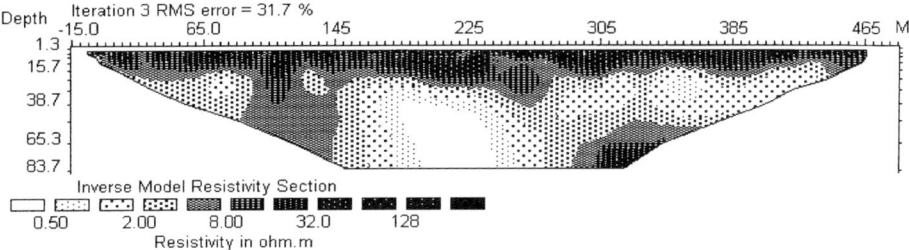

Figure 6 – Very low resistivity values underneath the landfill indicating possible salt water intrusion

.

These very low values (< 2 Ω m) indicate that the accumulations may be saline in nature. The landfill is approximately 3.5 km away from the coast, therefore the very low resistivity areas detected underneath the landfill could be sea water intrusion. The wastes are only about 5 m directly above it with the coastal alluvium separating them. Subsequent surveys conducted in the area adjacent to the landfill also show the same resistivity values with the very low values at about the same depths.

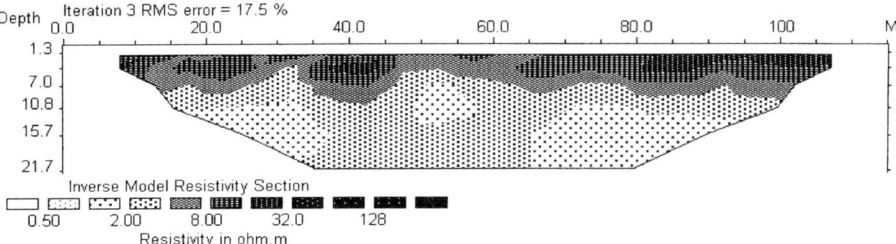

Figure 7 – Survey conducted 0.5 km away from the landfill show the same resistivity values occurring at relatively the same depths.

In order to ascertain whether the low resistivity area is caused by leachate or salt water intrusion, a survey was conducted in a nearby village, approximately 0.5 km from the landfill site. Once again, a low resistivity area within the same values and depth (4 – 6 m) as the other surveys was detected. Based on the depth and values of the low resistivity area, a probable conclusion is that the anomaly was caused by salt water intrusion underneath the area. This is because leachate migration would probably result in an increase in resistivity values as it moves further away from the landfill (due to dilution and filtering affects). This is not the case here as the values of the low anomalies were generally within the same range even with incresed distance from the landfill.

CONCLUSION

Lines 1 and 2 were successful in delineating the waste boundaries and detecting the bunds used to separate the wastes. These survey lines were conducted over an area with two layers of wastes differing in the decomposition level due to the age difference. The survey was successful in differentiating the two levels of wastes based on differing resistivity values. The newer wastes had resistivity values ranging from 16 to 32 Ω.m, while the older wastes below ranged between 4 and 8 Ω.m. The resistivities below these lines decreased with depth due to the higher level of decomposition with depth and the presence of marine clay as the landfill liner. Possible salt-water intrusion was detected underneath the landfill.

REFERENCES

1. Courtier, D.B. (1974). *Geology and mineral resources of the neighbourhood of Kulim, Kedah.* Geological Survey of Malaysia. Map Bulletin No. 3.
2. Daily, W. and Ramirez, A., (1995). *Electrical resistance tomography during in-situ trichloethylene remediation at the Savannah River site.* Journal of Applied Geophysics, 33, 239 - 249.
3. Dey, A. and Morrison, H.F. (1977).*Resistivity modelling for arbitrarily shaped two-dimensional structures.* Geophysical Prospecting, 27, 100 - 136.
4. Griffiths, D.H., Turnbull, J and Olayinka, A.L. (1990). *Two-dimensional resistivity mapping with a computer controlled array.* First Break, 8, 121-129.
5. Kamaludin, H. (1990). *A summary of quaternary geology investigations in Seberang Prai, Pulau Pinang and Kuala Kurau.* Geological Society of Malaysia Bulletin, 26, 47 - 53.
6. Loke, M.H. (1997). *Electrical Imaging surveys for environmental and engineering studies.* School of Physics, Universiti Sains Malaysia. p 1 - 17.
7. Loke, M.H.(1994). *The inversion of two dimensional resistivity data.* PhD thesis. School of Earth Sciences. University of Birmingham. United Kingdom.
8. Ramly, N.H. (1989). *Improvement of disposal sites using semi-aerobic reciculatory system.* Municipal Council of Seberang Prai. p 1 - 13.
9. Pokar, M and Loke, M.H. (1998) Electrical Tomographic Survey of a Landfill Site. Proceedings of the Environmental and Engineering Geophysical Society (European Section). Pp 123-126. Barcelona.

The Echline road construction project using sustainable design and construction techniques

Dr HL Robinson[(*)]**, Dr N Ghazireh and Mr H Jeffrey-Wright**
Tarmac Ltd, Technical Services, Wolverhampton, UK
(*) Visiting Professor at Liverpool John Moores University

ABSTRACT
The Scottish Executive Development Department, Trunk Roads - Design & Construction Division have funded the construction and testing of a trial section of experimental road pavement at Echline, South Queensferry, Edinburgh. The trial, completed in April 2000, had a clear objective to maximise the use of sustainable, in-situ, recycled and cold lay materials and minimise the overall construction energy requirements. The trial site will not receive traffic in the short to medium term, however the project has demonstrated that all the materials used can be manufactured and laid in a controlled manner. Over the first year the in-situ performance has been monitored with encouraging results, in keeping with that which would be expected based on previous experience. This paper will review the materials used in a flexible composite pavement design, the material production and pavement construction process and outline the testing regime in place to validate performance.

1. INTRODUCTION

Long-term projections of aggregate demand indicate that construction in the UK may need up to 7.3 – 7.9 billion tonnes over the next 15 years [17]. It is therefore important that the best use is made of all the nation's resources including a greater use of waste and recycled materials as well as more efficient use of primary aggregates. Waste and recycled materials already account for about 10% of the aggregates used in the UK and it is Government policy to increase this level of usage where this furthers the aims of materials conservation and environmental protection.

For most secondary materials the principal constraint associated with increasing use is the cost of transport from source to the end use location and also customer acceptance which is related to material performance and the need to maintain necessary standards of safety and quality assurance. To address these issues the Scottish Executive Development Department, Trunk Roads – Design and Construction Division have commissioned Tarmac Limited (then called Tarmac Quarry Products) to design, construct and assess the long term performance of a trial section of experimental road pavement at Echline located at South Queensferry in Edinburgh.

This paper summarises the innovative road design, the description and production of the proposed materials. A two-stage construction process was adopted in order to reduce the effect of severe weather conditions, particularly frost on the new mixes of recycled and low energy materials.

Geoenvironmental impact management, Thomas Telford, London, 2001.

2.   DESIGN PROPOSAL

The research trial consisted of an area 150metres long by 6metres wide with a 100metre long experimental section sandwiched between two turning areas constructed by conventional methods.   The technical design proposal submitted at tender stage addressed the following main criteria:

- Maximum use of recycled materials
- Maximum energy savings in road production
- Maximum conservation of primary materials
- Recycled materials: availability and sustainability
- Overall environmental benefit
- Road performance and durability

Whilst complying with the above criteria the trial section was designed to take a 100msa loading based on a design life of 40 years, although it is not expected to receive any traffic in the short to medium term.   A flexible composite pavement using by-product and recycled materials was proposed.

At design stage both short-term and long-term requirements have been considered as crucial to the pavement structure.   The short-term is mainly to ensure that each layer after construction will act as a platform for the construction of the next layer.   The long-term is to ensure the durability and performance of the road over its service life.   Considering the design constraints and the project requirements, a design has been proposed which optimises sustainable design and construction techniques.   The Transport Research Laboratory (TRL) in Scotland, independently justified the proposal.   The general layout of the design is illustrated in Figure 1.

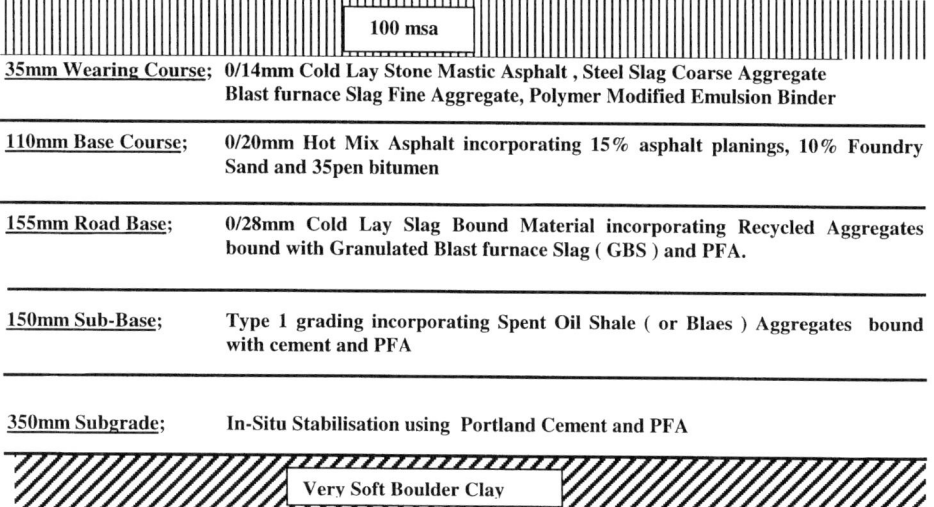

35mm Wearing Course;	**0/14mm Cold Lay Stone Mastic Asphalt , Steel Slag Coarse Aggregate Blast furnace Slag Fine Aggregate, Polymer Modified Emulsion Binder**
110mm Base Course;	**0/20mm Hot Mix Asphalt incorporating 15% asphalt planings, 10% Foundry Sand and 35pen bitumen**
155mm Road Base;	**0/28mm Cold Lay Slag Bound Material incorporating Recycled Aggregates bound with Granulated Blast furnace Slag (GBS) and PFA.**
150mm Sub-Base;	**Type 1 grading incorporating Spent Oil Shale (or Blaes) Aggregates bound with cement and PFA**
350mm Subgrade;	**In-Situ Stabilisation using Portland Cement and PFA**

note ; the overall pavement thickness (450mm) plus in-situ stabilisation of the sub-grade is considered to be adequate to protect the soil from frost damage, allowing for frost penetration of 500mm at worst in Edinburgh.

Figure 1. Pavement Design Layout

3.   MATERIALS RATIONALE
3.1   Wearing Course
The use of cold lay asphalt is established in France, Eire and Scandinavia with evidence of successful performance over the last 15 years [1], however cold lay emulsion asphalt is still relatively new to UK. In recent years, Tarmac has pioneered the development of a cold lay Stone Mastic Asphalt in association with Nynas UK [2,3]. A three year research programme culminated in a specification trial on the A52 at Rue Hill in Staffordshire in August 1997. Up until the site was inadvertently surface dressed in 2000, the material continued to perform well showing good texture retention and maintained stiffness, not dissimilar to hot lay SMA type materials. This despite the fact, cold lay materials require a curing period of typically 12 months to develop full stiffness, although experience shows they are capable of withstanding immediate trafficking. Cold lay wearing course is to a degree a seasonal product and should ideally be laid between March and September, however experience has shown that it can be laid in December and still perform adequately. The 14mm SMA Coldpave material laid at Echline is similar to that laid on the A52 except it used basic oxygen steel slag (BOS) as the coarse aggregate and blast furnace slag (BFS) as the fine aggregate. The UK steel industry generates 1.2 million tonnes per annum of steel slag at four sites; Teesside, Scunthorpe, Newport Gwent and Port Talbot. Some of this is used in asphalt or as surface dressing chippings , however the bulk of it is used in engineering projects around the steel works or dumped. Steel slag has reported polished stone values in the range 55 - 65, AAV results below 3 and flakiness less than 10, hence it is potentially a valuable sustainable resource of premium wearing course aggregate. Blast furnace slag (BFS) is also produced at the same four works with an annual output of 4 million tonnes most of which is used as a construction aggregate, regulated by BS 1047. BFS dust has been used in the SMA to counter the high density of the steel slag (typically 3.3) rendering the end product competitive in terms of its coverage rates.

The cold lay SMA was manufactured at Tarmac's Darlton plant in Derbyshire and transported northwards to the trial site. This was only for expediency since there was insufficient time to commission a local supply unit capable of supplying the required tonnage. Both steel slag and blast furnace slag are referred to in DMRB Vol.7 as being suitable for most pavement layers [4].

3.2   Base Course
The base course comprised a hot mix asphalt 20mm material complying with BS 4987 grading and binder content, incorporating 15% recycled asphalt planings, 10% foundry sand, and 50pen bitumen. The addition of recycled asphalt planings to hot mixes is established in other parts of the world, particularly the USA. Although it is allowed in UK, it is still not considered to be common practice.

The main British Standards make no reference to recycling because they are currently frozen in time during the development of European Standards, which will include extensive provision for recycling. The definitive UK guidance is contained in the Specification for Highway works. Provisions for monitoring compliance with the SHW requirements are included in the National Quality Assurance Scheme. The definitive statement in SHW is in clause 902.1: ' Reclaimed bituminous materials may be used in the production of bituminous roadbase, base course and wearing course'. This is qualified as subject to a number of detailed restrictions on HRA and dense base materials. The base course complied with the grading and binder content of the specification. A number of publications have discussed the use of recycled aggregates in asphalt [5] indicating end performance to be indistinguishable from materials using virgin aggregate. Tarmac as the industry's largest supplier of aggregates has taken a leading role in promoting the use of recycled materials in partnership with local authorities. In particular Tarmac has been involved in the reuse of foundry sand in asphalt which has been the subject of DETR development funding [6,7,8].

Tarmac's Highcraig plant produced the basecourse material. The recycled aggregates came from Tarmac's recently commissioned recycling depot at Addiewell, West Lothian. Foundry sand was supplied locally.

3.3   Roadbase
The roadbase comprised of a 0/28mm Slag Bound Material (SBM) incorporating recycled aggregates

Slag Bound Material (SBM) or otherwise known as *Grave-Laitier*, has been in use in France, Holland, Germany, Italy and Belgium for over 25 years. It's use in UK has to date been limited to occasional trials [9], however confidence in its performance has grown in the last few years to the extent where it will be included in the next amendment of the Specification For Highway Works. It is essentially a slow cementing 0/28mm continuously graded roadbase material, containing typically 15% granulated blast furnace slag (GBS) as the binder. The binder needs to be activated by a lime based catalyst present at around 1%. SBM behaves as an unbound granular material immediately after production, facilitating ease of handling and compaction. Strength gain is 70% complete after 3months and fully cured at 1 year, producing compressive strengths of 10 -15N ie equivalent to CBM3. It is important to recognise however that SBM materials are not prone to reflective cracking to the extent found with cement bound materials. This is attributed to their slow cementing behaviour ie micro cracks that form during the materials early life self heal unlike CBM materials where they can propagate leading to cracking. However, for the SBM layer transverse pre-cracking has been installed at 3m intervals using a vibrating plate with a blade welded on to the bottom. The joint was backfilled with bitumen emulsion prior to rolling.

The binder in SBM can be either GBS or a blend of GBS with PFA. Any suitable aggregate source can be used. The GBS chemistry is not dissimilar to cement and is remarkably consistent, in fact controlling the slag composition provides an indirect means of quality control on the iron production.

Today, SBM has been used to surface over 7000km of French national roads and is embodied in French design specifications covering the full range of traffic categories.

SBM can be produced on conventional readymix plants, asphalt plants or using mobile continuous mixers employed for making CBM or foamix type materials. Quality control focus is primarily on moisture content and grading to facilitate good compaction and hydration. The material is laid using conventional plant however a pneumatic tyred roller is preferred. Ultimate performance is affected by GBS reactivity and content, the catalyst type used and degree of compaction. Over recent years trials in UK have demonstrated good performance [9,10] culminating in March 1999 with a specification trial on the A485 Carmarthen by-pass link road supported by The Highways Agency and The Welsh Office. The trial is being monitored by TRL on behalf of the Highways Agency with 90 day core strengths indicating performance equivalent to CBM3 material.

Evidence from UK trials [11,12] suggests the material has sufficient load bearing capacity derived from good aggregate interlock to withstand early traffic loading. The post production shelf-life (typically 5days), provides logistical benefits and helps minimise waste. This is perceived to be one of the products key benefits compared to concrete which has to be placed within two hours of mixing and cannot be trafficked for seven days. Slag Bound Materials are also well suited to weak sub grades, able to adjust to the changing profile of the sub grade as it deforms under loading.

A European standard for Slag Bound Mixtures is nearing completion [13].

SBM also enables savings in carbon dioxide emissions associated with asphalt production ie 9 cubic metres per tonne.

3.4   Sub-Base
The sub-base comprised of locally available spent oil shale aggregate stabilised with 10% pfa and 5% Portland Cement to produce a type 1 graded material with a CBR > 30. Approximately 100 million tonnes of spent oil shale is available in the West Lothian area of Scotland. Recent research by TRL [14] suggests spent oil shale will provide an adequate sub-base particularly if stabilised using a hydraulic binder.

Burns (1978)[15] indicates that the addition of 5% or more of cement to spent oil shale sub-bases reduces the frost susceptibility to acceptable levels. Burns estimated that 0.5 to 1 Mt of pre-mixed spent oil shale and cement was used on the M8 and M9 motorways as sub-base. No reported failures due to heave of these material have been published or known. It is understood that the bings at Philpstoun were used for the M9 and those in the Uphall/Pumpherston area for the M8. Around 7% of cement was added to ensure a CBR >30%. The addition of PFA/Cement to the spent oil shale will reduce the frost susceptibility of this material to acceptable levels once adequate hydration has taken place.

3.5   In-Situ Stabilisation
The trial site ground condition indicated a weak naturally occurring foundation layer ie <5% CBR, hence locally available pfa/PC was used to stabilise the subgrade in-situ down to a depth of 350mm. The practice of stabilising soil with quicklime to form material suitable for use as capping and bulk fills has been common in the USA and Europe for decades. The Highways Agency also recognise the benefits of using this technique and included lime stabilisation of soils in the specification for Highway Works in 1986. To date its use in the UK has been sluggish. With the introduction of landfill tax, however, it's use is expected to grow. Soil stabilisation is primarily a quality controlled mix-in-place and compact technique that removes the need to import capping materials to site and also reduces transport movement on the public roads to and from the site.

4.   CONCLUSIONS
The road was successfully constructed with the wearing course containing 94% recycled aggregates, the base course 25% recycled aggregates, the roadbase 100% recycled aggregates / binder, whilst the sub-base contains 95% recycled aggregates.

The testing regime is designed to provide a strength development profile for all of the slow curing materials. The test data will enable the in-situ performance for each material to be cross referenced to similar materials laid elsewhere with known performance levels.

All of the materials are available in sufficient annual quantities to sustain the long term useage of each mixture. Table 1 summarises the availability of the secondary materials used in the construction.

Table 1: Available Tonnage of the Secondary Materials Used in the Construction

Material	Annual Production approx. tonnes (millions)	
Blast furnace Slag - air cooled & granulated	4.0	*source: BSC*
Steel Slag	1.2	*source:BSC*
Foundry Sand	1.3	*source: Castings Development Centre*
Recycled Asphalt Planings	> 1.0	*source:QPA*
PFA	> 5.0	*source: ScotAsh*
Spent Oil Shale	100.0	*stockpiled*

5. REFERENCES

1. 'Cold-lay asphalt thin surfacing', DETR Future Practice Report 85, January 1999

2. 'Cold-lay asphalt thin surfacing, An alternative solution for road maintenance', DETR Future Practice Profile 85, Energy Efficiency Best Practice Programme, January 1999.

3. H.L.Robinson, 'The Cold Road Ahead - Latest maintenance solutions may be cold rather than hot', Quarry Management, January 1999.

4. Conservation and the use of reclaimed materials in road construction and maintenance, DMRB Volume 7 , Section 1 , Part 2 HD 35/95.

5. J.F.Potter (TRL), J.Mercer (HA),'Full - Scale Performance Trials and Accelerated Testing of Hot-Mix Recycling in the UK, 8th International Conference on Asphalt Pavements, Seattle, Aug'97.

6.'Waste Sand Sets Scene For Profitable Partnership', DTI/DETR New Practice Case Study 95 at Precision Disc Castings Ltd and Tarmac Quarry Products Ltd, Environmental Technology Best Practice Programme.

7.J.G.Morley, 'External Recycling Of Foundry Waste Products', Castings Development Centre.

8. A.R.Hill, A.R.Dawson, 'Benefits and Difficulties of Using Foundry Sands and Other Secondary Materials in Road Construction', Foundry By-Products Conference, November 1997.

9. J.Kennedy,' Slag - Bound Mixtures, Use and Performance in Europe but particularly the UK', SCI Seminar, London, April 1998.

10. 'Use of Slag - bound material for roadbase construction, DETR New Practice Case Study 111, Energy Efficiency Best Practice Programme, January 1999.

11. H.L.Robinson,' Slag Bound Material (SBM) - For Flexible Composite Pavement Design', Proc.3rd European Symposium, Performance and Durability of Bituminous Materials and Hydraulic Stabilised Composites, Leeds, April 1999 pp 303-310.

12. H.L.Robinson, P.A.B.Acock, 'Low Energy Materials For Trench Reinstatement', Proc.3rd European Symposium, Performance and Durability of Bituminous Materials and Hydraulic Stabilised Composites, Leeds, April 1999 pp 465-476. 13. European Standard For Slag Bound Mixtures, Draft 19th,Pr EN 227 402, November 1998.

13. European Standard For Slag Bound Mixtures, Draft 19th,Pr EN 227 402, November 1998.

14. M.G.Winter, 'The Use of Spent Oil Shale in Earthwork Construction', Proc.2nd BGS Geo-environmental Conference, Thomas Telford, London, Sept 1999.

15. J.Burns (1978), 'The Use of Waste and Low Grade Materials in Road Construction: 6. Spent Oil Shale'.   TRRL Laboratory Report LR 818, Crowthorne: Transport Research Laboratory.

16. A Good Practice Guide on Energy Minimisation in Road Pavement Construction and Maintenance, Energy Efficiency Office DOE January 1997.

17. Department of the Environment (1994).'Sustainable Development: The UK Strategy'. DoE Command Paper CM2426, HMSO, London

The Long-Term Leaching Behaviour of a Pulverised Fuel Ash Mound

P. G. STUDDS* &   M. CROSS[#]

*Waterman Environmental   # W S Atkins, Leeds

Bradshaw House   The Calls

31 Waterloo Lane   Leeds UK LS2 7ES

Bramley

Leeds UK LS13 2JB

ABSTRACT: The long-term leaching behaviour of a pulverised fuel ash (PFA) disposal mound at a British coal-fired power station was investigated. Samples of weathered PFA, taken from boreholes extending through a PFA mound were taken from locations representing different constructional age bands within the mound (1976, 1980 & 1985). Small-disturbed samples taken at 1m depth intervals were analysed for natural moisture content and a range of chemical determinands. Permeability behaviour with change in depth from the mound surface was also investigated using undisturbed samples. Generally, the moisture content and degree of saturation were fairly constant with increasing depth within the mound and did not differ significantly from those at initial deposition and compaction. The permeability was relatively constant with depth, ranging from 1.23×10^{-7} ms^{-1} to 1.45×10^{-8} ms^{-1}, with a mean permeability of 5.43×10^{-8} ms^{-1}. There was no evidence of a saturated zone or potential wetting front within the mound. Arsenic, sulphur and boron were significant in relation to their toxicity and concentrations. Concentrations of boron, molybdenum, potassium and sodium showed a slight increase with depth, suggesting downward migration of leachate derived from PFA. Concentrations of all determinands within the silty clay located below the constructed drainage blanket beneath the mound were below the Environment Agency threshold levels. PFA leachate concentrations did not constitute an environmental hazard to groundwater beneath the mound.

key words: pulverised fuel ash (PFA), coal-fired power station, permeability; landfill; leachability, ground water contamination.

INTRODUCTION

Pulverised fuel ash (PFA) is formed during the combustion of coal in coal-fired power stations comprises mainly colourless, glassy spherical particles in the fine-sand-silt size range (Cabrera et al 1886). While much of the PFA produced in the UK is used for engineering purposes it is also often disposed to landfill. The disposal of PFA waste from Drax coal-fired power station (Selby, Yorkshire, National Grid Reference SE 655 277) commenced in 1974. Barlow Ash Disposal Mound, located at the north-east side of the power station, is roughly rectangular in shape (0.8 x 1.8 km) with an area of approximately 100 ha and a maximum height of 36.5 m above OS datum.

PFA has been recognised as a potential source of environmental contaminants due to the chemical composition of the leachate formed when exposed to rainwater (Ainsworth and Rai 1987; Fruchter et al. 1988, 1990; Eary et al. 1990; Mattigod et al. 1990). However, there have been few field based studies on PFA and these have mainly been concerned with PFA lagoons where leachates are characterised by rapid reactions of the soluble fractions of the surface associated elements in the PFA. Little information is available from field sites on the long-term weathering of PFA, which may be due in part to the difficulties of monitoring ash mounds (Carlson and Adriano, 1993). Studies of ash pond effluents show similar results to those of batch-column leaching tests, with Ca and S (present as SO_4^{2-}) as principle cationic and anionic constituents (Dreesen *et al.* 1977; Talbot *et al.* 1978; Simsiman *et al.* 1987). Groundwater samples bordering a PFA settling ponds showed increased concentrations of Ca, K, Fe, and SO_4^{2-} for the major elements, and As, B, Mn. Mo, Ni, Sr and Zn for the trace elements (Hardy 1981). Concentrations of As, B, Mo, and Se were also found to be elevated in effluent water from a PFA pond by Dreesen et al. (1977).

Simsiman et al. (1987) monitored a PFA mound for three years and observed large B, Na, and SO_4^{2-} plumes in the adjacent groundwater system.. Sakata (1988) reported Ca and SO_4^{2-} as major ions in the leachate of weathered PFA from ash mounds and also showed inorganic elements had infiltrated into the underlying soil. In addition to high concentrations of B Le-Seur and Drake (1987) detected As and Se over analytical detection limits, and Se exceeding 0.1 mgl^{-1} in the groundwater samples from shallow monitoring wells within a PFA landfill. The infiltration of As leachate into underlying soil and elevated concentrations of B, Mo, Mn and Sr in shallow groundwater were also reported by Rehage and Holcombe (1990).

UK based studies on the long term leaching of PFA under natural conditions at Barlow ash mound have been carried out by Lee and Spears (1995). Their research examined weathering reactions in the PFA based on the composition of porewaters and from the geochemistry and mineralogy of the PFA. The field investigations undertaken by Lee and Spears were limited and only concentrated on the top 5.0m of the ash mound.

This study follows on from the research carried out by Lee and Spears (1995) and examines the long term leaching behaviour of PFA in the field under natural weathering conditions. The investigation techniques employed aim to provide data on the chemical characteristics of leachate derived throughout the total depth profile of a PFA mound. The particular objectives of the investigation were to determine:

- the changing physical and chemical characteristics of PFA throughout the total depth within the mound;
- the contamination levels of leachate originating from various layers of PFA of different age deposited in the mound; and
- the potential impact on environmental waters, and groundwater in particular.

METHODS
Three 150mm diameter cable percussion boreholes were placed through the PFA mound, the furnace bottom ash (FBA) drainage blanket and into the superficial deposits below comprising glacial silty clays. Borehole 1 was bored to 26m below existing ground level (begl) and was located in part of the mound constructed in 1985-86, borehole 2 was bored to 23m begl and was located in part of the mound constructed in 1980-81 and borehole 3 was bored to 15.5m begl and was located in part of the mound constructed in 1976.

Small disturbed samples of PFA were taken at depth intervals of 1m from each borehole. Undisturbed (U100) samples were collected where appropriate. In total, 9 No. U100 samples and 23 No. small disturbed samples were collected from BH1, 6 No. U100 samples and 16 No. small disturbed samples from BH2 and 5 No. U100 samples and 15 No. small disturbed samples from BH3.

Moisture Content
Specimens were tested using the method specified in BS 1377: Part 2, 1990, Determination of Moisture Content.

Permeability Tests
Two methods were used to investigate the permeability of the PFA. The first method indirectly calculated the permeability of the PFA from the consolidation response upon application of a load increment. The second method was direct measurement using a triaxial cell constant head test.

Permeability Data from Consolidation tests (indirect measurement)
If a soil is laterally confined, fully saturated and the drainage and compression are one dimensional, the soil volume will decrease according to Terzaghi's theory (1943). The amount of settlement due to primary consolidation can be found using the coefficient of volume compressibility, m_v, and the time taken for a given percentage of primary consolidation by using the coefficient of consolidation, c_v. These two parameters can be used to give an estimate of the permeability of the soil. One-dimensional consolidation tests were performed on 10 No. undisturbed samples with specimen dimensions of approximately 75mm diameter by 20mm thick. Specimens were tested using the method specified in BS 1377: Part 5, 1990, Clause 3.

Permeability Data from Triaxial Cell Tests (direct measurement)
Six undisturbed U100 samples were collected from mound, extruded, trimmed and placed in the triaxial cell. Specimen dimensions were approximately 100mm diameter by 98mm long. Specimens were tested using the method specified in BS 1377: Part 6, 1990, Test 6 - Constant Head Test. A flexible wall type permeameter was chosen to reduce sidewall leakage. Permeability was measured when the inflow rate and outflow rates were within 5% of each other.

Leachability Tests
Leachability tests by extraction to DIN 38414 – S4 were carried out on 16 No PFA samples. Samples were selected for leachability testing from various depths within the PFA, representing different moisture contents found within the PFA depth profile. The Environment Agency approved method (NRA, 1994) was used for the leachate preparation, this involves mixing 1 part soil to 10 parts water. Leachability tests were also carried out on specimens collected from within the FBA drainage layer and from the natural silty clay located beneath the drainage blanket in each borehole.

Groundwater Analysis
Groundwater, sampled from the base of the mound in the FBA drainage blanket in BH1 and BH2, was analysed for a range of chemical determinands.

RESULTS
Moisture Content
Fig. 1 shows the results of percentage moisture content plotted against depth for samples taken from boreholes 1 to 3. The data displays an overall trend of slightly increasing moisture content with depth. As expected, higher moisture contents were found at depths relating to the FBA drainage layer at the base of the mound. The maximum water content recorded was 40%, although readings from samples of granular material with free draining water are likely to be subjective. Within the silty clay just below the drainage blanket, the moisture content dropped to an average of 24%.

The moisture content in PFA sampled from Borehole 1 increased from 11 to 16% over the first 14m into the mound. PFA is placed at the conditioned moisture content of 14-15%, which is similar to the moisture contents recorded in the first 14m of the mound. The slightly lower moisture contents at the surface of the mound are probably related to evapotranspiration processes and natural drainage in near surface horizons. A wetter layer, with moisture content of 20%, was detected at a depth of 17m. The water content increased below 20m corresponding to gravel material and eventually FBA observed below this depth. The mean moisture content through the PFA at BH1 was 14.6%.

The moisture content profile in Borehole 2 was more variable, ranging between 10 and 17%. The gravel material present below a depth of 20m was found to have a higher moisture content of 21%. The mean moisture content through the PFA at BH2 was 13.0%.

The moisture content profile in Borehole 3 was more variable, ranging between 13 and 19%. The drainage blanket in BH3 was quite definitive at a depth of 11.8 to 13.2m, which is reflected by the sharp increase in water content to 40%. The mean moisture content through the PFA at BH3 was 14.9%.

Permeability Data from Consolidation tests (indirect measurement)
Permeability data obtained from consolidation tests are given in Fig. 2. The permeability of the PFA was relatively constant ranging from 1.23×10^{-7} ms^{-1} to 1.45×10^{-8} ms^{-1}, with a mean permeability of 5.43×10^{-8} ms^{-1}.

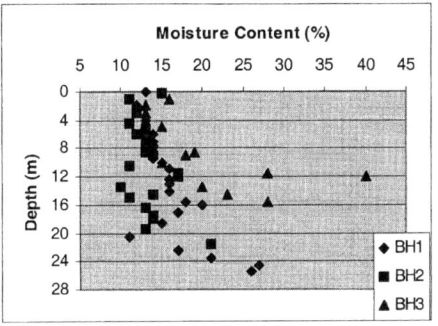

Figure 1 Moisture Content Profile through   Figure 2 Permeability against Depth
PFA Mound   (Data from all Boreholes)

Permeability Data from Triaxial Cell Tests (direct measurement)
Permeability data obtained from triaxial cell tests are given in Fig. 2. Generally the coefficient of permeability was approximately constant after five days of testing. The permeability of the PFA was relatively constant ranging from 9.5×10^{-8} ms^{-1} to 4.0×10^{-8} ms^{-1}, with a mean permeability of 7.25×10^{-8} ms^{-1}. From the data presented in Fig. 2 there appears to be a slight decrease in permeability with increasing depth. Such a trend would be expected in a PFA mound because the degree of consolidation would be greater in the deeper deposits and hence the size of the flow channels would be less. There does not appear to be any correlation between the age of the PFA and permeability.

Leachability Tests
The results of the leachability tests carried out on PFA samples from the three boreholes are summarised in Figs.3 to 9. The results of the analysis of leachate samples from PFA taken from BH1 are summarised in Table 1. On comparison with the Environment Agency's leachate quality threshold concentrations (NRA, 1994), S and As levels were found to exceed the threshold concentrations in most of the PFA leachate samples. This trend was observed in samples tested from all three boreholes.

Table 1. Chemical analyses of leachate from PFA sampled from BH1 (unit mg/l)

Depth	pH	Cl	S	Ca	Na	K	Mn	As	B	Cd	Cr	Cu	Pb	Zn	Se	Mo
EA Threshold	5.5 - 9.5	200	150					0.01	2	0.001	0.05	0.02	0.05	0.5	0.01	
1.0	7.8	<0.5	90	122	<0.1	1.5	<0.01	0.1	0.43	<0.005	<0.01	<0.01	<0.01	0.01	<0.01	0.01
2.0	8	<0.5	228	279	<0.1	3.6	<0.01	0.1	0.98	<0.005	<0.01	<0.01	<0.01	0.01	<0.01	0.06
3.0	9.1	<0.5	253	304	0.17	6.3	<0.01	0.07	3.4	<0.005	<0.01	<0.01	<0.01	<0.01	<0.01	0.1
6.5	9.1	<0.5	206	243	10	22	<0.01	0.07	1.9	<0.005	0.01	<0.01	<0.01	<0.01	<0.01	0.12
17.8	9.3	<0.5	240	266	27	25	0.05	0.04	1.8	<0.005	0.05	<0.01	<0.01	0.03	<0.01	0.69
23.5	8.5	<0.5	8.5	20	4.3	2.8	0.08	0.18	<0.01	<0.005	0.01	<0.01	<0.01	0.06	<0.01	0.03
24.6	8.6	<0.5	7.3	16	3.2	0.61	0.02	0.08	<0.01	<0.005	<0.01	<0.01	<0.01	<0.01	<0.01	0.04
25.5	8.3	<0.5	2.2	11	5.3	0.7	<0.01	0.03	<0.01	<0.005	<0.01	<0.01	<0.01	<0.01	<0.01	<0.01

Note: Shaded boxes denote concentration above EA leachate quality threshold.

For each element, graphs were plotted to show the variation of leached determinand concentration (mg/l) with depth. For several of the major species found in PFA, a trend was observed between leachate concentration and depth.

- The leached As profile on Figure 3 showed a trend of decreasing concentration with depth apart from an increase in concentration from PFA sampled from BH2 at a depth of 21.5m bgl.
- The concentration of B increased rapidly with depth within the first 3m of BH1, and then decreased with depth below 3m. Extremely low levels of B were detected in the samples from the drainage layer and the underlying silty clay (Fig. 4). Boron was found in most of the PFA samples and exceeded the EA threshold value in one sample.
- The Ca concentration increased over the first 3m, and then remained relatively constant with depth through the PFA (Fig. 5).
- Potassium levels generally increased rapidly with depth in all three boreholes. Concentrations appeared to become more constant below a depth of about 6.5m in BH1 and 11.5m in BH2. Extremely low levels of K were measured in the drainage layer and underlying silty clay (Fig. 6).
- Sodium concentrations in the PFA were found to increase with depth in Boreholes 1 and 2. However, an increase above the detection limits was not actually observed until a depth of 3m in BH1, and below 2.5m in BH2. Levels of Na in the FBA and silty clay were low (Fig. 7).
- The leached S profile with depth is shown in Fig. 8. The S concentration increased over the first 3m, and then remained relatively constant with depth through the PFA.
- Molybdenum levels appeared to increase with depth through the PFA in BH1 and BH2 (Fig. 9).

The trends observed for Mo, B, K and Na are not conclusive, but appear to indicate that substances are being leached from PFA in the upper layers and translocated downwards as pore-water slowly infiltrates under gravity.

Concentrations of S, As and B were found to be below the EA thresholds in the samples of silty clay located below the FBA drainage blanket. Levels of all other determinands tested fell below the EA thresholds. The age of the PFA sampled from the mound does not appear to significantly effect the concentration of leachate produced. Typically the leachate concentration was similar for all three age bands tested.

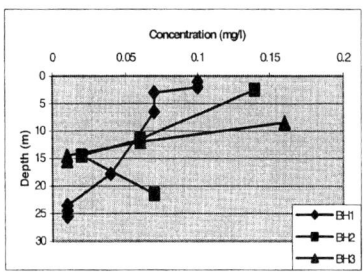

Figure 3 Arsenic Profile

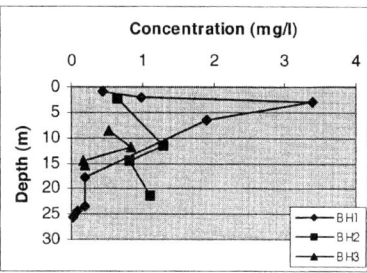

Figure 4 Boron Profile

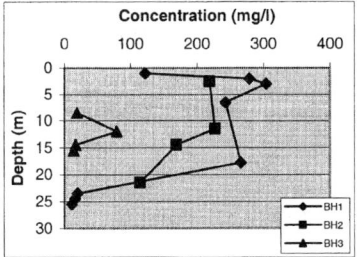

Figure 5 Calcium Profile

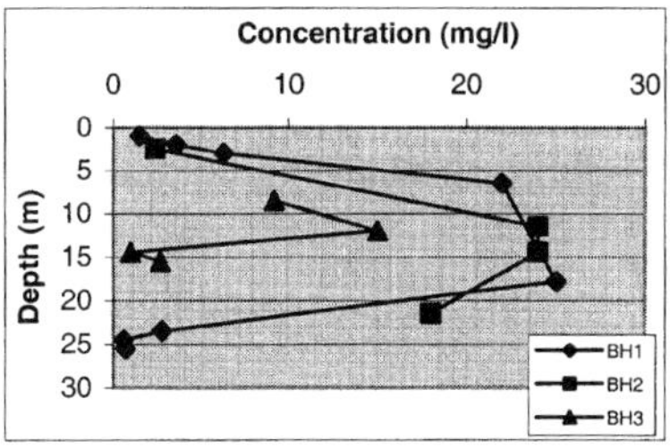

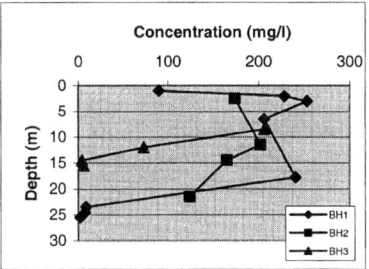

Figure 6 Potassium Profile

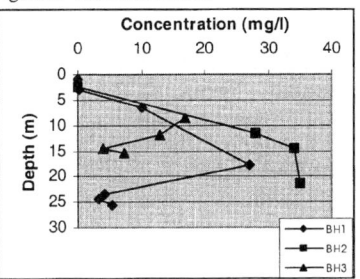

Figure 7 Sodium Profile

Figure 8 Sulphur Profile

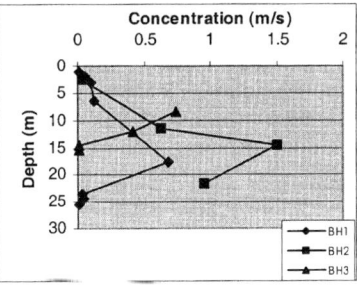

Figure 9 Molybdenum Profile

Groundwater Analysis

The results of the groundwater analysis are shown in Table 2. Contaminant concentrations were compared to the UK Drinking Water Standards as defined in The Water Supply (Water Quality) Regulations 1989. Concentrations of B and S exceeding the UK Drinking Water Standards were recorded in the groundwater.

Table 2. Groundwater Analysis units (μg/l)

Borehole	pH	SO₄	As	B	Cd	Cr	Cu	Pb	Ni	Zn	Se	Hg
UK Water Standards	5.5 - 9.5	250 mg/l	50	2000	5	50	3000	50	50	5000	10	1
BH1	8.1	1710	<0.75	4660	<1	<10	<10	<10	<10	<20	5.44	<0.1
BH2	7.4	947	2.79	994	<1	<10	<10	<10	<10	<20	<0.75	<0.1

Note: Shaded boxes denote concentration above UK Drinking Water Standards.

DISCUSSION

No evidence was found for the presence of a saturated zone or potential wetting front within the PFA ash mound. Typically, the moisture content increased slightly with depth. However, the highest moisture content recorded in the PFA (20%, BH1, 16m bgl) was still only slightly above the deposited PFA moisture content of 14-15% even though it had been there for at least 15 years. The mean moisture contents of the PFA sampled from boreholes 1, 2 and 3 were 14.6%, 13.0% and 14.9% respectively. Again these moisture contents are similar to the deposited PFA moisture content and suggest that the moisture content of the PFA does not change significantly with the age of the PFA. These results concur with Mitchell (1999) who observed that the degree of saturation within Barlow ash mound did not change significantly over time and concluded that there were very low flow rates of precipitated water through the mound.

Coefficients of permeability for PFA ranging from 1×10^{-6} to 1×10^{-7} ms^{-1} have been reported (Brown and Owens, 1984; Havukaine, 1987; Studds and Cousens, 1996). Slightly lower values of the order of 5.0 $\times 10^{-8}$ ms^{-1} were measured both by direct and indirect measurement. This corresponds to British Standards Institution (BS 8004, 1986) classification of 'Very low' permeability. Due to the low coefficients of permeability recorded it would be expected that the volume of water permeating through the mound would be low. Mitchell (1999) suggested that movement of water within the mound would be generated by pressure gradients associated with differing degrees of suction.

The results of the leachate analysis were compared with the findings of Lee and Spears (1995) who carried out similar analyses on samples of Drax PFA. Lee and Spears (1995) used a hand auger to obtain samples from four shallow boreholes on the Barlow ash mound bored to a maximum depth of 5.0m. Two of the boreholes (BH2 and BH3) were located in PFA deposited in 1978 but differ on slope position on the mound. The other two boreholes (BH1 and BH4) were from the oldest part of the mound where the PFA was placed in 1975. The PFA samples were collected at 0.3m depth intervals and were stored in sealed plastic bags in the refrigerator before the porewaters were extracted. Samples were centrifuged to extract porewaters. Samples were extracted at 3000 r.p.m. over a period of 60 minutes using the method of Edmunds and Bath (1976). In the laboratory, a Perkin Elmer M2100 atomic absorption spectrophotometer (AAS0 and Phillips PV8210 1.5m ICP-AES) was used to determine concentrations of cations in the porewater samples.

The conclusions of the PFA leachate testing carried out by Lee and Spears (1995) are provided below.

1. Nearly all cations and anions in porewaters from the Barlow ash mound show depth related trends. The fact that the concentrations mainly increase with depth indicates that the PFA is reactive and elements in the PFA are leached in contact with water. Confirmation of depth trends was obtained from changes in bulk composition of the PFA.
2. Significant concentrations of major and trace elements are present in porewaters. Among the elements in the leachate, B, Mo, As and Se might be of more concern in terms of their toxicity and concentrations.
3. Some clay minerals, including kaolinite and illite, were identified in weathered PFA from different depths in the Barlow ash mound. However, these clays were also detected in the imported topsoil covering the PFA. Although translocation of clay minerals has taken place some *in situ* formation of kaolinite within the PFA cannot be precluded.
4. Scanning electron microscope (SEM) analysis showed that no substantial changes were found in the surface texture of weathered PFA, except for the encrusted surface on the weathered ash.
5. Ca and SO_4^{2-} are the two main species in the porewaters and for most samples equilibrium calculations indicate saturation with respect to gypsum. Evidence of equilibrium in the porewater is provided by the attainment of constant concentrations in depth profiles.

It should be noted that Lee and Spears (1995) carried out chemical analyses on PFA samples collected from only the top 5.0m of the mound. The chemical analyses presented in this paper include PFA materials sampled throughout the entire depth within the mound. In relation to Lee and Spears conclusion 1, from the analyses presented in this paper the concentrations of all cations and anions do not all increase in concentration with depth to the base of the PFA mound. The concentration of B increased with depth in the first 3.0m of BH1 and then decreased in concentration with depth below 3.0m. Potassium levels increased with depth in all boreholes to approximately 6.5m bgl.; below this depth concentrations became more constant. Other substances detected in measurable concentrations, namely Ca and S, increased over the first 3.0m, and then remained relatively constant with depth through the PFA. The leachate concentrations of As decreased with depth in all three boreholes.

In relation to conclusion 2, only As, S and in one instance B (BH1, 3.0m) exceeded the EA leachate quality threshold.

No testing was carried out for the presence of clay minerals in this investigation. Scanning Electron Microscope (SEM) work undertaken by Faulkner (1998) on samples of near surface PFA samples did not find evidence for the presence of clay minerals. Faulkner examined Drax PFA and confirmed the presence of cemented PFA associated with pozzolanic activity and the corresponding formation of calcium hydroxides. Calcium hydroxides, which are low in solubility, would act as a cement, binding the PFA spheres together causing a reduction in pore space. This process may cause an increase in the strength of the PFA and possibly reduce the permeability of near surface PFA. Older PFA samples from the Barlow ash mound were shown to contain greater amounts of calcium hydroxide than younger PFA samples. Although a corresponding decrease in permeability with age of the PFA was not observed during this study.

CONCLUSIONS
- No evidence was found for the presence of a saturated zone or potential wetting front within the PFA ash mound.
- Environment Agency leachate quality threshold concentrations for sulphur, arsenic and boron were exceeded.
- Concentrations of boron, molybdenum, potassium and sodium generally showed a slight increase with depth, suggesting downward migration of leachate derived from PFA.
- Concentrations of boron and sulphate in groundwater from the drainage layer were found to exceed UK drinking water limits.
- The FBA drainage layer and associated perimeter cut-off drain around the mound appeared to be effective in reducing leachate movement downward into the silty clay.
- Concentrations of all determinands were below the Environment Agency threshold levels in the silty clay.

Acknowledgements: The authors would like to thank Mr. P. Faulkner and Mr. D. Mitchell for their permission to report some of their results and Mr. I. Fenton and Mr. R. Coombs of National Power for their assistance. The opinions expressed are those of the authors.

REFERENCES
Ainsworth, C. C. and Rai, D. 1987. *Chemical characterisation of fossil fuel wastes.* Electric Power Research Institute, Palo Alto, CA, EPRI Report EA-5321.

British Standards Institution, 1990. BS 1377, *Methods of test for soils for civil engineering purposes.* British Standards Institution, Milton Keynes.

Brown, J. and Owens, P. M., 1984. General and environmental aspects of ash disposal allied to land reclamation. *Proceedings of the 2nd International. Conference on Ash Technology and Marketing, London,* 619-626.

Cabrera, J. G., Hopkins, C. J., Wooley, G. R., Lee, R. F., Shaw, J., Plowman, C and Fox, H. 1986. Evaluation of the properties of British pulverised fuel ashes and their influence on the strength of concrete. *CANMET/ACI, Proceedings of the 2nd International. Conference on the use of fly ash, silica fume, slag and natural pozzolans in concrete, Madrid.*

Carlson, C. L. and Adriano, D. C. 1993. Environmental impacts of coal combustion residues. *Journal of Environmental Quality,* 22, 227-247.

Dreesen, D. R., Gladney, E. S., Owens, J. W., Perkins, B. L. Wienke, C. L. and Wangen, L. E. 1977. Comparison of levels of trace elements extracted from fly ash an level found in effluent waters from a coal fired power plant. *Environmental Science and Technology*, 10, 1017-1019.

Eary, L. E., Rai, D., Mattigod, S. V. and Ainsworth, C. C. 1990. Geochemical factors controlling the mobilisation of inorganic constituents from the fossil fuel combustion residue: II Review of the minor elements. *Journal of Environmental Quality*, 19, 202-214.

Edmunds, W. M. and Bath, A. H. 1976. Centrifuge extraction and chemical analysis of interstitial waters. *Environmental Science and Technology*, 10, 467-472.

Faulkner, P. 1998. *The development and characterisation of a PFA mound.* Unpublished MSc Thesis, Department of Earth Sciences, University of Leeds.

Fruchter, J. S., Rai, D. and Zachara, J. M. 1990. Identification of solubility-controlling solid phases in a large fly ash field lysimeter. *Environmental Science and Technology*, 24, 1173-1179.

Fruchter, J. S.' Rai, D., Zachara, J. M. and Schmidt, R. L. 1988. Leachate chemistry at Montour fly ash test cell. *EPRI Report EA-5922.* Palo Alto, California. Electric Power Research Institute.

Hardy, M. A. 1981. *Effects of coal-fly ash disposal on water quality in and around the Indiana dunes National lake-shore, Indiana.* US Geological Survey, Indianapolis, Water Resources Investigations, 81-106.

Havukaine, J. 1987. The utilisation of coal ash in earth works. *In: Reclamation, treatment and utilisation of coal mining wastes* (Ed. A. K. M. Rainbow), Elsevier, Amsterdam.

Lee, S. and Spears, D. A. 1995. The long term weathering of PFA and implications for groundwater pollution. *Quarterly Journal of Engineering Geology*, 28, S1-S15.

Le Seur, S. L. and Drake, E. E. 1987. Hydrogeology of an alkaline fly ash landfill in eastern Iowa. *Ground Water*, 25, 519-526.

Mattigod, S. V., Rai, D., Eary, L. E. and Ainsworth, C. C. 1990. Geochemical factors controlling the mobilisation of inorganic constituents from the fossil fuel combustion residue: Review of the major elements. *Journal of Environmental Quality*, 19, 188-201.

Mitchell, D. A. 1999. *A study of the physical and drainage characteristics of the Barlow Ash Mound, Drax Power Station.* Unpublished MSc. Thesis, Department of Civil Engineering, The University of Leeds.

National Rivers Authority, 1994. R&D Note 301, HMSO, London.

Rehage, J. A. and Holcombe, L. J., 1990. Environmental performance assessment of coal ash site: *Little Canada structural ash fill.* Electric Power Research Institute, Palo Alto, CA, EPRI-EN-6532.

Sakata, M. 1987. Movement and neutralisation of alkaline leachate at coal ash disposal. *Environmental Science and Technology*, 21, 771-777.

Simsiman, G. V.' Chesters, G. and Anderson, A. W. 1987. Effect of ash disposal ponds in groundwater at coal-fired power plant. *Water Research*, 21, 417-426.

Studds, P. G. and Cousens, T. W. 1996. Investigation of Alternative Materials for Landfill liners. DoE Contract No. 7/10/245, Unpublished, Department of Civil Engineering, The University of Leeds.

Talbot, R. W., Anderson, M. A. and Anders, W. A. 1978. Qualitative model of heterogeneous equilibria in a fly ash pond. *Environmental Science and Technology*, 12, 1056-1062.

Terzaghi K. 1994. *Theoretical Soil Mechanics.* Wiley, New York.

The Water Supply (Water Quality) Regulations 1989. UK Drinking Water Standards, HMSO, London

Hydrochemical properties and environmental isotopes of groundwater of the Quaternary – Pre Quaternary aquifers in El - Arish – Rafah area, northeast Sinai, Egypt.

By:
M. A. Tantawi
Geology Department, Faculty of Science, Minia University, 61519 El Minia, Egypt.

ABSTRACT

Twenty three groundwater samples were collected from El-Arish – Rafah area to study the quality of the water and identify the origin of recharge processes in the area of study, using hydrochemical techniques and environmental isotopes. The area of study is characterized by the presence of several water-bearing rock types. The mineral species that affect the chemical composition of the examined water points were evaluated using computer code (WATEQF). A great part of samples are saturated with dolomite and some other saturated with calcite. The water quality of the water samples varies from fresh to saline.

The stable isotope composition of the collected water points varied in a relatively wide range indicating a presence of different source of recharge in the area. The enrichment of sand dunes samples in isotopic content is either trace back to loss of water by evaporation or due to sea water contribution. The local paleowater plays an important role in recharging the water points especially in the Lower and Upper Cretaceous water samples. Moreover, the rain water reflects a significant contribution in providing the different aquifers in the area of investigation. A decreasing of salinity is recorded with increasing of stable isotope content reflecting the paleowater effects rather than evaporation.

INTRODUCTION:

The establishment of new settlement and land reclamation projects receive a great attention from the Egyptian government to release the artesian of over population. Throughout the history, the groundwater represent an important source of water for different purposes. Even now, the groundwater plays a critical role in satisfying the water requirements in many localities in Egypt. The utilized natural resources of Sinai Peninsula, particularly water resources are inadequate in relation to their actual potentials. Thus, serous development programs are required for ideal use of these vital resources. Isotopic and geochemical investigation are undertaken to study the quality of water and to identify the origin of recharge process in the area of study. For this purpose twenty three groundwater samples were collected for the present study (Fig. 1). The collected water samples were chemically and isotopically analyzed for major constituent and stable isotopes oxygen – 18 and deuterium.

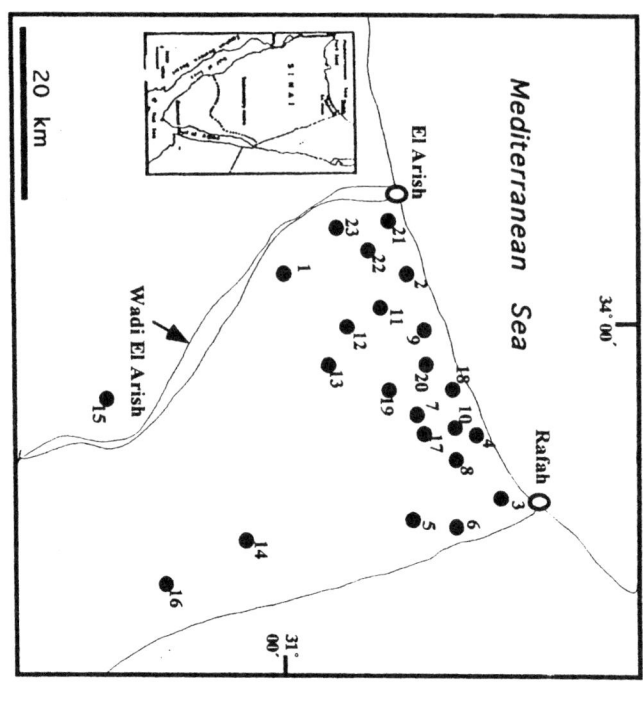

Fig. (1): Location map of the studied area and sampling sites.

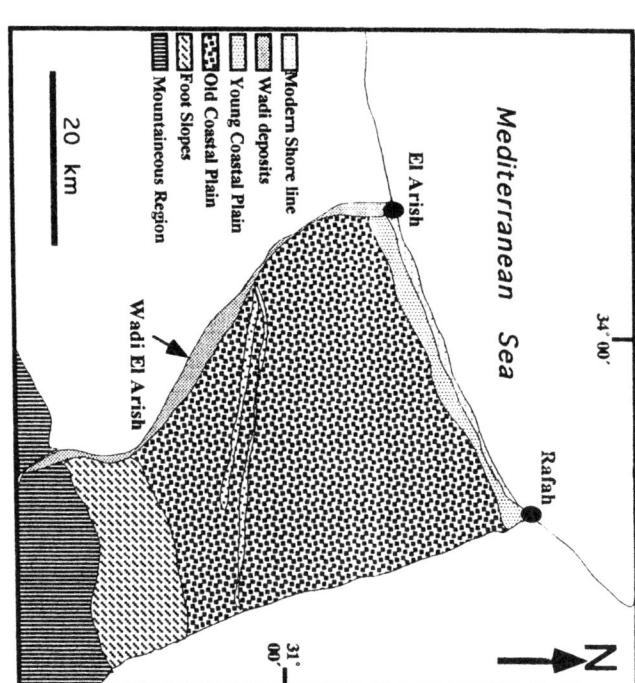

Fig. (2): Geomorphologic map of the studied area (after Taha 1968).

LOCATION AND CLIMATE:

The area of study occupies an area of about 900 km^2 in the northern part of Sinai Peninsula, between longitudes 33° 40$^\backslash$ and 34° 20$^\backslash$ and latitudes 30° 55$^\backslash$ and 31° 20$^\backslash$. It is characterized by arid to semi-arid climatic conditions. The main annual rainfall ranges between 90 mm and 30 mm, the maximum and minimum temperature are (34 °C and 18 °C) in the summer and (29 °C and 6 °C) in the winter. The mean monthly evaporation is about 5.2 mm and the mean relative humidity equal 60%.

GEOHYDROLOGIC SETTING:

The ground surface has a moderate relief (Fig. 2). It is bounded from the south by a mountainous region formed of a series of limy ridges separated by shallow synclinal basins (water collectors). A vast piedmont plain is found to the north. It is separated from the mountainous region by transitional zone of foot slopes underlain by soft Cretaceous and Eocene rocks. The piedmont plain comprises both old and young coastal belts with no sharp line of demarcation between them except for the sudden changes in the attitudes and altitudes of the land features which become parallel to the present coast near the shore line (Nasr, 1993). A series of foreshore dunes and sand sheets are formed in the coastal belts and sabkha soils are also developed near the shore in the area of low ground levels. Several wadies dissect the area of study and drain the water toward east in Wadi El Arish. These wadies are controlled by local tectonic and lithology.

The stratigraphic succession in the northeastern part of Sinai is entirely composed of sedimentary rock belonging to Mesozoic, Tertiary and Quaternary ages (Said, 1990). The Quaternary sediments cover most of the study area while Eocene and Upper Cretaceous are only exposed in southward (El-Shazly et al.; 1974).

The groundwater occurs in the Quaternary rocks in three aquifers; sand dunes, alluvial and calcareous sandstone (Kurkar or Fagra Formation).

1) The sand dunes aquifer has a local geographic extent in coastal strip; between El Sheikh Zoid and Rafah. It is formed under unconfined conditions and is hydraulically connected with the underlying alluvial aquifer. The water resources of this aquifer depend essentially on annual rainfall during winter time.

2) The alluvial aquifer have a wide distribution almost over the coastal plain and is particularly noticable along the courses of all the drainage lines dominating northeastern Sinai. This aquifer is the most productive one in northeast Sinai. It composed of alternating sands, clays and silts dominated by coarse gravels. The water of this aquifer appears in the of normal water table condition (fresh zone) floating on the sea water. The aquifer is recharged from the upward leakage from the underling aquifers (calcareous sandstone in delta Wadi El-Arish and the fissured limestone in Wadi Hassana), or by lateral leakage from the surrounding rocks and from local precipitation (Gomaa, 1984).

3) The calcareous sandstone aquifer (Kurkar or Fagra), occupies most part of the concerned area and largely distributed on the coastal belt from northern Sinai until Lebanon (Nasr, 1993). It is found under confined conditions and has a thickness varying from 5 to 50 m. The recharge of the Fagra aquifer takes place through direct precipitation on the exposed area of the formation in Palestine outside the country, or through the upward leakage from deeper aquifers and the lateral flow from adjacent aquifers.

The Pre-Quaternary rocks (Lower and Middle Eocene fractured limestone and marl intercalated with shale and Upper Cretaceous – Turonian limestone) are also developed as aquifers in the southern portion of the area of study. The Eocene aquifer is formed under unconfined while Upper Cretaceous is confined.

RESULTS AND DISCUSSION:

The results of the chemical and isotopic analysis as well as the Data of some reference points (local rainwater, Abd El Samie, 1995; Mediterranean Sea water, Awad et al., 1994 and Nubian Sandston water at Nekhel, Gat and Issar, 1974 are listed in table (1).

HYDROGEOCHEMISTRY:

A- Salinity:

The total concentration of dissolved salts in the area of study varies widely in the range from 420 to 6501 mg/L with an average of 2481 mg/L reflecting all the categories from fresh to saline. Calcareous sandstone aquifer (Fagra or Kurkar Formation) contains groundwater of significantly greater salinity ranging from 1865 to 5407 mg/L followed by alluvial aquifer (1300 – 4685 mg/L), while sand dunes one has lowest content (420 – 1695 mg/L) due to the lithic nature and the high infiltration rates. The relative increase of salinity of sample number 3 in sand dunes aquifer is accompanied with $MgCl_2$ salt which is developed on account of Na_2SO_4 and $NaHCO_3$ in salts composition. This could be due to contamination with sea water which invades under the effect of the high abstraction rate in Rafah area. The highest salinity is recorded in sample no. 16 (6501 mg/L) which collected from Eocene carbonate (Limestone).

b- Correlation Coefficient Matrix and Saturation index:

The major elements which highly correlate with total dissolved salts (Table 2 and Figs. 3, 4) and have a great bear on mineralization of the studied groundwater follow the order Cl, Na, Mg and SO_4. The correlation coefficients between TDS and previously elements are 0.90, 0.77, 0.63 and 0.54 respectively. The dissolution of evaporites that dominate the area could be explain the pattern of dissolved salts increasing as well as the positive correlation coefficient between the different ion pairs as follow:
Cl - Na c. c. = 0.81, Cl – Mg c. c. = 0.62 and SO_4– Na c. c. = 0.58

TDS	1							
Ca	0.19	1						
Mg	0.63	0.07	1					
Na	0.77	0.07	0.29	1				
K	0.07	0.03	0.09	0.10	1			
HCO₃	0.15	0.13	0.08	0.09	0.005	1		
Cl	0.90	0.15	0.62	0.81	0.14	0.12	1	
SO₄	0.54	0.18	0.21	0.58	0.13	0.002	0.41	1
	TDS	Ca	Mg	Na	K	HCO₃	Cl	SO₄

Table (2): Correlation Coefficient between TDS and different ions.

The mineral species that effect the chemical Composition of the examined groundwater were evaluated using computer code (WATEQF) program (Plummer et al., 1976). The activities of the ions are calculated from the analytical concentrations with attention to the effects of ionic strength and complex or ion pair formation. Table (3) represent the ion activity products of calcite, aragonite, dolomite, Gypsum and anhydrite. It shows that a great part of sample points are somewhat saturated with dolomite and some other with calcite. One sample (No. 11, El Mazer1) is saturated with anhydrite due to its enrichment of sulphate ion.

Table (1): Chemical and isotopic data of the collected groundwater samples in El Arish - Rafah area, Sinai, Egypt.

Serial Number	Well Name	Rock Type	pH	TDS	Units	Ca²⁺	Mg²⁺	Na⁺	K⁺	Total Cations	HCO₃⁻	Cl⁻	SO₄⁻	Total Anions	Ionic Balance	Delta ^{18}O‰	Delta D‰	d
1	El Arish Lehphin	Alluvium	7.4	2937	ppm	80	91	775	12		13	789	1038		1.69	-4.78	-26.00	12.24
					epm	4	7.58	33.70	0.31	45.58	0.21	22.23	21.63	44.07				
					epm%	9	16	74	1		1	50	49					
2	El Arish Airport	Alluvium	7.6	4685	ppm	193	125	1273	12		189	1858	1000		-0.36	-5.14	-28.90	12.22
					epm	9.65	10.42	55.35	0.31	75.73	3.10	52.34	20.83	76.27				
					epm%	13	14	75	0		4	69	27					
3	Rafah1	Sand Dunes	7.6	1695	ppm	100	73	378	7		134	705	209		2.38	-1.89	-9.90	5.22
					epm	5	6.08	16.44	0.18	27.70	2.20	19.86	4.35	26.41				
					epm%	18	22	59	1		8	75	17					
4	Rafah2	Sand Dunes	8.1	1210	ppm	33	28	315	5		315	345	210		-3.91	-0.95	-9.90	-2.30
					epm	1.65	2.33	13.70	0.13	17.81	5.16	9.72	4.38	19.26				
					epm%	9	13	77	1		27	50	23					
5	El Sheikh Zoed	Sand Dunes	7.4	582	ppm	31	19	180	14		325	177	45		0.27	-2.01	-11.10	4.98
					epm	1.55	1.58	7.83	0.36	11.32	5.33	4.99	0.94	11.26				
					epm%	14	14	69	3		47	44	9					
6	Gaica9	Calcareous Sandstone (Kurkar)	7.8	5407	ppm	167	174	1496	12		315	2371	864		-0.98	-4.82	-25.80	12.76
					epm	8.35	14.50	65.04	0.31	88.20	5.16	66.79	18	89.95				
					epm%	9	16	74	0		6	74	20					
7	Gaica8	Calcareous Sandstone (Kurkar)	7.6	3415	ppm	370	55	649	4		735	855	640		1.59	-4.37	-19.60	15.36
					epm	18.5	4.58	28.22	0.10	51.40	12.38	24.08	13.33	49.79				
					epm%	36	9	55	0		25	48	27					
8	Gaica7	Calcareous Sandstone (Kurkar)	7.8	1856	ppm	184	68	359	20		240	797	330		-3.53	-3.39	-15.20	11.92
					epm	9.2	5.67	15.61	0.51	30.99	3.93	22.45	6.88	33.26				
					epm%	30	18	50	2		12	67	21					
9	El Mokaddaba	Calcareous Sandstone (Kurkar)	7.7	3430	ppm	250	168	694	9		78	1232	990		0.26	-2.65	-8.40	12.80
					epm	12.5	14	30.17	0.23	56.90	1.28	34.7	20.63	56.61				
					epm%	22	25	53	0		2	61	37					
10	El Kharroba	Calcareous Sandstone (Kurkar)	7.4	1910	ppm	150	106	300	11		144	610	499		-0.49	-4.4	-29.10	6.10
					epm	7.5	8.83	13.04	0.28	29.65	2.36	17.18	10.40	29.94				
					epm%	25	30	44	1		8	57	35					
11	El Galawza	Calcareous Sandstone (Kurkar)	7.5	2069	ppm	185	118	323	10		204	570	720		-1.50	-3.68	-24.70	4.74
					epm	9.25	9.83	14.04	0.26	33.38	3.34	16.06	15	34.40				
					epm%	28	29	42	1		10	47	43					
12	El Maezer1	Calcareous Sandstone (Kurkar)	7.7	1993	ppm	141	95	272	10		130	339	643		3.80	-3.72	-22.50	7.26
					epm	7.05	7.92	11.83	0.26	27.06	2.13	9.55	13.40	25.08				
					epm%	26	29	44	1		9	38	53					
13	El Maezer2	Calcareous Sandstone (Kurkar)	7.6	2160	ppm	101	95	400	17		130	560	590		1.00	-3.36	-23.30	3.58
					epm	5.05	7.92	17.39	0.44	30.80	2.13	15.77	12.29	30.19				
					epm%	16	26	57	1		7	52	41					

Table (1): Cont.

Serial Number	Well Name	Rock Type	pH	TDS	Units	Ca++	Mg++	Na+	K+	Total Cations	HCO3-	Cl-	SO4--	Total Anions	Ionic Balance	Delta 18O‰	Delta D‰	d
14	El Amro	Upper Calcareous Limestone	7.9	3225	ppm	142	124	1297	21		410	1609	955					
					epm	7.1	10.33	56.39	0.54	74.36	6.72	45.32	19.90	71.94	1.65	-5.79	-37.8C	8.52
					epm%	9	14	76	1		9	63	28					
15	El Halal	Lower Calcareous Limestone Nubian Sandstone	7.5	2341	ppm	200	169	507	14		215	1098	480					
					epm	10	14.08	22.04	0.36	46.48	3.52	30.93	10.00	44.45	2.23	-6.15	-39.23	9.97
					epm%	22	30	47	1		8	70	22					
16	El Godirate	Lower Eocene Limestone	7.5	6501	ppm	148	450	1099	13		602	2502	686					
					epm	7.4	37.50	47.78	0.33	93.01	9.87	70.48	14.29	94.64	-0.87	-6.28	-28.80	21.44
					epm%	8	40	52	0		10	75	15					
17		Alluvium	8.2	1365	ppm	26	24	334	4		315	340	235					
					epm	1.3	2	14.52	0.1	17.92	5.16	9.58	4.9	19.64	-4.58	-1.54	-8.40	9.86
					epm%	7	11	81	1		26	49	25					
18	El Goora	Sand Dunes	8.3	420	ppm	40	18	52	3		172	78	38					
					epm	2	1.50	2.26	0.08	5.84	2.82	2.2	0.79	5.81	0.26	-0.67	-6.00	-0.64
					epm%	34	26	39	1		48	38	14					
19	Soliman	Calcareous Sandstone (Kurkar)	8.1	3160	ppm	75	68	900	5		415	910	570					
					epm	3.75	5.67	39.13	0.13	48.68	6.80	25.63	11.88	44.31	4.70	-3.85	-17.50	13.3
					epm%	8	12	80	0		15	58	27					
20	Hemida	Alluvium	8.2	1725	ppm	105	75	390	8		225	555	370					
					epm	5.25	6.25	16.96	0.21	28.67	3.69	15.63	7.71	27.03	2.94	-2.4	-12.00	7.2
					epm%	18	22	59	1		14	58	28					
21	Samarani	Alluvium	7.7	1300	ppm	120	45	270	5		165	508	217					
					epm	6	3.75	11.74	0.13	21.62	2.70	14.31	4.52	21.53	0.21	-3.3	-21.50	4.9
					epm%	28	17	54	1		13	66	21					
22	Salim	Alluvium	7.6	2295	ppm	185	162	435	9		415	942	357					
					epm	9.25	13.50	18.91	0.23	41.89	6.80	26.54	7.44	40.78	1.34	-3.5	-20.70	7.3
					epm%	22	32	45	1		17	65	18					
23	Omm Fattoh	Alluvium	7.4	2560	ppm	280	65	425	15		370	970	410					
					epm	14	5.42	18.48	0.38	38.28	6.07	27.32	8.54	41.93	-4.55	-4.52	-23.20	12.96
					epm%	37	14	48	1		15	65	20					
24	Rain Water															-3.4	-4.40	22.8
25	Nubian Sandstone (Nekhel)															-8.38	-54.50	12
26	Sea Water															1.77	9.70	-4.46

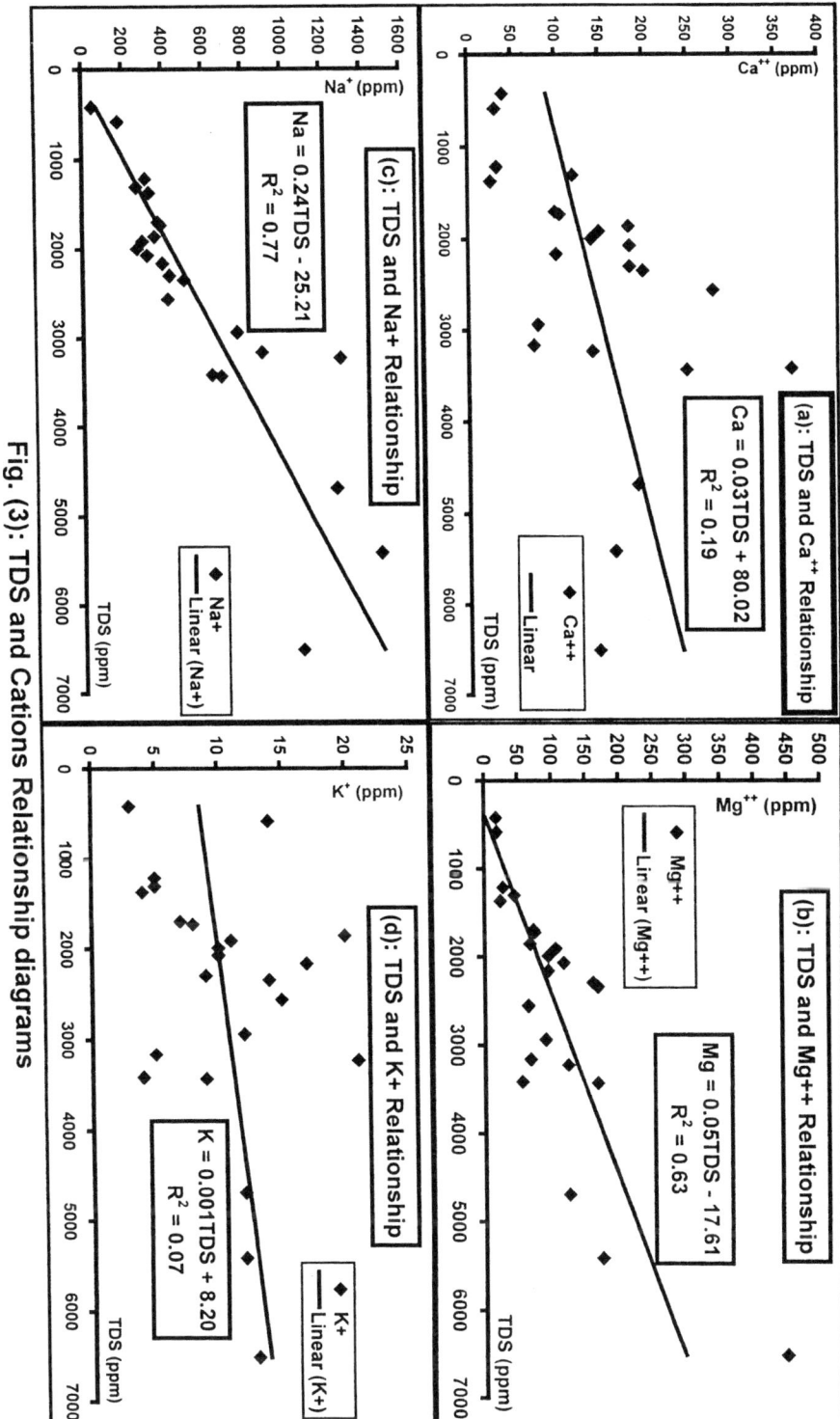

Fig. (3): TDS and Cations Relationship diagrams

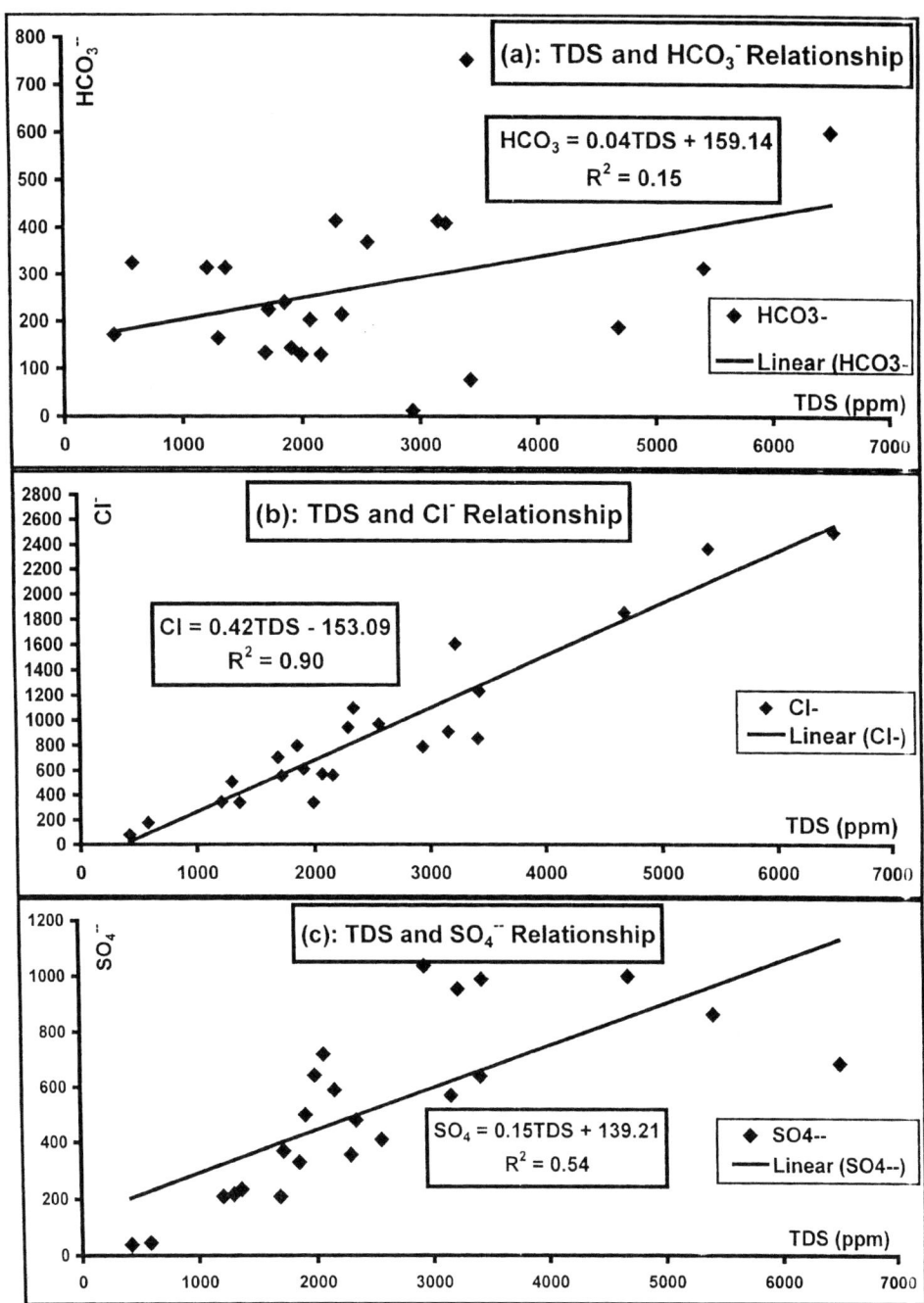

Fig, (4): TDS and Anions Relationship diagrams

Table (3): Hydrochemical Parameters, ion dominance and water type and saturation index of different minerals(WATEQF).

Serial No.	Well Name	Rock Type	r(Na+K)/rCl	rCa/rCl	rMg/rCl	rHCO3/rCl	rSO4/rCl	Hydrochemical Formula	Ion Dominance	Water Type	Calcite	Aragonite	Dolomite	Gypsum	Anhydrite
1	El Arish Lehphin	Alluvium	1.53	0.18	0.34	0.01	0.97	$Cl_{(50)}\,SO_{4(49)}\,HCO_{3(1)}$ / $(Na+K)_{(75)}\,Mg_{(16)}\,Ca_{(9)}$	Cl > SO₄ > HCO₃ / Na > Mg > Ca	Na - Cl	-0.40	-0.50	-0.40	-1.00	-1.22
2	El Arish Airport		1.06	0.20	0.20	0.06	0.40	$Cl_{(68)}\,SO_{4(27)}\,HCO_{3(4)}$ / $(Na+K)_{(73)}\,Mg_{(16)}\,Ca_{(13)}$	Cl > SO₄ > HCO₃ / Na > Mg > Ca	Na - Cl	0.30	0.16	0.77	-0.75	-0.97
17			1.53	0.18	0.21	0.54	0.51	$Cl_{(48)}\,HCO_{3(28)}\,SO_{4(25)}$ / $(Na+K)_{(82)}\,Mg_{(11)}\,Ca_{(7)}$	Cl > HCO₃ > SO₄ / Na > Mg > Ca	Na - Cl	-0.15	-0.30	-0.25	-1.02	-0.85
20	Hemida		1.10	0.34	0.24	0.24	0.49	$Cl_{(46)}\,HCO_{3(28)}\,SO_{4(25)}$ / $(Na+K)_{(60)}\,Mg_{(22)}\,Ca_{(7)}$	Cl > SO₄ > HCO₃ / Na > Mg > Ca	Na - Cl	-0.03	-0.02	-0.45	-0.75	-0.65
21	Samarani		0.83	0.42	0.26	0.19	0.32	$Cl_{(66)}\,SO_{4(21)}\,HCO_{3(13)}$ / $(Na+K)_{(55)}\,Mg_{(28)}\,Ca_{(17)}$	Cl > SO₄ > HCO₃ / Na > Mg > Ca	Na - Cl	0.05	-0.12	-0.37	-0.80	-0.70
22	Salim		0.72	0.35	0.51	0.26	0.28	$Cl_{(65)}\,SO_{4(20)}\,HCO_{3(15)}$ / $(Na+K)_{(46)}\,Mg_{(32)}\,Ca_{(22)}$	Cl > SO₄ > HCO₃ / Na > Mg > Ca	Na - Cl	-0.40	-0.50	-0.50	-1.20	-0.95
23	Omm Fattoh		0.69	0.51	0.20	0.22	0.31	$Cl_{(75)}\,SO_{4(17)}\,HCO_{3(16)}$ / $(Na+K)_{(48)}\,Mg_{(37)}\,Ca_{(14)}$	Cl > SO₄ > HCO₃ / Na > Ca > Mg	Na - Cl	-0.12	-0.35	-0.40	-0.76	-0.85
3	Rafah1	Sand Dunes	0.84	0.25	0.31	0.11	0.22	$Cl_{(75)}\,SO_{4(17)}\,HCO_{3(18)}$ / $(Na+K)_{(60)}\,Mg_{(22)}\,Ca_{(18)}$	Cl > SO₄ > HCO₃ / Na > Mg > Ca	Na - Cl	0.08	-0.06	0.36	-1.41	-1.63
4	Rafah2		1.42	0.17	0.24	0.53	0.45	$Cl_{(50)}\,HCO_{3(27)}\,SO_{4(23)}$ / $(Na+K)_{(78)}\,Mg_{(13)}\,Ca_{(9)}$	Cl > HCO₃ > SO₄ / Na > Mg > Ca	Na - HCO₃	-0.27	-0.21	-0.72	-1.13	-0.93
5	El Sheikh Zoed		1.64	0.31	0.32	1.07	0.19	$HCO_{3(47)}\,Cl_{(44)}\,SO_{4(8)}$ / $(Na+K)_{(72)}\,Mg_{(14)}\,Ca_{(14)}$	HCO₃ > Cl > SO₄ / Na > Ca > Mg	Na - HCO₃	-0.13	-0.27	-0.11	-2.35	-2.57
18	El Goora		1.06	0.91	0.68	1.28	0.36	$HCO_{3(46)}\,Cl_{(38)}\,SO_{4(14)}$ / $(Na+K)_{(40)}\,Ca_{(34)}\,Mg_{(26)}$	HCO₃ > Cl > SO₄ / Na > Ca > Mg	Na - HCO₃	-0.45	-0.58	-0.45	-1.95	-1.75

Table (3): Cont.

Serial No.	Well Name	Rock Type	rNa+K/rCl	rCa/rCl	rMg/rCl	rHCO3/rCl	rSO4/rCL	Hydrochemical Formula	Ion Dominance	Water Type	Calcite	Aragonite	Dolomite	Gypsum	Anhydrite
16	El Gudirate	Eocene Limestone	0.68	0.10	0.53	0.14	0.20	$Cl_{(75)} SO_{4(15)} HCO_{3(10)} / (Na+K)_{(52)} Mg_{(40)} Ca_{(8)}$	Cl > SO₄ > HCO₃	Na - Cl	0.95	0.81	2.70	-1.15	-1.37
15	El Halal	Calcareous Nubian Sandstone	0.72	0.32	0.46	0.11	0.32	$Cl_{(70)} SO_{4(22)} HCO_{3(8)} / (Na+K)_{(48)} Mg_{(30)} Ca_{(22)}$	Cl > SO₄ > HCO₃	Na - Cl	0.76	0.62	1.80	-0.95	-1.17
14	El Amro	Lower Calcareous Limestone	1.26	0.16	0.23	0.15	0.44	$Cl_{(63)} SO_{4(28)} HCO_{3(9)} / (Na+K)_{(77)} Mg_{(14)} Ca_{(9)}$	Cl > SO₄ > HCO₃	Na - Cl	0.39	0.25	1.07	-0.89	-1.11
19	Soliman	Upper Calcareous Limestone	1.53	0.15	0.22	0.27	0.46	$Cl_{(58)} SO_{4(27)} HCO_{3(15)} / (Na+K)_{(60)} Mg_{(12)} Ca_{(8)}$	Na > Mg > Ca	Na - Cl	0.70	0.61	1.25	-0.99	-1.18
13	El Maezer2	Calcareous Sandstone (Kurkar)	1.13	0.32	0.50	0.14	0.78	$Cl_{(52)} SO_{4(41)} HCO_{3(7)} / (Na+K)_{(56)} Mg_{(26)} Ca_{(18)}$	Na > Mg > Ca	Na - Cl	0.32	-0.16	0.27	-1.04	-1.26
12	El Maezer1		1.27	0.74	0.83	0.22	1.40	$SO_{4(47)} Cl_{(38)} HCO_{3(9)} / (Na+K)_{(45)} Mg_{(29)} Ca_{(26)}$	SO₄ > Cl > HCO₃	Na - SO₄	0.23	0.08	0.63	-0.87	1.09
11	El Galawza		0.89	0.58	0.61	0.21	0.93	$Cl_{(61)} SO_{4(37)} HCO_{3(2)} / (Na+K)_{(43)} Mg_{(29)} Ca_{(28)}$	Na > Mg > Ca	Na - Cl	0.32	0.18	0.79	-0.75	-0.97
10	El Kharroba		0.78	0.36	0.40	0.04	0.59	$Cl_{(57)} SO_{4(38)} HCO_{3(5)} / (Na+K)_{(45)} Mg_{(30)} Ca_{(25)}$	Na > Mg > Ca	Na - Cl	0.16	-0.12	0.49	-0.95	-1.16
9	El Mokaddaba		0.88	0.41	0.25	0.18	0.31	$Cl_{(67)} SO_{4(21)} HCO_{3(12)} / (Na+K)_{(52)} Ca_{(30)} Mg_{(18)}$	Na > Ca > Mg	Na - Cl	0.74	0.59	1.39	-0.61	-0.83
8	Gaica7		0.72	0.77	0.19	0.51	0.55	$Cl_{(57)} SO_{4(35)} HCO_{3(8)} / (Na+K)_{(53)} Ca_{(25)} Mg_{(22)}$	Na > Ca > Mg	Na - Cl	1.24	1.09	2.00	-1.02	-1.20
7	Gaica8		1.18	0.77	0.19	0.51	0.55	$Cl_{(46)} SO_{4(27)} HCO_{3(25)} / (Na+K)_{(55)} Ca_{(36)} Mg_{(9)}$	Na > Ca > Mg	Na - Cl	0.65	0.50	1.60	-0.58	-0.80
6	Gaica9		0.98	0.13	0.22	0.08	0.27	$Cl_{(74)} SO_{4(20)} HCO_{3(6)} / (Na+K)_{(74)} Mg_{(16)} Ca_{(10)}$	Na > Mg > Ca	Na - Cl	0.03	0.01	0.24	-0.92	-1.14

c) Ion Dominance and Water Type:

The major portion of samples (87%) has the following sequence of ion dominance and water type(Table 3):
Na > Mg > Ca / Cl > SO$_4$ > HCO$_3$ and sodium chloride water type in twenty samples. Sodium bicarbonate water type is recorded in two samples of sand dunes aquifer while sodium sulphate type is recognized in one sample of calcareous sandstone aquifer (El-Maezer1).
The Na – Cl / SO$_4$ ratio is negative and Cl – Na / Mg is positive in about 52% of the samples where MgCl$_2$, CaCl$_2$ and MgSO$_4$ salts are formed due to the dissolution of marine salts. Sodium increases in the rest of samples due to the leaching of terrestrial salts and / or cation exchange on the surface of the fine sediments.

d) Hydrochemical Classification:

The hydrochemical composition as indicated in Piper diagram (Piper, 1953) (Fig. 5) reflects a high stage of mineralization and a secondary salinity evaluation where salinity index Cl – SO$_4$ exceeds alkalies and alkalinities, and the most water points accordingly fall in the field number (4).
From table (3) we can distinguished that the groundwater in all aquifers of the study area are characterized by low ca^{++} than Mg^{++} and Na$^+$ which indicate that the ion exchange process between alkaline earth metals (Ca & Mg) and alkali metal (Na) adsorped on the rock surface. The anionic sequence in the most of the studied samples is Cl > SO$_4$ > HCO$_3$, but a noticable variation recorded in 50% of samples of sand dunes aquifer (sample No. 5 El-Sheikh Zoed and sample No. 18 El Goora) where HCO$_3$ > Cl > SO$_4$. One sample (No. 12, El Maezer 1) among the calcareous sandstone aquifer (Kurkar) has a different anionic dominance, where SO$_4$ exceeds Cl and HCO$_3$.
According to the hydrochemical coefficient (ion ratios, Table 3), the majority of the collected water of the different aquifer are of meteoric origin. The rest of the samples are of marine origin. The prevalence of Na$^+$ followed by Mg^{++} ions reflect the effect of the Mediterranean sea spray on the water type in the area.

ENVIRONMENTAL ISOTOPES:

The environmental isotopes oxygen-18 and deuterium are used as natural tracers in the water cycle to define the recharge sources, the mixing between different sources and the salinization processes.
The stable isotopic data of the studied groundwater and refernce samples are listed in table(1) in term of individual ratio (^{18}O / ^{16}O or D / ^{1}H) as well as their relationship referred d-excess (d = δD ‰ - 8 δ^{18}O‰). The water samples varied in relatively wide range in their isotopic composition.
The data for the groundwater samples in the concerned area are drown against the global meteoric water line (GMWL, Craig, 1961), Mediterranean precipitation water line (MPWL, Gat and Carmi, 1970) and paleowater line (PWL, Sonntag et al., 1979) (Fig. 6) to know the relationship with these lines and define the recharge sources.
The high spread of the values indicate the high inhomogeneity of the recharge conditions and the intermingling of water of different origin in the bulk of water. The positions of the data points in the conventional ^{18}O – D diagram (Fig. 6), can be highlights on this phenomena as follow:

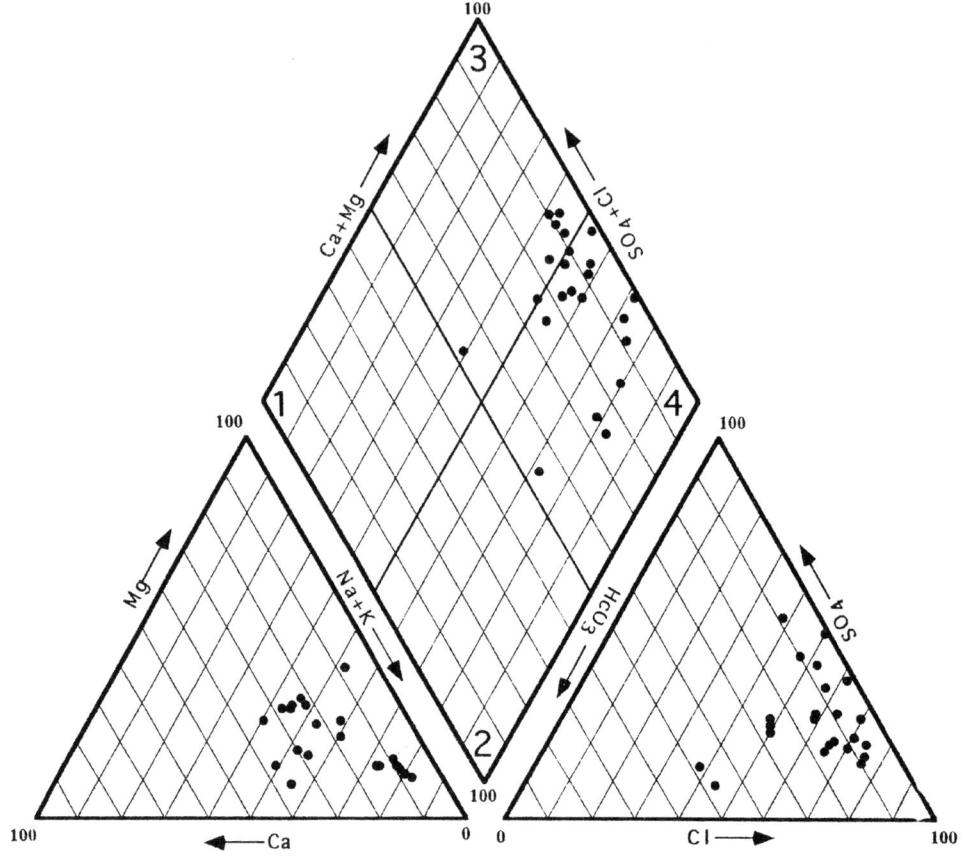

Fig. (5): Piper diagram

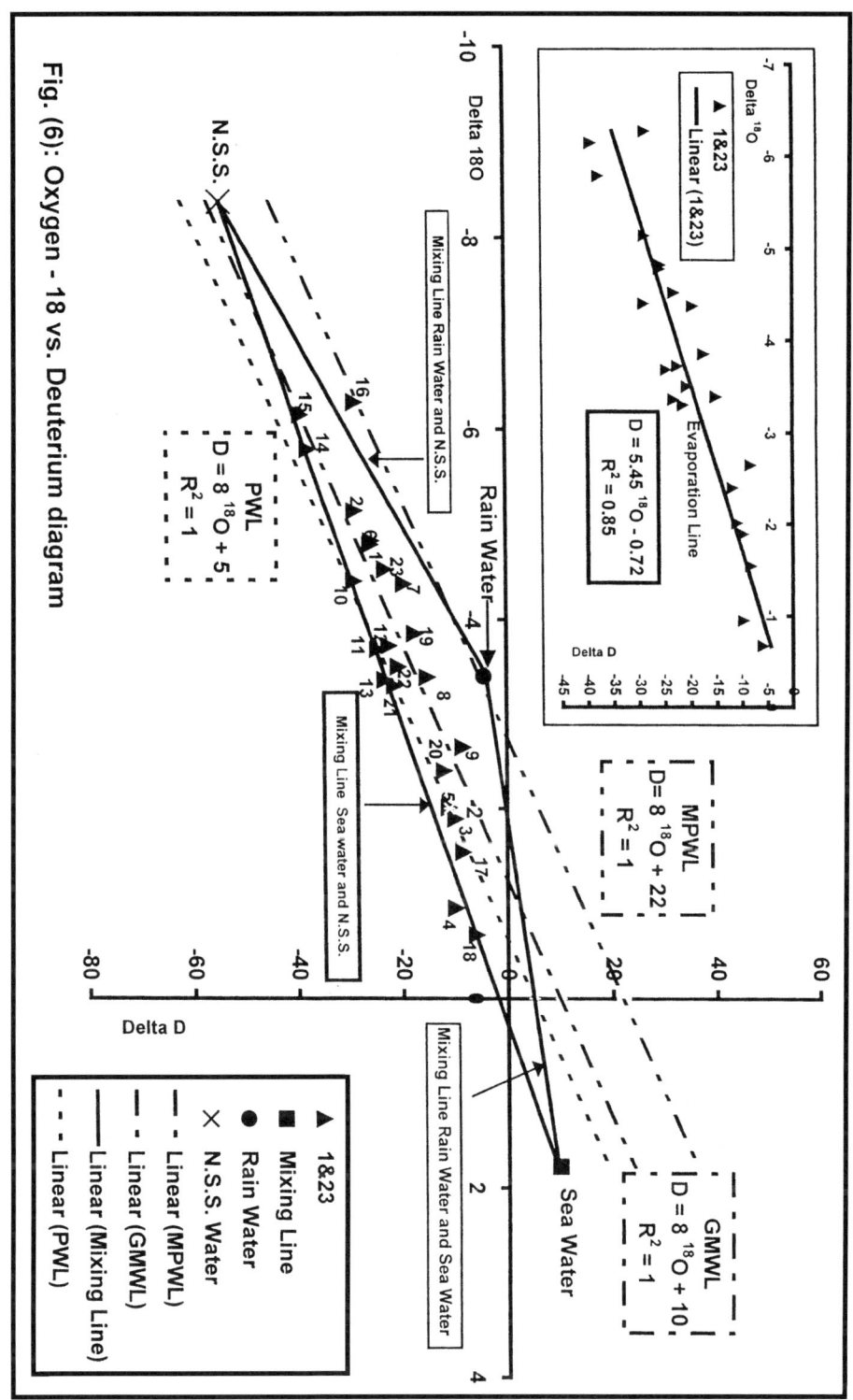

Fig. (6): Oxygen - 18 vs. Deuterium diagram

- The rain water of Rafah station (Abd El-Samie, 1995) falls very near to Mediterranean precipitation water line where the precipitation water is connected with air mass coming mainly from Europe and approached the region from west to northwest.
- The water points of the different aquifers distribute within the band of global meteoric water lines (δD‰ $=8$ $\delta^{18}O$‰ $+ 10$), PWL (δD‰ $= 8$ $\delta^{18}O$‰ $+ 5$) and MPWL (δD‰ $= 8$ $\delta^{18}O$‰ $+ 22$)} that have slopes equal 8 indicating the meteoric origin and equilibrium condensation of these waters. They show some scatter relative to Rafah rainwater (d-excess is generally low while ^{18}O and D content is more positive in sand dunes and more negative in the rest of samples) due to post depositional processes.
- The enrichment of sand dunes samples in isotopic content is either trace back to loss of water by evaporation or due to sea water contribution. The barren nature of this aquifer and the high abstraction rate can facilitate these effects.
- The isotopic composition of Pleistocene or early Holocene paleowater stands out quite notably relative to present day recharged aquifers; (the paleowater is represented by sample No. 25 (N.S.S. water in Nekhel area, Gat and Issar, 1974) which indicates the isotopic composition of rain water that supplied the Nubian Sandstone aquifer during pluvial times in the last cooling period. About 90 km north of Nekhel area, Nubian Sandstone is exposed in the erosion cirques of Gabel El-Halal anticline.
- All the collected samples (except sample No. 16) of the investigated area distribute inside the solid triangle of the mixing lines of rain water – sea water enriched end members and, paleowater depleted one. The relative position of the points with the three end members reflect the contribution of three sources of recharge (paleowater, rain water and sea water).
- The Lower and Upper Cretaceous groundwater samples No. (14, 15 and 16) lie very near the Nubian Sandstone water reflecting the high contribution of paleowater through fracture system combined with the recent day precipitation water.
- El-Godirate spring (No. 16) is the only sample that nearly have the same d-excess (21.44) as rain water (22.8) but its isotopic composition is more deleted due to high altitudes recharge.
The oxygen-18 content shows an inverse relation with d-excess (Fig. 7) due to evaporation. There is a high dispersion due to other effects such as mixing and leaching processes.
Salinity of the collected samples are drown versus oxygen-18 and deuterium (Figs. 8, 9). It showed that, the salinity decreases with the increasing of isotopic contents, reflecting the paleowater effects rather than evaporation.

CONCLUSION RECOMMENDATION:

. The total concentration of dissolved salts in the concerned area varies widely in the range from 420 to 6501 mg/L with an average of 2481 mg /L reflecting all the categories from fresh to saline.
. The major elements which highly correlate with the total dissolved salts and have a great bear of mineralization of the groundwater follow the order Cl, Na, Mg and SO_4, with a correlation coefficient equal 0.90, 0.77, 0.63 and 0.54 respectively.
.   The isotopic data reveal that the origin of groundwater rechargability is a mixture of three water sources: local rain water and run off, Mediterranean Sea water intrusion as well as contribution of Sinai paleowater which seeped to the aquifer via over pumping through faults and fractured present in the area.
. The salinity decreases with the increasing of environmental stable isotopes reflecting the paleowater effects rather than evaporation.

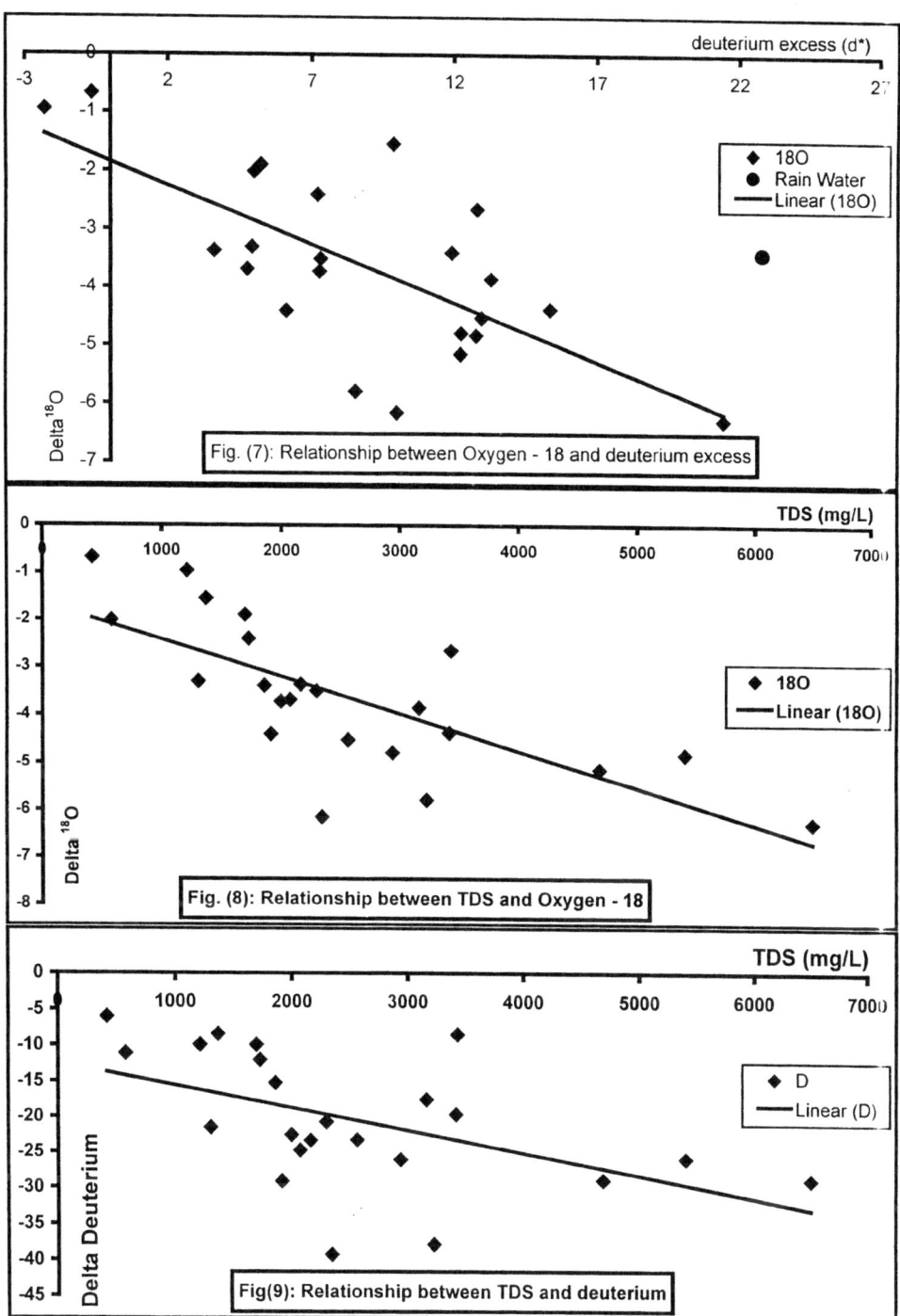

Fig. (7): Relationship between Oxygen - 18 and deuterium excess

Fig. (8): Relationship between TDS and Oxygen - 18

Fig(9): Relationship between TDS and deuterium

. Carbon-14 dating is required to study the contribution of paleowater in the rechargability process of groundwater in Sinai.

REFERENCES:

. Abd El Samie, S. G. (1995): Isotope and hydrochemical studies on the groundwater of Sinai Peninsula. Ph. D. Thesis, Ain Shams University, Faculty of Science, Egypt. 226p.

. Awad, M. A., Hammad, F. A., Aly, A. I. M. and Sadek, M. A. (1994): Use of environmental isotopes and hydrochemistry as indicators for the origin of groundwater resources in El Dabaa area, northwestern costal zone of Egypt. Envir. Geochem. And Health v. 16 no, 1, pp. 31 – 38.

. Craig, H. (1961): standard for reporting concentration of deuterium and Oxygen-18 in natural waters, Science 133, pp. 1833 – 1840.

. El-Shazly, E. M., Abdel Hady, M. A., El-Ghawaby, M. A., El-Kassas, I. A. El-Shazly, M. M., Salman, A. B. and El-Rakaiby, M. (1974): Geology of Sinai Peninsula from ERTS-1 Satellite Images " ASRT, Remote Sensing Project, Cairo, Egypt. 20p.

. Gat, J. R. and Carmi, I. (1970): Evalution of the isotopic composition of atmospheric water in the Mediterranean Sea area. Journal Geophysic, v. 75, pp. 3039 – 3048.

. Gat, J. R. and Issar, A. (1974): Desert isotope hydrology, water sources of the Sinai desert. Geochemica Cosmochimica ACTA, v. 38, pp. 1117 – 1131.

. Gomaa, M. A. (1984): Hydrological and hydrochemical studies on wadi El-arish, North Sinai, egypt. M. Sc. Thesis, Faculty of science, Zagazic University, Zagazic, Egypt.

. Nasr, I. M. (1993): Hydrogeological and geophysical studies in the northeastern part of Sinai Peninsula, Egypt. Ph. D. thesis, Faculty of Science, Al-Azhar University, Cairo, Egypt. 194p.

. Piper, A. M. (1953): A graphic procedure in the geochemical interpretation of water analysis. Am. Geophy. Union Trans.. V. 25 no. 6, Washington, D. C., pp. 914 – 928.

. Plummer, L. N., Jones, B. F. and Truesdell, A. H. (1976): " WATWQF – A Fortran IV version of WATEQ, a program for calculating chemical equilibrium of natural waters" U. S. Geological Survey, Nat. Tech. Infor. Surv., PB 261027, 21 p.

. Said, R. (1990): The geology of Egypt. A Balkeman, Roterdam, Bookfield.

. Sonntag, C., Klitzsch, E., Lonert, E. P., El-Shazly, E. M., Munnich, K. O., Junghans, Ch., Thorweihe, U., Weisstroffe, K. and Swaillem, F. (1979): Paleoclimatic information from deuterium and oxygen-18 in carbon-14 dated North Saharian groundwater. Groundwater formation in the past, Isotope Hydrology. Proceeding of Symposium Neuherberg 1978. pp. 569 – 581. IAEA, Vienna.

. Taha, A. A. (1968): Geology of water supplies of El-Arish – Rafah area, Northern Sinai, Egypt. M. sc. Thesis, Faculty of Science, cairo University, Cairo, Egypt.

Field investigations of slurry trench cut-off walls to control pollution migration

P TEDD. Building Research Establishment Ltd

INTRODUCTION

Slurry trench cut-off walls using self hardening cement bentonite are the most common form of in-ground vertical barrier in the UK for controlling lateral migration of pollution from landfill sites and contaminated land, with well over one hundred slurry walls having been constructed and some several kilometres long. The first use of a slurry trench cut-off wall in the UK for pollution control was adjacent to a landfill was in 1983. Jefferis (1997) provides an overview of the development of cement-bentonite cut-off walls in the UK and a National Specification for their construction as barriers to pollution migration was completed in 1999 (Doe et al, 1999).

The Building Research Establishment (BRE) began research into the performance of these cut-off walls in the early 1990s as concerns were being expressed about their long term performance, particularly in aggressive ground conditions. They investigated a number of cut-off walls (Tedd et al 1993) and started to develop methods for measuring in-situ permeability (Tedd et al 1995). In 1996, BRE in collaboration with Lattice Property Holdings (formally British Gas Properties) and Advantica (formally British Gas Research & Technology) began work on part of a disused gasworks site with the principal aim of assessing the in-situ performance of cement bentonite slurry trench cut-off walls in chemically aggressive ground. The ground conditions at the site consist of approximately 3m of fill, containing spent oxide, coal residues, carbon black and foul lime, overlying stiff low permeability clay. There are elevated levels of sulphates and other contaminants that could potentially react with the cement-bentonite.

Two test cells were constructed using slurry walls to isolate parts of the site in which the water levels were raised to create local hydraulic gradients across the walls of the test cells. Each test cell is 10m square in plan and 5m deep toeing into the underlying aquiclude as shown in Fig. 1. A high density polyethylene (HDPE) membrane was installed into one of the test cells. A nominal 0.5m layer of clay and HDPE membrane was placed over each cell to prevent rain infiltration. To monitor the performance, wells and piezometers have been installed inside and outside the test cells.

A number of isolated lengths of slurry wall were also constructed at the site for the purpose of assessing and developing in-situ techniques for measuring the properties, particularly permeability, of the cut-off wall material. Exhumation of a number of isolated sections of wall has been carried out and samples have been taken for laboratory permeability tests. Partial exhumation of the test cells has recently taken place, four and half years after their

Geoenvironmental impact management, Thomas Telford, London, 2001.

construction. These unique observations provide an insight into the behaviour of cement-bentonite walls in chemically aggressive ground.

Some of the initial findings from the test site and laboratory assessment of the cement-bentonite have been described by Tedd et al (1997 [a,b]). The objective of this paper is to describe the field investigations that have been undertaken at the test site and assess their value in determining the in-situ properties, particularly permeability, of cut-off walls. The chief criterion for design of the set slurry is to keep the permeability below 1×10^{-9} m/s and by this means it is intended to minimise the flow of contaminated ground water through the wall.

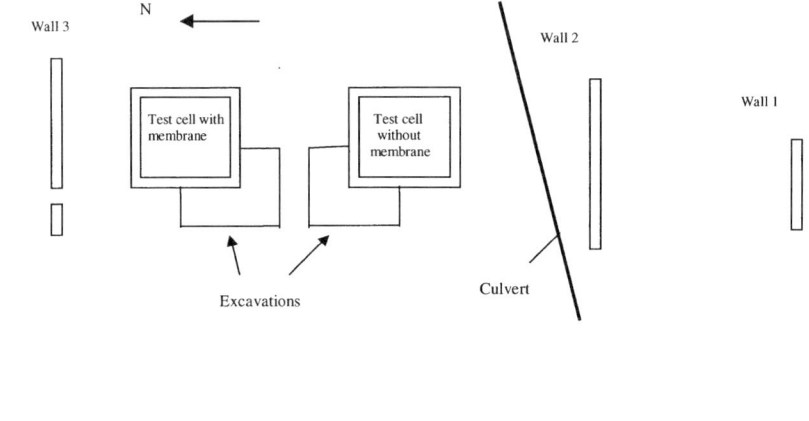

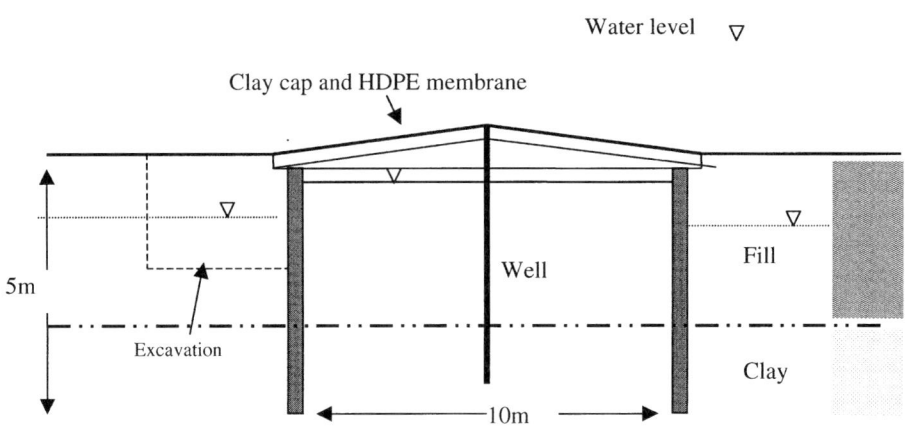

Fig. 1. Plan of test site and section through test cell showing partial excavations

IN-SITU TESTING OF THE CUT-OFF WALLS

There are currently no accepted methods for measuring in-situ permeability within the cut-off walls. Tedd et al (1995) conducted a comprehensive survey of existing methods. With any method where a hole is drilled into the set slurry there are concerns that drilling may damage the wall both due to the size of the hole and also the method of drilling. An assessment of the methods used at the test site is summarised in Table 1 and the results are compared with laboratory measured values. The walls used for the tests contain varying amounts of the surrounding ground and laboratory tests have shown this to affect both permeability and strength. Walls 1 and 2 were constructed in relatively uncontaminated ground whereas wall 3, described as contaminated, contained significant quantities of spent oxide, giving a higher permeability and lower strength material.

Cast-in piezometers

Standpipe piezometers with porous tips were installed in the walls at a number of locations when the slurry was two to three days old and still soft. Falling head permeability tests at 28 days and later in these piezometers gave very low values of permeability, less than 1×10^{-10} m/s. In a number of piezometers, the increased water level at the start of the test did not alter for a period of many months. It seems likely that the filter stone of these piezometers had become partially blocked by the fines in the setting cement-bentonite.

BRE Packer system

A number of in-situ tests have been undertaken using the BRE packer system (Tedd et al, 1995). This involves drilling a 50mm hole in the centre of the wall with a hand auger, and inserting a packer to isolate part of the hole, and undertaking a falling head permeability test. These test gave values of the order of 2×10^{-9} m/s for an uncontaminated wall (wall 1) and 5×10^{-9} m/s in the contaminated wall (wall 3) at 90 days

Piezocone

In reporting the use of the piezocone in a 1.5m wide cement-bentonite wall constructed in Italy, Manassero (1994) claimed that the piezocone provided a large number of tests in a short time and the information obtained included not only permeability, but also strength and deformation characteristics. To investigate the use of the method at the test site a number of profiles have been undertaken in the walls at the test site at 28 and 90 days, and 4 years. The work is described in detail by Tedd et al (1997[b]). Hardened cement-bentonite is a stiff, hard, brittle material that only exhibits plastic behaviour when loaded under drained conditions at high effective confining pressures. It was therefore envisaged that the use of a piezocone in cut-off walls could cause cracking. Good repeatability of results has been found in terms of cone resistance, sleeve friction and pore pressure. Cone resistance has been useful in identifying the general uniformity of the walls and the weaker contaminated slurry. Results have also identified very low strengths at the base of these test walls. Permeabilities derived from piezocone pore pressure dissipation tests were several orders of magnitude larger than laboratory measured values and BRE packer tests. It does not therefore appear that realistic and acceptable values of permeability can be obtained from piezocone tests in these brittle cement materials. Localised cracking in the more brittle material was suggested as a possible explanation but there was no evidence of any cracking during the subsequent excavation of the walls.

The use of the piezocone to identify defects in the wall needs careful consideration. The very low strength found at the base of these test walls will not necessarily occur in walls constructed though different ground conditions. Many cut-off walls have been constructed through sands and gravels resulting in much stronger material towards the base of the wall which would either be impossible to penetrate or be mistaken for the ground. Other cut-off walls have been constructed through peat where it is known that some pockets of peat were present but were not continuous across the wall. Piezocone profiles here could have indicated a defective wall where the cut-off wall is performing satisfactorily.

Table 1. Summary of in-situ methods of measuring properties of cement bentonite cut-off walls

Method	Range of permeability values m/s	Comment on measured permeability values	Comment on method
Piezometer	$< 1 \times 10^{-10}$	Very low values probably due to filter stones blocked	Installed during construction. Discrete measurement at single depth
BRE packer	1 to 5×10^{-9}	Possibly higher than actual values due to leakage passed packer	Post construction installation. Multiple level measurement. Difficulties in making hole where gravel is included in the wall
Piezocone	1×10^{-7}	Significantly higher than other measurements possibly due to localised leakage	Doubt over permeability measurements. Continuous profile of cone resistance and therefore possible use in identifying defects. Good repeatability of cone resistance. Pushing in cone does not crack the wall.
Self- boring permeameter	1×10^{-7} to 10^{-8}	Significantly higher than other measurements possibly due to localised leakage	Doubt over permeability measurements. Frequent measurements in same location. Also measures strength and stiffness.
Laboratory tests samples cast in tubles	1×10^{-9} to 10^{-10}	Value depends on amount of contamination in sample and duration of test	Samples not representative of actual wall material due to stress and ground water conditions, and any local discontinuities.
Laboratory prepared from block samples	1 to 9×10^{-10} 1×10^{-9} to 6×10^{-11}	Uncontaminated Contaminated Lower values from below water table	Confining conditions during test different to in-situ conditions.

Note: all the values above are were measured at 90 days or later after construction and therefore are not significantly affected by age of the sample

As most walls in the UK tend to be 0.6m wide, there is a risk of damage to the wall and the possibility of the cone coming out of the side of the wall particularly with the deeper walls. It is concluded that the application of piezocone testing for routine quality assurance of slurry cut-off walls is unlikely to find acceptance in the UK.

Self-boring permeameter

Most recently, Cambridge Insitu and Cambridge University undertook pressuremeter and in-situ permeability tests on wall 2 at the test site using a modified Cambridge Self Boring Pressuremeter (SBP). The SBP was developed in the 1970's to carry out minimal disturbance in-situ tests on soils to obtain strength, stiffness and in-situ horizontal stress. The instrument, which is 90mm diameter, bores itself vertically into the ground to the selected depth using a rotating cutting head and water flush to remove the soil. A membrane behind the cutting head is inflated to carry out horizontal loading of the test cavity to determine the in-situ properties.

The modified SBP allows water to be pumped to the cutting head at a constant rate using a flow pump while monitoring the water pressure, thus allowing the permeability to be determined. Fifteen permeability tests were undertaken using the SBP down to a depth of 4.1m. Values of permeability ranged between 1×10^{-7} and 1×10^{-8} m/s. These values are an order of magnitude larger than those measured using the BRE packer system in the same length of wall and considerably larger than those measured in the laboratory on cast samples.

It is considered that the values of permeability measured by SBP are not representative of the actual permeability of the wall. There are several reasons why the field tests may be giving misleading results:

- A fault in the wall at the location of the tests or preferential flow paths through the wall caused by surrounding soil entering the slurry during construction, however this was not indicated in the post testing excavation of the test wall.
- Vertical leakage from the cavity up the sides of the instrument. This is a common problem with in-situ permeability tests but checks carried out on site could not find evidence of leakage.
- The tests were affected by disturbance caused to the material by the SBP.

Five pressuremeter tests were undertaken giving values of undrained shear strength and shear modulus. Values of undrained shear strength varied between 270 and 440 kPa, the same order of magnitude as those values determined from laboratory unconfined compression tests. Initial shear moduli values from the SBP were between 30 and 50 MPa whereas unload reload values are typically between 100 and 150 MPa. Youngs moduli values from unconfined compression tests and drained consolidated drained triaxial tests gave values typically between 250 and 370 MPa i.e. of the same order of magnitude as the unload reload values. Further details of the test are reported by Ratnam et al (2001).

SAMPLING THE CEMENT-BENTONITE FROM CUT-OFF WALL

It is the usual practice in the UK to adopt a performance specification for the construction of slurry trench cut-off walls in which the properties of the set slurry are specified in accordance with the National Specification. To demonstrate compliance with the specified permeability of the set material in the trench, samples of the fluid slurry are taken from the trench, cast in plastic tubes, stored under water and tested in a triaxial permeability apparatus at some specified time after sampling. Values obtained for the slurry used at the test site are summarised in Table 1.

Various methods have been attempted to obtain samples from the wall once the slurry had set such that the material cured in the wall can be tested. Experience at a number of sites has shown that the quality of the samples obtained by rotary coring is very dependent on the amount and type of ground that is contained in the slurry. The drilling action causes the set slurry to re- slurrify and gravel in the slurry results in significant amounts of the material being eroded, making it unsuitable for most types of testing. At the test site coring by dynamic continuous sampling was used to avoid the problem of erosion from rotary coring. Concerns that this method would damage the comparatively brittle material were confirmed with all the samples being shattered and not suitable for strength or permeability testing.

During the exhumation of the walls 1 and 3, block samples were taken from the wall from above and below the water table, and subsequently trimmed for triaxial permeability tests. These tests are unique as they measure the permeability of material cured under site conditions as distinct from samples cast in tubes and cured in the laboratory. The permeability tests were carried out in accordance with the ICE Specification on slurry walls and the results are summarised below in Table 1. All the measured permeabilities were nearly 1×10^{-9} m/s or lower. At both walls, the permeabilities of the samples taken from depth, below the water table were significantly lower than for the one near the surface. It is possible that this difference is due to the different curing conditions or possibly some structure in the samples associated with drying near the top of the all. Of particular importance is the very low permeability below the water table where flow would occur.

EXHUMATION OF CUT-OFF WALLS
The isolated walls
Exhumation of cut-off walls in-service is rarely undertaken unless a problem has been identified. Exhumation provides a unique opportunity of assessing the quality and likely performance of a slurry cut-off wall system. Three of the walls at the test site have been exhumed; one in relatively uncontaminated ground, one in slightly contaminated ground and one in very contaminated ground.

The exhumations established that the walls were continuous except towards the top of wall 3 in the contaminated ground where some very tight oxidised fissures were present to a depth of approximately 1.4m depth. All the exposed walls had a minimum width 0.6m and generally there was very little overbreak except neat the tops of the walls where it was up to 200mm. The depth of the filter cake was only a few millimetres.

In contrast to wall 1 which was in relatively uncontaminated ground, there was distinctive change in the material with depth of wall 3 in the contaminated ground, going from the typical dark blue near the surface to a dark grey due to the effect of the contamination ground being mixed into the slurry. Below about 1m depth, the outer 100 mm of the wall adjacent to the ground appeared to have more material mixed into it, otherwise the wall was reasonably uniform across its width. The exhumation of this cut-off wall has shown no signs of any obvious deterioration during the 3 years that the wall has been in place.

The ground water level adjacent to the wall was initially at 1.3m but was lowered to 3m by pumping from the adjacent well. The pH of the ground water is 3.7 and contains high levels of sulphate. The ground conditions consisted of black ash, gravels and layers of varying thickness of spent oxide, generally near the surface.

All the holes formed by the piezocone were open down to the bottom of the wall, about 5m, and were the same diameter (45 mm) as the piezocone. The clay beneath the wall had squeezed into the hole. There was no sign of any cracking in the cement-bentonite wall around the hole formed by the piezocone. This exercise illustrates that despite appearing to be a brittle, non-plastic material at low confining stresses, cement-bentonite can be displaced in a piezocone test without causing cracking of the surrounding material.

Partial exhumation of the test cells
The objective of this work was to determine if any obvious leakage paths exist through the walls that could account for the decrease in levels inside the test cells. The ground water level within the test cells has been maintained higher than outside. Another objective was to determine the extent of any chemical reaction with surrounding ground and ground water that could eventually lead to long term deterioration and reduced performance of the walls. Excavations were made on the outside of the test cells at the corners and included approximately 5m of each wall, as shown in Fig. 1. Corners were included as leakage was considered more likely there.

Generally the outside walls forming the test cells were reasonably regular with little overbreak except at the corners where it was generally less than 200mm. There was no evidence of leakage except for three very small weeps at approximately 1m depth on part of one wall forming the test cell without the membrane These minor weeps are possibly "cold joints" associated with topping up the trench after the majority of the slurry had settled. The top of the slurry had typically settled 0.6m 24 hours after placing, after which it was topped up. This is a potential problem on short lengths of wall. The was no evidence of leakage at the corners of the test cells.

Although the minor weeps appear unsatisfactory it is unlikely that the water level would be so close to the top of slurry trench cut-off walls in most working situations. The lowest leak was 0.5m below the top of the cut-off wall. Concerns that the flows through the cement bentonite could lead to erosion and increase in flow are not substantiated by laboratory experiments that have shown that relatively fast flowing water (3.6m/s) through cement bentonite does not lead to erosion and increase in leakage.

Where the wall had been in contact with the spent oxide there had been a reaction. The surface was a soft white oxidised material over a soft black material, some 20mm thick, similar to that seen in laboratory tests in which samples of cement-bentonite had been immersed in the site leachate. These limited excavations have generally provided confidence in the performance of cement-bentonite cut-off walls in very aggressive ground. There are no obvious discontinuities in the walls leading to large leakage. The minor weeps towards the top of the walls of the test cell without a membrane suggests it would be prudent to include a membrane as part of the detail if very high water tables are expected.

CONCLUSIONS
There are limitations with all the methods used to measure discrete in-situ permeabilities. Both the piezocone and SBP give measured permeability values orders of magnitude larger than would be expected from laboratory tests or from the BRE packer system. The BRE packer system appears the most satisfactory for measuring discrete permeabilities in cement-bentonite cut-off walls and has been used on a number of occasions.

Exhumations of the cut-off walls, two in uncontaminated ground and one in chemically aggressive ground, have shown them to be continuous and, on the evidence of laboratory permeability tests from block samples, to have a low permeability down to the level of the aquiclude. There were no signs of any obvious deterioration of the cement bentonite during the 3 years that the walls have been in contact with aggressive ground water. In contrast, unconfined samples of cement-bentonite immersed in ground water from the borehole adjacent to the walls disintegrated in less than one year. Laboratory immersion tests are often used to assess the chemical compatibility and durability of the set cement bentonite in contaminated ground water. It can be concluded that such immersion tests provide a simple method of demonstrating that a specific mix of cement-bentonite is durable when no visible reaction takes place. However, when a reaction does take place, less severe testing may still indicate that the durability is adequate.

In-situ studies have revealed satisfactory performance of the cement-bentonite walls during the four and half year period of monitoring. Full exploitation of the study has yet to be achieved by complete exhumation and examination of the slurry walls at the test site.

ACKNOWLEDGEMENTS
This work has been funded by the Department of the Environment. Transport and the Regions, Lattice Property Holdings (LPH). The support of S Wallace of LPH and I R Holton and R Chown of BRE in undertaking the work is gratefully acknowledged.

REFERENCES
Doe G, Jefferis S A and Tedd P (1999). Specification for the construction of slurry trench cut-off walls as barriers to pollution migration. Thomas Telford.

Jefferis S A (1997). The origins of the slurry trench and a review of cement-bentonite cut-off walls in the UK. International Containment Technology Conference. St Petersburg, Florida. pp52-61.

Manassero M.(1994).Hydraulic Conductivity Assessment of Slurry Wall Piezocone Tests. Journal of Geotechnical Engineering. ASCE Vol.120,No 10, pp1725-1746.

Ratnam S, Soga K, Mair R J, Whittle R and Tedd P (2001). An insitu permeability measurement technique for cut-off walls using the Cambridge self boring pressure meter. Proceedings of the IV ICSMGE, Istanbul.

Tedd P, Holton I R, Butcher A P, Wallace S and Daly P J (1997a). Investigation of the performance of cement-bentonite cut-off walls in aggressive ground at a disused gasworks site. Proc. of the 1st Int. Containment Technology Conf. St Petersburg, Florida, USA.pp 125-132. Also published in Land Contamination & Reclamation Vol5, No 3, pp 217-223

Tedd P, Butcher A P and Powell J J M (1997b). Assessment of the piezocone to measure the in-situ properties of cement bentonite cut-off walls. Proc. of Conf. on Contaminated ground: fate of pollutants and remediation. Cardiff. Thomas Telford, pp 48-55.

Tedd P, Quarterman R S T and Holton I R (1995) Development of an instrument to measure in-situ permeability of slurry trench cut-off walls. 4th Int. Symp. on Field Measurements in Geomechanics. Bergamo,Italy. pp441-446.

Tedd P, Paul V and Lomax C (1993). Investigation of an eight year old slurry trench wall. Green'93, Int.Symp on Waste Disposal by Landfill. Bolton Institute. Balkema 1995.pp581-590.